Advances in Statistical Modeling and Inference

Essays in Honor of Kjell A Doksum

SERIES IN BIOSTATISTICS

Series Editor: Heping Zhang *(Yale University School of Medicine, USA)*

Vol. 1 Development of Modern Statistics and Related Topics
— In Celebration of Prof. Yaoting Zhang's 70th Birthday
edited by Heping Zhang & Jian Huang

Vol. 2 Contemporary Multivariate Analysis and Design of Experiments
— In Celebration of Prof. Kai-Tai Fang's 65th Birthday
edited by Jianqing Fan & Gang Li

Vol. 3 Advances in Statistical Modeling and Inference
— Essays in Honor of Kjell A. Doksum
edited by Vijay Nair

Series in Biostatistics – Vol. 3

Advances in Statistical Modeling and Inference

Essays in Honor of Kjell A Doksum

Editor

Vijay Nair

University of Michigan, USA

World Scientific

NEW JERSEY • LONDON • SINGAPORE • BEIJING • SHANGHAI • HONG KONG • TAIPEI • CHENNAI

Published by

World Scientific Publishing Co. Pte. Ltd.
5 Toh Tuck Link, Singapore 596224
USA office: 27 Warren Street, Suite 401-402, Hackensack, NJ 07601
UK office: 57 Shelton Street, Covent Garden, London WC2H 9HE

British Library Cataloguing-in-Publication Data
A catalogue record for this book is available from the British Library.

Series in Biostatistics — Vol. 3
ADVANCES IN STATISTICAL MODELING AND INFERENCE
Essays in Honor of Kjell A Doksum

ISBN-13 978-981-270-369-9
ISBN-10 981-270-369-1

Printed in Singapore by World Scientific Printers (S) Pte Ltd

Preface

This volume covers a broad range of contemporary topics in statistical modeling and inference. There have been many exciting developments in the field of Statistics over the last quarter century, stimulated by the rapid advances in computing and data-measurement technologies. The increased computing power and availability of large datasets have led to considerable new research in flexible modeling, semiparametric and nonparametric methods, and computationally-intensive techniques. These developments have allowed us to move away from parametric techniques that rely on restrictive assumptions to much more flexible methods for modeling and analysis of data. There is also extensive use of simulation and Monte Carlo techniques for doing statistical inference. This book provides an overview of some of these advances as well as description of new research in methodology and theory.

There are 32 chapters written by leading international researchers on a broad range of topics: semiparametric methods, transformation models, nonparametric regression, rank-based inference, mixture models, survival and reliability analysis, Bayesian inference, resampling methods, and inference under constraints. Researchers, graduate students, as well as practitioners will find the volume to be useful.

The book was prepared in honor of Professor Kjell A. Doksum to celebrate his 65th birthday. Authors include Kjell's friends, colleagues, and former students from all over the world. In recognition of Kjell's passion for soccer, the volume begins with a chapter by Brillinger that deals with modeling ordinal data and application to Norwegian soccer.

Part 1 covers topics in survival analysis. Aalen and Gjessing provide a modern perspective on the role of stochastic processes in modeling random phenomena in survival and event history analysis. Jewell examines the correspondence between complex survival models and categorical regression models for polytomous data, generalizing earlier connections between binary regression models and survival analysis. Borgan and Langholz develop methods for assessing the goodness-of-fit for sample risk set data using martingale residuals.

Part 2 deals with reliability techniques and related applications. Singpurwalla examines relationships between reliability/survival analysis and the mathematical theory of finance and uses them to characterize asset pricing formula, model interest rates, etc. Block, Dugas, and Samaniego study new applications of the "system-signature" concept to reliability analysis of system lifetimes. Li and Shaked review generalizations of the total-time-on-test transforms, stochastic orders based on these transforms, and their applications.

Part 3 includes five chapters on advances in semiparametric methods. Beran develops a very flexible modeling and estimation strategy, called Adaptive Shrinkage on Penalty bases, for discrete two-way layouts and studies its large-sample behavior. You and Jiang consider varying-coefficient partially-linear models and propose a penalized spline-based least-squares estimation methodology for serially correlated data. In related work, Chong, Wang, and Zhu develop a semi-linear index model for flexible dimension reduction that incorporates both discrete and continuous predictors. Chaudhuri explores extensions of semiparametric single-index models for multivariate lifetime data and related inference methods. The chapter by Samarov and Tsybakov proposes and examines a data-based method for selecting the best estimator from a collection of arbitrary density estimators.

Part 4 is concerned with the related area of transformation models. Klassen considers a general class of semiparametric transformation models and determines the semiparametric information for the Euclidean parameter in the model. Scheike also considers semiparametric transformation models and examines the modified partial likelihood estimators in this context. Taylor and Liu examine the effect of embedding a standard model in a larger family of models indexed by an additional parameter and discuss parameter interpretations, variance inflation, predictions, and so on.

The three chapters in Part 5 cover topics in nonparametric regression. Müller examines smooth nonparametric estimation of conditional moments and correlation functions and proposes a general linear unbiased estimation scheme. Hallin, Jureckova and Koul study the asymptotic properties of rank score statistics for regression and serial autoregression. Støve and Tjøstheim develop a new convolution smoother for nonparametric regression that outperforms standard kernel estimators.

Part 6 deals with clustering and mixture models. Zou, Yandell, and Fine review gene mapping in the context of semiparametric and nonparametric inference for mixture models and discuss estimation and model selection issues. Lau and Lo consider model-based clustering and Bayesian nonparametric methods for mixture models, and develop Monte Carlo methods using a weighted Chinese restaurant process for inference. James describes a class of species-sampling mixture models that can be derived from Doksum's neutral-to-the-right processes.

The two chapters in Part 7 develop Bayesian nonparametric inference for quantiles using Dirichlet process priors. Johnson and Sim obtain an asymptotic expansion for the posterior distribution of a percentile with a leading normal term. Hjort and Petrone develop Bayesian inference for the quantile function and related quantities such as the Lorenz curve and Doksum's shift function.

Part 8 is concerned with rank-based methods. Aaberge studies empirical rank-dependent family of inequality measures, motivated by applications to modeling income distributions. Zheng and Lo develop a modified Kendall rank-order test for evaluating repeatability of studies with very large data sets when only a small proportion of "interesting or important" objects. Miura uses rank statistics of the geometric Brownian motion to define some new concepts in finance and studies their stochastic properties.

Part 9 covers inference based on Monte Carlo and resampling methods. Lindqvist and Taraldsen review and extend a general approach for Mote Carlo computations of conditional expectations given a sufficient statistic. Kong, McCullagh, Meng, and Nicolae explore the use of a likelihood-based theory for Monte Carlo integration, following up on their earlier work. Schweder studies confidence nets, a family of nested confidence regions, and uses bootstrapping to get product confidence nets for high-dimensional parameters.

Part 10 deals with topics in constrained inference. Koenker and Mizera examine a unified approach to density estimation using total variation regularization and develop methods that are capable of identifying features such as sharp peaks. Fan and Zhang study the bounded normal mean problem and develop a better approximation to the minimax risk. The final chapter by Rojo and Batún-Cutz examines estimation of symmetric distributions under a peakedness constraint.

The volume spans a broad range of areas in statistical modeling and inference. It is worth noting that Professor Kjell Doksum has made significant contributions to all of these topics.

Several people have made important contributions during the preparation of this volume. First, I want to thank the authors for their enthusiastic support for a volume to honor Kjell Doksum and for their patience during the editing and publication process. The initial plans for a festschrift for Kjell were conceived in collaboration with Dorota Dabrowska. I am greatly indebted to Dorota for the tremendous amount of time and effort she invested on the original project. I wish we could have completed it. Thanks are also due to the many referees for their help with the reviewing process. Anupap (Paul) Somboonsavatdee did an extraordinary job with the technical aspects of editing and proof-reading this volume. He really rescued me! Sheela Nair, Aijun Zhang, Mary Ann King, and Matthew Linn also helped out with editing at various points. I am grateful to Joan Fujimura and Teresa Doksum for background information, pictures, and encouragement.

I have known Kjell Doksum for more than 30 years as a teacher, mentor, and, best of all, friend. On behalf of his friends, colleagues, and students, I am pleased to dedicate this volume in Kjell's honor and to celebrate his 65-th birthday!

Vijay Nair
Ann Arbor, MI, USA
August 15, 2006

Contents

Kjell Doksum: Family, Career, and Contributions

Kjell Doksum: May, 2006

Family Background

Kjell Doksum was born in Sandefjord, Norway on July 20, 1940. His family had gone there from Oslo to get away from the ravages of World War II. Kjell has many hair-raising stories to tell of his first few years under the Nazi occupation. The family soon returned to Oslo, and Kjell grew up in an apartment across from the Frydenlund brewery, in downtown Oslo close to the Bislett stadium, where he watched and developed his passion for soccer. He attended Vestheim gymnasium (high school) and moved to the U.S after graduation.

Kjells father, Filip Doksum (born Filip Karlsen), was a mathematics teacher in Oslo. His mother Elise Olsen died when Kjell was four years old. Kjells stepmother Astrid helped raise him. Kjell has two brothers. Older brother Olav lives in Tomter, Norway, and worked for the intelligence branch of the Norwegian defense department. Younger brother Sigmund

Kjell with daughters: Kathryn, Teresa, and Margrete

lives in Birkerød, Denmark, and is a poet and novelist.

Kjell is married to Joan Fujimura, a sociology professor at the University of Wisconsin, Madison. Their daughters are Kathryn Doksum of Newport, Oregon, Teresa Doksum of Stoneham, Massachusetts, and Margrete Doksum (deceased). Kathryn is a certified public accountant, and Teresa is a health services researcher. Kjell and Joan have four lovely grandchildren: Matthew, Kevin, Emma, and Calvin.

Why Statistics?

Kjells original career plan and lifelong passion was soccer. He started playing in the streets of Oslo. Had it worked out, he would have been a professional soccer player. But, alas, his skills on the cobblestones did not transfer to soccer played on grass fields. Nevertheless, Kjell has been an avid soccer fan and played recreational soccer until the ripe age of 64. Generations of students, faculty, and visitors at Berkeley, Stanford, Madison, and at other campuses Kjell has spent time will fondly recall his presence on the soccer field and his organization of soccer teams, games, and parties.

His second career was fishing, which lasted all of two weeks. In the Spring of 1959, Kjell was about to graduate from high school in Oslo. He had not paid much attention in high school and had no plans to attend a university. He was drafted into the Norwegian military and was supposed to report to bootcamp up by the North Pole on July 1st. But in April, his aunt and uncle from San Diego visited Oslo and invited Kjell to come to

Berkeley soccer team in the mid-1970s: Front row, from left – Kjell is second and Vijay Nair is fourth

Kjell and Joan's wedding, Berkeley, 1987: Right to left – Kjell, Joan, Teresa, Kathryn, and Margrete

San Diego. Kjell saw this as a clear sign that he should move to California.

All of Kjell's relatives and their friends in San Diego were fishermen, so Kjell had no choice. He spent two weeks on a 36 foot boat, fishing albacore tuna off the coast of California in July 1959. But something unusual

Kjell (right) with Chuck Bell (left) and other friends

Kjell (far left background) with Jerzey Neyman (jacket and tie), David Blackwell (bow tie), and Elizabeth Scott (far right) and others at Berkeley

happened. Southern California was hit by the only July storm in recorded history. It lasted two weeks and produced waves bigger than the boat. Kjell was seasick every minute of those two weeks. It was clear that he is not cut out to be a fisherman. So he signed up to become a student at San Diego

At the Olympics soccer game between Norway and Brazil, 1996 (score 2-2): Right to left: Kjell, Joan, Teresa, and Kathryn

State College (SDSC).

After about a year in college, Kjell got a telegram saying he had been drafted into – this time – the American military and he was to go to bootcamp in Los Angeles the following day. Kjell showed the telegram to the professor he knew best, the chair of the Math department. Professor Smith said, "You can either go to bootcamp or become a math major." Again, the choice was clear – so Kjell became a math major.

How did he end up in Statistics? The story starts with the international club at SDSC. The club's talent show was very popular, mainly because of a Middle-Eastern math student who performed an Arabian Harem Dance with minimum wardrobe. (At the end of her performance, she would invariably have a wardrobe malfunction.) One day, the dancer/math student cornered Kjell and started grilling him about his social life. Upon learning of its non-existence, she tried to arrange a blind date for him with her Norwegian roommate. When Kjell went to the address, no one answered. After a few minutes he started to leave, but then a math professor (Dr. Gindler) came out of the next apartment and asked what he was doing there. When Kjell explained, he offered him a ride home. As they were driving, he asked Kjell about his plans for Spring break. Kjell said he was scrubbing floors to make money to pay for room and board and nonresident tuition. So Gindler introduced him to the statistician Dr. C. B. Bell, who had an NSF

After Berkeley-Stanford soccer game to celebrate Kjell's 65th Birthday, October 28, 2005 – Right to left (Kjell, Emma, Calvin, Teresa, Friend Tom, Matthew, Kathryn, Vijay, and Friend Debbie)

grant and support for students. Kjell was at another crossroad: Should he scrub floors or study Statistics? Again, the optimal decision was clear. That is how Kjell Doksum ended up as a statistician.

Career and Research Contributions

Kjell received his Masters from SDSC in 1963 and became a Ph.D. student in the Berkeley Statistics department. He finished his Ph.D. in 1965 with Erich Lehmann, spent a year as a post-doc with Professor Bell in Paris, then returned to Berkeley as an assistant professor in 1966.

Most of Kjell's academic life was spent at the University of California-Berkeley Statistics Department where he became associate professor in 1973 and full professor in 1978. He took early retirement from Berkeley in 2002 and has been Professor at the Statistics Department in the University of Wisconsin, Madison since 2003. Kjell has also visited the Universities of Paris, Oslo, Trondheim, Harvard, Hitotsubashi in Tokyo, and Columbia as well as the Bank of Japan in Tokyo.

Kjell has made pioneering contributions to statistical theory, methodology and applications. He has worked on randomization methods, nonparametric and rank-based inference, survival and reliability analysis, semiparametric techniques and transformation models, probability measures, and Bayesian inference. Almost all of these areas are covered in the various chapters in this volume.

His early work, joint with C. B. Bell starting in 1964, was on the use of randomization in statistical inference. The idea was to replace subsets of the data from an experiment by standard normal random samples placed in the same order as the data subset. These Gaussian randomized tests were asymptotically as efficient as the classical tests they were derived from under normality and more efficient in non-normal cases. They also have connections with Monte Carlo and bootstrapping methods.

Much later, he revisited this topic when he proposed a Monte Carlo approach to rank (partial) likelihood methods for semiparametric models and developed its properties. In this approach, the data in a likelihood are replaced by random samples put in the same order as the original data, then a Monte Carlo average of these randomized likelihoods that estimate the partial likelihood is computed.

Kjell and his collaborators have studied the asymptotic power of rank tests for nonparametric classes of alternatives. They developed asymptotically minimax tests for 0-1 loss functions for alternatives separated from the null hypthesis by a certain distance. In particular, they found minimax tests for the two-sample, matched pair, and independence problems as well as for reliability and life-testing problems.

In the early 1970's, Kjell introduced the concept of a shift function and developed inference procedures. This is a general measure of the difference between populations and is closely related to the population version of quantile-quantile plots. In joint work with Sievers, he developed a general class of simultaneous confidence regions for the shift function that are useful in formal model selection. In related work, he introduced measures of location and symmetry and developed simultaneous confidence procedures for inference. His contributions to simultaneous inference also include confidence procedures for nonparametric regression curves.

In one of his most cited papers, Kjell proposed and developed the properties of "neutral-to-the-right" processes that are very useful in nonparametric Bayesian inference. He introduced a general class of probabilities on the class of all probability measures and showed that the posterior distribution given a sample is also in the same class. The Dirichlet process is a special case of these neutral processes. In other work related to Bayesian inference, he and Lo showed that conditioning on robust estimates produced consistent Bayes estimates even when the original Bayes estimates based on all the data are not.

Kjell has also made seminal contributions, jointly with Bickel, to transformation models and semiparametric inference. Their work has provided deep insights into the statistical properties of procedures based on transformed data. In joint work with Bjerve and others, he introduced and analyzed the concept of local correlation in the nonparametric regression framework and, with Samarov, a global measure of correlation in the nonparametric framework with multiple covariates. The latter is a nonparametric version of "R-squared" and provides a measure of explanatory power in a nonparametric context. He has studied, jointly with Chaudhuri and Samarov, quantile regression methods to investigate general relationships between a response and covariates.

His research also covers reliability and survival analysis. He has introduced and studied, with Høyland , new classes of degradation models for reliability and life testing. His work in survival analysis includes modeling time of infection (joint with Normand) and graphical methods for checking treatment effects and model assumptions with censored survival data (joint with Dabrowska and others).

Kjell has supervised over 20 PhD students from all over the world. He has also contributed to the statistics education of a large number of other students who have taken courses from him or have used his text book with Peter Bickel *Mathematical Statistics: Basic Ideas and Selected Topics* (Bickel and Doksum 1977, 2001, 2007, Pearson Prentice Hall).

Kjell was Vice-Chair of the Statistics Department (1987-88) and Assistant Dean of the College of Letters and Science (1978-80). He has also provided extensive service to the statistical profession. He has served on the editorial boards of (and as guest editor for) the Journal of the American Statistical Association, Scandinavian Journal of Statistics, Life Data Analysis, and Sankya. He has been Executive Secretary of the Institute of Mathematical Statistics. Together with Ingram Olkin and Bruce Trumbo, he played a role in the founding of the IMS journal *Statistical Science.*

Kjell is a Fellow of the Institute of Mathematical Statistics, Fellow of the American Statistical Association, and Elected Member of the International Statistical Institute. He is an Elected Foreign Member of The Royal Norwegian Society of Sciences and Letters, Trondheim.

Reminiscences of a 40-year Friendship

Peter J. Bickel
Department of Statistics
University of California, Berkeley, CA

Kjell and I have run on parallel courses in many ways. First, we were both students of Erich Lehmann at Berkeley, obtaining our PhDs within 2 years of each other: 1963 and 1965. We both followed Hodges and Lehmann's research campaign in nonparametric and robust statistics in our theses and a little after. We both stayed on in the Berkeley faculty for many years.

From left to right: Peter Bickel, Juliet Shaffer, Erich Lehmann, and Kjell Doksum

We started collaborating early on with a paper (1969) applying Le Cam's contiguity ideas in testing for constant failure rate. Our temperaments were well suited from the beginning: I going off into vague generalities with great ease, and Kjell bringing us back to concrete examples. We dared venture out of our theoretical cocoon together in 1981, challenging George Box and David Cox on their analysis of transformations. We were vigorously slapped down for speaking of effects on unspecified scales. In retrospect I, at least, admit our conceptual error, but still maintain that, in pretending that variability in estimation of the scale and of the effect on that scale have nothing to do with each other, our distinguished colleagues and their defenders erred just as much. I can't resist stating the correct conclusion noted by our friend Bill van Zwet. One must always talk of joint estimation of scale and effects on the scale revealing the compatibility of possibly quite variable scales and differing effects on these.

Passing from these old battles, I want to focus on our largest and, I think, most effective collaboration: our textbook, Mathematical Statistics, in its first (1976) edition and second (2001) edition for which volume I has appeared and Volume II is being readied for publication in 2007. In both endeavors, Kjell and I were sufficiently different on one score but not sufficiently so on another. I have already mentioned the counterpoint between generality and concreteness between us, which has served us well. The point on which we are more similar (and not good) is carelessness. I'm by far the worse sinner there but Kjell isn't innocent either, as readers of the first printings of both editions have learned to their and our sorrow!

Kjell and I have both passed the 65-year mark. I have every expectation that our friendship and collaboration will continue and BD Edition Three, not to speak of Volume II of the second edition, will all see the light of day. If I had a glass, I'd raise a toast to Kjell. Many happy returns!

PART 0

Statistics and Soccer

Chapter 1

MODELLING SOME NORWEGIAN SOCCER DATA

David R. Brillinger
Statistics Department
University of California, Berkeley, CA, U.S.A.

E-mail: brill@stat.berkeley.edu

Results of Norwegian Elite Division soccer games are studied for the year 2003. Previous writers have modelled the number of goals a given team scores in a game and then moved on to evaluating the probabilities of a win, a tie and a loss. However in this work the probabilities of win, tie and loss are modelled directly. There are attempts to improve the fit by including various explanatories.

Key words: Binary data; Empirical process; Football; Generalized linear model; Ordinal data; Residuals; Soccer.

1 Introduction

Kjell Doksum has been a steady contributor to the theory and practice of nonparametric statistics and soccer. In former case he has studied the quantile function, probability plotting and, what is most pertinent to this article, the introduction of randomness to ease analyses. In the latter case he has potted lots of goals during his lifetime.

Previous studies have modelled the number of goals a team of interest scores in a soccer game as a function of explanatories such as site of game, opponent, and FIFA rating. References include Lee (1997), Dyte and Clarke (2000), Karlis and Ntzoufras (2000, 2003a, 2003b) and references therein. In our work the respective probabilities of win (W), tie (T), and loss (L) are modeled directly and are examined as a functions of possible explanatories. A reason for employing W, T, L is the thought that the ultimate purpose of a game is to decide a winner. It is felt that the response W, T, L better represents this event than the number of goals scored. The latter may be inflated by a team's "giving up" or be deflated by a team's moving to a defensive strategy.

To an extent the approach is that taken in Brillinger (1996) for hockey data. The sections of the paper after the Introduction are: Some previous soccer modeling, Norwegian soccer, Ordinal data, Results, Assessing fit, Another model, Uses, Extensions, Discussion and summary.

2 Some previous soccer modelling

There has been previous work on modelling the number of goals scored by each team in a game. For example Lee (1997) employs independent Poissons for the number of home and away goals, with

$$E\{home\ goals\ by\ team\ i\} = exp\{\alpha + \Delta + \beta_i + \gamma_j\}$$

$$E\{away\ goals\ by\ team\ i\} = exp\{\alpha + \gamma_i + \beta_j\}$$

respectively, where Δ represents the home effect, β_i refers to team i playing at home and γ_i refers to team i playing away, and j refers to any arbitrary team. On the other hand Dyte and Clarke (2000) employ the expected value

$$exp\{\alpha + \beta U_i + \gamma V_j + \Delta\}$$

where U_i is i's FIFA rating, V_j is j's and Δ is again a home team effect. Karlis and Ntzoufras (2000, 2000a, 2000b) employ the Poisson and bivariate Poisson in their work. In each case the model is used to determine resultant win, tie, loss from the goals scored. In this paper the focus is on the win-tie-loss result as the basic response.

Panaretos(2002) adopts a "game viewpoint" employing explanatories such as: fouls committed, off-sides, and shots on goal. Brillinger (2005) employs mutual information as a measure of the strength of association of the effect of playing at home and the number of goals scored for various premier leagues around the world.

3 Norwegian soccer

Table 1 lists the Norwegian Elite Division teams for the 2003 season, and Figure 1 is a map displaying their locations. One notes the two northerly teams and wonders whether travel and weather might not play important roles in their games. Also listed are identifiers for the map showing locations. (The reason for switching the last five teams from numbers to letters is to have less overprinting in the figure.) The teams are listed in the table in order of their final standings for 2003.

Table 1 The 2003 Elitserien teams. The identifier provides their location on the map of Figure 1. The teams are in the order of their 2003 finish

Team identifier	*Team*
1	Rosenborg
2	Bodo-Glimt
3	Stabaek
4	Odd-Grenland
5	Viking
6	Brann
7	Lillestrom
8	Sogndal
9	Molde
10(a)	Lyn
11(b)	Tromso
12(c)	Valerenga
13(d)	Aalesund
14(e)	Bryne

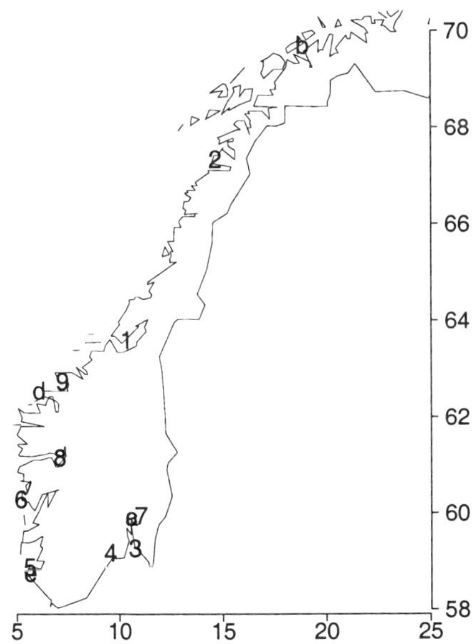

Figure 1 Locations of the 2003 Elitserien Teams. Table 1 lists the team names corresponding to the identifiers. The x-axis is longitude east and the y-axis latitude.

The Elitserien has 14 teams, each playing all the others, home and away. There were 182 games in the 2003 season and the season goes on 26 weeks with a break in the summer. There are 7 games each week and the design is balanced. The data employed in the analyses came from the url http://www.soccerway.com/national/norway/tippeligaen/2003/round-1/results

Table 2 The results of the first week's games as an illustration

Home	Visitor	Result
Rosenborg	Valerenga	1 - 0
Lillestrom	Bodo/Glimt	1 - 0
Aalesund	Tromso	2 - 3
Viking	Bryne	3 - 0
Sogndal	Stabek	2 - 1
Odd_Grenland	Molde	1 - 0
Lyn	Brann	0 - 0

To show the character of the original data the first week's results are displayed in Table 2. Table 3 provides the final 2003 season results. The left hand columns give the at-home results and the right hand the away for the complete season.

Table 3 The season's results for 2003. The left hand columns are home games and the right hand columns away games

Identifier	W	T	L	W	T	L
1	9	2	2	10	2	1
2	7	2	4	7	3	3
3	6	4	3	5	5	3
4	6	4	3	5	1	7
5	6	3	4	3	7	3
6	7	1	5	3	6	4
7	7	4	2	3	3	7
8	7	4	2	2	4	7
9	6	2	5	3	2	8
10(a)	4	3	6	4	3	6
11(b)	4	4	5	4	1	8
12(c)	4	5	4	2	5	6
13(d)	4	5	4	3	2	8
14(e)	7	1	5	0	0	13

4 Ordinal data

This section motivates and lays out the analysis approach taken in the paper.

4.1 *The cut-point approach*

The random variables of principal concern are ordinal-valued namely *loss, tie, win.* These will be denoted by

$$0,\ 1,\ 2$$

respectively.

A number of different models have been proposed for the analysis of ordinal data. These include: continuation ratio (see Fienberg (1980)), stereotype (see Andersen (1984)) and grouped continuous (see McCullagh and Nelder (1989)). This last is the one employed in the analyses presented.

The approach to be followed starts by supposing that there exists a latent variable, Λ, whose value in some sense represents the difference in strengths of two teams in a game. It further assumes the existence of cutpoints θ_1 and θ_2 such that

$$Y = 0 \text{ if } \Lambda < \theta_1,\quad Y = 1 \text{ if } \theta_1 < \Lambda < \theta_2 \text{ and } Y = 2 \text{ if } \theta_2 < \Lambda$$

so for example

$$Prob\{Y = 1\} = F_\Lambda(\theta_2) - F_\Lambda(\theta_1) \tag{1}$$

where F_Λ is the c.d.f. of Λ. In practice the choice of F_Λ is sensibly based on the subject matter of the problem. The *complimentary loglog* link corresponds to situations in which of an internal variate crosses a threshold. It may be based on an extreme value distribution. In the present context this may be reasonable, with a win for a particular team resulting from the team members putting out maximum efforts to exceed those of the opponent. What is basic though is that its choice makes standard generalized linear model programs available via the Pregibon trick.

Thc extreme value distribution of the first type is given by

$$Prob\{\Lambda \le \eta\} = 1 - exp\{-e^\eta\},\ -\infty < \eta < \infty \tag{2}$$

One can write

$$log(-log(1 - Prob\{\Lambda \le \lambda\})) = \lambda$$

and see the appearance of the *cloglog* link. Pregibon (1980) noted that one could employ standard statistical packages in analyses of such multinomial data when one proceeded via conditional probabilities. Here the distributions involved in the modelling are $Prob\{Y = 2\}$ and $Prob\{Y = 1|Y \neq 2\}$.

Explanatory variables, x, may be introduced directly by writing

$$\Lambda = \mathrm{E} + \beta' x$$

where E has the standard extreme value distribution. Now (1) becomes

$$F_{\mathrm{E}}(\theta_2 - \beta' x) - F_{\mathrm{E}}(\theta_1 - \beta' x).$$

4.2 *Some formulas*

To begin consider $Prob\{Y = 2\}$, as opposed to $Prob\{Y \neq 2\}$, and $Prob\{Y = 1|Y \neq 2\}$, as opposed to $Prob\{Y = 0|Y \neq 2\}$. The response is binary in each case. In the work the following parametrization will be employed,

$$Prob\{Y = 2\} = 1 - exp\{-e^{\eta-\theta_2}\}$$

$$Prob\{Y = 1|Y \neq 2\} = 1 - exp\{-e^{\eta-\psi}\} \qquad (3)$$

with $\eta = \beta' x$. The other probabilities of interest may be obtained from these. The advantage of this parametrization is that θ_2, ψ and β may be estimated directly via the function glm() of R and S. See Pregibon (1980) and McCullagh and Nelder (1989). The pertinent material is in McCullagh and Nelder (1989) on page 170. One sees there a multinomial probability mass function being represented as the product of binomials. One follows that representation in setting up the response and explanatory matrices for glm.

Now the basic probabilities are parameterized as

$$Prob\{Y = 2\} = 1 - exp\{-e^{\eta-\theta_2}\}$$

$$Prob\{Y = 1\} = exp\{-e^{\eta-\theta_2}\} - exp\{-e^{\eta-\theta_1}\}$$

$$Prob\{Y = 0\} = exp\{-e^{\eta-\theta_1}\}$$

This fits in with (3) via the connection

$$e^{-\psi} = e^{-\theta_1} - e^{-\theta_2}.$$

4.3 *The setup*

Suppose that:

β_i is the "strength" of team i when playing at home

and

γ_i is the "weakness" of team i when playing away

These will be assumed constant.

Now consider the model

$$Prob\{i\ wins\ at\ home\ playing\ j\} \ = \ 1 \ - \ exp\{-e^{\beta_i+\gamma_j-\theta_2}\}$$

and

$$Prob\{i\ loses\ at\ home\ playing\ j\} \ = \ exp\{-e^{\beta_i+\gamma_j-\theta_1}\} \qquad (4)$$

with the probability of a tie 1 minus the sum of these two.

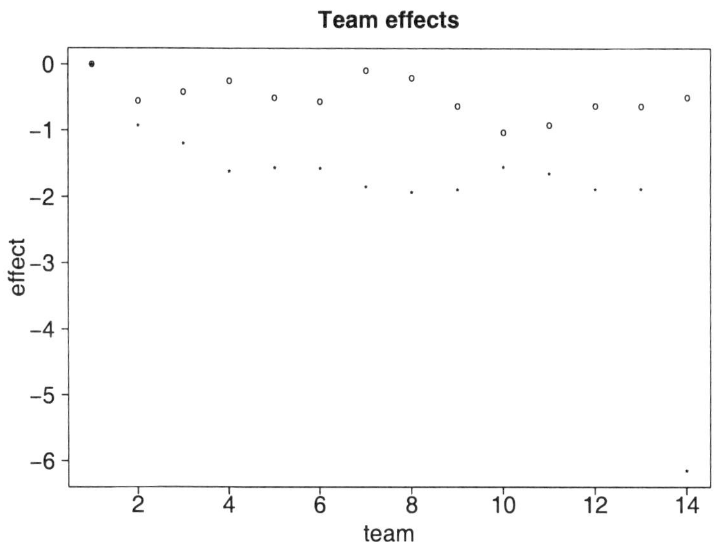

Figure 2 The estimated home effects, $\hat{\beta}_i$, are denoted "o" and the away, $\hat{\gamma}_i$, are "*".

In the fitting the results of the individual games will be assumed statistically independent. The fixed effects are meant to handle the connections amongst teams.

5 Results

The parametrization employed is (3). The estimation method is maximum likelihood. Figure 2 shows the resulting $\hat{\beta}_i$ and $\hat{\gamma}_j$ of (4). The values have been anchored by setting $\hat{\beta}_1,\ \hat{\gamma}_1 \ = \ 0$. One sees $\hat{\gamma}_{14}$ sitting near -6.15 in an attempt to get to $-\infty$ following losing all its away games. The residual

Table 4 Fitted values. The left hand columns refer to home games and the right hand to away. The fitted values are the number of games, 13, times the fitted probability

Team	W	T	L	W	T	L
1	8.35	3.10	1.55	10.10	1.52	1.38
2	5.96	3.26	3.78	6.73	2.95	3.32
3	6.40	3.25	3.36	5.56	3.31	4.13
4	6.91	3.20	2.89	3.73	3.65	5.62
5	5.96	3.21	3.83	3.87	3.63	5.50
6	5.76	3.19	4.06	3.83	3.63	5.54
7	7.44	3.13	2.43	2.79	3.67	6.55
8	6.96	3.21	2.83	2.44	3.61	6.95
9	5.41	3.14	4.45	2.45	3.60	6.95
10(a)	4.31	2.78	5.91	3.74	3.67	5.59
11(b)	4.60	2.90	5.51	3.36	3.68	5.96
12(c)	5.44	3.14	4.41	2.47	3.60	6.92
13(d)	5.42	3.14	4.44	2.47	3.61	6.92
14(e)	5.41	3.50	4.09	0.00	0.00	13.00

deviance is 334.8 with $182 - 26 - 2 = 154$ degrees of freedom. The degrees of freedom here and later in the paper are those as if the model were fitted directly, i.e. the Pregibon trick was not employed. The away performances values stand out.

Table 4 gives the fitted wins-ties-losses for home and away. These number. Figure 3 plots fitted versus actual. One notes that the fitting definitely picks up Bryne losing all its away games. One also sees a clustering about the diagonal line in Figure 3.

6 Assessing fit

There is an issue of how to assess the fit of an ordinal response model. The link function may be checked by nonparametric regression, see Figure 11 in Brillinger et al (1980). Figure 4 shows the kernel estimate based on the data $(\hat{\eta}_i, y_i)$ where $\hat{\eta}_i$ is the fitted linear predictor and y_i is the observed Bernouli value. The smooth curve is the extreme value cumulative distribution function. The two follow each other.

6.1 *Chi-squared statistics*

It was indicated that the residual deviance of model (4) was 334.8 with 154 degrees of freedom, but the interpretation must be made with care. Further, one cannot simply interpret a chi-squared statistic based on the values of

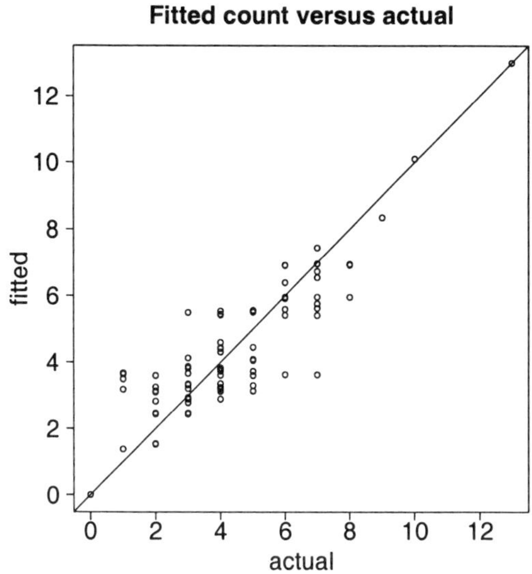

Figure 3 Fitted counts of wins, ties, losses against corresponding actual.

Tables 3 and 4 because the entries are negatively correlated following the competitive character of the variates - one wins, another loses.

6.2 *Uniform residuals*

> "The idea is to obtain randomized rank-sum statistics for the independence, randomness, k-sample and two factor problems analogous to the statistics of ... others." "... one essentially replaces the original data ... by a random sample ... known to have distribution ... advantage ... of having a continuous distribution ..."

Bell and Doksum (1965)

In the case of a continuous variate, Y, the random variable $F(Y)$ has a uniform distribution, see Fisher (1932). Supposing the distribution depends on an unknown parameter θ with estimate $\hat{\theta}$, the $\hat{U} = F(Y|\hat{\theta})$ may be anticipated to have an approximate uniform distribution. The variates $\hat{U}_i = F(Y_i|\hat{\theta})$ were employed in Brillinger and Preisler (1983) to examine the overall fit of a model. They are an aid in various nonstandard cases, such as for random effect models.

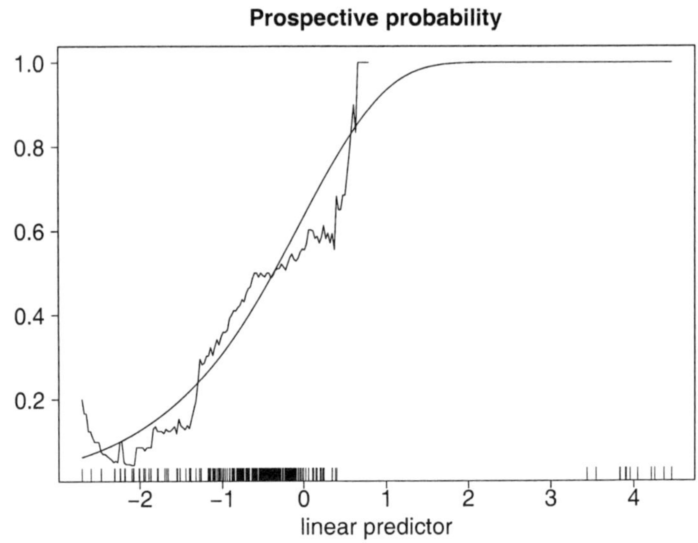

Figure 4 Prospective probabilities. The smooth curve is expression (2). There is a rug plot giving the linear predictor values.

In the present case the response employed is binary, $Y = 0,\ 1$ so various of the classical model effect procedures appear not particularly effective. In this binary case uniform residuals may be computed as follows.

Suppose

$$Prob\{Y = 1|explanatories\} = \pi$$

and that U_1 and U_2 denote independent uniforms on the intervals $(0, 1 - \pi)$, $(1 - \pi, 1)$, respectively . Then the variate

$$U = U_1(1 - Y) + U_2 Y \tag{5}$$

has a uniform distribution on the interval $(0, 1)$. An effect of constructing these values is that the data that are 1's will become spread out in the upper interval and those that were 0's in the lower.

In the null case $E\{U\} = 1/2$ whereas when

$$Prob\{Y = 1|explanatories\} = \pi_0$$

then $E\{U\} = (1 + \pi - \pi_0)/2$.

In practice one has $\hat{\pi}$ an estimate of π and forms

$$\hat{U} = \hat{U}_1(1 - Y) + \hat{U}_2 Y$$

where $\hat{U}_1$ and $\hat{U}_2$ are uniform on $(0, 1 - \hat{\pi})$ and $(1 - \hat{\pi}, 1)$ respectively. When an estimate of $\hat{\pi}$ is employed, we refer to $\hat{U}$, as a uniform residual.

One can equally employ normal residuals, $\Phi^{-1}(\hat{U}_i)$. Working with these has the advantage of spreading the values out in a familiar manner in the null case. We refer to $\Phi^{-1}(\hat{U})$ as a normal residual. Doksum (1966) uses the term "normal deviate" in the situation referred to at the beginning of this section. Various traditional residual plots may now be constructed using the $\hat{U}$ or $\Phi^{-1}(\hat{U})$, e.g. normal probability plots involving the normal residual, $\Phi^{-1}(\hat{U})$ versus an appropriate normal quantile and of $\Phi^{-1}(\hat{U})$ versus explanatories.

Discrete response cases were considered in Brillinger (1996) and Dunn and Smyth (1996).

6.2.1 *Results*

The normal residuals, $\Phi(\hat{U})$ were computed fitting the model (4) as discussed in Section 4. The idea is that they should have an approximate $N(0, 1)$ distribution if the model is fitting well.

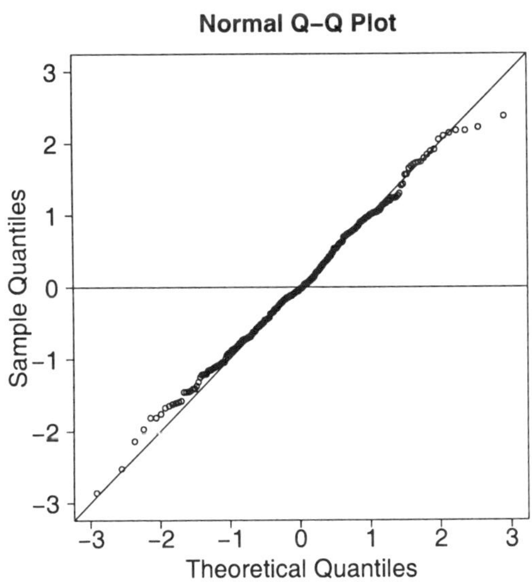

Figure 5 A normal probability plot of the "normal residuals".

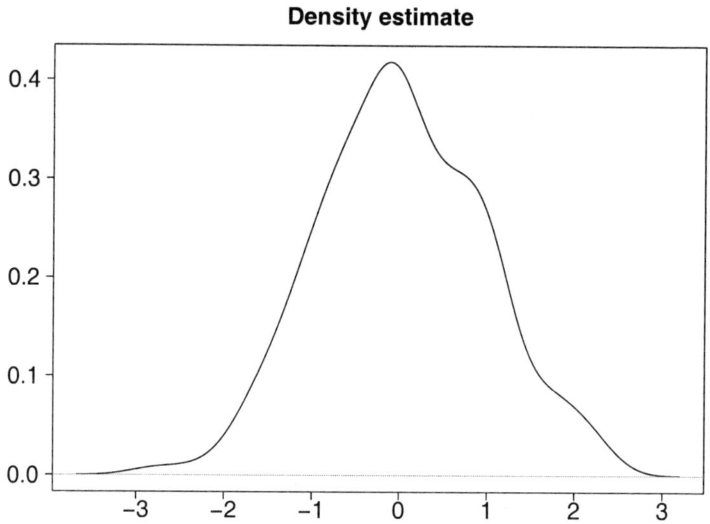

Figure 6 A kernel density estimate of the residuals.

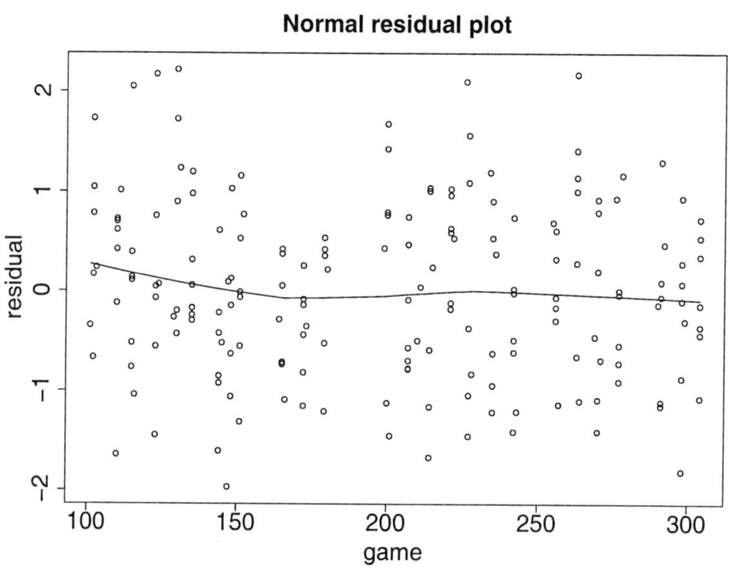

Figure 7 Normal residuals versus day of game for the home wins. A loess line has been added.

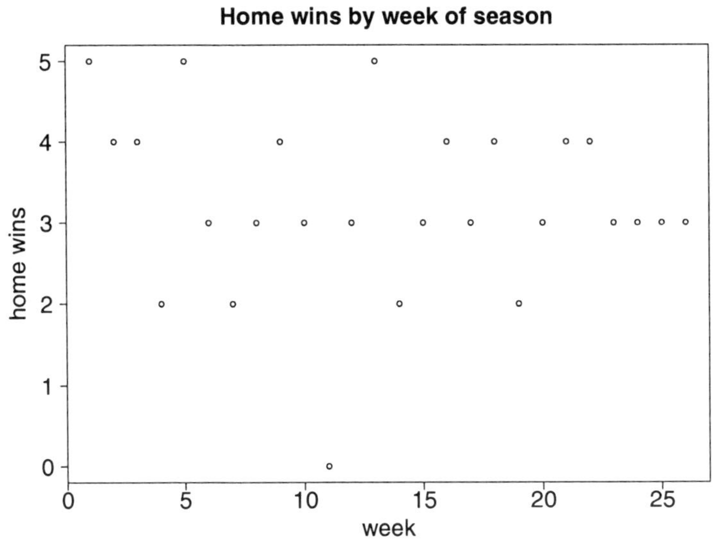

Figure 8 Actual number of wins versus week into season. There are typically 7 games a week.

The plots are shown in Figures 5 and 6. There are some indications of asymmetry. Next consideration turns to seeking to improve the fit by including possible explanatory variables. Available variables include: day of year, distance between cities, and results of the preceding game for the teams. Figure 7 provides a plot of normal residuals vs. day in year. A loess line, see Cleveland, Grosse and Shyu (1992) has been added. There is an indication of departure in the earlier part of the season. Figure 8 seeks to confirm this. It plots weekly totals of home wins by week of the season. There is a trend downwards as the season progresses. Figure 9 plots the residuals against the distance between the towns involved in the game. There is an indication of a bowing upwards.

An analysis of deviance was carried out fitting for the pair of teams in the game the explanatories of: the home team, the away team, day of game, distance between the towns involved and the results (W, T or L) of the teams' preceding game. The results are presented in Table 5. One sees that the visiting team and their previous week's result appear most important. The final deviance is 297.99 on 182-32-2 = 148 degrees of freedom. The degrees of freedom are less than in the previous case because, in order to include the previous week's result, a week of data must be dropped.

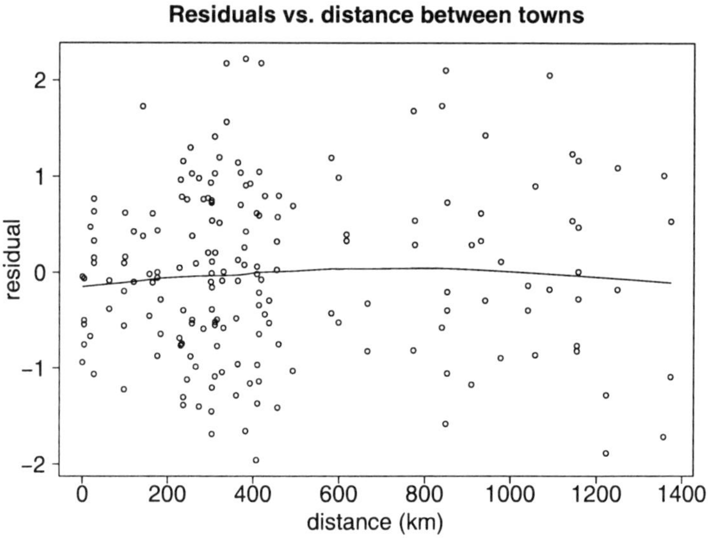

Figure 9 Normal residuals versus distance between towns. A loess line has been added.

Table 5 Analysis of deviance table for the inclusion of explanatories successively

Term	Df	Deviance change
Home team	13	8.76
Away team	13	42.59
Previous home	2	1.77
Previous away	2	4.26
Day	1	2.33
Distance	1	1.00

7 Another model

A simpler model is next considered. Let δ_i denote the strength of team i whether playing at home or away, i.e. assume $\beta = \gamma$.

In the computations expressions (4) are replaced by

$$Prob\{i \ wins \ at \ home \ playing \ j\} = 1 - exp\{-e^{\delta_i - \delta_j - \theta_2}\}$$

and

$$Prob\{i \ loses \ at \ home \ playing \ j\} = exp\{-e^{\delta_i - \delta_j - \theta_1}\}. \tag{6}$$

The fitted values, $\hat{\delta}_i$, are given in Figure 10.

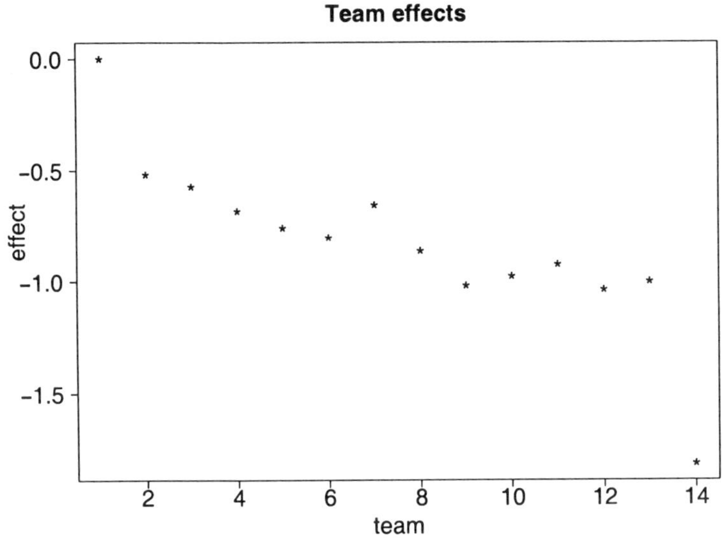

Figure 10 The $\hat{\delta}_i$, of the model (6), are indicated by "*".

One sees in the figure that $\hat{\delta}_1$, the champion's strength, is particularly large while $\hat{\delta}_{14}$, the lowest team's is particularly small. This result is consistent with Figure 2.

The residual deviance of the model (6) is 357.36 with 182-14-2 = 166 degrees of freedom to be compared with the previous 334.8 with 154 degrees of freedom.

8 Uses

The fitted models obtained may be put to some uses. For example one could run Monte Carlos to estimate the probability of each team being champion or of being relegated, as in Lee (1997). Alternately one could examine the effects of a switch from 2 to 3 points for a win, again via simulation, if every thing else remained fixed.

Further, one could use the fitted models to assess various betting strategies. In that connection one referee mentioned that one could fit the model to say the first 20 week's data and then see how well that model predicts the next week's results. The other referee brought up the idea of using the model to rank the teams, but backed away because the design was completely balanced. I can add that if some explanatory with teeth could be found to include in the model, then ranking could proceed. The referee also added that if desired one could fit a single home advantage parameter for all the teams.

9 Extensions

A study was made of the effect of handling omitted variables by including additive random effects in the linear predictor. The resulting model corresponds to a different link function, for example the inverse link function becomes

$$1 - \int exp\{-e^{\eta+\sigma z}\}\phi(z)dz$$

instead of (2) if the effects are assumed to be $IN(0, \sigma^2)$. The resulting $\hat{\sigma}$ turned out to be near 0.

Harville (2003) and Stern (2004) are recent papers concerned with related problems for other sports and might be consulted by the interested reader.

10 Discussion and summary

Conditioning was employed to take the ordinal-valued case to a pair of conditionally independent cases. The advantage was that standard statistical packages became available for the analyses.

One could have modeled the goals and then obtained W-T-L results afterwards but the choice was made to try something different.

The fine away performance of Rosenborg and poor away performance of Bryne are perhaps the most notable features noted.

Acknowledgement

I thank Phil Spector and Ryan Lovett for help with the computing. I thank the two referees for important comments that improved the paper. The research was supported in part by NSF Grants DMS-02-03921 and DMS-05-04162.

Kjell, jeg snakker svaert lite norsk, men takk venn.

References

1. ANDERSEN, J. A. (1984). Regression of ordered categorical values. *J. Royal Statist. Soc.* B 46, 19-35.
2. BELL, C. B. AND DOKSUM, K. A. (1965). Some new distribution-free statistics. *Ann. Math. Statist.* 36, 203-214.
3. BRILLINGER, D. R., UDIAS, A. AND BOLT, B. A. (1980). A probability model for regional focal mechanism solutions. *Bull. Seismol. Soc. Amer.* 70, 149-170.
4. BRILLINGER, D. R. (1996). An analysis of an ordinal-valued time series, pp. 73-87 in *Athens Conf. on Applied Probability and Series Analysis. Volume II: Time Series Analysis.* Lecture Notes in Statistics, vol. 115. Springer-Verlag, New York.
5. BRILLINGER, D. R. (2005). Some data analyses using mutual information. *Brazilian J. Prob. and Statist.* 18, 163-183.
6. BRILLINGER, D. R. AND PREISLER, H. K. (1983). Maximum likelihood estimation in a latent variable problem. Pp. 31-65 in *Studies in Econometrics, Time Series and Multivariate Statistics.* (Eds. S. Karlin, T. Amemiya and L.A. Goodman). Academic Press, New York.
7. CLEVELAND, W.S., GROSSE, E. AND SHYU, W.M. (1992). Local regression models. Pp. 309-376 in *Statistical Models in S.* Eds. J. M. Chambers and T. J. Hastie. Wadsworth, Pacific Grove.
8. DOKSUM, K. A. (1966). Distribution-free statistics based on normal deviates in analysis of variance. *Rev. Inter. Statist. Inst.* 34, 376-388.
9. DOKSUM, K. A. (1974). Empirical probability plots and statistical inference for nonlinear models in the two-sample case. *Ann. of Statistics* 2, 267- 277.
10. DUNN, P. K. AND SMYTH, G. K. (1996). Randomized quantile residuals. *J. Computational and Graphical Statistics* 5, 236-244.
11. DYTE, D. AND CLARKE, S. R. (2000) A ratings based Poisson model for World Cup soccer simulation. *J. Operational Research Soc.* 51, 993-998.
12. FIENBERG, S. E. (1980). *The Analysis of Cross-classified Data.* MIT Press, Cambridge.
13. HARVILLE, D. (2003). The selection or seeding of college basketball or football teams for postseason competition. J. American Statistical Assoc. 98, 17-27.
14. KARLIS D. AND NTZOUFRAS J. (2000). On modelling soccer data. *Student* 3, 229-245.
15. KARLIS, D. AND NTZOUFRAS, J. (2003a) Analysis of sports data Using bivariate Poisson models *The Statistician* 52, 381-393.
16. KARLIS, D. AND NTZOUFRAS, J. (2003b) Bayesian and non-Bayesian analysis of soccer data using bivariate Poisson regression models. *16th Panhelenic Conference in Statistics.* Kavala, April 2003.

17. Lee, A. J. (1997). Modelling scores in the Premier League: is Manchester United *really* the best? *Chance*, 15-19.

18. McCullagh, P. and Nelder, J. (1989). *Generalized Linear Models.* Chapman and Hall, London.

19. Panaretos, V. (2002). A statistical analysis of the European soccer champion league. *Proc. Joint Statistics Meeting.*

20. Pearson, E. S. (1950). On questions raised by the combination of tests based on discontinuous distributions. *Biometrika* 37, 383-398.

21. Pregibon, D. (1980). Discussion of paper by P. McCullagh. *J. Royal Statist. Soc. B* 42, 139.

22. Stern, H. (2004). Statistics and the college football championship. *The American Statistician* 58, 179-185.

PART 1
Survival Analysis

Chapter 2

STOCHASTIC PROCESSES IN SURVIVAL ANALYSIS

Odd O. Aalen and Håkon K. Gjessing

Dept. of Biostatistics, Institute of Basic Medical Sciences
University of Oslo, Oslo, NORWAY

Norwegian Institute of Public Health
Oslo, NORWAY

E-mails: o.o.aalen@medisin.uio.no & hakon.gjessing@fhi.no

The objects studied in survival and event history analysis are stochastic phenomena developing over time. It is therefore natural to use the highly developed theory of stochastic processes. We argue that this theory should be used more in event history analysis. Some specific examples are treated: Markov chains, martingale-based counting processes, birth type processes, diffusion processes and Lévy processes. Some less well known applications are given, with the internal memory of the process as a connecting issue.

Key words: Counting process; Markov chain; Diffusion process; Lévy process; Exploding process; Frailty; Survival analysis.

1 Introduction

The objects studied in survival and event history analysis are stochastic phenomena developing over time. It seems rather obvious that the large body of theory of stochastic processes should have a bearing on the statistical theory. There is probably a wide agreement on this view, but the connection between survival and analysis and stochastic processes should be made much stronger than has so far been generally accepted.

Important parts of stochastic process theory can in fact be connected to survival analysis. These include

- Markov chains

- birth processes
- counting processes and martingale theory
- Wiener processes and more general diffusion processes
- Lévy processes

We shall have a look at these types of processes with a view as to their applicability in survival analysis.

Indeed, the work of Kjell Doksum contains several examples on the use of stochastic processes. He has written a number of papers where an underlying risk process has been modeled as a Wiener process, see e.g. Doksum & Høyland (1992). Furthermore, Doksum was a pioneer in the nonparametric Bayesian approach to survival analysis and introduced the "neutral to the right" prior distributions, which means that the cumulative hazard rates are in fact Lévy processes (Doksum, 1974).

There are two important general aspects of survival analysis which are connected to the use of stochastic processes. One is the issue of *time*. The common regression method in survival analysis, the proportional hazards or Cox regression, is based on an assumption of proportionality. This in effect decouples the statistical analysis from the development over time, implicitly assuming that no changes take place when time passes. Time is relegated to a nuisance parameter instead of being in fact *the* major parameter of survival data. This has numerous implications for the actual practice of survival analysis, since it de-emphasizes the fact that changes over time, e.g. in the effect of covariates, are likely to occur and should be examined and understood. We believe that the time aspect should play a much more central role in survival analysis.

The second general aspect is whether survival and event history data should be analyzed just as they present themselves, or whether one should try to look behind the data even though it may be speculative. The major tradition, like in most of statistics, is very pragmatic. One computes Kaplan-Meier survival curves and runs regression analyses, which are straightforward analyses of the actual observed data. However, there are also attempts at looking below the surface. One example is frailty theory, based on the recognition that some individuals have higher risk than others. Such models will rarely be identifiable for univariate survival data, but nevertheless it may yield considerable insight to speculate about the frailty effects that may be present. We believe that models which allow fruitful speculations on underlying mechanisms should be applied much more than is presently the case. Indeed, speculation is part of the scientific creativity, and statisticians should contribute to this aspect too.

Two types of models for underlying mechanisms will be presented, namely first-passage models and extended frailty models (with stochastic hazard rates). An issue running through several of the models studied is the question of the memory of the process; how well does it remember the past. The distinction between short-term and long-term memory is important in stochastic process theory, and plays a special role in the extended frailty models presented below. Models with long range dependence often result in distributions that are sub-exponential, that is, an eventually declining hazard rate, see e.g. Samorodnitsky (2002).

Why are stochastic processes of importance in survival analysis? The main reason is of course that event histories and the associated covariate histories develop over time, and the theory of stochastic processes is our tool of analyzing such development. The processes mimic some underlying structure, maybe in a superficial fashion. The models are usually not correct descriptions of the phenomena in great detail. They are rather some kind of coarse analogues that may still yield important insights.

However, stochastic processes have a function beyond more or less appropriate mimicking of event and covariate histories. An almost unavoidable aspect of event history observations is the occurrence of censoring. The martingale concept, and associated stochastic integrals, are ideally suited for handling censored processes. This is due to the martingale property being invariant to certain operations that would destroy more classical relations like independence.

2 Probabilistic or statistical assumptions

The common stochastic processes fall into two categories, those that are amenable to detailed probabilistic calculations, and those where the assumed structure gives rise to results of a more conceptual nature. A typical example of the former ones are the Markov processes where probabilistic calculations can be carried out precisely because of the Markov property. An example of the second category would be the martingales, where the basic results are less computational. For example, a major result is the invariance of the martingale property under optional stopping and stochastic integration.

From a statistical point of view, one will sometimes need the probabilistic computational power of a stochastic process. However, quite often this is not relevant. In many cases the dependence, say, on past observations can be arbitrarily complex, that is far from Markov or similar assumptions, and with a possible long-term memory. The important thing is the dependence of the model on statistical parameters, which needs to be tractable. This statistical dependence is an entirely different matter than probabilistic dependence. Then martingale results guarantee (possibly approximate) unbiasedness and asymptotic normality and produces variance formulas.

3 Stochastic processes modelling observed data

3.1 *Markov chains*

The Markov chain is in many ways the simplest type of stochastic process and has long played a major role in biostatistical modelling. In fact, the simplest Markov chain in survival analysis is the competing risks model which goes back to Bernoulli in 1760 and his assessment of the importance of smallpox on mortality. A more recent example of major importance is the Armitage-Doll model for the development of cancer (i.e. carcinogenesis), which appeared in 1954 (Armitage and Doll, 1954). This so-called multi-stage model, is really a simple Markov chain which describes how a cell moves through a number of different stages before becoming cancerous. The model has been a considerable inspiration for understanding the development of cancer.

It is interesting to note that the Armitage-Doll model is quite primitive from a biological point of view. It is rather doubtful that the changes of a cell really constitute a Markov process on a set of well-defined states. Above all, the model ignores cell division and cell death, that is, the whole dynamic process taking place in the tissue. Nevertheless, the model has been quite important, with 555 citations as of 2004. A number of further developments of this model have arisen, incorporating issues like cell division and cell death. Still these models are of Markov type, being related to branching processes. An interesting example is the model of Portier et al (2000). These approaches really constitute complex survival models since the aim is to compute the distribution of time to malignancy, that is the cancer incidence. Hence they demonstrate the application of complex stochastic processes in survival analysis.

A general difficulty with the Markov process is the basic Markov assumption which may appear unrealistic in many cases. For instance, one may not actually believe in the lack of memory property (conditional on the present state). It is important to note, however, that the Markov property may be much less of a restriction than one thinks. This point has been made by Datta and Satten (2001) and Glidden (2002). In fact, the basic Markov tool of multiplying transition matrices often has a validity beyond the Markov framework. Basically, the multiplication of transition matrices is simply a description of the movements of individuals on the chain and does not necessarily depend on probabilistic assumptions. This is connected to the fact that the Markov assumption is really made on the level of individuals. The statistical estimation, on the other hand, is usually taking place on a more aggregate level. One often does not follow individuals, but the estimation is merely dependent on the numbers of individuals present in the various states at any given time, and the transitions that occur for them. Hence, much estimation for Markov chains will have a broader validity than one might think.

3.2 *Counting processes*

The Markov chain assumption implies a highly specific stochastic framework. The details are specified in such a way that explicit probabilistic calculations can be made. For many statistical purposes, however, one is not dependent on such detailed calculations. Rather, the main thing is how the statistical parameters enter the model, and whether features like incomplete observation, often termed censoring in the survival case, can be incorporated. The counting process structure (Aalen, 1978) takes care of precisely these issues; for event histories the basic observations are counts of transitions or events that take place over time. A very fruitful model for such counting processes is defined by considering the intensity processes given the entire past.

Consider for now just a single counting processes $N(t)$, and its intensity process $\lambda(t)$. The intensity process generalizes the intensity of a Poisson process, by letting the intensity be a function of happenings in the past. The interesting probabilistic feature here is not how $\lambda(t)$ depends on the past in a detailed fashion, as it would be for a Markov chain, but the fact that

$$N(t) - \int_0^t \lambda(s)ds$$

is a (local) martingale. This is a very different kind of assumption, it does not allow explicit calculation of probabilities like the Markov assumption does, but it has other properties, like the fact that the martingale property is preserved under stochastic integration of predictable processes. Censoring may be represented as a stochastic integral with respect to a censoring process, and so one has that the all-important martingale property is preserved under the fundamental operation of censoring. In addition, the martingale property in many cases implies asymptotic normality. Usually, asymptotic theory is associated with some underlying independence structure, or a modification of this. The martingale property is a more robust assumption which achieves more or less the same.

The second basic assumption concerns how the parameters enter the model, and a common formulation is the multiplicative intensity model

$$\lambda(t) = Y(t)\alpha(t)$$

where $\alpha(t)$ is the statistical part in a parametric or non-parametric version, while $Y(t)$ is an observable quantity. Again, $Y(t)$ may have an arbitrarily complicated dependence on the past (with some qualification to be discussed below). No assumption of the Markovian or any similar type is of relevance here. The martingale property enters into the estimation and testing since it forms the basis for proving unbiasedness, for computing and estimating variances, and for asymptotic theory.

In the counting process theory there is no difficulty with a long term memory in the process. The martingale property is compatible with complex dependence on the past. It may be useful to model such dependence more explicitly, however, and we shall illustrate this in the next section.

3.2.1 *Counting beyond 1*

In spite of counting process theory now having an almost 30 years history in event history analysis, the actual applications have been very limited in scope. Mostly, the individual counting processes have been counting at most a single event each, which have then been aggregated into a larger counting process. The multivariate survival data, where each individual experiences several events or where related groups of individuals are lumped together, have only to a slight extent been included in the theory. Rather, the multivariate data have been handled by mixed models of the frailty type (see e.g. Hougaard, 2000), which are certainly useful, but which, for instance, have little ability to include changes over time. What has been missing from this picture are individual intensity processes which depend in a more detailed fashion on the past of the individuals. Again the type of dependence is not of the probabilistic type, but of the statistical type. What matters is the type of dependence on statistical parameters, the probabilistic dependence may, in principle, be arbitrarily complex and have long-term memory.

One possibility is to define a model with dynamic covariates. The dynamic covariates may be quantities like the number of previous events for the individual (or for the group), or the time since last event. Numerous possibilities exist along these lines and can be alternatives to the mixed (or frailty) models. Dynamic models have been suggested by several authors (Kalbfleisch and Prentice, 2002; Aalen et al, 2004; Peña and Hollander, 2004; Gandy and Jensen, 2004; Miloslavsky, Keleş and van der Laan, 2004).

The existence of dynamic models follows from a general theorem for submartingales, namely the Doob-Meyer decomposition which states, essentially, that any submartingale can be decomposed into a martingale and a compensator. Since a counting process is a submartingale, there is essentially always an intensity process however the counting processes comes about. For instance, there might be an underlying random effects, or frailty model, of a possibly complex and general nature, nevertheless the whole thing can be reformulated by intensity processes depending on the past.

Simple examples of dynamic models are the Cox type model (e.g. Kalbfleisch and Prentice, 2002):

$$\lambda_i(t) = \alpha(t)\exp(\beta N_i(t-)) \tag{1}$$

or the additive model (Aalen et al, 2004):

$$\lambda_i(t) = \alpha(t) + \beta(t) N_i(t-) \tag{2}$$

where the index i refers to individual process i, and $N_i(t-)$ are the number of occurrences prior to time t in this process. Methods for statistical analysis of such models may be found in Aalen et al (2004) and Fosen et al, (2005). Here we shall focus on a particular aspect of these models which is related to stochastic process theory.

3.2.2 *Dynamic models and explosion*

When introducing more complex dependence on the past into the model, as illustrated above, one has to be careful. Actually, what one is doing is to specify stochastic processes with a particular dynamic structure, e.g. similar to birth processes. Then the question arises whether the process in question is well defined. It turns out that in this respect the two models defined in equations (1) and (2) behave very differently.

Considering first the model (2), then it is clear that this defines a birth process with immigration. If $\alpha(t)$ and $\beta(t)$ are constant, then such a process is well defined and even has explicit solutions. Certainly the process is well defined even for time-varying parameters under weak conditions, and so there is no conceptual difficulty with the additive model.

The Cox type model (1), on the other hand, may run into difficulties. When defining dynamic models one should be aware of the phenomenon of "explosion". A large intensity process leads to many new events in a short interval of time. These events are fed back into the intensity process through the contribution of $N_i(t-)$. This again leads to even more events and eventually the process explodes, i.e. $N_i(t) \to \infty$ when $t \to \tau^-$, where τ is a random time which is finite with positive probability. Such processes are sometimes called "dishonest" processes; Cox and Miller (1965, p. 163) point out that for instance by defining the intensity process as $\lambda_i(t) = N_i(t-)^2$ one gets a dishonest process. Clearly, this also creates potential problems for applying model (1) where the growth in the intensity process due to increasing number of events is stronger than for the square function. In general, let us define the intensity by

$$\lambda_i(t) = h(N_i(t-), t)$$

for some non-negative function $h(x,t)$ of two arguments. The question is what kind of functions h will lead to a well-defined process N_i, and which functions will cause explosions. Mathematically, this is related to the Lipschitz condition used in differential equations theory (Birkhoff and Rota, 1989). Typically, a local Lipschitz condition in the first argument of h guarantees a unique solution up to a certain point in time, but does not exclude the possibility of explosion at a later time. A global Lipschitz condition, however, sets a growth restriction on the behavior of h, and guarantees a unique non-explosive solution. Similar conditions exist for stochastic differential equations driven by Wiener or Poisson processes, that is, including the counting processes considered here (Protter, 1990). Additive processes usually satisfy global Lipschitz conditions, whereas exponentially based processes only satisfy local conditions. In general one has to be careful when defining dynamic models to ensure that they are actually well defined, and this may be a non-trivial issue.

Another general criterion for models to be non-explosive is the Feller condition which can be used on processes where the intensity is only dependent on the number of previous occurrences, i.e.: $\lambda_i(t) = h(N_i(t-))$ for a nonnegative function $h(x)$ (not depending on t). Then non-explosiveness on finite intervals is guaranteed if and only if

$$\sum_{i=1}^{\infty} \frac{1}{h(i)} \text{ diverges,}$$

see, for instance, Allen (2003). The intuitive justification for this criterion is that when the process is in state i its intensity is constant until the next jump, and thus the expected time in that state is $1/h(i)$. The above sum then represents the total expected time used spent in all states. If this quantity converges it seems clear that the process moves faster and faster to new states, and that it will explode in finite time. The Feller condition immediately holds for a linear function $h(\cdot)$, so the linear or additive model will be a safe choice. For a quadratic function $h(i) = i^2$, on the other hand, it is clear that the above sum converges, and there will be explosion with a positive probability. Also for an exponential form $\lambda_i(t) = \exp(\beta N_i(t-))$ it is clear that $\sum_{i=1}^{\infty} \exp(-\beta i)$ converges for all positive β, implying once more a positive probability for explosion on finite time intervals.

The situation is more difficult when h also depends on t. For instance, one may have a Cox model of the type $\lambda_i(t) = \alpha(t)\exp(\beta N_i(t-)/t)$, i.e. $h(x,t) = \alpha(t)\exp(\beta x/t)$. This is a sensible model since it implies that it is the average number of events per time unit that is of importance. We will return to this model shortly, after some general considerations.

Assume that h is a convex function in the first argument. By counting process theory and Jensen's inequality we have

$$\mathrm{E}N_i(t) = \mathrm{E}\int_0^t h(N_i(s-), s)ds = \int_0^t \mathrm{E}h(N_i(s-), s)ds \geq \int_0^t h(\mathrm{E}N_i(s-), s)ds.$$

Consider a function $f(t)$ satisfying

$$f(t) = \int_0^t h(f(s), s)ds. \tag{3}$$

A general comparison theorem for differential equations can be applied (Birkhoff & Rota, 1989, Chapter 1, Section 11). From the theorem it follows that $\mathrm{E}N_i(t) \geq f(t)$. Hence, we may solve (3) and study whether the solution is explosive. Differentiating the equation we get

$$f'(t) = h(f(t), t), \tag{4}$$

with initial condition $f(0) = 0$. If the solution to this differential equation explodes, then the expectation of the process will explode. (The opposite direction is more complex; the process may explode even though the equation above has a non-explosive solution).

Note that the solution to (4) is just what is often termed the deterministic solution, as opposed to the stochastic solution determined by a counting process with intensity process $\lambda_i(t) = h(N_i(t-), t)$. The relationship between deterministic and stochastic solutions is of much interest in areas like population dynamics and the mathematical theory of epidemics (see e.g. Allen, 2003). Often the relationship between stochastic and deterministic solutions is close, but this is not always the case.

Let us first consider the Cox type model $\lambda_i(t) = \alpha(t)\exp(\beta N_i(t-))$ defined in (1). The special case of (4) relevant here is

$$f'(t) = \alpha(t)\exp(\beta f(t)).$$

A solution to this equation with initial condition $f(0) = c$ yields

$$f(t) = -\frac{1}{\beta}\log(e^{-\beta c} - \beta\int_0^t \alpha(s)ds).$$

If $\beta \leq 0$ there is no explosion. For $\beta > 0$ we see that the deterministic solution explodes if $\int_0^t \alpha(s)ds$ reaches $e^{-\beta c}/\beta$ at some finite time. In particular, for $c = 0$ this means that if $\beta\int_0^t \alpha(s)\, ds = 1$ has a solution with t finite, $f(t)$ explodes.

We shall consider in somewhat more detail the special case of (4) corresponding to the Cox type model $\lambda_i(t) = \alpha\exp(\beta N_i(t-)/t)$. We will assume that the process starts at time 1 (just to avoid the singularity at 0), and that $EN(1) = c \geq 0$ is the initial value. The relevant equation is

$$f'(t) = \alpha \exp(\beta f(t)/t), \quad t \geq 1,$$

with $f(1) = c$. When $\beta \leq 0$ no explosion occur, so we will focus on $\beta > 0$. Aided by Mathematica (Wolfram, 1999), or by substituting $a(t) \stackrel{\text{def}}{=} f(t)/t$ and separating the equation, we find the following implicit solution:

$$\int_c^{\frac{f(t)}{t}} \frac{1}{g(u)}\, du = \log(t), \tag{5}$$

where $g(u) \stackrel{\text{def}}{=} \alpha e^{\beta u} - u$, $u \geq 0$. The solution fulfills the initial condition $f(1) = c$. From (5) one may decide whether $f(t)$ explodes or not, by observing that we cannot necessarily solve this equation for $f(t)$ for all values of α, β and t. For some combinations of α and β the left hand side may remain bounded as $t \to \infty$.

To analyze equation (5) in detail, consider the auxiliary function $g(u)$, the denominator of the integrand. We have $g'(u) = \alpha\beta e^{\beta u} - 1$ and $g''(u) = \alpha\beta^2 e^{\beta u}$. Note that $g(0) = \alpha > 0$. Since g'' is strictly positive g is convex and has a unique minimum, which we denote u_0. By setting $g'(u_0) = 0$ we find that $u_0 = -\log(\alpha\beta)/\beta$ and $g(u_0) = 1/\beta - u_0$. There are now three possibilities:

(1) $g(u) > 0$ for all $u \geq 0$. Then the integrand of (5) is non-singular and $\int_c^\infty 1/g(u)\, du < \infty$. For large enough t (5) cannot have a solution for f, and explosion occurs.
(2) $g(u_0) = 0$, i.e. g is tangential to the x-axis. Then $1/g(u)$ has a non-integrable singularity at $u = u_0$. If $0 \leq c < u_0$ there will be no explosion, if $c > u_0$ an explosion will occur. In the very special case $c = u_0$ the solution (5) is not valid but is replaced by the simple solution $f(t) = ct$, and no explosion.
(3) $g(u_0) < 0$. Then g has two zeros u_1 and u_2, $u_1 < u_2$, and the integrand has non-integrable singularities at these two values. Accordingly, if $0 \leq c < u_1$ or $u_1 < c < u_2$ there is no explosion. If $c > u_2$ the solution explodes. If $c = u_1$ or $c = u_2$ there is no explosion, as above.

In conclusion: There is explosion if g has no zero, or if c is larger than the largest zero of g. This translates into saying that there is an explosion in finite time if either $g(u_0) > 0$ or ($g(c) > 0$ and $g'(c) > 0$), i.e. that $\alpha\beta > e^{-1}$ or ($\alpha e^{\beta c} > c$ and $\alpha\beta e^{\beta c} > 1$). Note in particular that when the starting level c is large enough, then the second condition is necessarily fulfilled. As a numerical illustration, put $c = 1$ and $\alpha = 1$. Then explosion in finite time occurs if $\beta > e^{-1} = 0.368$.

4 Stochastic processes modelling underlying developments

In statistics one often assumes the existence of unobserved random variables or processes. For instance, the latent variables of mixed models are useful conceptual and practical tools. The corresponding concept in survival analysis is termed frailty models. Thinking in terms of frailty is useful even in cases where the frailties are completely unobservable (Aalen, 1994). We shall present a considerable extension of the frailty model below.

Another example of unobserved, latent concepts in survival analysis are the underlying processes in first-passage time models. Here the time of the event in question is assumed to correspond to a process crossing a certain level. In particular, models of this nature are of importance when one attempts to understand the shape of a hazard rate, see Aalen and Gjessing (2001). Here we shall focus on the concept of quasi-stationarity.

4.1 *First-passage time models. Quasi-stationarity*

One may think of the occurrence of an event as resulting when some underlying risk process crosses a certain limit. This has been suggested by a number of authors, including Doksum (Doksum and Høyland, 1992), but we shall here focus on an aspect of these models which is not well known.

Consider a number of independent individuals moving on some (unobserved) state space. The space is divided in two parts, the transient space prior to the event occurring, and the absorbing space the entrance into which means that the event in question has occurred. Clearly the population of individuals on the transient space will diminish as more and more become absorbed. However, in many cases a phenomenon termed quasi-stationarity will occur, that is, the expected distribution of individuals on the transient space will converge to a limiting distribution. This is not a stationary distribution since absorption of probability mass takes place all the time, and so it is termed quasi-stationary.

4.1.1 *Quasi-stationarity for diffusion processes*

We shall consider a first-passage time for a diffusion process, which leads to a useful and different view of the hazard rate. Consider a Markovian diffusion process $X(t)$ on the positive half line with zero as the absorbing state, and let the event time be defined as $T = \inf_{t \geq 0}\{t : X(t) = 0\}$, the time until absorption. Let $\varphi_t(x)$ be the density on the transient state space, i.e. $P(X(t) \in dx)$ at time t, $x > 0$, and let $\sigma^2(x)$ and $\mu(x)$ be the variance and drift diffusion coefficients respectively. The evolution forward in time is described by Kolmogorov's forward equation (Karlin and Taylor, 1981,

p. 220):

$$\frac{\partial}{\partial t}\varphi_t(x) = \frac{1}{2}\frac{\partial^2}{\partial x^2}\left[\sigma^2(x)\varphi_t(x)\right] - \frac{\partial}{\partial x}\left[\mu(x)\varphi_t(x)\right].$$

Assume that the process is in a quasi-stationary state. Then one can write $\varphi_t(x) = e^{-\theta t}\psi(x)$, where θ is the constant hazard rate and $\psi(x)$ is the quasi-stationary distribution. Insertion into the above equation yields

$$-\theta\psi(x) = \frac{1}{2}\frac{\partial^2}{\partial x^2}\left[\sigma^2(x)\psi(x)\right] - \frac{\partial}{\partial x}\left[\mu(x)\psi(x)\right].$$

This is an eigenvalue equation which in some instances can be solved explicitly for the quasi-stationary distribution and the corresponding constant hazard rate θ.

Consider the process prior to quasi-stationarity, and let θ_t denote the hazard rate of the time to absorption. Let $\psi_t(x) = P(X(t) \in dx | X(t) > 0)$ denote the density on transient space, conditioned on non-absorption, so that $\int_0^\infty \psi_t(x)\, dx = 1$. We can write

$$\varphi_t(x) = \exp(-\int_0^t \theta_s\, ds)\, \psi_t(x)$$

for the connection between the non-conditioned and conditioned densities. The following result holds under suitable regularity assumptions:

$$\theta_t = \frac{\sigma^2(0)}{2}\psi_t'(0)$$

that is, the hazard rate is proportional to the slope of the normalized density at zero (Aalen and Gjessing, 2001), see Figure (1). Note that $\psi_t(x)$ can be considered the risk distribution of survivors in the context of survival analysis.

Hence this diffusion model gives a different representation of the hazard rate than the common one. The derivative of $\psi_t(x)$ at time zero depends on the absorption and the diffusion in the process. We can say that the shape of the hazard rate is created in a balance between two forces: the attraction of the absorbing state and the general diffusion within the transient space. It turns out that the various common shapes of hazard rates occur naturally depending on how the starting distribution on the transient state space relates to the *quasi-stationary distribution*. Simplifying quite a bit, one could say that the shape of the hazard rate depends on the distance between the starting point, or starting distribution, and the state of absorption. A great distance leads to an increasing hazard rate, an intermediate distance

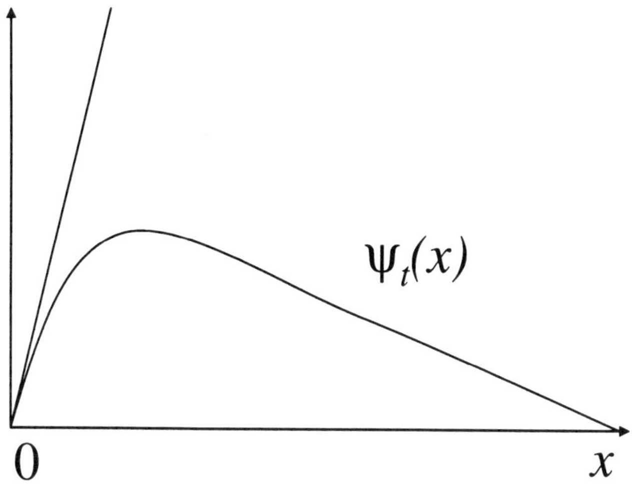

Figure 1 *The hazard rate of time to absorption is proportional to the derivative at 0 of the distribution of survivors.*

leads to a hazard rate that is first increasing and then declining, and a small distance leads to an (essentially) decreasing hazard rate.

We shall illustrate this by considering the special case of a Wiener process with drift and absorption in state 0. In this case the variance and drift coefficients σ^2 and μ are independent of x. We assume $\mu < 0$, that is, a drift towards zero. The first-passage time till absorption given start in a fixed state follows the well known inverse Gaussian distribution.

It is known that in this case there is a whole class of quasi-stationary distributions. One of those is the limiting distribution of survivors given a fixed initial state, sometimes called the canonical quasi-stationary distribution. This is given by the following gamma distribution:

$$\phi_{\mathrm{can}}(x) = \frac{\mu^2}{\sigma^4}\, x \exp\left(-\frac{\mu\, x}{\sigma^2}\right)$$

and yields a constant hazard of absorption equal to $\theta_0 = (\mu/\sigma)^2/2$. The more general quasi-stationary distributions are given by:

$$\phi(x;\theta) = \frac{\theta}{\eta}\left(\exp\left(-\frac{\mu-\eta}{\sigma^2}x\right) - \exp\left(-\frac{\mu+\eta}{\sigma^2}x\right)\right)$$

for $0 < \theta < \theta_0$, where we define $\eta = \sqrt{\mu^2 - 2\,\sigma^2\,\theta}$. Here the constant hazard of absorption equals θ.

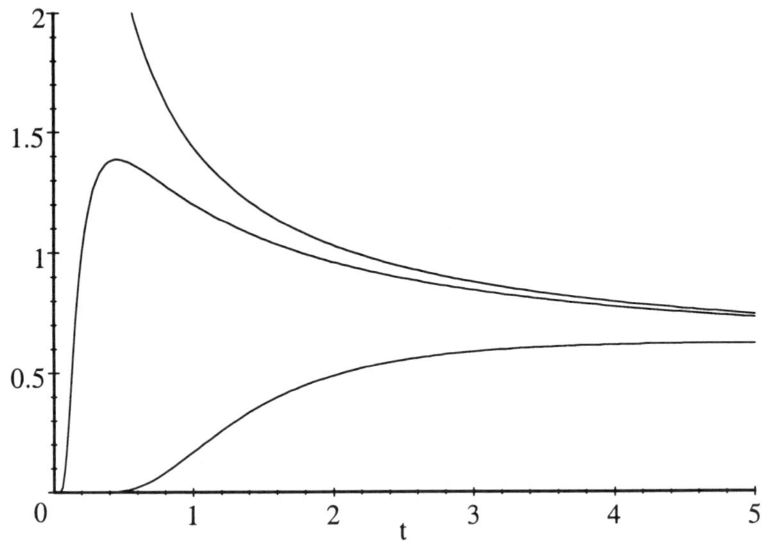

Figure 2 *Hazard rates for time to absorption when process starts out in c=0.2 (upper curve), c=1 (middle curve) and c=3 (lower curve). In all cases $\mu = 1$ and $\sigma^2 = 1$.* (Reprinted from Aalen and Gjessing (2001) by permission of the authors).

Starting out at some level c it is well known that the time to absorption is determined by the inverse Gaussian distribution. The shape of the hazard rate of this distribution for various values of c is shown in Figure 2. The values of c are placed at the beginning of the quasi-stationary distribution, close to the mode of the distribution, and in its tail. We see the following from the figure: If c is close to zero compared to the quasi-stationary distribution one gets, essentially, a decreasing hazard rate; a value of c far from zero gives essentially an increasing hazard rate; while an intermediate value of c yields a hazard that first increases and then decreases. The wording "essentially" is used here because the continuous nature of the model and the non-compact state space yield hazard rates that will, strictly speaking, always increase to a maximum and then decrease, but for c small or large they can be seen as just decreasing or just increasing for most practical purposes.

This relationship is a quite general phenomenon (Aalen and Gjessing, 2001, 2003). It explains the various shapes of the hazard rate in an alternative fashion to that derived from e.g. frailty considerations.

4.2 An extended frailty model: Frailty as a stochastic process

A basic fact of life is that individuals are dissimilar. From a medical viewpoint there is considerable variation in the risk of developing various diseases, and in the prognosis for patients. This variation may be due to genetics, lifestyle or other factors. Some of these factors may be controlled in a statistical analysis, while others are unknown. Such unknown factors are usually counted as a part of the error terms, but when considering processes developing over time, selection effects and artefacts may appear as a result of the unexplained variation (see e.g. Aalen, 1994). In particular, the shape of hazard functions may be strongly influenced, so that, for instance, the observed hazard may be pulled down to increase much more slowly, or even decrease, compared to what would have been observed in a homogenous group.

In the standard frailty model, frailty is assumed to be given at time zero, and to follow an individual throughout life. No changes in frailty takes place. This is clearly a gross simplification, and it might be interesting to try models which are more flexible. From a biological point of view one would think that some aspects of frailty are given early in life and stays with the individual throughout life, as for instance genetic factors. Other aspects of frailty may be determined by more or less random developments and events happening later in life, i.e. the general stresses of life.

One flexible generalized frailty model is discussed by Gjessing et al (2003), where frailty is generated by a stochastic process. More precisely, we consider frailty distributions defined by a nonnegative Lévy process $Z(t)$ (also called a "subordinator") the Laplace transform of which is given by the Lévy–Khintchine formula

$$L(c;t) = E\exp\{-cZ(t)\} = \exp\{-t\Phi(c)\}.$$

The function $\Phi(c)$ is called the Laplace exponent of the Lévy process. The family of Lévy processes contains a number of important special cases, like compound Poisson processes, gamma processes, stable processes etc. In fact, all nonnegative Lévy processes are limits of compound Poisson processes.

To consider processes with a varying "rate", define the nonnegative deterministic rate function $r(t)$ with integral $R(t) = \int_0^t r(u)\,du$, and let $Z(R(t))$ be the time-transformed subordinator. Conditional on Z, we define our hazard rate processes h as

$$h(t) = \lambda(t)\int_0^t a(u, t-u)\,dZ(R(u)), \tag{6}$$

where $a(u, t-u)$ is a nonnegative weight function, and $\lambda(t)$ is a basic hazard rate.

A number of results have been proven for this model in Gjessing et al (2003). An important quantity is the population hazard rate $\mu(t)$, by which we mean the average hazard rate among survivors at a given time. If T is the lifetime of an individual, we define $\mu(t) = E[h(t)|T > t]$. In our model we may derive the following expressions for the population survival and hazard functions:

$$S(t) = \exp(-\int_0^t \Phi(b(u,t))\, r(u)\, du)$$

$$\mu(t) = \lambda(t) \int_0^t \Phi'(b(u,t))\, a(u, t-u)\, r(u)\ du, \tag{7}$$

where $b(u,t) = \int_u^t \lambda(s)\, a(u, s-u)\, ds$.

Another important result concerns quasi-stationarity. We shall make the following conditions:

(1) $\lambda(\infty) = \lim_{t\to\infty} \lambda(t)$, $r(\infty) = \lim_{t\to\infty} r(t)$ and $a(\infty, v) = \lim_{t\to\infty} a(t, v)$ all exist and are finite.
(2) $E[Z(1)] < \infty$ and $a(t, v) \leq \tilde{a}(v)$ for some function $\tilde{a}$ with $\int_0^\infty \tilde{a}(s)\, ds < \infty$.

Under these conditions, a quasi-stationary distribution exists for $h(t)$, conditional on $T > t$, as $t \to \infty$. This implies that $\mu(t)$ converges to a limit.

We shall consider now some important special cases. The first is a formulation of the standard frailty model, by which we mean that frailty is determined at the beginning, and then not changing later on. To fit in the present framework, we let frailty be generated over some initial finite time interval. Our next model generalizes this frailty model by letting the weight function $a(u, v)$ be only dependent on the first argument u. This means that frailty contributions are accumulated along the way and nothing is forgotten. In fact, both these models preserves the memory of previous frailty.

Our third model is a moving average formulation defined by letting $a(u, v)$ be only dependent on the second argument v. Under an integrability assumption on the weight function, the moving average model implies that the past is gradually forgotten, and quasi-stationarity is achieved as indicated above.

In fact the memory of past events is a major issue in frailty models. In medicine and biology properties determined by genetics would be expected to have a long term effect, while many other events throughout life may have a more limited effect and be "forgotten" over time. It is reason to believe that the standard frailty models, with all frailty placed at time zero and perfectly remembered, exaggerates the effects of frailty. We shall give examples illustrating this issue.

4.2.1 *Special case 1: Standard frailty model*

We assume $a(t,v) \equiv 1$. Let $r(t)$ be equal to ρ up to time T and 0 after this time, and assume that $\lambda(t)$ is equal to 0 up to time T. From the general model (6) it follows that the hazard process equals

$$h(t) = \lambda(t)\, Z(\rho T), \quad t \geq 0.$$

The population hazard rate is $\mu(t) = \rho\,\lambda(t)\,\Phi'(\Lambda(t))$, $t \geq 0$, where $\Lambda(t) \stackrel{\text{def}}{=} \int_0^t \lambda(s)\,ds$. We recognize the hazard rate of the standard frailty model, where the frailty distribution is generated by a Lévy process, as are almost all common frailty distributions. For instance, the very broad PVF distributions described in Hougaard (2000) are distributions of Lévy processes.

4.2.2 *Special case 2: Cumulative frailty model*

Let $a(t,v) \equiv a_1(t)$ depend only on the *first* argument, let $r(t) \equiv \lambda(t) \equiv 1$. We have

$$h(t) = \int_0^t a_1(u)\,dZ(u),$$

so that frailty is accumulated as time passes. This is a reasonable model for a situation where frailty is gradually building up throughout life and all previous frailty contributions are remembered.

For an explicit result, consider the gamma Lévy process with Laplace exponent $\Phi(c) = \rho\{\log(\nu + c) - \log\nu\}$. Assume $r(t) \equiv \lambda(t) \equiv 1$ and $a_1(t) = a/t$. Then $\mu(t)$ can be computed from (7) to yield:

$$\mu(t) = \rho\frac{\log a - \log\nu}{a - \nu}$$

which is seen to be a constant rate.

In general, however, it seems that $\mu(t)$ will eventually decrease. Examples are given below.

4.2.3 *Special case 3: Moving average frailty model*

Let $a(t,v) \equiv a_2(v)$ depend only on the *second* argument, let $r(t) \equiv \lambda(t) \equiv 1$. We have

$$h(t) = \int_0^t a_2(t-u)\,dZ(u) = \int_0^t a_2(v)\,dZ(t-v)$$

which is seen to be a moving average process. Note that here the past will be gradually forgotten if $a(v)$ decreases over time. Define $A(v) = \int_0^v a_2(u)\,du$. Then $b(u,t) = A(t-u)$ and

$$S(t) = \exp(-\int_0^t \Phi(A(v))\,dv) \quad \text{and} \quad \mu(t) = \Phi(A(t)).$$

Note that $\mu(t)$ is increasing. It is clear that if either Φ is bounded (i.e. Z is compound Poisson,) or $A(\infty) \stackrel{\text{def}}{=} \lim_{t\to\infty} A(t) < \infty$, then $\lim_{t\to\infty} \mu(t) = \Phi(A(\infty)) < \infty$, so that the hazard converges to a limit, which means that there is a quasi-stationary distribution for the hazard of survivors.

Let us now consider the special case $a_2(v) = ae^{-\kappa v}$. Then $A(v) = a(1-e^{-\kappa v})/\kappa$ and we can prove that

$$\begin{aligned} \mu(t) &= \Phi(A(t)), \\ Var[h(t)|T > t] &= a\Phi'(0) - a_2(t)\Phi'(A(t)) - \kappa\Phi(A(t)) \\ &= \mu'(0) - \mu'(t) - \kappa\mu(t). \end{aligned}$$

4.2.4 *Special case 4: Frailty model with no memory*

Here we shall assume that the frailty equals instantaneous jump of the Lévy process. Assume that $a(t,v)$ depends only on the argument v, and that it equals the Dirac delta function in this argument. Then it can be proven that

$$\mu(t) = r(t)\,\Phi(\lambda(t)).$$

Clearly, this is a model where there is no recollection of past frailty. Notice that frailty nevertheless has an influence on the shape of the hazard rate, since in general $\mu(t)$ will have a different shape from $\lambda(t)$.

4.2.5 *Example*

We shall give illustrations for special cases 2 and 3. Assume that $a_1(t) = \exp(-t)$ and $a_2(v) = \exp(-v)$. Thus both weight functions are exponentially decreasing, but with the difference that the first one starts from time 0 and weights the process forward in time, while the second one starts from the present time and weights the process backwards in time. We consider two different forms for the basic hazard rate, namely a constant one ($\lambda(t) = 1$) and an increasing one ($\lambda(t) = t^2$). We use the Gamma process as the driving frailty process. Population hazard rates for these models are shown in Figures 3 and 4.

One sees that shapes of the population hazard rates for the cumulative frailty model (left panels in the figures) is very different from the basic hazard rate. After reaching a top, the population hazard rate turns down

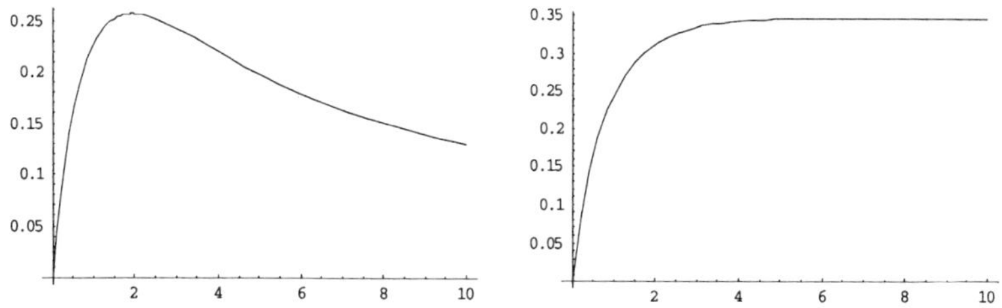

Figure 3 *Population hazard rates for cumulative frailty model (left panel) and moving average frailty model (right panel). Constant basic hazard ($\lambda(t) = 1$). Gamma frailty with shape parameter 0.5 and scale parameter 1; $a_1(t) = e^{-t}$; $a_2(v) = e^{-v}$.*

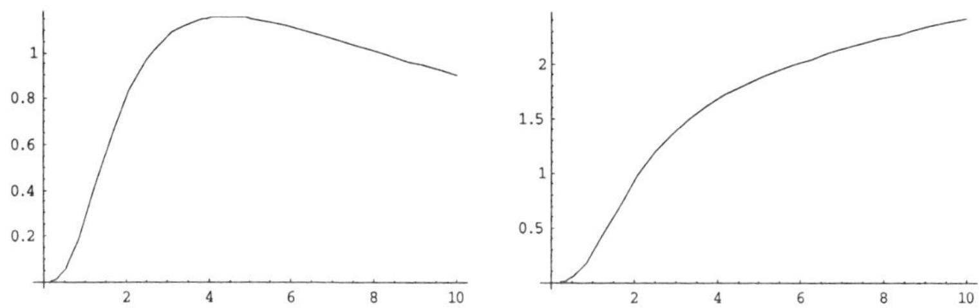

Figure 4 *Population hazard rates for cumulative frailty model (left panel) and moving average frailty model (right panel). Increasing basic hazard ($\lambda(t) = t^2$). Gamma frailty with shape parameter 0.5 and scale parameter 1; $a_1(t) = e^{-t}$; $a_2(v) = e^{-v}$.*

in both cases. This feature is similar to what is often seen in standard frailty models and is due to the long term memory of the cumulative frailty model.

The right panels show the population hazard rates for the moving average model. In Figure 3 the population hazard converges to a constant, hence assuming eventually the same shape as the basic hazard rate. This is due to quasi-stationarity. In Figure 4 the population has a continuous increase, but much more slowly than the basic hazard rate t^2.

We see clearly that the effects of frailty on the hazard rates depend strongly on the degree of memory in the frailty process. The very strong effects often seen in standard frailty models is due to the long-term memory of such models and might be considerably weaker in frailty models with less memory.

A general result on asymptotic constancy of the hazard rate, due to quasi-stationarity, was given already by Keilson (1966). As he points out, the phenomenon is closely related to the process losing its memory when time passes. This is true for Markov processes and regenerative processes, and also for the model presented here. If the process preserves the memory of early events, the hazard will generally not become constant. Hence, the asymptotic behavior of hazard rates depends on whether previous effects are retained in the system, and our example illustrates this.

5 Conclusion

We have illustrated how theory from stochastic processes may illuminate various aspects of survival analysis. Both practical statistical methods and useful qualitative insights may be derived.

References

1. AALEN, O. O. (1978). Nonparametric inference for a family of counting processes. *The Annals of Statistics* **6**, 701-726.
2. AALEN, O. O. (1994). Effects of frailty in survival analysis. *Stat. Methods Med. Res.* 3 227-43.
3. AALEN, O. O. AND GJESSING, H. (2001). Understanding the shape of the hazard rate. *Statistical Science,* 16, 1-22.
4. AALEN, O.O AND GJESSING, H. K. (2003). A look behind survival data: Underlying processes and quasi-stationarity. In: Lindqvist, B. H., Doksum, K. A. (Eds.), *Mathematical and Statistical Methods in Reliability*, pp. 221–34. World Scientific Publishing, Singapore.
5. AALEN, O. O., FOSEN, J., WEEDON-FEKJAER, H., BORGAN, O. and HUSEBYE, E. (2004). Dynamic analysis of multivariate failure time data. *Biometrics*, 60, 764-773.
6. ALLEN, L. J. S. (2003). *Stochastic processes with applications in biology,* Pearson Prentice Hall.
7. ARMITAGE, P. AND DOLL, R. (1954) The age distribution of cancer and a multi-stage theory of carcinogenesis. *Br. J. Cancer*, 8, 1-15.
8. BIRKHOFF, G. AND ROTA, G.-C. (1989). *Ordinary Differential Equations.* Wiley, New York.
9. COX, D. R. AND MILLER, H. D. (1965). *The Theory of Stochastic Processes.* Methuen, London.
10. DATTA, S. AND SATTEN, G. A. (2001). Validity of the Aalen–Johansen estimators of stage occupation probabilities and Nelson–Aalen estimators of integrated transition hazards for non-Markov models. *Statistics & Probability Letters*, 55, 403 – 411
11. DOKSUM, K. A. (1974). Tailfree and neutral random probabilities and their posterior distributions. *Ann. Probab.* 2, 183-201.

12. DOKSUM K. A. AND HØYLAND A. (1992). Models for variable-stress accelerated life testing experiments based on Wiener processes and the inverse gaussian distribution. *Technometrics*, 34: 74-82.

13. FOSEN, J., BORGAN, O., WEEDON-FEKJAER, H. and AALEN, O. O. (2005). Path analysis for survival data with recurrent events. Unpublished manuscript.

14. GANDY, A. AND JENSEN, U. (2004). A Nonparametric Approach to Software Reliability, *Applied Stochastic Models in Business and Industry*, **20**, 3-15.

15. GJESSING H. K., AALEN O. O. AND HJORT N. L. (2003). Frailty models based on Lévy processes. *Advances in Applied Probability.* 35(2), 532–550.

16. GLIDDEN, D. V. (2002). Robust Inference for Event Probabilities with Non-Markov Event Data, *Biometrics*, 58, 361-368.

17. HOUGAARD, P. (2000). *Analysis of Multivariate Survival Data,* Springer-Verlag, New York.

18. KALBFLEISCH, J. D. AND PRENTICE, R. L. (2002). *The Statistical Analysis of Failure Time Data.* Wiley, New Jersey.

19. KARLIN, S. AND TAYLOR, H. M. (1981). *A Second Course in Stochastic Processes.* Academic Press, New York.

20. KEILSON, J. (1966). A limit theorem for passage times in ergodic regenerative processes. *Ann. Math. Statist.* 37, 866-870.

21. MILOSLAVSKY, M., KELEŞ, S. AND VAN DER LAAN, M. J. (2004). Recurrent events analysis in the presence of time-dependent covariates and dependent censoring, *J. R. Statist. Soc. B*, 66, 239–257.

22. PEÑA, E. A. AND HOLLANDER, M. (2004). Models for Recurrent Phenomena in Survival Analysis and Reliability. In: *Mathematical Reliability: An Expository Perspective*, edited by T. Mazzuchi, N. Singpurwalla and R. Soyer, pp. 105-123, Kluwer.

23. PORTIER, C. J, SHERMAN, C. D. AND KOPP-SCHNEIDER, A. (2000). Multistage stochastic models of the cancer process: A general theory for calculating tumor incidence. *Stochastic Environmental Research and Risk Assessment,* 14, 173-179.

24. PROTTER, P. (1990). *Stochastic Integration and Differential Equations: A New Approach.* Springer-Verlag, Berlin.

25. SAMORODNITSKY, G. (2002). Long range dependence, heavy tails and rare events. Lecture Notes Maphysto, Centre for Mathematical Physics and Stochastics, Aarhus.

26. WOLFRAM, S. (1999). *The Mathematica book.* 4th edition. Wolfram Media/ Cambridge University Press.

Chapter 3

CORRESPONDENCES BETWEEN REGRESSION MODELS FOR COMPLEX BINARY OUTCOMES AND THOSE FOR STRUCTURED MULTIVARIATE SURVIVAL ANALYSES

Nicholas P. Jewell

Division of Biostatistics, School of Public Health
University of California, Berkeley, CA 94720, U.S.A.

E-mail: jewell@stat.berkeley.edu

Doksum and Gasko (1990) described a one-to-one correspondence between regression models for binary outcomes and those for continuous time survival analyses. This correspondence has been exploited heavily in the analysis of current status data (Jewell and van der Laan 2004), Shiboski (1998)). Here, we explore similar correspondences for complex survival models and categorical regression models for polytomous data. We include discussion of competing risks and progressive multi-state survival random variables.

Key words: Competing risk; Current status data; Multinomial logit model; Proportional odds; Sequential logit model.

1 Introduction

Consider a continuous time survival response T that is measured on an individual with a known p-dimensional set of covariates $\mathbf{Z} = (Z_1, \ldots, Z_p)$. A survival regression model focuses on the relationship between the (conditional) distribution function $F_{\mathbf{z}}$, of T, given $\mathbf{Z} = \mathbf{z}$, and $\mathbf{z}$. Examples of such models include the Cox model (Cox 1972) and the proportional odds model (Bennett 1983); in each of these, the model can be made fully parametric if both the regression relationship and a baseline version of F are parametrically described, semi-parametric if only one of these is parametrically modeled, or nonparametric if both are only loosely specified.

At a fixed value t, a binary characteristic is defined by the event $T < t$ which occurs with probability $p_t = \Pr(T < t)$. A survival time regression model for T automatically induces a binary regression model relating $p_{t;\mathbf{z}} = \Pr(T < t|\mathbf{Z} = \mathbf{z})$ to both t and $\mathbf{z}$. Doksum and Gasko (1990) examine this correspondence in detail for several familiar survival models including those noted above. One simple example is the proportional odds model for which

$$F_{\mathbf{z}}(t) = \frac{e^{\alpha(t)+\beta\mathbf{z}}}{1+e^{\alpha(t)+\beta\mathbf{z}}} \tag{1}$$

where $\alpha(t)$ is a non-decreasing function with $\alpha(0) = -\infty, \alpha(\infty) = \infty$, and where β is a p-dimensional vector of regression coefficients. With this model

$$\log \frac{p_{t;\mathbf{z}}}{1-p_{t;\mathbf{z}}} = \alpha(t) + \beta\mathbf{z},$$

a logistic regression model in $\mathbf{z}$ with 'intercept' $\alpha(t)$. As noted, if F_0 is assumed to follow a parametric model (for example, the log-logistic distribution, Kalbfleisch and Prentice 2002, Chapter 2.2.6), then this logistic model is also fully parametric (with $\alpha(t) = a + b\log t$ in the log-logistic distribution case with a and $b > 0$ suitably chosen).

This correspondence between survival time and binary outcome regression models has been heavily exploited in the analysis of current status data. Current status observation refers to a form of incompleteness in that the data consists of independent observations on the random variable $(C, \Delta = I(T < C), \mathbf{Z})$ instead of $(T, \mathbf{Z})$. Here, C can be random or deterministic, and is usually referred to as the *monitoring time*; it is typically assumed that C is independent of T (conditionally on $\mathbf{Z}$ in the regression setting) and is uninformative. The variable Δ indicates the *current status* of an individual at time C, namely whether $T < C$ or not. A review of various forms and examples of current status data is given in Jewell and van der Laan (2004) and the references contained therein.

As follows from the work of Doksum and Gasko (1990), a regression model for the (unobserved) T immediately leads to a binary regression model for the observed binary outcome Δ with C included as an additional covariate along with $\mathbf{Z}$. For example, with the proportional odds model for T in (1), the model for Δ is given by the logistic regression

$$\log \frac{p_{c;\mathbf{z}}}{1-p_{c;\mathbf{z}}} = \alpha(c) + \beta\mathbf{z}, \tag{2}$$

where $p_{c;\mathbf{z}} = \Pr(\Delta = 1|C = c, \mathbf{Z} = \mathbf{z})$, and $\alpha(c)$ is necessarily non-decreasing in c with $\alpha(0) = -\infty$ and $\alpha(\infty) = \infty$.

There are two primary properties of this correspondence of regression models for current status data that make it particularly useful. First, for many standard survival models, the effects of C and $\mathbf{Z}$ are additive in the binary regression setting when the appropriate link function is used. This is illustrated by the proportional odds model when a logistic link is used for Δ as shown in (2). A similar property holds for the proportional hazards model with the complementary log-log link for Δ (see Jewell and van der

Laan 2004). Second, the parameter β in the binary regression model for the observed Δ has an immediate interpretation in the survival regression model for the unobserved T. For example, in the logistic regression model (2), the regression coefficient β_k is nothing more than the log odds ratio of failure by time t, associated with a unit increase in the kth component of $\mathbf{Z}$ (holding other component variables constant), as given by the original proportional odds model (1). Although this assumes no interaction terms in $\mathbf{Z}$, the ideas and interpretations immediately generalize to more complex models. For more detail on the application of the proportional odds model in the context of current status data see Rossini and Tsiatis (1996).

The purpose of this article is to examine analogous correspondences between regression models for more complex survival data and their current status counterparts. We pay specific attention to two settings: (i) competing risks survival models which naturally lead to unordered polytomous current status outcomes (Section 2), and (ii) progressive multi-state survival models, corresponding to ordinal categorical current status outcomes (Section 3). While such correspondences naturally extend to these more complex settings, we show that the attractive features of additivity in C and $\mathbf{Z}$, and interpretability of regression coefficients, only can be guaranteed with additional assumptions, at least in the models considered here. In Section 4, we therefore briefly discuss the advantages (and disadvantages) of modeling marginal distributions separately using the simpler survival and binary regression connections for standard current status data.

2 Competing Risks Survival Models and Polytomous Regression Models

Competing risks survival data arise in situations where, in addition to observations on the failure time T, there is also information on a categorical variable J which takes on the values $1, \ldots, m$, and represents the cause or type of failure at time T. It is standard to assume that all failures are associated with one and only one value of J. The joint distribution of the random variable (T, J) is of primary interest. See Crowder (2001) for a recent treatment of the topic.

The cause-specific hazard function for cause $J = 1, \ldots, m$, (Kalbfleisch and Prentice 2002) is defined by

$$\lambda_j(t) = \lim_{h \to 0} h^{-1} \text{Pr}[t \leq T < t + h, J = j | T \geq t].$$

Related to these cause-specific hazards are the sub-distribution functions of primary interest given by

$$F_j(t) = \text{Pr}(T < t, J = j), \qquad j = 1, \ldots, m$$

with the overall survival function then

$$S(t) = 1 - \sum_{j=1}^{m} F_j(t).$$

Note that the cause-specific density function

$$f_j(t) = \lim_{h \to 0} h^{-1}\Pr[t \le T < t+h, J = j],$$

is the derivative of F_j. Finally, these functions are related through

$$f_j(t) = \lambda_j(t)F(t)$$

where $F(t) = 1 - S(t) = F_1(t) + \cdots + F_m(t)$.

We now introduce the covariate $\mathbf{Z}$ into the notation, writing for example

$$F_j(t; \mathbf{z}) = \Pr(T < t, J = j | \mathbf{Z} = \mathbf{z}),$$

for $j = 1, \ldots, m$, with

$$F_0(t; \mathbf{z}) = \Pr(T \ge t | \mathbf{Z} = \mathbf{z}) = S(t; \mathbf{z}).$$

Before further discussion of regression models, it will be helpful to introduce an alternative description of the joint distribution of (T, J). For each j, let α_j be a non-decreasing function on $[0, \infty)$ for which $\alpha_j(0) = -\infty$ and $\alpha_j(\infty) = \infty$. Further, assume that these m functions are commensurate in the sense that the functions

$$\frac{e^{\alpha_j(t)}}{1 + \sum_{k=1}^{m} e^{\alpha_k(t)}} \tag{3}$$

are non-decreasing, for $j = 1, \ldots, m$. Then

$$F_j(t) = \frac{e^{\alpha_j(t)}}{1 + \sum_{k=1}^{m} e^{\alpha_k(t)}} \tag{4}$$

define sub-distribution functions for m competing risks. Note that solving (4) yields the inverse relationships

$$\alpha_j(t) = \log\left[\frac{F_j(t)}{S(t)}\right] \tag{5}$$

for $j = 1, \ldots, m$. We can thus characterize the joint distribution of (T, J) equally well in terms of either $\{\alpha_1, \ldots, \alpha_m\}$ or $\{F_1, \ldots, F_m\}$, with the appropriate constraints on either set of functions.

We are now in a position to describe a natural regression model for $F_1, \ldots, F_m$. For each $j = 1, \ldots, m$ and each covariate value $\mathbf{z}$ we write

$$F_j(t; \mathbf{z}) = \frac{e^{\alpha_j(t) + \beta_j \mathbf{z}}}{1 + \sum_{k=1}^{m} e^{\alpha_k(t) + \beta_k \mathbf{z}}}, \tag{6}$$

where β_j is a $1 \times p$ vector of regression coefficients. This model introduces the key additive separation of the effects of t and z on the sub-distribution functions that we noted was valuable in the standard setting. We refer to this model as the *proportional odds model with competing risks* as it generalizes the model of the same name in the single risk setting (Bennett 1983). Before proceeding further, however, for (6) to describe a set of sub-distribution functions, we need the functions $\{\alpha_j^* = \alpha_j(t) + \beta_j \mathbf{z} : j = 1, \ldots, m\}$ to satisfy the constraints, described by (3), assuming that $\{\alpha_j : j = 1, \ldots, m\}$ do. Trivially $\{\alpha_j^*; j = 1, \ldots, m\}$ possess the same limits as $\{\alpha_j; j = 1, \ldots, \alpha_m\}$ at both 0 and ∞. In considering the constraints (3), we consider the case where $m = 2$ for simplicity.

Differentiating (3) with respect to t shows that (3) is equivalent to

$$\alpha_1'(t) + e^{\alpha_2(t)}[\alpha_1'(t) - \alpha_2'(t)] \geq 0,$$
$$\alpha_2'(t) + e^{\alpha_1(t)}[\alpha_2'(t) - \alpha_1'(t)] \geq 0,$$

for all t. Therefore, for (6) to correspond to a survival model for competing risks for any value of β and $\mathbf{z}$, we need

$$\alpha_1'(t) + e^{a_2} e^{\alpha_2(t)}[\alpha_1'(t) - \alpha_2'(t)] \geq 0,$$
$$\alpha_2'(t) + e^{a_1} e^{\alpha_1(t)}[\alpha_2'(t) - \alpha_1'(t)] \geq 0,$$

for all t and any value of $a_1 = \beta_1 \mathbf{z}$ and $a_2 = \beta_2 \mathbf{z}$. Without further restrictions on α_1 and α_2, this holds if and only if $\alpha_1'(t) - \alpha_2'(t) = 0$ for all t. Noting that $\alpha_1'(t) = [F_1 S]^{-1}[f_1 - f_1 F_2 + f_2 F_1]$ (where $F_1(t) = F_1(t; \mathbf{0})$, etc), with an analogous expression for $\alpha_2'(t)$, it follows that

$$\alpha_1'(t) - \alpha_2'(t) = \frac{f_1}{F_1} - \frac{f_2}{F_2} = \left(\log \frac{F_1}{F_2}\right)'.$$

Thus $\alpha_1'(t) - \alpha_2'(t) = 0$ is equivalent to F_1 and F_2 being proportional, in turn, equivalent to proportionality of the two cause-specific hazard functions λ_1 and λ_2.

In sum, we have shown that the proportional odds regression model (6) only yields proper sub-distribution functions F_j for all values of β and $\mathbf{z}$ if the cause-specific hazards are proportional for all values of $\mathbf{z}$, a very restrictive condition. With this assumption, however, the parameters α_j and β_j have specific interpretations as follows. First, it follows from (5) that, for individuals at the baseline level of $\mathbf{Z} = \mathbf{0}$, $\alpha_j(t)$ is just the log odds, at time t, that a failure of type j has occurred as against no failure. Further, from (6) it follows that

$$\frac{F_j(t; \mathbf{z})}{S(t; \mathbf{z})} = e^{\alpha_j(t) + \beta_j \mathbf{z}},$$

so that the kth component of the regression coefficient β_j is the log odds of failure by time t, due to cause j, as against no failure, associated with a unit increase in the kth component of $\mathbf{Z}$ (holding other component variables constant). This is the case at all values of t. Similarly, note that

$$\frac{F_j(t;\mathbf{z})}{F_k(t;\mathbf{z})} = e^{\alpha_j(t)-\alpha_k(t)}e^{(\beta_j-\beta_k)\mathbf{z}},$$

showing that the log odds of failure by time t due to cause j, as against failure by time t due to cause k, is linear in $\mathbf{z}$ with slope $\beta_j - \beta_k$, again true for all t.

Given the restriction of proportional cause-specific hazards, why is the model (6) appealing in the first place? The answer is in its relationship to a regression model for a polytomous outcome generated by a current status observation scheme. Specifically, suppose that, for each individual, information on survival status, and, if relevant, cause of failure, is available only at a single time C. Thus, the observed data can be represented as (C, Δ), where $\Delta = 0$ if $T \geq C$, $\Delta = j$ if $T < C$ with $J = j$, for $1 \leq j \leq m$. It is therefore assumed that if an individual is known to have failed at the observation time C, the cause of failure is also available. As before, we assume that the monitoring time C is independent of T and is uninformative.

Note first that the distribution of Δ is related to that of (T, J) simply as follows:

$$\Pr(\Delta = j|C) = F_j(C), \tag{7}$$

for $j = 1, \ldots, m$ with $\Pr(\Delta = 0|C) = S(C)$.

For a fixed C, it is natural to consider a regression model which links the distribution of Δ to covariates $\mathbf{Z}$. A natural model is the *multinomial logistic model* which describes the dependence of $\Pr(\Delta = j|C, \mathbf{Z} = \mathbf{z})$ on the explanatory variables. In particular, the model states that

$$\Pr(\Delta = j|C, \mathbf{Z} = \mathbf{z}) = \frac{e^{\alpha_j + \beta_j \mathbf{z}}}{1 + \sum_{k=1}^{m} e^{\alpha_k + \beta_k \mathbf{z}}}, \quad j = 1, \ldots, m, \tag{8}$$

with necessarily

$$\Pr(\Delta = 0|C, \mathbf{Z} = \mathbf{z}) = \frac{1}{1 + \sum_{k=1}^{m} e^{\alpha_k + \beta_k \mathbf{z}}}.$$

See, for example, McCullagh and Nelder (1989), Chapter 5.2.4.

Extending this model to allow for varying C, while ensuring additivity of effects of C and $\mathbf{Z}$, immediately suggests replacing α_j with $\alpha_j(C)$ in (8), where the functions α_j satisfy the constraints given in (3) along with appropriate limits. Through (7) and (8), this immediately corresponds to the proportional odds model (6) for (T, J). As a consequence of our analysis of (6), this shows that we can only 'properly' use the multinomial logistic model for current status competing risks data, with additive effects of C and the covariates, if we are willing to assume that the underlying cause-specific hazards are proportional. Even in this restrictive case, it is important to note that practical issues remain for joint estimation of α and $\beta_1, \ldots, \beta_m$, particularly when α is treated nonparametrically.

2.1 *The Proportional Hazards Model*

Extending the ubiquitous Cox proportional hazards model (Cox 1972), the proportional hazards model for competing risks (Crowder 2001, Chapter 1.4.1; Kalbfleisch and Prentice 2002, Chapter 8.12) specifies that the conditional cause-specific hazard functions satisfy

$$\lambda_j(t; \mathbf{z}) = \lambda_{0j}(t) e^{\beta_j \mathbf{z}}, \tag{9}$$

for $j = 1, \ldots, m$, where λ_{0j} is the baseline cause-specific hazard function for cause j for individuals with $\mathbf{z} = \mathbf{0}$. This should not be confused with the assumption of proportional cause-specific hazards, at any fixed value of $\mathbf{Z}$, that we discussed earlier in Section 2, and that we return to briefly below.

It is of interest to determine the form of polytomous regression model that the proportional hazards model for (T, J), given in (9), induces on current status observations (C, Δ). First, without covariates, note that

$$\Pr(\Delta = j | C) = \int_0^C \lambda_j(u) \exp\left[-\int_0^u \left(\sum_{k=1}^m \lambda_k(t)\right) dt\right] du.$$

Now introducing the covariates $\mathbf{Z}$, under (9), we have

$$\Pr(\Delta = j | C, \mathbf{Z} = \mathbf{z}) = \int_0^C \lambda_{0j}(u) e^{\beta_j \mathbf{z}} \prod_{k=1}^m \exp\left[-\int_0^u \lambda_{0k}(u) e^{\beta_k \mathbf{z}} dt\right] du. \tag{10}$$

This explicitly links the proportional hazards model for (T, J) to a multinomial regression model for the current status observation (C, Δ), albeit a rather cumbersome one. In particular, there appears to be no convenient link function which separates the right hand side of (10) into additive effects for C and $\mathbf{z}$. It is plausible that further assumptions might lead to a simpler relation than (10). Suppose, for example, we now additionally assume proportional cause-specific hazard functions, so that, in particular, $\lambda_{0j}(t) = a_j \lambda_0(t)$ for all t and $j = 1, \ldots, m$, where the a_j's are positive constants and λ_0 is an unspecified hazard function. Then (10) simplifies to

$$\begin{aligned}
&\Pr(\Delta = j|C, \mathbf{Z} = \mathbf{z}) \\
&= a_j e^{\beta_j \mathbf{z}} \int_0^C \lambda_0(u) \prod_{k=1}^m \exp\left[e^{-a_k e^{\beta_k \mathbf{z}}} \int_0^u \lambda_0(t)dt\right] du \\
&= \frac{a_j e^{\beta_j \mathbf{z}}}{\sum_{k=1}^m a_k e^{\beta_k \mathbf{z}}} \left[1 - \exp\left\{\left(-\sum_{k=1}^m a_k e^{\beta_k \mathbf{z}}\right) \int_0^C \lambda_0(t)dt\right\}\right]. \quad (11)
\end{aligned}$$

Note that, for simplicity, we can absorb the constants $a_1, \ldots, a_m$ into the regression terms so long as a constant is included in $\mathbf{Z}$, yielding

$$\Pr(\Delta = j|C, \mathbf{Z} = \mathbf{z}) = \frac{e^{\beta_j \mathbf{z}}}{\sum_{k=1}^m e^{\beta_k \mathbf{z}}} \left[1 - \exp\left\{\left(-\sum_{k=1}^m e^{\beta_k \mathbf{z}}\right) \int_0^C \lambda_0(t)dt\right\}\right],$$

where we adjust our definition and interpretation of $\beta_1, \ldots, \beta_m$. However, the main point is that the effects of C and $\mathbf{z}$ remain inextricably linked in (11), even when further restrictions are placed on the shape of λ_0. The closest analogue, arising from (11), to the univariate correspondence of the proportional hazards model to a complementary log-log regression model for Δ, is that

$$\begin{aligned}
\log\left[-\log\left\{\Pr(T > C|J = j, C, \mathbf{z})\right\}\right] &= \log\left[-\log\left\{1 - \frac{F_j(C;\mathbf{z})}{F_j(\infty;\mathbf{z})}\right\}\right] \\
&= \log\left(\sum_{k=1}^m e^{\beta_k \mathbf{z}}\right) + \log \Lambda_0(C), \quad (12)
\end{aligned}$$

where Λ_0 is the integrated hazard function associated with λ_0. Unfortunately, $\Pr(T > C|J = j, C, \mathbf{z})$, in the left hand side of (12), does not obviously correspond to any (conditional) expectation of an observable random variable with current status data (except where the cause of failure is also observed for those for whom the failure event has not occurred at time C). Even then, the right hand side, while showing additivity of the effects for C and $\mathbf{z}$, does not yield a simple linear term in $\mathbf{z}$ when $m > 1$.

In sum, although the proportional hazards model for competing risks data necessarily induces a multinomial regression model for the categorical data produced by current status observation, the resulting model does not simply correspond to a recognizable multinomial regression model which might allow the use of existing software (possibly adapted to allow for monotonicity constraints in the nonparametric case). Similarly, application of a 'standard' generalized linear model for nominal multinomial outcomes to current status observations of competing risks data cannot be simply interpreted in terms of an underlying proportional hazards model even with additional restrictive assumptions.

2.2 *Mixture Models for Competing Risks*

Larson and Dinse (1985) suggested a mixture model for competing risks data which, in its simplest form, is as follows. First, a multinomial logistic regression model is assumed for $F_j(\infty; \mathbf{z})$, the fraction of all eventual events from cause j, so that

$$F_j(\infty; \mathbf{z}) = \frac{e^{\alpha_j \mathbf{z}}}{\sum_{k=1}^{m} e^{\alpha_k \mathbf{z}}},$$

for some set of regression coefficients $\alpha_1, \ldots, \alpha_m$, where a constant term is included in $\mathbf{Z}$, and for identifiability we assume, for example, $\alpha_1 = 0$. The second part of the model specifies regression relationships for the conditional distribution functions $H(T; J)$ that determine properties of event times associated with each specific cause. In particular, a proportional hazards model for these distribution functions yields $1 - H(t; J = j, \mathbf{Z} = \mathbf{z}) = \exp\left(\Lambda_j(t) e^{\beta_j \mathbf{z}}\right)$ for some set of integrated hazard functions $\Lambda_j, j = 1, \ldots, m$, so that

$$\Pr(\Delta = j | C, \mathbf{Z} = \mathbf{z}) = \frac{e^{\alpha_j \mathbf{z}}}{\sum_{k=1}^{m} e^{\alpha_k \mathbf{z}}} \left[1 - \exp\left(-e^{\beta_j \mathbf{z}} \int_0^C \lambda_0(t) dt\right)\right]. \quad (13)$$

Note the similarity with (11). Again we can rewrite (13) to obtain the analogue of (12), namely

$$\begin{aligned} \log\left[-\log\left\{\Pr(T > C | J = j, C, \mathbf{z})\right\}\right] &= \log\left[-\log\left\{1 - \frac{F_j(C; \mathbf{z})}{F_j(\infty; \mathbf{z})}\right\}\right] \\ &= \beta_j \mathbf{z} + \log \Lambda_0(C). \end{aligned}$$

This yields additive effects for C and $\mathbf{z}$, and now a linear term in $\mathbf{z}$ on the right hand side, but, of course, suffers from the same drawback as (12) in that the left hand side does not correspond to the (conditional) expectation of an observable random variable with current status data.

3 Progressive Multi-State Survival Models and Ordinal Polytomous Regression Models

We now turn to generalizations of a simple survival random variable in a quite different direction. Suppose interest focuses on a finite state survival process where individuals have to successively progress through each of $m + 1$ states over time. The illness-death model is a special case of this scenario with $m = 2$. Specifically, let $X(t)$ be a counting process with m jump times denoted by the random variables $T_1, \ldots, T_m$, where necessarily $T_1 \le T_2 \le \cdots \le T_m$. We wish to understand the joint distribution, F, of $(T_1, \ldots, T_m)$ and the influence of explanatory variables on its properties.

We focus here solely on models for the marginal distributions of F, denoted by $F_1, \ldots, F_m$ since only these marginals are identifiable from current status data. One immediate consequence of this is that the constraint $\Pr(T_1 \leq \cdots \leq T_m) = 1$ does not imply a stronger constraint on the marginals other than that $F_1 \geq \cdots \geq F_m$. This follows, since for any set of marginal distributions $F_1, \ldots, F_m$ with $F_1 \geq \cdots \geq F_m$, there exists an m-dimensional distribution with $\Pr(T_1 \leq \cdots \leq T_m) = 1$ that has marginals $F_1, \ldots, F_m$. To keep things simple, we also assume throughout that $F_1, \ldots, F_m$ are all continuous.

As in Section 2 we first consider the scenario absent covariates, and introduce a useful parameterization of F. For $j = 1$, let $\alpha_1(t)$ be defined by

$$\alpha_1(t) = \log\left[\frac{1 - F_1(t)}{F_1(t)}\right]. \tag{14}$$

Now consider the conditional probabilities of T_j, given T_{j-1}, for $j > 1$. In particular, define

$$\alpha_j(t) = \log\left[\frac{G_j(t)}{(1 - G_j(t))}\right], \tag{15}$$

where

$$G_j(t) = \Pr(T_j \geq t | T_{j-1} < t) = \frac{F_{j-1}(t) - F_j(t)}{F_{j-1}(t)} = 1 - \frac{F_j(t)}{F_{j-1}(t)}, \tag{16}$$

for $1 < j \leq m$. We can solve (14)–(16) for F_j, giving

$$F_j = \prod_{k=1}^{j} \frac{1}{1 + e^{\alpha_k}}. \tag{17}$$

The functions α_j re-express the marginal distributions $F_1, \ldots, F_m$, and necessarily have to satisfy appropriate conditions for (17) to yield proper distribution functions. The conditions for α_1 are straightforward in that $\alpha_1(0) = \infty$, $\alpha_1(\infty) = -\infty$ and α_1 is non-increasing. For α_j with $j > 1$, the constraints are more complex. Formally, $\alpha_j(\infty) = -\infty$, and the α_js possess the mutual properties that the functions $\prod_{k=1}^{j} \frac{1}{1+e^{\alpha_k}}$ are all non-decreasing for $j = 1, \ldots, m$. For example, with $j = 2$, this requires that

$$\alpha_1' e^{\alpha_1} + \alpha_2' e^{\alpha_2} + (\alpha_1' + \alpha_2') e^{\alpha_1 + \alpha_2} \leq 0, \tag{18}$$

with analogous conditions for the other α_js for $j > 2$. Note that, along with the conditions on α_1, α_2 being non-increasing is a sufficient condition for (18); in general, α_j being non-increasing for all j implies proper distribution functions F_j (along with the appropriate limit conditions). However, it is not necessary that α_j be non-increasing. For example, with $m = 2$—the standard nonparametric illness-death model—α_2 being non-increasing is equivalent to F_2/F_1 being non-decreasing. However, suppose that, for small t, progression to illness (the first transition) is immediately followed by the second transition (to death), but for large t, there is a much longer gap between the two transitions. Then, initially F_2/F_1 is close to 1 and then decreases as t gets larger.

We now introduce regression effects of $\mathbf{Z}$ on each of $T_1, \ldots, T_m$. In principal, we cannot simply postulate separate unlinked regression models for each of $T_1, \ldots, T_m$ in turn, as this may lead to violations of the stochastic ordering of $T_1, \ldots, T_m$ for certain values of regression coefficients and/or $\mathbf{Z}$. Suppose, alternatively, that we focus on the effects of $\mathbf{Z}$ on the functions $\alpha_1, \ldots, \alpha_m$, and assume that these are linear,

$$\alpha_j(t; \mathbf{Z} = \mathbf{z}) = \alpha_j(t; \mathbf{Z} = \mathbf{0}) + \beta_j \mathbf{z}, \tag{19}$$

or equivalently,

$$F_j(t; \mathbf{Z} = \mathbf{z}) = \prod_{k=1}^{j} \frac{1}{1 + e^{\alpha_k(t;\mathbf{z})}} = \prod_{k=1}^{j} \frac{1}{1 + e^{\alpha_k(t;\mathbf{Z}=\mathbf{0}) + \beta_k \mathbf{z}}}, \tag{20}$$

for $j = 1, \ldots, m$. For (20) to correspond to proper distribution functions, it is necessary that the constraining conditions, exemplified by (18)—when $\mathbf{Z} = \mathbf{0}$— imply that the same conditions hold for $\alpha_j(t; \mathbf{Z} = \mathbf{z})$ in (19). However, this is not guaranteed for all values of β_j and $\mathbf{z}$ except in particular circumstances. One such is the additional assumption that $\alpha_j(t; \mathbf{Z} = \mathbf{0})$ is non-increasing for all j, or equivalently that $F_j(t; \mathbf{Z} = \mathbf{0})/F_{j-1}(t; \mathbf{Z} = \mathbf{0})$ is non-decreasing in t for $j > 1$. This additional condition implies that the regression model (20) always yields a set of proper distribution functions $F_j(t; \mathbf{Z})$ for all j, β_j, and any value of $\mathbf{Z}$.

We call the model (20) a *proportional odds model* for $T_1, \ldots, T_m$ because of the interpretation of the regression coefficient vectors β_j. Note that a unit increase in the kth component of $\mathbf{Z}$ (holding other components fixed) increases the log odds of being in state j, conditional on being in state j or higher, by β_{jk}, the kth component of β_j. As in the other cases we have studied, the functions $\alpha_j(t; \mathbf{Z} = \mathbf{0})$ determine the shape of the baseline distribution functions $F_j(t; \mathbf{Z})$ for $\mathbf{Z} = \mathbf{0}$, $j = 1, \ldots, m$.

We now relate these ideas to current status observation on $T_1, \ldots, T_m$, at a monitoring time C. Here, the observed data can be represented as $Y = (C, \Phi)$, where $\Phi = j$ if $T_{j-1} < C \leq T_j$ for $j = 1, \ldots, m+1$, where $T_0 \equiv 0$ and $T_{m+1} \equiv \infty$. As before, we assume that observation times are independent of $T_1, \ldots, T_m$, and are uninformative.

For a fixed C, we again focus on models for $p_{j;\mathbf{z}} = \Pr(\Phi = j | \mathbf{Z} = \mathbf{z})$. Note that, suppressing the dependence on $\mathbf{z}$ for the moment, $p_1 = \Pr(T_1 \geq C) = 1 - F_1(C)$, $p_{m+1} = \Pr(T_m < C) = F_m(C)$, and

$$p_j(C) = \Pr(T_{j-1} < C \leq T_j) = F_{j-1}(C) - F_j(C), \tag{21}$$

for $j = 2, \ldots, m$. Note that $p_{j+1}(C) + \cdots + p_{m+1}(C) = F_j(C)$ for $j = 1, \ldots, m$.

A natural regression model here is the so-called *sequential logit model* for ordinal categorical data that is defined by logistic regression models for the sequential probabilities $p_{j;\mathbf{z}}/(p_{j;\mathbf{z}} + \cdots + p_{m+1;\mathbf{z}})$. In terms of log odds, this yields

$$\log \frac{p_{j;\mathbf{z}}}{p_{j+1;\mathbf{z}} + \cdots + p_{m+1;\mathbf{z}}} = \alpha_j + \beta_j \mathbf{z}, \qquad (22)$$

for $j = 1, \ldots, m$. This is also referred to as the *continuation ratio logit model*; see, for example, Agresti (2002, Chapter 7.4.3).

We now want to incorporate varying monitoring times C, again with the idea of assuming that the effects of C are additive to those of the covariates. This is achieved by assuming that only the intercept terms α_j depend on C, and not the slope coefficients β_j, in (22). The final model is therefore

$$\log \frac{p_{j;\mathbf{z}}(C)}{p_{j+1;\mathbf{z}}(C) + \cdots + p_{m+1;\mathbf{z}}(C)} = \alpha_j(C) + \beta_j \mathbf{z}. \qquad (23)$$

Using (21), the model (23) therefore corresponds exactly with the proportional odds model (20). The consequence again is that the sequential logistic model for ordered multi-state current status data, with additive effects of C and the covariates, corresponds with the proposed proportional odds model for $T_1, \ldots, T_m$ so long as the intercept functions in C satisfy the constraints induced by the functions in (20) being non-decreasing, as discussed earlier, and the associated limit conditions.

The situation is therefore somewhat more satisfying than in the competing risks situation where the multinomial logistic model for current status data, with additive effects, implied that the underlying competing risks model is only proper if the intercept functions have identical derivatives (corresponding to the restrictive condition of proportional cause-specific hazards). With ordered multi-state current status data, the sequential logistic model (23) corresponds to any set of marginal distributions for $T_1, \ldots, T_m$, albeit with cumbersome monotonicity conditions on the intercept functions. As previously noted, the simple conditions that α_j be non-increasing for all j may be more useful in practice, but requires the additional assumption that the distribution functions $F_j(t; \mathbf{z})/F_{j-1}(t; \mathbf{z})$ are non-decreasing in t for $j = 2, \ldots, m$.

The regression model (20) has been previously suggested in an example concerning transitions of women from a disease-free state, to onset of preclinical fibroids, to diagnosis of fibroids (i.e. $m = 2$) in Dunson and Baird (2001), as part of a richer data structure where T_2 is often observed directly (for the single group setting for such data, see van der Laan, Jewell and Petersen 1997). Although Dunson and Baird (2001) developed the model in an ad hoc fashion, they also invoked the assumption that F_2/F_1 be non-decreasing to simplify semiparametric estimation strategies, arguing that this assumption is reasonable in the fibroid example. For previous work on current status data for multi-state stochastic processes in the single group setting, see Jewell and van der Laan (1995,1997) and van der Laan and Jewell (2003).

We note here that there is an obvious alternative sequential logistic model which focuses on conditional probabilities in the alternative 'direction' from (23). Specifically, we could sequentially use a logistic model for the probabilities $p_{j;\mathbf{z}}/(p_{1;\mathbf{z}} + \cdots + p_{j;\mathbf{z}})$ for $j = 1, \ldots, m+1$ which is linear in $\mathbf{z}$ with an additive term in C. In analogous fashion this leads to the regression model

$$S_j(t; \mathbf{Z} = \mathbf{z}) = \prod_{k=j}^{m} \frac{1}{1 + e^{\gamma_k(t;\mathbf{Z}=\mathbf{0}) + \beta_k \mathbf{z}}}, \tag{24}$$

where the new intercept functions $\gamma_k(t; \mathbf{Z} = \mathbf{0})$ again determine the shape of the baseline distribution functions $F_j(t; \mathbf{Z})$. This proportional odds model again requires appropriate constraints on the functions $\gamma_1, \ldots, \gamma_m$ for (24) to yield proper survival functions. Although the model (24) differs from (20) there is no *a priori* reason to prefer one over the other.

4 Unlinked Regression Models for Current Status Data

In the competing risks and multi-state survival scenarios of Sections 2–3, we avoided the use of simple unlinked regression models for the sub-distribution functions in the former case, and the marginal distribution functions in the latter, since the use of such may not lead to a proper joint distribution function. However, as we have now explored, correspondences between a full data regression model and a multivariate binary regression model for incomplete current status observations are not as straightforward as in the univariate setting, at least when additive effects of the monitoring time and covariates are desired. Further, Jewell, van der Laan and Henneman (2003) show that, in the competing risks setting, smooth functionals of the sub-distribution functions can be efficiently estimated—asymptotically—using separate unlinked nonparametric maximum likelihood estimators of the individual sub-distribution functions. The advantage of this approach is that the unlinked estimators are much simpler than the full nonparametric maximum likelihood estimator while they retain consistency. A similar result was established for nonparametric estimators of the marginal distributions for finite multi-state counting processes in van der Laan and Jewell (2003). This suggests that there may be little or no asymptotic precision gained by estimating regression relationships jointly rather than separately, and that the simpler estimators may, in fact, outperform, the more complex simultaneous modeling investigated in Sections 2–3 with small or moderate sample sizes. While this opinion is speculative and remains to be more fully addressed elsewhere, both in theory and simulations, we give a brief outline of this strategy here.

4.1 *Competing Risks Models*

We continue to use the notation of Section 2. Recall that current status data is represented by (C, Δ), where $\Delta = 0$ if $T \geq C$, $\Delta = j$ if $T < C$ with $J = j$, for $1 \leq j \leq m$. Define the observed binary random variables $\Psi_j = 1$ if $\Delta = j$ and $\Psi_j = 0$ otherwise. Note that

$$E(\Psi_j | C, \mathbf{Z} = \mathbf{z}) = F_j(C; \mathbf{z}), \tag{25}$$

for $1 \leq j \leq m$. Thus, taking each j separately, (25) allows construction of a regression model for $F_j(t; \mathbf{z})$ in correspondence with a binary regression model for Ψ_j as for standard univariate current status data. For example, a logistic regression model for Ψ_j with covariates $\mathbf{Z}$ leads to a proportional odds relationship between $\mathbf{Z}$ and $F_j(t; \mathbf{z})$ as in (1), the only difference being that $F_j(\infty; \mathbf{z})$ may be less than 1 so that the corresponding incidence function $\alpha_j(t)$ potentially has a finite limit at ∞.

The advantage to using these separate models is their simplicity, with the consequence that they can be fit using standard software for univariate current status data, leading to semi-parametric estimators $\hat{F}_j(t; \mathbf{z})$ for $j = 1, \ldots, m$ and any t and $\mathbf{z}$. The disadvantage, as previously noted, is that, even though $\hat{F}_j(t; \mathbf{z})$ is non-decreasing in t for any fixed value of $\mathbf{z}$ as desired, $\sum_{j=1}^{m} \hat{F}_j(t; \mathbf{z})$ may exceed 1 for some values of t and $\mathbf{z}$, violating the requirement that $F(t; \mathbf{z}) = \sum_{j=1}^{m} F_j(t; \mathbf{z})$ is a distribution function. This, however, may not be a major drawback in large samples as the estimator $\sum_{j=1}^{m} \hat{F}_j(t; \mathbf{z})$ will consistently estimate the true $F(t; \mathbf{z})$ so long as appropriate semiparametric estimation procedures are used for the separate regression models.

A slight variant on this strategy can be described as follows. First, we use standard univariate current status regression methods to yield an estimator $\hat{F}(t; \mathbf{z})$, based on the observations $(C_i, (\Psi)_i)$ where $\Psi = \sum_{j=1}^{m} \Psi_j$ indicates only whether the outcome event has occurred by time C without regard to failure type.

Now, for each j, consider the constructed variable $W_j = F(C)\Psi_j$, and note that $E(W_j | C, \Psi = 1) = F_j(C)$. This suggests using current status type regression techniques (that is, isotonic dependence on C and additive linear dependence on $\mathbf{Z}$ with an appropriate link function) for the constructed outcomes $(W_j)_i = \hat{F}(C_i; \mathbf{z}_i)(\Psi_j)_i$ against C_i, using only observations with $(\Psi)_i = 1$, that is, observations where an event of any type has occurred by the monitoring time. This yields estimators $\hat{F}_j(t; \mathbf{z})$ for each j.

While this approach still does not guarantee estimators $\hat{F}_j(t; \mathbf{z})$ which sum to less than 1, this may be somewhat less likely than the first unlinked method since, for each j, the constructed outcomes $(W_j)_i$ are smaller than the respective outcomes $(\Psi_j)_i$ for the previous estimators. In the single sample setting, this approach is related to the full nonparametric maximum likelihood estimator of $F_1, \ldots, F_m$—see Jewell et al. (2003).

4.2 *Multi-State Survival Models*

With the notation of Section 3, current status data is given by (C, Φ), where $\Phi = j$ if $T_{j-1} < C \leq T_j$ for $j = 1, \ldots, m+1$, where $T_0 \equiv 0$ and $T_{m+1} \equiv \infty$. In this setting define $\Psi_j = 1$ if $\Phi > j$, and $\Psi_j = 0$ otherwise. Note that

$$E(\Psi_j | C, \mathbf{Z} = \mathbf{z}) = F_j(C; \mathbf{z}), \tag{26}$$

for $1 \leq j \leq m$. Thus, we can separately estimate marginal regression models for $F_j(t; \mathbf{z})$ for each j using univariate current status methods on the data (C, Ψ_j). Again, the advantages of this approach are simplicity, use of standard current status methods only, and direct regression modeling of the marginal distributions, presumably the primary relationships of interest. But once more, although estimates of $F_j(t; \mathbf{z})$ obtained in this way are each distribution functions they are not guaranteed to be stochastically ordered, as required by the structure of the data. Again, this is unlikely to be a serious problem in large samples for similar reasons to those discussed with competing risks data.

Finally, there are variants to this approach similar to the one suggested in Section 4.1 for competing risks data. For example, suppose we obtain the estimator $\hat{F}_1(t; \mathbf{z})$ using the data on Ψ_1 as described. Now, consider the constructed variable $W_2 = F_1(C)\Psi_2$, where again it immediately follows that $E(W_2 | C, \Psi_1 = 1) = F_2(C)$. As before, this suggests using current status regression techniques for the constructed outcomes $(W_2)_i = \hat{F}_1(C_i; \mathbf{z}_i)(\Psi_2)_i$, against C_i, using only observations with $(\Psi_1)_i = 1$, thereby yielding an estimator $\hat{F}_2(t; \mathbf{z})$. This process then is repeated to yield estimators $\hat{F}_3(t; \mathbf{z})$, $\hat{F}_4(t; \mathbf{z})$, and so on. Again this approach does not guarantee stochastic ordering of the estimated marginals of F, although it may be more likely since, for each j, the constructed outcomes $(W_j)_i$ are smaller than the respective outcomes $(\Psi_j)_i$ for the previous estimators (and smaller than $\hat{F}_{j-1}(C_i; \mathbf{z}_i)$).

5 Motivating Examples

We briefly describe illustrations of competing risks and multi-state survival data where the need for practical regression models for current status data motivated the development in the earlier sections. In the competing risks case, Krailo and Pike (1983) discuss data from the National Center for Health Statistics' Health Examination Survey, originally analysed by MacMahon and Worcester (1966). In particular, they focus on the menopausal history of 3,581 female respondents from 1960-1962 who provided cross-sectional information on their age and their menopausal status. For those who had experienced menopause, further retrospective information on the exact age when their periods stopped was deemed unreliable by McMahon and Worcester because of extreme digit preference. Thus, Krailo and Pike (1983) concentrated on the simple current status information on menopausal status, in addition to the response on whether menopause had occurred due to an operation or not. Thus natural and operative menopause provide the two causes of 'failure' (here, menopause) in the context of competing risks. Jewell et al. (2003) analyze this current status data with a nonparametric model. To extend these 'one-sample' models to allow for regression effects requires the kinds of models introduced in Section 2.

This example suggests interesting extensions to simple current status observation of competing risks data. According to MacMahon and Worcester (1966), the original data from the Health Examination Survey contained reliable information about the exact age at operative menopause, despite the concerns about information about age at natural menopause. This raises the problem of estimation of regression models for the subdistribution functions F_1 and F_2 in the case where exact times of failures are observed when a failure due to the first risk has occurred before the observation time but where only current status information is available regarding failures due to the second risk. Jewell et al. (2003) consider this problem in the 'one sample' case.

We now turn briefly to examples of regression based on current status observation of a multi-state survival process, namely the onset and diagnosis of uterine fibroids. The compound 2,3,7,8-tetrachlorodibenzo-*p*-dioxin, commonly known as TCDD or dioxin, is a toxic hydrocarbon and environmental contaminant. It has a half-life of approximately 8 years in humans and, in addition to being a carcinogen, has been shown to disrupt endocrine pathways. On July 10, 1976, an explosion at a chemical plant in Seveso, Italy, exposed local residents to the highest known environmental dioxin levels in a residential area of about 18 km^2 around the plant. A number of health assessments were launched soon after the explosion and many blood samples were collected from residents with sera stored for subsequent analyses. The Seveso Womens' Health Study (SWHS) was initiated in 1996, assembling a historical cohort of more than 500 women who were under 40 years of age at the time of the explosion, who were resident in the most heavily exposed areas, and who had sufficient stored sera from the period 1976–1980 available for analysis. Individual level of dioxin exposure was evaluated using the stored sera. For a detailed description of the study see Eskenazi et al. (2000).

Uterine fibroids are noncancerous growths in the uterus, commonly referred to as fibroids. Although uterine fibroids may be present in up to 75% of all women, about a half of these women do not have symptoms. Symptoms, leading to a diagnosis, may develop slowly over a period of several years or rapidly over a period of several months and may include abnormal menstrual bleeding, pelvic pain and pressure and urinary problems. During the period 1996–98 eligible women—still menstruating—in the SWHS were interviewed and received a transvaginal ultrasound, a screening instrument that can detect the presence of fibroids in women without symptoms. Prior diagnosis of fibroids was determined at interview and medical records used to calculate the age at diagnosis. With age as the time scale of interest, all women included in the analysis contributed current status data on onset of the disease with medical records potentially providing exact ages at diagnosis where this had occurred. If only the prior existence of a diagnosis of fibroids is known, then the data structure corresponds with what is envisioned in Section 3 where the monitoring time corresponds with age at screening. Here, regression effects may focus on dioxin exposure information although other covariate effects may also be of substantial interest. As in the case of the competing risks example, right-censored information on the age at diagnosis at the time of screening provides an interesting variant to the 'pure' current status form of data structure considered in Section 3. van der Laan et al. (1997) consider a 'one sample' version of this kind of data structure. Dunson and Baird (2001) consider a regression model in this context, with their approach also applied to the analysis of fibroids data arising from a National Institute of Environmental Health Sciences cross-sectional study of the premenopausal incidence of uterine fibroids. The primary covariate of interest in their regression analysis was race. Young and Jewell (2006) compare Dunson and Baird's (2001) model to an extension of the approach of van der Laan et al. (1997) to the regression setting using data examples and simulations.

Multi-state examples occur in quite different contexts than disease progression. For example, in cross-sectional life/sexual history surveys questions are often asked about the number of distinct sexual partners experienced by the respondent by their age at survey. Similarly, employment history questionnaires may focus on the number of distinct employment (or unemployment) experiences of the respondent. Often, with such data, there may be little or no information on the exact ages where a respondent transitions between 'states' that describe the current cumulative number of partners or experiences. This therefore produces current status data of exactly the sort considered in Section 3. Although such data precludes study of association between the time spent in various states, there is often still considerable interest in investigating and comparing marginal regression models for times until specified transitions.

6 Discussion

We have considered correspondences between regression models for multinomial outcomes and various multivariate survival models that extend those developed by Doksum and Gasko (1990) in a univariate setting. While this suggests some useful regression survival models that can be identified from current status observation, the correspondences are not generally straightforward. This motivates the simpler approach of examining several unlinked univariate regression models as suggested in Section 4. However, there are a wider range of multinomial models that can be considered here so that this should only be considered as a preliminary investigation. Doksum and Gasko (1990) also consider correspondences with linear transformation models. It is natural to consider extensions of these ideas to the multivariate setting in which multivariate survival regression models correspond to multivariate binary anlaogues. Space does not permit further discussion of results in this area and details will appear elsewhere. It is important to note that several approaches to multivariate current status data with a common monitoring time have already appeared (Wang and Ding 2000, Dunson and Dinse 2002, Ding and Wang 2004, Jewell, van der Laan and Lei 2005).

References

1. AGRESTI, A. (2002). *Categorical Data Analysis.* 2nd ed. Wiley, New York

2. BENNETT, S. (1983). Analysis of survival data by the proportional odds model. *Statistics in Medicine* **2**, 273–7.

3. COX, D.R. (1972). Regression models and life tables (with discussion). *Journal of the Royal Statistical Society, Series B* **34**, 187–220.

4. CROWDER, M.J. (2001). *Classical Competing Risks.* Chapman & Hall, New York.

5. DING, A.A. AND WANG, W. (2004). Testing independence for bivariate current status data. *J. Amer. Statist. Assoc.* **99**, 145–55.

6. DOKSUM, K.A. AND GASKO, M. (1990). On a correspondence between models in binary regression analysis and in survival analysis. *International Statistical Review* **58**, 243–52.

7. DUNSON, D.B. AND BAIRD, D.D. (2001). A flexible parametric model for combining current status and age at first diagnosis data. *Biometrics* **57**, 396–403.

8. DUNSON, D.B. AND DINSE, G.E. (2002). Bayesian models for multivariate current status data with informative censoring. *Biometrics* **58**, 79–88.

9. ESKENAZI, B., MOCARELLI, P., WARNER, M., SAMUELS, S. VERCELLINI, P., OLIVE, D., NEEDHAM, L., PATTERSON, D. AND BRAMBILLA, P. (2000). Seveso Women's Health Study: a study of the effects of 2,3,7,8-tetrachlorodibenzo-*p*-dioxin on reproductive health. *Chemosphere* **40**, 1247–53.

10. JEWELL, N.P. AND VAN DER LAAN, M. (1995). Generalizations of current status data with applications. *Lifetime Data Analysis* **1**, 101–9.

11. JEWELL, N.P. AND VAN DER LAAN, M. (1997). Singly and doubly censored current status data with extensions to multi-state counting processes. In *Proceedings of First Seattle Conference in Biostatistics* Lin, D-Y. ed., Springer Verlag, 171–84.

12. JEWELL, N.P., VAN DER LAAN, M. AND X. LEI (2005). Bivariate current status data with univariate monitoring times. *Biometrika* **92**, 847–62.

13. JEWELL, N.P., VAN DER LAAN, M. AND HENNEMAN, T. (2003). Nonparametric estimation from current status data with competing risks. *Biometrika* **90**, 183–97.

14. JEWELL, N.P. AND VAN DER LAAN, M. (2004). Current status data: Review, recent developments and open problems. In *Advances in Survival Analysis*, Handbook in Statistics #23, 625–42, Elsevier, Amsterdam.

15. KALBFLEISCH, J.D. AND PRENTICE, R.L. (2002). *The Statistical Analysis of Failure Time Data.* 2nd ed. Wiley, New York.

16. KRAILO, M.D. AND PIKE, M.C. (1983). Estimation of the distribution of age at natural menopause from prevalence data. *Am. J. Epidemiol.* **117**, 356–61.

17. LARSON, M.G. AND DINSE, G.E. (1985). A mixture model for the regression analysis of competing risks data. *Applied Statistics* **34**, 201–11.

18. MCCULLAGH, P. AND NELDER, J.A. (1989). *Generalized Linear Models.* 2nd ed. Chapman & Hall, New York.

19. MACMAHON, B. AND WORCESTER, J. (1966). Age at menopause, United States 1960–1962. *National Center for Health Statistics; Vital and Health Statistics, Series 11: Data from the National Health Survey, no. 19* Washington, DC: DHEW Publication no. (HSM) 66–1000.

20. Rossini, A. and Tsiatis, A.A. (1996). A semiparametric proportional odds regression model for the analysis of current status data. *J. Amer. Statist. Assoc.* **91**, 713–21.

21. Shiboski, S.C. (1998). Generalized additive models for current status data. *Lifetime Data Analysis* **4**, 29–50.

22. van der Laan, M. and Jewell, N. P. (2003). Current status data and right-censored data structures when observing a marker at the censoring time. *Annals of Statistics*, **31**, 512–35.

23. van der Laan, M., Jewell, N.P. and Petersen, D. (1997). Efficient estimation of the lifetime and disease onset distribution. *Biometrika* **84**, 539–54.

24. Wang, W. and Ding, A.A.(2000). On assessing the association for bivariate current status data. *Biometrika* **87**, 879–93.

25. Young, J.G. and Jewell, N.P. (2006). Regression analysis of a disease onset using diagnosis data. Preprint.

Chapter 4

USING MARTINGALE RESIDUALS TO ASSESS GOODNESS-OF-FIT FOR SAMPLED RISK SET DATA

Ørnulf Borgan and Bryan Langholz

Department of Mathematics
University of Oslo, Oslo, NORWAY

Department of Preventive Medicine
University of Southern California, Los Angeles, CA, U.S.A.

E-mails: borgan@math.uio.no & langholz@usc.edu

Standard use of Cox's regression model and other relative risk regression models for censored survival data requires collection of covariate information on all individuals under study even when only a small fraction of them die or get diseased. For such situations risk set sampling designs offer useful alternatives. For cohort data, methods based on martingale residuals are useful for assessing the fit of a model. Here we introduce grouped martingale residual processes for sampled risk set data, and show that plots of these processes provide a useful tool for checking model-fit. Further we study the large sample properties of the grouped martingale residual processes, and use these to derive a formal goodness-of-fit test to go along with the plots. The methods are illustrated using data on lung cancer deaths in a cohort of uranium miners.

Key words: Chi-squared test; Cohort sampling; Counter-matching; Counting process; Cox's regression model; Martingale; Matching; Nested case-control study; Relative risk regression; Survival analysis.

1 Introduction

Cox regression is central to modern survival analysis, and it is the method of choice when one wants to assess the influence of risk factors and other covariates on mortality or morbidity. A number of methods, both graphical methods and formal tests, have been proposed to assess the goodness-of-fit of Cox's model; see e.g. the recent textbooks by Hosmer and Lemeshow (1999), Klein and Moeschberger (2003), and Therneau and Grambsch (2000).

One important tool for checking the fit of Cox's regression model is the martingale residuals introduced by Barlow and Prentice (1988). Therneau, Grambsch and Fleming (1990) proposed to use a smoothed plot of these residuals versus a covariate as a means to detect its correct functional form, while Grambsch, Therneau and Fleming (1995) suggested a similar, improved plot; see Section 5.7 in Therneau and Grambsch (2000) for a review and further discussion. Another approach was taken by Aalen (1993). In the context of his additive model (see Aalen 1989), he proposed to plot martingale residual processes, aggregated over groups of individuals, versus time as an omnibus procedure to check the fit of a model. Aalen's idea was implemented for Cox's regression by Grønnesby and Borgan (1996), who also derived a formal goodness-of-fit test to go along with the graphical procedure.

The commonly used methods for inference in Cox's regression model, including the methods for goodness-of-fit, require collection of covariate information on all individuals under study. This may be very expensive in large epidemiologic cohort studies of a rare disease. Risk set sampling designs, where covariate information is collected for all failing individuals (cases), but only for a sample of the non-failing ones (controls) then offer useful alternatives which may drastically reduce the resources that need to be allocated to a study for data collection and checking.

In the present paper we use the counting process framework of Borgan, Goldstein and Langholz (1995) to generalize the martingale residual processes to sampled risk set data. In this context it does not seem feasible to obtain graphical procedures analogous to the smoothed martingale residual plot of Therneau et al. (1990) or the related plot of Grambsch et al. (1995). However, we may still generalize the grouped martingale residual processes plots of Grønnesby and Borgan (1996) and the accompanying goodness-of-fit test. In doing this we will not restrict ourselves to Cox's regression model, but consider a general class of relative risk regression models.

The outline of the paper is as follows. In Section 2 we introduce the class of relative risk regression models, describe the type of failure time data considered for the cohort, and review how the cohort data may be formulated by means of counting processes. Then we outline how the martingale residuals and grouped martingale residual processes follow naturally from the counting process formulation. Section 3 is devoted to risk set sampling. We first introduce the general framework for risk set sampling of Borgan et al. (1995), describe how it specializes for simple random and counter-matched sampling, and review methods for inference for sampled risk set data. Then we outline how sampled risk set data can be described by processes counting jointly the occurrence of failures and the sampling of controls, and we use this counting process formulation to generalize the grouped martingale residual processes and accompanying goodness-of-fit test of Grønnesby and Borgan (1996) to sampled risk set data. An illustration for a study of lung cancer death in a cohort of uranium miners is provided in Section 4, while proofs are collected in Section 5. In Section 6 we briefly explain how the results extend to matched risk set sampling designs, while some concluding comments are given in the final Section 7. Throughout the paper we will without further references use standard results for counting processes (e.g. Andersen et al. 1993).

2 Cohort data

We consider a cohort of n individuals, and denote by $\alpha(t; \mathbf{z}_i)$ the hazard rate at time t for an individual i with vector of covariates $\mathbf{z}_i(t) = (z_{i1}(t), \ldots, z_{ip}(t))^{\mathsf{T}}$. Here the time-variable t may be age, time since employment, or some other time-scale relevant to the problem at hand, where we throughout assume that $t \in (0, \tau]$ for a given terminal study time τ. A covariate may be time-fixed or time-dependent; in the latter case its value at time t is assumed to be known "just before" time t, i.e., the covariate is assumed to be predictable. We assume that the covariates of individual i are related to its hazard rate by the relative risk regression model

$$\alpha(t; \mathbf{z}_i) = c(\boldsymbol{\beta}_0, \mathbf{z}_i(t))\, \alpha_0(t). \tag{1}$$

Here $c(\boldsymbol{\beta}_0, \mathbf{z}_i(t))$ is a relative risk function, $\boldsymbol{\beta}_0 = (\beta_{01}, \ldots, \beta_{0p})^{\mathsf{T}}$ is a vector of regression coefficients describing the effect of the covariates, while the baseline hazard rate $\alpha_0(t)$ is left unspecified. Throughout we use $\boldsymbol{\beta}_0$ to denote the vector of true regression coefficients, while we use $\boldsymbol{\beta}$ as an argument in the partial likelihood and similar quantities. We normalize the relative risk function by assuming $c(\boldsymbol{\beta}_0, \mathbf{0}) = 1$. Thus $\alpha_0(t)$ corresponds to the hazard rate of an individual with all covariates identically equal to zero. For the exponential relative risk function $c(\boldsymbol{\beta}_0, \mathbf{z}_i(t)) = \exp(\boldsymbol{\beta}_0^{\mathsf{T}} \mathbf{z}_i(t))$, formula (1) gives the usual Cox regression model. Other possibilities include the linear relative risk function $c(\boldsymbol{\beta}_0, \mathbf{z}_i(t)) = 1 + \boldsymbol{\beta}_0^{\mathsf{T}} \mathbf{z}_i(t)$ and the excess relative risk model $c(\boldsymbol{\beta}_0, \mathbf{z}_i(t)) = \prod_{j=1}^{p}(1 + \beta_{0j}\, z_{ij}(t))$.

The individuals in the cohort may be followed over different periods of time, i.e., our observations may be subject to left-truncation and right censoring. It is a fundamental assumption throughout that the left truncation and right censoring are independent in the sense that the additional knowledge of which individuals have entered the study or have been censored before any time t do not carry information on the risks of failure at t; see Sections III.2-3 in Andersen et al. (1993) and Sections 1.3 and 6.2 in Kalbfleisch and Prentice (2002) for a general discussion on the concept of independent censoring.

We let $t_1 < t_2 < \cdots$ be the times when failures are observed and, assuming that there are no tied failures, denote by i_j the individual who fails at t_j. The risk set $\mathcal{R}_j$ is the collection of all individuals who are under observation "just before" time t_j. In particular the case i_j is a member of $\mathcal{R}_j$. Then the vector of regression parameters in (1) is estimated by $\widehat{\boldsymbol{\beta}}$, the value of $\boldsymbol{\beta}$ maximizing Cox's partial likelihood, while the cumulative baseline hazard rate $A_0(t) = \int_0^t \alpha_0(u)\mathrm{d}u$ is estimated by the Breslow estimator

$$\widehat{A}_0(t) = \sum_{t_j \le t} \frac{1}{\sum_{l \in \mathcal{R}_j} c(\widehat{\boldsymbol{\beta}}, \mathbf{z}_l(t_j))},$$

e.g. Section VII.2 in Andersen et al. (1993).

In order to define the martingale residuals, we first need to review some basic facts on counting processes, (cumulative) intensity processes and martingales. To this end, introduce the processes

$$N_i(t) = \sum_{t_j \le t} I\{i_j = i\}; \qquad i = 1, 2, \ldots, n; \tag{2}$$

counting the number of observed events for individual i in $(0, t]$ (which is 0 or 1 for survival data). The intensity processes λ_i of the counting process N_i is given heuristically by $\lambda_i(t)\mathrm{d}t = P(\mathrm{d}N_i(t) = 1 \,|\, \mathcal{H}_{t-})$, where $\mathrm{d}N_i(t)$ is the increment of N_i over the small time interval $[t, t+\mathrm{d}t)$, and $\mathcal{H}_{t-}$ denotes all information available to the researcher "just before" time t. Then by (1) and the independent censoring assumption,

$$\lambda_i(t) = Y_i(t)\, \alpha(t; \mathbf{z}_i) = Y_i(t)\, c(\boldsymbol{\beta}_0, \mathbf{z}_i(t))\, \alpha_0(t), \tag{3}$$

with $Y_i(t)$ a left-continuous at risk indicator for individual i. Thus $\mathcal{R}(t) = \{i \,|\, Y_i(t) = 1\}$ is the risk set at time t, and $n(t) = |\mathcal{R}(t)|$ is the number at risk "just before" time t. Note that $\mathcal{R}_j = \mathcal{R}(t_j)$.

Corresponding to λ_i, we define the cumulative intensity process

$$\Lambda_i(t) = \int_0^t \lambda_i(u)\, \mathrm{d}u = \int_0^t Y_i(u)\, c(\boldsymbol{\beta}_0, \mathbf{z}_i(u))\, \alpha_0(u)\, \mathrm{d}u. \tag{4}$$

By standard results on counting processes, it then follows that $M_i(t) = N_i(t) - \Lambda_i(t)$; $i = 1, 2, \ldots, n$; are local square integrable martingales. If we insert the maximum partial likelihood estimator $\widehat{\boldsymbol{\beta}}$ for $\boldsymbol{\beta}_0$ and the increment $\mathrm{d}\widehat{A}_0(u)$ of the Breslow estimator for $\alpha_0(u)\mathrm{d}u$ in (4), we get the estimated cumulative intensity processes

$$\widehat{\Lambda}_i(t) = \int_0^t Y_i(u)\, c(\widehat{\boldsymbol{\beta}}, \mathbf{z}_i(u))\, \mathrm{d}\widehat{A}_0(u) = \sum_{t_j \le t} \frac{Y_i(t_j)\, c(\widehat{\boldsymbol{\beta}}, \mathbf{z}_i(t_j))}{\sum_{l \in \mathcal{R}_j} c(\widehat{\boldsymbol{\beta}}, \mathbf{z}_l(t_j))},$$

and the *martingale residual processes* $\widehat{M}_i(t) = N_i(t) - \widehat{\Lambda}_i(t)$. Evaluating these processes at the terminal study time τ, we obtain the *martingale residuals* $\widehat{M}_i = \widehat{M}_i(\tau)$ first considered by Barlow and Prentice (1988).

Following Aalen (1993), Grønnesby and Borgan (1996) considered the *grouped* martingale residual processes, obtained by aggregating the individual martingale residual processes $\widehat{M}_i(t)$ over groups of individuals. Specifically, assume that we have some grouping of the individuals, typically based on the values of one or two covariates, and denote the groups by $J = 1, \ldots, G$. We will allow the grouping of the individuals to depend on time. Thus an individual may move from one group to another as time passes, as will often be the case when the grouping is performed on the basis of one or more time-dependent covariates. It is a prerequisite, however, that the information used for grouping at time t is available "just before" time t, i.e., the grouping must be based on the "history" $\mathcal{H}_{t-}$. Then, if we denote by $\mathcal{J}(u)$ the set of all individuals who belong to group J at time u, the group J martingale residual process takes the form

$$\widehat{M}_J(t) = \int_0^t \sum_{i \in \mathcal{J}(u)} \mathrm{d}\widehat{M}_i(u) = N_J(t) - \sum_{t_j \le t} \frac{\sum_{i \in \mathcal{R}_j \cap \mathcal{J}(t_j)} c(\widehat{\boldsymbol{\beta}}, \mathbf{z}_i(t_j))}{\sum_{l \in \mathcal{R}_j} c(\widehat{\boldsymbol{\beta}}, \mathbf{z}_l(t_j))}. \quad (5)$$

Here $N_J(t) = \int_0^t \sum_{i \in \mathcal{J}(u)} \mathrm{d}N_i(u)$ is the observed number of failures in group J in $(0, t]$, while the last term on the right-hand side of (5) is an estimate of the expected number of failures in that group if the relative risk regression model (1) holds true. In Section 4 we illustrate how a plot of the grouped martingale residual processes provides a useful tool for checking the fit of the model.

For the special case of an exponential relative risk function, Grønnesby and Borgan (1996) studied the large sample properties of the grouped martingale residual processes. The corresponding results for a general relative risk function, may be obtained as a special case of the results for sampled risk set data given in Section 3.5 below. There we also derive a formal goodness-of-fit test based on the grouped martingale residual processes.

3 Risk set sampling designs

In Sections 3.4 and 3.5 below we will see how martingale residuals may be defined for risk set sampling designs. Before we do that, however, we will review the framework for risk set sampling of Borgan et al. (1995) and generalize some of their results to the situation with a general relative risk function.

3.1 *A model for risk set sampling*

For risk set sampling one selects, whenever a failure occurs, a (typically small) number of controls for the failing individual. The set consisting of these controls together with the failing individual (the case) is called a *sampled risk set.* In order to describe in general terms how the sampling of controls is performed, we need to introduce the "cohort and sampling history" $\mathcal{F}_{t-}$, which contains information about events in the cohort (i.e. $\mathcal{H}_{t-}$) as well as on the sampling of controls, up to, but not including, time t. Based on the parts of this history that are available to the researcher, one decides on a sampling strategy for the controls. Such a strategy may be described in probabilistic terms as follows. Let $\mathcal{P}$ be the power set of $\{1, 2, \ldots, n\}$, i.e. the set of all subsets of $\{1, 2, \ldots, n\}$, and let $\mathcal{P}_i = \{\mathbf{r} : \mathbf{r} \in \mathcal{P},\, i \in \mathbf{r}\}$. Then, given $\mathcal{F}_{t-}$, if an individual i fails at time t, we select the set $\mathbf{r} \in \mathcal{P}_i$ as our sampled risk set with (known) probability $\pi_t(\mathbf{r} \,|\, i)$. Thus, if $Y_i(t) = 1$, then $\pi_t(\mathbf{r} \,|\, i)$ is a probability distribution over sets $\mathbf{r} \in \mathcal{P}_i$. For notational convenience we let $\pi_t(\mathbf{r} \,|\, i) = 0$ whenever $Y_i(t) = 0$.

It turns out to be useful to have a factorization of the sampling probabilities $\pi_t(\mathbf{r} \,|\, i)$. To this end we introduce

$$\pi_t(\mathbf{r}) = n(t)^{-1} \sum_{l \in \mathbf{r}} \pi_t(\mathbf{r} \,|\, l), \tag{6}$$

and note that

$$\sum_{\mathbf{r} \in \mathcal{P}} \pi_t(\mathbf{r}) = n(t)^{-1} \sum_{l=1}^{n} \sum_{\mathbf{r} \in \mathcal{P}_l} \pi_t(\mathbf{r} \,|\, l) = n(t)^{-1} \sum_{l=1}^{n} Y_l(t) = 1.$$

Thus $\pi_t(\mathbf{r})$ is a probability distribution over sets $\mathbf{r} \in \mathcal{P}$. We also introduce

$$w_i(t, \mathbf{r}) = \frac{\pi_t(\mathbf{r} \,|\, i)}{\pi_t(\mathbf{r})}, \tag{7}$$

and get the factorization

$$\pi_t(\mathbf{r} \,|\, i) = w_i(t, \mathbf{r})\, \pi_t(\mathbf{r}). \tag{8}$$

Note that the above framework allows the sampling probabilities to depend in an arbitrary way on events in the past, i.e., on events that are contained in $\mathcal{F}_{t-}$. The sampling probabilities may, however, not depend on events in the future. For example, one may not exclude as a potential control for a current case an individual that subsequently fails. Also note that the selection of controls is done independently at the different failure times, so that subjects may serve as controls for multiple cases, and cases may serve as controls for other cases that failed when the case was at risk. A basic assumption throughout is that not only the truncation and censoring, but also the sampling of controls, are independent in the sense that the additional knowledge of which individuals have entered the study, have been censored or have been selected as controls before any time t do not carry information on the risks of failure at t.

3.2 *Two common sampling designs*

The most common risk set sampling design is simple random sampling; the classical *nested case-control design* (Thomas 1977). For this design, if individual i fails at time t, one selects $m-1$ controls by simple random sampling from the $n(t)-1$ non-failing individuals at risk. In probabilistic terms the design is given by

$$\pi_t(\mathbf{r} \mid i) = \binom{n(t)-1}{m-1}^{-1} I\{\,|\mathbf{r}| = m, \, \mathbf{r} \subset \mathcal{R}(t)\}$$

for any set $\mathbf{r} \in \mathcal{P}_i$. Here the factorization (8) applies with

$$\pi_t(\mathbf{r}) = \binom{n(t)}{m}^{-1} I\{\,|\mathbf{r}| = m, \, \mathbf{r} \subset \mathcal{R}(t)\}\,; \quad \mathbf{r} \in \mathcal{P};$$

$$w_i(t, \mathbf{r}) = \frac{n(t)}{m} I\{i \in \mathbf{r}\}. \tag{9}$$

To select a simple random sample, the only piece of information needed from $\mathcal{F}_{t-}$ is the at risk status of the individuals. Often, however, some additional information is available for all cohort members, e.g., a surrogate measure of the exposure of main interest may be available for everyone. Langholz and Borgan (1995) have developed an "exposure" stratified design which makes it possible to incorporate such information into the sampling process in order to obtain a more informative sample of controls. For this design, called *counter-matching*, one applies the additional piece of information from $\mathcal{F}_{t-}$ to classify each individual at risk into one of say, S, strata. We denote by $\mathcal{R}_s(t)$ the subset of the risk set $\mathcal{R}(t)$ which belongs to stratum s, and let $n_s(t) = |\mathcal{R}_s(t)|$ be the number at risk in this stratum just before time t. If individual i fails at t, we want to sample our controls such that the sampled risk set contains a prespecified number m_s of individuals from each stratum s; $s = 1, \ldots, S$. This is obtained as follows. Assume that the failing individual i belongs to stratum $s(i)$. Then for $s \neq s(i)$ one samples randomly without replacement m_s controls from $\mathcal{R}_s(t)$. From the case's stratum $s(i)$ only $m_{s(i)} - 1$ controls are sampled. The failing individual i is, however, included in the sampled risk set so this contains a total of m_s from each stratum. Even though it is not made explicit in the notation, we note that the classification into strata may be time-dependent. A crucial assumption, however, is that the information on which the stratification is based has to be known "just before" time t.

In probabilistic terms, counter-matched sampling may be described as follows. For any set $\mathbf{r} \in \mathcal{P}_i$ which is a subset of $\mathcal{R}(t)$ and satisfies $|\mathbf{r} \cap \mathcal{R}_s(t)| = m_s$ for $s = 1, \ldots, S$, we have

$$\pi_t(\mathbf{r} \mid i) = \left\{ \binom{n_{s(i)}(t) - 1}{m_{s(i)} - 1} \prod_{s \neq s(i)} \binom{n_s(t)}{m_s} \right\}^{-1}.$$

For counter-matched sampling the factorization (8) applies with

$$\pi_t(\mathbf{r}) = \left\{ \prod_{s=1}^{S} \binom{n_s(t)}{m_s} \right\}^{-1} I(|\mathbf{r} \cap \mathcal{R}_s(t)| = m_s;\ s = 1, \ldots, S); \quad \mathbf{r} \in \mathcal{P};$$

$$w_i(t, \mathbf{r}) = \frac{n_{s(i)}(t)}{m_{s(i)}} I\{i \in \mathbf{r}\}.$$

Other sampling designs for the controls are discussed in Borgan et al. (1995) and Langholz and Goldstein (1996). Note that also the full cohort study is a special case of our general framework in which the full risk set is sampled with probability one, i.e., $\pi_t(\mathbf{r} \mid i) = I\{\mathbf{r} = \mathcal{R}(t)\}$ for all $i \in \mathcal{R}(t)$, and $\pi_t(\mathbf{r} \mid i) = 0$ otherwise.

3.3 *Inference for sampled risk set data*

As in Section 2 we denote by $t_1 < t_2 < \cdots$ the times when failures are observed, and let i_j be the individual who fails at t_j. As described above, the sampled risk set $\widetilde{\mathcal{R}}_j$ is selected according to a sampling distribution $\pi_{t_j}(\mathbf{r} \mid i_j)$ specified by the researcher, and it consists of the case i_j and its controls. Covariate information is collected on the cases and their controls, but are not needed for the other individuals in the cohort. It was shown by Borgan et al. (1995) that from sampled risk set data one may estimate the vector of regression parameters in (1) by $\widehat{\boldsymbol{\beta}}$, the value of $\boldsymbol{\beta}$ maximizing the partial likelihood

$$L(\boldsymbol{\beta}) = \prod_{t_j} \frac{c(\boldsymbol{\beta}, \mathbf{z}_{i_j}(t_j)) w_{i_j}(t_j, \widetilde{\mathcal{R}}_j)}{\sum_{l \in \widetilde{\mathcal{R}}_j} c(\boldsymbol{\beta}, \mathbf{z}_l(t_j)) w_l(t_j, \widetilde{\mathcal{R}}_j)}. \tag{10}$$

We note that (10) is similar to the full cohort partial likelihood. In fact, the full cohort partial likelihood is the special case of (10) in which the entire risk set is sampled with probability one and all weights are unity. Note that for simple random sampling, the weights (9) are the same for all individuals and hence cancel from (10) giving the partial likelihood of Oakes (1981).

The maximum partial likelihood estimator $\widehat{\boldsymbol{\beta}}$ enjoys similar large sample properties as ordinary maximum likelihood estimators. Specifically $\widehat{\boldsymbol{\beta}}$ is approximately multinormally distributed around the true parameter vector $\boldsymbol{\beta}_0$ with a covariance matrix that may be estimated as $\mathcal{I}(\widehat{\boldsymbol{\beta}})^{-1}$, the inverse of the expected information matrix

$$\mathcal{I}(\widehat{\boldsymbol{\beta}}) = \sum_{t_j} \left\{ \frac{\mathbf{S}^{(2)}_{\widetilde{\mathcal{R}}_j}(\widehat{\boldsymbol{\beta}}, t_j)}{S^{(0)}_{\widetilde{\mathcal{R}}_j}(\widehat{\boldsymbol{\beta}}, t_j)} - \left(\frac{\mathbf{S}^{(1)}_{\widetilde{\mathcal{R}}_j}(\widehat{\boldsymbol{\beta}}, t_j)}{S^{(0)}_{\widetilde{\mathcal{R}}_j}(\widehat{\boldsymbol{\beta}}, t_j)} \right)^{\otimes 2} \right\}. \tag{11}$$

Here

$$S^{(0)}_{\widetilde{\mathcal{R}}_j}(\widehat{\boldsymbol{\beta}}, t_j) = \sum_{l \in \widetilde{\mathcal{R}}_j} c(\widehat{\boldsymbol{\beta}}, \mathbf{z}_l(t_j))\, w_l(t_j, \widetilde{\mathcal{R}}_j), \tag{12}$$

$$\mathbf{S}^{(1)}_{\widetilde{\mathcal{R}}_j}(\widehat{\boldsymbol{\beta}}, t_j) = \sum_{l \in \widetilde{\mathcal{R}}_j} \dot{\mathbf{c}}(\widehat{\boldsymbol{\beta}}, \mathbf{z}_l(t_j))\, w_l(t_j, \widetilde{\mathcal{R}}_j), \tag{13}$$

$$\mathbf{S}^{(2)}_{\widetilde{\mathcal{R}}_j}(\widehat{\boldsymbol{\beta}}, t_j) = \sum_{l \in \widetilde{\mathcal{R}}_j} \frac{\dot{\mathbf{c}}(\widehat{\boldsymbol{\beta}}, \mathbf{z}_l(t_j))^{\otimes 2}}{c(\widehat{\boldsymbol{\beta}}, \mathbf{z}_l(t_j))}\, w_l(t_j, \widetilde{\mathcal{R}}_j),$$

where $\dot{\mathbf{c}}(\boldsymbol{\beta}, \mathbf{z}_i(t)) = \partial c(\boldsymbol{\beta}, \mathbf{z}_i(t))/\partial\boldsymbol{\beta}$, and $\mathbf{v}^{\otimes 2}$ of a column vector $\mathbf{v}$ equals the matrix $\mathbf{v}\mathbf{v}^\intercal$. The main steps in the proofs of these properties for the situation with a general relative risk function are given in Section 5.1. For the special case of Cox's regression model, detailed proofs are provided by Borgan et al. (1995).

3.4 *Counting process formulation and martingale residuals*

To derive the partial likelihood (10) and study the asymptotic properties of the maximum partial likelihood estimator, Borgan et al. (1995) expressed the sampled risk set data by means of the processes

$$N_{(i,\mathbf{r})}(t) = \sum_{j \geq 1} I\{t_j \leq t, (i_j, \widetilde{\mathcal{R}}_j) = (i, \mathbf{r})\} \tag{14}$$

counting the observed number of failures for individual i in $(0, t]$ with associated sampled risk set $\mathbf{r}$. These counting processes are also key for deriving the martingale residual processes for sampled risk set data.

From the counting processes $N_{(i,\mathbf{r})}(t)$ we may aggregate over sets $\mathbf{r} \in \mathcal{P}_i$ to recover the counting process (2) registering the observed failures for the ith individual, i.e., $N_i(t) = \sum_{\mathbf{r} \in \mathcal{P}_i} N_{(i,\mathbf{r})}(t)$. In a similar manner we may for a set $\mathbf{r} \in \mathcal{P}$ aggregate over individuals $i \in \mathbf{r}$ to obtain the process

$$N_{\mathbf{r}}(t) = \sum_{i \in \mathbf{r}} N_{(i,\mathbf{r})}(t) = \sum_{j \geq 1} I\{t_j \leq t, \widetilde{\mathcal{R}}_j = \mathbf{r}\} \tag{15}$$

counting the number of times in $(0, t]$ the sampled risk set equals the set $\mathbf{r}$.

The assumption that not only truncation and censoring, but also the sampling of controls, are independent ensures that the intensity processes of the counting processes N_i are given by (3), not only w.r.t. the "cohort history" $\mathcal{H}_{t-}$, but also w.r.t. the "cohort and sampling history" $\mathcal{F}_{t-}$. From this and (8) it follows that the intensity processes $\lambda_{(i,\mathbf{r})}(t)$ of the counting processes (14) take the form

$$\lambda_{(i,\mathbf{r})}(t) = \lambda_i(t)\pi_t(\mathbf{r} \mid i) = Y_i(t)c(\boldsymbol{\beta}_0, \mathbf{z}_i(t))w_i(t, \mathbf{r})\pi_t(\mathbf{r})\alpha_0(t). \tag{16}$$

Therefore by general results for counting processes

$$M_{(i,\mathbf{r})}(t) = N_{(i,\mathbf{r})}(t) - \Lambda_{(i,\mathbf{r})}(t) \tag{17}$$

with

$$\Lambda_{(i,\mathbf{r})}(t) = \int_0^t Y_i(u)c(\boldsymbol{\beta}_0, \mathbf{z}_i(u))w_i(u, \mathbf{r})\pi_u(\mathbf{r})\alpha_0(u)du \tag{18}$$

are local square integrable martingales. As for cohort data, we will insert estimates for $\boldsymbol{\beta}_0$ and $\alpha_0(u)du$ in (18) to obtain estimated cumulative intensity processes $\widehat{\Lambda}_{(i,\mathbf{r})}(t)$. For $\boldsymbol{\beta}_0$ we insert the maximum partial likelihood estimator $\widehat{\boldsymbol{\beta}}$, and for $\alpha_0(u)du$ we insert $d\widehat{A}_{0\mathbf{r}}(t)$, where

$$\widehat{A}_{0\mathbf{r}}(t) = \sum_{t_j \leq t, \widetilde{\mathcal{R}}_j = \mathbf{r}} \frac{1}{\sum_{l \in \mathbf{r}} c(\widehat{\boldsymbol{\beta}}, \mathbf{z}_l(t_j))w_l(t_j, \mathbf{r})\pi_{t_j}(\mathbf{r})}. \tag{19}$$

Thus we get the estimated cumulative intensity processes

$$\widehat{\Lambda}_{(i,\mathbf{r})}(t) = \int_0^t Y_i(u)c(\widehat{\boldsymbol{\beta}}, \mathbf{z}_i(u))w_i(u, \mathbf{r})\pi_u(\mathbf{r})d\widehat{A}_{0\mathbf{r}}(u)$$

and the corresponding *martingale residual processes*

$$\widehat{M}_{(i,\mathbf{r})}(t) = N_{(i,\mathbf{r})}(t) - \widehat{\Lambda}_{(i,\mathbf{r})}(t). \tag{20}$$

The martingale residual processes (20) are of little use in their own right, in fact most of them will be identically equal to zero. But they provide the building blocks for the grouped martingale residual processes for sampled risk set data.

3.5 *Grouped martingale residual processes and a chi-squared goodness-of-fit test*

As in Section 2, we assume that we have a grouping of the individuals into G groups, and denote by $\mathcal{J}(u)$ the set of all individuals who belong to group J at time u; $J = 1, \ldots, G$. Then the group J martingale residual process for sampled risk set data corresponding to (5) is given by

$$\begin{aligned}\widehat{M}_J(t) &= \int_0^t \sum_{i \in \mathcal{J}(u)} \sum_{\mathbf{r} \in \mathcal{P}_i} \mathrm{d}\widehat{M}_{(i,\mathbf{r})}(u) \\ &= \int_0^t \sum_{i \in \mathcal{J}(u)} \mathrm{d}N_i(u) - \int_0^t \sum_{\mathbf{r} \in \mathcal{P}} \sum_{i \in \mathbf{r} \cap \mathcal{J}(u)} \mathrm{d}\widehat{\Lambda}_{(i,\mathbf{r})}(u),\end{aligned}$$

which may be rewritten as

$$\widehat{M}_J(t) = N_J(t) - \sum_{t_j \le t} \frac{\sum_{i \in \widetilde{\mathcal{R}}_j \cap \mathcal{J}(t_j)} c(\widehat{\boldsymbol{\beta}}, \mathbf{z}_i(t_j))\, w_i(t_j, \widetilde{\mathcal{R}}_j)}{\sum_{l \in \widetilde{\mathcal{R}}_j} c(\widehat{\boldsymbol{\beta}}, \mathbf{z}_l(t_j))\, w_l(t_j, \widetilde{\mathcal{R}}_j)} \tag{21}$$

with $N_J(t) = \int_0^t \sum_{i \in \mathcal{J}(u)} \mathrm{d}N_i(u)$. As for cohort data, these grouped martingale residual processes may be interpreted as observed minus expected number of events in the given groups.

In Section 5.2 we note that, if we could have used the true value $\boldsymbol{\beta}_0$ instead of its estimate $\widehat{\boldsymbol{\beta}}$ in (21), then the grouped martingale residual processes would have been martingales. However, since the regression coefficients have to be estimated, the grouped martingale residual processes are only approximately martingales. In Section 5.2 we also show that, properly normalized, the vector of grouped martingale residual processes $(\widehat{M}_1, \ldots, \widehat{M}_G)^\top$ converges weakly to a mean zero multivariate Gaussian process. Further the covariance between $\widehat{M}_I(s)$ and $\widehat{M}_J(t)$ can be estimated by

$$\widehat{\sigma}_{IJ}(s,t) = \widehat{\phi}_{IJ}(0, s \wedge t, \widehat{\boldsymbol{\beta}}) - \widehat{\boldsymbol{\psi}}_I(0, s, \widehat{\boldsymbol{\beta}})^\top \mathcal{I}(\widehat{\boldsymbol{\beta}})^{-1}\, \widehat{\boldsymbol{\psi}}_J(0, t, \widehat{\boldsymbol{\beta}}), \tag{22}$$

where

$$\widehat{\phi}_{IJ}(s_1, s_2, \widehat{\boldsymbol{\beta}}) = \sum_{s_1 < t_j \le s_2} \frac{S^{(0)}_{\widetilde{\mathcal{R}}_j I}(\widehat{\boldsymbol{\beta}}, t_j)}{S^{(0)}_{\widetilde{\mathcal{R}}_j}(\widehat{\boldsymbol{\beta}}, t_j)} \left\{ \delta_{IJ} - \frac{S^{(0)}_{\widetilde{\mathcal{R}}_j J}(\widehat{\boldsymbol{\beta}}, t_j)}{S^{(0)}_{\widetilde{\mathcal{R}}_j}(\widehat{\boldsymbol{\beta}}, t_j)} \right\} \tag{23}$$

with δ_{IJ} a Kronecker delta, and

$$\widehat{\boldsymbol{\psi}}_J(s_1, s_2, \widehat{\boldsymbol{\beta}}) = \sum_{s_1 < t_j \le s_2} \left\{ \frac{\mathbf{S}^{(1)}_{\widetilde{\mathcal{R}}_j J}(\widehat{\boldsymbol{\beta}}, t_j)}{S^{(0)}_{\widetilde{\mathcal{R}}_j}(\widehat{\boldsymbol{\beta}}, t_j)} - \frac{S^{(0)}_{\widetilde{\mathcal{R}}_j J}(\widehat{\boldsymbol{\beta}}, t_j)\, \mathbf{S}^{(1)}_{\widetilde{\mathcal{R}}_j}(\widehat{\boldsymbol{\beta}}, t_j)}{S^{(0)}_{\widetilde{\mathcal{R}}_j}(\widehat{\boldsymbol{\beta}}, t_j)^2} \right\}. \tag{24}$$

Here $S^{(0)}_{\widetilde{\mathcal{R}}_j J}(\widehat{\boldsymbol{\beta}}, t_j)$ and $\mathbf{S}^{(1)}_{\widetilde{\mathcal{R}}_j J}(\widehat{\boldsymbol{\beta}}, t_j)$ are given by expressions similar to (12) and (13), but with the summation restricted to individuals $l \in \widetilde{\mathcal{R}}_j \cap \mathcal{J}(t_j)$.

As will be illustrated in Section 4, a plot of the grouped martingale residual processes is a useful tool for assessing the fit of the relative risk regression model (1). In addition the grouped martingale residual processes may be used to derive formal goodness-of fit tests. In Section 7 we briefly discuss different possible goodness-of-fit tests. Here we restrict our attention to a simple chi-squared test based on a comparison of observed and expected number of events in the G groups in K disjoint time intervals. To this end let $0 = a_0 < a_1 < \cdots < a_{K-1} < a_K = \tau$ be a partitioning of the study time interval, and introduce (for $H = 1, 2, \ldots, K$ and $J = 1, 2, \ldots, G$)

$$\widehat{M}_{HJ} = \widehat{M}_J(a_H) - \widehat{M}_J(a_{H-1}) = O_{HJ} - E_{HJ}. \tag{25}$$

Here $O_{HJ} = N_J(a_H) - N_J(a_{H-1})$ is the observed number of events in group J in time interval H, while

$$E_{HJ} = \sum_{a_{H-1} < t_j \le a_H} \frac{\sum_{i \in \widetilde{\mathcal{R}}_j \cap \mathcal{J}(t_j)} c(\widehat{\boldsymbol{\beta}}, \mathbf{z}_i(t_j))\, w_i(t_j, \widetilde{\mathcal{R}}_j)}{\sum_{l \in \widetilde{\mathcal{R}}_j} c(\widehat{\boldsymbol{\beta}}, \mathbf{z}_l(t_j))\, w_l(t_j, \widetilde{\mathcal{R}}_j)}$$

is the corresponding expected number under model (1). The martingale residual processes (21) sum to zero at any given time t. To derive a chi-squared goodness-of-fit test, we therefore disregard the contribution from one of the groups, say the first group, and consider the $K(G-1)$-vector $\widehat{\mathbf{M}}$ with elements $\widehat{M}_{HJ}$ for $H = 1, 2, \ldots, K$; $J = 2, 3, \ldots, G$. By the large sample distributional results for the grouped martingale residual processes summarized in connection with (22), it follows that $\widehat{\mathbf{M}}$ is approximately mean zero multinormally distributed in large samples when model (1) holds true. Its covariance matrix may be estimated by the matrix $\widehat{\boldsymbol{\Sigma}} = \{\widehat{\sigma}_{LI,HJ}\}$ with elements

$$\begin{aligned}\widehat{\sigma}_{LI,HJ} &= \widehat{\mathrm{Cov}}(\widehat{M}_{LI}, \widehat{M}_{HJ}) \\ &= \delta_{LH}\, \widehat{\phi}_{IJ}(a_{H-1}, a_H, \widehat{\boldsymbol{\beta}}) - \widehat{\boldsymbol{\psi}}_I(a_{L-1}, a_L, \widehat{\boldsymbol{\beta}})^{\mathsf{T}} \mathcal{I}(\widehat{\boldsymbol{\beta}})^{-1} \widehat{\boldsymbol{\psi}}_J(a_{H-1}, a_H, \widehat{\boldsymbol{\beta}});\end{aligned}$$

$H, L = 1, 2, \ldots, K$; $J, I = 2, 3, \ldots, G$; where δ_{LH} is a Kronecker delta. Therefore a goodness-of-fit test may be based on the statistic $\chi^2 = \widehat{\mathbf{M}}^{\mathsf{T}} \widehat{\boldsymbol{\Sigma}}^{-1} \widehat{\mathbf{M}}$, which is approximately chi-squared distributed with $K(G-1)$ degrees of freedom in large samples when model (1) holds true.

Large sample results for the grouped martingale residual processes and the goodness-of-fit test for full cohort data, are the special cases of the above results in which the sampled risk set equals the full risk set with probability one and all weights are unity. In particular for cohort data with exponential relative risk function and only one time interval (i.e. $K = 1$), the test statistic χ^2 specializes to the goodness-of-fit statistic of Grønnesby and Borgan (1996).

May and Hosmer (2004) showed how the test of Grønnesby and Borgan (1996) can be obtained as the score test for the addition of categorical grouping variables. A similar result holds here as well. More specifically, consider the extension of model (1) where an individual i who belongs to group J at time $t \in (a_{H-1}, a_H]$ has a hazard rate of the form

$$\alpha(t; \mathbf{z}_i) = c(\boldsymbol{\beta}_0, \mathbf{z}_i(t))\, e^{\gamma_{HJ}}\, \alpha_0(t); \tag{26}$$

$J = 2, 3, \ldots, G$. Then by some straightforward, but tedious algebra along the lines of Appendix A in May and Hosmer (2004) one may show that the goodness-of-fit statistic χ^2 is algebraically equivalent to the score test for the hypothesis that all the additional $K(G-1)$ parameters γ_{HJ} in (26) are equal to zero.

4 An illustration

To illustrate the use of the grouped martingale residual processes and the accompanying goodness-of-fit test, we will use data on lung cancer death among a cohort of uranium miners from the Colorado Plateau. The cohort was assembled to study the effects of radon exposure and smoking on lung cancer risk and has been described in detail in earlier publications; e.g. Hornung and Meinhardt (1987). The cohort consists of 3,347 Caucasian male miners recruited between 1950 and 1960 and was traced for mortality outcomes through December 31, 1982, by which time 258 lung cancer deaths were observed. Exposure data include radon exposure, in working level months (WLM), and smoking histories, in number of packs of cigarettes (20 cigarettes per pack) smoked per day. We consider age as the basic time scale and summarize radon and smoking data into cumulative exposures lagged by two years. Thus we consider the covariates $\mathbf{z}(t) = (z_{i1}(t), z_{i2}(t))^{\top}$, where $z_{i1}(t)$ is cumulative radon exposure measured in working level months (WLM) up to two years prior to age t, and $z_{i2}(t)$ is cumulative smoking in number of packs smoked up to two years prior to t. Although covariate information is available on all cohort subjects, in order to illustrate the methods we selected simple random and counter-matched samples with three controls per case. These data sets are denoted 1:3 simple random and counter-matched samples, respectively. The 23 tied failure times were broken randomly so that there was only one case per risk set. Counter-matching was based on radon exposure grouped into four strata according to the quartiles of the cumulative radon exposure for the cases (Langholz and Goldstein 1996, Section 5), and one control was sampled at random from each stratum except the one of the case.

As has been the case in previous analyzes of these data (cf. Langholz

Table 1 Observed and expected number of lung cancer deaths.

Exposure group[a]	Observed numbers	Expected numbers 1:3 simple	Expected numbers 1:3 counter-matched
		Below 60 years of age	
Group I	30	30.7	35.5
Group II	39	45.9	48.4
Group III	81	73.4	66.1
		Above 60 years of age	
Group I	27	27.7	25.3
Group II	45	36.1	36.9
Group III	36	44.2	45.8

a) Group I: below 500 WLMs; group II: 500–1500 WLMs; group III: above 1500 WLMs.

and Goldstein 1996 and their references), the excess relative risk model was used. Thus the hazard rate for miner i is assumed to take the form

$$\alpha(t; \mathbf{z}_i) = [1 + \beta_{01}\, z_{i1}(t)]\,[1 + \beta_{02}\, z_{i2}(t)]\,\alpha_0(t). \tag{27}$$

For the 1:3 simple random data, the estimated radon excess relative risk (with standard error) is $\widehat{\beta}_1 = 0.556$ (0.215) per 100 WLMs cumulative radon exposure, while the smoking excess relative risk is $\widehat{\beta}_2 = 0.276$ (0.093) per 1000 packs of cigarettes smoked. For the 1:3 counter-matched data, the estimates become $\widehat{\beta}_1 = 0.420$ (0.137) and $\widehat{\beta}_2 = 0.205$ (0.068).

Figure 1 shows the grouped martingale residual processes (21) for both data sets when the individuals are aggregated over groups defined by cumulative radon exposure (group I: below 500 WLMs; group II: 500–1500 WLMs; group III: above 1500 WLMs), while Table 1 summarizes the observed and expected number of lung cancer deaths in the three radon exposure groups for ages below and above 60 years. From the plots and the table it is seen that more lung cancer deaths than expected occur in the high exposure group (group III) below the age of 60 years, while fewer cases than expected occur above this age, the pattern being most pronounced for the counter-matched data. The chi-squared goodness-of-fit statistic with $2(3-1) = 4$ degrees of freedom based on the observed and expected numbers of Table 1 takes the values 10.5 and 14.2, respectively, for the 1:3 simple random sample and the 1:3 counter-matched sample, with corresponding P-values 0.032 and 0.007. Thus our analysis shows that the excess relative risk model (27), where the effect of radon depends linearly on cumulative exposure, is too simplistic.

The lack of fit is further illustrated in Table 2 for the 1:3 simple random sample. The table shows relative risks within radon exposure categories for individuals below and above 60 years of age, as well as the relative risks

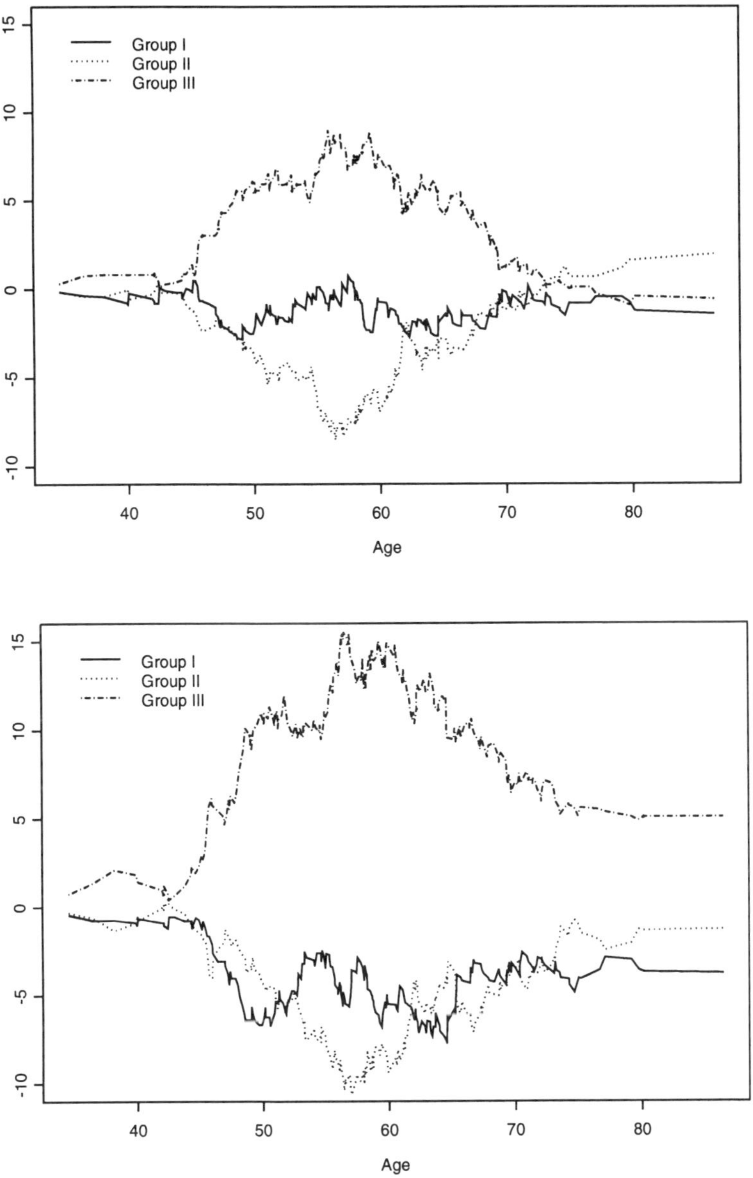

Figure 1 Grouped martingale residual processes for the uranium miners based on a 1:3 simple random sample (upper panel) and a 1:3 counter-matched sample (lower panel). Grouping is done according to cumulative radon exposure: Group I: below 500 WLMs; group II: 500–1500 WLMs; group III: above 1500 WLMs.

Table 2 Relative risks within categories of cumulative radon exposure by age at lung cancer death and relative risks predicted by the excess relative risk model (27). 1:3 simple random sampling.

Radon exposure category[a]	Mean exposure within category[b]	Relative risks for categorical model	Relative risks[c] predicted by model (27)
		Below 60 years of age	
Group I	180	1	1
Group II	896	2.35	2.99
Group III	2885	10.91	8.51
		Above 60 years of age	
Group I	187	1	1
Group II	923	3.89	3.01
Group III	3034	5.61	8.76

a) Group I: below 500 WLMs; group II: 500–1500 WLMs; group III: above 1500 WLMs.

b) Mean among controls.

c) Computed at category mean, normalized to mean of first category.

predicted by the excess relative risk model (27). Prior to age 60 years, lung cancer mortality rates increase faster than linear with radon exposure level while after age 60, the dose response is quite a bit slower than linear. There are a number of possible ways one might choose to accommodate this pattern of rates. One could simply accommodate the variation in a model that allows for changing shape of the dose response curve with age. But, since miners tended to experience the larger exposures at earlier ages, the observed change in exposure response curve shape with age may well be due to the time since exposure. Thus, a biologically appealing approach would be to summarize the exposure history in a way that accounts for the time since exposure (latency) and, perhaps, rate of exposure. In fact, it has been found that latency effects are a significant component in describing radiation exposure and lung cancer risk in the Colorado Plateau miners, e.g. Lubin et al. (1994) and Langholz et al. (1999). It is, however, beyond the scope of this paper to pursue such alternative models. Here we are content with the above illustration of how the grouped residual process plots play a useful role by identifying model lack of fit and by suggesting candidate changes that may yield a better fitting model.

5 Outline of proofs for sampled risk set data

In this section, we give an outline of the proofs of the large sample properties of the maximum partial likelihood estimator $\widehat{\boldsymbol{\beta}}$ and the grouped martingale residual processes for sampled risk set data when we have a general relative risk function. Formal proofs may be written out along the lines of Borgan et al. (1995), who give detailed proofs of the large sample properties of the maximum partial likelihood estimator for the special case of an exponential relative risk function.

5.1 *Large sample properties of* $\widehat{\boldsymbol{\beta}}$

The estimator $\widehat{\boldsymbol{\beta}}$ is the solution to $\mathbf{U}(\boldsymbol{\beta}) = \mathbf{0}$, where $\mathbf{U}(\boldsymbol{\beta}) = \partial \log L(\boldsymbol{\beta})/\partial\boldsymbol{\beta}$ is the vector of score functions, and $L(\boldsymbol{\beta})$ is the partial likelihood (10). Using counting process notation, the vector of score functions may be expressed as

$$\mathbf{U}(\boldsymbol{\beta}) = \sum_{\mathbf{r}\in\mathcal{P}}\sum_{i\in\mathbf{r}}\int_0^\tau \left\{ \frac{\dot{\mathbf{c}}(\boldsymbol{\beta},\mathbf{z}_i(u))}{c(\boldsymbol{\beta},\mathbf{z}_i(u))} - \frac{\mathbf{S}_{\mathbf{r}}^{(1)}(\boldsymbol{\beta},u)}{S_{\mathbf{r}}^{(0)}(\boldsymbol{\beta},u)} \right\} \mathrm{d}N_{(i,\mathbf{r})}(u),$$

where τ is the terminal study time, and

$$S_{\mathbf{r}}^{(0)}(\boldsymbol{\beta},u) = \sum_{l\in\mathbf{r}} Y_l(u)\, c(\boldsymbol{\beta},\mathbf{z}_l(u))\, w_l(u,\mathbf{r}), \tag{28}$$

$$\mathbf{S}_{\mathbf{r}}^{(1)}(\boldsymbol{\beta},u) = \sum_{l\in\mathbf{r}} Y_l(u)\, \dot{\mathbf{c}}(\boldsymbol{\beta},\mathbf{z}_l(u))\, w_l(u,\mathbf{r}). \tag{29}$$

Further the observed partial information matrix $\mathbf{I}(\boldsymbol{\beta}) = -\partial\mathbf{U}(\boldsymbol{\beta})/\partial\boldsymbol{\beta}^{\mathsf{T}}$ becomes

$$\mathbf{I}(\boldsymbol{\beta}) = \sum_{\mathbf{r}\in\mathcal{P}}\sum_{i\in\mathbf{r}}\int_0^\tau \frac{\partial}{\partial\boldsymbol{\beta}^{\mathsf{T}}} \left\{ \frac{\mathbf{S}_{\mathbf{r}}^{(1)}(\boldsymbol{\beta},u)}{S_{\mathbf{r}}^{(0)}(\boldsymbol{\beta},u)} - \frac{\dot{\mathbf{c}}(\boldsymbol{\beta},\mathbf{z}_i(u))}{c(\boldsymbol{\beta},\mathbf{z}_i(u))} \right\} \mathrm{d}N_{(i,\mathbf{r})}(u). \tag{30}$$

If we evaluate the score function at $\boldsymbol{\beta}_0$, we find by some straightforward algebra [using (17) and (18)]:

$$\mathbf{U}(\boldsymbol{\beta}_0) = \sum_{\mathbf{r}\subset\mathcal{P}}\sum_{i\in\mathbf{r}}\int_0^\tau \left\{ \frac{\dot{\mathbf{c}}(\boldsymbol{\beta}_0,\mathbf{z}_i(u))}{c(\boldsymbol{\beta}_0,\mathbf{z}_i(u))} - \frac{\mathbf{S}_{\mathbf{r}}^{(1)}(\boldsymbol{\beta}_0,u)}{S_{\mathbf{r}}^{(0)}(\boldsymbol{\beta}_0,u)} \right\} \mathrm{d}M_{(i,\mathbf{r})}(u). \tag{31}$$

Here the integrands are predictable processes. Thus the score function is a sum of (vector-valued) stochastic integrals when evaluated at the true value of the regression coefficients. If, on the right hand side of (31), we replace the upper limit of integration by t, we get a stochastic process. This stochastic process is a martingale with a predictable variation process that evaluated at τ becomes

$$\langle \mathbf{U}(\boldsymbol{\beta}_0)\rangle(\tau) = \sum_{\mathbf{r}\in\mathcal{P}}\sum_{i\in\mathbf{r}}\int_0^\tau \left\{ \frac{\dot{\mathbf{c}}(\boldsymbol{\beta}_0,\mathbf{z}_i(u))}{c(\boldsymbol{\beta}_0,\mathbf{z}_i(u))} - \frac{\mathbf{S}_{\mathbf{r}}^{(1)}(\boldsymbol{\beta},u)}{S_{\mathbf{r}}^{(0)}(\boldsymbol{\beta},u)} \right\}^{\otimes 2} \lambda_{(i,\mathbf{r})}(u)\mathrm{d}u.$$

Using (16), we get after some straightforward algebra that

$$\langle \mathbf{U}(\boldsymbol{\beta}_0) \rangle (\tau) = \sum_{\mathbf{r} \in \mathcal{P}} \int_0^{\tau} \mathbf{V}_{\mathbf{r}}(\boldsymbol{\beta}_0, u)\, S_{\mathbf{r}}^{(0)}(\boldsymbol{\beta}, u)\, \pi_u(\mathbf{r})\, \alpha_0(u) \mathrm{d}u, \tag{32}$$

where

$$\mathbf{V}_{\mathbf{r}}(\boldsymbol{\beta}, u) = \frac{\mathbf{S}_{\mathbf{r}}^{(2)}(\boldsymbol{\beta}, u)}{S_{\mathbf{r}}^{(0)}(\boldsymbol{\beta}, u)} - \left(\frac{\mathbf{S}_{\mathbf{r}}^{(1)}(\boldsymbol{\beta}, u)}{S_{\mathbf{r}}^{(0)}(\boldsymbol{\beta}, u)} \right)^{\otimes 2} \tag{33}$$

with $S_{\mathbf{r}}^{(0)}(\boldsymbol{\beta}, u)$ and $\mathbf{S}_{\mathbf{r}}^{(1)}(\boldsymbol{\beta}, u)$ given by (28) and (29), respectively, and

$$\mathbf{S}_{\mathbf{r}}^{(2)}(\boldsymbol{\beta}, u) = \sum_{l \in \mathbf{r}} Y_l(u) \frac{\dot{\mathbf{c}}(\boldsymbol{\beta}, \mathbf{z}_l(u))^{\otimes 2}}{c(\boldsymbol{\beta}, \mathbf{z}_l(u))} w_l(u, \mathbf{r}).$$

If we insert $\mathrm{d}N_{(i,\mathbf{r})}(u) = \lambda_{(i,\mathbf{r})}(u)\mathrm{d}u + \mathrm{d}M_{(i,\mathbf{r})}(u)$ [cf. (17)] and use (16) in (30), we find after some algebra that the observed information matrix evaluated at $\boldsymbol{\beta}_0$ may be decomposed as

$$\begin{aligned} \mathbf{I}(\boldsymbol{\beta}_0) &= \langle \mathbf{U}(\boldsymbol{\beta}_0) \rangle (\tau) \\ &+ \sum_{\mathbf{r} \in \mathcal{P}} \sum_{i \in \mathbf{r}} \int_0^{\tau} \frac{\partial}{\partial \boldsymbol{\beta}^{\mathsf{T}}} \left\{ \frac{\mathbf{S}_{\mathbf{r}}^{(1)}(\boldsymbol{\beta}_0, u)}{S_{\mathbf{r}}^{(0)}(\boldsymbol{\beta}_0, u)} - \frac{\dot{\mathbf{c}}(\boldsymbol{\beta}_0, \mathbf{z}_i(u))}{c(\boldsymbol{\beta}_0, \mathbf{z}_i(u))} \right\} \mathrm{d}M_{(i,\mathbf{r})}(u). \end{aligned}$$

Thus, at the true value of the vector of regression coefficients, the observed information matrix equals the predictable variation process of the score function plus a stochastic integral.

By the martingale central limit theorem, we may now show, under suitable regularity conditions, that $n^{-1/2}\mathbf{U}(\boldsymbol{\beta}_0)$ converges weakly to a multinormal distribution with mean zero and a covariance matrix $\boldsymbol{\Sigma}_{\boldsymbol{\beta}}$ that is the limit in probability of $n^{-1} \langle \mathbf{U}(\boldsymbol{\beta}_0) \rangle (\tau)$. We may also show that both $n^{-1}\mathbf{I}(\boldsymbol{\beta}_0)$ and $n^{-1}\mathbf{I}(\widehat{\boldsymbol{\beta}})$ converge in probability to $\boldsymbol{\Sigma}_{\boldsymbol{\beta}}$. From these results the large sample properties of $\widehat{\boldsymbol{\beta}}$ follow in the usual way. The main steps in the derivations are as follows. Since $\widehat{\boldsymbol{\beta}}$ is the solution to the score equation $\mathbf{U}(\widehat{\boldsymbol{\beta}}) = \mathbf{0}$, a Taylor expansion of the score equation around $\boldsymbol{\beta}_0$ gives $\mathbf{0} = \mathbf{U}(\widehat{\boldsymbol{\beta}}) \approx \mathbf{U}(\boldsymbol{\beta}_0) - \mathbf{I}(\boldsymbol{\beta}_0)(\widehat{\boldsymbol{\beta}} - \boldsymbol{\beta}_0)$. From this we obtain

$$\sqrt{n}\left(\widehat{\boldsymbol{\beta}} - \boldsymbol{\beta}_0\right) \approx \left(n^{-1}\mathbf{I}(\boldsymbol{\beta}_0)\right)^{-1} n^{-1/2}\mathbf{U}(\boldsymbol{\beta}_0) \approx \boldsymbol{\Sigma}_{\boldsymbol{\beta}}^{-1}\, n^{-1/2}\mathbf{U}(\boldsymbol{\beta}_0), \tag{34}$$

and it follows that $\sqrt{n}(\widehat{\boldsymbol{\beta}} - \boldsymbol{\beta}_0)$ converges weakly to a multinormal distribution with mean zero and covariance matrix $\boldsymbol{\Sigma}_{\boldsymbol{\beta}}^{-1}\boldsymbol{\Sigma}_{\boldsymbol{\beta}}\boldsymbol{\Sigma}_{\boldsymbol{\beta}}^{-1} = \boldsymbol{\Sigma}_{\boldsymbol{\beta}}^{-1}$. Thus, in large samples, $\widehat{\boldsymbol{\beta}}$ is approximately multinormally distributed around $\boldsymbol{\beta}_0$ with covariance matrix $n^{-1}\boldsymbol{\Sigma}_{\boldsymbol{\beta}}^{-1}$.

In order to estimate the covariance matrix of $\widehat{\boldsymbol{\beta}}$ we may use $\mathbf{I}(\widehat{\boldsymbol{\beta}})^{-1}$, the inverse of the observed information, or we may use the inverse of the (estimated) expected information matrix. The (estimated) expected information is obtained from (32) by inserting $\widehat{\boldsymbol{\beta}}$ for $\boldsymbol{\beta}_0$ and the increment

$$\mathrm{d}\widehat{A}_{0\mathbf{r}}(u) = \frac{\mathrm{d}N_{\mathbf{r}}(u)}{S_{\mathbf{r}}^{(0)}(\widehat{\boldsymbol{\beta}}, u)\pi_u(\mathbf{r})}$$

of the Breslow type estimator (19) for $\alpha_0(u)\mathrm{d}u$ to get

$$\mathcal{I}(\widehat{\boldsymbol{\beta}}) = \sum_{\mathbf{r}\in\mathcal{P}} \int_0^{\tau} \mathbf{V}_{\mathbf{r}}(\widehat{\boldsymbol{\beta}}, u)\, \mathrm{d}N_{\mathbf{r}}(u), \tag{35}$$

where $\mathbf{V}_{\mathbf{r}}(\boldsymbol{\beta}, u)$ is given by (33). This justifies (11) of Section 3.3. By (35) and (30) we note that while the expected information matrix depends only on quantities that are aggregates over each sampled risk set, the observed information matrix depends specifically on the covariates of the cases. Therefore the expected information matrix tends to be the most stable of the two, and it is the one we recommend. For Cox's regression model the observed and expected information matrices coincide.

5.2 *Large sample properties of the grouped martingale residual processes*

We will derive similar large sample properties for the grouped martingale residuals for sampled risk set data as those of Grønnesby and Borgan (1996) for Cox regression with cohort data. To this end we first note that the grouped martingale residual processes (21) may be given as

$$\widehat{M}_J(t) = \sum_{\mathbf{r}\in\mathcal{P}} \int_0^t \sum_{i\in\mathbf{r}\cap\mathcal{J}(u)} \mathrm{d}N_{(i,\mathbf{r})}(u) - \sum_{\mathbf{r}\in\mathcal{P}} \int_0^t \frac{S_{\mathbf{r}J}^{(0)}(\widehat{\boldsymbol{\beta}}, u)}{S_{\mathbf{r}}^{(0)}(\widehat{\boldsymbol{\beta}}, u)}\, \mathrm{d}N_{\mathbf{r}}(u), \tag{36}$$

where $S_{\mathbf{r}}^{(0)}(\boldsymbol{\beta}, u)$ is given by (28) and

$$S_{\mathbf{r}J}^{(0)}(\boldsymbol{\beta}, u) = \sum_{l\in\mathbf{r}\cap\mathcal{J}(u)} Y_l(u) c(\boldsymbol{\beta}, \mathbf{z}_l(u)) w_l(u, \mathbf{r}).$$

We also note that the intensity process of the counting process $N_{\mathbf{r}}(t)$ given by (15) takes the form

$$\lambda_{\mathbf{r}}(t) = \sum_{i\in\mathbf{r}} \lambda_{(i,\mathbf{r})}(t) = S_{\mathbf{r}}^{(0)}(\boldsymbol{\beta}_0, t)\, \pi_t(\mathbf{r})\, \alpha_0(t), \tag{37}$$

where we have used (16) and (28) to get the last equality. We also introduce the martingales

$$M_{\mathbf{r}}(t) = \sum_{i\in\mathbf{r}} M_{(i,\mathbf{r})}(t) = N_{\mathbf{r}}(t) - \int_0^t \lambda_{\mathbf{r}}(u)\mathrm{d}u. \tag{38}$$

Then, using (17), (18), (28), and (36) – (38), we find after some straightforward algebra that the normalized grouped martingale residual processes may be decomposed as

$$n^{-1/2}\widehat{M}_J(t) = \tag{39}$$

$$X_J^*(t) - n^{-1/2}\sum_{\mathbf{r}\in\mathcal{P}}\int_0^t \left\{ \frac{S_{\mathbf{r}J}^{(0)}(\widehat{\boldsymbol\beta},u)}{S_{\mathbf{r}}^{(0)}(\widehat{\boldsymbol\beta},u)} - \frac{S_{\mathbf{r}J}^{(0)}(\boldsymbol\beta_0,u)}{S_{\mathbf{r}}^{(0)}(\boldsymbol\beta_0,u)} \right\} \mathrm{d}N_{\mathbf{r}}(u),$$

where

$$X_J^*(t) = n^{-1/2}\sum_{\mathbf{r}\in\mathcal{P}}\sum_{i\in\mathbf{r}}\int_0^t \left\{ \delta_{iJ}(u) - \frac{S_{\mathbf{r}J}^{(0)}(\boldsymbol\beta_0,u)}{S_{\mathbf{r}}^{(0)}(\boldsymbol\beta_0,u)} \right\} \mathrm{d}M_{(i,\mathbf{r})}(u)$$

with $\delta_{iJ}(u) = 1$ if $i \in \mathcal{J}(u)$, i.e. if individual i belongs to group J at time u, and $\delta_{iJ}(u) = 0$ otherwise. Note that $X_J^*(t)$ is a stochastic integral, and hence itself a martingale. Thus the grouped martingale residual processes would have been martingales if we could use the true value $\boldsymbol\beta_0$ instead of its estimate $\widehat{\boldsymbol\beta}$ in (36).

We now take a closer look at the last term in (39). By a Taylor series expansion, one may show that this term is asymptotically equivalent to

$$-n^{-1}\sum_{\mathbf{r}\in\mathcal{P}}\int_0^t \frac{\partial}{\partial\boldsymbol\beta^{\mathsf T}} \left\{ \frac{S_{\mathbf{r}J}^{(0)}(\boldsymbol\beta_0,u)}{S_{\mathbf{r}}^{(0)}(\boldsymbol\beta_0,u)} \right\} \mathrm{d}N_{\mathbf{r}}(u)\,\sqrt{n}\left(\widehat{\boldsymbol\beta} - \boldsymbol\beta_0\right).$$

Now, using (37) and (38), the latter expression may be shown to be asymptotically equivalent to $-\boldsymbol\psi_J(0,t,\boldsymbol\beta_0)^{\mathsf T}\sqrt{n}\,(\widehat{\boldsymbol\beta} - \boldsymbol\beta_0)$, where $\boldsymbol\psi_J(s_1,s_2,\boldsymbol\beta_0)$ is the uniform (in s_1 and s_2) limit in probability of

$$n^{-1}\sum_{\mathbf{r}\in\mathcal{P}}\int_{s_1}^{s_2} \left\{ \frac{\mathbf{S}_{\mathbf{r}J}^{(1)}(\boldsymbol\beta_0,u)}{S_{\mathbf{r}}^{(0)}(\boldsymbol\beta_0,u)} - \frac{S_{\mathbf{r}J}^{(0)}(\boldsymbol\beta_0,u)\,\mathbf{S}_{\mathbf{r}}^{(1)}(\boldsymbol\beta_0,u)}{S_{\mathbf{r}}^{(0)}(\boldsymbol\beta_0,u)^2} \right\} S_{\mathbf{r}}^{(0)}(\boldsymbol\beta_0,u)\pi_u(\mathbf{r})\alpha_0(u)\mathrm{d}u, \tag{40}$$

and

$$\mathbf{S}_{\mathbf{r}J}^{(1)}(\boldsymbol\beta,u) = \sum_{l\in\mathbf{r}\cap\mathcal{J}(u)} Y_l(u)\dot{\mathbf{c}}(\boldsymbol\beta,\mathbf{z}_l(u))w_l(u,\mathbf{r}).$$

Further, using (31) and (34), one may show that $\sqrt{n}\,(\widehat{\boldsymbol\beta} - \boldsymbol\beta_0)$ is asymptotically equivalent to $\boldsymbol\Sigma_{\boldsymbol\beta}^{-1}\,\mathbf{X}^{**}(\tau)$, where

$$\mathbf{X}^{**}(t) = n^{-1/2}\sum_{\mathbf{r}\in\mathcal{P}}\sum_{i\in\mathbf{r}}\int_0^t \left\{ \frac{\dot{\mathbf{c}}(\boldsymbol\beta_0,\mathbf{z}_i(u))}{c(\boldsymbol\beta_0,\mathbf{z}_i(u))} - \frac{\mathbf{S}_{\mathbf{r}}^{(1)}(\boldsymbol\beta_0,u)}{S_{\mathbf{r}}^{(0)}(\boldsymbol\beta_0,u)} \right\} \mathrm{d}M_{(i,\mathbf{r})}(u)$$

Combining all this, we get from (39) that, for $J = 1,\ldots,G$, the normalized martingale residual processes $n^{-1/2}\,\widehat{M}_J(t)$ are asymptotically equivalent (as stochastic processes in t) to

$$X_J(t) = X_J^*(t) - \boldsymbol\psi_J(t,\boldsymbol\beta_0)^{\mathsf T}\boldsymbol\Sigma_{\boldsymbol\beta}^{-1}\,\mathbf{X}^{**}(\tau). \tag{41}$$

Now $X_J^*(t)$ and $\mathbf{X}^{**}(t)$ are linear combinations of stochastic integrals, and hence themselves martingales. For given groups $I, J = 1, \ldots, G$ we find after some algebra that the predictable (co)variation process of the first of these martingales takes the form

$$\langle X_I^*, X_J^* \rangle(t) = \tag{42}$$

$$n^{-1} \sum_{\mathbf{r} \in \mathcal{P}} \int_0^t \frac{S_{\mathbf{r}I}^{(0)}(\boldsymbol{\beta}_0, u)}{S_{\mathbf{r}}^{(0)}(\boldsymbol{\beta}_0, u)} \left\{ \delta_{IJ} - \frac{S_{\mathbf{r}J}^{(0)}(\boldsymbol{\beta}_0, u)}{S_{\mathbf{r}}^{(0)}(\boldsymbol{\beta}_0, u)} \right\} S_{\mathbf{r}}^{(0)}(\boldsymbol{\beta}_0, u) \pi_u(\mathbf{r}) \alpha_0(u) du,$$

with $\delta_{IJ} = 1$ if $I = J$, and $\delta_{IJ} = 0$ otherwise, while $\langle X_J^*, \mathbf{X}^{**} \rangle(t)$ equals (40) and $\langle \mathbf{X}^{**} \rangle(\tau) = \langle \mathbf{U}(\boldsymbol{\beta}_0) \rangle(\tau)$ is given by (32). If, on the right hand side of (42), we integrate over $(s_1, s_2]$ instead of $(0, t]$, one may show that the resulting integral converges uniformly (in s_1 and s_2) to a limit function $\phi_{IJ}(s_1, s_2, \boldsymbol{\beta}_0)$, say. By (41) and the above results we may now conclude, using the martingale central limit theorem, that the normalized vector of grouped martigale residual processes $n^{-1/2}(\widehat{M}_1, \ldots, \widehat{M}_G)^\top$ converges weakly to a mean zero Gaussian process $\mathbf{U} = (U_1, \ldots, U_G)^\top$. The (I, J)-th entry of the covariance matrix $\boldsymbol{\Sigma}(s, t) = \mathrm{E}\{\mathbf{U}(s)^\top \mathbf{U}(t)\}$ between $\mathbf{U}(s)$ and $\mathbf{U}(t)$ becomes

$$\begin{aligned} \sigma_{IJ}(s, t) &= \mathrm{Cov}(U_I(s), U_J(t)) \\ &= \phi_{IJ}(0, s \wedge t, \boldsymbol{\beta}_0) - \boldsymbol{\psi}_I(0, s, \boldsymbol{\beta}_0)^\top \boldsymbol{\Sigma}_{\boldsymbol{\beta}}^{-1} \boldsymbol{\psi}_J(0, t, \boldsymbol{\beta}_0), \end{aligned} \tag{43}$$

where $\boldsymbol{\psi}_J(s_1, s_2, \boldsymbol{\beta}_0)$ and $\phi_{IJ}(s_1, s_2, \boldsymbol{\beta}_0)$ are defined just above (40) and just below (42), respectively. For estimation of the covariances (43), we may estimate $\boldsymbol{\Sigma}_{\boldsymbol{\beta}}$ consistently by $1/n$ times the expected information matrix (35). Further, using (37) and (38), one may prove that $\phi_{IJ}(s_1, s_2, \boldsymbol{\beta}_0)$ can be estimated uniformly (in s_1 and s_2) consistently by $1/n$ times (23), while $\boldsymbol{\psi}_J(s_1, s_2, \boldsymbol{\beta}_0)$ can be estimated uniformly consistently by $1/n$ times (24). Combining this, it follows that the asymptotic covariances (43) may be estimated uniformly consistently by $n^{-1}\widehat{\sigma}_{IJ}(s, t)$, where $\widehat{\sigma}_{IJ}(s, t)$ is given by (22) in Section 3.4.

6 Matched risk set sampling

In order to keep the presentation simple, we have so far considered the relative risk regression model (1), where the baseline hazard rate is assumed to be the same for all individuals in the cohort. Sometimes this may not be reasonable, e.g., to control for the effect of one or more confounding factors, one may want to adopt a stratified version of (1) where the baseline hazard differs between (possibly time-dependent) population strata generated by the confounders. The regression coefficients are, however, assumed the same across these strata. Thus the hazard rate of an individual i from population stratum h is assumed to take the form

$$\alpha(t; \mathbf{z}_i) = c(\boldsymbol{\beta}_0, \mathbf{z}_i(t))\, \alpha_{0h}(t). \tag{44}$$

When the stratified proportional hazards model (44) applies, the sampling of controls should be restricted to those at risk in the same population stratum as the case. We say that the controls are *matched* by the stratification variable. In particular for simple random sampling, if an individual in population stratum h fails at time t, one selects at random $m-1$ controls from the $n^{(h)}(t) - 1$ non-failing individuals at risk in this population stratum. Similarly one may combine matching and counter-matching by selecting the controls among those in the sampling strata used for counter-matching who belong to the population stratum of the case. Note the distinction between the population strata, which form the basis for stratification in (44), and the sampling strata used for the counter-matched sampling of the controls.

In general, matched risk set sampling may be described as follows. Given $\mathcal{F}_{t-}$, if an individual i in population stratum h fails at time t, we select our sampled risk set according to a probability distribution $\pi_t(\mathbf{r} \mid i)$ over sets $\mathbf{r}$ that contain i and where *all individuals in* $\mathbf{r}$ *belong to population stratum* h at time t. (Note that the sampling distribution will depend on the population stratum h of the failing individual, even though this is not made explicit in the notation.) For such sampling distributions we have the factorization $\pi_t(\mathbf{r} \mid i) = w_i(t, \mathbf{r})\, \pi_t(\mathbf{r})$, where $\pi_t(\mathbf{r})$ is given by (6) with $n(t)$ replaced by $n^{(h)}(t)$, and $w_i(t, \mathbf{r})$ is obtained from (7) as before. In particular for matched risk set sampling with simple random sampling of the controls, the weights are $w_i(t, \mathbf{r}) = [n^{(h)}(t)/m] I\{i \in \mathbf{r}\}$ for individuals in population stratum h, while for matched risk set sampling with counter-matched sampling of the controls the weights are $w_i(t, \mathbf{r}) = [n^{(h)}_{s(i)}(t)/m_{s(i)}] I\{i \in \mathbf{r}\}$. Here $s(i)$ denotes the sampling stratum of individual i, while $n^{(h)}_s(t)$ is the number of individuals at risk "just before" time t in population stratum h who belong to sampling stratum s.

The general theory of Sections 3 and 5 goes through almost unchanged for matched risk set sampling. In particular the partial likelihood (10) and the formula (11) for the expected information matrix apply without modification provided one uses the appropriate weights as just described. Also the expressions (21) and (22) for the grouped martingale residuals and their estimated covariances, as well as the chi-squared goodness-of-fit test derived from these expressions, remain valid for matched risk set sampling.

In order to prove these extensions of the results of Sections 3 and 5, we have to consider the processes $N^{(h)}_{(i,\mathbf{r})}(t)$, counting the observed number of failures for individual i with associated sampled risk set $\mathbf{r}$ *while being a member of population stratum* h, and their associated (cumulative) intensity processes and martingales. The proofs follow step by step the arguments of Sections 3 and 5, and we omit the details.

7 Discussion

We have shown how plots of grouped martingale residual processes and the accompanying chi-squared goodness-of-fit test provide useful tools for checking the fit of relative risk regression models based on sampled risk set data. However, a number of questions remain to be better understood in relation with these methods.

To use the methods one has to define a (possibly time-dependent) grouping of the individuals, and it is then a question how this best can be done. If the grouping is based on current covariate values, one has to decide which covariates to use for the grouping and how the cut points should be chosen. Another option is to follow the approach of Grønnesby and Borgan (1996) and group the individuals according to their values of the estimated relative risks $c(\widehat{\boldsymbol{\beta}}, \mathbf{z}_i(t))$. As these depend on $\widehat{\boldsymbol{\beta}}$, such a grouping will violate our assumption that the grouping at time t should only depend on information available "just before" time t. We conjecture, however, that the large sample distributions of the grouped martingale residual processes and the accompanying chi-squared goodness-of-fit test can still be used as approximations in large samples, but simulation studies are needed to investigate this further.

A useful feature of the grouped martingale residual process plots is that they show how deviations from the model may change over time. For instance, in the uranium miners example, we saw how the highest radon exposure group had more observed lung cancer deaths than expected for ages below 60 years and fewer thereafter. Such deviations give useful hints as to how the model may be modified to obtain a better fit. However, a better understanding is needed on how various deviations from the relative risk regression model (1) will turn up in the plots.

For the special case of an exponential relative risk function, one may use standard software for Cox regression to maximize the partial likelihood (10), formally treating the label of the sampled risk sets as a stratification variable in the Cox regression and including the $\log w_l(t_j, \widetilde{\mathcal{R}}_j)$ as offsets in the model. The package Epicure fits a wide variety of relative risk functions $c(\boldsymbol{\beta}, \mathbf{z}_i(t))$ and was used to estimate the parameters for the uranium miners data in Section 4. But available statistical packages are in general not able to perform all the computations needed for the grouped martingale residual process plots and accompanying chi-squared goodness-of-fit test, and the computations in Section 4 were done in separate programs written by the authors for SAS and for S-Plus. However, for the special case of Cox's regression model, the extended model (26) becomes a Cox model as well, and our chi-squared goodness-of-fit test can be computed as the score test for the addition of categorical grouping variables using standard software for Cox regression.

Our chi-squared goodness-of-fit test is based on a comparison of observed and expected number of failures in cells obtained by partitioning the space of covariates and time. This is in line with the test suggested by Schoenfeld (1980) for Cox's regression model with cohort data. In fact, apart from details in the estimation of covariances, Schoenfeld's test is the special case of ours in which the relative risk function is exponential and the entire risk set is sampled with probability one.

In order to use our chi-squared test, one has to decide on a grouping according to both covariates and time. As an alternative one may group only according to covariates and consider the maximum value of the chi-squared statistic over a time interval $[\tau_1, \tau_2] \subset (0, \tau]$. P-values for such a supremum type test statistic should be obtainable by the simulation approach of Lin, Wei and Ying (1993) based on the asymptotic representation (41) of the grouped martingale residual processes. It should even be possible to avoid the grouping according to covariates by using the individual martingale residual processes $\widehat{M}_{(i,\mathbf{r})}(t)$ [cf. (20)] to derive cumulative sum of martingale-based residuals along the line of Lin et al. (1993).

Acknowledgements

Part of the work of Ørnulf Borgan was done during sabbatical leaves at the Department of Medical Epidemiology and Biostatistics, Karolinska Institutet, Stockholm, in the fall of 2004, and at the School of Mathematics and Statistics, University of Newcastle upon Tyne, in the spring of 2005. Borgan acknowledges the departments for their hospitality and for providing the best working facilities, while the University of Oslo and Johan and Mimi Wessmann's foundation are acknowledged for their economic support which made these stays possible. The work of Bryan Langholz was supported by grant CA-42949 from the United States National Cancer Institute. The paper was finalized as a part of the project "Statistical Analysis of Complex Event History Data" at the Centre for Advanced study at the Norwegian Academy of Science and Letters in Oslo the academic year 2005/2006.

References

1. AALEN, O. O. (1989). A linear regression model for the analysis of life times. *Statist. Med.* **8**, 907-925.
2. AALEN, O. O. (1993). Further results on the non-parametric linear regression model in survival analysis. *Statist. Med.* **12**, 1569–1588.
3. ANDERSEN, P. K., BORGAN, Ø., GILL, R. D., AND KEIDING, N. (1993). *Statistical models based on counting processes.* Springer Series in Statistics. Springer-Verlag, New York.
4. BARLOW, W. E. AND PRENTICE, R. L. (1988). Residuals for relative risk regression. *Biometrika* **75**, 65–74.
5. BORGAN, Ø., GOLDSTEIN, L., AND LANGHOLZ, B. (1995). Methods for the analysis of sampled cohort data in the Cox proportional hazards model. *Ann. Statist.* **23**, 1749–1778.
6. GRAMBSCH, P. M., THERNEAU, T. M., AND FLEMING, T. R. (1995). Diagnostic plots to reveal functional form for covariates in multiplicative intensity models. *Biometrics* **51**, 1469–1482.
7. GRØNNESBY, J. K. AND BORGAN, Ø. (1996). A method for checking regression models in survival analysis based on the risk score. *Lifetime Data Anal.* **2**, 315–328.
8. HORNUNG, R. AND MEINHARDT, T. (1987). Quantitative risk assessment of lung cancer in U. S. uranium miners. *Health Physics* **52**, 417–30.
9. HOSMER, JR., D. W. AND LEMESHOW, S. (1999). *Applied survival analysis.* Wiley Series in Probability and Statistics: Texts and References Section. John Wiley & Sons Inc., New York.
10. KALBFLEISCH, J. D. AND PRENTICE, R. L. (2002). *The statistical analysis of failure time data.* Wiley Series in Probability and Statistics. Wiley-Interscience [John Wiley & Sons], Hoboken, NJ.
11. KLEIN, J. P. AND MOESCHBERGER, M. L. (2003). *Survival Analysis. Techniques for Censored and Truncated Data. 2nd ed..* Springer Verlag, New York.
12. LANGHOLZ, B. AND BORGAN, Ø. (1995). Counter-matching: A stratified nested case-control sampling method. *Biometrika* **82**, 69–79.
13. LANGHOLZ, B. AND GOLDSTEIN, L. (1996). Risk set sampling in epidemiologic cohort studies. *Statist. Sci.* **11**, 35–53.
14. LANGHOLZ, B., THOMAS, D. C., XIANG, A., AND STRAM, D. (1999). Latency analysis in epidemiologic studies of occupational exposures: Application to the Colorado Plateau uranium miners cohort. *Amer. J. Indust. Med.* **35**, 246–256.
15. LIN, D. Y., WEI, L. J., AND YING, Z. (1993). Checking the Cox model with cumulative sums of martingale-based residuals. *Biometrika* **80**, 557–572.

16. LUBIN, J., BOICE, J., EDLING, C., HORNUNG, R., HOWE, G., KUNZ, E., KUSIAK, R., MORRISON, H., RADFORD, E., SAMET. J., TIRMARCHE, M., WOODWARD. A., XIANG, Y., AND PIERCE, D. (1994). Radon and lung cancer risk: A joint analysis of 11 underground miners studies. NIH Publication 94-3644. U.S. Department of Health and Human Services, Public Health Service, National Institutes of Health. Bethesda, MD.

17. MAY, S. AND HOSMER, D. W. (2004). A cautionary note on the use of the Grønnesby and Borgan goodness-of-fit test for the Cox proportional hazards model. *Lifetime Data Anal.* **10**, 283–291.

18. OAKES, D. (1981). Survival times: aspects of partial likelihood. *Internat. Statist. Rev.* **49**, 235–264.

19. PRENTICE, R. L. AND SELF, S. G. (1983). Asymptotic distribution theory for Cox-type regression models with general relative risk form. *Ann. Statist.* **11**, 804–813.

20. SCHOENFELD, D. (1980). Chi-squared goodness-of-fit tests for the proportional hazards regression model. *Biometrika* **67**, 145–153.

21. THERNEAU, T. M. AND GRAMBSCH, P. M. (2000). *Modeling survival data: extending the Cox model.* Statistics for Biology and Health. Springer-Verlag, New York.

22. THERNEAU, T. M., GRAMBSCH, P. M., AND FLEMING, T. R. (1990). Martingale-based residuals for survival models. *Biometrika* **77**, 147–160.

23. THOMAS, D. C. (1977). Addendum to: Methods of cohort analysis: Appraisal by application to asbestos mining. By F. D. K. Liddell, J. C. McDonald and D. C. Thomas. *J. Roy. Statist. Soc. Ser. A* **140**, 469–491.

PART 2
Reliability Techniques

Chapter 5

RELIABILITY AND SURVIVAL IN FINANCIAL RISK

Nozer D. Singpurwalla
Department of Statistics
The George Washington University, Washington, DC, U.S.A.
Email: nozer@gwu.edu

The aim of this paper is to create a platform for developing an interface between the mathematical theory of reliability and the mathematics of finance. This we are able to do because there exists an isomorphic relationship between the survival function of reliability, and the asset pricing formula of fixed income investments. This connection suggests that the exponentiation formula of reliability theory and survival analysis be reinterpreted from a more encompassing perspective, namely, as the *law of a diminishing resource.* The isomorphism also helps us to characterize the asset pricing formula in non-parametric classes of functions, and to obtain its crossing properties. The latter provides bounds and inequalities on investment horizons. More generally, the isomorphism enables us to expand the scope of mathematical finance and of mathematical reliability by importing ideas and techniques from one discipline to the other. As an example of this interchange we consider interest rate functions that are determined up to an unknown constant so that the set-up results in a Bayesian formulation. We may also model interest rates as "shot-noise processes", often used in reliability, and conversely, the failure rate function as a Lévy process, popular in mathematical finance. A consideration of the shot noise process for modelling interest rates appears to be new.

Key words: Asset pricing; Bayesian analysis; Failure rate; Interest rate; Non-parametric classes; Risk-free bond; Shot-noise process; Zero coupon bond.

1 Introduction

An area that has recently experienced an outburst of activity in the mathematical sciences is what is known as "financial risk analysis". However reliability theory, and survival analysis, which are some of the stalwart tools of risk analysis, have played little or no role in mathematical finance. There are many plausible causes behind the absence of a synergism between these two fields. One reason could be that the term "financial risk" needs to be better articulated and defined. Whereas mathematical finance has benefitted from topics in stochastic processes, statistical inference, and probabilistic modelling, the constructive role that reliability theory is able to play here remains to be exploited. The aim of this paper is to point out some avenues via which the above can be done. To do so, we use the well known asset pricing formula of a fixed income instrument (like a risk-free zero coupon bond) as a "hook". Underlying this formula is the use of an unknown (future) interest rate function. This can be a deterministic function or the realization of a stochastic process. We liken the interest rate function to a (deterministic or stochastic) failure rate function, and then using results from reliability theory explore its consequences on asset pricing. Among these consequences are bounds and inequalities on the investment horizon.

Associated with any failure rate function is a survival function. As currently interpreted, this function encapsulates the risk of failure of an item over time. By risk we mean here probability of failure. Since a risk-free zero coupon bond cannot (by definition) default, the survival function that results from looking at the interest rate as a failure rate cannot encapsulate the risk of a bond's default. This dilemma motivates us to seek alternate, more global, ways of interpreting the survival function. Our view is that the survival function be viewed as one that describes the phenomenon of a diminishing resource over time. In mathematical finance, this resource is a bond's present value; in reliability theory this resource is an item's "hazard potential" [cf. Singpurwalla (2004)] With the above perspective on a survival function the failure rate can be seen as the rate at which the item's hazard potential gets depleted, and the interest rate as the rate at which a bond's present value shrinks. This interpretation of the interest rate appears to be new and can be seen as one of the merits of the isomorphism between survival analysis and mathematical finance.

The remainder of this paper is organized as follows. In Section 2 we give an overview of the derivation of the asset pricing formula under both constant and varying interest rates and point out the relationship between this formula and the survival function of reliability theory. The material of this section is standard and can be found, for example, in Ross (1999). In

Section 3 we present arguments that attempt to give a unifying perspective for the present value and the survival functions. Drawing from an analogy in physics about the decay of radioactive material, we claim that the two formulae in question describe the law of diminishing resource. In Section 4 we invoke several ideas and results from reliability theory to characterize present value functions into non-parametric classes and show how some of these results can be exploited for practical purposes. Section 5 pertains to a discussion of interest rate functions with unknown parameters, or as realizations of stochastic processes. In both cases we draw upon known results in reliability with a view of enhancing the state of the art in mathematical finance. Section 6 concludes the paper with some pointers for future research.

2 Asset Pricing of Risk Free Bonds: An Overview

The material of this section is for the benefit of those working in reliability and survival analysis whose familiarity with the various instruments of finance may be limited. The focus here is the derivation of the asset pricing formula for a risk free bond assuming a deterministic and known interest rate function. The section ends by pointing out the isomorphism between the asset pricing formula and the exponentiation formula of reliability and survival analysis.

A risk free *zero coupon bond* pays, with certainty, the buyer of the bond—the *bondholder*— \$1 at time T after the time of purchase; T is known as the *holding period* (of the buyer) or *maturity*. The bondholder purchases the bond at some calendar time t at a price $P(t,T)$, known as the *present value* at t. Clearly, $P(T,T) = 1$, and $P(t,T)$ decreases in T. Risk-free bonds are generally issued by governments and do not default because governments can always honor payments by "printing" their currency. The present value $P(t,T)$ depends on what the bond holder and the *bond issuer* think of the interest rate that will prevail during the period $(t, t+T]$.

2.1 *Interest Rates and Present Value Analysis*

To keep matters simple, suppose that an amount P is borrowed now, at time $t = 0$, for a period T with the understanding that at time T the amount returned is $P + rP = P(1+r)$. The amount P is known as the *principal* and r the *simple interest* rate per time T. When T is taken to be one year, r is the simple annual interest rate, and the compounding of interest is once per year. If the interest rate is compounded semi-annually, then the amount paid at the end of the year is $P(1+r/2)^2$. In this case r is called the *nominal* interest rate. If the compounding is done n times per year then the amount paid at the end of the year is $P(1+r/n)^n$ and with *continuous compounding* the amount paid at year's end is $P \lim_{n\to\infty}(1+r/n)^n = Pe^r$.

With present value analysis we consider the reverse of the above process. Specifically, what should P be at time $t = 0$ so that at the end of the i-th period of compounding the amount paid (or *payoff*) is V, supposing that the nominal interest rate is r? it is easy to see that the principal $V(1+r)^{-i}$ would yield V at time i. The quantity $V(1+r)^{-i}$ is known as the *present value* at time $t = 0$ of the payoff V at time $t = i$.

2.1.1 *Present Value Under Varying Interest Rates*

Suppose that the nominal interest rate changes with time continuously, as $r(s)$, $s \geq 0$. The quantity $r(s)$ is called the *spot* (or *instantaneous*) interest rate at s. Consequently, an amount x invested at time s becomes $x(1+r(s)h)$ at time $s+h$—approximately—assuming that h is small. Let $D(T)$ denote the amount one has at time T if one invests one monetary unit at time 0. Then for h small and interest rate $r(s)$, $0 \leq s \leq T$

$$D(s+h) \approx D(s)(1+r(s)h),$$

or that the rate of change of the amount at time s is

$$\frac{D(s+h) - D(s)}{h} \approx D(s)r(s).$$

Taking the limit as $h \downarrow 0$, we have

$$\lim_{h\downarrow 0} \frac{D(s+h) - D(s)}{h} = D(s)r(s),$$

or that

$$r(s) = \frac{D'(s)}{D(s)},$$

where $D'(s)$ is the derivative of $D(s)$ at s, assuming it exists at any s.

Integrating over s from $[0, T]$, we have

$$\log(D(T)) - \log(D(0)) = \int_0^T r(s)ds.$$

Since $D(0) = 1$, the above can be written as

$$D((T))^{-1} = \exp\left[-\int_0^T r(s)ds\right].$$

But $(D(T))^{-1}$ is $P(0,T)$, the present value at time 0 of a bond that pays one monetary unit at time T. Thus in general we have the relationship

$$P(t,T) = \exp\left[-\int_{t}^{T+t} r(s)ds\right], \tag{1}$$

where$P(t,T)$ is the present value, at time t, of a risk free bond yielding one monetary unit at time $t+T$, under a continuously changing interest rate $r(s)$, s≥ 0. If we let $R(t,T)$ denote the exponent of the expression for $P(t,T)$, then

$$P(t,T) = \exp(-R(t,T)).$$

The average of the spot interest rate $r(s)$ is

$$\widetilde{R}(t,T) = \frac{1}{T}\int_{t}^{T+t} r(s)ds; \tag{2}$$

it is called the *yield curve.*

2.2 *Isomorphism with the Survival Function*

Mathematically, Equation (1) is identical to the *exponentiation formula* of reliability theory and survival analysis with $r(s)$ as the failure rate function, and $P(t,T)$ as the survival function. Observe that $P(t,0) = 1$ and $P(t,T)$ is a decreasing function of T, which asymptotes to 0 as T increases to infinity. Similarly $R(t,T)$ can be identified with the *cumulative failure* (or hazard) rate at T, and $\widetilde{R}(t,T)$—yield curve—with the *failure rate average.*

As two special cases, suppose that $r(s) = r$, a constant, for $s \geq t$, or that $r(s) = \alpha r(rs)^{\alpha-1}$, for $s \geq t$ and some constant $\alpha \geq 1$. Then $P(t,T) = \exp(-r(T-t))$ in the first case, and $P(t,T) = \exp(-r(T-t)^{\alpha})$ in the second. These present value functions would correspond to the exponential and the Weibull survival functions, respectively.

3 Re-interpreting the Present Value and Survival Functions

In what follows, we set $t = 0$, so that $P(t,T)$ becomes $P(0,T) \stackrel{\text{def}}{=} P(T)$, $R(0,T) \stackrel{\text{def}}{=} R(T)$, and $\widetilde{R}(0,T) \stackrel{\text{def}}{=} \widetilde{R}(T)$. In the context of reliability and survival analysis the interpretation of $P(T)$ as a survival function, $r(s)$ as the failure rate function, and $\widetilde{R}(T)$ as the failure rate average have an intuitive import that is embedded in the context of ageing and wear. How can one justify looking at $P(T)$, the present value function as a survival function and the interest rate $r(s)$ as a hazard function? Alternatively put, how can one see the relationships

$$P(T) = \exp\left[-\int_0^T r(s)ds\right], \tag{3}$$

and

$$R(T) = \int_0^T r(s)ds, \tag{4}$$

from the perspective of hazard, risk and failure, especially since risk-free bonds do not default? More specifically, we may ask if there is a common theme—different from the ones in reliability and finance—that drives the likes of Equations (3) and (4)? The aim of this section is to show that there is indeed a common theme that is able to provide meaning to the above equations in a unified manner. This common theme causes us to look at the exponentiation formula of Equation (3) as encapsulating the phenomenon of a depleting resource. However, in order to do so, we need to first revisit the derivation of the exponentiation formula from first principles. The material that follows is standard and found in Barlow and Proschan (1975).

To keep our notation distinct, let X denote the time to failure of an item and let $F(x) = Pr(X \leq x)$. Suppose that $F(x)$ is absolutely continuous so that its derivative $\frac{dF(x)}{dx} \stackrel{\text{def}}{=} f(x)$ exists (almost everywhere). We now consider

$$Pr(x < X \leq x + dx | X > x) = \frac{F(x + dx) - F(x)}{\overline{F}(x)},$$

where $\overline{F}(x) = 1 - F(x)$. If we divide both sides of the above expression by dx, we get a rate in the sense that

$$\frac{1}{\overline{F}(x)} \frac{F(x + dx) - F(x)}{dx}$$

is the rate at which $F(x)$ increases at x, multiplied by $(\overline{F}(x))^{-1}$. Taking the limit as $dx \downarrow 0$, we have

$$\lim_{dx \downarrow 0} \frac{F(x + dx) - F(x)}{\overline{F}(x)dx} = \frac{f(x)}{\overline{F}(x)} \stackrel{\text{def}}{=} h(x). \tag{5}$$

The right hand side of Equation (5) is defined as the *failure* (or *hazard*) rate function, denoted here as $h(x)$. The qualifier "failure" is added because the function $F(x)$ whose rate of increase is being discussed represents the probability of failure by x. A motivation for referring to $h(x)$ as a rate has been given above. Namely, it is the rate at which the distribution function $F(x)$ increases in x. The exponentiation formula of Equation (3) is an immediate consequence of the relationship $h(x) = f(x)/\overline{F}(x)$.

It is important to note that the development above is not contingent on the fact that $F(x)$ necessarily be a probability distribution function. All that we require is for $F(x)$ to be absolutely continuous with respect to Lebesgue measure, and that for $h(x)$ to be non-negative $F(x)$ be non decreasing. To underscore this point, and also to pave the path for looking at the asset pricing formula from the point of risk and reliability, we turn to a scenario from physics, a scenario that does not involve failure nor does it involve the probability of failure. What we have in mind is the decay of radioactivity (as a function of time) of certain materials, say carbon 14. But before we do so, it is useful to note that Equation (5) may also be written as

$$\lim_{dx \downarrow 0} \frac{1}{\overline{F}(x)} \frac{\overline{F}(x+dx) - \overline{F}(x)}{dx} = -h(x),$$

so that $-h(x)$ encapsulates the rate at which $\overline{F}(x)$ decreases in x.

3.1 *The Exponentiation Formula as the Law of a Diminishing Resource*

Turning to the problem of radioactive decay, it has been claimed that for certain materials the amount of radioactivity decreases exponentially over time, so that if $H(t)$ denotes the level of radioactivity at time t, then $H(t) = \exp(-\lambda t)$, for some $\lambda > 0$. Note that $H(t)$ is absolutely continuous and behaves like a survival function. The rate at which this function decreases is $-\lambda \exp(-\lambda t)$, and so now our analogue of $h(t)$ is $\lambda \exp(-\lambda t)/H(t) = \lambda$, a constant. The exponentiation formula of Equation (3) holds here as well, though it does not have the interpretation used in reliability. Our position here is that the exponentiation formula is ubiqutious in any scenario involving an absolutely continuous monotonically decreasing function, the interpretation of the function being context dependent. In reliability, it is the item's survival function; in radioactivity it is the amount of radioactivity that is remaining, and in finance it is the present value at any time T.

3.1.1 *Interest Rate as a Proportion Loss in Present Value*

In the context of reliability, the quantity $h(x)dx$ is, approximately, the conditional probability of failure at x. In the context of radioactive decay, λdt is the proportion of radioactive loss in the time interval t, $t + dt$. This interpretation will hold irrespective of the functional form of $H(t)$. The interpretation has a broader ramification in the sense that when $P(T)$ denotes the present value at time T and $r(s)$ is the interest rate, then $r(s)ds$ is the proportion loss of present value at time s in the interval s, $s + ds$. Thus one may liken the interest rate as a form of a hazard or risk posed to the present value of the function vis a vis its failure to maintain a particular value at any time. We now have at hand a point of view that unites the failure rate function and the interest rate function.

Our theme of interpreting interest rate as a proportion loss in present value has a synergetic effect in reliability. Specifically, since the survival function $\overline{F}(x)$ decreases in x from $\overline{F}(0) = 1$, the exponentiation formula of Equation (3) can be seen as a law which prescribes life-times as a consequence of some diminishing resource, with $\overline{F}(0) = 1$ interpreted as an item's initial resource. This resource gets depleted over time, with the proportion depleted at x being of the form $h(x)dx$. The amount of resource at x is given by the exponentiation formula of Equation (3).

Thus to recap, the well known exponentiation formula of reliability and survival can also be seen as a law governing a depletion of a resource, with the proportion loss at x governed by the failure[interest] rate $h(x)[r(x)]$. This interpretation is a consequence of the isomorphism between the survival and present value functions. We have now established a platform for discussing financial risk from the point of view of more traditional tools of risk analysis, namely, reliability theory and survival analysis. In what follows we show how this common platform enables us to import some ideas and notions from the latter to the former, and vice versa.

4 Characterizing Present Values Under Monotone Interest Rates

This section is mainly directed towards those working in mathematical finance. Its aim is to describe the qualitative behavior of the present value function $P(T)$ when the underlying interest function $r(s)$, $s \leq T$, or the yield curve $\widetilde{R}(T)$, is monotonic (increasing or decreasing) in T. By increasing (decreasing) we mean non-decreasing (non-increasing); thus a constant interest rate function is both increasing and decreasing. When a bond is issued, the precise nature of the interest rate that will prevail during the life of the bond will not be known. However, one can speculate its general nature as being edging upwards or downwards depending on ones view about the strength of the economy. Thus the objective here is to characterize the behavior of $P(T)$ when the interest rate function, or the yield curve is monotonic but not precisely known. The practical motivation for characterizing present value functions will become clear in what follows. For now it suffices to say that such characterizations facilitate a comparison with present value functions under constant interest rate functions and enable one to obtain bounds and inequalities for investment horizons. The exercise here parallels that in reliability theory wherein comparison against the exponential survival function has proved to be valuable.

4.1 *Non-parametric Classes of Present Value Functions*

By a non parametric class of present value functions, we mean a class of functions whose precise form is unknown (i.e. they are not parametrically defined) but about which some general features can be specified.

Definition 1. The present value function $P(T)$ is defined to be IIR (DIR)—for increasing (decreasing) interest rate—if for each $\tau \geq 0$, $P(T+\tau)/P(T)$ is decreasing (increasing) in $T \geq 0$.

A consequence of Definition 1 is that when $P(T)$ is absolutely continuous the interest rate function $r(T)$ is increasing (decreasing) in T. Conversely, when $r(T)$ is increasing (decreasing) in T, $P(T)$ is IIR (DIR). When $r(t) = \lambda$, a constant greater than 0, $P(T) = \exp(-\lambda T)$, which is both IIR and DIR. All present value functions that display the IIR (DIR) property constitute a class that we label "IIR (DIR) class".

Interest rate functions are often not monotonic even though they may reflect a tendency to edge upwards. They may contain aberrations (or kinks) that are not too severe, in the sense that their average is monotone. In other words, whereas $r(T)$ is not monotone, the yield curve $\widetilde{R}(T)$ is monotone. To bring this feature into play we introduce

Definition 2. The present value function $P(T)$ is defined to be IAIR (DAIR)—for increasing (decreasing) average interest rate—if $-[\log P(T)]/T$ is increasing (decreasing) in $T \geq 0$.

A consequence of Definition 2 is that $P(T)$ IAIR (DAIR) is tantamount to $\widetilde{R}(T)$ increasing (decreasing) in $T \geq 0$. Analogous to IIR (DIR) class, we define the IAIR (DAIR) class as a collection of functions $P(T)$ that display the IAIR (DAIR) property. Verify that the IAIR class, denoted $\{IAIR\}$, encompass the IIR class—denoted $\{IIR\}$—so that $\{IIR\} \subseteq \{IAIR\}$. Similarly $\{DIR\} \subseteq \{DAIR\}$.

A further generalization of Definitions 1 and 2, a generalization whose merits will be pointed out later, is obtained via Definition 3 below.

Definition 3. The present value function $P(T)$ is said to display a NWO (NBO)—for new worse (better) than old—property if for each τ, $T \geq 0$,

$P(T+\tau) \leq (\geq)\, P(T)P(\tau)$.

It can be shown—details omitted [cf. Barlow and Proschan (1975)] – that

$$\{IIR\} \subseteq \{IAIR\} \subseteq \{NWO\},$$

and

$$\{DIR\} \subseteq \{DAIR\} \subseteq \{NBO\},$$

where the $\{NWO\}$ and the $\{NBO\}$ classes contain all present value functions that display the NWO and NBO property, respectively.

4.1.1 *Financial Interpretation of NBO (NWO) Feature*

Consider the case of equality in Definition 3. Now

$$P(T+\tau) = P(T)P(\tau), \tag{6}$$

and the above relationship holds if and only if $P(T) = \exp(-\lambda T)$, for some $\lambda \geq 0$ and $T \geq 0$. The interest rate function underlying this form of the present value function is $r(s) = \lambda$. Equation (6) also implies that

$$\frac{P(T) - P(T+\tau)}{P(T)} = 1 - P(\tau),$$

and since $P(0) = 1$, the above relationship can also be written as

$$\frac{P(T) - P(T+\tau)}{P(T)} = \frac{P(0) - P(\tau)}{P(0)}. \tag{7}$$

Because $P(T)$ is a decreasing function of T, the left hand side of Equation (7) describes the proportion loss in present value during a time interval $[0, \tau]$ at the time T, whereas the right hand side describes the proportion loss in the same time interval, but at time 0. This is an *analogue of the memoryless property* of the exponential distribution in the context of finance. Its practical consequence is that under a constant interest rate function, there is no reason to prefer one investment horizon over another, so long as the holding period is the same.

We now consider the case of strict inequality. Suppose that $P(T)$ is NWO, so that

$$P(T+\tau) < P(T)P(\tau),$$

and as a consequence

$$\frac{P(T) - P(T+\tau)}{P(T)} < \frac{P(0) - P(\tau)}{P(0)}. \tag{8}$$

This means that under Equation (8) the proportion loss in present value at some time $T > 0$ is always less than the proportion loss at time 0. Vice-versa when $P(T)$ is NBO and the inequality above is reversed. To a bondholder, the greater the drop in present value, the more attractive is the bond. Consequently, for $P(T)$'s that are NWO, an investment for any fixed holding period that is made early on in the life of the bond is more attractive than one (for the same holding period) that is made later on. In the IIR or the IAIR case, the above claim makes intuitive sense because the aforementioned properties are a manifestation of increasing interest rates and increasing yield curves, and $\{IIR\} \subseteq \{IAIR\} \subseteq \{NWO\}$. A similar claim can be made in the case of $P(T)$ that is NBO.

It is of interest to note that our definition of NWO and NBO is a **reverse** of that used in reliability theory, namely, the NBU and NWU classes. This makes sense, because a decrease of the present value function is a consequence of an earned resource (namely interest) whereas the decrease of the survival function is a consequence of a depleted resource.

4.2 *Present Value Functions that are Log Concave and PF$_2$*

Suppose that the present value function $P(T)$ belong to one of the several non-parametric classes introduced in Section 4.1, and suppose that the spot interest rate at time of issue of bond is $\lambda > 0$. Were the interest rate over the investment horizon T to remain a constant at λ, then the present value function should be of the form $\exp(-\lambda T)$, $T \geq 0$. The purpose of this section is to compare $P(T)$ and $\exp(-\lambda T)$. Such a comparison could provide new insights about desirable asset pricing investment horizons. To do so, we need to introduce the notions of log concavity and Polya Frequency Functions of Order 2 – abbreviated PF$_2$. These notions have turned out to be useful in reliability theory.

Definition 4. A function $h(x)$, $-\infty < x < \infty$ is said to be PF$_2$ if: $h(x) \geq 0$ for $-\infty < x < \infty$, and

$$\begin{vmatrix} h(x_1 - y_1) & h(x_1 - y_2) \\ h(x_2 - y_1) & h(x_2 - y_2) \end{vmatrix} \geq 0$$

for all $-\infty < x_1 < x_2 < \infty$ and $-\infty < y_1 < y_2 < \infty$, or equivalently $\log h(x)$ is concave on $(-\infty, +\infty)$, or equivalently for fixed $\Delta > 0$, $h(x + \Delta)/h(x)$ is decreasing in x for $a \leq x \leq b$, where

$$a = \inf_{h(y)>0} y \quad \text{and} \quad b = \sup_{h(y)>0} y.$$

The above equivalencies are given in Barlow and Proschan (1975, p.76). Log concavity and PF_2 enable us to establish crossing properties of $P(\bullet)$.

To start with, suppose that $P(\bullet)$ is IIR (DIR). Then, from Definition 1 we have that for each $\tau \geq 0$, $P(T+\tau)/P(T)$ is decreasing (increasing) in $T \geq 0$. As a consequence we have:

Claim 1: $P(\bullet)$ IIR is equivalent to $P(\bullet)$ being both log-concave and PF_2.

Since $P(\bullet)$ IIR is equivalent to an increasing interest rate function $r(\bullet)$, and vice-versa, the essence of Claim 4.1 is that increasing interest rate functions lead to log-concave present value functions. What is the behavior of $P(\bullet)$ if instead of the interest rate function being increasing it is the yield curve that is increasing? More generally, suppose that $P(\bullet)$ is IAIR (DAIR). Then, $P^{1/T} \downarrow (\uparrow) T$, for $T \geq 0$; see Definition 4.2. Consequently we have

Claim 2: *$P(\bullet)$ IAIR (DAIR) implies that for all $T \geq 0$ and any α, $0 < \alpha < 1$,*

$$P(\alpha T) \geq (\leq) P^{\alpha}(T). \tag{9}$$

To interpret Equation (9), let $Q(T) = 1/P(T)$. Then $Q(T)$ is the amount received at time T for every unit of money invested at time $T = 0$. Consequently taking reciprocals in Equation (4.4), we have

$$Q(T/2) \leq (\geq)(Q(T))^{1/2}.$$

Thus, here again, long investment horizons yield more bang for a buck than short horizons when the yield curve is monotonic increasing, and vice-versa when the yield curve is monotone decreasing. Claim 2 prescribes how the investment horizon scales.

To explore the crossing properties of present value functions that are IAIR (DAIR), we introduce

Definition 5. A function $h(x)$, $0 \leq x \leq \infty$ is said to be star-shaped if $h(x)/x$ is increasing in x. Otherwise, it is said to be anti star-shaped. Equivalently, $h(x)$ is star-shaped (anti star-shaped), if for all α, $0 \leq \alpha \leq 1$,

$$h(\alpha x) \leq (\geq) \alpha h(x).$$

It is easy to verify that any convex function passing through the origin is star-shaped. [cf. Barlow and Proschan (1975, p.90)]

Since $P(\bullet)$ IAIR (DAIR) implies — see Definition 2 — that $-[\log P(T)]/T$ is increasing (decreasing) in $T \geq 0$, it now follows that

Claim 3: *$P(\bullet)$ IAIR (DAIR) implies that $T(\widetilde{R}(T))$ is star-shaped (anti star-shaped).*

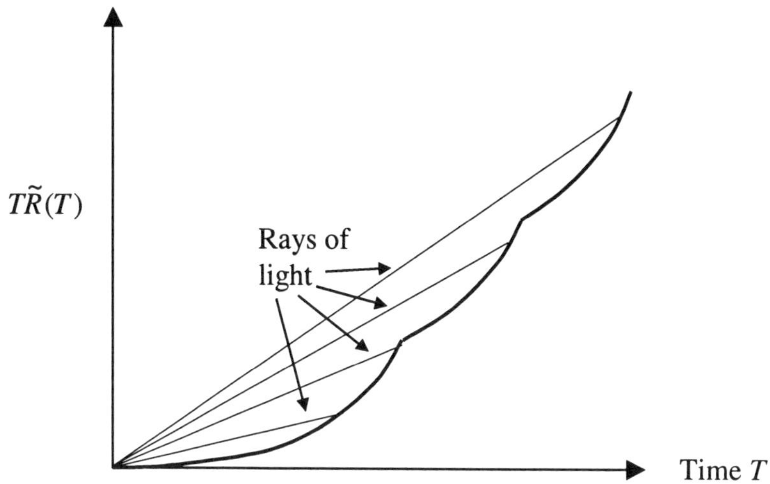

Figure 1 Star-Shapedness of $T(\tilde{R}(T))$ when $P(\cdot)$ is IAIR

Recall that $\widetilde{R}(T)$ is the yield curve. The star-shapedness property, illustrated above, is useful for establishing Theorem 4.1 which gives bounds on $P(\bullet)$. The essence of the star-shapedness property is that there exists a point from which a ray of light can be drawn to all points of the star-shaped function $T(\widetilde{R}(T)) = \int_0^T r(u)du$, with the origin as the point from which the rays of light can be drawn.

It is clear from an examination of Figure 1, that a star-shaped function can cross a straight line from the origin at most once, and that if it does so, it will do it from below. Thus we have

Theorem 1. *The present value function $P(\bullet)$ is IAIR (DAIR) iff for $T \geq 0$ and each $\lambda > 0$, $(P(T) - \exp(-\lambda T))$, has at most one change of sign, and if a change of sign actually occurs, it occurs from $+$ to $-$ (from $-$ to $+$).*

A formal proof of this theorem is in Barlow and Proschan (1975, p. 90). Its import is that the present value function under a monotonically increasing yield curve will cross the present value function under a constant interest rate λ—namely $\exp(-\lambda T)$—at most once, and that if it does cross it will do so from above. The reverse is true when the yield curve decreases monotonically.

Figure 2 illustrates the aforementioned crossing feature for the case of $P(\bullet)$ IAIR, showing a crossing at some time T^*. In general, T^* is unknown;

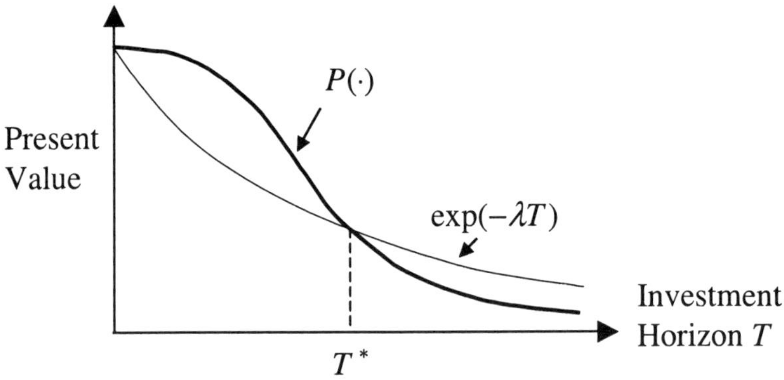

Figure 2 Crossing Properties of an IAIR Present Value Function

it will be known only when a specific functional form is assumed for $P(\bullet)$.

The essence of Figure 2 is that when the yield curve is predicted to be monotone increasing, and having a spot interest rate $\lambda > 0$ at $T = 0$, then the investment horizon should be at least T^*. Investment horizons smaller than T^* will result in smaller total yields than those greater than T^*. The investment horizon of T^* is an equilibrium point.

The illustration of Figure 2 assumes that $P(T)$ and $\exp(-\lambda T)$ cross, whereas Theorem 1 asserts that there is at most one crossing. Thus we need to explore the conditions under which a crossing necessarily occurs and the point at which the crossing occurs. That is, we need to find T^*, assuming that $T^* < \infty$. For this, we need to introduce the notion of "star-ordering".

Definition 6. Let $F(T) = 1 - P(T)$ and $G(T) = (1 - e^{-\lambda T})$, for $\lambda > 0$ and $T \geq 0$. Clearly, $F(0) \equiv G(0) = 0$.Then $F(T)$ is said to be star-ordered with respect to $G(T)$, written $F \underset{*}{<} G$, if $G^{-1}[F(T)]$is star-shaped; i.e. $G^{-1}[F(T)]/T$ is increasing in $T \geq 0$.

With the above definition in place, we have the following as a theorem. It is compiled from a collection of results on pages 107-110 in Barlow and Proschan (1975).

Theorem 2. *Let* $F \underset{*}{<} G$. *Then*

i) $P(\bullet)$ *is IAIR, and*

ii) *$P(T)$ crosses* $\exp(-\lambda T)$ *at most once, and from above, as* $T \uparrow \infty$, *for each* $\lambda > 0$. *Furthermore if* $\int_0^\infty P(u)du = 1/\lambda$, *then*

iii) *A single crossing must occur, and* T^*, *the point at which the crossing occurs is greater than* $1/\lambda$. *Finally a crossing will necessarily occur at* $T^* = 1/\lambda$, *if*

iv) *$P(u)$ is DIR and*

$$\int_0^\infty P(u)du = 1/\lambda.$$

Under *iv)* above, the interest rate is monotonically decreasing; in this case the investment horizon should be no more than T^*.

In parts *ii)* and *iv)* of Theorem 2, we have imposed the requirement that

$$\int_0^\infty P(u)du = 1/\lambda. \tag{10}$$

How must we interpret the condition of Equation (10)? To do so, we appeal to the isomorphism of Section 2. Since $P(u)$ behaves like a survival function, with $P(0) = 1$ and $P(T)$ decreasing in T, we may regard T as a random variable with distribution function $(1 - P(\bullet))$. Consequently, the left hand side of Equation (10) is the expected value of T. With this as an interpretation, we may regard the investment horizon as an unknown quantity whose distribution is prescribed by the present value function, and whose mean is $1/\lambda$.

5 Present Value Functions Under Stochastic Interest Rates

The material of Section 4 was based on the premise that whereas the spot interest rate over the holding period of a bond is unknown, its general nature—a monotonic increase or decrease—can be speculated. Such speculations may be meaningful for small investment horizons; over the long run interest rates cannot be assumed to be monotonic. In any case, the scenario of Section 4 pertains to the case of deterministic but partially specified interest rates. In this section we consider the scenario of interest rate functions that are specified up to some unknown constants, or are the realization of a stochastic process. An analogue of the above two scenarios in reliability theory is a consideration of hazard functions that are stochastic about which much has been written. A recent overview is given by Yashin and Manton (1997).

5.1 *Interest Rate Functions with Random Coefficients*

Recall [see Equation (3)] the exponentiation formula for the present value function under a specified interest rate function $r(s)$, $s \geq 0$, as

$$P(T) = \exp(-R(T)), \tag{11}$$

where $R(T)$ is the cumulative interest rate function. Suppose now that $r(s)$, $s \geq 0$ cannot be precisely specified. Then the $R(T)$ of Equation (11) becomes a random quantity. Let $\pi[R(T)]$ describe our uncertainty about $R(T)$ for any fixed $T \geq 0$. We require that $\pi(\bullet)$ be assessed and specified. Thus our attention now centers around assessing $P(T;\pi)$, the present value function when $\pi[R(T)]$ can be specified for any desired value of T. In other words, $P(T;\pi)$ refers to the fact that the present value function depends on π. In what follows we shall show that

$$P(T;\pi) = E_\pi[\exp(-R(T))], \tag{12}$$

where E_π denotes the expectation with respect to $\pi\,(\bullet)$. To see why, we may use a strategy used in reliability theory which which begins by noting that the right-hand side of Equation (11) can also be written as

$$\exp(-R(T)) = Pr(X \geq R(T)),$$

where X is a random variable whose distribution function is a unit exponential. Consequently when $R(T)$ is random

$$\begin{aligned} P(T;\pi) &= \int_0^\infty Pr(X \geq R(T)|R(T))\pi[R(T)]dR(T) \\ &= \int_0^\infty \exp(-R(T))\pi[R(T)]dR(T) \\ &= E_\pi[\exp(-R(T))]. \end{aligned}$$

Thus in order to obtain the present value function for any investment horizon T, when we are uncertain about interest rate function over the horizon $[0,T]$, all we need do is specify our uncertainty about the cumulative interest rate at T, via $\pi[R(T)]$. What is noteworthy here is that the functional form of $R(T)$, $T \geq 0$ does not matter. All that matters is the value of $R(T)$.

5.1.1 *Consideration of Special Cases*

As an illustration of how we may put Equation (12) to work, suppose that $r(s) = \lambda$, $s \geq 0$, but that λ is unknown. This means that at time 0^+, the spot interest rate is to take some value λ, $\lambda \geq 0$ that is unknown at time 0 when the bond is purchased and that the interest rate is to remain constant over the life of the bond.

Suppose further that our uncertainty about λ is described by a gamma distribution with scale parameter α and a shape parameter β. Then $U \stackrel{\text{def}}{=} \lambda T$ has a density at u of the form

$$\pi\left(u;\alpha,\beta\right) = \frac{\exp(-\alpha u)\alpha^{\beta}u^{\beta-1}}{T^{\beta}\Gamma\left(\beta\right)},$$

from which it follows that the present value function is

$$P(T;\alpha,\beta) = \left(\frac{\alpha}{T+\alpha}\right)^{\beta}, \tag{13}$$

which is of the same form as the survival function of a Pareto distribution. In reliability, such functions are a consequence of doing a Bayesian analysis of lifetimes.

The argument carries forward to a higher level of sophistication wherein one assigns a prior to the survival function itself, the classic examples being the *Dirichlet process prior* of Ferguson [cf. Ferguson, Phadia and Tiwari (1992)], the *Tailfree and Neutral to the Right Process priors* of Doksum (1974), and the *Beta process priors* of Hjort (1990). Invoking the above ideas in the context of financial risk analysis could lead to interesting possibilities.

It can be verified that the present value function of Equation (13) belongs to the DIR class of functions of Definition 11. For this class we are able to provide an upper bound on $P(T)$; see Theorem 3 below. The implication for this theorem is that for scenarios of the type considered here, short investment horizons are to be preferred over long ones.

Theorem 3. **[Barlow and Proschan [1], p.116]** *If P(T) is DIR with mean μ, then*

$$P(T;\mu) \leq \begin{cases} \exp\left(-T/\mu\right), & \textit{for } T \leq \mu, \\ \frac{\mu}{T}e^{-1}, & \textit{for } T \geq \mu; \end{cases} \tag{14}$$

this bound is sharp.

The dark line of Figure 3 illustrates the behavior of this bound. It shows that the decay in present value for time horizons smaller than μ is greater than the decay in present value for time horizons greater than μ.

The dotted line of Figure 3 shows the behavior of the upper bound had its decay been of the form $\exp(-T/\mu)$ for all values of T. Clearly investment horizons greater than μ would not be of advantage to a holder of the bond.

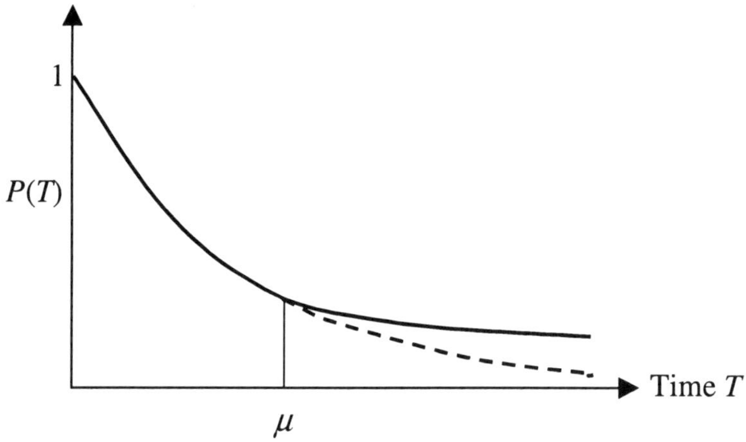

Figure 3 Upper Bound on $P(T)$ when $P(T)$ is DIR

For the special case considered here, namely λ unknown with its uncertainty described by $\pi(\lambda;\alpha,\beta)$, $P(T;\alpha,\beta) = (\alpha/(T+\alpha))^{\beta}$. Were $P(T;\alpha,\beta)$ be interpreted as a survival function, then the μ of Theorem 3 would be of the form

$$\mu = \int_0^\infty \left(\frac{\alpha}{T+\alpha}\right)^{\beta} dT = \frac{\alpha}{\beta - 1};$$

it exists if $\beta > 1$. Consequently, under this $P(T;\alpha,\beta)$ the investment horizon should not exceed $\alpha/(\beta-1)$.

Recall that were λ to be known with certainty, $P(T)$ would be $\exp(-\lambda T)$, $\lambda > 0$, $T \geq 0$, and that there would be no restrictions on the investment horizon so that a bond holder could choose any value of T as an investment horizon. With λ unknown, the net effect is to choose shorter investment horizons, namely those that are at most $\alpha/(\beta-1)$. A similar conclusion can also be drawn in the case wherein $\pi(\lambda;\alpha,\beta)$ be a uniform over $[\alpha,\beta]$. It can be verified that in the uniform case

$$P(T;\alpha,\beta) = \frac{e^{-T\alpha} - e^{-T\beta}}{T(\beta-\alpha)},$$

and that $P(T;\alpha,\beta)$ is again DIR.

Whereas the above conclusions regarding uncertainty about $r(s)$, $s \geq 0$ causing a lowering of the investment horizon have been made based on a consideration of a special case, namely $r(s) = \lambda$, $\lambda > 0$, $s \geq 0$, the question arises about the validity of this claim, were $r(s)$ to be any other function of s, say $r(s) = \alpha\lambda(\lambda s)^{\alpha-1}$, for some $\lambda > 0$, and $\alpha > 0$. When α is assumed known, and uncertainty about λ is described by $\pi(\lambda; \bullet)$, then Equation (12) would be a scale mixture of exponentials and by Theorem 4.7 of Barlow and Proschan (1975, p.103), it can be seen that $P(T; \bullet)$ is DIR, so that Theorem 5.1 comes into play and the inequalities of Equation (14) hold. Thus once again, uncertainty about λ causes a lowering of the investment horizon. Indeed, the essence of Theorem 3 will always hold if the cumulative interest rate $R(T)$ is such that any function of T does not entail unknown parameters.

5.2 *Interest Rates as the Realization of a Stochastic Process*

In this section we consider the case of interest rates that are the realization of a stochastic process. A consideration of stochastic processes for describing interest rate function is not new to the literature in mathematical finance. Indeed much has been written and developed therein; so much so, that some of the results can be profitably imported for use in reliability theory, wherein a consideration of stochastic failure rate functions has proven to be of value [cf. Singpurwalla (1995)]. One such example, is to describe the failure rate function by a Lévy process and to explore the hitting time of this process to a random threshold so that survival function can be introduced; the details are in Singpurwalla (2004).

The focus of this section, however, is to describe the use of a shot-noise process for modelling interest rates and to explore its consequences on the present value function. A use of the shot-noise process for describing the failure rate function has been considered by Singpurwalla and Youngreen (1993). Given below is the adaptation of this process for describing the interest rate function and some justification as to why this could be a meaningful thing to do.

We start by first noting that when the interest rate function is the realization of a stochastic process, say $\{r(s); s \geq 0\}$, then as a consequence of an argument on "randomized stopping times" by Pitman and Speed (1973), the present value function $P(T)$ is of the form $E[\exp(-R(T)]$. Here $\{R(T); T \geq 0\}$ is the *cumulative interest rate process* with $R(T) = \int_0^T r(u)du$, and as in Equation (5.2) the expectation is with respect to the distribution of $R(T)$. Clearly, an evaluation of $P(T)$ would be dependent on the ease with which $E[\exp(-R(T))]$ can be computed. With that in mind, we consider below as a special case a shot-noise process for $\{r(s); s \geq 0\}$.

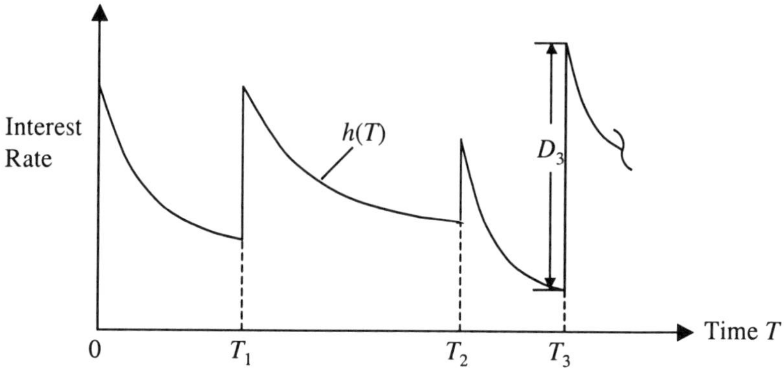

Figure 4 Sample Path of a Shot-Noise Process

5.2.1 *The Shot-Noise Process for Interest Rates*

The shot-noise process of physics is an attractive model for describing the fluctuations of the interest rate function. Our rationale for doing so is that interest rates take an upward jump when certain deleterious economic events occur. Subsequent to their upward jump, the interest rates tend to come down—or even remain constant—until the next deleterious event occurs. In Figure 4 the deleterious events are shown to occur at times $T_1, T_2, T_3, \ldots$. Such events are assumed to occur at random and are governed by say a Poisson process with rate m, $m > 0$. The amount by which the interest rate jumps upward at time T_i is supposed to be random; let this be denoted by a random variable D_i. Finally, suppose that the rate at which the interest rate decays is governed by a function, $h(s)$, $s \geq 0$; this function is called the *attenuation function*. Then, it is easy to see that for any time $T \geq 0$,

$$r(T) = \sum_{i=1}^{\infty} D_i h(T - T_i),$$

with $h(u) = 0$ whenever $u < 0$.

In what follows, we suppose that the T_i's and the D_i's are serially and contemporaneously independent. We also suppose that the D_i's are identically distributed as a random variable D.

If $D = d$, a constant, and if the attenuation function is of the form $h(u) = (1+u)^{-1}$—i.e. the interest rate decays slowly, then it can be shown that the present value function takes the form

$$P(T; m) = \exp(-mT)(1+T)^m. \tag{15}$$

If, on the contrary, D has an exponential distribution with scale parameter b, and $h(u) = \exp(-au)$—that is, the interest rate decays exponentially, then

$$P(T; m, a, b) = \exp\left(-\frac{mbT}{1+ab}\right)\left(\frac{1+ab-\exp(-aT)}{ab}\right)^{mb/(1+ab)}. \quad (16)$$

The $P(T)$ of Equation (15) is the survival function of a Pareto distribution. If in Equation (16) we set $a = b = 1$, and $m = 2$, then a change of time scale from T to $\exp(T)$ would result in the present value function having the form of the survival function of a beta distribution on $(0, 1)$ with parameters 1 and 2.

Thus to summarize, the consideration of a shot-noise process for the interest rate function results in some interesting forms of the present value function. A possible drawback of describing the interest rate by a shot-noise process is that except for the random times at which the interest rate shoots up by a random amount, the process is essentially deterministic.

6 Summary, Conclusions, and Future Work

Equations (13) through (16) were originally obtained in the context of reliability under dynamic environments. The isomorphism of Section 2 has enabled us to invoke them in the context of finance, and what is given in Section 5 barely scratches the surface. Much more can be done along these lines. For example, a hierarchical modelling of interest rate is one possibility. Another possibility, and one that is motivated by work of Dykstra and Laud (1981) is to describe the cumulative interest rate by a gamma process or to look at the present value functions as Dirichlet or neutral to the right processes. Another possibility, and one that is motivated by the enormous literature in survival analysis is to model interest rates as a function of covariates and markers. The Markov Additive Process of Cinlar (1972) presents an opportunity for doing the above. The purpose of this paper is mainly to open the door to other possibilities by creating a suitable platform, which we feel has been done.

But, as correctly pointed out by a referee, our discussion here has been one-sided. We have pointed out how results in reliability and survival analysis can be brought to bear on mathematical finance. It would be a folly not to acknowledge that the reverse can also be true. Indeed, this is something that has already been done by us [see Singpurwalla (2004)], where we capitalize on the several results on hitting times of stochastic processes – such as the Lévy – that can be used to generate new families of survival functions for items experiencing dynamic environments.

Acknowledgements

Research supported by Grants DAAD 19-01-1-0502 under a MURI and DAAD 19-02-01-0195, The U. S. Research Office. Some detailed comments by a referee and several conceptual issues raised by Professor Dabrowska have helped sharpen this paper. My thanks to both.

References

1. Barlow, R. E. and Proschan, F. (1975). *Statistical Theory of Reliability and Life Testing.* Holt, Rinehart and Winston, Inc., New York.
2. Cinlar, E. (1972). Markov additive processes. *II. Z. Wahrsch. Verw. Gebiete* **24** 94–121.
3. Doksum, K. A. (1974). Tailfree and neutral random probabilities and their posterior distributions". *Ann. Prob.* **2** 183–201.
4. Dykstra, R. L. and Laud, P. W. (1981). A Bayesian nonparametric approach to reliability. *Ann. Statist.* **9** 356–367.
5. Ferguson, T. S., Phadia, E. G. and Tiwari, R. C. (1992). Bayesian nonparametric inference. *Current Issues in Statistical Inference: Essays in Honor of D. Basu.* **17** 127–150.
6. Hjort, N. L. (1990). Nonparametric Bayes estimators based on beta processes in models for life history data. *Ann. Statist.* **18** 1259–1294.
7. Pitman, J. W. and Speed, T. P. (1973). A note on random times. *Stochastic Process. Appl.* **1** 369–374.
8. Ross, S. M. (1999). *An Introduction to Mathematical Finance.* Cambridge University Press, U. K.
9. Singpurwalla, N. D. (1995). Survival in dynamic environments. *Statist. Sci.* **10** 86–103.
10. Singpurwalla, N. D. (2004). The hazard potential of items and individuals. *Technical Report GWU/IRRA/TR-00/2.* The George Washington University.
11. Singpurwalla, N. D. and Youngreen M.A. (1993). Multivariate distributions induced by dynamic environments. *Scand. J. Statist.* **20** 251–261.
12. Yashin, A. I. and Manton, K. G. (1997). Effects of unobserved and partially observed covariate processes on system failure: a review of models and estimation strategies. *Statist. Sci.* **12** 20–34.

Chapter 6

SIGNATURE-RELATED RESULTS ON SYSTEM LIFETIMES

Henry W. Block, Michael R. Dugas, and Francisco J. Samaniego

Department of Statistics
University of Pittsburgh, Pittsburgh, PA, U.S.A.

Department of Statistics
University of California, Davis, CA, U.S.A

E-mails: hwb@stat.pitt.edu, mike@dugas.com & fjsamaniego@ucdavis.edu

The performance (lifetime, failure rate, etc.) of a coherent system in iid components is completely determined by its "signature" and the common distribution of its components. A system's signature, defined as a vector whose i^{th} element is the probability that the system fails upon the ith component failure, was introduced by Samaniego (1985) as a tool for indexing systems in iid components and studying properties of their lifetimes. In this paper, several new applications of the signature concept are developed for the broad class of mixed systems, that is, for stochastic mixtures of coherent systems in iid components. Kochar, Mukerjee and Samaniego (1999) established sufficient conditions on the signatures of two competing systems for the corresponding system lifetimes to be stochastically ordered, hazard-rate ordered or likelihood-ratio ordered, respectively. Partial results are obtained on the necessity of these conditions, but all are shown not to be necessary in general. Necessary and sufficient conditions (NASCs) on signature vectors for each of the three order relations above to hold are then discussed. Examples are given showing that the NASCs can also lead to information about the precise number and locations of crossings of the systems' survival functions or failure rates in $(0, \infty)$ and about intervals over which the likelihood ratio is monotone. New results are established relating the asymptotic behavior of a system's failure rate, and the rate of convergence to zero of a system's survival function, to the signature of the system.

Key words: Survival function; Failure rate; Coherent system; Mixed system; Stochastic ordering; Hazard rate ordering; Likelihood ratio ordering; Temporal asymptotics.

1 Introduction

Characterizing the relationship between the design of a system of interest and that system's performance is an important problem in Reliability Theory. Historically, the tools available for studying such problems have been rather sparse. While a system's structure function, which expresses the state (i.e. the success or failure) of a system in terms of the states of its components, fully characterizes a system's design, it has proven to be an awkward tool when applied to the individual or comparative study of system performance. The notion of the "signature" of a coherent system (a monotone system in which every component is relevant), introduced by Samaniego (1985), provided some fresh possibilities in this area. For the sake of clarity, we mention here that the signature vector of such a system is an n-dimensional probability vector whose i^{th} element is the probability that the system fails upon the ith component failure. For an overview of system signatures and their applications, see Boland and Samaniego (2004).

The performance of individual components of n-component systems is typically characterized by the *cumulative distribution function* F of the lifetime involved or by equivalent functions such as the *survival function* $\overline{F} = 1 - F$, the *density function* f or the *failure rate* $r(t) = f(x)/(\overline{F}(x))$, the functions f and r being well defined when F is absolutely continuous. (See, for example, Barlow and Proschan (1981) for further details.) For coherent systems whose components have independent, identically distributed (iid) lifetimes, Samaniego (1985) established useful representations of the system's distribution, density and failure rate in terms the system's signature vector. We will briefly review these representations below. Further, we will review the "preservation theorems" obtained by Kochar, Mukerjee and Samaniego (KMS) (1999) showing that certain properties of system signatures will ensure similar properties for system lifetimes.

This paper has several purposes. One is to investigate the extent to which the sufficient conditions in KMS (1999) are in fact necessary. In brief, we find that they are indeed both necessary and sufficient for very small systems (e.g., when $n = 3$) but that the conditions are not necessary in general. Necessary and sufficient conditions on signatures for various orderings of system lifetimes are then obtained. Another line of investigation followed in the present paper is the study of the limiting behavior of a system. Specifically, we are interested in describing the asymptotic behavior of a system's failure rate, and the rate of convergence of its survival function to zero (both as $t \to \infty$) in terms of the system's signature. Ratios of failure rates and survival functions of competing systems are also studied. In the case of the asymptotics of failure rates, our goal is to explore possible extensions of recently established results by Block, Li and Savits (2003) where conditions which determine the asymptotic behavior of the failure rate of the mixture of lifetime distributions are identified. One of their principal results is that if the failure rate of the component lifetimes have limits, the failure rate of the mixture converges to the limit of the strongest component. The present work differs from earlier studies in that our focus and our results are based on signatures of coherent or mixed systems.

The distribution F_T of the lifetime T of a system in iid components is completely determined by its signature and the underlying component distribution F. An especially useful tool obtained in Samaniego (1985) was a representation for the failure rate r_T of the lifetime T of a coherent system with iid components in terms of the system's signature vector and the underlying component lifetime distribution F. We will display that representation, as well as Samaniego's (1985) representations of $\overline{F}_T$ and f_T, in the next section, and we'll exploit them in various ways in the sequel.

We will provide the basic background needed in the present study in Section 2. In Section 3, we will examine the questions of whether, or when, the sufficient conditions of KMS (1999) for various stochastic relationships between two system lifetimes are also necessary. We show that the answer is "yes" in low dimensional problems, and give examples showing that such necessity is not true in general. In Section 4, necessary and sufficient conditions on system signatures are identified for the lifetime distributions of two mixed systems in n i.i.d. components to be stochastically ordered, hazard-rate ordered or likelihood-ratio ordered. An interesting byproduct of these results is the ability to determine the number and location of crossings of the failure rates or survival functions of the lifetimes of two mixed systems. We are also able to identify intervals over which the likelihood ratio is monotone. These latter insights facilitate the definitive comparison of systems in finite intervals of interest, for example, in the interval $(0, T^*)$, where T^* is the mission time of the systems in question.

In Section 5, we obtain a new result on the asymptotic behavior of the failure rates of mixed systems based on coherent systems in n iid components. We show that the answer depends on the largest index of the signature vector's non-zero elements and the limit of the underlying common component failure rate. Our results in Section 5 include asymptotic comparisons of system failure rates in which the systems have their own signatures and underlying failure rates. Results on the rate of convergence to zero of the survival function of individual mixed systems, and the comparative rates of convergence for the survival functions of two mixed systems are also obtained.

2 Sufficient Conditions for the Comparison of System Life

In this section, we give background results on signatures from Samaniego (1985), on the notion of mixed systems from Boland and Samaniego (2004) and on the comparison of system lifetimes from Kochar, Mukerjee and Samaniego (1999). We first give a formal definition of the signature of a coherent system in iid components. Numerous examples of system signatures are given in the two papers cited above.

Definition 1. The signature of a coherent system with n iid component lifetimes is the probability vector $\mathbf{s} = (s_1, s_2, ..., s_n)$, where s_i is the probability the system fails upon the ith component failure.

The computation of an n-component system's signature typically involves combinatorial arguments for counting the number of permutations of the indexes of the n component failure times $\{X_1, X_2, \ldots, X_n\}$ which result in system failure upon a given (say the ith) ordered component failure time $X_{(i)}$. For example, the three-component system in which component 1 is arranged in series with a parallel system in components 2 and 3 will fail upon the first component failure if and only if $X_1 < min\{X_2, X_3\}$, an event that occurs with probability $1/3$ under an iid assumption. If this event does not occur, the system will necessarily fail upon the second component failure. Thus, the signature of this system is $(1/3, 2/3, 0)$. The signatures of the four remaining coherent systems of order three are easily found to be $(0, 2/3, 1/3)$, $(1, 0, 0)$, $(0, 1, 0)$ and $(0, 0, 1)$. The signature of the "bridge system" featured in Barlow and Proschan (1981, p. 9) and elsewhere is $\mathbf{s} = (0, 1/5, 3/5, 1/5, 0)$.

Consider a coherent system with n iid components with survival distribution $\overline{F}$, density function f and failure rate r. The following representations of $\overline{F}_T$, f_T and r_T, the corresponding survival function, density and failure rate of the system lifetime T, in terms of the system's signature $\mathbf{s}$, are given in Samaniego (1985):

$$\overline{F}_T(t) = \sum_{j=0}^{n-1}\Big(\sum_{i=j+1}^{n} s_i\Big)\binom{n}{j}(F(t))^j(\overline{F}(t))^{n-j}, \tag{1}$$

$$f_T(t) = \sum_{i=0}^{n-1}(n-i)s_{i+1}\binom{n}{i}(F(t))^i(\overline{F}(t))^{n-i}r(t), \tag{2}$$

and

$$r_T(t) = \frac{\sum\limits_{i=0}^{n-1}(n-i)s_{i+1}\binom{n}{i}(F(t))^i(\overline{F}(t))^{n-i}}{\sum\limits_{i=0}^{n-1}\Big(\sum\limits_{j=i+1}^{n} s_j\Big)\binom{n}{i}(F(t))^i(\overline{F}(t))^{n-i}}\, r(t). \tag{3}$$

In applications of these representations in the sequel, we will typically utilize forms of these expressions which are written as functions of the ratio

$$G(t) = \frac{F(t)}{\overline{F}(t)}.$$

Specifically,

$$\overline{F}_T(t) = (\overline{F}(t))^n \sum_{j=0}^{n-1} (\sum_{i=j+1}^{n} s_i) \binom{n}{j} (G(t))^j, \tag{4}$$

$$f_T(t) = (\overline{F}(t))^n \sum_{i=0}^{n-1} (n-i) s_{i+1} \binom{n}{i} (G(t))^i r(t) \tag{5}$$

and

$$r_T(t) = \frac{\sum\limits_{i=0}^{n-1} (n-i) s_{i+1} \binom{n}{i} (G(t))^i}{\sum\limits_{i=0}^{n-1} (\sum\limits_{j=i+1}^{n} s_j) \binom{n}{i} (G(t))^i} r(t). \tag{6}$$

The notion of "mixed systems" was introduced in Boland and Samaniego (2004), and will be utilized here in various ways. A mixed system is simply a stochastic mixture of several coherent systems, and can be physically realized through a randomization process that selects a coherent system at random according to predetermined probabilities. Mixed systems are most easily understood by picturing an available collection of coherent systems (perhaps all of them) of order n, where all n components in each available system have iid lifetimes distributed according to a common F. A mixed system may be implemented through a randomization process that picks one of these systems at random according to a fixed mixing distribution. Repeated use of the mixed system will result in the use of different coherent systems, each arising with a relative frequency tending toward the probability the mixing distribution assigns to that system. The signature of a mixed system is clearly the corresponding mixture of the signatures of the systems involved. For example, for $n = 3$, the 50-50 mixture of a series and a parallel system results in a mixed system with signature (1/2, 0, 1/2). Intuitively, since the series system is selected with probability 1/2, the chances that the mixed system fails upon the first component failure must also be 1/2.

Mixed systems are not arbitrary mathematical artifacts which simply serve to expand the collection of coherent systems of a given size. Because they can be easily realized physically through a randomization process, mixed systems represent a viable possibility in selecting a system for use. Their utility in reliability analysis is much the same as that of randomized rules in decision theory; indeed, there are circumstances in which the best choice of system, relative to a fixed criterion function, is a non-degenerate mixture of two or more coherent systems. An immediate example is the problem of finding an optimal system design in problems in which the criterion function depends on both a system's cost and its performance. Dugas and Samaniego (2005) demonstrate that certain mixed systems are optimal relative to specific criterion functions based on cost and performance. The representation results above, and all of the results obtained in the sequel, are established for the class of mixed systems. As a necessary byproduct, these results hold for coherent systems, which of course can be viewed as degenerate mixtures.

Suppose $\mathbf{s}$ is an arbitrary n-dimensional probability vector. Then there exists a mixed system with $\mathbf{s}$ as its signature. For example, the mixture of systems which fail with probability one upon the kth component failure (the so called k-out-of-n systems) according to the probabilities in $\mathbf{s}$ is a mixed system with signature $\mathbf{s}$. Mixed systems are thus indexed by the class of all n-dimensional probability vectors. Several of our examples in the sequel will involve mixed systems.

For the reader's convenience, we give below brief descriptions of the three stochastic relationships (stochastic, hazard rate and likelihood ratio ordering) which the developments in this and the following two sections utilize. Throughout the paper, "increasing" is taken to mean "nondecreasing". Given the random variables X_1 and X_2, discrete or continuous, with corresponding distributions F_1 and F_2, $X_1 \leq_{st} X_2$ if $\overline{F}_1(x) \leq \overline{F}_2(x)$ for all x. We write $X_1 \leq_{hr} X_2$ if the ratio of survival functions $\overline{F}_2(x)/\overline{F}_1(x)$ is increasing in x. Finally, $X_1 \leq_{lr} X_2$ if the ratio $f_2(x)/f_1(x)$ is increasing in x, where f_i represents the density or probability mass function of X_i. For Y_1 and Y_2 with distributions G_1 and G_2, the notation $Y_1 \leq Y_2$ and $G_1 \leq G_2$ will be used interchangeably.

The preservation results below are proven by Kochar, Mukerjee and Samaniego (1999) for coherent systems, but as stated below, they hold more broadly, with the same basic proofs, for arbitrary mixed systems.

Proposition 1. *Let $\mathbf{s_1}$, $\mathbf{s_2}$ be signatures of two mixed systems based on coherent systems with n iid components, and let T_1, T_2 be the corresponding system lifetimes. Then*

(a): if $\mathbf{s_1} \leq_{st} \mathbf{s_2}$, then $T_1 \leq_{st} T_2$;
(b): if $\mathbf{s_1} \leq_{hr} \mathbf{s_2}$, then $T_1 \leq_{hr} T_2$;
(c): if $\mathbf{s_1} \leq_{lr} \mathbf{s_2}$, then $T_1 \leq_{lr} T_2$.

We will now examine these ordering conditions on $\mathbf{s}_1$ and $\mathbf{s}_2$ more closely.

3 On the Necessity of the KMS Ordering Conditions

In this section, we examine the necessity of the sufficient conditions on signatures which are given in Proposition 1. We show that the conditions are not necessary in general. Some special cases in which necessity holds are noted.

The converse of Proposition 1(a) holds for n=2 and n=3. We show this below for n=3.

Proposition 2. *Let T_1 and T_2 be the lifetimes of two mixed systems of order $n = 3$. If $T_1 \leq_{st} T_2$, then $\mathbf{s}_1 \leq_{st} \mathbf{s}_2$.*

Proof. From (4) for $n = 3$, for $k = 1, 2$

$$\overline{F}_{T_k}(t) = (\overline{F}(t))^3 \sum_{j=0}^{2} (\sum_{i=j+1}^{3} s_{ki}) \binom{3}{j} (G(t))^j.$$

Thus $\overline{F}_{T_1}(t) \leq \overline{F}_{T_2}(t)$ implies

$$0 \leq (\overline{F}(t))^3 \sum_{j=0}^{2} (\sum_{i=j+1}^{3} s_{2i} - \sum_{i=j+1}^{3} s_{1i}) \binom{3}{j} (G(t))^j$$

which yields the inequality

$$0 \leq (\sum_{i=2}^{3} s_{2i} - \sum_{i=2}^{3} s_{1i}) + (s_{23} - s_{13}) G(t). \tag{7}$$

Letting $t \to \infty$ and $t \to 0$ shows that the two differences in (7) are positive, conditions that are equivalent to $\mathbf{s}_1 \leq_{st} \mathbf{s}_2$. □

A similar result holds for hazard rate ordering. The converse of Proposition 1(b) holds for small n; this is true, in particular, for $n = 2$ and $n = 3$. We show this for $n = 3$.

Proposition 3. *Let T_1 and T_2 be the lifetimes of two mixed systems of order $n = 3$. If $T_1 \leq_{hr} T_2$, then $\mathbf{s}_1 \leq_{hr} \mathbf{s}_2$.*

Proof. Hazard rate ordering for the two system lifetimes implies that

$$\frac{\overline{F}_{T_2}(t)}{\overline{F}_{T_1}(t)}$$

is increasing in t or, equivalently,

$$r_{T_2}(t) \leq r_{T_1}(t) \tag{8}$$

for all t. Employing the representation in (6), one may show that $\mathbf{s}_1 \leq_{hr} \mathbf{s}_2$ is implied by the inequality in (8), as the latter, after making the substitution $s_{i1} = 1 - s_{i2} - s_{i3}$, $i = 1, 2$, can be reduced algebraically, for all $t > 0$, to the inequality:

$$\begin{aligned} &3(s_{12}s_{23} - s_{13}s_{22})(G(t))^3 + [2(s_{23} - s_{13}) + 3(s_{12}s_{23} - s_{13}s_{22})](G(t))^2 \\ &+[(s_{22} + s_{23} - s_{12} - s_{13}) + 2(s_{23} - s_{13})]G(t) + (s_{22} + s_{23} - s_{12} - s_{13}) \\ &\quad > 0. \end{aligned} \tag{9}$$

The constant term and the coefficient of $G(t)^3$ in (9) must be greater than zero for (9) to hold for all t. Thus, $1 < (s_{22}+s_{23})/(s_{12}+s_{13})$ and $s_{22}/s_{12} < s_{23}/s_{13}$. But this implies that the coefficients of $(G(t))^2$ and $G(t)$ are greater than 0. Together, these conditions are equivalent to $\mathbf{s}_1 \leq_{hr} \mathbf{s}_2$. □

The following example shows that the converses of Propositions 1(a) and 1(b) do not hold for arbitrary n. The algebraic details are omitted.

Example 1. Consider two mixed systems with signatures $\mathbf{s_1} = (.1, .1, .8, 0)$ and $\mathbf{s_2} = (0, .3, .1, .6)$. Then $s_{11} + s_{12} = .2 < s_{21} + s_{22} = .3$, violating both $\mathbf{s_1} \leq_{st} \mathbf{s_2}$ and $\mathbf{s_1} \leq_{hr} \mathbf{s_2}$. However, $T_1 \leq_{hr} T_2$.

Finally, we examine the necessity of the condition on signatures in Proposition 1(c) for the likelihood ratio ordering of system lifetimes. First, we note that, as in the situations above, the necessity holds when $n = 3$.

Proposition 4. *Let T_1 and T_2 be the lifetimes of two mixed systems of order $n = 3$. If $T_1 \leq_{lr} T_2$, then $\mathbf{s}_1 \leq_{lr} \mathbf{s}_2$.*

Proof. Employing (5), the likelihood ratio ordering for the two system lifetimes implies that

$$\frac{f_{T_2}(t)}{f_{T_1}(t)} = \frac{\sum_{i=0}^{2}(3-i)s_{2,i+1}\binom{3}{i}(G(t))^i}{\sum_{i=0}^{2}(3-i)s_{1,i+1}\binom{3}{i}(G(t))^i} \tag{10}$$

is increasing in t. By taking the derivative of the expression in (10) with respect to $G(t)$, one obtains a condition equivalent to the ratio in (10) being increasing, namely,

$$G(t)^2(s_{12}s_{23} - s_{13}s_{22}) + G(t)(s_{11}s_{23} - s_{13}s_{21}) + (s_{11}s_{22} - s_{12}s_{21}) > 0 \tag{11}$$

for all $t > 0$. The coefficients of $G(t)^2$ and $G(t)^0$ in (11) must be greater than zero. But this implies that the coefficient of $G(t)$ is greater than 0. Together, these conditions imply $\mathbf{s}_1 \leq_{lr} \mathbf{s}_2$. □

Proposition 4 is not true for general n, as the following example shows.

Example 2. Consider two mixed systems with signatures $\mathbf{s_1} = (.3, .3, .4, 0)$ and $\mathbf{s_2} = (0, .4, .4, .2)$. $\mathbf{s_1}$ is not smaller than $\mathbf{s_2}$ in the likelihood ratio ordering since $4/3 = s_{22}/s_{12} > s_{23}/s_{13} = 1$. However, $T_1 <_{lr} T_2$.

4 Necessary and Sufficient Conditions for Stochastic Relationships between Systems

As is apparent from Section 3, the characterization of differences between two coherent or mixed systems with iid components in terms of properties of the system's respective signatures is a nontrivial matter. While the various ordering conditions on signatures of Section 2 (Proposition 1) are sufficient to imply corresponding orderings of the system lifetimes, we have seen that such conditions tend not to be necessary. In this section, we provide an affirmative answer to the question: can necessary and sufficient conditions (NASC) be identified in any problems of practical interest? These characterization results may be found in Block, Dugas and Samaniego (2006). The original treatment is augmented here by the addition of formal arguments establishing Theorem 3 and the addition of an example of the extension of that theorem to problems in which the monotonicity of likelihood ratios varies in two distinct intervals.

Theorem 1. *Let* $\mathbf{s_1}$ *and* $\mathbf{s_2}$ *be the signatures of two arbitrary mixed systems based on coherent systems in* n *iid components, with the same component lifetime distribution, and let* T_1 *and* T_2 *denote the system lifetimes. Then* $T_1 \leq_{st} T_2$ *if and only if*

$$g(x) \geq 0 \quad \text{for all} \quad x \geq 0, \tag{12}$$

where

$$g(x) = \sum_{j=0}^{n-1} \binom{n}{j} \sum_{i=j+1}^{n} (s_{2i} - s_{1i}) x^j \quad \text{for} \quad x \geq 0. \tag{13}$$

While condition (12) is a complex statement concerning the relationship between the two system signatures involved, it is an NASC and thus no essential simplification is possible. The condition is, however, computationally simple since it reduces to the problem of finding the minimum of a continuous function over a bounded interval.

Theorem 2. *Let* $\mathbf{s_1}$ *and* $\mathbf{s_2}$ *be the signatures of two arbitrary mixed systems based on coherent systems in* n *iid components, having the same component* *lifetime distribution, and let T_1 and T_2 be the respective system lifetimes. Then $T_1 \leq_{hr} T_2$ if and only if*

$$h_1(x) - h_2(x) \geq 0 \quad \textit{for all} \quad x \geq 0, \tag{14}$$

where h_j represents the rational function

$$h(x) = \frac{\sum_{i=0}^{n-1} (n-i) s_{i+1} \binom{n}{i} x^i}{\sum_{i=0}^{n-1} \left(\sum_{j=i+1}^{n} s_j \right) \binom{n}{i} x^i}, \tag{15}$$

with $\mathbf{s} = \mathbf{s_j}$, $j = 1, 2$.

While condition (14) is mathematically complex, after cross multiplying in the inequality $h_1(x) \geq h_2(x)$, checking (14) reduces to verifying that a certain polynomial of degree $2n - 3$ is nonnegative for all $x \geq 0$.

Let us now consider the case of likelihood ratio ordering between two system lifetimes. In this case, it follows from (5) that $T_1 \leq_{lr} T_2$ if and only if for all $t \geq 0$, the rational function

$$\frac{(\overline{F}(t))^n \sum_{i=0}^{n-1} (n-i) s_{2,i+1} \binom{n}{i} (G(t))^i r(t)}{(\overline{F}(t))^n \sum_{i=0}^{n-1} (n-i) s_{1,i+1} \binom{n}{i} (G(t))^i r(t)} \tag{16}$$

is increasing in $t > 0$. Define the polynomial m as follows:

$$m(x) = \sum_{i=0}^{n-1} (n-i) s_{i+1} \binom{n}{i} x^i, \tag{17}$$

where $\mathbf{s}$ is an n-dimensional probability vector (or signature). The desired result then follows:

Theorem 3. *Let $\mathbf{s_1}$ and $\mathbf{s_2}$ be the signatures of two arbitrary mixed systems based on coherent systems in n iid components, both with the same component lifetime distribution, and let T_1 and T_2 be the respective system lifetimes. Then $T_1 \leq_{lr} T_2$ if and only if the rational function*

$$\frac{m_2(x)}{m_1(x)} \tag{18}$$

is increasing in $x \geq 0$, where $m_j(x)$ is given by $m(x)$ in (17) with $\mathbf{s} = \mathbf{s_j}$, $j = 1, 2$.

Our main interest in Theorems 1–3 was the development of NASC for the ordering of two system failure times. However, when the ordering of these lifetimes does not hold, the NASCs in these theorems suggest the possibility of identifying distinct **intervals** of time in which the orderings do hold. In all three situations, the type of domination one is interested in may be found to fail when one concentrates on the whole real line, but it might nonetheless hold in an important interval such as $(0, T^*)$, where T^* is the mission time of the systems of interest.

We close with an example of two systems in iid components for which both the survival functions and the failure rates of two competing systems cross exactly once and whose likelihood ratio changes monotonicity. In each case, the changes can be identified to occur at a specific quantile of the common component lifetime distribution F. In all three of these illustrations, the two systems to be compared are the same, namely, the three-component systems having signatures $\mathbf{s}_1 = (1/2, 0, 1/2)$ and $\mathbf{s}_2 = (0, 1, 0)$ respectively. The first system results from selecting a series or a parallel system at random, each with probability 1/2, while the second system is simply a 2-out-of-3 system (i.e., fails upon the second component failure).

Example 3. From Theorem 1, it follows that the two survival functions will cross at the time $t_0 = F^{-1}(1/2)$, which leads to the conclusion that the 2-out-of-3 system is as good as or better than the mixed system if and only if $t \le t_0$. For details, see Block, Dugas, and Samaniego (2006).

Example 4. A comparison of the failure rates of these same two systems would proceed as follows. It is shown in Block, Dugas and Samaniego (2006) that the functions h_1 and h_2 in (15) cross exactly once, as do the failure rates of the two systems involved. The crossing of the two failure rates occurs at time $t_o = F^{-1}(1/4)$. It follows that the 2-out-of-3 system has a smaller failure rate than the mixed system for t such that $0 \le t < F^{-1}(1/4)$ and has a larger failure rate than the mixed system if $t > F^{-1}(1/4)$.

The following example describes the behavior of the likelihood ratio for the two systems above.

Example 5. To compare the likelihood ratio of the two systems, we calculate the polynomials m_1 and m_2 in this problem from (17).

$$m_1(x) = 1.5x^2 + 1.5 \tag{19}$$

and

$$m_2(x) = 6x \tag{20}$$

The derivative of the ratio $m_2(x)/m_1(x)$ has just one positive root, namely $x = G(t) = 1$. When $x < 1$, the ratio of interest is increasing, while it is decreasing when $x > 1$. Since $F(t) = .5$ when $x = 1$, the median failure time of the component distribution F proves to be the pivot around which the likelihood ratio ordering reverses for these two systems. Thus, the ratio of the likelihoods of the 2-out-of-3 system to the series-parallel mixture is increasing for $t < F^{-1}(1/2)$ and is decreasing thereafter.

5 Asymptotics for Failure Rates and Survival Functions of General Mixed Systems

The results of this section concern the asymptotic behavior of the failure rate and the survival function of the lifetime T of an arbitrary mixed system based on a collection of coherent systems in n iid components. Our first theorem demonstrates that the asymptotic failure rate of the system is a particular multiple of the failure rate of an individual component. This result can also be derived from a result in Block Li and Savits (2003). The proof presented here, however, is remarkably direct and shows quite clearly the utility and power of the representation of failure rates via system signatures. The argument requires no adjustments when considering arbitrary mixed systems, as the argument applies with equal force to coherent systems or stochastic mixtures of them. We also present two new results on the rates of convergence to zero of the survival functions of lifetimes of arbitrary mixed systems.

Theorem 4. *Let T be the lifetime of a mixed system based on a set of coherent systems in n iid components, each component having a common failure rate $r(t)$. Assume that $\lim_{t\to\infty} r(t) = r$ $(0 \le r \le \infty)$ exists. If the system has failure rate $r_T(t)$ and signature $\mathbf{s} = (s_1, s_2, ..., s_n)$, then*

$$\lim_{t\to\infty} r_T(t) = (n - k^* + 1)r \tag{21}$$

where $k^* = \max\{i|s_i > 0\}$.

Proof. Dividing numerator and denominator of equation (6) by $[G(t)]^{k^*-1}$ and letting $t \to \infty$, we obtain

$$r_T(t) = \frac{o(1) + (n - k^* + 1)s_{k^*}\binom{n}{k^*-1}}{o(1) + s_{k^*}\binom{n}{k^*-1}} r \to (n - k^* + 1)r,$$

as $t \to \infty$, where $r = \lim_{t\to\infty} r(t)$ and $r(t)$ is the common failure rate of the components involved in the mixed system. □

Remark 1. Let $r_j(t)$ be the failure rate of mixed system with signature $\mathbf{s}_j$ and let $k_j^* = \max\{i|s_{ji} > 0\}$, $j = 1, 2$. If $k_1^* \le k_2^*$, then

$$\lim_{t\to\infty} \frac{r_1(t)}{r_2(t)} = \frac{(n - k_1^* + 1)}{(n - k_2^* + 1)} \ge 1.$$

Remark 2. Note that if $\mathbf{s_1} \leq_{st} \mathbf{s_2}$, then $k_1^* \leq k_2^*$.

Remark 3. Regarding the behavior of $r_T(t)$ for t near zero, the following conclusion immediately follows from (6). If T is the lifetime of a mixed system with signature $\mathbf{s}$, then $\lim_{t\to 0^+} r_T(t) = ns_1 r(0)$, where $r(0) = \lim_{t\to 0^+} r(t)$.

Remark 4. It follows from Remarks 1 and 3 above that if $s_{11} > s_{21}$ and $k_2^* \leq k_1^*$, then the failure rates of the corresponding mixed systems cross at least once.

The limiting behavior of the survival function of the lifetime of a coherent or mixed system has not attracted much attention from the reliability community. Of course all these functions tend to zero, but precise results concerning their rates of convergence have heretofore been unavailable. Simple results, such as the fact that the survival function of a parallel system in iid components tends to zero much more slowly than the survival function of an individual component, and that the opposite is true for series systems, are trivially proven and are intuitively known by virtually all users of such systems. But what can be said about a general mixed system? Using signatures, the following two results provide definitive answers about rates of convergence of survival functions to zero in the general case. We note that, at the other extreme, the case of a series system, the ratio $\overline{F}_T(t)/(\overline{F}(t))^n$ is constant in t, so that the limiting result in Theorem 5 is automatic.

Theorem 5. *Let T be the lifetime of a mixed system with signature $\mathbf{s}$ based on a set of coherent systems in n iid components. Let F be the common lifetime distribution of the components. Then*

$$\frac{\overline{F}_T(t)}{[\overline{F}(t)]^{n-k^*+1}} \to \binom{n}{k^*-1} s_{k*},$$

where $k^* = \max\{i|s_i > 0\}$.

Proof. Using the representation in (1), the mixed system has lifetime distribution,

$$\overline{F}_T(t) = \sum_{j=0}^{k^*-1} \left(\sum_{i=j+1}^{n} s_i \right) \binom{n}{j} (F(t))^j (\overline{F}(t))^{n-j}$$

Dividing the above quantity by $[\overline{F}(t)]^{n-k^*+1}$ and letting $t \to \infty$, one obtains $\binom{n}{k^*-1} s_{k*}$ as the limit. □

The result above provides a number of interesting insights. Since the rate of convergence of $\overline{F}_T$ to zero is only affected by the largest index of a positive element of the signature vector, it is apparent that mixed systems can be creatively used to achieve the limiting behavior of T resembling that of a parallel system. Indeed, the survival function of any system design for which s_n is positive achieves the same basic rate of convergence to zero as that of the parallel system (which achieves the best possible rate). We see from the theorem above that all systems corresponding to the same value of $k^* = \max\{i|s_i > 0\}$ have the same rate of convergence but that the survival functions, for large t, are ordered, with larger values of s_{k^*} corresponding to larger survival probabilities for t sufficiently large.

Theorem 6. *Let T_1 and T_2 be the lifetimes of two mixed system, with signatures $\mathbf{s_1}$ and $\mathbf{s_2}$ respectively, each based on a set of coherent systems in n iid components. Let F be the common lifetime distribution of the components. Then, with $k_i^* = \max\{j|s_{ij} > 0\}$ for $i = 1, 2$, the following limits obtain:*

$$\frac{\overline{F}_{T_2}(t)}{\overline{F}_{T_1}(t)} \to \frac{s_{2k^*}}{s_{1k^*}} \qquad \text{if} \qquad \mathrm{k}_1^* = \mathrm{k}_2^* = \mathrm{k}^* \tag{22}$$

and

$$\frac{\overline{F}_{T_2}(t)}{\overline{F}_{T_1}(t)} \to \infty(0) \qquad \text{if} \qquad \mathrm{k}_1^* < (>) \mathrm{k}_2^* \tag{23}$$

Proof. Using the representation in (4), we obtain

$$\frac{\overline{F}_{T_2}(t)}{\overline{F}_{T_1}(t)} = \frac{\sum_{j=0}^{k_2^*-1} \left(\sum_{i=j+1}^{n} s_{2i} \right) \binom{n}{j} (G(t))^j}{\sum_{j=0}^{k_1^*-1} \left(\sum_{i=j+1}^{n} s_{1i} \right) \binom{n}{j} (G(t))^j}$$

Since $G(t) \to \infty$ as $t \to \infty$, the result follows by dividing both the numerator and the denominator by $(G(t))^{\max(k_1^*, k_2^*)-1}$. □

References

1. Barlow, R. E. and Proshan, F. (1981) *Statistical Theory of Reliability*, Silver Springs, MD: To Begin With Press
2. Block,H. W., Dugas, M. R. and Samaniego, F. J. (2006) "Characterizations of the Relative Behavior of Two Coherent Systems via Properties of their Signature Vectors", in *Advances in Distribution Theory, Order Statistics and Inference* (N. Balakrishnan, Editor) Basel, Switzerland: Birkhauser Verlag

3. BLOCK, H. W., LI, Y. AND SAVITS, T. (2003) "Initial and final behavior of the failure rate functions for mixtures and systems", *J. Appl. Prob.*, 40, 721-740.

4. BOLAND, P. AND SAMANIEGO, F. J. (2004) "The signature of a coherent system and it's applications in reliability", in *Mathematical Reliability: An Expository Perspective* (R. Soyer, T. Mazzuchi, and N. Singpurwalla, Editors), New York: Kluwer Academic Press, 1 - 29.

5. DUGAS, M. R. AND SAMANIEGO, F. J. (2005) "On optimal system designs under reliability and economics constraints", *Proceedings of the 10th Army Conference on Applied Statistics*, Aberdeen Proving Ground, MD: Army Research Laboratory Reports.

6. KOCHAR, S., MUKERJEE, H. AND SAMANIEGO, F. J. (1999) "The signature of a coherent system and its application to comparisons among systems", *Naval Research Logistics* 46, 507-523.

7. SAMANIEGO, F. J. (1985) "On closure of the IFR class under formation of coherent systems", *IEEE Trans Reliab*, R-34, 69-72.

Chapter 7

STOCHASTIC ORDERS BASED ON TOTAL TIME ON TEST TRANSFORM

Xiaohu Li and Moshe Shaked

Department of Mathematics
Lanzhou University, Gansu, CHINA

Department of Mathematics
University of Arizona, Tucson, AZ, U.S.A.

E-mails: xhli@lzu.edu.cn & shaked@math.arizona.edu

In this article we review recent work on generalizations of the total time on test transform, and on stochastic orders that are based on these generalizations. Applications in economics, statistics, and reliability theory, are described as well.

Key words: Proportional hazard; Reliability theory; Economics; Coherent system; Distortion function; Insurance premium.

1 Introduction

Let X be a nonnegative random variable with a continuous distribution function F and mean $\mu \leq \infty$. Denote the corresponding survival function by $\overline{F} = 1 - F$. The total time on test (TTT) transform T_X of X, is defined by

$$T_X(p) = \int_0^{F^{-1}(p)} \overline{F}(x)\, dx, \quad p \in (0,1), \tag{1}$$

where F^{-1} is the right-continuous inverse of F. The TTT transform is a theoretical version of the empirical TTT transform that is often used in statistical reliability theory. Roughly speaking, $T_X(p)$ gives the average time that an item spends on a test if the test is terminated when a fraction p of all the items on the test fail.

Barlow and Doksum (1972) introduced and studied a generalization of the TTT transform, $T_X^{(G)}$, defined by

$$T_X^{(G)}(p) = \int_0^{F^{-1}(p)} g[G^{-1}F(x)]\,dx, \quad p \in (0,1), \tag{2}$$

where G is an absolutely continuous distribution function with $G(0) = 0$, and $g = G'$ is the corresponding density function. When G is the unit mean exponential distribution, $T_X^{(G)}$ reduces to T_X. Barlow and Doksum (1972) noticed that the generalized TTT transform simplifies the study of the TTT transform. We will later see that it also has useful interpretations and applications in actuarial science, as well as in reliability theory.

An even more general transform has been introduced and studied in Li and Shaked (2004). Let $\mathfrak{H}$ denote the set of all functions h such that $h(u) > 0$ for $u \in (0,1)$, and $h(u) = 0$ for $u \notin [0,1]$. For $h \in \mathfrak{H}$, define the transform $\widetilde{T}_X^{(h)}$ by

$$\widetilde{T}_X^{(h)}(p) = \int_0^{F^{-1}(p)} h(F(x))\,dx, \quad p \in (0,1). \tag{3}$$

Obviously $T_X^{(G)}$ is a special case of $\widetilde{T}_X^{(h)}$. In fact,

$$T_X^{(G)} = \widetilde{T}_X^{(gG^{-1})}. \tag{4}$$

Example 1. (Reliability theory and statistics). Let X be a nonnegative random variable with survival function $\overline{F}$. For $\theta > 0$, let $X(\theta)$ denote a random variable with survival function $(\overline{F})^\theta$. In the theory of statistics, $(\overline{F})^\theta$ is often referred to as the *Lehmann's alternative.* In reliability theory terminology, different $X(\theta)$'s have *proportional hazards.* If $\theta < 1$ then $X(\theta)$ is the lifetime of a component with lifetime X which is subjected to *imperfect repair* procedure where θ is the probability of minimal (rather than perfect) repair (see Brown and Proschan (1983)). If $\theta = n$, where n is a positive integer, then $(\overline{F})^n$ is the survival function of $\min\{X_1, X_2, \ldots, X_n\}$ where $X_1, X_2, \ldots, X_n$ are independent copies of X; that is, $(\overline{F})^n$ is the survival function of a series system of size n where the component lifetimes are independent copies of X. For $\theta < 1$, choosing h to be

$$h(u) = (1-u)^\theta, \quad u \in [0,1], \tag{5}$$

it is seen that

$$\widetilde{T}_X^{(h)}(p) = \int_0^{F^{-1}(p)} h(F(x))\,dx = \int_0^{F^{-1}(p)} (\overline{F})^\theta(x)\,dx$$

and this gives, in the reliability theory setting, the expected total time on test of an item that is maintained with an imperfect repair procedure, and which is tested until its pth quantile lifetime or until its first perfect repair, whichever comes first. Equivalently, in the theoretical statistics setting, for $\theta > 0$ with h as in (5), $\widetilde{T}_X^{(h)}(p)$ gives the expected total time on test of an item under the Lehmann's alternative, if the item runs under the null hypothesis (that is, when $\theta = 1$) until it fails or until it reaches its pth quantile lifetime, whichever occurs first. Finally, when $\theta = n$, where n is a positive integer, then $\widetilde{T}_X^{(h)}(p)$ gives the expected total time on test of a series system of independent and identical components, which runs until it fails or until it reaches the pth quantile lifetime of its components, whichever occurs first.

When $h(u) = u^\theta$, a similar interpretation can be given to $\widetilde{T}_X^{(h)}(p)$ in the setting of proportional reversed hazard rates; see, for example, Di Crescenzo (2000) for a study of this model.

Example 2. (Actuarial science and insurance). Let X be the loss of an insurance contract. Let F be the distribution function of X. Then the expected loss is

$$E[X] = \int_0^\infty \overline{F}(x)\, dx.$$

The expected loss can be used as the net premium paid for the insurance contract. In order to give more loss weight to higher risks, a distortion pricing principle is sometimes used in practice. Let ψ be a *distortion function*; that is, $\psi : [0,1] \to [0,1]$ is an increasing concave function such that $\psi(0) = 0$ and $\psi(1) = 1$. The *distortion pricing principle* that is based on ψ states that

$$\rho_\psi(X) = \int_0^\infty \psi(\overline{F}(x))\, dx \tag{6}$$

is the premium paid for the insurance contract. See, Wang (1996), Wang, Young, and Panjer (1997), Hurlimann (1998), and Wu and Wang (2003) for more details.

Let $\psi \in \mathfrak{H}$ be a distortion function, and define h_ψ by $h_\psi(u) = \psi(1-u)$, $u \in [0,1]$. Then $\widetilde{T}_X^{(h_\psi)}(p)$ has an important and interesting interpretation in actuarial science. Suppose that an insurer accepts a random risk X that has a distribution function F. Suppose further that the insurer arranges a reinsurance for this risk determined by a retention level ℓ; that is, the reinsurer will pay any excess of a claim above the level ℓ (in other words, ℓ may be thought of as a deductible); see, for instance, Waters (1983). If ℓ is determined as the pth quantile of the claim distribution, then $\widetilde{T}_X^{(h_\psi)}(p)$ is the expected claim loss below the retention level; this is an important quantity for the insurer.

Example 3. (Reliability theory). We now describe another appearance of $\widetilde{T}_X^{(h)}$ in reliability theory. In order to do that, we need to recall the definition of the reliability function of a coherent system. Consider a coherent system (see Barlow and Proschan (1975)) and suppose that each of its components works with probability u, independently of each other. If the probability that the system works is $\phi(u)$, then $\phi : [0,1] \to [0,1]$ is called the *reliability function* of the system. If the lifetimes of the components are independent and identically distributed with survival function $\overline{F}$, then the survival function of the system lifetime, X say, is given by

$$\overline{F}_X(x) = \phi(\overline{F}(x)), \quad x \geq 0. \tag{7}$$

Thus, if $h \in \mathfrak{H}$ is such that $h(1-\cdot)$ is a reliability function of a coherent system then $\widetilde{T}_X^{(h)}(p)$ gives the expected total time on test of that coherent system, when it has independent and identical components, and which runs until it fails or until it reaches the pth quantile lifetime of its components, whichever occurs first. This observation generalizes the discussion about the series system in Example 1.

In this paper, 'increasing' stands for 'nondecreasing', and 'decreasing' stands for 'nonincreasing.'

2 Stochastic Orders Based on Transforms

Each of the transforms that are described in (1)–(3) can be used to define a stochastic order. Let X and Y be two nonnegative random variables with distribution functions F and K, and survival functions $\overline{F}$ and $\overline{K}$, respectively. Let T_X be as defined in (1), and let T_Y be similarly defined, with K replacing F. If

$$T_X(p) \leq T_Y(p), \quad p \in (0,1),$$

then, according to Kochar, Li, and Shaked (2002), X is said to be smaller than Y in the TTT transform order (denoted as $X \leq_{\text{ttt}} Y$ or as $F \leq_{\text{ttt}} K$). For $h \in \mathfrak{H}$, let $\widetilde{T}_X^{(h)}$ be as defined in (3), and let $\widetilde{T}_Y^{(h)}$ be similarly defined, with K replacing F. If

$$\widetilde{T}_X^{(h)}(p) \leq \widetilde{T}_Y^{(h)}(p), \quad p \in (0,1), \tag{8}$$

then, according to Li and Shaked (2004), X is said to be smaller than Y in the generalized TTT transform order with respect to h (denoted as $X \leq_{\text{ttt}}^{(h)} Y$ or as $F \leq_{\text{ttt}}^{(h)} K$). The transform that is described in (2) can also be used to define a stochastic order. However since the study of this order is similar to the study of the order $\leq_{\text{ttt}}^{(h)}$, we will not study it on its own. In this section we review some basic properties of the orders $\leq_{\text{ttt}}$ and $\leq_{\text{ttt}}^{(h)}$, and some of their consequences.

2.1 *The TTT Transform Order*

In this subsection we review some basic properties of the order $\leq_{\text{ttt}}$. Most of the results below, and their proofs, can be found in Kochar, Li, and Shaked (2002).

Let X and Y be two nonnegative random variables with distribution functions F and K, and survival functions $\overline{F}$ and $\overline{K}$, respectively.

A simple sufficient condition for the order $\leq_{\text{ttt}}$ is the usual stochastic order:

$$X \leq_{\text{st}} Y \Longrightarrow X \leq_{\text{ttt}} Y, \tag{9}$$

where $X \leq_{\text{st}} Y$ means that $\overline{F}(x) \leq \overline{K}(x)$ for all $x \geq 0$ (see, for example, Shaked and Shanthikumar (1994, Section 1.A)). In order to verify (9) we just notice that if $X \leq_{\text{st}} Y$ then $F^{-1}(p) \leq G^{-1}(p)$ for all $p \in (0,1)$. Another way to verify (9) is indicated after Theorem 3 below.

Using the fact that, for any nonnegative random variable X and for any $a > 0$, we have

$$T_{aX}(p) = aT_X(p), \quad p \in (0,1),$$

it is easy to see that, for any two nonnegative random variables X and Y, we have

$$X \leq_{\text{ttt}} Y \Longrightarrow aX \leq_{\text{ttt}} aY \text{ for any } a > 0. \tag{10}$$

The implication (10) may suggest that if $X \leq_{\text{ttt}} Y$ then $\phi(X) \leq_{\text{ttt}} \phi(Y)$ for any increasing function ϕ. But this is not true. What is true, however, is that the order $\leq_{\text{ttt}}$ is preserved under increasing concave transformations.

Theorem 1. *Let X and Y be two continuous random variables with interval supports, and with 0 being the common left endpoint of their supports. Then, for any increasing concave function ϕ, such that $\phi(0) = 0$, we have*

$$X \leq_{\text{ttt}} Y \Longrightarrow \phi(X) \leq_{\text{ttt}} \phi(Y).$$

If $\phi : [0,\infty) \to (-\infty,\infty)$ is an increasing concave function such that $\phi(0) \neq 0$, then the function $\phi(\cdot) - \phi(0) : [0,\infty) \to [0,\infty)$ satisfies the conditions in Theorem 1. It follows, for random variables X and Y as described in Theorem 1, that

$$X \leq_{\text{ttt}} Y \Longrightarrow E\phi(X) \leq E\phi(Y) \tag{11}$$

for every increasing concave function ϕ on $[0,\infty)$, provided the expectations exist. That is, for random variables X and Y as described in Theorem 1, we have

$$X \leq_{\text{ttt}} Y \Longrightarrow X \leq_{\text{icv}} Y, \tag{12}$$

where $\leq_{\text{icv}}$ denotes the increasing concave order (see, for example, Shaked and Shanthikumar (1994, Section 3.A)).

The order $\leq_{\text{ttt}}$ is also closed under the formation of series systems.

Theorem 2. *Let $X_1, X_2, \ldots, X_n$ be a collection of independent and identically distributed nonnegative random variables, and let $Y_1, Y_2, \ldots, Y_n$ be another collection of independent and identically distributed nonnegative random variables. Then*

$$X_1 \leq_{\text{ttt}} Y_1 \Longrightarrow \min\{X_1, X_2, \ldots, X_n\} \leq_{\text{ttt}} \min\{Y_1, Y_2, \ldots, Y_n\}.$$

2.2 *The Generalized TTT Transform Orders*

In this subsection we review some basic properties of the orders $\leq_{\text{ttt}}^{(h)}$. Most of the results below, and their proofs, can be found in Li and Shaked (2004).

Let X and Y be two nonnegative random variables with distribution functions F and K, and survival functions $\overline{F}$ and $\overline{K}$, respectively.

First we note that the order $\leq_{\text{ttt}}$ is a member in the class of orders $\leq_{\text{ttt}}^{(h)}$, $h \in \mathfrak{H}$; it is obtained by letting h be the function $h(u) = 1 - u$ on $[0, 1]$.

We also note that the usual stochastic order $\leq_{\text{st}}$ is a member of this class of orders; it is obtained when h is a constant function on $[0, 1]$. In order to see this, suppose that $h(u) = c$, $u \in [0, 1]$, for some $c > 0$. Then $X \leq_{\text{ttt}}^{(h)} Y$ means

$$\widetilde{T}_X^{(h)}(p) \leq \widetilde{T}_Y^{(h)}(p), \quad p \in (0, 1);$$

that is,

$$\int_0^{F^{-1}(p)} h(F(x))\, dx \leq \int_0^{K^{-1}(p)} h(K(x))\, dx, \quad p \in (0, 1).$$

The latter inequality is the same as

$$F^{-1}(p) \leq K^{-1}(p), \quad p \in (0, 1);$$

that is, $X \leq_{\text{st}} Y$.

Another common order that is a member of the class of orders $\leq_{\text{ttt}}^{(h)}$ is obtained by letting h be the function $h(u) = u$ on $[0, 1]$. Then $X \leq_{\text{ttt}}^{(h)} Y$ means

$$\int_0^{F^{-1}(p)} F(x)\, dx \leq \int_0^{K^{-1}(p)} K(x)\, dx, \quad p \in (0, 1).$$

This defines the so-called *location independent riskier* order (denoted as $X \leq_{\text{lir}} Y$ or $F \leq_{\text{lir}} G$); this order was introduced in Jewitt (1989) and was further studied in Fagiuoli, Pellerey, and Shaked (1999) and in Kochar, Li, and Shaked (2002).

A useful relationship among the orders $\leq_{\rm ttt}^{(h)}$ is given in the next theorem.

Theorem 3. *Let X and Y be two random variables with continuous distribution functions, having 0 as the common left endpoint of their supports. Let $h_1, h_2 \in \mathfrak{H}$. Suppose that*

$$h_2(u)/h_1(u) \text{ is decreasing on } (0,1).$$

Then

$$X \leq_{\rm ttt}^{(h_1)} Y \Longrightarrow X \leq_{\rm ttt}^{(h_2)} Y.$$

For example, the implication (9) easily follows from Theorem 3. Some more relationships between the usual stochastic order $\leq_{\rm st}$ and the orders $\leq_{\rm ttt}^{(h)}$ follow easily from Theorem 3 and are described next.

Theorem 4. *Let X and Y be two nonnegative random variables with continuous distribution functions, having 0 as the common left endpoint of their supports. Let $h \in \mathfrak{H}$.*

(a) *If h is decreasing on $(0,1)$ then $X \leq_{\rm st} Y \Longrightarrow X \leq_{\rm ttt}^{(h)} Y$.*
(b) *If h is increasing on $(0,1)$ then $X \leq_{\rm ttt}^{(h)} Y \Longrightarrow X \leq_{\rm st} Y$.*

The usual stochastic order is a useful tool that yields important inequalities (see, for example, Shaked and Shanthikumar (1994) or Müller and Stoyan (2002); Theorem 4(b) can be used to identify this order. On the other hand, in instances in which the usual stochastic order is known to hold, Theorem 4(a) gives conditions under which the generalized TTT transform order holds, and the inequalities that the latter yields can then be applied.

The following result is a part of Proposition 1 of Bartoszewicz (1986) although his notation is different than the notation in the present paper. It shows that the dispersive order yields a particular form of the generalized TTT transform order. Recall that a random variable X is said to be smaller than another random variable Y in the dispersive order (denoted as $X \leq_{\rm disp} Y$) if $K^{-1}(p) - F^{-1}(p)$ is increasing in $p \in (0,1)$, where F and K are the distribution functions of X and Y, respectively; see, for example, Shaked and Shanthikumar (1994, Section 2.B).

Theorem 5. *Let X and Y be two absolutely continuous random variables with 0 being the common left endpoint of their supports. Let G be any absolutely continuous distribution function on $[0,\infty)$ with density function g. Then*

$$X \leq_{\rm disp} Y \Longrightarrow X \leq_{\rm ttt}^{(gG^{-1})} Y.$$

It is of interest to point out the contrast between Theorem 4(a) and Theorem 5. Taking $h = gG^{-1}$, the conclusions of Theorem 4(a) and of Theorem 5 are the same. For random variables X and Y that have the same left endpoint of support, it is known that $X \leq_{\text{disp}} Y \Longrightarrow X \leq_{\text{st}} Y$. Thus the assumption $X \leq_{\text{st}} Y$ in Theorem 4(a) is weaker than the assumption in Theorem 5. However, in Theorem 5 nothing is assumed about G, whereas in Theorem 4(a), we obtain the implication only for some G's; that is, for G's with decreasing densities.

An extension of Theorem 1 is the following result.

Theorem 6. *Let X and Y be two continuous random variables with interval supports, and with 0 being the common left endpoint of their supports. Let $h \in \mathfrak{H}$ be decreasing on $[0,1]$. Then, for any increasing concave function ϕ, such that $\phi(0) = 0$, we have*

$$X \leq_{\text{ttt}}^{(h)} Y \Longrightarrow \phi(X) \leq_{\text{ttt}}^{(h)} \phi(Y).$$

Following the argument that leads to (12), we see that as a consequence of Theorem 6 we get, for random variables X and Y as described there, that

$$X \leq_{\text{ttt}}^{(h)} Y \Longrightarrow X \leq_{\text{icv}} Y,$$

whenever $h \in \mathfrak{H}$ is decreasing on $[0,1]$.

Let G be an absolutely continuous distribution function with $G(0) = 0$, and let $g = G'$ be the corresponding density function. From (4) it follows that if we take h in (8) to be gG^{-1}, then the order becomes a pointwise comparison of generalized TTT transforms (see (2)). Let us recall the definition of the convex transform order. Let X and Y be two nonnegative random variables with continuous distribution functions F and K, respectively, and with supports $[0,a)$ and $[0,b)$, respectively, for some finite or infinite constants a and b. Then X is said to be smaller than Y in the convex transform order (denoted as $X \leq_{\text{c}} Y$ or $F \leq_{\text{c}} K$) if $K^{-1}F$ is convex. The order $\leq_{\text{c}}$ is discussed, for example, in Shaked and Shanthikumar (1994, Section 3.C). Note that $X \leq_{\text{c}} Y$ is denoted in Fernandez-Ponce, Kochar, and Muñoz-Perez (1998) as $X \overset{\text{IFR}}{\succeq} Y$. The following result gives a condition, by means of the convex transform order $\leq_{\text{c}}$, under which a pointwise comparison of generalized TTT transforms with respect to one distribution, G_1, say, implies the pointwise comparison of generalized TTT transforms with respect to another distribution, G_2, say.

Proposition 1. *Let X and Y be two nonnegative random variables with continuous distribution functions, having 0 as the common left endpoint of their supports. Let G_1 and G_2 be two absolutely continuous distribution* *functions with supports* $[0, a)$ *and* $[0, b)$ *for some finite or infinite constants* a *and* b, *and density functions* g_1 *and* g_2. *If* $X \le_{\text{ttt}}^{(g_1 G_1^{-1})} Y$, *and if* $G_1 \le_{\text{c}} G_2$, *then* $X \le_{\text{ttt}}^{(g_2 G_2^{-1})} Y$.

Proof. Note that $G_1 \le_{\text{c}} G_2$ means that $g_2 G_2^{-1}(u)/g_1 G_1^{-1}(u)$ is decreasing on $[0, 1]$. The stated result thus follows from Theorem 3. □

Recall that a nonnegative random variable Z is IFR (increasing failure rate) if $Z \le_{\text{c}} E(1)$, where $E(1)$ denotes a unit mean exponential random variable. Also, a nonnegative random variable Z is DFR (decreasing failure rate) if $Z \ge_{\text{c}} E(1)$. From Proposition 1 we obtain the following corollary.

Corollary 1. *Let* G_1 *with support of the form* $[0, a)$ *be an* IFR *distribution function, and let* G_2 *with support* $[0, \infty)$ *be a* DFR *distribution function. Let* X *and* Y *be two nonnegative random variables with continuous distribution functions, having* 0 *as the common left endpoint of their supports. Then*

(a) $X \le_{\text{ttt}}^{(g_1 G_1^{-1})} Y \Longrightarrow X \le_{\text{ttt}} Y$.
(b) $X \le_{\text{ttt}} Y \Longrightarrow X \le_{\text{ttt}}^{(g_2 G_2^{-1})} Y$.
(c) $X \le_{\text{ttt}}^{(g_1 G_1^{-1})} Y \Longrightarrow X \le_{\text{ttt}}^{(g_2 G_2^{-1})} Y$.

3 Some Applications

3.1 *Reliability Theory and Statistics*

Let X be a nonnegative random variable with survival function $\overline{F}$. For $\theta > 0$, as in Example 1, let $X(\theta)$ denote a random variable with survival function $(\overline{F})^\theta$. Also, let Y be another nonnegative random variable with survival function $\overline{K}$, and let $Y(\theta)$ denote a random variable with survival function $(\overline{K})^\theta$. Define the function $h^{(\theta)}$ by $h^{(\theta)}(u) = (1-u)^\theta$, $u \in (0, 1)$. Li and Shaked (2004) proved that

$$X(\theta) \le_{\text{ttt}} Y(\theta) \Longleftrightarrow X \le_{\text{ttt}}^{(h^{(\theta)})} Y. \tag{13}$$

As a consequence they obtained the following result.

Proposition 2. *Let* X *and* Y *be two nonnegative random variables with continuous distribution functions, having* 0 *as the common left endpoint of their supports. Let* $X(\theta)$ *and* $Y(\theta)$ *be as described above.*

(a) *If* $\theta > 1$ *then* $X \le_{\text{ttt}} Y \Longrightarrow X(\theta) \le_{\text{ttt}} Y(\theta)$.
(b) *If* $\theta < 1$ *then* $X(\theta) \le_{\text{ttt}} Y(\theta) \Longrightarrow X \le_{\text{ttt}} Y$.

Proof. Note that

$$X \leq_{\rm ttt} Y \Longleftrightarrow X \leq_{\rm ttt}^{(h^{(1)})} Y. \tag{14}$$

Now, if $\theta > 1$ then $h^{(\theta)}/h^{(1)}$ is decreasing on $(0,1)$, and part (a) follows from Theorem 3, (13), and (14). On the other hand, if $\theta < 1$ then $h^{(1)}/h^{(\theta)}$ is decreasing on $(0,1)$, and part (b) follows, again, from Theorem 3, (13), and (14). □

If we take $\theta = n$ in Proposition 2, where n is a positive integer, we obtain Theorem 5.1(a) of Kochar, Li, and Shaked (2002); that is, we see that if the lifetimes of the (identical) components of one series system, are comparable to the lifetimes of the (identical) components of another series system, with respect to the order $\leq_{\rm ttt}$, then the two system lifetimes are also comparable with respect to the order $\leq_{\rm ttt}$, where n is the size of the series systems.

3.2 *Actuarial Science and Insurance*

Let X be the loss of an insurance contract (see Example 2), and let F be its distribution function. Let $\psi \in \mathfrak{H}$ be a distortion function (again, see Example 2), and define h_ψ, as in that example, by $h_\psi(u) = \psi(1-u)$, $u \in [0,1]$. Then the premium paid for the insurance contract, with respect to ψ (see (6)), is

$$\rho_\psi(X) = \int_0^\infty \psi(\overline{F}(x))\,dx = \int_0^\infty h_\psi(F(x))\,dx.$$

Let Y be the loss of another insurance contract, and let K be its distribution function. Then the premium paid for that contract, with respect to ψ, is

$$\rho_\psi(Y) = \int_0^\infty h_\psi(K(x))\,dx.$$

It follows that

$$X \leq_{\rm ttt}^{(h_\psi)} Y \Longrightarrow \rho_\psi(X) \leq \rho_\psi(Y);$$

that is, the order $\leq_{\rm ttt}^{(h_\psi)}$ gives a sufficient condition for a comparison of premiums with respect to ψ.

The orders $\leq_{\rm ttt}^{(h)}$ can also be used to compare premiums of two insurance contracts X and Y, with respect to two distortion functions ψ_1 and ψ_2. We use Theorem 3. Note that ψ_2/ψ_1 is increasing on $(0,1)$ if, and only if, h_{ψ_2}/h_{ψ_1} is decreasing on $(0,1)$. Therefore, by Theorem 3, if ψ_2/ψ_1 is increasing on $(0,1)$ then

$$X \leq_{\rm ttt}^{(h_{\psi_1})} Y \Longrightarrow X \leq_{\rm ttt}^{(h_{\psi_2})} Y.$$

For example, take $\psi_2(u) = u$. Then $h_{\psi_2}(u) = 1 - u$, $u \in [0,1]$, and the order $\le_{\rm ttt}^{(h_{\psi_2})}$ is just the order $\le_{\rm ttt}$. Since ψ_1 is concave (and increasing, and satisfies $\psi_1(0) = 0$) it follows that ψ_1 is anti-starshaped; that is, $\psi_1(u)/u$ is decreasing on $(0,1)$. In other words, ψ_2/ψ_1 is increasing on $(0,1)$. Thus, for any distortion function ψ_1 we have

$$X \le_{\rm ttt}^{(h_{\psi_1})} Y \Longrightarrow X \le_{\rm ttt} Y.$$

In words, if a loss X is smaller than a loss Y with respect to any distortion, then it is also smaller without any distortion.

3.3 *Reliability Theory*

Let X be the lifetime of a coherent system with reliability function ϕ (see Example 3). Suppose that the lifetimes of the components are independent and identically distributed with distribution function F and survival function $\overline{F}$. Define h_ϕ by $h_\phi(u) = \phi(1-u)$, $u \in [0,1]$. Note that $h_\phi \in \mathfrak{H} \Longleftrightarrow \phi \in \mathfrak{H}$. Using (7) it is seen that the expected lifetime of the system is

$$E[X] = \int_0^\infty \phi(\overline{F}(x))\, dx = \int_0^\infty h_\phi(F(x))\, dx.$$

Let Y be the lifetime of the same coherent system, but now suppose that the lifetimes of the components are independent and identically distributed with distribution function K. Then the expected lifetime of the system is

$$E[Y] = \int_0^\infty h_\phi(K(x))\, dx.$$

It follows that

$$X \le_{\rm ttt}^{(h_\phi)} Y \Longrightarrow E[X] \le E[Y];$$

that is, the order $\le_{\rm ttt}^{(h_\phi)}$ gives a sufficient condition for a comparison of the expected lifetimes of the same coherent system, but with different component lifetime distributions.

An interesting preservation result is stated next.

Proposition 3. *Let X and Y be two random lifetimes with continuous distribution functions, having 0 as the common left endpoint of their supports. Let ϕ be the reliability function of a coherent system. If*

$$\frac{\phi(u)}{u} \quad \text{is increasing on } (0,1), \tag{15}$$

then

$$X \le_{\rm ttt} Y \Longrightarrow X \le_{\rm ttt}^{(h_\phi)} Y. \tag{16}$$

Proof. Let ϕ_1 be the reliability function of the elementary system consisting of one component; that is, $\phi_1(u) = u$, $u \in [0,1]$. Note that

$$X \le_{\text{ttt}} Y \Longleftrightarrow X \le_{\text{ttt}}^{(h_{\phi_1})} Y. \tag{17}$$

Now, (15) is equivalent to the condition that h_ϕ / h_{ϕ_1} is decreasing on $(0,1)$. Thus, in light of (17), we see that (16) follows from Theorem 3. □

Let F and K denote, respectively, the distribution functions of X and Y in Proposition 3. The right hand side of (16) can be written as

$$\int_0^{F^{-1}(p)} \overline{F}_S(x)\,dx \le \int_0^{G^{-1}(p)} \overline{F}_T(x)\,dx, \quad p \in (0,1),$$

where $\overline{F}_S$ and $\overline{F}_T$ are defined, for $x \ge 0$, as

$$\overline{F}_S(x) = \phi(\overline{F}(x)) \quad \text{and} \quad \overline{F}_T(x) = \phi(\overline{K}(x));$$

that is, $\overline{F}_S$ is the survival function of the lifetime of the coherent system when the components lifetime distribution is F (see (7)), and $\overline{F}_T$ is the survival function of the lifetime of the coherent system when the components lifetime distribution is K. This means that if a coherent system repeatedly runs a test that is terminated when a fraction p of all the components on test fail, then the average time that the system with lifetime S spends on test is smaller than the average time that the system with lifetime T spends on test. This may be a useful observation in some life testing procedures.

Examples of coherent systems with reliability functions that satisfy (15) are described in Li and Shaked (2004).

Acknowledgements

We thank a referee for a careful reading of the manuscript and for valuable comments.

References

1. Barlow, R. E. and Doksum, K. A. (1972). Isotonic tests for convex orderings. *Proceedings of the Sixth Berkeley Symposium on Mathematical Statistics and Probability I* (edited by L. M. Le Cam, J. Neyman, and E. L. Scott) 293–323.

2. Barlow, R. E. and Proschan, F. (1975). *Statistical Theory of Reliability and Life Testing: Probability Models.* Holt, Rinehart and Winston, New York.

3. Bartoszewicz, J. (1986). Dispersive ordering and the total time on test transformation. *Statistics and Probability Letters* **4** 285–288.
4. Brown, M. and Proschan, F. (1983). Imperfect repair. *Journal of Applied Probability* **20** 851–859.
5. Di Crescenzo, A. (2000). Some results on the proportional reversed hazards model. *Statistics and Probability Letters* **50** 313–321.
6. Fagiuoli, E., Pellerey, F. and Shaked, M. (1999). A characterization of the dilation order and its applications. *Statistical Papers* **40** 393–406.
7. Fernandez-Ponce, J. M., Kochar, S. C., and Muñoz-Perez, J. (1998). Partial orderings of distributions based on right-spread functions. *Journal of Applied Probability* **35** 221–228.
8. Hurlimann, W. (1998). On stop-loss order and the distortion pricing principle. *ASTIN Bulletin* **28** 119–134.
9. Jewitt, I. (1989). Choosing between risky prospects: The characterization of comparative statics results, and location independent risk. *Management Science* **35** 60–70.
10. Kochar, S. C., Li, X. and Shaked, M. (2002). The total time on test transform and the excess wealth stochastic orders of distributions. *Advances in Applied Probability* **34** 826–845.
11. Li, X. and Shaked, M. (2004). A general family of univariate stochastic orders. *Journal of Statistical Planning and Inference*, to appear.
12. Müller, A. and Stoyan, D. (2002). *Comparison Methods for Stochastic Models and Risks*, John Wiley & Sons, New York.
13. Shaked, M. and Shanthikumar, J. G. (1994). *Stochastic Orders and Their Applications*, Academic Press, Boston.
14. Wang, S. (1996). Premium calculation by transforming the layer premium density. *ASTIN Bulletin* **26** 71–92.
15. Wang, S. S., Young, V. R. and Panjer, H. H. (1997). Axiomatic characterization of insurance prices. *Insurance: Mathematics and Economics* **21** 173–183.
16. Waters, H. R. (1983). Some mathematical aspects of reinsurance. *Insurance: Mathematics and Economics* **2** 17–26.
17. Wu, X. and Wang, J. (2003). On characterization of distortion premium principle. *ASTIN Bulletin* **33** 1–10.

PART 3
Semiparametric Methods

Chapter 8

ADAPTIVELY DENOISING DISCRETE TWO-WAY LAYOUTS

Rudolf Beran

Department of Statistics
University of California, Davis, CA 95616, USA

E-mail: beran@wald.ucdavis.edu

The unrestricted least squares estimator for the means of a two-way layout is usually inadmissible under quadratic loss and the model of homoscedastic independent Gaussian errors. In statistical practice, this least squares estimator may be modified by fitting hierarchical submodels and, for ordinal factors, by fitting polynomial submodels. ASP, an acronym for **A**dapative **S**hrinkage on **P**enalty bases, is an estimation (or denoising) strategy that chooses among submodel fits and more general shrinkage or smoothing fits to two-way layouts without assuming that any submodel is true. ASP fits distinguish between ordinal and nominal factors; respect the ANOVA decomposition of means into overall mean, main effects, and interaction terms; and are designed to reduce risk substantially over the unrestricted least squares estimator. Multiparametric asymptotics, in which the number of factor-level pairs tends to infinity, and numerical case studies both support the methodology.

Key words: Estimated risk; Penalized least squares; Bi-monotone shrinkage; Bi-flat shrinkage; Annihilator matrix.

1 Introduction

A fundamental data type in the sciences, engineering, and informatics is the discrete two-way layout. Instances include the data recorded in agricultural field trials, in DNA microassays, in digital imaging, and in other settings where regression or ANOVA are established tools. The factors in a two-way layout may be ordinal or nominal. The levels of an ordinal factor are real-values that indicate at least order and possibly more. The levels of a nominal factor are pure labels that convey no ordering information.

In devising trustworthy fits to discrete layouts, it is essential not to make strong unsupported assumptions concerning how the mean response at a grid point depends on the factor levels associated with that grid-point. On the other hand, the unrestricted least squares estimator tends to overfit the means to a two-way layout, especially when there is little replication. A formal theoretical statement of this difficulty is Stein's (1956) inadmissibility result for the least squares fit to a one-way layout with independent identically distributed Gaussian errors. Better estimation (or denoising) techniques for two-way layouts with unrestricted means rely on biased estimators, such as those generated by the statistical regularization methods of this paper.

The acronym ASP stands for **A**dapative **S**hrinkage on **P**enalty bases. An ASP fit to a discrete two-way layout is constructed in three stages:

1) Devise a candidate class of constrained penalized least squares (PLS) estimators whose *three* quadratic penalty terms express tentative notions about the two main effects and the interactions in the means of the two-way layout.

2) Estimate the risk of each candidate estimator under a general model on the means that does *not* assume any of the prior notions in step one.

3) Define the ASP fit to be a candidate fit that minimizes estimated risk under the general model—the adaptive aspect of the procedure.

Wood (2000) treated penalized least squares with multiple quadratic penalties. The present paper differs from his work by constructing the three penalty terms to address the possible unimportance or smoothness of some main effects or interactions; by using estimated risk under a general model rather than cross-validation to select penalty weights and terms; by treating bi-monotone shrinkage strategies more general than penalized least squares; and by developing asymptotics for ASP estimators in large two-way layouts.

An ASP fit is a biased estimator that trades bias against variance so as to achieve, approximately, the lowest quadratic risk attained over the class of candidate estimators for the means of the two-way layout. Multiparametric asymptotics, in which the total number of cells in the two-way layout tends to infinity, indicate that the estimated risk of an ASP estimator is a trustworthy approximation to its risk (Section 4). This asymptotic analysis relies on results that were developed by Beran and Dümbgen (1998) for abstract shrinkage estimators and were applied by Beran (2000, 2002) to one-way layouts.

The smoothing spline literature seeks to estimate a mean function that is deemed to be a function of *continuous* ordinal factors, using observations typically made at a scattered set of factor levels (cf. Wahba 1990, Wahba et al. 1995, Heckman and Ramsay 2000, Lin 2000). This scattered set may be viewed as an incomplete subset of the smallest discrete grid that contains it. The discrete fitting techniques of this paper may be compared and contrasted with the spline literature as follows:

a) Our aim is to estimate well a possibly large *discrete* array of means rather than a smooth mean surface. Risks of competing estimators are evaluated under a model that puts *no* restrictions on the unknown means.

b) The factors affecting the discrete means in the two-way layout can be either ordinal or nominal or one of each. For both factors ordinal, ASP fits rely on discrete splines that are akin to continuous smoothing splines. For both factors nominal, ASP yields multiple-shrinkage estimators very close to those of Stein (1966).

Tukey (1977) experimented with certain smoothing algorithms for fitting one- and higher-way layouts with ordinal factors. In ordinal one-way layouts where wavelet bases provide a sparse representation of the means, Donoho and Johnstone (1995) used adaptive shrinkage through soft-thresholding. Beran and Dümbgen (1998) proposed and studied adaptive symmetric linear estimators that perform monotone shrinkage relative to a fixed orthonormal basis. ASP estimators for two-way layouts with either ordinal or nominal factors can be represented canonically as a closed set of bi-monotone shrinkage estimators acting on a tensor product basis determined by the three penalty terms (Section 2).

This paper considers a complete balanced two-way layout in which the first factor has p_1 distinct levels, the second factor has p_2 distinct levels, and q observations are taken at each combination of factor levels. Without loss of generality in the theory, we take $q = 1$. This corresponds to using the averages over replications in place of the raw observations. Subscripting is arranged so that, for an ordinal factor, the factor levels are a strictly increasing function of subscript. The statistical model used in all risk calculations is

$$y_{ij} = \mu(s_{1i}, s_{2j}) + \epsilon_{ij} \qquad 1 \leq i \leq p_1,\ 1 \leq j \leq p_2, \tag{1}$$

where the $\{y_{ij}\}$ are the (averaged) observations, the $\{s_{ki}\}$ are the levels of factor k, and the errors $\{\epsilon_{ij}\}$ are independent, identically distributed $N(0, \sigma^2)$ random variables. *Both the function μ and the variance σ^2 are unknown.* If factor k is ordinal, then $s_{k1} < s_{k2} < \ldots < s_{kp_k}$. For notational simplicity, we usually write m_{ij} instead of $\mu(s_{1i}, s_{2j})$.

Let M denote the $p_1 \times p_2$ matrix with elements $\{m_{ij}\}$. The Frobenius matrix norm $|\cdot|$ is defined by $|C|^2 = \mathrm{tr}(C'C) = \mathrm{tr}(CC')$. The normalized quadratic loss and corresponding risk of any estimator $\hat{M}$ of M is

$$L(\hat{M}, M) = (p_1 p_2)^{-1} |\hat{M} - M|^2, \qquad R(\hat{M}, M, \sigma^2) = \mathrm{E}L(\hat{M}, M). \tag{2}$$

The unrestricted least squares estimator of M is the matrix Y with elements $\{y_{ij}\}$. It has risk σ^2. This least squares estimator underlies classical analysis of variance for the two-way layout but is less useful in fitting response surfaces or analyzing a digital image. Indeed, Stein (1956) proved that it is inadmissible whenever $p_1 p_2 \geq 3$.

1.1 *Penalized least squares and submodels*

Estimators of M that may dominate least squares are suggested by the following class of penalized least squares estimators. For $k = 1$ or 2, define the $p_k \times 1$ unit vector $u_k = p_k^{-1/2}(1, 1, \ldots, 1)'$. Let A_k be any matrix with p_k columns such that $A_k u_k = 0$. Examples of such *annihilator* matrices that have additional useful properties are presented in Section 1.2 and are treated more fully in Section 3. Let $A = (A_1, A_2)$ and let $\nu = (\nu_1, \nu_2, \nu_{12})$ be any vector in $[0, \infty]^3$. The *candidate penalized least squares* (PLS) *estimator* of M is defined to be

$$\hat{M}_{PLS}(\nu, A) = \underset{M}{\operatorname{argmin}}\, S(M, \nu, A), \tag{3}$$

where

$$S(M, \nu, A) = |Y - M|^2 + \nu_1 |A_1 M u_2|^2 + \nu_2 |u_1' M A_2'|^2 + \nu_{12} |A_1 M A_2'|^2. \tag{4}$$

The three penalty terms in (4) are designed to measure departures in M from certain submodels. The unrestricted model for the mean matrix of the two-way layout is

- *Full Model:* $M = \gamma_0 u_1 u_2' + \gamma_1 u_2' + u_1 \gamma_2' + \Gamma_{12}$, where γ_0 is a scalar, γ_k is a $p_k \times 1$ vector such that $u_k' \gamma_k = 0$, and Γ_{12} is a $p_1 \times p_2$ matrix such that $u_1' \Gamma_{12} = 0$, $\Gamma_{12} u_2 = 0$ (cf. Scheffé 1959).

The vanishing of one or more penalty terms indicates when M satisfies a designated submodel of the full model.

- *Additive Model:* This is the submodel of the full model for which $\Gamma_{12} = 0$. For this submodel, the penalty term $|A_1 M A_2'|^2$ vanishes.
- *Row-effects Model:* This is the submodel of the full model for which $\gamma_2 = 0$ and $\Gamma_{12} = 0$. For this submodel, the penalty terms $|A_1 M A_2'|^2$ and $|u_1' M A_2'|^2$ both vanish.
- *Column-effects Model:* This is the submodel of the full model for which $\gamma_1 = 0$ and $\Gamma_{12} = 0$. For this submodel, the penalty terms $|A_1 M A_2'|^2$ and $|A_1 M u_2|^2$ both vanish.
- *Constant Model:* This is the submodel of the full model for which each $\gamma_k = 0$ and $\Gamma_{12} = 0$. For this submodel, the penalty terms $|A_1 M A_2'|^2$, $|u_1' M A_2'|^2$, and $|A_1 M u_2|^2$ all vanish.

Thus, if ν_{12} is very large, the candidate PLS estimator will fit a nearly additive model. If ν_1 and ν_{12} are both large, the PLS estimator will fit a nearly row-effects model. If ν_2 and ν_{12} are both large, the candidate PLS estimator will fit a a nearly column-effects model. Finally, if every component of ν is large, the candidate PLS estimator will fit a nearly constant model. Further properties of the PLS fit will depend on the precise choice of A_k and will be discussed in Sections 2 and 3. The value of ν and hence the extent of submodel fitting will be chosen to minimize estimated risk of the candidate PLS estimator.

1.2 *Examples of annihilators and fits*

The following examples introduce suitable choices of the annihilator matrices A_k and corresponding ASP fits to data.

Example 1. Two ordinal factors. Estimating a response surface or denoising a digital image deals with responses indexed by two ordinal factors. Vague prior information may suggest that the mean function $\mu(s_{1i}, s_{2j})$ behaves locally like a polynomial function of the factor levels. Suppose that each set of factor levels is equally spaced. To have the PLS estimator favor a fit that is locally polynomial of degree $r-1$ in the levels of the first factor and of degree $c-1$ in the levels of the second factor, we take A_1 and A_2 to be, respectively, the r-th and c-th difference operators of column dimensions p_1 and p_2 respectively. More explicitly, consider the $(p-1) \times p$ matrix $\Delta(p) = \{\delta_{i,j}\}$ in which $\delta_{i,i} = 1$, $\delta_{i,i+1} = -1$ for every i and all other entries are zero.

Define

$$D_1(p) = \Delta(p), \qquad D_d(p) = \Delta(p-d+1)D_{d-1} \quad \text{for } 2 \le d \le p-1. \tag{5}$$

The annihilators just mentioned are $A_1 = D_r(p_1)$ and $A_2 = D_c(p_2)$ respectively.

Subplot (1,1) of Figure 1 displays a greyscale plot of a 70×50 two-way layout with one observation per cell. The artificial data was obtained by adding pseudo-random Gaussian white noise to the means in subplot (2,1). Section 3.2 provides mathematical details for this example. Both factors are ordinal. Subplot (3,1) gives an adaptive PLS estimator that uses the second difference annihilator for each factor. This ASP estimator recovers major features of the true means far more clearly than the unrestricted least squares estimator, which coincides with the raw data in subplot (1,1). The fitting errors displayed in subplot (3,2) are the difference between the ASP estimator and the true mean matrix. The fitting errors appear homogeneously random except at the central dip. Sections 2, 3.1, and 3.2 develop

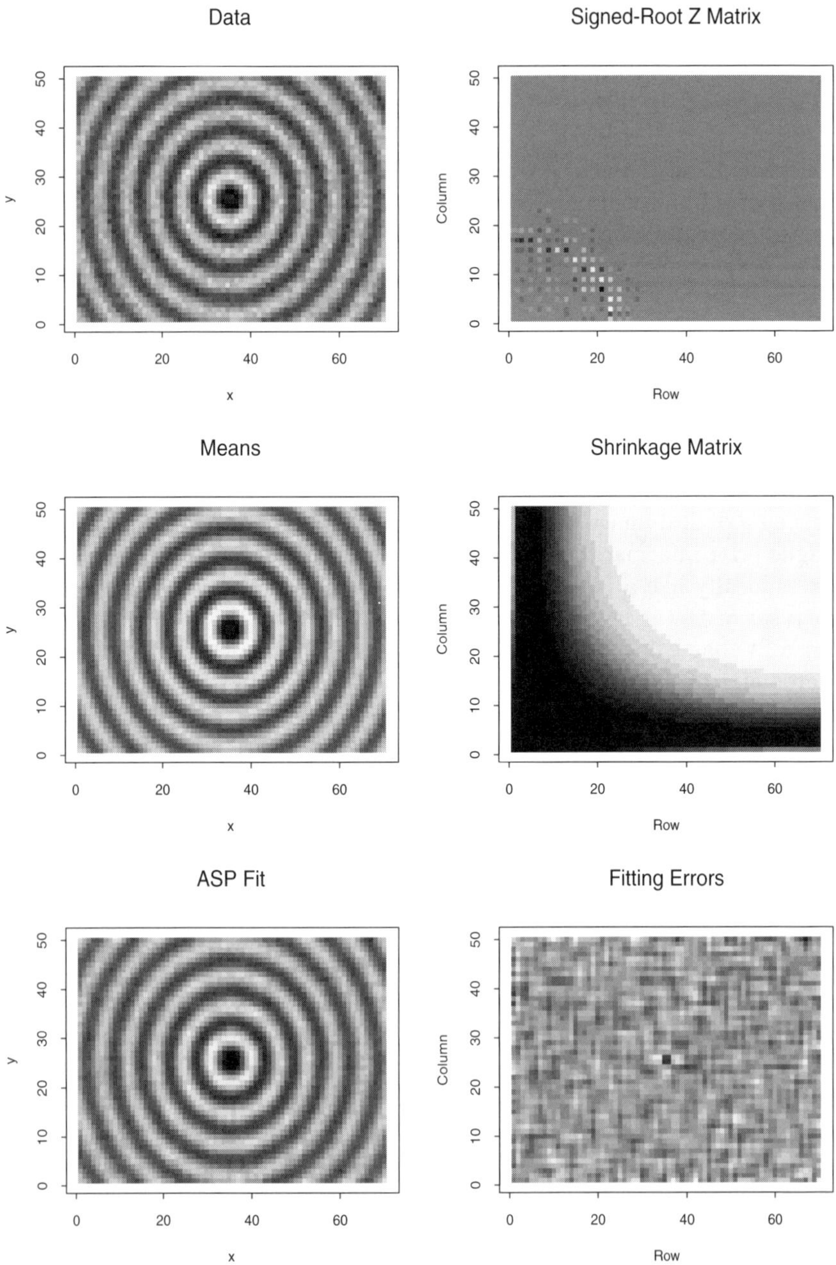

Figure 1 ASP fit and diagnostics for the artificial data. Both factors are ordinal. Both annihilators are second difference.

this example by discussing adaptive choice of r, c and the penalty weights and by presenting more general constructions of annihilator matrices.

Example 2. Two nominal factors. Classical analysis of variance deals with such data. When both factors are purely nominal, permutation of the subscripts (labels) should not affect the estimator of M. The matrix

$$A_k = I_{p_k} - u_k u_k' \tag{6}$$

is an annihilator that is invariant under permutations of row and column labels. We call it the flat annihilator for reasons that will become clear in Section 3.3. When both A_k are flat annihilators, the corresponding candidate PLS estimators are equivariant under permutations of row and column labels.

Subplot (1,1) of Figure 2 displays a linearly interpolated 6×8 two-way layout with one observation by cell. The data comes from p. 238 of Anderson and Bancroft (1952) and is reprinted on p. 138 of Scheffé (1959). Cell (i, j) in the layout reports the amount of cooking fat number j that is absorbed in baking a batch of donuts on day i of the experiment. Both factors in this example are treated as nominal. Subplot (1,2) gives an interpolated adaptive PLS estimator that uses the flat annihilator (6) for each factor. The cross-sections in the second row of Figure 2 show how this ASP fit, unlike the unrestricted least squares fit, recovers near-additivity in the dependence of fat-absorption on the day and the oil used. Sections 2 and 3.3 develop this example.

Example 3. One nominal and one ordinal factor. Classical analysis of covariance deals with such data. If the first factor is nominal with equally spaced levels while the second factor is ordinal, we may take $A_1 = I_{p_1} - u_1 u_1'$ and $A_2 = D_c(p_2)$. The resulting candidate PLS estimators are equivariant under permutations of the levels of the first factor, shrink the least squares estimator for the main effects of the first factor, and favor a fit that is locally of degree $c - 1$ in the levels of the second factor.

Subplot (1,1) of Figure 3 displays a linearly interpolated 52×3 two-way layout with one observation by cell. The data comes from Chatterjee et al. (1995). Cell (i, j) in the layout reports the grape yield harvested in year j from row i of a vineyard with 52 rows. Vineyard row in this example is an ordinal factor. Harvest year is treated as a nominal factor because weather and viticulture can vary considerably from year to year. Subplot (1,2) gives an interpolated adaptive PLS estimator that uses the third difference annihilator for the ordinal factor and the flat annihilator for the nominal factor. The cross-sections in the second row of Figure 3 show how this ASP

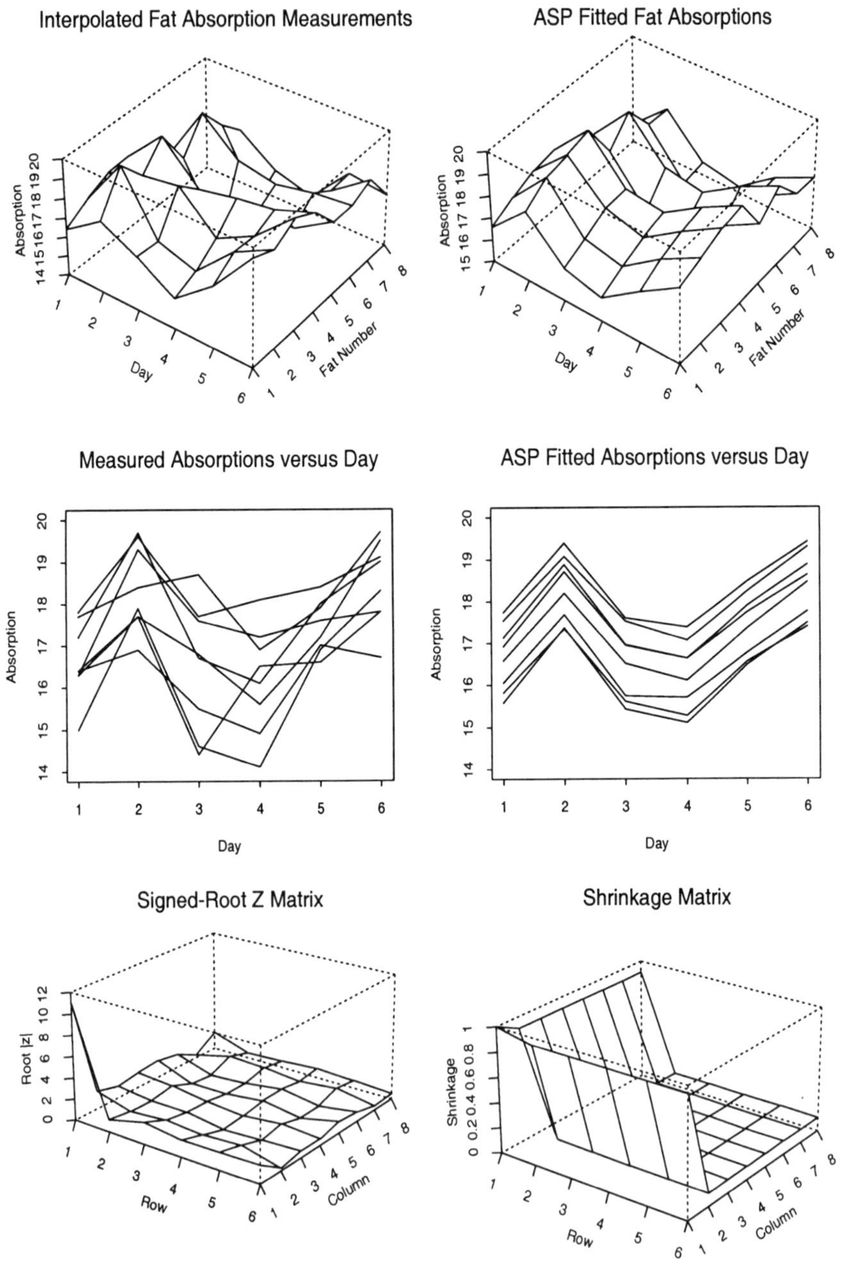

Figure 2 ASP fit and diagnostics for the data on fat-absorption by donuts. The factors day and fat number are both nominal. Both annihilators are flat.

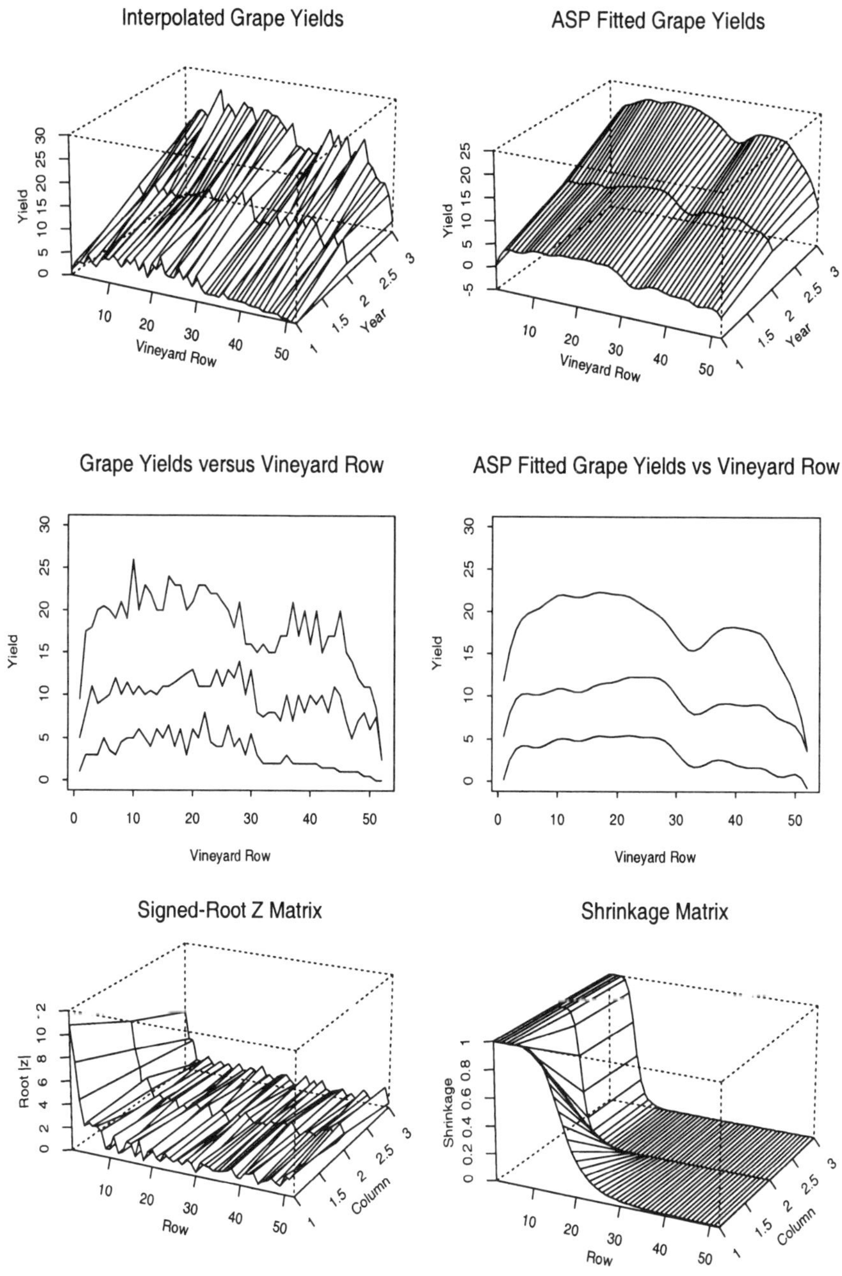

Figure 3 ASP fit and diagnostics for the data on vineyard grape yields. The factor vineyard row is ordinal while the factor year is nominal. The annihilators are respectively third difference and flat.

fit, more clearly than the unrestricted least squares fit, brings out leading features of the grape yield over the three harvest years. Among these is a dip in yield at and near vineyard row 33. Sections 2 and 3.4 develop this example.

1.3 *Outline of the paper*

Section 2 defines ASP estimators in several stages. Candidate PLS estimators are expressed in canonical form with respect to an orthogonal basis determined by the three penalty terms in (4). This representation suggests larger classes of candidate estimators that contain the candidate PLS estimators and have the mathematical advantage of forming closed convex sets. ASP estimators are then defined as estimators that minimize estimated risk over the class of candidate estimators being considered. A theorem develops conditions under which estimated risk is a trustworthy surrogate for the unknown risk as the number of cells $p_1 p_2$ tends to infinity. It is shown that ASP estimators will greatly dominate unrestricted least squares estimators when the selected penalty basis is economical.

Section 3 discusses algorithmic aspects, including how to devise appropriate annihilator matrices that express vague prior information about M and how to minimize the estimated risk when the factors are both ordinal or both nominal or mixed. In the case of two nominal factors, a simple closed form solution exists that essentially coincides with an estimator proposed by Stein (1966). Section 4 provides proofs of theorems stated in Sections 2 and 3.

The ASP methodology generalizes to k-way layouts. The motivating PLS estimator uses a separate penalty term for each of the $2^k - 1$ main effects and interactions in the ANOVA decomposition of the means.

2 Defining ASP Estimators

We start by studying the form and risk of candidate PLS estimators in terms of a canonical penalty basis for the regression space.

2.1 *Candidate PLS estimators and penalty bases*

The PLS criterion may be vectorized as follows. Let $y = \text{vec}(Y) = \{\{y_{ij} \colon 1 \le i \le p_1\}, 1 \le j \le p_2\}$, the column vector obtained by sequentially stacking the columns of Y with first column on top and last column at the bottom. Similarly, let $m = \text{vec}(M)$. Then, from (4),

$$S(M, \nu, A) = |y - m|^2 + \nu_1 m'(u_2 u_2' \otimes A_1' A_1) m + \nu_2 m'(A_2' A_2 \otimes u_1 u_1') m \\ + \nu_{12} m'(A_2' A_2 \otimes A_1' A_1) m. \quad (7)$$

Let $\hat{m}_{PLS}(\nu, A) = \text{vec}(\hat{M}_{PLS}(\nu, A))$, the right side being defined through (3). By calculus,

$$\hat{m}_{PLS}(\nu, A) = [I_{p_1 p_2} + \nu_1(u_2 u_2' \otimes A_1' A_1) + \nu_2(A_2' A_2 \otimes u_1 u_1') \\ + \nu_{12}(A_2' A_2 \otimes A_1' A_1)]^{-1} y. \quad (8)$$

This expression for the candidate PLS estimator can be simplified to reveal its essential structure. Suppose that the $p_k \times p_k$ symmetric matrix $A_k' A_k$ has the spectral decomposition $A_k' A_k = U_k \Lambda_k U_k'$, where the eigenvector matrix satisfies $U_k U_k' = U_k' U_k = I_{p_k}$ and the diagonal matrix $\Lambda_k = \text{diag}\{\lambda_{ki}\}$ gives the ordered eigenvalues $0 = \lambda_{k1} \leq \lambda_{k2} \leq \ldots \leq \lambda_{kp_k}$. This eigenvalue ordering, the reverse of the customary, is adopted here because the eigenvectors associated with the smallest eigenvalues play the greatest role in determining the numerical value and risk of the candidate PLS estimator. Because the annihilator A_k satisfies $A_k u_k = 0$, the eigenvalue λ_{k1} is necessarily zero and has u_k as corresponding eigenvector. Thus, the first column of U_k is u_k. It follows from this discussion that

$$(A_2' A_2 \otimes A_1' A_1)(U_2 \otimes U_1) = U_2 \Lambda_2 \otimes U_1 \Lambda_1 = (U_2 \otimes U_1)(\Lambda_2 \otimes \Lambda_1).$$

Consequently,

$$A_2' A_2 \otimes A_1' A_1 = (U_2 \otimes U_1)(\Lambda_2 \otimes \Lambda_1)(U_2 \otimes U_1)' \quad (9)$$

gives a spectral decomposition of the symmetric matrix on the right side.

The $p_k \times p_k$ matrix $u_k u_k'$ is symmetric, idempotent, has eigenvalue 1 associated with the eigenvector u_k, and has eigenvalue 0 repeated $p_k - 1$ times. Let $E_k = \text{diag}\{e_{ki}\}$ denote the $p_k \times p_k$ diagonal matrix that has 1 in the $(1,1)$ cell and zeroes elsewhere. Because u_k is the first column of U_k, we may write $u_k u_k' - U_k E_k U_k'$, a spectral decomposition of the left-hand side. As in the preceding paragraph,

$$u_2 u_2' \otimes A_1' A_1 = (U_2 \otimes U_1)(E_2 \otimes \Lambda_1)(U_2 \otimes U_1)', \\ A_2' A_2 \otimes u_1 u_1' = (U_2 \otimes U_1)(\Lambda_2 \otimes E_1)(U_2 \otimes U_1)'. \quad (10)$$

Combining (8), (9) and (10) yields

$$\hat{m}_{PLS}(\nu, A) = U[I_{p_1 p_2} + \nu_1(E_2 \otimes \Lambda_1) + \nu_2(\Lambda_2 \otimes E_1) + \nu_{12}(\Lambda_2 \otimes \Lambda_1)]^{-1} U' y \quad (11)$$

for $U = U_2 \otimes U_1$. Let

$$f_{ij}(\nu) = [1 + \nu_1 \lambda_{1i} e_{2j} + \nu_2 e_{1i} \lambda_{2j} + \nu_{12} \lambda_{1i} \lambda_{2j}]^{-1}. \quad (12)$$

The matrix inverse in (11) is a diagonal matrix whose main diagonal is the vector $f(\nu) = \{\{f_{ij}(\nu)\colon 1 \le i \le p_1\}\colon 1 \le j \le p_2\}$. Let $z = (U_2 \otimes U_1)'y$. Then

$$\hat{m}_{PLS}(\nu, A) = (U_2 \otimes U_1)\,\mathrm{diag}\{f(\nu)\}z. \tag{13}$$

The columns of $U_2 \otimes U_1$ constitute the *penalty basis* generated by the annihilator matrices A_1 and A_2. Equation (13) shows that the PLS candidate estimator maps the data-vector y into its coefficient vector z with respect to the penalty basis, then shrinks z through componentwise multiplication by $f(\nu)$, then maps the result back to the original basis.

To obtain a compact matrix expression for the candidate PLS estimator of M, define the matrix $F(\nu) = \{f_{ij}(\nu)\}$. Because $\lambda_{k1} = 0$ and $e_{ki} = 0$ if $i \ge 2$, expression (12) is equivalent to

$$F(\nu) = \begin{pmatrix} 1 & (1+\nu_2\lambda_{22})^{-1} & \dots & (1+\nu_2\lambda_{2p_2})^{-1} \\ (1+\nu_1\lambda_{12})^{-1} & (1+\nu_{12}\lambda_{12}\lambda_{22})^{-1} & \dots & (1+\nu_{12}\lambda_{12}\lambda_{2p_2})^{-1} \\ \vdots & \vdots & \ddots & \vdots \\ (1+\nu_1\lambda_{1p_1})^{-1} & (1+\nu_{12}\lambda_{1p_1}\lambda_{22})^{-1} & \dots & (1+\nu_{12}\lambda_{1p_1}\lambda_{2p_2})^{-1} \end{pmatrix}. \tag{14}$$

Let $Z = U_1'YU_2$ and let $F(\nu).Z$ denote the componentwise product of the two matrices. Equation (13) is equivalent to

$$\hat{M}_{PLS}(\nu, A) = U_1[F(\nu).Z]U_2'. \tag{15}$$

Note that the least squares estimator Y of M is the special case of (15) when every component of $F(\nu)$ equals 1, or equivalently, when $\nu_1 = \nu_2 = \nu_{12} = 0$.

2.2 *Candidate shrinkage estimators and their risks*

A *shrinkage class* $\mathcal{F}$ consists of $p_1 \times p_2$ matrices $F = \{f_{ij}\}$ such that $0 \le f_{ij} \le 1$ for every i and j. The associated *candidate shrinkage estimator* of M is defined to be

$$\hat{M}(F, A) = U_1[F.Z]U_2'. \tag{16}$$

In this paper we will consider the following shrinkage classes, each of which is inspired by (14) and each of which generates a class of candidate shrinkage estimators for M:

The *Unrestricted* shrinkage class $\mathcal{F}_U$ consists of all $p_1 \times p_2$ shrinkage matrices with elements in $[0, 1]$.

The *PLS* shrinkage class $\mathcal{F}_{PLS}$ is the subset of shrinkage matrices defined in (14) by $\{F(\nu)\colon \nu \in [0, \infty]^3\}$. The candidate PLS estimator $\hat{M}_{PLS}(\nu, A)$ described in (15) coincides with $\hat{M}(F(\nu), A)$ in the notation (16).

The *Bi-Flat* shrinkage class $\mathcal{F}_{BF}$ is the subset of $\mathcal{F}_U$ defined by $f_{ij} = 1$ if $i = j = 1$; $= c_1$ if $j = 1, i \geq 2$; $= c_2$ if $i = 1, j \geq 2$; and $= c_{12}$ if $i \geq 2, j \geq 2$, where c_1, c_2 and c_{12} are any constants in $[0,1]$. This is the specialization of PLS shrinkage obtained when $\lambda_{ki} = 1$ for $i \geq 2$ and every k.

The *Submodel* shrinkage class $\mathcal{F}_{SM}$ is the subset of $\mathcal{F}_{BF}$ in which the possible values of c_1, c_2, and c_{12} are restricted to either 0 or 1. This shrinkage class is suggested by classical techniques for choosing a hierarchical submodel.

The *Monotone Score* shrinkage class $\mathcal{F}_{MS}$ is the subset of $\mathcal{F}_U$ defined by $f_{ij}(\lambda_{1i}, \lambda_{2j}) = 1$ if $i = j = 1$; $= g_1(\lambda_{1i})$ if $j = 1, i \geq 2$; $= g_2(\lambda_{2j})$ if $i = 1, j \geq 2$; and $= g_{12}(\lambda_{1i}\lambda_{2j})$ if $i \geq 2, j \geq 2$, where g_1, g_2, g_{12} are any functions nonincreasing in their arguments.

The *Bi-Monotone* shrinkage class $\mathcal{F}_{BM}$ is the subset of $\mathcal{F}_U$ defined by $f_{11} = 1$; $\{f_{i1} \colon i \geq 2\}$ is nonincreasing in i; $\{f_{1j} \colon j \geq 2\}$ is nonincreasing in j; and $\{f_{ij} \colon i, j \geq 2\}$ is nonincreasing in i for each fixed j and nonincreasing in j for each fixed i.

The *Bi-Nested* shrinkage class $\mathcal{F}_{BN}$ is the subset of $\mathcal{F}_{BM}$ such that: $f_{11} = 1$; each f_{i1} is either c_1 or 0 for $i \geq 2$; each f_{1j} is either c_2 or 0 for $j \geq 2$; and each f_{ij} is either c_{12} or 0 for $i \geq 2$ and $j \geq 2$. Here c_1, c_2 and c_{12} are any constants in $[0,1]$.

The *Flat* $\times$ *Monotone* shrinkage class $\mathcal{F}_{F \times M}$ is the subset of $\mathcal{F}_U$ defined by $f_{ij} = 1$ if $i = j = 1$; $= c$ if $j = 1, i \geq 2$; $= g_j$ if $i = 1, j \geq 2$; and $= h_j$ if $i \geq 2, j \geq 2$, where c is any constant in $[0,1]$ and $\{g_j\}$, $\{h_j\}$ are each any nonincreasing sequence.

Evidently, $\mathcal{F}_{SM} \subset \mathcal{F}_{BF} \subset \mathcal{F}_{PLS} \subset \mathcal{F}_{MS} \subset \mathcal{F}_{BM}$. Also $\mathcal{F}_{BF} \subset \mathcal{F}_{BN} \subset \mathcal{F}_{BM}$ and $\mathcal{F}_{F \times M} \subset \mathcal{F}_{BM}$. With the exception of $\mathcal{F}_{PLS}$ (in general), $\mathcal{F}_{SM}$, and $\mathcal{F}_{BN}$, these shrinkage classes are closed convex subsets of $\mathcal{F}_U$. PLS, monotone score, bi-nested and bi-monotone shrinkage are useful when both factors are ordinal. Bi-flat shrinkage, a specialization of PLS, is useful when both factors are nominal. PLS and Flat $\times$ monotone shrinkage are useful when the row factor is nominal while the column factor is ordinal. However, the unrestricted shrinkage class $\mathcal{F}_U$ does not generate low risk ASP estimators. These matters will be developed in the remainder of the paper.

Let $f = \text{vec}(F)$. The risk of the candidate estimator $\hat{M}(F, A)$ defined in (16) may be expressed simply through the penalty basis representation of $\hat{m}(f, A) = \text{vec}(\hat{M}(F, A))$. Let $\xi = \text{E}(z) = (U_2 \otimes U_1)'m$ and $\hat{\xi}(f) = \text{diag}\{f\}z$. Then

$$\hat{m}(f, A) = (U_2 \otimes U_1)\hat{\xi}(f), \quad m = (U_2 \otimes U_1)\xi.$$

The normalized quadratic loss (2) thus reduces to

$$L(\hat{M}(F,A),M) = (p_1p_2)^{-1}|\hat{m}(f,A) - m|^2 = (p_1p_2)^{-1}|\hat{\xi}(f) - \xi|^2. \quad (17)$$

For any vector x, let ave(x) denote the average of its components. From (17), the risk of candidate shrinkage estimator $\hat{M}(F,A)$ is

$$R(\hat{M}(F,A),M,\sigma^2) = r(f,A,\xi^2,\sigma^2),$$

where

$$r(f,A,\xi^2,\sigma^2) = \text{ave}[f^2\sigma^2 + (1-f)^2\xi^2]. \quad (18)$$

Multiplication of vectors on the right side of (18) is done componentwise, as in the S language.

2.3 *Estimated risks and ASP estimators*

If the risk function (18) were known, we would seek an *oracle* estimator of M—the candidate estimator that minimizes risk over the class of shrinkage vectors and the class of annihilator matrices under consideration. This oracular strategy is usually unavailable. Instead, we will estimate the risk function from the data, then choose the candidate estimator that minimizes estimated risk. The result is called an ASP estimator of M.

The risk function (18) contains two quantities, σ^2 and ξ^2, that are usually unknown. The sampling scheme and the ordinal or nominal character of the factors both influence methods for estimating σ^2. Basic possibilities include:

Replicated layout. Fundamental in this setting is the least squares estimator of σ^2, the normalized residual sum of squares in the ANOVA table for the two-way layout.

One observation per combination of factor levels. If the penalty basis is economical in the sense that the coefficients $\{\xi_{ij} : q_1 < i \le p_1, q_2 < j \le p_2\}$ are close to zero, then the high-component estimator is

$$\hat{\sigma}^2 = [(p_1 - q_1)(p_2 - q_2)]^{-1} \sum_{i>q_1}^{p_1} \sum_{j>q_2}^{p_2} z_{ij}^2. \quad (19)$$

The classical pooled interaction estimator of ANOVA, suitable when the means approximately follow the additive model described in the Introduction, is equivalent to (19) with $q_1 = q_2 = 1$.

For the variance estimator (19), $\mathrm{E}(\hat{\sigma}^2 - \sigma^2)^2$ converges to zero if and only if $(p_1 - q_1)(p_2 - q_2)$ tends to infinity and the sum of squared biases $[(p_1 - q_1)(p_2 - q_2)]^{-1} \sum_{i>q_1}^{p_1} \sum_{j>q_2}^{p_2} \xi_{ij}^2$ tends to zero as p_1p_2 tends to infinity. When the number of replications is greater than one but not large enough to make the least squares estimator of variance accurate, it may be useful to combine it with a pooled interaction estimator. Robust analogs of these variance estimators are the medians of the respective sets of $\{|z_{ij}|\}$ divided by $\Phi^{-1}(.75)$. Here Φ^{-1} denotes the quantile function of the standard normal distribution.

Having devised a variance estimator $\hat{\sigma}^2$, we may estimate ξ^2 by $z^2 - \hat{\sigma}^2$ and hence the risk function $r(f, A, \xi^2, \sigma^2)$ by

$$\hat{r}(f, A) = \text{ave}[\hat{\sigma}^2 f^2 + (1-f)^2(z^2 - \hat{\sigma}^2)] = \text{ave}[(f-\hat{g})^2 z^2] + \hat{\sigma}^2 \,\text{ave}(\hat{g}),$$

where $\hat{g} = (z^2 - \hat{\sigma}^2)/z^2$. Apart from considerations entering into the estimation of σ^2, this equation is an application of the Stein (1981) unbiased estimator of risk or of the risk estimator that underlies Mallow's (1973) discussion of C_p.

For fixed annihilator pair A and shrinkage class $\mathcal{F}$, the *shrinkage-adaptive* estimator is defined to be $\hat{M}(\hat{F}, A)$, where

$$\hat{f} = \text{vec}(\hat{F}) = \underset{f \in \mathcal{F}}{\text{argmin}}\, \hat{r}(f, A) = \underset{f \in \mathcal{F}}{\text{argmin}}\, \text{ave}[(f - \hat{g})^2 z^2]. \tag{20}$$

Computation of $\hat{F}$ is a weighted least squares problem that will be discussed further in Section 3 because the details depend upon the shrinkage class. When clarity requires, we will add a subscript to $\hat{F}$ to indicate the shrinkage class being used.

In general, ASP estimators involve adaptation over annihilators as well as over the shrinkage vector. Let $\mathcal{A}$ be a class of of annihilator pairs. The ASP estimator of M determined by annihilator class $\mathcal{A}$ and shrinkage class $\mathcal{F}$ is defined to be $\hat{M}(\hat{A}, \hat{F})$, where

$$(\hat{f}, \hat{A}) = \underset{A \in \mathcal{A}, f \in \mathcal{F}}{\text{argmin}}\, \hat{r}(f, A). \tag{21}$$

The following theorem gives conditions under which shrinkage adaptation to minimize estimated risk approximately minimizes true risk as $p_1 p_2$ tends to infinity. Section 4 gives the proof, which draws on abstract results for shrinkage estimators established by Beran and Dümbgen (1998).

Theorem 1. *Fix the annihilator pair A and let $\mathcal{F}$ be a subset of $\mathcal{F}_{DM}$ that is closed in $[0,1]^{p_1 p_2}$. In particular, $\mathcal{F}$ can be any shrinkage class listed in Section 2.2 other than $\mathcal{F}_U$. Suppose that $\hat{\sigma}^2$ is consistent in that, for every $a > 0$ and $\sigma^2 > 0$,*

$$\lim_{p_1 p_2 \to \infty} \sup_{\text{ave}(\xi^2) \le \sigma^2 a} \text{E}|\hat{\sigma}^2 - \sigma^2| = 0. \tag{22}$$

a) Let $V(f)$ denote either the loss $L(\hat{M}(F, A), M)$ or the estimated risk $\hat{r}(f, A)$. Then for every annihilator pair A, every $t > 0$, and every $\sigma^2 > 0$,

$$\lim_{p_1 p_2 \to \infty} \sup_{\text{ave}(\xi^2) \le \sigma^2 a} \text{E} \sup_{f \in \mathcal{F}} |V(f) - R(\hat{M}(F, A), M, \sigma^2)| = 0. \tag{23}$$

b) If $\hat{f} = \mathrm{vec}(\hat{F}) = \mathrm{argmin}_{f \in \mathcal{F}} \hat{r}(f, A)$, *then*

$$\lim_{p_1 p_2 \to \infty} \sup_{\mathrm{ave}(\xi^2) \le \sigma^2 a} |R(\hat{M}(\hat{F}, A), M, \sigma^2) - \min_{f \in \mathcal{F}} R(\hat{M}(F, A), M, \sigma^2)| = 0. \tag{24}$$

c) For W *equal to either* $L(\hat{M}(\hat{F}, A), M)$ *or* $R(\hat{M}(\hat{F}, A), M, \sigma^2)$,

$$\lim_{p_1 p_2 \to \infty} \sup_{\mathrm{ave}(\xi^2) \le \sigma^2 a} \mathrm{E}|\hat{r}(\hat{f}, A) - W| = 0. \tag{25}$$

d) Let $\#(\mathcal{A})$ *denote the cardinality of the class* $\mathcal{A}$. *If* $\#(\mathcal{A}) \cdot \min\{p_1^{-1/2}, p_2^{-1/2}\}$ *and* $\#(\mathcal{A}) \cdot \mathrm{E}|\hat{\sigma}^2 - \sigma^2|$ *both converge to zero as* $p_1 p_2 \to \infty$, *then convergences (23) to (25) to hold for the ASP estimator* $\hat{M}(\hat{F}, \hat{A})$, *defined in (21).*

Because $\max\{p_1, p_2\} \le p_1 p_2 \le [\max\{p_1, p_2\}]^2$, the condition $p_1 p_2 \to \infty$ is equivalent to $\max\{p_1, p_2\} \to \infty$. By part a, the loss, risk and estimated risk of a candidate estimator converge together asymptotically. Uniformity of this convergence over the shrinkage class $\mathcal{F}$ makes the estimated risk of a candidate estimators a trustworthy surrogate for its true risk or loss. By part b, the risk of the shrinkage adaptive-estimator $\hat{M}(\hat{F}, A)$ converges to that of the best candidate estimator. Thus, when $\mathcal{F}$ is a closed subset of $\mathcal{F}_{BM}$, shrinkage adaptation works as intended. This covers every shrinkage class defined in Section 2.2 except $\mathcal{F}_U$. Moreover, because the unrestricted least squares estimator is one of the candidate estimators indexed by these shrinkage classes, its asymptotic risk is at least as large as that of the best-shrinkage adaptive estimator. In practice, the risk of the best shrinkage-adaptive estimator is often much smaller than that of the unrestricted least squares estimator and this is the point. Part c shows that the loss, risk, and plug-in estimated risk of an adaptive estimator converge together asymptotically. Part d preserves these conclusions for ASP estimators in which the cardinality of the annihilator class is finite or slowly growing in the sense described.

The pleasant properties stated in Theorem 1 break down when the shrinkage class is $\mathcal{F}_U$. Then the estimator $\hat{M}(\hat{F}, A)$ is dominated by the least squares estimator Y (see Beran and Dümbgen 1998, p. 1829). Adaptation works when the shrinkage class is not too large, in a sense made precise in Section 4.

3 Annihilators and Algorithms

This section treats methods for constructing annihilator matrices and algorithms for minimizing estimated risk so as to construct ASP estimators. Case studies illustrate what ASP estimators can accomplish on data.

3.1 *Role of basis economy*

The following discussion motivates techniques for selecting annihilator matrices. Let $\Xi = \{\xi_{ij}\} = U_1' M U_2$, so that $\xi = \text{vec}(\Xi)$. Heuristically, a penalty basis is economical if all components outside the upper left corner of Ξ are close to zero. In that case, we need only to identify and estimate from the data the relatively few non-zero components of ξ, estimating the remaining components by zero. The quadratic risk then accumulates small squared biases from ignoring the nearly zero components of ξ but does not accumulate the many variance terms that would arise in attempting to estimate these unbiasedly.

An idealized formulation of basis economy facilitates mathematical analysis of how economy affects estimation risk in the two-way layout. Let $\mathcal{S}$ denote the set of all subsets of $\{(i,j)\colon 1 \le i \le p_1, 1 \le j \le p_2\}$. For given subset $S \in \mathcal{S}$, let $F(S) = \{f_{ij}(S)\}$ where $f_{ij}(S) = 1$ or 0 according to whether or not $(i,j) \in S$. Define

$$\mathcal{S}_0 = \{S \in \mathcal{S}\colon F(S) \in \mathcal{F}_{BM}\}.$$

For every $S \in \mathcal{S}_0$, every $a > 0$, and every $\sigma^2 > 0$, consider the projected ball

$$B(a, S, \sigma^2) = \{\xi \in R^{p_1 p_2}\colon \text{ ave}(\xi^2) \le \sigma^2 a \text{ and } \xi_{ij} = 0 \text{ for } (i,j) \notin S\}.$$

Formally, we will say that the penalty basis associated with the annihilator pair A is *economical* if $\xi \in B(a, S, \sigma^2)$ for some finite $a > 0$ and $\#(S)$, the cardinality of S, is small relative to $p_1 p_2$. Though this formulation is too simple to serve as a complete definition of basis economy, it yields the following quantitative result that shows how basis economy affects the risk of estimators of M.

Theorem 2. *Suppose that* $S \in \mathcal{S}_0$ *and*

$$\lim_{p_1 p_2 \to \infty} (p_1 p_2)^{-1} \#(S) = b.$$

Then, for every $a > 0$ *and every* $\sigma^2 > 0$*, the asymptotic minimax quadratic risk over all estimators of* M *is*

$$\lim_{p_1 p_2 \to \infty} \inf_{\hat{M}} \sup_{\xi \in B(a,S,\sigma^2)} R(\hat{M}, M, \sigma^2) = \sigma^2[ab/(a+b)]. \tag{26}$$

The bi-monotone shrinkage-adaptive estimator $\hat{M}(\hat{F}_{BM}, A)$ *satisfies*

$$\lim_{p_1 p_2 \to \infty} \sup_{\xi \in B(a,S,\sigma^2)} R(\hat{M}(\hat{F}_{BM}, A), M, \sigma^2) = \sigma^2[ab/(a+b)]. \tag{27}$$

The same holds for the bi-nested shrinkage-adaptive estimator $\hat{M}(\hat{F}_{BN}, A)$.

This theorem reveals substantially more than formal asymptotic minimaxity of the bi-monotone shrinkage-adaptive estimator $\hat{M}(\hat{F}_{BM}, A)$. When $b \in [0,1]$ is close to zero—in which case the penalty basis is highly economical—the right side of (27) is much smaller than the risk σ^2 of the unrestricted least squares estimator of M. To the extent that the PLS and other shrinkage-adaptive estimators defined in Section 2 approximate $\hat{M}(\hat{F}_{BM}, A)$, their performance also benefits strongly from economy of the basis.

3.2 *Both factors ordinal*

The ideal choice of penalty basis $U_2 \otimes U_1$ would have its first basis vector proportional to the unknown mean vector m so that only the first component of ξ would be nonzero. Though unrealizable, this ideal selection suggests that prior information or conjecture about m should be exploited in devising the annihilator matrices A_k that generate the penalty basis. The following discussion relates prior notions about the local behavior of the mean function μ in (1) to constructions of A_1 and A_2 for two ordinal factors.

Let $t = (t_1, t_2, \ldots, t_p)$ denote the levels of an ordinal factor, where p may be either p_1 or p_2. Let $g_0, g_1, \ldots, g_{d-1}$ be a given set of real-valued functions defined on the real line such that $g_0 \equiv 1$. We will construct a sparse matrix $B_d = B_d(t,p)$ to annihilate functions that behave locally like a linear combination of the $\{g_h\colon 0 \le h \le d-1\}$. For each i such that $1 \le i \le p-d$, let $\mathcal{G}_i$ denote the subspace of R^{d+1} that is spanned by the d vectors $\{(g_h(t_i), \ldots, g_h(t_{i+d}))\colon 0 \le h \le d-1\}$. Assume that the dimension of $\mathcal{G}_i$ is d. This condition is satisfied, for instance, when $g_h(t_i) = t_i^h$. Define the $(p-d) \times p$ local annihilator matrix $B_d = \{b_{ij}\}$ as follows: In the i-th row of B_d, the subvector $\{b_{ij}\colon i \le j \le i+d\}$ is the unit vector in R^{d+1}, unique up to sign, that is orthogonal to $\mathcal{G}_i$. The remaining elements of B_d are zero.

Theorem 3. *Let $\bar{g}_h = (g_h(t_1), g_h(t_2), \ldots, g_h(t_p))'$. Each row vector of the local annihilator matrix B_d has unit length and*

$$B_d \bar{g}_h = 0 \quad \text{for } 0 \le h \le d-1.$$

Proof: The definition of B_d ensures that its rows have unit length and that

$$\sum_{j=1}^{p} b_{ij} g_h(t_j) = \sum_{j=i}^{i+d} b_{ij} g_h(t_j) = 0 \quad \text{for } 0 \le h \le d-1.$$

Of frequent utility is the *local polynomial* annihilator, which is obtained by setting $g_h(t_i) = t_i^h$ in the foregoing definition of B_d. If we conjecture that the unknown mean function $\mu(s_{1i}, s_{2j})$ behaves locally like a polynomial of degree $r-1$ in the first ordinal factor and like a polynomial of degree $c-1$ in the second ordinal factor, we would take the annihilators that generate the penalty basis to be

$$A_1 = B_r(s_1, p_1), \qquad A_2 = B_c(s_2, p_2),$$

where $s_1 = (s_{11}, \ldots, s_{1p_1})'$ and $s_2 = (s_{12}, \ldots, s_{1p_2})'$. When the factor levels are equally spaced, the local polynomial annihilator B_d becomes a scalar multiple of the d-th difference matrix defined in display (5) of Example 1.

We turn next to the computation of shrinkage-adaptive estimators for the case of two ordinal factors once the annihilators A have been fixed.

Penalized least squares. From Section 2.2 and definition (20), the shrinkage-adaptive PLS estimator is

$$\hat{M}(F(\hat{\nu}), A) = U_1[F(\hat{\nu}).Z]U_2', \tag{28}$$

where

$$\hat{\nu} = \underset{\nu \in [0,\infty]^3}{\text{argmin}} \ \text{ave}[(f(\nu) - \hat{g})^2 z^2]. \tag{29}$$

Let $\hat{g}_{ij} = (z_{ij}^2 - \hat{\sigma}^2)/z_{ij}^2$. Because of (14), equation (29) is equivalent to

$$\hat{\nu}_1 = \underset{\nu_1 \in [0,\infty]}{\text{argmin}} \sum_{i=2}^{p_1} [(1 + \nu_1 \lambda_{1i})^{-1} - \hat{g}_{i1}]^2 z_{i1}^2,$$

$$\hat{\nu}_2 = \underset{\nu_2 \in [0,\infty]}{\text{argmin}} \sum_{j=2}^{p_2} [(1 + \nu_2 \lambda_{2j})^{-1} - \hat{g}_{1j}]^2 z_{1j}^2,$$

$$\hat{\nu}_{12} = \underset{\nu_{12} \in [0,\infty]}{\text{argmin}} \sum_{i=2}^{p_1} \sum_{j=2}^{p_2} [(1 + \nu_{12} \lambda_{1i} \lambda_{2j})^{-1} - \hat{g}_{ij}]^2 z_{ij}^2. \tag{30}$$

Calculation of $\hat{\nu} = (\hat{\nu}_1, \hat{\nu}_2, \hat{\nu}_{12})$ thus amounts to solving three nonlinear, weighted least squares problems, each of which can be treated with minimization algorithms for a function of a single variable.

Example 1 (continued) : The data matrix for this example, displayed in Figure 1, is constructed as $Y = M + E$ with $p_1 = 70$ and $p_2 = 50$. The components of the error matrix E are pseudo-random independent Gaussian with mean 0 and standard deviation $\sigma = .15$. The mean matrix M has components $m_{ij} = \mu[i-(p_1+1)/2, j-(p_2+1)/2)]$ for $1 \le i \le p_1, 1 \le j \le p_2$, where $\mu(u, v) = 2t^{-1/4}\sin(t)$ with $t = \sqrt{u^2 + v^2}$. The penalty basis is generated by using a second difference annihilator for each factor.

Subplot (1,2) in Figure 1 explores the economy of this basis empirically by plotting the transformed components $\{|z_{ij}|^{1/2}\}$ of Z as surrogates for the likewise transformed components of Ξ. The transformation reduces the vertical range and makes it easier to see what is happening when z_{ij} is close to zero. The components outside the upper right corner of the matrix Z are relatively small, supporting the conclusion that the chosen penalty basis is economical here. The nature of this economy motivates using variance estimator (19) with $q_1 = 29$ and $q_2 = 24$. The estimated risk of the least squares estimator is then .0220, in good agreement with the actual risk $.15^2$. The estimated risk .0084 of the shrinkage-adaptive PLS estimator—the ASP estimator—is much smaller than that of the unrestricted least squares estimator. In this example, reduction of estimated risk accompanies visually better recovery of the response surface or image. The actual loss incurred by the ASP estimator is .0082, in approximate agreement with the estimated risk. This is to be expected from part c of Theorem 7.

The shrinkage matrix that defines the adaptive PLS estimator is shown in subplot (2,2). Shrinkage of the interaction coefficients $\{z_{ij} : i \geq 2, j \geq 2\}$ increases pronouncedly with i and j, though is less dramatic than the shrinkage of the main-effect coefficients $\{z_{i1} : i \geq 2\}$ and $\{z_{1j} : j \geq 2\}$. The shrinkage matrix reflects the non-additivity of the true means in this example. Experimenting with d-th difference annihilators of orders 1 through 4 did not reduce estimated risk below that achieved with second differences.

Bi-monotone shrinkage. The shrinkage-adaptive BM estimator is

$$\hat{M}(\hat{F}_{BM}, A) = U_1[\hat{F}_{BM}.Z]U_2',$$

where $\text{vec}(\hat{F}_{BM}) = \hat{f}_{BM}$ and

$$\hat{f}_{BM} = \underset{f \in \mathcal{F}_{BM}}{\text{argmin}}\, \text{ave}[(f - \hat{g})^2 z^2]. \tag{31}$$

Consider the generic decomposition

$$\sum_{i=1}^{p_1}\sum_{j=1}^{p_2} a_{ij}^2 = a_{11}^2 + \sum_{i=2}^{p_1} a_{i1}^2 + \sum_{j=2}^{p_2} a_{1j}^2 + \sum_{i=2}^{p_1}\sum_{j=2}^{p_2} a_{ij}^2. \tag{32}$$

To evaluate (31), we may proceed as follows:

a) Decompose the left side of (31) into minimizations of three separate sums formed as in (32). Minimize each of these sums *without* the constraint that $f \in [0,1]^p$. Weighted isotonic regression with the pool adjacent violators (PAV) algorithm accomplishes this for the first two sums. Iterative use of the PAV algorithm handles the third sum. Roberston, Wright and Dykstra (1988) describe both algorithms. Bril et al. (1984) provide a Fortran implementation of the latter. Burdakow et al. (2004) review more efficient algorithms for isotonic regression in several variables.

b) Then each component of $\hat{f}_{BM}$ is the positive part of the unconstrained minimizer found in part a. An extension of the argument in Section 5 of Beran and Dümbgen (1998) establishes this point.

Bi-nested shrinkage. The components of $\hat{f}_{BN} = \operatorname{argmin}_{f \in \mathcal{F}_{BN}} \operatorname{ave}[(f - \hat{g})^2 z^2]$ are given by $\hat{f}_{BN,ij} = 1(\hat{f}_{BM,ij} \geq 1/2)$ or may be found directly by finite search.

Monotone score shrinkage. The shrinkage-adaptive MS estimator is

$$\hat{M}(\hat{F}_{MS}, A) = U_1[\hat{F}_{MS}.Z]U_2',$$

where $\operatorname{vec}(\hat{F}_{MS}) = \hat{f}_{MS}$ and

$$\hat{f}_{MS} = \operatorname*{argmin}_{f \in \mathcal{F}_{MS}} \operatorname{ave}[(f - \hat{g})^2 z^2].$$

The components $\{\hat{f}_{ij}\}$ of the matrix $\hat{F}_{MS}$ may be found as follows:

First step. Set $\hat{f}_{11} = 1$.

Second step. Let $w = \{z_{i1} \colon 2 \leq i \leq p_1\}$ and let $\hat{h} = (w^2 - \hat{\sigma}^2)/w^2$. Let $\mathcal{K} = \{k \in R^q \colon k_1 \geq k_2 \geq \ldots \geq k_q\}$, where $q = p_1 - 1$. Find $\hat{k} = \operatorname{argmin}_{k \in \mathcal{K}} \operatorname{ave}[(k - \hat{h})^2 w^2]$, using an algorithm for weighted isotonic least squares such as the PAV. Set $\hat{f}_{i1} = \max\{\hat{k}_{i-1}, 0\}$ for $2 \leq i \leq p_1$.

Third step. Repeat the second step, letting $w = \{z_{1j} \colon 2 \leq j \leq p_2\}$ and $q = p_2 - 1$. Having found $\hat{k}$, set $\hat{f}_{1j} = \max\{\hat{k}_{j-1}, 0\}$ for $2 \leq j \leq p_2$.

Fourth step. Let $y = \operatorname{vec}(\{z_{ij} \colon 2 \leq i \leq p_1, 2 \leq j \leq p_2\})$, let $q = (p_1 - 1)(p_2 - 1)$, and let $v = \operatorname{vec}(\{\lambda_{1i}\lambda_{2j} \colon 2 \leq i \leq p_1, 2 \leq j \leq p_2\})$ be the vector of corresponding scores. Suppose first that these scores contain no ties. Let ρ denote the rank vector of v and define the q dimensional vector w through $w_{\rho_i} = y_i$. Repeat the second step using these definitions of w and q. Having found $\hat{k}$, define the vector $\hat{n}$ to have i-th component $\max\{\hat{k}_{\rho_i}, 0\}$. Let $\hat{N} = \{\hat{n}_{ij}\}$ be the $(p_1 - 1) \times (p_2 - 1)$ matrix such that $\hat{n} = \operatorname{vec}(\hat{N})$. Set $\hat{f}_{ij} = \hat{n}_{i-1,j-1}$ for $2 \leq i \leq p_1, 2 \leq j \leq p_2$. In the presence of ties among the components of v, we pool the corresponding components of y^2 in constructing w^2 and reduce q accordingly.

Monotone score shrinkage, a special case of bi-monotone shrinkage, has the computational advantage that the PAV algorithm converges in a finite number of steps.

3.3 *Both factors nominal*

As was noted after (6), the flat annihilator $A_k = I_{p_k} - u_k u_k'$ is invariant under permutations of row and column labels. This makes A_1 and A_2 suitable for defining candidate PLS estimators when both factors are nominal.

Let U_k denote any orthogonal matrix whose first column is the vector u_k. We may write $U_k = (u_k, C_k)$, where $u_k' C_k = 0$ and $C_k' C_k = I_{p_k - 1}$. The columns of C_k are any set of orthonormal contrasts in R^{p_k}. The matrix A_k is symmetric and idempotent. The eigenvalues of $A_k' A_k$ are $\lambda_{k1} = 0$, $\lambda_{k2} = \ldots \lambda_{kp_k} = 1$ and the columns of the matrix U_k defined above give corresponding eigenvectors.

PLS and bi-flat shrinkage. It follows from Section 2.2 that the class of candidate PLS estimators generated by the flat annihilators coincides with the class of BF candidate estimators $\hat{M}(c, A) = U_1[F.Z]U_2'$ for $F \in \mathcal{F}_{BF}$ through the correspondence $c_1 = (1+\nu_1)^{-1}$, $c_2 = (1+\nu_2)^{-1}$ and $c_{12} = (1+\nu_{12})^{-1}$.

The adaptive BF estimator $\hat{M}(\hat{c}, A)$ has components

$$\begin{aligned}
\hat{m}_{ij}(\hat{c}, A) = y_{\cdot\cdot} &+ \left[1 - \frac{(p_1-1)\hat{\sigma}^2}{\sum_{i=1}^{p_1}\sum_{j=1}^{p_2}(y_{i\cdot} - y_{\cdot\cdot})^2}\right]_+ (y_{i\cdot} - y_{\cdot\cdot}) \\
&+ \left[1 - \frac{(p_2-1)\hat{\sigma}^2}{\sum_{i=1}^{p_1}\sum_{j=1}^{p_2}(y_{\cdot j} - y_{\cdot\cdot})^2}\right]_+ (y_{\cdot j} - y_{\cdot\cdot}) \\
&+ \left[1 - \frac{(p_1-1)(p_2-1)\hat{\sigma}^2}{\sum_{i=1}^{p_1}\sum_{j=1}^{p_2}(y_{ij} - y_{i\cdot} - y_{\cdot j} + y_{\cdot\cdot})^2}\right]_+ (y_{ij} - y_{i\cdot} - y_{\cdot j} + y_{\cdot\cdot}).
\end{aligned}$$

because, by calculus, $\hat{c} = (\hat{c}_1, \hat{c}_2, \hat{c}_{12})$ with

$$\begin{aligned}
\hat{c}_1 &= \operatorname*{argmin}_{c_1 \in [0,\infty]} \sum_{i=2}^{p_1} [c_1 - \hat{g}_{i1}]^2 z_{i1}^2 = \left[1 - (p_1-1)\hat{\sigma}^2 / \sum_{i=2}^{p_1} z_{i1}^2\right]_+, \\
\hat{c}_2 &= \operatorname*{argmin}_{c_2 \in [0,\infty]} \sum_{j=2}^{p_2} [c_2 - \hat{g}_{1j}]^2 z_{1j}^2 = \left[1 - (p_2-1)\hat{\sigma}^2 / \sum_{j=2}^{p_2} z_{1j}^2\right]_+, \\
\hat{c}_{12} &= \operatorname*{argmin}_{c_{12} \in [0,\infty]} \sum_{i=2}^{p_1}\sum_{j=2}^{p_2} [c_{12} - \hat{g}_{ij}]^2 z_{ij}^2 = \left[1 - (p_1-1)(p_2-1)\hat{\sigma}^2 / \sum_{i=2}^{p_1}\sum_{j=2}^{p_2} z_{ij}^2\right]_+.
\end{aligned}$$

Through different reasoning, Stein (1966, p. 358) obtained an estimator akin to this for the case when $\hat{\sigma}^2$ is an independent least squares estimator of variance. In that setting, Stein refined the right side of $\hat{M}(\hat{c}, A)$ slightly—subtracting 2 from the three factors (p_1-1), (p_2-1) and $(p_1-1)(p_2-1)$—so as to reduce risk in estimating M. The effects of his modification decrease to vanishing as p_1 and p_2 increase. Devising such improvements to the other adaptive estimators treated in this paper is an open question.

Example 2 (continued) : A flat annihilator for each nominal factor generates the penalty basis, whose dramatic empirical economy is revealed by subplot (3,1) in Figure 2. The nature of this economy motivates using variance estimator (19) with $q_1 = q_2 = 3$. The shrinkage matrix that defines the adaptive PLS or BF estimator is shown in subplot (3,2). Shrinkage of the main effects is slight while shrinkage of the interactions is great, making this ASP fit nearly additive as noted earlier. The estimated risk of the least squares estimator is .4616 while the considerably smaller estimated risk of the ASP estimator is .1522. In this example, reduction of estimated risk accompanies clarification of how fat absorption depends on fat number and day.

Submodel shrinkage. The components of $\hat{f}_{SM} = \operatorname{argmin}_{f \in \mathcal{F}_{SM}} \operatorname{ave}[(f - \hat{g})^2 z^2]$ are given by $\hat{f}_{SM,ij} = 1(\hat{f}_{BF,ij} \geq 1/2)$ or may be found directly by finite search.

3.4 *One nominal and one ordinal factor*

Suppose that the first factor is nominal while the second factor is ordinal. A suitable pair of annihilators is then $A_1 = I_{p_1} - u_1 u_1'$ as in Section 3.3 and $A_2 = B_c(s_2, p_2)$ as in Section 3.2.

Penalized least squares. Here the shrinkage-adaptive PLS estimator is a specialization of (28) and (30) that is obtained by setting $\lambda_{11} = 0$, $\lambda_{12} = \ldots = \lambda_{1p_1} = 1$. The value of $\hat{\nu}_1$ is given by

$$(1 + \hat{\nu}_1)^{-1} = \left[1 - (p_1 - 1)\hat{\sigma}^2 / \sum_{i=2}^{p_1} z_{i1}^2\right]_+.$$

Calculating $\hat{\nu}_2$ and $\hat{\nu}_{12}$ amounts to minimizing the respective nonlinear weighted least squares criteria in (30).

Example 3 (continued): A third difference annihilator for the ordinal factor and a flat annihilator for the nominal factor generate the penalty basis, whose empirical economy is revealed by subplot (3,1) in Figure 3. The nature of this economy motivates using variance estimator (19) with $q_1 = 24$ and $q_2 = 0$. The shrinkage matrix that defines this ASP estimator is shown in subplot (3,2). Shrinkage is negligible for the main effects of the nominal factor but is pronounced for the higher order coefficients, whether main effect or interaction, of the ordinal factor. Strong shrinkage of the highest order interaction coefficients makes the ASP fit roughly additive, as seen in subplot (2,2). The estimated risk of the least squares estimator is 1.8751 while the much smaller estimated risk of the ASP estimator is .3918. In this example, reduction of estimated risk accompanies greater understanding of how grape yield depends on row number and year.

Experimenting with d-th difference annihilators of orders one through four on the row factor does not reduce estimated risk below that achieved with the third difference annihilator. If the factor year is treated as ordinal rather than nominal, the first difference annihilator on that factor best controls estimated risk of PLS candidate estimators. However, the corresponding ASP fit virtually coincides with that obtained in the preceding paragraph.

Flat × monotone shrinkage. Larger than the PLS class is the shrinkage class $\mathcal{F}_{F\times M}$ defined previously. The shrinkage-adaptive F × M estimator is

$$\hat{M}(\hat{F}_{F\times M}, A) = U_1[\hat{F}_{F\times M}.Z]U_2',$$

where $\text{vec}(\hat{F}_{F\times M}) = \hat{f}_{F\times M}$ and

$$\hat{f}_{F\times M} = \underset{f\in\mathcal{F}_{F\times M}}{\text{argmin}} \text{ ave}[(f-\hat{g})^2 z^2].$$

The components $\{\hat{f}_{ij}\}$ of the matrix $\hat{F}_{F\times M}$ may be found as follows:

First step. Set $\hat{f}_{11} = 1$.

Second step. For $i \geq 2$, set $\hat{f}_{i1} = \left[1 - (p_1 - 1)\hat{\sigma}^2/\sum_{i=2}^{p_1} z_{i1}^2\right]_+$.

Third step. Let $w = \{z_{1j} \colon 2 \leq j \leq p_2\}$ and let $\hat{h} = (w^2 - \hat{\sigma}^2)/w^2$. Let $\mathcal{K} = \{h \in R^q \colon k_1 \geq k_2 \geq \ldots \geq k_q\}$, where $q = p_2 - 1$. Find $\hat{k} = \text{argmin}_{k\in\mathcal{K}} \text{ave}[(k-\hat{h})^2 w^2]$, using an algorithm for weighted isotonic least squares. Set $\hat{f}_{1j} = \max\{\hat{k}_{j-1}, 0\}$ for $2 \leq j \leq p_2$.

Fourth step. Letting $w^2 = \{\sum_{i=2}^{p_1} z_{ij}^2 \colon 2 \leq j \leq p_2\}$ and $q = p_2 - 1$, find $\hat{k}$ as in the third step. Set $\hat{f}_{ij} = \max\{\hat{k}_{j-1}, 0\}$ for $2 \leq i \leq p_1$, $2 \leq j \leq p_2$.

4 Multiparametric Asymptotics

Adaptation works when estimated risk converges to actual risk uniformly over the class of candidate estimators. Empirical process theory provides sufficient conditions for such uniform convergence. For our purpose, the richness of a shrinkage class $\mathcal{F} \subset \mathcal{F}_U$ is characterized through the covering number $J(\mathcal{F})$ that is defined as follows. For any probability measure Q on the set $T = \{(i,j) \colon 1 \leq i \leq p_1, 1 \leq j \leq p_2\}$, consider the pseudo-distance $d_Q(f,g) = [\int (f-g)^2 dQ]^{1/2}$ on $[0,1]^T$. For every positive u, let

$$N(u, \mathcal{F}, d_Q) = \min\{\#\mathcal{F}_0 \colon \mathcal{F}_0 \subset \mathcal{F}, \inf_{f_0\in\mathcal{F}_0} d_Q(f_0, f) \leq u \quad \forall f \in \mathcal{F}\}.$$

Let

$$N(u, \mathcal{F}) = \sup_Q N(u, \mathcal{F}, d_Q),$$

where the supremum is taken over all probabilities on T. Define

$$J(\mathcal{F}) = \int_0^1 [\log N(u, \mathcal{F})]^{1/2} du.$$

Important in proving Theorem 1 is the fact $J(\mathcal{F}_{BM}) = O(\min\{p_1^{1/2}, p_2^{1/2}\})$, which follows from Example 5 on p. 1832 of Beran and Dümbgen (1998) and implies

$$(p_1 p_2)^{-1/2} J(\mathcal{F}_{BM}) = O(\min\{p_1^{-1/2}, p_2^{-1/2}\}). \tag{33}$$

In particular, because $\max\{p_1, p_2\} \leq p_1 p_2 \leq [\max\{p_1, p_2\}]^2$, the right side of (33) tends to zero as $p_1 p_2 \to \infty$.

Proof of Theorem 1: *Part a.* By Theorem 1 in Beran and Dümbgen (1998), there exists a finite constant C such that

$$\mathrm{E} \sup_{f \in \mathcal{F}} |V(f) - R(\hat{M}(F, A), M, \sigma^2)| \leq C \left[J(\mathcal{F}) \frac{\sigma^2 + \sigma\sqrt{\mathrm{ave}(\xi^2)}}{\sqrt{p_1 p_2}} + \mathrm{E}|\hat{\sigma}^2 - \sigma^2| \right].$$

Limit (23) follows from this, the inclusion of $\mathcal{F}$ in $\mathcal{F}_{BM}$, (33), and (22).

Parts b and c. In analogy to $\hat{f} = \mathrm{argmin}_{f \in \mathcal{F}} \hat{r}(f, A)$, let

$$\tilde{f} = \underset{f \in \mathcal{F}}{\mathrm{argmin}}\, r(f, A, \xi^2, \sigma^2).$$

Then $\min_{f \in \mathcal{F}} R(\hat{M}(F, A), M, \sigma^2) = r(\tilde{f}, A, \xi^2, \sigma^2)$. Let $\tilde{F}$ be the shrinkage matrix vectorized by $\tilde{f}$. We first show that (23) implies

$$\lim_{p_1 p_2 \to \infty} \sup_{\mathrm{ave}(\xi^2) \leq \sigma^2 a} \mathrm{E}|W - r(\tilde{f}, A, \xi^2, \sigma^2)| = 0, \tag{34}$$

where W can be $L(\hat{M}(\hat{F}, A), M)$ or $L(\hat{M}(\tilde{F}, A), M)$ or $\hat{r}(\hat{f}, A)$.

Indeed, (23) with $V(f) = \hat{r}(f, A)$ entails

$$\lim_{p_1 p_2 \to \infty} \sup_{\mathrm{ave}(\xi^2) \leq \sigma^2 a} \mathrm{E}|\hat{r}(\hat{f}, A) - r(\tilde{f}, A, \xi^2, \sigma^2)| = 0,$$

$$\lim_{p_1 p_2 \to \infty} \sup_{\mathrm{ave}(\xi^2) \leq \sigma^2 a} \mathrm{E}|\hat{r}(\hat{f}, A) - r(\hat{f}, A, \xi^2, \sigma^2)| = 0.$$

Hence, (34) holds for $W = \hat{r}(\hat{f}, A)$ and

$$\lim_{p_1 p_2 \to \infty} \sup_{\mathrm{ave}(\xi^2) \leq \sigma^2 a} \mathrm{E}|r(\hat{f}, A, \xi^2, \sigma^2) - r(\tilde{f}, A, \xi^2, \sigma^2)| = 0. \tag{35}$$

On the other hand, (23) with $V(f) = L(\hat{M}(F, A), M)$ gives

$$\lim_{p_1 p_2 \to \infty} \sup_{\mathrm{ave}(\xi^2) \leq \sigma^2 a} \mathrm{E}|L(\hat{M}(\hat{F}, A), M) - r(\hat{f}, A, \xi^2, \sigma^2)| = 0,$$

$$\lim_{p_1 p_2 \to \infty} \sup_{\mathrm{ave}(\xi^2) \leq \sigma^2 a} \mathrm{E}|L(\hat{M}(\tilde{F}, A), M) - r(\tilde{f}, A, \xi^2, \sigma^2)| = 0.$$

These limits together with (35) establish the remaining two cases of (34).

The limits (24) and (25) are immediate consequences of (34).

Part d. This conclusion follows by combining the separate results for $\mathcal{F} = \mathcal{F}_{MS}$ and $\mathcal{F} = \mathcal{F}_{ST}$.

Proving Theorem 2 requires a preliminary result. Let $\mathcal{E} = \{c \in R^{p_1 p_2} : c_i \in [1, \infty], 1 \le i \le p_1 p_2\}$. For every $c \in \mathcal{E}$, define the ellipsoid

$$E(a, c, \sigma^2) = \{\xi \in R^{p_1 p_2} : \text{ ave}(c\xi^2) \le \sigma^2 a\}.$$

When $\xi \in E(a, c, \sigma^2)$ and $c_i = \infty$, it is to be understood that $\xi_i = 0$ and $c_i^{-1} = 0$. Let

$$\begin{aligned} \xi_0^2 &= \sigma^2[(\alpha/c)^{1/2} - 1]_+, \\ g_0 &= \xi_0^2/(\sigma^2 + \xi_0^2) = [1 - (c/\alpha)^{1/2}]_+, \end{aligned} \tag{36}$$

where α is the unique positive number such that $\text{ave}(c\xi_0^2) = \sigma^2 a$. Define

$$\tau(a, c, \sigma^2) = r(f, A, \xi^2, \sigma^2) = \sigma^2 \text{ ave}[\xi_0^2/(\sigma^2 + \xi_0^2)]. \tag{37}$$

Evidently, $\tau(a, c, \sigma^2) \in [0, \sigma^2]$ for every $a > 0$ and every $c \in \mathcal{E}$.

The following theorem, specialized from the argument of Pinsker (1980), establishes that the linear estimator $g_0 z$ is typically asymptotically minimax among all estimators of ξ.

Theorem 4. *Suppose that* $\liminf_{p_1 p_2 \to \infty} \tau(a, c, \sigma^2) > 0$. *Then,*

$$\lim_{p_1 p_2 \to \infty} \Big[\inf_{\hat{\xi}} \sup_{\xi \in E(a,c,\sigma^2)} (p_1 p_2)^{-1} \mathrm{E}|\hat{\xi} - \xi|^2 - \tau(a, c, \sigma^2)\Big] = 0 \tag{38}$$

and

$$\lim_{p_1 p_2 \to \infty} \Big[\sup_{\xi \in E(a,c,\sigma^2)} (p_1 p_2)^{-1} \mathrm{E}|g_0 z - \xi|^2 - \tau(a, c, \sigma^2)\Big] = 0. \tag{39}$$

Proof of Theorem 2: Limit (26) is the specialization of (38) when $c_{ij} = 1$ for $(i, j) \in S$ and is infinite otherwise. In that case, $\lim_{p_1 p_2 \to \infty} \tau(a, c, \sigma^2) = \sigma^2[ab/(a + b)]$ by specialization of (36) and (37).

The coefficients of g_0 are $g_{0,ij} = [1 - \alpha^{-1/2}]_+$ for $(i, j) \in S$ and are zero otherwise. By the definition of $\mathcal{S}$, $g_0 \in \mathcal{F}_{BN} \subset \mathcal{F}_{BM}$. Consequently the oracle estimator $\tilde{f} z$ that is defined by (34) when $\mathcal{F}$ is either $\mathcal{F}_{BN}$ or $\mathcal{F}_{BM}$ satisfies

$$\sup_{\xi \in E(a,c,\sigma^2)} (p_1 p_2)^{-1} \mathrm{E}|\tilde{f} z - \xi|^2 \le \sup_{\xi \in E(a,c,\sigma^2)} (p_1 p_2)^{-1} \mathrm{E}|g_0 z - \xi|^2.$$

From this, (39), and the preceding evaluation of $\tau(a, c, \sigma^2)$,

$$\lim_{p_1 p_2 \to \infty} \sup_{\xi \in E(a,c,\sigma^2)} \mathrm{E}|\tilde{f} z - \xi|^2 = \sigma^2[ab/(a + b)]. \tag{40}$$

Limit (27) follows from (40) and limit (24) in Theorem 1.

Acknowledgments

This research was supported in part by National Science Foundation Grant DMS 0404547.

References

1. ANDERSON, R. L. AND BANCROFT, T. A. (1952). *Statistical Theory in Research.* McGraw-Hill, New York.
2. BERAN, R. (2000). REACT scatterplot smoothers: superefficiency through basis economy. *J. Amer. Statist. Assoc.* **63** 155–171.
3. BERAN, R. (2002). Improving penalized least squares through adaptive selection of penalty and shrinkage. *Ann. Inst. Statist. Math.* **54** 900–917.
4. BERAN, R. AND DÜMBGEN, L. (1998). Modulation of estimators and confidence sets. *Ann. Statist.* **26** 1826–1856.
5. BRIL, G., DYKSTRA, R. L., PILLERS, C., and ROBERTSON, T. (1984). Isotonic regression in two variables. *J. R. Statist. Soc. (C)* **33** 352–357.
6. BURDAKOW, O., GRIMVALL, A. AND HUSSIAN, M. (2004). A generalized PAV algorithm for monotonic regression in several variables. In *Compstat Proceedings in Computational Statistics, 16th Symposium.* (J. Antoch, ed.) 761–767. Physica-Verlag, Heidelberg.
7. CHATTERJEE, S., HANDCOCK, M. S., and SIMONOFF, J. S. (1995). *A Casebook for a First Course in Statistics and Data Analysis.* Wiley, New York.
8. DONOHO, D. L. AND JOHNSTONE, I. M. (1995). Adapting to unknown wavelet shrinkage. *J. Amer. Statist. Soc.* **90** 1200-1224.
9. HECKMAN, N. E. AND RAMSAY, J. O. (2000). Penalized regression with model-based penalties. *Can. J. Statist.* **28** 241–258.
10. LIN, YI (2000). Tensor Product Space ANOVA Fits. *Ann. Statist.* **28** 734–755.
11. MALLOWS, C. L. (1973). Some comments on C_p. *Technometrics* **15** 661–676.
12. PINSKER, M. S. (1980). Optimal filtration of square-integrable signals in Gaussian noise. *Problems Inform. Transmission.* **16** 120–133.
13. ROBERTSON, T., WRIGHT, F. T. AND DYKSTRA, R. L. (1988). *Order Restricted Statistical Inference.* Wiley, New York.
14. SCHEFFÉ, H. (1959). *The Analysis of Variance.* Wiley, New York.
15. STEIN, C. (1956). Inadmissibility of the usual estimator for the mean of a multivariate normal distribution. In *Proc. Third Berkeley Symp. Math. Statist. Probab.* (J. Neyman, ed.) 197–206. Univ. California Press, Berkeley.

16. Stein, C. (1966). An approach to the recovery of inter-block information in balanced incomplete block designs. In *Festschrift for Jerzy Neyman.* (F. N. David, ed.) 351–364. Wiley, New York.

17. Tukey, J. W. (1977). *Exploratory Data Analysis.* Addison-Wesley, Reading MA.

18. Wahba, G. (1990). *Spline Models for Observational Data.* SIAM, Philadelphia.

19. Wahba, G., Wang Y., Gu, C., Klein, R. and Klein, B. (1995). Smoothing spline ANOVA for exponential families with application to the Wisconsin epidemiological study of diabetic retinopathy. *Ann. Statist.* **23** 1868-1895.

20. Wood, S. N. (2000). Modelling and smoothing parameter estimation with multiple quadratic penalties. *J. Roy. Statist. Soc. (B)* **62** 413-428.

Chapter 9

INFERENCES FOR VARYING-COEFFICIENT PARTIALLY LINEAR MODELS WITH SERIALLY CORRELATED ERRORS

Jihong You and Jiancheng Jiang

Department of Biostatistics
University of North Carolina, Chapel Hill, NC, U.S.A.

Department of Mathematics and Statistics
University of North Carolina, Charlotte, NC, U.S.A.

E-mails: jyou@bios.unc.edu & jjiang1@uncc.edu

Varying-coefficient partially linear (VCPL) models are very useful tools. This chapter focuses on inferences for the VCPL model when the errors are serially correlated and modeled as an AR process. A penalized spline least squares (PSLS) estimation is proposed based on the penalized spline technique. This approach is then improved by a weighted PSLS estimation. We investigate the asymptotic theory under the assumption that the number of knots is fixed, though potentially large. The weighted PSLS estimators of all parameters are shown to be $\sqrt{n}$-consistent, asymptotically normal and asymptotically more efficient than the un-weighted ones. The proposed method can be used to make simultaneous inference for the parametric and nonparametric components by virtue of the sandwich formula for the joint covariance matrix. Simulations are conducted to demonstrate the finite sample performance of the proposed estimators. A real data analysis is used to illustrate the application of the proposed method.

Key words: Varying-coefficient; Partially linear; Serial correlation; Penalized spline; Asymptotic normality.

1 Introduction

Parametric regression models provide powerful tools for analyzing practical data when the models are correctly specified, but may suffer from large modeling biases if the structures of models are misspecified. As an alternative, nonparametric smoothing eases the concerns on modeling biases.

However, the nonparametric method is hampered by the so-called "curse of dimensionality" in multivariate settings (see for example Stone 1985, Hastie and Tibshirani 1990, and Fan and Gijbels 1996). One of the methods for attenuating this difficulty is to model covariate effects via a partially linear structure, a combination of linear and nonparametric parts. It retains nice features of both the parametric and the nonparametric regression models, which includes partially linear regression model (see Engle et al. 1986),partially nonlinear regression model (see Andrews 1996), single-index regression model (see Ichimura 1993, Delecroix, Härdle and Hristache 2003), varying-coefficient partially linear regression model (see Fan, Yao and Cai 2003), and so on. The varying-coefficient partially linear regression models are useful tools for modeling the relationship between the response and its covariates, while addressing possible interaction among the covariates (see Fan and Huang 2005). A general varying-coefficient partially linear regression model has the following form

$$Y = \boldsymbol{\beta}^T \mathbf{X} + \boldsymbol{\alpha}^T(U)\mathbf{Z} + \varepsilon, \tag{1}$$

where Y is the response, $\mathbf{X}, U$ and $\mathbf{Z}$ are the regressors, $\boldsymbol{\beta} = (\beta_1, \ldots, \beta_q)^T$ is a vector of p-dimensional unknown parameters, $\boldsymbol{\alpha}(\cdot) = (\alpha_1(\cdot), \ldots, \alpha_q(\cdot))^T$ is a vector of unknown functions, ε is the random error and the superscript T denotes the transpose of a vector or matrix.

Model (1) permits the interaction between the covariates U and $\mathbf{Z}$ in such a way that a different level of covariate U is associated with a different linear model. This allows one to examine the extent to which the effect of covariate $\mathbf{Z}$ varies over different levels of the variable U. When $\boldsymbol{\beta} = 0$, model (1) reduces to the varying-coefficient regression model widely studied in the literature (see, for example, Hastie and Tibshirani (1993), Carroll, Ruppert and Welsh (1998), Fan and Zhang (1999), Xia and Li (1999), Brumback and Rice (1998), Hoover, Rice, Wu and Yang (1998), and Huang, Wu and Zhou (2002) among others).

The model (1) has been studied by several authors. Zhang, Lee and Song (2002) developed a procedure for estimation of the linear part and the nonparametric part. Li, Huang, Li and Fu (2002) considered a local least squares estimation for model (1). Fan and Huang (2005) studied a generalized likelihood ratio test based on a profile least-squares estimation. Zhou and You (2004) employed a wavelet method for estimating model (1).

However, the previous results for model (1) focused only on the assumption that the errors are i.i.d. In practice, the independence assumption may be inappropriate. For example, when data were recorded over time, such as daily exchange rates, it is likely that the current response values are correlated with their past ones. Ignoring the dependence of errors may deteriorate the efficiency of estimators. It is a good practice to model the dependence structure via a stationary process, for instance, an AR process. In this paper we employ the following model:

$$\varepsilon_i = \psi_1 \varepsilon_{i-1} + \cdots + \psi_d \varepsilon_{i-d} + e_i, \qquad (2)$$

where ψ's are unknown parameters, $\{e_i\}$ is a sequence of i.i.d. random variables with mean zero and variance σ_e^2. It should be noted that the AR process is usually sufficient for modeling serially correlated errors because an MA or an ARMA process can be well approximated by an AR process (see Brockwell and Davis 1991).

To our knowledge, there is no formal research work for model (1) and (2) in the literature. We here focus on inference for model (1) and (2). A penalized spline least squares (*PSLS*) estimation will be proposed for the parametric and nonparametric components based on the penalized spline technique. This approach is then improved by a weighted *PSLS* estimation. We investigate the asymptotic theory under the assumption that the number of knots is fixed, though potentially large. It is shown that the estimators of all parameters are $\sqrt{n}$ consistent and asymptotically normal. Moreover, we show that theoretically and empirically the proposed weighted estimators of the parametric and nonparametric components are asymptotically more efficient than the un-weighted ones. The asymptotic covariate matrix of the estimators is of sandwich form. A consistent estimator is proposed based on the sandwich formula. In addition, we consider the choice of the smoothing parameters. The advantages of the proposed estimation method are manifold. The numerical implementation of the estimators is fast and stable. The joint asymptotic normality of the estimators of all parameters facilitates simultaneous inferences for the parametric and nonparametric components.

This paper is organized as follows. In Section 2 we first introduce an un-weighted *PSLS* estimator of the parametric and nonparametric components. Based on the un-weighted estimators, we fit the error structure and then construct a weighted *PSLS* estimator. The asymptotic properties of the proposed estimators are investigated. In Section 3 we discuss implementation details of the proposed approach related to the penalty terms and selection of the number and placements of knots. Simulation studies are conducted in Section 4. A practical data example is analyzed in Section 5. Section 6 concludes. The proofs of the main results are collected in Appendix.

2 Penalized Spline Least Squares Estimation

For theoretic study, we consider only the fixed design points model. The proposed method can be adapted to the case of random design points with a conceptually straightforward extension, conditional on the observed covariates.

According to Ruppert and Carroll (1997), Ruppert and Carroll (2001), and Ruppert (2002), the unknown univariate function $\alpha_s(\cdot)$ can be approximated by a penalized spline

$$\alpha_s(u) = \delta_{s0} + \delta_{s1}u + \ldots + \delta_{sm_s}u^{m_s} + \sum_{k=1}^{\kappa_s} \delta_{s,m_s+k}(u - \vartheta_{sk})_+^{m_s}, \quad (3)$$

where, for any number u, u_+ equals u if u is positive and equals 0 otherwise, m_s is the spline degree and $\{\vartheta_{sk}\}_{k=1}^{\kappa_s}$ are spline knots. Different degrees and knots could be used for different functions $\alpha_s(\cdot)$'s. However, the number and location of knots are not crucial and the smoothness can easily be controlled by a single smoothing parameter λ_s. For different components $\alpha_s(\cdot)$'s, we allow different smoothing parameters λ_s's. Here the truncated power basis is used for notational simplicity.

Let $\boldsymbol{\delta}_s = (\delta_{s0}, \delta_{s1}, \ldots, \delta_{s,m_s+\kappa_s})^T$ and

$$\mathbf{B}_s(u) = (1, u, \ldots, u^{m_s}, (u - \vartheta_{s1})_+^{m_s}, \ldots, (u - \vartheta_{s\kappa_s})_+^{m_s}).$$

Then the mean function of the model (1) can be written as

$$\mathbf{X}_i^T\boldsymbol{\beta} + \sum_{s=1}^{q} Z_{si}(\mathbf{B}^T(U_i)\boldsymbol{\delta}_s),$$

where Z_{si} is the sth element of $\mathbf{Z}_i$. Denote by

$$\boldsymbol{\theta} = (\boldsymbol{\beta}^T, \boldsymbol{\delta}_1^T, \ldots, \boldsymbol{\delta}_q^T)^T$$

and $\boldsymbol{\delta} = (\boldsymbol{\delta}_1^T, \ldots, \boldsymbol{\delta}_q^T)^T$. Then the *PSLS* estimator of $\boldsymbol{\theta}$ minimizes

$$Q_{n,\lambda}(\boldsymbol{\theta}) = \frac{1}{n}(\mathbf{Y} - \mathbf{X}\boldsymbol{\beta} - \Xi\boldsymbol{\delta})^T(\mathbf{Y} - \mathbf{X}\boldsymbol{\beta} - \Xi\boldsymbol{\delta}) + \sum_{s=1}^{q} \lambda_s\boldsymbol{\delta}_s^T\boldsymbol{\Omega}_s\boldsymbol{\delta}_s \quad (4)$$

where $\Xi = (\Xi_1, \ldots, \Xi_n)^T$, and $\Xi_i = \mathbf{Z}_i\mathbf{D}(U_i)$ with $\mathbf{D}(u) = \text{blcokdiag}\{\mathbf{B}_1(u), \cdots, \mathbf{B}_q(u)\}$ as a block diagonal matrix with q rows and $\sum_{s=1}^q(m_s + 1 + k_s)$ columns. The coefficients λ's are penalty parameters for $\alpha(\cdot)$'s, $\boldsymbol{\Omega}$'s are appropriate semi-definite symmetric matrices. A common choice for $\boldsymbol{\Omega}_s$ is

$$\boldsymbol{\delta}^T\boldsymbol{\Omega}_s\boldsymbol{\delta} = \int_{\min(U_i)}^{\max(U_i)} [\alpha_s''(u)]^2 du,$$

which yields the usual quadratic integral penalty (see Ruppert 2002).

By (4) we can obtain a closed form of $\widehat{\boldsymbol{\theta}}_n$, which is equal to

$$\widehat{\boldsymbol{\theta}}_n = [(\mathbf{X}, \Xi)^T(\mathbf{X}, \Xi) + \text{blockdiag}(\mathbf{0}_{p\times p}, n\boldsymbol{\Lambda}\boldsymbol{\Omega})]^{-1}(\mathbf{X}, \Xi)^T\mathbf{Y},$$

where $\boldsymbol{\Lambda} = \text{blockdiag}\{\boldsymbol{\lambda}_1^T, \cdots, \boldsymbol{\lambda}_q^T\}$ with $\boldsymbol{\lambda}_s$ being a $(1+m_s+k_s)\times 1$ vector of each component λ_s, and the penalty matrix $\boldsymbol{\Omega}$ is block diagonal with the s-th block being $\boldsymbol{\Omega}_s$.

2.1 *Asymptotic Development*

There are two kinds of asymptotic approaches that one can use for the penalized splines. The first one is to let the number of basis functions grow asymptotically as well as a penalty decay asymptotically in a certain order. The second one is to consider fixed-knot penalized splines.

The first kind of asymptotics is ideal for providing an insight into the studied problem. Unfortunately, there are few results on this topic, even for a univariate penalized spline model. The main hurdle lies in that the penalized splines involve two different features: the number of basis functions and the penalty. They jointly determine the property of a spline. Recently, Hall and Opsomer (2005) showed that the penalized splines can achieve the optimal nonparametric rate in a univariate penalized spline regression by taking a white-noise model representation and by assuming the spline estimator as an integral over a continuously varying set of basis functions, subject to a penalty. However, this white-noise presentation ignores the effect of the number of basis functions on the properties of the estimator. Therefore, the first kind of asymptotic approach is far from being completely solved.

The second kind of asymptotics has been adopted by many authors, e.g. Gray (1994), Wand (1999), Yu and Ruppert (2002), Yu and Ruppert (2004), Jarrow, Ruppert, and Yu (2004), Wu and Yu (2004), and Carroll et al. (2004), among others. Assuming a fixed but potentially large number of knots, the penalized splines enjoy most of the parametric asymptotic properties, where all the parameters can be jointly estimated at a $\sqrt{n}$-rate. The large number of knots though fixed allows flexible fits with a roughness penalty to avoid over-fitting. Ruppert (2002) found that the properties of the penalized spline estimators are relatively insensitive to the choices of basis functions, given that enough of them are used and the bias due to spline approximation is negligible compared to the variance.

We will investigate the second asymptotic properties of the proposed estimators under the assumption that the number of knots is fixed, though potentially large. Before presenting the asymptotic properties of $\hat{\boldsymbol{\theta}}_n$, we make the following assumptions.

(A1) $\lim_{n\to\infty}(\mathbf{X},\Xi)^T(\mathbf{X},\Xi) = \mathbf{\Sigma}$, *where* $\mathbf{\Sigma}$ *is a* $[p+\sum_{s=1}^q(m_s+1+\kappa_s)] \times [p+\sum_{s=1}^q(m_s+1+\kappa_s)]$ *positive definite matrix.*

(A2) $\psi(\zeta) = 1-\psi_1\zeta-\cdots-\psi_d\zeta^d \neq 0$ *for all* ζ *such that* $|\zeta| \leq 1$ *and* $\{e_i\}$ *is a sequence of i.i.d. random variables with mean zero and variance* σ_e^2.

Let $\lambda_n = \max(\lambda_1,\ldots,\lambda_q)$. The following theorems establish the asymptotic properties of the *PSLS* estimator $\widehat{\boldsymbol{\theta}}_n$.

Theorem 1. *Suppose that assumptions (A1) and (A2) hold. If the smoothing parameter satisfies* $\lambda_n = o(1)$, *then* $P(\lim_{n\to\infty} \widehat{\boldsymbol{\theta}}_n = \boldsymbol{\theta})$.

Theorem 2. *Suppose that assumptions (A1) and (A2) hold. If the smoothing parameter satisfies* $\lambda_n = o(n^{-1/2})$, *then*

$$\sqrt{n}(\widehat{\boldsymbol{\theta}}_n - \boldsymbol{\theta}) \xrightarrow{D} N\left(0, \boldsymbol{\Sigma}^{-1}\boldsymbol{\Sigma}_1\boldsymbol{\Sigma}^{-1}\right),$$

where

$$\boldsymbol{\Sigma}_1 = \lim_{n\to\infty} \frac{1}{n}(\mathbf{X}, \Xi)^T \mathbf{V}(\mathbf{X}, \Xi) \quad \textit{and} \quad \mathbf{V} = (E(\varepsilon_i\varepsilon_j))_{i,j=1}^n.$$

As pointed out in Yu and Ruppert (2002), Yu and Ruppert (2004), and Carroll, et al. (2004), the working assumption here is that the true function is a spline. More precisely, the spline parameter $\boldsymbol{\delta}$ should be called as the best projection of the true smooth function on the spline space.

Note that the estimator $\widehat{\boldsymbol{\theta}}_n$ does not take the serial correlation in (2) into the account. Therefore, it may not be asymptotically efficient. However, it is a consistent estimator. This can be used to build an improved estimator by fitting the model (2).

2.2 *Fitting the Error Model*

In this section we will estimate the autoregressive coefficients $\boldsymbol{\psi} = (\psi_1, \ldots, \psi_d)^T$ and σ_e^2 in the error structure (2) based on the residuals from the model (1).

Yule-Walker's equation implies that if ε_i's were observable, one could estimate $\boldsymbol{\psi}$ and σ_e^2 by means of

$$\tilde{\boldsymbol{\psi}}_n = \widetilde{\boldsymbol{\Gamma}}^{-1}\tilde{\boldsymbol{\gamma}} \text{ and } \tilde{\sigma}_e^2 = \tilde{\gamma}(0) - \tilde{\boldsymbol{\gamma}}^T\widetilde{\boldsymbol{\Gamma}}^{-1}\tilde{\boldsymbol{\gamma}},$$

where $\tilde{\boldsymbol{\psi}}_n = (\tilde{\psi}_1, \ldots, \tilde{\psi}_d)^T$, $\widetilde{\boldsymbol{\Gamma}} = (\tilde{\gamma}(i-j))_{i,j=1}^d$, $\tilde{\boldsymbol{\gamma}} = (\tilde{\gamma}(1), \ldots, \tilde{\gamma}(d))'$ and

$$\tilde{\gamma}(h) = \sum_{i=1}^{n-h} \varepsilon_i\varepsilon_{i+h}/n.$$

Since the ε_i's can not be observed, a natural method for estimating the parameters is to use the residuals from the model (1) as their pseudo-observations:

$$\widehat{\varepsilon}_i = Y_i - \mathbf{X}_i^T\widehat{\boldsymbol{\beta}}_n - \mathbf{Z}_i^T\mathbf{D}(U_i)\widehat{\boldsymbol{\delta}}_n, \quad i = 1, \ldots, n.$$

This results in our estimators:

$$\widehat{\boldsymbol{\psi}}_n = \widehat{\boldsymbol{\Gamma}}^{-1}\widehat{\boldsymbol{\gamma}} \text{ and } \widehat{\sigma}_e^2 = \widehat{\gamma}(0) - \widehat{\boldsymbol{\gamma}}^T\widehat{\boldsymbol{\Gamma}}^{-1}\widehat{\boldsymbol{\gamma}},$$

where $\widehat{\mathbf{\Gamma}} = (\widehat{\gamma}(i-j))_{i,j=1}^d$, $\widehat{\boldsymbol{\gamma}} = (\widehat{\gamma}(1), \ldots, \widehat{\gamma}(d))^T$ and $\widehat{\gamma}(h) = n^{-1}\sum_{i=1}^{n-h} \widehat{\varepsilon}_i \widehat{\varepsilon}_{i+h}$.

The following theorems summarize the asymptotic properties of $\widehat{\boldsymbol{\psi}}_n$ and $\widehat{\sigma}_e^2$.

Theorem 3. *Suppose that assumptions (A1) and (A2) hold. If the smoothing parameter satisfies $\lambda_n = o(1)$ and $E[e_1^4] < \infty$, then $\widehat{\boldsymbol{\psi}}_n$ and $\widehat{\sigma}_e^2$ are strongly consistent estimators of $\boldsymbol{\psi}$ and σ_e^2, respectively.*

Theorem 4. *Suppose that assumptions (A1) and (A2) hold. If the smoothing parameter satisfies $\lambda_n = o(n^{-1/2})$ and $E[e_1^4] < \infty$, then*

$$\sqrt{n}(\widehat{\boldsymbol{\psi}}_n - \boldsymbol{\psi}) \xrightarrow{D} N(0, \sigma_e^2 \mathbf{\Gamma}^{-1}),$$

where $\mathbf{\Gamma}$ is the covariance matrix $(\gamma(i-j))_{i,j=1}^d$ of $\{\varepsilon_i\}$. Moreover,

$$\sqrt{n}(\widehat{\sigma}_e^2 - \sigma_e^2) \xrightarrow{D} N(0, Var(e_1^2)).$$

Theorems 3 and 4 show that the estimators of $\boldsymbol{\psi}$ and σ_e^2 based on the residuals $\widehat{\varepsilon}_1, \ldots, \widehat{\varepsilon}_n$ are asymptotically equivalent to those based on the actual errors $\varepsilon_1, \ldots, \varepsilon_n$.

2.3 *Weighted Penalized Spline Least Squares Estimation*

Like the weighted least squares estimation, the fitted error structure enables one to construct the weighted *PSLS* estimator of $\boldsymbol{\theta}$ by minimizing

$$Q_{n,\lambda^w}^w(\boldsymbol{\theta}) = \frac{1}{n}(\mathbf{Y} - \mathbf{X}\boldsymbol{\beta} - \boldsymbol{\Xi}\boldsymbol{\delta})\widehat{\mathbf{V}}^{-1}(\mathbf{Y} - \mathbf{X}\boldsymbol{\beta} - \boldsymbol{\Xi}\boldsymbol{\delta}) + \sum_{s=1}^{q} \lambda_s^w \boldsymbol{\delta}_s^T \boldsymbol{\Omega}_s \boldsymbol{\delta}_s, \quad (5)$$

where

$$\widehat{\mathbf{V}}^{-1} = \widehat{\sigma}_e^{-2}(\mathbf{I} + \widehat{\psi}_{n1}\mathbf{J} + \cdots + \widehat{\psi}_{nd}\mathbf{J}^d)^T(\mathbf{I} + \widehat{\psi}_{n1}\mathbf{J} + \cdots + \widehat{\psi}_{nd}\mathbf{J}^d)$$

with $\mathbf{J} = \begin{pmatrix} 0 & \mathbf{I}_{n-1} \\ \mathbf{0} & \mathbf{0} \end{pmatrix}$ being an $n \times n$ matrix. The weighted *PSLS* estimator of $\boldsymbol{\theta}$ admits the following closed form:

$$\widehat{\boldsymbol{\theta}}_n^w = \left[(\mathbf{X}, \boldsymbol{\Xi})^T \widehat{\mathbf{V}}^{-1}(\mathbf{X}, \boldsymbol{\Xi}) + \text{blockdiag}(\mathbf{0}_{p\times p}, n\boldsymbol{\Lambda}^w \boldsymbol{\Omega})\right]^{-1} (\mathbf{X}, \boldsymbol{\Xi})^T \widehat{\mathbf{V}}^{-1}\mathbf{Y},$$

where $\boldsymbol{\Lambda}^w$ is defined in the same way as $\boldsymbol{\Lambda}$ but with λ_s replaced by λ_s^w. Put $\lambda_n^w = \max(\lambda_1^w, \cdots, \lambda_q^w)$. The following results reveal that $\widehat{\boldsymbol{\theta}}_n^w$ is asymptotically more efficient than $\widehat{\boldsymbol{\theta}}_n$.

Theorem 5. *Suppose that assumptions (A1) and (A2) hold. If the smoothing parameters satisfy $\max(\lambda_n, \lambda_n^w) = o(1)$ and $E[e_1^4] < \infty$, then $\widehat{\boldsymbol{\theta}}_n^w$ is a strongly consistent estimator of $\boldsymbol{\theta}$.*

Theorem 6. *Suppose that assumptions (A1) and (A2) hold. If the smoothing parameters satisfy* $\max(\lambda_n, \lambda_n^w) = o(n^{-1/2})$ *and* $E[e_1^4] < \infty$*, then*

$$\sqrt{n}(\widehat{\boldsymbol{\theta}}_n^w - \boldsymbol{\theta}) \xrightarrow{D} N\left(0, \boldsymbol{\Sigma}_2^{-1}\right) \quad as \quad n \to \infty$$

where $\boldsymbol{\Sigma}_2 = \lim_{n\to\infty} n^{-1}(\mathbf{X}, \boldsymbol{\Xi})^T \mathbf{V}^{-1}(\mathbf{X}, \boldsymbol{\Xi})$ *provided that the limit exists.*

Remark 1.
Since $\mathbf{W} \equiv \mathbf{V}^{-\frac{1}{2}}(\mathbf{X}, \boldsymbol{\Xi})\left((\mathbf{X}, \boldsymbol{\Xi})^T \mathbf{V}^{-1}(\mathbf{X}, \boldsymbol{\Xi})\right)^{-1}(\mathbf{X}, \boldsymbol{\Xi})^T \mathbf{V}^{-\frac{1}{2}}$ is an idempotent matrix with rank $p + \sum_{s=1}^q (m_s + 1 + \kappa_s)$,

$$\begin{aligned}
&\boldsymbol{\Sigma}^{-1}\boldsymbol{\Sigma}_2\boldsymbol{\Sigma}^{-1} - \boldsymbol{\Sigma}_3^{-1} \\
&= \left\{ \left[\lim_{n\to\infty} \frac{1}{n}(\mathbf{X}, \boldsymbol{\Xi})^T(\mathbf{X}, \boldsymbol{\Xi})\right]^{-1} \lim_{n\to\infty} \frac{1}{n}(\mathbf{X}, \boldsymbol{\Xi})^T \mathbf{V}(\mathbf{X}, \boldsymbol{\Xi}) \right. \\
&\quad \times \left. \left[\lim_{n\to\infty} \frac{1}{n}(\mathbf{X}, \boldsymbol{\Xi})^T(\mathbf{X}, \boldsymbol{\Xi})\right]^{-1} \right\} - \lim_{n\to\infty} \frac{1}{n}(\mathbf{X}, \boldsymbol{\Xi})^T \mathbf{V}^{-1}(\mathbf{X}, \boldsymbol{\Xi}) \\
&= \lim_{n\to\infty} \left\{ n[(\mathbf{X}, \boldsymbol{\Xi})^T(\mathbf{X}, \boldsymbol{\Xi})]^{-1}(\mathbf{X}, \boldsymbol{\Xi})^T \mathbf{V}^{\frac{1}{2}}(\mathbf{I} - \mathbf{W})\mathbf{V}^{\frac{1}{2}}(\mathbf{X}, \boldsymbol{\Xi}) \right. \\
&\quad \times \left. [(\mathbf{X}, \boldsymbol{\Xi})^T(\mathbf{X}, \boldsymbol{\Xi})]^{-1} \right\} \\
&\geq 0.
\end{aligned}$$

This implies that $\widehat{\boldsymbol{\theta}}_n^w$ is asymptotically more efficient than $\widehat{\boldsymbol{\theta}}_n$ in the sense that $\widehat{\boldsymbol{\theta}}_n^w$ has a smaller asymptotic covariance matrix.

Theorem 6 shows that the asymptotic covariance matrix admits a sandwich formula, which enables one to construct a consistent estimator of $\boldsymbol{\Sigma}_2$. In fact, define

$$\widehat{\boldsymbol{\Sigma}}_2 = \frac{1}{n}(\mathbf{X}, \boldsymbol{\Xi})^T \widehat{\mathbf{V}}^{-1}(\mathbf{X}, \boldsymbol{\Xi}),$$

then $\widehat{\boldsymbol{\Sigma}}_2$ is a consistent estimator of $\boldsymbol{\Sigma}_2$.

Theorem 7. *Suppose that assumptions (A1) and (A2) hold. If the smoothing parameters satisfy* $\max(\lambda_n, \lambda_n^w) = o(1)$ *and* $E(e_1^4) < \infty$*, then* $\widehat{\boldsymbol{\Sigma}}_2$ *is a consistent estimator of* $\boldsymbol{\Sigma}_2$.

Theorems 6 and 7 can be used to make joint inferences for the parametric component $\boldsymbol{\beta}$ and spline coefficients. For example if one wants to test the null hypothesis $H_0 : \alpha_s(u) \equiv 0$ for some $s \in \{1, \cdots, q\}$, which is equivalent to test $H_0 : \boldsymbol{\delta}_s = 0$. More generally, one can test the hypothesis $H_0 : \mathbf{C}\boldsymbol{\theta} = 0$ based on the following corollary, where $\mathbf{C}$ is a known $f \times \{p + \sum_{s=1}^q (m_s + 1 + \kappa_s)\}$ matrix with rank f.

Corollary 1. *Suppose that assumptions (A1) and (A2) hold. If the smoothing parameters satisfy* $\max(\lambda_n, \lambda_n^w) = o(n^{-1/2})$ *and* $E[e_1^4] < \infty$, *then under the null hypothesis* H_0

$$n(\mathbf{C}\widehat{\boldsymbol{\theta}}_n^w)^T \left(\mathbf{C}^T \widehat{\boldsymbol{\Sigma}}_2^{-1} \mathbf{C}\right)^{-1} (\mathbf{C}\widehat{\boldsymbol{\theta}}_n^w) \xrightarrow{D} \chi_f^2,$$

where χ_f^2 *is the chi-square distribution with* f *degrees of freedom.*

If one is interested in the parametric component $\boldsymbol{\beta}$, then the upper left $p \times p$ block sub-matrix of $\widehat{\boldsymbol{\Sigma}}_2$ can be used to calculate the test statistic. A joint confidence region for a set of parametric components can similarly be constructed.

3 Choice of Smoothing Parameters

Selection of the smoothing parameter is essential in nonparametric regression. The estimators $\widehat{\boldsymbol{\theta}}_n$ and $\widehat{\boldsymbol{\theta}}_n^w$ depend on the parameters $\boldsymbol{\lambda} = (\lambda_1, \ldots, \lambda_p)$ and $\boldsymbol{\lambda}^w = (\lambda_1^w, \ldots, \lambda_p^w)$. To describe such dependence, we denote by $\widehat{\boldsymbol{\theta}}(\boldsymbol{\lambda}) = \widehat{\boldsymbol{\theta}}_n$ and $\widehat{\boldsymbol{\theta}}^w(\boldsymbol{\lambda}^w) = \widehat{\boldsymbol{\theta}}_n^w$. Motivated by Yu and Ruppert (2002), and Wu and Yu (2004), we propose to select the smoothing parameter $\boldsymbol{\lambda}$ by minimizing the generalized cross validation score

$$\mathrm{GCV}(\boldsymbol{\lambda}) = \frac{\left[\mathbf{Y} - (\mathbf{X}, \boldsymbol{\Xi})\widehat{\boldsymbol{\theta}}(\boldsymbol{\lambda})\right]^T \left[\mathbf{Y} - (\mathbf{X}, \boldsymbol{\Xi})\widehat{\boldsymbol{\theta}}(\boldsymbol{\lambda})\right]}{(1 - n^{-1}\mathrm{tr}(\mathbf{S}(\boldsymbol{\lambda})))^2},$$

where the numerator is the model averaged squared residual. The trace of the smoothing matrix $\mathbf{S}(\boldsymbol{\lambda})$, often called the *degree of freedom* of the fit (see Hastie and Tibshirani 1990), can be calculated as

$$\mathrm{tr}(\mathbf{S}(\boldsymbol{\lambda})) = \mathrm{tr}\left\{[(\mathbf{X}, \boldsymbol{\Xi})^T(\mathbf{X}, \boldsymbol{\Xi}) + n\boldsymbol{\Lambda}\boldsymbol{\Omega}]^{-1} (\mathbf{X}, \boldsymbol{\Xi})^T(\mathbf{X}, \boldsymbol{\Xi})\right\}.$$

Similarly, we choose the smoothing parameter $\boldsymbol{\lambda}^w$ by minimizing the generalized cross validation score

$$\mathrm{GCV}(\boldsymbol{\lambda}^w) = \frac{\left[\mathbf{Y} - (\mathbf{X}, \boldsymbol{\Xi})\widehat{\boldsymbol{\theta}}^w(\boldsymbol{\lambda}^w)\right]^T \left[\mathbf{Y} - (\mathbf{X}, \boldsymbol{\Xi})\widehat{\boldsymbol{\theta}}^w(\boldsymbol{\lambda}^w)\right]}{(1 - n^{-1}\mathrm{tr}(\mathbf{S}^w(\boldsymbol{\lambda})))^2},$$

where

$$\mathrm{tr}(\mathbf{S}^w(\boldsymbol{\lambda}^w)) = \mathrm{tr}\left\{\left[(\mathbf{X}, \boldsymbol{\Xi})^T\widehat{\mathbf{V}}^{-1}(\mathbf{X}, \boldsymbol{\Xi}) + n\boldsymbol{\Lambda}^w\boldsymbol{\Omega}\right]^{-1} (\mathbf{X}, \boldsymbol{\Xi})^T\widehat{\mathbf{V}}^{-1}(\mathbf{X}, \boldsymbol{\Xi})\right\}.$$

With q penalty parameters λ_s in the model (1), a full grid search algorithm for $\boldsymbol{\lambda}$ might be not practical computationally. However, the two-step GCV algorithm proposed by Ruppert and Carroll (2001) can be used to reduce the burden of calculation. We will use the algorithm in simulations. See the afore-mentioned paper for details.

One of the outstanding features of the penalized splines is that the selection of the number of knots as well as knot location is no longer crucial, where the complicated knot selection problem is reduced to the choice of a single smoothing parameter λ_s. Some authors, e.g. Ruppert (2002), observed that the value of the number of knots κ_s is not too important, provided it is large enough. As in Wu and Yu (2004), we can simply choose approximately $\min(n/40, 40)$ knots. Our numerical study shows that it works well. Given a fixed number of knots, we recommend the knots be placed at equally-spaced sample quantiles of the index u.

4 Simulations

In this section we carry out simulations to demonstrate the finite sample performances of the proposed estimators. The two estimators, the unweighted and weighted, will be compared to illustrate the efficiency of the weighted *PSLS* estimation.

The data are generated from the following varying-coefficient partially linear regression model

$$y_i = x_{1i}\beta_1 + x_{2i}\beta_2 + z_i\alpha(u_i) + \varepsilon_i, i = 1, \ldots, n,$$

where x_{1i}'s are i.i.d. $N(0,1)$, x_{2i}'s are i.i.d. Bernoulli(0.45), z_i's are i.i.d. $N(0,1)$ and u_i's are i.i.d. $U(0,1)$; the parameters are set as $\beta_1 = 1.5$ and $\beta_2 = 2$, $\alpha(u)$ is taken as $2\sin(2\pi u)$; ε_i's satisfy

$$\varepsilon_i = \theta_1 \varepsilon_{i-1} + e_i, \quad i = 1, \cdots, n.$$

We consider different levels of correlation with $\theta_1 = 0.7, 0.5, 0.3$, 0, -0.3, -0.5 and -0.7 and two kind of sample sizes with $n = 200, 400$. In each case the number of simulated realizations is $1,000$ and the values of the x_{1i}, x_{2i}, z_i and u_i are generated only once.

For the coefficient functions, we assess the estimator $\widehat{\alpha}(\cdot)$ via the Square-Root of Averaged Squared Errors (RASE):

$$\text{RASE} = \left[n^{-1} \sum_{i=1}^{n} \{\widehat{\alpha}(u_i) - \alpha(u_i)\}^2 \right]^{1/2}.$$

For a given sample size, the estimated biases and the mean and standard deviation of the RASEs are calculated. The results are listed in Tables 1 and 3. Figure 1 gives the averaged penalized splines estimate of the function $\alpha(\cdot)$ and the corresponding 2.5% and 97.5% quantiles among 1,000 simulations, and the estimated 95% confidence intervals based on the normal approximation in Theorems 2 and 6. Tables 1 and 3 and Figure 1 demonstrate that the true coefficient functions and the averages of their estimators virtually overlay, which indicates that there is little bias. Further, the confidence intervals are very close to the quantile bands. This again confirms similar findings for the variance estimates in Yu and Ruppert (2002). Moreover, Tables 1 and 3 and Figure 1 also show that the weighted *PSLS* estimator improves the un-weighted *PSLS* estimator.

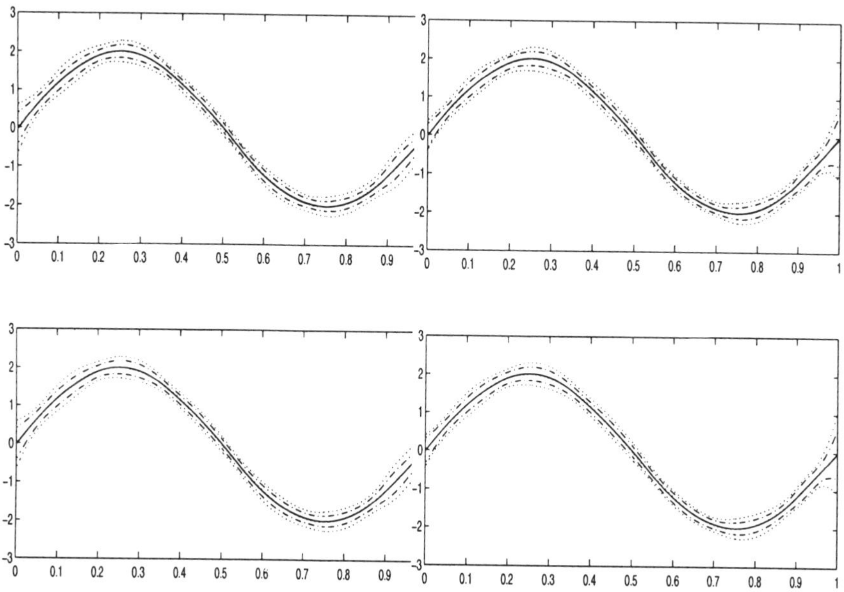

Figure 1 The estimators of the nonparametric component and its confidence intervals Left panel: $n = 400$ and $\theta_1 = 0.7$. Right panel: $n = 400$ and $\theta_1 = -0.7$. (a) - from the Monte Carlo simulation and (b) - the asymptotic confidence band. Dotted: the un-weighted *PSLS* estimator; dash-dotted: weighted *PSLS* estimator.

For the parameters (β_1, β_2), we compare three estimators: the un-weighted *PSLS* estimator $\widehat{\beta}_n$, the weighted *PSLS* estimator $\widehat{\beta}_n^w$, and the ideal weighted *PSLS* estimator $\tilde{\beta}_n^w$ with $\mathbf{V}$ being known. The estimators and their standard deviations (Std) were evaluated along with the average of the estimated standard error (SE) for the estimators. The coverage probability (CP) of the 95% confidence intervals for β was also calculated based on the normal approximation. In addition, we calculated the relative efficiency (RE) of the estimators with respect to the ideal weighted *PSLS* estimator in terms of the ratio of mean squared errors. These re-

Table 1 The estimators of the nonparametric function with $n = 200$.

	$\widehat{\alpha}(\cdot)$			$\widehat{\alpha}^w(\cdot)$			$\tilde{\alpha}^w(\cdot)$		
	RASE	Std(RASE)	Bias	RASE	Std(RASE)	Bias	RASE	Std(RASE)	Bias
$\theta_1 = 0.7$	0.206	0.054	-0.001	0.127	0.032	0.001	0.127	0.032	0.001
$\theta_1 = 0.5$	0.202	0.053	0.001	0.160	0.042	0.002	0.160	0.042	0.002
$\theta_1 = 0.3$	0.199	0.054	-0.001	0.182	0.049	-0.001	0.182	0.049	-0.001
$\theta_1 = 0$	0.221	0.063	-0.004	0.222	0.064	-0.004	0.221	0.063	-0.004
$\theta_1 = -0.3$	0.209	0.052	-0.001	0.193	0.048	-0.001	0.192	0.048	-0.001
$\theta_1 = -0.5$	0.209	0.056	-0.001	0.168	0.044	-0.001	0.168	0.044	-0.001
$\theta_1 = -0.7$	0.190	0.055	-0.005	0.119	0.032	-0.000	0.119	0.032	-0.000

Table 2 The estimators of the parametric components with $n = 200$.

		β_1					β_2				
		Mean	Std	SE	CP	RE	Mean	Std	SE	CP	RE
$\theta_1 = 0.7$	$\widehat{\beta}_n$	1.498	0.079	0.075	0.935	3.294	1.990	0.192	0.185	0.932	4.716
	$\widehat{\beta}_n^w$	1.499	0.043	0.047	0.958	1.005	1.994	0.090	0.091	0.952	1.046
	$\tilde{\beta}_n^w$	1.499	0.043	0.046	0.956	1.000	1.994	0.088	0.090	0.956	1.000
$\theta_1 = 0.5$	$\widehat{\beta}_n$	1.501	0.064	0.067	0.958	1.596	1.997	0.148	0.149	0.958	1.837
	$\widehat{\beta}_n^w$	1.500	0.051	0.055	0.969	1.005	1.999	0.109	0.111	0.957	1.011
	$\tilde{\beta}_n^w$	1.500	0.050	0.054	0.966	1.000	1.999	0.109	0.111	0.949	1.000
$\theta_1 = 0.3$	$\widehat{\beta}_n$	1.495	0.070	0.068	0.951	1.232	1.996	0.131	0.129	0.943	1.236
	$\widehat{\beta}_n^w$	1.496	0.064	0.063	0.950	1.013	1.996	0.119	0.116	0.937	1.025
	$\tilde{\beta}_n^w$	1.496	0.063	0.062	0.950	1.000	1.995	0.118	0.116	0.941	1.000
$\theta_1 = 0$	$\widehat{\beta}_n$	1.498	0.073	0.074	0.952	1.000	1.992	0.109	0.107	0.944	1.000
	$\widehat{\beta}_n^w$	1.498	0.073	0.074	0.949	1.009	1.992	0.109	0.107	0.945	1.004
	$\tilde{\beta}_n^w$	1.498	0.073	0.074	0.951	1.000	1.992	0.109	0.107	0.942	1.000
$\theta_1 = -0.3$	$\widehat{\beta}_n$	1.500	0.073	0.074	0.954	1.227	1.996	0.100	0.100	0.956	1.185
	$\widehat{\beta}_n^w$	1.501	0.066	0.067	0.953	1.007	1.995	0.092	0.093	0.960	1.013
	$\tilde{\beta}_n^w$	1.501	0.066	0.067	0.952	1.000	1.995	0.091	0.092	0.957	1.000
$\theta_1 = -0.5$	$\widehat{\beta}_n$	1.501	0.068	0.070	0.956	1.640	1.996	0.090	0.088	0.950	1.514
	$\widehat{\beta}_n^w$	1.501	0.054	0.056	0.953	1.011	1.997	0.073	0.074	0.953	1.000
	$\tilde{\beta}_n^w$	1.501	0.053	0.056	0.956	1.000	1.997	0.073	0.073	0.951	1.000
$\theta_1 = -0.7$	$\widehat{\beta}_n$	1.500	0.084	0.081	0.936	2.675	2.000	0.081	0.082	0.949	2.569
	$\widehat{\beta}_n^w$	1.501	0.051	0.053	0.953	1.004	2.000	0.050	0.052	0.951	1.003
	$\tilde{\beta}_n^w$	1.501	0.051	0.052	0.952	1.000	2.000	0.050	0.051	0.950	1.000

sults are summarized in Tables 2 and 4. Based on Tables 2 and 4 we make the following observations: (i) all three methods yield unbiased estimates; (ii) the proposed variance estimators are consistent; (iii) the nominal 95% confidence intervals based on the proposed standard errors provide good coverages for the cases studied; (iv) the weighted *PSLS* estimator and the ideal weighted estimator perform almost equally well, and both of them have smaller standard deviations than the un-weighted *PSLS* estimator. This improvement becomes greater when the absolute value of the autore-

Table 3 The estimators of the nonparametric function with $n = 400$.

	$\widehat{\alpha}(\cdot)$			$\widehat{\alpha}^w(\cdot)$			$\tilde{\alpha}^w(\cdot)$		
	RASE	Std(RASE)	Bias	RASE	Std(RASE)	Bias	RASE	Std(RASE)	Bias
$\theta_1 = 0.7$	0.131	0.034	0.002	0.080	0.020	0.002	0.080	0.020	0.002
$\theta_1 = 0.5$	0.133	0.034	-0.001	0.106	0.026	-0.001	0.106	0.026	-0.001
$\theta_1 = 0.3$	0.138	0.035	-0.004	0.126	0.032	-0.003	0.126	0.032	-0.003
$\theta_1 = 0$	0.153	0.040	-0.0000	0.154	0.040	-0.000	0.153	0.040	-0.000
$\theta_1 = -0.3$	0.140	0.035	0.002	0.129	0.033	0.002	0.129	0.033	0.002
$\theta_1 = -0.5$	0.141	0.037	-0.002	0.111	0.0289	-0.001	0.111	0.029	-0.001
$\theta_1 = -0.7$	0.135	0.038	-0.002	0.080	0.0203	-0.002	0.0802	0.020	-0.002

Table 4 The estimators of the parametric components with $n = 400$.

		β_1					β_2				
		Mean	Std	SE	CP	RE	Mean	Std	SE	CP	RE
$\theta_1 = 0.7$	$\widehat{\beta}_n$	1.4994	0.0480	0.0486	0.9530	2.4530	1.9933	0.1359	0.1337	0.9420	5.1452
	$\widehat{\beta}_n^w$	1.501	0.030	0.031	0.956	1.000	1.996	0.060	0.061	0.951	1.016
	$\tilde{\beta}_n^w$	1.501	0.030	0.031	0.956	1.000	1.996	0.059	0.061	0.952	1.000
$\theta_1 = 0.5$	$\widehat{\beta}_n$	1.500	0.054	0.053	0.939	1.680	1.998	0.098	0.103	0.953	1.929
	$\widehat{\beta}_n^w$	1.499	0.041	0.041	0.950	1.002	1.997	0.071	0.073	0.956	1.005
	$\tilde{\beta}_n^w$	1.499	0.041	0.041	0.953	1.000	1.997	0.071	0.073	0.958	1.000
$\theta_1 = 0.3$	$\widehat{\beta}_n$	1.499	0.052	0.052	0.945	1.162	1.996	0.090	0.088	0.946	1.209
	$\widehat{\beta}_n^w$	1.498	0.048	0.047	0.946	1.001	1.997	0.082	0.079	0.942	1.010
	$\tilde{\beta}_n^w$	1.498	0.048	0.047	0.947	1.000	1.996	0.082	0.079	0.943	1.000
$\theta_1 = 0$	$\widehat{\beta}_n$	1.498	0.052	0.051	0.944	1.000	2.004	0.074	0.073	0.948	1.000
	$\widehat{\beta}_n^w$	1.498	0.052	0.051	0.942	0.999	2.003	0.074	0.073	0.944	0.999
	$\tilde{\beta}_n^w$	1.498	0.052	0.051	0.944	1.000	2.004	0.074	0.073	0.947	1.000
$\theta_1 = -0.3$	$\widehat{\beta}_n$	1.499	0.049	0.051	0.958	1.149	2.003	0.066	0.067	0.953	1.150
	$\widehat{\beta}_n^w$	1.498	0.045	0.047	0.954	0.993	2.001	0.061	0.063	0.955	1.002
	$\tilde{\beta}_n^w$	1.498	0.046	0.046	0.952	1.000	2.001	0.061	0.062	0.951	1.000
$\theta_1 = -0.5$	$\widehat{\beta}_n$	1.502	0.048	0.049	0.946	1.626	1.995	0.065	0.065	0.950	1.681
	$\widehat{\beta}_n^w$	1.501	0.038	0.039	0.965	1.002	1.996	0.050	0.052	0.945	1.001
	$\tilde{\beta}_n^w$	1.501	0.038	0.039	0.962	1.000	1.996	0.050	0.051	0.945	1.000
$\theta_1 = -0.7$	$\widehat{\beta}_n$	1.498	0.049	0.049	0.947	2.900	2.000	0.069	0.068	0.948	2.870
	$\widehat{\beta}_n^w$	1.498	0.029	0.031	0.969	1.004	1.999	0.040	0.041	0.956	0.999
	$\tilde{\beta}_n^w$	1.498	0.029	0.030	0.968	1.000	1.999	0.040	0.040	0.953	1.000

gressive coefficient increases.

To explore the sensitivity to the selection of λ in estimation of the finite parameters, we calculated the estimators for different λ's over a large range. Table 5 reports the estimated parameters along with some related statistics under moderate correlation of the error. It is evident that the estimated values are quite robust against the selection of λ. This supports the conclusion in Theorems 1 and 6 that the estimators of the finite parameters are $\sqrt{n}$-consistent for a large range of the smoothing parameters λ_n and λ_n^w.

Table 5 The estimators of the parametric components with $n = 400$ and different λ.

			β_1					β_2				
			Mean	Std	SE	CP	RE	Mean	Std	SE	CP	RE
$\theta_1 = 0.5$	$\lambda = n^{-0.5}$	$\widehat{\beta}_n$	1.504	0.046	0.049	0.950	1.685	2.018	0.103	0.104	0.944	2.009
		$\widehat{\beta}_n^w$	1.499	0.036	0.039	0.973	0.998	2.015	0.073	0.076	0.951	1.035
		$\tilde{\beta}_n^w$	1.499	0.036	0.038	0.969	1.000	2.015	0.072	0.076	0.951	1.000
	$\lambda = n^{-0.8}$	$\widehat{\beta}_n$	1.493	0.048	0.048	0.947	1.482	2.012	0.103	0.103	0.945	2.000
		$\widehat{\beta}_n^w$	1.492	0.039	0.039	0.944	1.000	2.013	0.073	0.073	0.948	1.011
		$\tilde{\beta}_n^w$	1.492	0.039	0.039	0.944	1.000	2.013	0.072	0.073	0.951	1.000
	$\lambda = n^{-1.0}$	$\widehat{\beta}_n$	1.502	0.053	0.051	0.941	1.550	2.003	0.098	0.101	0.956	1.961
		$\widehat{\beta}_n^w$	1.505	0.042	0.041	0.946	1.001	2.005	0.070	0.072	0.955	1.004
		$\tilde{\beta}_n^w$	1.505	0.042	0.041	0.940	1.000	2.005	0.069	0.072	0.959	1.000
	$\lambda = n^{-1.2}$	$\widehat{\beta}_n$	1.499	0.055	0.054	0.946	1.676	1.999	0.104	0.104	0.946	2.002
		$\widehat{\beta}_n^w$	1.501	0.043	0.043	0.953	1.005	1.998	0.073	0.073	0.951	1.007
		$\tilde{\beta}_n^w$	1.501	0.043	0.043	0.953	1.000	1.998	0.073	0.073	0.948	1.000
$\theta_1 = -0.5$	$\lambda = n^{-0.5}$	$\widehat{\beta}_n$	1.507	0.049	0.050	0.950	1.635	2.008	0.063	0.063	0.955	1.407
		$\widehat{\beta}_n^w$	1.504	0.038	0.042	0.966	1.006	2.015	0.051	0.054	0.953	1.002
		$\tilde{\beta}_n^w$	1.504	0.038	0.041	0.960	1.000	2.015	0.051	0.052	0.949	1.000
	$\lambda = n^{-0.8}$	$\widehat{\beta}_n$	1.502	0.047	0.047	0.951	1.632	1.991	0.067	0.066	0.944	1.628
		$\widehat{\beta}_n^w$	1.504	0.037	0.037	0.946	1.002	1.994	0.052	0.052	0.944	0.998
		$\tilde{\beta}_n^w$	1.504	0.037	0.037	0.947	1.000	1.994	0.052	0.052	0.943	1.000
	$\lambda = n^{-1.0}$	$\widehat{\beta}_n$	1.495	0.049	0.048	0.938	1.646	2.001	0.060	0.061	0.949	1.573
		$\widehat{\beta}_n^w$	1.496	0.038	0.038	0.950	1.000	2.003	0.048	0.050	0.963	0.999
		$\tilde{\beta}_n^w$	1.496	0.038	0.038	0.946	1.000	2.003	0.048	0.049	0.963	1.000
	$\lambda = n^{-1.2}$	$\widehat{\beta}_n$	1.501	0.048	0.047	0.948	1.519	1.995	0.062	0.062	0.942	1.686
		$\widehat{\beta}_n^w$	1.502	0.039	0.038	0.942	1.002	1.995	0.048	0.049	0.954	1.003
		$\tilde{\beta}_n^w$	1.502	0.038	0.038	0.943	1.000	1.995	0.047	0.049	0.954	1.000

5 Real Data Analysis

We now illustrate the application of the proposed method in stock market. In finance and security analysis, the risk of an individual stock is often measured by its (standardized) regression slope against a market index. If this slope is greater than 1, the change in the stock price is expected to be more than that in the index and thus the stock is considered to be more risky. The data set we considered consists of the daily closing prices of the common stock price of Microsoft during the first ten months of year 2000 and the Standard & Poor's (S&P) 100 index for the same time period. Cui, Zhu and He (2002) employed the following purely parametric regression to model the relationship between the common stock price and the S&P100 index:

$$y_i = \beta_0 I(i \leq 64) + \beta_1 I(i > 64) + x_i\beta_3 + \varepsilon_i, \quad i = 1, \cdots, 206, \tag{6}$$

where y_i is the common stock price at i-th day divided by the price on the first day and x_i denotes the change in the S&P100 index.

Applying the least squares technique, we calculated the parameter estimators as $\bar{\beta}_{n0} = -0.2934$, $\bar{\beta}_{n1} = -0.6160$, $\bar{\beta}_{n2} = 1.2775$, $\bar{\rho}_n = 0.8168$ and $\bar{\sigma}_n^2 = 0.0030$. The corresponding error variances were calculated as $0.0709, 0.0723$ and 0.0734, respectively. Moreover, by fitting the estimated error structure, we obtained the weighted least squares estimator $\bar{\beta}_{n0}^w = -0.0488$, $\bar{\beta}_{n1}^w = -0.2667$ and $\bar{\beta}_{n3}^w = 0.9600$. The corresponding error variances were computed to be $0.0017, 0.0016$ and 0.0017, respectively.

To test if the regression slope changes over time, we fit the dataset using the semiparametric regression model:

$$y_i = \beta_0 I(i \leq 64) + \beta_1 I(i > 64) + x_i \alpha(u_i) + \varepsilon_i, \quad i = 1, \cdots, 206. \tag{7}$$

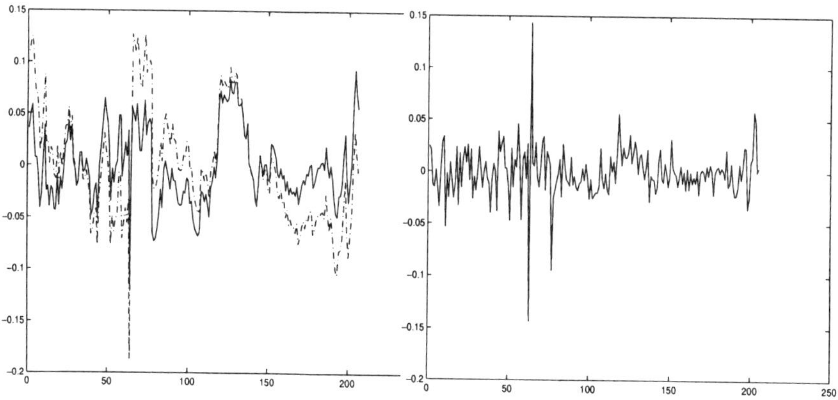

Figure 2 The residual plots. Left panel: dash-dotted - the residuals from model (6), solid -the residuals from model (7); right panel: the residuals from the AR(1) process.

Applying the *PSLS* estimation technique, we got the un-weighted *PSLS* estimator $\widehat{\beta}_{n0} = -0.3049$, $\widehat{\beta}_{n1} = -0.4863$. Correspondingly, the autoregressive coefficient was calculated as $\widehat{\rho}_n - 0.7410$ and the error variance $\widehat{\sigma}_n^2 = 0.0013$. The error variances of $\widehat{\beta}_{n0}$ and $\widehat{\beta}_{n1}$ were estimated as 0.0322 and 0.0307, respectively. Moreover, the weighted *PSLS* estimators were calculated as $\widehat{\beta}_{n0}^w = -0.2429$, $\widehat{\beta}_{n1}^w = -0.3034$, The corresponding error variances of $\widehat{\beta}_{n0}^w$ and $\widehat{\beta}_{n1}^w$ were computed to be 0.0144 and 0.0139, respectively.

Figure 2 shows the residuals of the models (6) and (7). Obviously, from Figure 2, we can see that the model (7) has smaller fitted errors. Figure 2 also shows the residuals after fitting the AR(1) process to $\{\varepsilon_i\}$ for the model (7). There is no significant difference between the residuals and the white noise, which validates the specification of the AR(1) model in this case. Figure 3 shows the estimators of $\alpha(\cdot)$ based on the unweighted *PSLS*

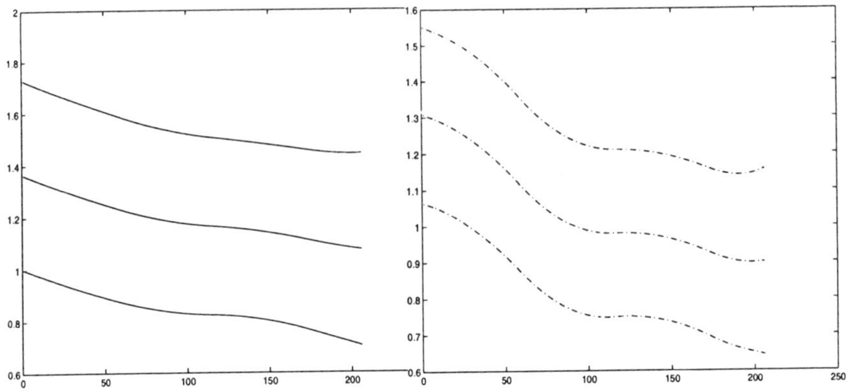

Figure 3 The estimated curves $\widehat{\alpha}(\cdot)$ with 95% confidence intervals. Left panel: the un-weighted *PSLS* estimator; right panel: the weighted *PSLS* estimator.

and the weighted *PSLS* methods and the corresponding confidence regions. Obviously, the regression slope is decreasing over the time. This implies the risk of this stock is decreasing.

6 Concluding Remarks

The varying-coefficient partially linear regression model provides a useful tool for statistical modeling. In this paper we have theoretically and empirically studied the statistical inference for the model when the errors are serially correlated and modeled as an AR process. We proposed a weighted *PSLS* estimator. The asymptotic properties were investigated under the assumption that the number of knots is fixed, though potentially large. We showed that the proposed estimators of all parameters are $\sqrt{n}$-consistent, and asymptotically normally distributed. The efficiency of the weighted *PSLS* estimator was demonstrated in comparison to several other estimators. Simultaneous inference procedures were proposed for both components of the model based on the sandwich formula of the joint covariance matrix in Theorem 6. The methodology was applied to daily stock price data of Microsoft.

The success of the weighted *PSLS* estimator depends on the specification of the order d of autoregressive model. A mis-specification of the parameter d may generally deteriorate the efficiency of the weighted *PSLS* estimator, although the resulting estimator is consistent. Fortunately, the AIC criterion for model selection can be employed to rapidly identify the value of d, based on the residuals from the model (1).

Interesting topics for further studies include extending our results to ARMA error structures, and to nonlinear time series error structures such as the ARCH and GARCH models.

7 Appendix. Proof of the Main Results

To facilitate the proofs of the theorems in the previous sections, we first present the following lemma.

Lemma 1. *For the autoregressive process $\{\varepsilon_i\}$ defined in (2), let $\{c_i\}_{i=1}^n$ be a sequence of real numbers and let $u(\cdot)$ be a function defined on integers such that*

$$\lim_{n\to\infty} \frac{1}{n}\sum_{i=1}^{n-j} c_i c_{i+|j|} = u(j), \quad j = 0, \pm 1, \pm 2, \ldots.$$

Assume that $0 < v = \sum_{j=-\infty}^{\infty} u(j)\gamma(j) < \infty$ with $\gamma(j) = E(\varepsilon_1\varepsilon_{1+j})$. Then

$$\left(\sum_{i=1}^n c_i^2\right)^{-1/2} \sum_{i=1}^n c_i\varepsilon_j \xrightarrow{D} N\left(0, [u(0)]^{-1}v\right) \quad \text{as } n\to\infty.$$

Proof. It is straightforward to prove this result by applying arguments contained in the proof of Theorem 6.3.4 in Fuller (1976) and Proposition 2.2 of Huber (1973)

Proof of Theorem 1. By the definition of $\widehat{\boldsymbol{\theta}}_n$, we have

$$\begin{aligned}\widehat{\boldsymbol{\theta}}_n &= \left[(\mathbf{X},\Xi)^T(\mathbf{X},\Xi) + \text{blockdiag}(\mathbf{0}_{p\times p}, n\boldsymbol{\Lambda}\boldsymbol{\Omega})\right]^{-1}(\mathbf{X},\Xi)^T\mathbf{Y}\\ &= \boldsymbol{\theta} + \left[(\mathbf{X},\Xi)^T(\mathbf{X},\Xi) + \text{blockdiag}(\mathbf{0}_{p\times p}, n\boldsymbol{\Lambda}\boldsymbol{\Omega})\right]^{-1}(\mathbf{X},\Xi)^T\boldsymbol{\varepsilon}.\end{aligned}$$

Note that $(\mathbf{A}+a\mathbf{B})^{-1} = \mathbf{A}^{-1} - a\mathbf{A}^{-1}\mathbf{B}\mathbf{A}^{-1} + O(a^2)$. It follows that

$$\left[n^{-1}(\mathbf{X},\Xi)^T(\mathbf{X},\Xi) + \text{blockdiag}(\mathbf{0}_{p\times p}, \boldsymbol{\Lambda}\boldsymbol{\Omega})\right]^{-1} = \lim_{n\to\infty} n^{-1}(\mathbf{X},\Xi)^T(\mathbf{X},\Xi) + o(1).$$

Moreover, a simple algebra shows that the sth element of $n^{-1}(\mathbf{X},\Xi)^T\boldsymbol{\varepsilon}$ is of order $o(1)$ almost surely. The result of the theorem follows.

Proof of Theorem 2. Using the same argument as in Theorem 1, we have

$$\sqrt{n}(\widehat{\boldsymbol{\theta}}_n - \boldsymbol{\theta}) = \sqrt{n}\left[(\mathbf{X},\Xi)^T(\mathbf{X},\Xi)\right]^{-1}(\mathbf{X},\Xi)^T\boldsymbol{\varepsilon} + o_p(1).$$

Applying Lemma 1, for any nonzero p-vector ζ we have

$$\frac{1}{\sqrt{n}}\zeta^T(\mathbf{X},\Xi)^T\boldsymbol{\varepsilon} \xrightarrow{D} N(0, \zeta^T\boldsymbol{\Sigma}_1\zeta) \quad \text{as } n\to\infty,$$

where $\mathbf{\Sigma}_1$ is defined in Theorem 2. By Slutsky theorem the proof is complete.

Proof of Theorem 3. It suffices to show that

$$\widehat{\gamma}(h)-\gamma(h)=\frac{1}{n}\sum_{i=1}^{n-h}\varepsilon_i\varepsilon_{i+h}-\gamma(h)+o(1)\quad \text{a.s.},$$

where h is a positive integer. It can be rewritten as

$$\begin{aligned}\widehat{\gamma}(h)-\gamma(h)=&\frac{1}{n}\sum_{i=1}^{n-h}(\widehat{\varepsilon}_i-\varepsilon_i)(\widehat{\varepsilon}_{i+h}-\varepsilon_{i+h})+\frac{1}{n}\sum_{i=1}^{n-h}(\widehat{\varepsilon}_i-\varepsilon_i)\varepsilon_{i+h}\\&+\frac{1}{n}\sum_{i=1}^{n-h}(\widehat{\varepsilon}_{i+h}-\varepsilon_{i+h})\varepsilon_i+\frac{1}{n}\sum_{i=1}^{n-h}(\varepsilon_i\varepsilon_{i+h}-\gamma(h)).\end{aligned}$$

By the definition of $\widehat{\varepsilon}_i$ and Theorem 1, it can be shown that

$$\begin{aligned}&\frac{1}{n}\sum_{i=1}^{n-h}(\widehat{\varepsilon}_i-\varepsilon_i)\varepsilon_{i+h}=o(1)\quad \text{a.s.,}\qquad \text{and}\\&\frac{1}{n}\sum_{i=1}^{n-h}(\widehat{\varepsilon}_i-\varepsilon_i)(\widehat{\varepsilon}_{i+h}-\varepsilon_{i+h})=o(1),\quad \text{a.s.}\end{aligned}\tag{8}$$

Thus, the theorem follows.

Proof of Theorem 4. According to Theorem 8.1.1 of Rockwell and Davis (1989) and the proof of Theorem 3, it is easy to complete the proof of Theorem 4.

Proof of Theorem 5. Let

$$\tilde{\boldsymbol{\theta}}_n^w=\left[(\mathbf{X},\Xi)^T\mathbf{V}^{-1}(\mathbf{X},\Xi)+\text{blockdiag}(\mathbf{0}_{p\times p},n\mathbf{\Lambda}^w\mathbf{\Omega})\right]^{-1}(\mathbf{X},\Xi)^T\mathbf{V}^{-1}\mathbf{Y}$$

where

$$\mathbf{V}^{-1}=\sigma_e^{-2}(\mathbf{I}+\psi_{n1}\mathbf{J}+\cdots+\psi_{nd}\mathbf{J}^d)^T(\mathbf{I}+\psi_{n1}\mathbf{J}+\cdots+\psi_{nd}\mathbf{J}^d)$$

with $\mathbf{J}$ defined in Section 2.3. Using the same argument as in Theorem 1, one can show that $\tilde{\boldsymbol{\theta}}_n^w$ is a strongly consistent estimator of $\boldsymbol{\theta}$.

Note that $(\mathbf{A}+a\mathbf{B})^{-1}=\mathbf{A}^{-1}-a\mathbf{A}^{-1}\mathbf{B}\mathbf{A}^{-1}+O(a^2)$. It suffices to show

$$\frac{1}{n}\left((\mathbf{X},\Xi)^T\widehat{\mathbf{V}}^{-1}(\mathbf{X},\Xi)-(\mathbf{X},\Xi)^T\mathbf{V}^{-1}(\mathbf{X},\Xi)\right)=o(1)\quad \text{a.s.,}\tag{9}$$

$$\frac{1}{n}\left((\mathbf{X},\Xi)^T\widehat{\mathbf{V}}^{-1}\boldsymbol{\varepsilon}-(\mathbf{X},\Xi)^T\mathbf{V}^{-1}\boldsymbol{\varepsilon}\right)=o(1)\quad \text{a.s..}\tag{10}$$

Since

$$\max_{1\le i\le n}\sum_{j=1}^{n}|\gamma(i-j)|\le 2\sum_{l=0}^{\infty}|\gamma(l)|\le 2\sigma_e^2\left(\sum_{k=0}^{\infty}|\theta_k|\right)^2=O(1),$$

where $\theta(\zeta) = 1 + \theta_1\zeta + \cdots = 1/\psi(\zeta)$ with $\psi(z) = 1 + \psi_1\zeta + \cdots + \psi_d\zeta^d$, there exists a constant c_1 such that $\lambda_{\max}(\mathbf{V}) < c_1$. In addition, since

$$\mathbf{V}^{-1} = \sigma^{-2}(\mathbf{I} + \psi_1\mathbf{J} + \cdots + \psi_d\mathbf{J}^d)^T(\mathbf{I} + \psi_1\mathbf{J} + \cdots + \psi_d\mathbf{J}^d)$$

there exists a constant c_2 such that $\lambda_{\min}(\mathbf{V}) > c_2 > 0$. Moreover, by the definition of $\widehat{\mathbf{V}}$ and the strong consistency of ψ_i, when n is large enough,

$$0 < c_2 < \lambda_{\min}(\widehat{\mathbf{V}}) \leq \lambda_{\max}(\widehat{\mathbf{V}}) < c_1 \text{ a.s..}$$

This together with Assumption 1 leads to (9) and (10).

Proof of Theorem 6. By the proof of Theorem 5, we have

$$\sqrt{n}(\widehat{\boldsymbol{\theta}}_n^w - \boldsymbol{\theta}) = \sqrt{n}(\tilde{\boldsymbol{\theta}}_n^w - \boldsymbol{\theta}) + o_p(1).$$

In addition, using the same argument as in the proof of Theorem 2, we obtain

$$\sqrt{n}(\widehat{\boldsymbol{\theta}}_n^w - \boldsymbol{\theta}) = \sqrt{n}\left[(\mathbf{X}, \Xi)^T\mathbf{V}^{-1}(\mathbf{X}, \Xi)\right]^{-1}(\mathbf{X}, \Xi)^T\mathbf{V}^{-1}\varepsilon + o_p(1).$$

Applying Lemma 1, for any nonzero p-vector ζ we have

$$\frac{1}{\sqrt{n}}\zeta^T(\mathbf{X}, \Xi)^T\mathbf{V}^{-1}\varepsilon \xrightarrow{D} N(0, \zeta^T\boldsymbol{\Sigma}_2\zeta) \quad \text{as } n \to \infty$$

where $\boldsymbol{\Sigma}_2$ is defined in Theorem 6. By Slutsky theorem, the proof is complete.

Proof of Theorem 7. It follows from Theorem 3.

Acknowledgement The authors are grateful to two referees for their insightful comments and suggestions which greatly improved the presentation of this paper.

References

1. Andrews, D.W. K. (1996). Nonparametric kernel estimation for semiparametric models. *Econometric Theory*, **11**, 560-596.

2. Brockwell, P. J. and Davis, R. A. (1991). *Time series: theory and methods.* Second edition. Springer Series in Statistics. Springer-Verlag, New York.

3. Brumback, B. and Rice, J. A. (1998). Smoothing spline models for the analysis of nested and crossed samples of curves (with discussion). *Jour. Amer. Statist. Assoc.*, **93**, 961-994.

4. Carroll, R. J., Ruppert, D. and Welsh, A. H. (1998) Local estimating equations. *Jour. Amer. Statist. Assoc.*, **93**, 214–227.

5. Carroll, R. J., Ruppert, D., Crainiceanu, C., Tosteson, T. and Karagas, M. (2004). Nonlinear and nonparametric regression and instrumental variables. *Jour. Amer. Statist. Assoc.*, **99**, 736-750.

6. CUI, H., HE, X. AND ZHU, L. (2002). On regression estimators with denoised variables. *Statist. Sinica.*, **12**, 1191-1205.

7. DELECROIX, M., HÄRDLE, W. and HRISTACHE, M. (2003). Efficient estimation in conditional single-index regression. *J. Multivariate Anal.*, **86**, 213–226.

8. ENGLE, R. F., GRANGER, W. J., RICE, J. and WEISS, A. (1986). Semiparametric estimates of the relation between weather and electricity sales. *Jour. Amer. Statist. Assoc.*, **80**, 310-319.

9. FAN, J. AND GIJBELS, I. (1996). *Local Polynomial Modelling and Its Applications.* Chapman and Hall, London.

12. FAN, J. AND HUANG, T. (2005). Profile Likelihood Inferences on semiparametric varying-coefficient partially linear models. *Bernoulli*, **11**, 1031-1057.

10. FAN, J. AND ZHANG, W. (1999). Statistical estimation in varying coefficient models. *Ann. Statist.*, **27**, 1491-1518.

11. FAN, J., YAO, Q. AND CAI, Z. (2003). Adaptive varying-coefficient linear models. *J. Roy. Stat. Soc. Ser. B.* **65**, 57–80.

13. GRAY, R. J. (1994). Spline-based tests in survival analysis. *Biometrics* , **50**, 640-652.

14. HALL, P. AND OPSOMER, J. D. (2005). Theory for penalized spline regression. *Biometrika*, **92**, 105-118.

15. HASTIE, T. J. AND TIBSHIRANI, R. J. (1990), *Generalized Additive Models,* New York: Chapman and Hall.

16. HASTIE, T. J. AND TIBSHIRANI, R. J. (1993). Varying-coefficient models. *J. Roy. Statist. Soc. B*, **55**, 757-796.

17. HOOVER, D. R., RICE, J. A., WU, C. O. AND YANG, L. P. (1998). Nonparametric smoothing estimates of time-varying coefficient models with longitudinal data. *Biometrika*, **85**, 809-822.

18. HUANG, J., WU, C. O. AND ZHOU, L. (2002). Varying-coefficient model and basis function approximations for the analysis of repeated measurements. *Biometrika*, **89**, 809-822.

19. ICHIMURA, H. (1993). Semiparametric least squares(SLS) and weighted SLS estimation of single-index models. *J. Econometrics*, **58**, 71-120.

20. JARROW, R., RUPPERT, D. AND YU, Y. (2004). Estimating the term structure of corporate debt with a semiparametric penalized spline model, *Jour. Amer. Statist. Assoc.*, **99**, 57-66.

21. LI, Q., HUANG, C., LI, D. AND FU, T. (2002). Semiparametric smooth coefficient models. *J. Business and Econ. Statist.*, **3**, 412-422.

22. RUPPERT, D. (2002). Selecting the number of knots for penalized splines. *Journal of Computational and Graphical Statistics*, **11**, 735-757.

23. RUPPERT, D. AND CARROLL, R. (1997). Penalized regression splines. Manuscript, School of Operations Research and Industrial Engineering, Cornell University, USA.

24. RUPPERT, D. AND CARROLL, R. (2001). Spatially-adaptive penalties for spline fitting. *Australian and New Zealand Journal of Statistics*, **42**, 205-223.

25. STONE, C. J. (1985), "Additive Regression and Other Nonparametric Models," *The Annals of Statistics*, **13**, 689 – 705.

25. WAND, M. P. (1999). On the optimal amount of smoothing in penalized spline regression. *Biometrika*, **86**, 936-940.

26. WU, Z. AND YU, Y. (2004). Single-Index varying coefficient models with dependent data. *Manuscript*, Department of Quantitative Analysis and Operations Management. University of Cincinnati, U.S.A.

27. XIA, Y. AND LI, W. K. (1999). On the estimation and testing of functional-coefficient linear models. *Statistica Sinica*, **9**, 737-757.

28. YU, Y. AND RUPPERT, D. (2002). Penalized spline estimation for partially linear single-index models. *Jour. Amer. Statist. Assoc.*, **97**, 1042-1054.

29. YU, Y. and RUPPERT, D. (2004). Root-n consistency of penalized spline estimator for partially linear single-index models under general Euclidean space. *Statist. Sinica*, **14**, 449-455.

30. ZHANG, W., LEE, S. Y. AND SONG, X. (2002). Local polynomial fitting in semivarying coefficient models. *J. Multivariate Anal.*, **82**, 166-188.

31. ZHOU, X. AND YOU, J. (2004). Wavelet estimation in varying-coefficient partially linear regression models. *Statist. Probab. Lett.*, **68**, 91-104.

Chapter 10

SEMI-LINEAR INDEX MODEL WHEN THE LINEAR COVARIATES AND INDICES ARE INDEPENDENT

Yun Sam Chong, Jane-Ling Wang, and Lixing Zhu

William E. Wecker Associates, Inc.
Novato, CA, U.S.A.

Department of Statistics,
University of California, Davis, CA, U.S.A.

Department of Mathematics
Hong Kong Baptist University, Hong Kong, CHINA

E-mails: chong@mail.wecker.com, wang@wald.ucdavis.edu & lzhu@hkbu.edu.hk

This chapter develops a flexible dimension-reduction model that incorporates both discrete and continuous covariates. Under this model, some covariates, $\boldsymbol{Z}$, are related to the response variable, Y, through a linear relationship, while the remaining covariates, $\boldsymbol{X}$, are related to Y through k indices which depend only on $\boldsymbol{X}'\mathbf{B}$ and some unknown function g of $\boldsymbol{X}'\mathbf{B}$. To avoid the curse of dimensionality, k should be much smaller than p. This is often realistic as the key features of a high dimensional variable can often be extracted through a low-dimensional subspace. We develop a simple approach that separates the dimension reduction stage to estimate $\mathbf{B}$ from the remaining model components when the two covariates $\boldsymbol{Z}$ and $\boldsymbol{X}$ are independent. For instance, one can apply any suitable dimension reduction approach, such as the average derivative method, projection pursuit regression or sliced inverse regression, to get an initial estimator for $\mathbf{B}$ which is consistent at the $\sqrt{n}$ rate, and then estimate the regression coefficient of $\boldsymbol{Z}$ and the link function g through a profile approach such as partial regression. All three estimates can be refined by iterating the procedure once. Such an approach is computationally simple and yields efficient estimates for both parameters at the $\sqrt{n}$ rate. We provide both theoretical proofs and empirical evidence.

Keywords: Partial regression; Single-index; Nonparametric smoothing; Dimension reduction; Projection pursuit regression; Sliced inverse regression.

1 Introduction

Kjell Doksum has made seminal contributions to *dimension reduction methods*. This includes work on transformation models [Doksum (1987), Dabrowska and Doksum (1988a), Dabrowska and Doksum (1988b), and Doksum and Gasko (1990)]. Another line of related research is the average derivative estimator (ADE) method [Doksum and Samarov (1995) and Chaudhuri, and Doksum and Samarov (1997)], where the average derivative approach is shown to be a promising dimension reduction tool. All these papers employ semiparametric models to accomplish the dimension reduction goal and to explore inference for the parametric components.

Our objective here is to explore the dimension reduction topic through a particular semiparametric model. We show that in a simple and special situation, efficiency for the parametric estimators can easily be achieved by various dimension reduction tools. The model is motivated by the fact that many dimension-reduction methods, such as projection pursuit regression (PPR), average derivative estimation method (ADE), and sliced inverse regression (SIR), assume implicitly that the predictors are continuous variables and will not work well when some of the predictors are discrete. One solution to this problem is the use of a semiparametric model where it is assumed that the response variable, Y, has a parametric relationship with some q-dimensional covariates $\boldsymbol{Z}$ (some of which may be discrete), but a nonparametric relationship with other p-dimensional covariates $\boldsymbol{X}$. If $\boldsymbol{X}$ is of high dimension, additional dimension reduction is needed and the most common approach is to assume that all information contained in $\boldsymbol{X}$ about Y is carried through a few, say k, indices. More specifically, we assume:

$$Y = \boldsymbol{Z}'\boldsymbol{\theta} + g(\boldsymbol{X}'\mathbf{B}) + e. \tag{1}$$

The function g (we call it the link function) and the $p \times k$ matrix $\mathbf{B}$ describe the dimension-reduction model through which Y and $\boldsymbol{X}$ are related, $\boldsymbol{\theta}$ is the vector of parameters describing the linear relationship between Y and $\boldsymbol{Z}$, and e is an error term. Model (1) is a semi-linear model with k indices and will be abbreviated as **SLIM** (semi-linear indices model) hereafter. Dimension reduction is accomplished because k is usually much smaller than the dimension p of $\boldsymbol{X}$. In addition, the other covariate vector $\boldsymbol{Z}$ is related to Y through a linear relation. When the link function g is unknown, the matrix $\mathbf{B}$ is not identifiable but the linear subspace spanned by it is identifiable. We thus assume hereafter that the column vectors of $\mathbf{B}$ are all of unit length with nonnegative first components.

The special case $p = 1$ has vast appeal to econometricians and is called the "partial linear model" [Engle, Granger, Rice and Weiss (1986), Heckman (1986), Rice (1986), Denby (1986), Chen(1988), Speckman (1988),

Severini and Staniswalis (1994), Bhattacharya and Zhao (1997), Hamilton and Troung (1997), Mammen and van de Geer (1997) among others]. Model (1) also includes another popular model when $q = 0$, $k = 1$, but p might be larger than 1, in which case Y and $\boldsymbol{X}$ are related through a single dimension reduction direction called the index and the resulting model is called the "single index model" in the economics literature [Stoker (1989), Härdle, Hall and Ichimura (1993), Chiou and Müller (1998, 1999), and Stute and Zhu (2005)]. In contrast to partial linear models and single index models, where hundreds of papers appeared in the literature, the results are sparse for the **SLIM** model in (1). Carroll, Fan, Gijbels, and Wand (1997) were the first to explore this topic, focusing on the case $k = 1$ with a single index. Their methods sometimes encounter numerical difficulties and this was noticed independently by Chong (1999) and Yu and Ruppert (2002). Yu and Ruppert (2002) circumvented the problem by assuming that (in addition to $k = 1$) the link function g lies in a known, finite-dimensional spline space, yielding a flexible parametric model. The approach in Chong (1999) is different and completely nonparametric, employing a local polynomial smoother to estimate g. Moreover, the number of indices k is not restricted to be 1 or even known, and is being estimated along the way.

We consider in this paper that a random sample of n observations are collected, and use $\mathbf{y} = (y_1, \ldots, y_n)'$ to denote the vector of observed responses and

$$\mathbf{Z} = \begin{bmatrix} z_{11} & \cdots & z_{1q} \\ \vdots & & \vdots \\ z_{n1} & \cdots & z_{nq} \end{bmatrix} \text{ and } \mathbf{X} = \begin{bmatrix} x_{11} & \cdots & x_{1p} \\ \vdots & & \vdots \\ x_{n1} & \cdots & x_{np} \end{bmatrix}$$

to represent the observed values of $\boldsymbol{Z}$ and $\boldsymbol{X}$, with the first subscript representing the observation number and the second subscript representing the position in the array of variables. Restating equation (1) to reflect the observations we obtain,

$$\mathbf{y} = \mathbf{Z}\boldsymbol{\theta} + g(\mathbf{XB}) + \boldsymbol{\epsilon},$$

where $\boldsymbol{\epsilon} = (e_1, \ldots, e_n)'$ is the $n \times 1$ vector of observed errors.

Our goal is to show that simple and non-iterative algorithms are available when the two covariates $\boldsymbol{Z}$ and $\boldsymbol{X}$ are independent of each other, and the procedures yield efficient estimators for parameters $\boldsymbol{\theta}$ and $\boldsymbol{\beta}$. The independence assumption could be fulfilled, for instance, in clinical trials when $\boldsymbol{Z}$ represents the type of treatments and patients are assigned to treatments randomly. Other examples of independent $\boldsymbol{Z}$ and $\boldsymbol{X}$ are plentiful such as in social studies where participants are assigned to different groups randomly and $\boldsymbol{Z}$ represents the group indicators. Specifically in this paper, we show that our procedures provide adaptive estimates for $\mathbf{B}$, in the sense that the asymptotic variance of the estimator for $\mathbf{B}$ is equal to that of an estimator that assumes a known link function. We also show that $\boldsymbol{\theta}$ can be estimated as efficiently as when g and $\mathbf{B}$ are both known. Finally, we illustrate our procedures through simulation studies and show that they compare favorably with the procedure in Carroll et al. (1997). We note here that the algorithms in Section 2 were first reported in the Ph.D. thesis of Chong (1999), and have not been published previously. Moreover, the asymptotic results reported here are different from and more general than those obtained in Chong (1999). For instance, Theorem 2 in Section 2 is for a different and better estimator than the one in Theorem 2 of Chong (1999), and Theorem 3 in Section 2 is completely new.

2 Main Results

Hereafter, we assume that $\boldsymbol{X}$ and $\boldsymbol{Z}$ are independent. Consequently, we may consider $\boldsymbol{Z}'\boldsymbol{\theta}$ of equation (1) to be part of the error term and use only the values of Y and $\boldsymbol{X}$ to obtain an estimate of $\mathbf{B}$. The theorem below shows that we can obtain a $\sqrt{n}$-consistent estimate for $\mathbf{B}$ when we apply the sliced inverse regression (SIR) method in Li (1991) to Y and $\boldsymbol{X}$, if the following linear condition is satisfied:

$$\text{for any } \boldsymbol{b} \in \Re^p, E(\boldsymbol{X}'\boldsymbol{b}|\boldsymbol{X}'\mathbf{B}) \text{ is linear in } \boldsymbol{X}'\boldsymbol{\beta}_1, \ldots, \boldsymbol{X}'\boldsymbol{\beta}_k. \tag{2}$$

Theorem 1. *Under condition (2), $E(\boldsymbol{X}|Y) - E(\boldsymbol{X}) \propto \boldsymbol{\Sigma}_x\mathbf{B}\boldsymbol{a}^*$ for some $\boldsymbol{a}^* \in \Re^k$, where $\boldsymbol{\Sigma}_x$ is the covariance matrix of $\boldsymbol{X}$, and "$\propto$" stands for "proportional to".*

The proof is similar to the one in Li (1991) and follows from $E(\boldsymbol{X}|Y) = E(E(\boldsymbol{X}|\boldsymbol{X}'\mathbf{B}, \boldsymbol{Z}'\boldsymbol{\theta}, e)|Y) = E(E(\boldsymbol{X}|\boldsymbol{X}'\mathbf{B})|Y)$, where the last equality follows from the fact that $\boldsymbol{X}$ is independent of $\boldsymbol{Z}$ and e. It then follows that $E(\boldsymbol{X}|Y) - E(\boldsymbol{X}) = E(\boldsymbol{\Sigma}_x\mathbf{B}\boldsymbol{a}^*|Y)$ for some $\boldsymbol{a}^* \in \Re^k$. Details of the proof will not be presented here as they can be found in Section 5.3 of Chong (1999). We have thus shown that SIR, proposed in Li (1991) and reviewed in Chen and Li (1998), can be employed to estimate $\mathbf{B}$. Variants of SIR, such as SIR II [Li (1991)], SAVE [Cook and Weisberg (1991)], PHD [Li (1992)] etc., are also feasible in case SIR fails. All these SIR based procedures are simple as they do not involve smoothing and separate the dimension reduction stage from the model fitting stage. Li (1991) and Zhu and Ng (1995) states that SIR yields a $\sqrt{n}$-consistent estimate for $\mathbf{B}$, when $\boldsymbol{Z}$ is not present in model. Zhu and Fang (1996) used kernel estimation, where the $\sqrt{n}$-consistency also holds. We show in *Theorem 3* that these results continue to hold when $\boldsymbol{Z}$ is independent of $\boldsymbol{X}$ as in our setting.

By the same token, the average derivative method (ADE) can also be applied under certain smoothness conditions as described in Härdle and Stoker (1989) and Samarov (1993). The resulting estimate would be $\sqrt{n}$ consistent like SIR and it has the same advantage as SIR (or its variants) in that it separates the dimension reduction stage from the model fitting step. A relevant method, the Outer Product of Gradients estimation proposed by Xia, Tong, Li and Zhu (2002) can also be applied. While SIR relies on the linear conditional mean design condition (2), it is simpler to implement than ADE, as the latter involves the estimation of the derivative of g. These two different approaches compliment each other as dimension reduction tools.

If the additivity assumption is satisfied in the projection pursuit regression (PPR) model in Friedman and Stuetzle (1981), one can also employ the PPR-estimators for $\mathbf{B}$, which were shown to be $\sqrt{n}$-consistent in Hall (1989). Hristache, Juditsky and Spokoiny (2001) provided a new class of $\sqrt{n}$-consistent estimators. These projection pursuit type estimators typically yield more efficient initial estimators for $\mathbf{B}$ than SIR or ADE since PPR utilizes the additive model structure and attempts to estimate $\mathbf{B}$ iteratively while estimating the unknown link function. However, ADE and SIR (or its variants) have the advantage that they rely on no model assumption, separate the dimension reduction stage from model fitting, and are thus computationally simpler and more robust than the PPR approach.

2.1 *Estimation of θ*

There are two ways to estimate $\boldsymbol{\theta}$:

(1) *Procedures starting with dimension reduction*: Start with a dimension-reduction procedure to obtain an estimate $\hat{\mathbf{B}}$ for $\mathbf{B}$ and then follow the steps of the partially linear model, using $\boldsymbol{X}'\hat{\mathbf{B}}$ instead of the unknown $\boldsymbol{X}'\mathbf{B}$ to estimate $\boldsymbol{\theta}$.

(2) *Procedures starting with initial estimation of the linear component*: Because $\boldsymbol{Z}$ and $\boldsymbol{X}$ are independent, linear regression of Y on $\boldsymbol{Z}$ will yield a consistent estimate of $\boldsymbol{\theta}$. The linear regression procedure is computationally simple, so we may start with this initial estimate of $\boldsymbol{\theta}$ and use it to improve the dimension-reduction step above.

When using the partial linear model to estimate $\boldsymbol{\theta}$, there are two common approaches based on either partial splines, as in Wahba (1984), or the partial regression proposed independently by Denby (1986) and Speckman (1986). The partial regression method is a profile approach so it is also referred to as the profile estimator in the literature. Simulation results in Chapter 6 of Chong (1999) suggest that the two procedures provide numerically equivalent estimators under the independence assumption, but the partial spline procedure might be biased when the independence assumption is violated as demonstrated in Rice (1986). There is thus no advantage to employ the partial spline procedure in our setting and we recommend the use of the partial regression estimator, even though partial spline estimators would be fine when $\boldsymbol{Z}$ and $\boldsymbol{X}$ are independent as reported in Heckman (1986).

The partial regression stage involves a smoothing method to estimate the unknown link function g. The choice of smoother can be subjective; we employed the local polynomial smoother due to its appealing properties as reported in Fan (1993). This results in a linear smoother in the sense that we may construct a smoothing matrix $\mathbf{S}$ such that $\mathbf{Su}$ represents the result of smoothing a vector of generic observations $\mathbf{u}$ using the linear smoother $\mathbf{S}$. Details are given in Appendix B. Below we use the partial regression procedure to estimate $\boldsymbol{\theta}$ and provide the algorithms for each of the approaches above.

Algorithm for Procedure 1 which begins with dimension reduction:

(i) Apply a dimension-reduction procedure to $\mathbf{X}$ and $\mathbf{y}$ to obtain an estimate $\hat{\mathbf{B}}$ of $\mathbf{B}$.
(ii) Use $\mathbf{X}\hat{\mathbf{B}}$ to obtain a smoothing matrix $\mathbf{S}$.
(iii) Take $\hat{\boldsymbol{\theta}} = (\mathbf{Z}'(\mathbf{I}-\mathbf{S})'(\mathbf{I}-\mathbf{S})\mathbf{Z})^{-1}\mathbf{Z}'(\mathbf{I}-\mathbf{S})'(\mathbf{I}-\mathbf{S})\mathbf{y}$ to be the estimate for $\boldsymbol{\theta}$.

Algorithm for Procedure 2 which starts with an initial estimator of the linear component:

(i) Apply least squares to $\mathbf{Z}$ and $\mathbf{y}$ to obtain an initial estimate $\hat{\boldsymbol{\theta}}_0$ of $\boldsymbol{\theta}$.
(ii) Apply a dimension-reduction procedure to $\mathbf{X}$ and $\mathbf{y}-\mathbf{Z}'\hat{\boldsymbol{\theta}}_0$ to obtain an estimate $\hat{\mathbf{B}}$ of $\mathbf{B}$.
(iii) Use $\mathbf{X}\hat{\mathbf{B}}$ to obtain a smoothing matrix $\mathbf{S}$.
(iv) Take $\hat{\boldsymbol{\theta}}_1 = (\mathbf{Z}'(\mathbf{I}-\mathbf{S})'(\mathbf{I}-\mathbf{S})\mathbf{Z})^{-1}\mathbf{Z}'(\mathbf{I}-\mathbf{S})'(\mathbf{I}-\mathbf{S})\mathbf{y}$ to be the revised estimate for $\boldsymbol{\theta}$.

Simulation results in Section 3 suggest that Procedure 2 which uses the residuals to perform the dimension reduction step is slightly more efficient than Procedure 1. We thus present the asymptotic distribution of $\hat{\boldsymbol{\theta}}_1$ based on Procedure 2 only. Note that, following Theorem 1 or the discussions afterwards at the end of Section 1, many initial $\sqrt{n}$-consistent estimators of $\mathbf{B}$ exist. We thus make such an assumption in the following theorem.

Let $\boldsymbol{Z} = (Z_1, \cdots, Z_q)'$ be centered at 0.

Theorem 2. *Under conditions (1)–(11), listed in Appendix A, and* $\|\hat{\mathbf{B}} - \mathbf{B}\| = O_P(n^{-1/2})$,

$$\sqrt{n}(\hat{\boldsymbol{\theta}}_1 - \boldsymbol{\theta}) \Rightarrow N(0, \mathbf{E}^{-1}\sigma^2)$$

for the partial regression estimate in Procedure 2 using residuals, where

$$\mathbf{E} = \begin{bmatrix} E(Z_1^2) & \cdots & E(Z_1 Z_q) \\ \vdots & & \vdots \\ E(Z_1 Z_q) & \cdots & E(Z_q^2) \end{bmatrix},$$

and $E(Z_i) = 0$ *for all* i.

In other words, when we start with a $\sqrt{n}$-consistent estimate for $\mathbf{B}$, the resulting $\hat{\boldsymbol{\theta}}_1$ is consistent for $\boldsymbol{\theta}$ with the same efficiency as an estimate that we would obtain if we knew $\mathbf{B}$ and g. This illustrates the adaptiveness of $\hat{\boldsymbol{\theta}}_1$; no iteration is required and many $\sqrt{n}$-consistent estimators for $\mathbf{B}$ exist.

2.2 *Estimation of* B *and* g

In addition to $\boldsymbol{\theta}$, we may also be interested in estimating $\mathbf{B}$ and g. Although both procedures in Section 2.1 involve the estimation of $\mathbf{B}$, we will generally want a more refined estimate. For instance, after obtaining $\hat{\boldsymbol{\theta}}_1$, we can obtain a revised estimate $\hat{\mathbf{B}}_2$ for $\mathbf{B}$, and then an estimate for g by smoothing $Y - \boldsymbol{Z}'\hat{\boldsymbol{\theta}}_1$ on $\boldsymbol{X}'\hat{\mathbf{B}}_2$.

(i) Apply a dimension-reduction procedure to $\mathbf{X}$ and $\mathbf{y} - \mathbf{Z}\hat{\boldsymbol{\theta}}_1$ to obtain a revised estimate $\hat{\mathbf{B}}_2$ of $\mathbf{B}$.
(ii) Use $\mathbf{X}\hat{\mathbf{B}}_2$ to obtain a smoothing matrix $\mathbf{S}$.
(iii) Let $\hat{g}(\mathbf{X}\hat{\mathbf{B}}_2) = \mathbf{S}(\mathbf{y} - \mathbf{Z}\hat{\boldsymbol{\theta}}_1)$ be the estimate for g.

Next, we present the asymptotic results for the parameters defining the indices. Theorem 3 below demonstrates optimality in the sense that the asymptotic variance of $\hat{\mathbf{B}}_2$ is equal to the nonlinear least squares estimator that is obtained when the link function $g(\cdot)$ is known and when the linear part $\boldsymbol{Z}'\boldsymbol{\theta}$ is absent in the model. That is, the impact of nonparametric estimation of $g(\cdot)$ and the linear part $\boldsymbol{Z}'\boldsymbol{\theta}$ is negligible asymptotically. Let

$$\mathbf{W} = \int \Big\{\boldsymbol{X} - E(\boldsymbol{X}|\boldsymbol{X}'\mathbf{B})\Big\}\Big\{\boldsymbol{X} - E(\boldsymbol{X}|\boldsymbol{X}'\mathbf{B})\Big\}'(g'(\boldsymbol{X}'\mathbf{B}))^2 f_{\boldsymbol{X}}(\boldsymbol{X})d\,\boldsymbol{X},$$

and $\mathbf{W}^-$ denotes its generalized inverse, where $f_{\boldsymbol{X}}$ is the density function of the p-dimensional vector, $\boldsymbol{X}$.

Theorem 3. *Under conditions (1) - (11) stated in Appendix A, and in addition for $h = O(n^{-1/(4+k)})$ and any unit-vector $u \neq \mathbf{B}$, we have*

$$n^{1/2}u'(\hat{\mathbf{B}} - \mathbf{B}) \Longrightarrow N(0, u'\sigma^2(\mathbf{W}^-)\mathbf{u}).$$

2.3 *Iterated estimate of $\boldsymbol{\theta}$*

While the estimator for $\boldsymbol{\theta}$ in Section 2.1 is already asymptotically efficient, it might be improved in the finite sample case by iterating the algorithm. For instance, following the steps of the previous section and after obtaining estimates for $\mathbf{B}$ and g, one can use partial regression to obtain the revised estimate for $\boldsymbol{\theta}$.

- Let the partial regression estimate $\hat{\boldsymbol{\theta}}_2 = (\mathbf{Z}'(\mathbf{I}-\mathbf{S})'(\mathbf{I}-\mathbf{S})\mathbf{Z})^{-1}\mathbf{Z}'(\mathbf{I}-\mathbf{S})'(\mathbf{I}-\mathbf{S})\mathbf{y}$ be the revised estimate for $\boldsymbol{\theta}$.

The simulation studies in Section 3 indicate some improvement by adding one iteration.

3 Simulations

In this section we check the numerical performance of the procedures in Section 2 and compare it to the **GPLSIM** algorithm of Carroll, Fan, Gijbels, and Wand (1997). We consider two different models: a linear model,

$$Y = 2 + \boldsymbol{X}'\mathbf{B} + \boldsymbol{Z\theta} + 0.3\ e, \tag{3}$$

and a quadratic model

$$Y = (2 + \boldsymbol{X}'\mathbf{B})^2 + \boldsymbol{Z\theta} + 0.3\ e. \tag{4}$$

In each model, $\boldsymbol{\theta}$ is a scalar with value 1. The variable $\boldsymbol{Z}$ will be a single binary variable with values 0 and 1, and $\boldsymbol{Z} = 1$ with probability 1/2. The e's are standard normal, and $\mathbf{B}$ is the vector $(0.75, 0.5, -0.25, -0.25, 0.25)'$. The $\boldsymbol{X}$'s are standard multivariate normal, with mean $(0,0,0,0,0)'$ and covariance $\mathbf{I}_5$. Thus, in these simulations the assumption on the distribution of $\boldsymbol{X}$ is satisfied for both projection pursuit regression and sliced inverse regression, and we focus on these two dimension reduction methods. The average derivative method can also be used at additional computational cost. Although the simulations shown here use $\boldsymbol{X}$ with independent components, we also ran simulations that have a correlation structure on $\boldsymbol{X}$. The results for those simulations were not much different from those shown here.

We ran $N = 100$ simulations each on the linear model and the quadratic model, and the sample size is $n = 100$ in both cases. The link function is estimated with a local linear smoother as defined in Appendix B, and the bandwidths are chosen by a generalized cross-validation method. Here the generalized cross-validation procedure may be preferred due to its computational advantage over the least squares cross validation method. Simulation results not reported here, which can be found in Chapter 6 of Chong (1999), show that the two-cross validation methods yielded very similar results. We thus only report the findings based on generalized cross-validation [Craven and Wahba (1979)], defined in equation (17) of Appendix B.

The performance of the two types of partial regression estimators, with and without an initial estimate of $\boldsymbol{\theta}$, were compared using two types of dimensions reduction tools, the PPR and SIR with 2, 5, 20, and 20 elements per slice. The results of estimating $\boldsymbol{\theta}$ for the linear and quadratic model are reported in Table 1 and Table 2 respectively.

We find that, as expected, PPR generally outperforms SIR, but only slightly. With only one iteration, the estimators in section 2.3 are nearly as efficient as the one with $\mathbf{B}$ known regardless of which dimension reduction method has been employed. Iteration helps the estimator without an initial estimate of $\boldsymbol{\theta}$ much more than the one with an initial estimate. This suggests also that further iteration will not improve the estimation of $\boldsymbol{\theta}$ much . We also compared the performance of the dimension reduction estimators in Section 2.2, but due to space limitation, these results are not reported here. Details of additional simulations can be found in Chong (1999).

In both simulations we tried to compare our approach with the **GPLSIM** algorithm in Carroll et. al (1997) but were unable to obtain any meaningful results for their procedure due to computational difficulties, triggered possibly by the relatively high dimension of $\boldsymbol{X}$. The minimization in the **GPLSIM** is now for a five-dimensional vector $\mathbf{B}$ and a scalar $\boldsymbol{\theta}$, whereas it is for a three-dimensional $\mathbf{B}$ and a scalar $\boldsymbol{\theta}$ in the simulation model (5) in that paper. We thus instead adopt the simulation model presented in that article. The simulation has $n = 200$ with $N = 100$ simulations based on the model

$$Y_i = \sin(\pi \frac{\mathbf{B}\boldsymbol{X}_i - A}{B - A}) + \boldsymbol{\theta}\boldsymbol{Z}_i + e_i, \tag{5}$$

with $A = \sqrt{3}/2 - 1.645/\sqrt{12}$, $B = \sqrt{3}/2 + 1.645/\sqrt{12}$, $\boldsymbol{X}_i$ distributed as a uniform variable on the cube $[0,1]^3$, $\boldsymbol{Z}_i = 0$ for i odd, $\boldsymbol{Z}_i = 1$ for i even, and $e_i \sim N(0, \sigma^2 = 0.01)$. The parameters are $\mathbf{B} = (1/\sqrt{3}, 1/\sqrt{3}, 1/\sqrt{3})'$ and $\boldsymbol{\theta} = 0.3$.

Since the design is nearly symmetric, we do not use the SIR procedure here (Li (1991)) and only PPR was employed to estimate $\mathbf{B}$. Again, when implementing the program for **GPLSIM** we encountered difficulties. The

Table 1 *Estimates of* $\boldsymbol{\theta}$ *in the linear model (3)* with (two columns in the right panel) or without (two columns in the central panel) initial estimates for $\boldsymbol{\theta}$. The fourth and sixth columns are the resulting estimates for $\boldsymbol{\theta}$ after one iteration of the estimates on the third and fifth columns respectively. The subscript for SIR in the first column stands for the number of slices used.

		no initial $\hat{\boldsymbol{\theta}}$	iteration with no initial $\hat{\boldsymbol{\theta}}$	with initial $\hat{\boldsymbol{\theta}}$	iteration with initial $\hat{\boldsymbol{\theta}}$
PPR	Bias	-0.0411	-0.0029	-0.0013	-0.0007
	SD	0.0623	0.0581	0.0603	0.0592
	MSE	0.00557	0.00338	0.00363	0.0035
SIR_2	Bias	-0.0459	-0.005	-0.0058	-0.0024
	SD	0.0708	0.0606	0.0618	0.061
	MSE	0.00712	0.0037	0.00385	0.00373
SIR_5	Bias	-0.0447	-0.0047	-0.0008	-0.0007
	SD	0.0621	0.0606	0.064	0.0623
	MSE	0.00585	0.00369	0.0041	0.00388
SIR_{10}	Bias	-0.0423	-0.0034	-0.0023	-0.0003
	SD	0.0655	0.06	0.0624	0.0603
	MSE	0.00608	0.00361	0.0039	0.00364
SIR_{20}	Bias	-0.0441	-0.0032	0.0006	-0.001
	SD	0.065	0.0612	0.065	0.0599
	MSE	0.00617	0.00376	0.00423	0.00358
B known	Bias	0.0025	0.0025		
	SD	0.0564	0.0564		
	MSE	0.00318	0.00318		

initial estimates for both **B** and $\boldsymbol{\theta}$ seem crucial, so we decided to use our estimates of **B** and $\boldsymbol{\theta}$ in Sections 2.1 and 2.2 as the initial estimates for the **GPLSIM** procedure, and then iterate our procedure once to make both procedures comparable as **GPLSIM** utilizes the same initial estimates. We used only procedure 1 in this simulation since the results in Tables 1 and 2 show no benefit using the initial estimator for $\boldsymbol{\theta}$ in procedure 2 if we iterate once for both estimates of **B** and $\boldsymbol{\theta}$ using our procedures. The results for the three procedures, (a) our estimator in Section 2.1, (b) **GPLSIM** using our estimator as the initial estimator in its iterated algorithm, and (c) one iteration of our procedure as described in Section 2.3, are reported in the last three rows of Table 3. For comparison with the optimal procedure, we also include in the first row the nonlinear least squares (NLS) procedure available in S-PLUS, which use the true link function g. Relative efficiencies with respect to this optimal procedure are reported for all three procedures in the last column of Table 3. To save computing time, we used the same bandwidth (based on generalized cross-validation) for the iteration as for the first partial regression step. The **GPLSIM** procedure uses a plug-in method for estimating the bandwidth. The results in Table 3 suggest that the iterated PPR estimators outperform those from **GPLSIM** slightly. Note that both approaches utilize the PPR estimators in the second row as initial estimators. Our procedures are computationally much more stable and simpler than **GPLSIM**.

Table 2 *Estimates of $\boldsymbol{\theta}$ as in Table except that the quadratic model (4) is used.*

		no initial $\hat{\boldsymbol{\theta}}$	iteration with no initial $\hat{\boldsymbol{\theta}}$	with initial $\hat{\boldsymbol{\theta}}$	iteration with initial $\hat{\boldsymbol{\theta}}$
PPR	Bias	-0.0466	-0.0043	-0.005	-0.0009
	SD	0.0625	0.0586	0.0786	0.059
	MSE	0.00608	0.00345	0.0062	0.00348
SIR_2	Bias	-0.0385	-0.0002	-0.002	0.0043
	SD	0.0915	0.0767	0.1011	0.077
	MSE	0.00985	0.00588	0.0102	0.00595
SIR_5	Bias	-0.0391	-0.004	-0.0017	-0.0002
	SD	0.0731	0.0731	0.0974	0.0707
	MSE	0.00688	0.00536	0.00949	0.005
SIR_{10}	Bias	-0.0425	-0.0022	-0.0049	0.0001
	SD	0.086	0.0815	0.1021	0.0808
	MSE	0.0092	0.00665	0.0105	0.00653
SIR_{20}	Bias	-0.04	-0.0068	-0.0062	-0.0078
	SD	0.0853	0.0887	0.0993	0.0885
	MSE	0.00887	0.00791	0.0099	0.00789
B known	Bias	0.0006	0.0006		
	SD	0.0571	0.0571		
	MSE	0.00326	0.00326		

4 Conclusions

We have demonstrated that when $\boldsymbol{X}$ and $\boldsymbol{Z}$ are independent, the estimation of the dimension-reduction direction $\mathbf{B}$ is straightforward and much simpler algorithms than those in the literature are available. Consequently, the problem to estimate the linear parameter, $\boldsymbol{\theta}$, is equivalent to the corresponding problem in the partially linear model, in the sense that the same efficiency can be attained as in the partial linear model which assumes a known $\mathbf{B}$. In addition, we show that the indices, $\mathbf{B}$, in the semiparametric index components can also be estimated optimally. The theoretical results presented in Theorems 2 and 3 here improve upon those in Carroll et. al (1977), where the asymptotic distributions of both estimates for $\mathbf{B}$ and $\boldsymbol{\theta}$ were derived under the additional stringent assumption that those estimators are already known to be $\sqrt{n}$-consistent. We show that such an assumption can be dispensed with when $\boldsymbol{X}$ and $\boldsymbol{Z}$ are independent.

Table 3 *Comparison of our procedures with **GPLSIM** using the model in Carroll et. al (1997), where the second row corresponding to NLS gives the results of the nonlinear least square estimates when the link function g is known. The third row marked by PPR corresponds to the results from our procedure 1 in section 2.1 when PPR is used to estimate* **B**. *This estimate is also used as the initial estimator for **GLPSIM** (reported in the fourth row) and the iterated PPR in section 2.3 (reported in the fifth row). The last column gives the relative efficiency of NLS to the other three estimates.*

Estimate of $\boldsymbol{\theta}$	Mean	SD	MSE	Relative efficiency
NLS	0.3007	0.0106	0.000114	1
PPR	0.2945	0.0229	0.000556	4.89
GPLSIM	0.3053	0.0165	0.000302	2.66
PPR- iterated	0.3	0.0165	0.000273	2.4

Acknowledgments

Wang's research was supported in part by NSF grants DMS-0406430 and DMS-98-03627. Zhu's research was supported by the grants (HKU7181/02H and HKU7060/04P) from The Research Grants Council of Hong Kong.

APPENDIX A: Proofs

We first present the assumptions for the theorems. When $k > 1$, a product kernel with marginal kernels each satisfying the assumed conditions (7)-(11) below should be employed.

(1) $E(e) = 0$, $Var(e) = \sigma^2 < \infty$.
(2) $E(\boldsymbol{Z}) = 0$, $E(\|\boldsymbol{Z}\|^2) < \infty$.
(3) $h = \text{const} \cdot n^{-a}$, where $0 < a < \frac{1}{k+2}$.
(4) g is twice differentiable, with the second derivative bounded and continuous.
(5) The density function, $f_{\boldsymbol{X}} : \Re^p \to \Re$, of the p-dimensional random vector $\boldsymbol{X}$ is twice differentiable with the second derivative bounded and continuous.
(6) $f_{\boldsymbol{X}}$ is bounded away from zero.

(7) K is Lipschitz continuous on the real line.
(8) K has support $[-1, 1]$.
(9) $K(u) \geq 0$ for all u and $\int_{-1}^{1} K(u)du = 1$.
(10) $\int_{-1}^{1} uK(u)du = 0$.
(11) $\int_{-1}^{1} u^2 K(u)du = M_K \neq 0$.

Remark: Assumptions (1) and (2) are necessary conditions for the asymptotic normality of an estimator. Assumption (3) is commonly used in nonparametric estimation. Assumptions (4) and (5) are also common conditions. Assumptions (5) and (6) imply that the distribution of $\boldsymbol{X}$ has bounded support and that $f_{\boldsymbol{X}}$ is bounded from above. With assumption (4), we can also conclude that g is bounded from above. These conditions are used to avoid boundary effects when a nonparametric smoother is employed to construct an estimator of a nonparametric regression function. All conditions on the kernel function are commonly used in the literature. Therefore, the imposed conditions are mild.

Without loss of generality and for simplicity, we will focus our proof on the single-index model with $k = 1$, although the proof can be extended to multiple-indices models. For this simplification, we will use a vector $\boldsymbol{\beta}$ instead of the matrix $\mathbf{B}$ to describe the relationship between Y and $\boldsymbol{X}$ when $k = 1$. This is a partially linear single-index model, given by the equation

$$Y = \boldsymbol{Z}'\boldsymbol{\theta} + g(\boldsymbol{X}'\boldsymbol{\beta}) + e.$$

We divide the tedious proof of Theorem 2 into five Lemmas. Suppose we observe the data $(Y_j, \boldsymbol{X}_j, \boldsymbol{Z}_j)$, $j = 1, \ldots, n$, where $\boldsymbol{X}_j \in \Re^p$ and $\boldsymbol{Z}_j \in \Re^q$. Consequently, $\boldsymbol{\beta} \in \Re^p$ and $\boldsymbol{\theta} \in \Re^q$. For $i_1 = 0, 1$, $i_2 = 0, 1, 2$, and any $\boldsymbol{\beta}^* \in \Re^p$ define

$$\xi_j^{i_1,i_2}(\mathbf{x}, \boldsymbol{\beta}^*) = Y_j^{i_1} K_h((\boldsymbol{X}_j - \mathbf{x})'\boldsymbol{\beta}^*)((\boldsymbol{X}_j - \mathbf{x})'\boldsymbol{\beta}^*)^{i_2}$$

$$\alpha_n^{i_1,i_2}(\mathbf{x}, \boldsymbol{\beta}^*) = n^{-1} \sum_{j=1}^{n} [\xi_j^{i_1,i_2}(\mathbf{x}, \boldsymbol{\beta}^*) - E(\xi_j^{i_1,i_2}(\mathbf{x}, \boldsymbol{\beta}^*))].$$

Lemma 1. *Under conditions (1)-(8), for $i_1 = 0, 1$, $i_2 = 0, 1, 2$, and $\|\boldsymbol{\beta}^* - \boldsymbol{\beta}\| = O(n^{-\frac{1}{2}})$,*

$$\sup_{\mathbf{x} \in \Re^p} \sup_{\boldsymbol{\beta}^* : \|\boldsymbol{\beta}^* - \boldsymbol{\beta}\| = O(n^{-1/2})} \sqrt{nh} |\alpha_n^{i_1,i_2}(\mathbf{x}, \boldsymbol{\beta}^*) - \alpha_n^{i_1,i_2}(\mathbf{x}, \boldsymbol{\beta})| \xrightarrow{P} 0. \quad (6)$$

Proof. In order to use arguments such as those provided in the proof of Theorem II. 37 in Pollard (1984, pages 34-35), we first show that for any $\epsilon > 0$, $\mathbf{x} \in \Re^p$, $\boldsymbol{\beta}^* \in \Re^p$, $\|\boldsymbol{\beta}^* - \boldsymbol{\beta}\| = O(n^{-\frac{1}{2}})$, and large n,

$$P\left(\sqrt{nh} |\alpha_n^{i_1,i_2}(\mathbf{x}, \boldsymbol{\beta}^*) - \alpha_n^{i_1,i_2}(\mathbf{x}, \boldsymbol{\beta})| > \frac{\epsilon}{2}\right) \leq \frac{1}{2}. \quad (7)$$

This is in preparation to apply the symmetrization approach in Pollard (1984, pages 14-16). The left-hand side of (7) is equal to

$$P\left(|\alpha_n^{i_1,i_2}(\mathbf{x},\boldsymbol{\beta}^*) - \alpha_n^{i_1,i_2}(\mathbf{x},\boldsymbol{\beta})| > \frac{\epsilon}{2\sqrt{nh}}\right),$$

which is less than or equal to

$$\frac{4nh}{\epsilon^2}E\{[\alpha_n^{i_1,i_2}(\mathbf{x},\boldsymbol{\beta}^*) - \alpha_n^{i_1,i_2}(\mathbf{x},\boldsymbol{\beta})]^2\}$$

by Chebychev's inequality. We now prove that this value is less than or equal to 1/2. Define

$$A(\boldsymbol{\beta}) = K_h((\boldsymbol{X}-\mathbf{x})'\boldsymbol{\beta})((\boldsymbol{X}-\mathbf{x})'\boldsymbol{\beta})^{i_2}$$
$$A(\boldsymbol{\beta}^*) = K_h((\boldsymbol{X}-\mathbf{x})'\boldsymbol{\beta}^*)((\boldsymbol{X}-\mathbf{x})'\boldsymbol{\beta}^*)^{i_2}.$$

Recalling the definition of $\alpha_n^{i_1,i_2}$ and the independence of $\xi_j^{i_1,i_2}$, an elementary calculation yields that

$$\begin{aligned}\frac{4nh}{\epsilon^2}E\{[\alpha_n^{i_1,i_2}(\mathbf{x},\boldsymbol{\beta}^*) - \alpha_n^{i_1,i_2}(\mathbf{x},\boldsymbol{\beta})]^2\} &= \frac{4h}{\epsilon^2}Var[\xi^{i_1,i_2}(\mathbf{x},\boldsymbol{\beta}^*) - \xi^{i_1,i_2}(\mathbf{x},\boldsymbol{\beta})] \\ &\le \frac{4h}{\epsilon^2}E\{Y^{2i_1}[A(\boldsymbol{\beta}^*) - A(\boldsymbol{\beta})]^2\}. \qquad (8)\end{aligned}$$

If $i_1 = 0$, then $Y^{2i_1} = 1$. Let M_{f_X} be the upper bound of $f_{\boldsymbol{X}}$. Define $B(\boldsymbol{\beta}^*) = K((\boldsymbol{t}-\mathbf{x})'\boldsymbol{\beta}^*/h)$, $B(\boldsymbol{\beta}) = K((\boldsymbol{t}-\mathbf{x})'\boldsymbol{\beta}/h)$, and let $T = \{\boldsymbol{t} \in \Re^p : B(\boldsymbol{\beta}^*) > 0 \text{ or } B(\boldsymbol{\beta}) > 0\}$. Then the right hand side of (8) is equal to

$$\begin{aligned}&\frac{4h}{\epsilon^2}\int_{\boldsymbol{t}\in T}\left[\frac{1}{h}B(\boldsymbol{\beta}^*)((\boldsymbol{t}-\mathbf{x})'\boldsymbol{\beta}^*)^{i_2} - \frac{1}{h}B(\boldsymbol{\beta})((\boldsymbol{t}-\mathbf{x})'\boldsymbol{\beta})^{i_2}\right]^2 f_{\mathbf{X}}(\boldsymbol{t})d\boldsymbol{t} \\ &= \frac{4h^{2i_2}}{h\epsilon^2}\int_{\mathbf{x}+h\boldsymbol{u}\in T}[K(\boldsymbol{u}'\boldsymbol{\beta}^*)(\boldsymbol{u}'\boldsymbol{\beta}^*)^{i_2} - K(\boldsymbol{u}'\boldsymbol{\beta})(\boldsymbol{u}'\boldsymbol{\beta})^{i_2}]^2 f_{\mathbf{X}}(\mathbf{x}+h\boldsymbol{u})d(h\boldsymbol{u}) \\ &\le \frac{4h^{2i_2}h^p}{h\epsilon^2}CM_{f_X}\int_U(\boldsymbol{u}'(\boldsymbol{\beta}^*-\boldsymbol{\beta}))^2 d\boldsymbol{u}, \text{ where } U = \{\boldsymbol{u}\in\Re^p : \mathbf{x}+h\boldsymbol{u}\in T\} \\ &= \frac{4h^{2i_2}h^p}{h\epsilon^2}CM_{f_X}O(h^{-p})O(h^{-2}n^{-1}) = \frac{h^{2i_2}}{\epsilon^2}O\left(\frac{1}{nh^3}\right). \qquad (9)\end{aligned}$$

Then, if $i_1 = 1$, with the independence of $\boldsymbol{X}, \boldsymbol{Z}$ and e, (8) is equal to

$$\begin{aligned}\frac{4h}{\epsilon^2}E\left\{Y^2\left[A(\boldsymbol{\beta}^*) - A(\boldsymbol{\beta})\right]^2\right\} &= \frac{8h}{\epsilon^2}E\left\{g^2(\boldsymbol{X}'\boldsymbol{\beta})\left[A(\boldsymbol{\beta}^*) - A(\boldsymbol{\beta})\right]^2\right\} \\ &+\frac{8h}{\epsilon^2}\left[E(\boldsymbol{Z}'\boldsymbol{\theta})^2 + E(e^2)\right]\times E\left\{\left[A(\boldsymbol{\beta}^*) - A(\boldsymbol{\beta})\right]^2\right\}. \qquad (10)\end{aligned}$$

The second term of (10) has the same order as (8), because $E(\boldsymbol{Z}'\boldsymbol{\theta})^2+E(e^2)$ is bounded. Similar to (9), the first term of (10) can be bounded as follows:

$$\frac{8}{h\epsilon^2}\int_{\boldsymbol{t}\in T} g^2(\boldsymbol{t}'\boldsymbol{\beta})[K_h((\boldsymbol{t}-\mathbf{x})'\boldsymbol{\beta}^*)((\boldsymbol{t}-\mathbf{x})'\boldsymbol{\beta}^*/h)^{i_2} - \{K_h((\boldsymbol{t}-\mathbf{x})'\boldsymbol{\beta}/h)$$
$$\times ((\boldsymbol{t}-\mathbf{x})'\boldsymbol{\beta})^{i_2}\}]^2 f_{\boldsymbol{X}}(\boldsymbol{t})d\boldsymbol{t} \;=\; \frac{h^{2i_2}}{\epsilon^2}O\left(\frac{1}{nh^3}\right).$$

Therefore, the left-hand side of (7) has order $O(1/(nh^3\epsilon^2))$ and is less than or equal to $\frac{1}{2}$ when n is large enough. We have thus proved (7). This inequality ensures the use of symmetrization arguments. The next step is to show that conclusion (6) is the maximum value of an empirical process indexed by a VC class of functions. Hereafter, we will suppress the i_1, i_2 superscripts.

Let $\mathcal{F}_n = \{f_{n,\mathbf{x},\boldsymbol{\beta}^*}(\cdot,\cdot) : \|\mathbf{x}\| \le C \text{ and } \|\boldsymbol{\beta}^*\| \le A\}$ be a class of functions indexed by $\mathbf{x}$ and $\boldsymbol{\beta}^*$ consisting of

$$f^{i_1,i_2}_{n,\mathbf{x},\boldsymbol{\beta}^*}(y,\boldsymbol{t}) = y^{i_1}[K((\boldsymbol{t}-\mathbf{x})'\boldsymbol{\beta}^*/h)((\boldsymbol{t}-\mathbf{x})'\boldsymbol{\beta}^*)^{i_2} - K((\boldsymbol{t}-\mathbf{x})'\boldsymbol{\beta}/h)((\boldsymbol{t}-\mathbf{x})'\boldsymbol{\beta})^{i_2}].$$

Therefore, the left-hand side of (6) is equal to $\sqrt{nh}\sup_{f\in\mathcal{F}_n}\left|\sum_{j=1}^n f(Y_j,\boldsymbol{X}_j)\right|$. Note that

$$f^{i_1,i_2}_{n,\mathbf{x},\boldsymbol{\beta}^*}(Y_j,\boldsymbol{X}_j) = h\cdot(\xi^{i_1,i_2}_j(\mathbf{x},\boldsymbol{\beta}^*) - \xi^{i_1,i_2}_j(\mathbf{x},\boldsymbol{\beta})),$$

and

$$A = \|\boldsymbol{\beta}\| + O(n^{-1/2}), \quad \text{since } \|\boldsymbol{\beta}^* - \boldsymbol{\beta}\| = O(n^{-1/2}).$$

We next show that $\mathcal{F}_n$ is a VC class of functions. That is, for any n,

$$N_1(\epsilon_n, P_n, \mathcal{F}_n) \le (\mathit{const.})\cdot n^w, \text{ for some } w,$$

where $N_1(\epsilon_n, P_n, \mathcal{F}_n)$ is the minimum m of the set functions F° consisting of functions $\{f_1^\circ,\ldots,f_m^\circ\}$, each in $\mathcal{F}_n$, such that

$$\min_{i\in 1,\ldots,m} n^{-1}\sum_{j=1}^n |f(Y_j,\boldsymbol{X}_j) - f_i^\circ(Y_j,\boldsymbol{X}_j)| < \epsilon_n \text{ for every } f\in\mathcal{F}_n. \tag{11}$$

The proof goes as follows: For each set F° satisfying (11) and for each f_i° in there, we can find a pair $(\boldsymbol{s}_i,\boldsymbol{\beta}_i)$ such that $f_i^\circ(y,\boldsymbol{t}) \equiv f_{n,\boldsymbol{s}_i,\boldsymbol{\beta}_i}(y,\boldsymbol{t})$. Then

$$\begin{aligned}
&|f^{i_1,i_2}_{n,\mathbf{x},\boldsymbol{\beta}^*}(Y,\boldsymbol{X}) - f^{i_1,i_2}_{n,\boldsymbol{s}_i,\boldsymbol{\beta}_i}(Y,\boldsymbol{X})|\\
=&|Y^{i_1}[K((\boldsymbol{X}-\mathbf{x})'\boldsymbol{\beta}^*/h)((\boldsymbol{X}-\mathbf{x})'\boldsymbol{\beta}^*)^{i_2}\\
&-K((\boldsymbol{X}-\boldsymbol{s}_i)'\boldsymbol{\beta}_i/h)((\boldsymbol{X}-\boldsymbol{s}_i)'\boldsymbol{\beta}_i)^{i_2}\\
&\quad + K((\boldsymbol{X}-\boldsymbol{s}_i)'\boldsymbol{\beta}/h)((\boldsymbol{X}-\boldsymbol{s}_i)'\boldsymbol{\beta})^{i_2} - K((\boldsymbol{X}-\mathbf{x})'\boldsymbol{\beta}/h)((\boldsymbol{X}-\mathbf{x})'\boldsymbol{\beta})^{i_2}]|\\
\le&\frac{|Y^{i_1}|h^{i_2}}{h}M(|(\boldsymbol{X}-\mathbf{x})'\boldsymbol{\beta}^* - (\boldsymbol{X}-\boldsymbol{s}_i)'\boldsymbol{\beta}_i| + |(\boldsymbol{X}-\boldsymbol{s}_i)'\boldsymbol{\beta} - (\boldsymbol{X}-\mathbf{x})'\boldsymbol{\beta}|)
\end{aligned}$$

$$
\begin{aligned}
&=\frac{|Y^{i_1}|h^{i_2}}{h}M(|\boldsymbol{X}'(\boldsymbol{\beta}^* - \boldsymbol{\beta}_i) - \mathbf{x}'\boldsymbol{\beta}^* + \boldsymbol{s}_i'\boldsymbol{\beta}_i| + |(\mathbf{x} - \boldsymbol{s}_i)'\boldsymbol{\beta}|)\\
&\le\frac{|Y^{i_1}|h^{i_2}}{h}M(|\boldsymbol{X}'(\boldsymbol{\beta}^* - \boldsymbol{\beta}_i)| + |(\boldsymbol{s}_i - \mathbf{x})'\boldsymbol{\beta}^*| + |\boldsymbol{s}_i'(\boldsymbol{\beta}_i - \boldsymbol{\beta}^*)| + |(\mathbf{x} - \boldsymbol{s}_i)'\boldsymbol{\beta}|)\\
&\le\frac{|Y^{i_1}|h^{i_2}}{h}M(|\boldsymbol{X}'(\boldsymbol{\beta}^* - \boldsymbol{\beta}_i)| + \|\boldsymbol{s}_i - \mathbf{x}\|A + C\|\boldsymbol{\beta}_i - \boldsymbol{\beta}^*\| + \|\mathbf{x} - \boldsymbol{s}_i\|\|\boldsymbol{\beta}\|),
\end{aligned}
$$

for some constant M. Since K has a compact support, for large n, $n^{-1}\sum_{j=1}^n |Y_j^{i_1}|\|\boldsymbol{X}_j\|$ and $n^{-1}\sum_{j=1}^n |Y_j^{i_1}|$ are bounded by a constant with probability one. For all $\mathbf{x}$ with $\|\mathbf{x}\| < C$ and all $\boldsymbol{\beta}^*$ with $\|\boldsymbol{\beta}^*\| < A$,

$$
n^{-1}\sum_{j=1}^n |f^{i_1,i_2}_{n,\mathbf{x},\boldsymbol{\beta}^*}(Y_j,\boldsymbol{X}_j) - f^{i_1,i_2}_{n,\boldsymbol{s}_i,\boldsymbol{\beta}_i}(Y_j,\boldsymbol{X}_j)| \le (\text{const})\frac{h^{i_2}}{h}(\|\boldsymbol{\beta}^* - \boldsymbol{\beta}_i\| + \|\boldsymbol{s}_i - \mathbf{x}\|).
$$

That is, for any two functions in $\mathcal{F}_n$, the distance only relates to the distances of $\boldsymbol{\beta}$ and $\mathbf{x}$ and h^{i_2}/h. Letting $\epsilon_n = \frac{\epsilon}{8}\sqrt{\frac{h}{n}}$, we thus have

$$
N_1(\epsilon_n, P_n, \mathcal{F}_n) \le (\text{const})\cdot\frac{1}{\epsilon^2}\left(\frac{n}{h^3}\right)^p = (\text{const})\cdot n^w,
$$

where we let $w = p(1+3a)$. Let the constant be $M/2$. This means that $\mathcal{F}_n$ is a VC-class of functions.

Next, we invoke similar arguments that are used to prove Theorem II.37 (Pollard,1984, p.35) or those of Zhu (1993) and arrive at

$$
\begin{aligned}
&P(\sup_{\mathbf{x}}\sup_{\boldsymbol{\beta}^*}\sqrt{nh}|\alpha_n(\mathbf{x},\boldsymbol{\beta}^*) - \alpha_n(\mathbf{x},\boldsymbol{\beta})| > \epsilon)\\
\le& Mn^w E\left(\max_{i\in\{1,\ldots,N_1\}}\exp\left(\frac{-\frac{n}{2}(\frac{\epsilon^2}{8^2}\cdot\frac{h}{n})}{n^{-1}\sum_{j=1}^n(f_{n,\boldsymbol{s}_i,\boldsymbol{\beta}_i}(Y_j,\boldsymbol{X}_j))^2}\right)\right)\\
=& Mn^w E\left(\max_{i\in\{1,\ldots,N_1\}}\exp\left(\frac{-\epsilon^2/128}{n^{-1}h^{-1}\sum_{j=1}^n(f_{n,\boldsymbol{s}_i,\boldsymbol{\beta}_i}(Y_j,\boldsymbol{X}_j))^2}\right)\right)\\
\le& Mn^w\left(\sup_{\mathbf{x}}\sup_{\boldsymbol{\beta}^*}\exp\left(\frac{-\epsilon^2/128}{n^{-1}h\sum_{j=1}^n(\xi_j(\mathbf{x},\boldsymbol{\beta}^*) - \xi_j(\mathbf{x},\boldsymbol{\beta}))^2}\right)\right)\\
=& Mn^w\exp\left(\frac{-\epsilon^2/128}{O(1/nh^3)}\right)\to 0,
\end{aligned}
$$

because $n^{-1}h\sum_{j=1}^n(\xi_j(\mathbf{x},\boldsymbol{\beta}^*) - \xi_j(\mathbf{x},\boldsymbol{\beta}))^2$ has the same order as $hE((\xi(\mathbf{x},\boldsymbol{\beta}^*) - \xi(\mathbf{x},\boldsymbol{\beta}))^2)$, which in turn has the same order as (8). Therefore, $P(\sup_{\mathbf{x}}\sup_{\boldsymbol{\beta}^*}\sqrt{nh}|\alpha_n(\mathbf{x},\boldsymbol{\beta}^*) - \alpha_n(\mathbf{x},\boldsymbol{\beta})| > \epsilon) \to 0$, for any given ϵ and relation (6) holds. □

Lemma 2. *Under conditions (1)-(8), for* $i_1 = 0, 1$, $i_2 = 0, 1, 2$, *and* $\|\boldsymbol{\beta}^* - \boldsymbol{\beta}\| = O(n^{-\frac{1}{2}})$,

$$\sup_{\mathbf{X}\in\Re^p} \sup_{\boldsymbol{\beta}^*:\|\boldsymbol{\beta}^*-\boldsymbol{\beta}\|=O(n^{-1/2})} |n^{-1}\sum_{j=1}^{n}[Y_j^{i_1}A(\boldsymbol{\beta}^* - E(Y_j^{i_1}A(\boldsymbol{\beta}^*))]| = O_P(1/\sqrt{nh}).$$

Proof. Similar arguments as used in the proof of Lemma 1 apply. Readers are referred to Chong (1999) for details. □

Lemma 3. *Under conditions (1)-(8), for* $i_2 = 0, 1, 2$ *and* $\|\boldsymbol{\beta}^* - \boldsymbol{\beta}\| = O(n^{-\frac{1}{2}})$,

$$\sup_{\mathbf{X}\in\Re^p} \sup_{\boldsymbol{\beta}^*:\|\boldsymbol{\beta}^*-\boldsymbol{\beta}\|=O(n^{-1/2})} |n^{-1}\sum_{j=1}^{n} \boldsymbol{Z}_j'(\hat{\boldsymbol{\theta}} - \boldsymbol{\theta})A(\boldsymbol{\beta}^*)| = O_P(\frac{1}{n\sqrt{h}}). \quad (12)$$

Proof. Note that $\boldsymbol{Z}_j'(\hat{\boldsymbol{\theta}} - \boldsymbol{\theta}) = \sum_{i=1}^{q} Z_{ji}(\hat{\theta}_i - \theta_i)$. The left-hand side of (12) is equal to

$$\begin{aligned}
&\sup_{\mathbf{X}} \sup_{\boldsymbol{\beta}^*} |n^{-1}\sum_{j=1}^{n}[\sum_{i=1}^{q} Z_{ji}(\hat{\theta}_i - \theta_i)]A(\boldsymbol{\beta}^*)| \\
\le &\sum_{i=1}^{q} \sup_{\mathbf{X}} \sup_{\boldsymbol{\beta}^*} |n^{-1}\sum_{j=1}^{n} Z_{ji}(\hat{\theta}_i - \theta_i)A(\boldsymbol{\beta}^*))| \\
\le &\sum_{i=1}^{q} |\hat{\theta}_i - \theta_i| \sup_{\mathbf{X}} \sup_{\boldsymbol{\beta}^*} |n^{-1}\sum_{j=1}^{n} Z_{ji}A(\boldsymbol{\beta}^*)|. \qquad (13)
\end{aligned}$$

Since $q \ll n$, the order of (13) is the same as the order of a single term of the first summation. Without loss of generality, we may take $q = 1$. Because $\hat{\boldsymbol{\theta}}$ is obtained through a least squares regression of Y and $\boldsymbol{Z}$, we have $|\hat{\boldsymbol{\theta}} - \boldsymbol{\theta}| = O_P(n^{-1/2})$. Note that $\boldsymbol{Z}_j$'s are independent of the $\boldsymbol{X}_j$'s. Similar arguments as in the proof of Lemma 1 can be applied again to yield

$$\sup_{\mathbf{X}} \sup_{\boldsymbol{\beta}^*} |n^{-1}\sum_{j=1}^{n} \boldsymbol{Z}_j A(\boldsymbol{\beta}^*)| = O_P(1/\sqrt{nh}).$$

For details see Chong (1999). □

Lemma 4. *Under conditions (4)-(11), letting* $E_{i_1,i_2} = E(Y^{i_1}A(\boldsymbol{\beta}^*))$, *we have*

$$E_{i_1,i_2} = \begin{cases} f(\mathbf{x}'\boldsymbol{\beta}) + O(h^2), & i_1 = 0, i_2 = 0 \\ \Big[E(\boldsymbol{Z}'\boldsymbol{\theta}) + g(\mathbf{x}'\boldsymbol{\beta})\Big] f(\mathbf{x}'\boldsymbol{\beta}) + O(h^2), & i_1 = 1, i_2 = 0 \\ O(h^2), & i_1 = 0, i_2 = 1 \\ O(h^2), & i_1 = 1, i_2 = 1 \\ O(h^2), & i_1 = 0, i_2 = 2. \end{cases}$$

Also, uniformly over $\mathbf{x} \in \Re^p$, $\|\boldsymbol{\beta}^* - \boldsymbol{\beta}\| = O(n^{-1/2})$,

$$
\begin{aligned}
|E(K_h((\boldsymbol{X}-\mathbf{x})'\boldsymbol{\beta}^*)) - f(\mathbf{x}'\boldsymbol{\beta})| &= O(h^2+n^{-1/2}),\\
|E(YK_h((\boldsymbol{X}-\mathbf{x})'\boldsymbol{\beta}^*)) - \Big(E(\boldsymbol{Z}'\boldsymbol{\theta}) + g(\mathbf{x}'\boldsymbol{\beta})\Big) f(\mathbf{x}'\boldsymbol{\beta})| &= O(h^2+n^{-1/2}),\\
|E(K_h((\boldsymbol{X}-\mathbf{x})'\boldsymbol{\beta}^*)(\boldsymbol{X}-\mathbf{x})'\boldsymbol{\beta}^*)| &= O(h^2),\\
|E(YK_h((\boldsymbol{X}-\mathbf{x})'\boldsymbol{\beta}^*)(\boldsymbol{X}-\mathbf{x})'\boldsymbol{\beta}^*)| &= O(h^2), \ \textit{and}\\
|E(K_h((\boldsymbol{X}-\mathbf{x})'\boldsymbol{\beta}^*)((\boldsymbol{X}-\mathbf{x})'\boldsymbol{\beta}^*)^2)| &= O(h^2).
\end{aligned}
$$

Proof. The proof follows from Taylor expansions. See Lemma 5 of Chong (1999). □

Now define $\hat{g}(\mathbf{x}'\boldsymbol{\beta}^*) = \mathbf{S}_{\mathbf{X}'\boldsymbol{\beta}^*}\boldsymbol{Y}$, where the smoothing matrix $\mathbf{S}$ is based on the variables, $\boldsymbol{X}'_1\beta^*, \cdots, \boldsymbol{X}'_n\beta^*$ is defined in Appendix B and the subscript on $\mathbf{S}$ denotes the point to which the smoother is applied.

Lemma 5. *Under conditions (1)–(11),*

$$
\begin{aligned}
\sup_{\mathbf{x}\in\Re^p}\ \sup_{\boldsymbol{\beta}^*:\|\boldsymbol{\beta}^*-\boldsymbol{\beta}\|=O(n^{-1/2})} & |\hat{g}(\mathbf{x}'\boldsymbol{\beta}^*) - E(\boldsymbol{Z}'\boldsymbol{\theta}) - g(\mathbf{x}'\boldsymbol{\beta})|\\
&= O_P\left(\frac{1}{\sqrt{nh}} + h^2\right). \qquad (14)
\end{aligned}
$$

Proof. Let $S_{n,i_2}(\mathbf{x},\boldsymbol{\beta}^*) = n^{-1}\sum_{j=1}^n K_h((\boldsymbol{X}_j-\mathbf{x})'\boldsymbol{\beta}^*)((\boldsymbol{X}_j-\mathbf{x})'\boldsymbol{\beta}^*)^{i_2}$. Then

$$
\begin{aligned}
S_0 &= \alpha_n^{0,0}(\mathbf{x},\boldsymbol{\beta}^*) + E(K_h((\boldsymbol{X}-\mathbf{x})'\boldsymbol{\beta}^*))\\
&= O_P(1/\sqrt{nh}) + f(\mathbf{x}'\boldsymbol{\beta}) + O(h^2 + 1/\sqrt{n}),
\end{aligned}
$$

$$
\begin{aligned}
S_1 &= \alpha_n^{0,1}(\mathbf{x},\boldsymbol{\beta}^*) + E(K_h((\boldsymbol{X}-\mathbf{x})'\boldsymbol{\beta}^*)(\boldsymbol{X}-\mathbf{x})'\boldsymbol{\beta}^*)\\
&= O_P(1/\sqrt{nh}) + O(h^2),
\end{aligned}
$$

$$
\begin{aligned}
S_2 &= \alpha_n^{0,2}(\mathbf{x},\boldsymbol{\beta}^*) + E(K_h((\boldsymbol{X}-\mathbf{x})'\boldsymbol{\beta}^*)((\boldsymbol{X}-\mathbf{x})'\boldsymbol{\beta}^*)^2)\\
&= O_P(1/\sqrt{nh}) + O(h^2),
\end{aligned}
$$

by Lemmas 2 and 4. Applying these two lemmas again, we obtain

$$
\begin{aligned}
& n^{-1}\sum_{j=1}^n \xi_j^{1,0}(\mathbf{x},\boldsymbol{\beta}^*)\\
&= \alpha_n^{1,0}(\mathbf{x},\boldsymbol{\beta}^*) + E(YK((\boldsymbol{X}-\mathbf{x})'\boldsymbol{\beta}^*))\\
&= O_P(1/\sqrt{nh}) + \Big(E(\boldsymbol{Z}'\boldsymbol{\beta}) + g(\mathbf{x}'\boldsymbol{\beta})\Big) f(\mathbf{x}'\boldsymbol{\beta}) + O(h^2 + \sqrt{n}), \ \text{and}
\end{aligned}
$$

$$n^{-1}\sum_{j=1}^{n}\xi_j^{1,1}(\mathbf{x},\boldsymbol{\beta}^*)=\alpha_n^{1,1}(\mathbf{x},\boldsymbol{\beta}^*)+E(YK((\boldsymbol{X}-\mathbf{x})'\boldsymbol{\beta}^*)(\boldsymbol{X}_j-\mathbf{x})'\boldsymbol{\beta}^*)$$
$$=O_P(1/\sqrt{nh})+O(h^2).$$

The bounds for the above five equations are uniform over $\mathbf{x}$ and $\boldsymbol{\beta}^*$. We have

$$\hat{g}(\mathbf{x}'\beta^*)=\frac{n^{-1}\sum_{i=1}^{n}\xi_i^{1,0}(\mathbf{x},\boldsymbol{\beta}^*)S_2-n^{-1}\sum_{i=1}^{n}\xi_i^{1,1}(\mathbf{x},\boldsymbol{\beta}^*)S_1}{S_0S_2-S_1^2},$$

so

$$\begin{aligned}&\hat{g}(\mathbf{x}'\boldsymbol{\beta}^*)-E(\boldsymbol{Z}'\boldsymbol{\theta})-g(\mathbf{x}'\boldsymbol{\beta})\\
&=\frac{(\Big(E(\boldsymbol{Z}'\boldsymbol{\theta})+g(\mathbf{x}'\boldsymbol{\beta})\Big)f(\mathbf{x}'\boldsymbol{\beta})+O_P(h^2+\frac{1}{\sqrt{nh}}))S_2-O_P(h^2+\frac{1}{\sqrt{nh}})S_1}{(f(\mathbf{x}'\boldsymbol{\beta})+O_P(h^2+\frac{1}{\sqrt{nh}}))S_2-S_1^2}\\
&\quad-\frac{\Big(E(\boldsymbol{Z}'\boldsymbol{\theta})+g(\mathbf{x}'\boldsymbol{\beta})\Big)((f(\mathbf{x}'\boldsymbol{\beta})+O_P(h^2+\frac{1}{\sqrt{nh}}))S_2-S_1^2}{(f(\mathbf{x}'\boldsymbol{\beta})+O_P(h^2+\frac{1}{\sqrt{nh}}))S_2-S_1^2}\\
&=\frac{(O_P(h^2+\frac{1}{\sqrt{nh}}))^2}{O_P(h^2+\frac{1}{\sqrt{nh}})}=O_P\left(h^2+\frac{1}{\sqrt{nh}}\right),\end{aligned}$$

with bounds uniform over $\mathbf{x}$ and $\boldsymbol{\beta}^*$, showing (14). □

Proof of Theorem 2. Define $\mathbf{E}_n=\widehat{Cov}(\boldsymbol{Z})$, the sample covariance matrix of $\boldsymbol{Z}$, and $\widehat{Cov}(\boldsymbol{Z},Y)$ the sample covariance between $\mathbf{Z}$ and Y. Further, let

$$\mathbf{g}=\begin{bmatrix}g(\boldsymbol{X}_1'\boldsymbol{\beta})\\ \vdots\\ g(\boldsymbol{X}_n'\boldsymbol{\beta})\end{bmatrix}\quad\text{and}\quad\hat{\mathbf{g}}=\begin{bmatrix}\hat{g}(\boldsymbol{X}_1'\hat{\boldsymbol{\beta}})\\ \vdots\\ \hat{g}(\boldsymbol{X}_n'\hat{\boldsymbol{\beta}})\end{bmatrix},\quad\text{then}$$

$$\begin{aligned}\hat{\boldsymbol{\theta}}&=(\mathbf{E}_n)^{-1}\widehat{Cov}(\boldsymbol{Z},((\mathbf{I}-\mathbf{S})\boldsymbol{Y})\\
&=(\mathbf{E}_n)^{-1}(\mathbf{Z}-\bar{\mathbf{Z}})'(\boldsymbol{Y}-\bar{\boldsymbol{Y}}-(\hat{\mathbf{g}}-\mathbf{g})\\
&=\mathbf{Z}'(\mathbf{g}+\mathbf{Z}\boldsymbol{\theta}+\boldsymbol{\epsilon}-\hat{\mathbf{g}})\\
&=\boldsymbol{\theta}+(\mathbf{Z}'\mathbf{Z})^{-1}\mathbf{Z}'\boldsymbol{\epsilon}+(\mathbf{Z}'\mathbf{Z})^{-1}\mathbf{Z}'(\mathbf{g}-\hat{\mathbf{g}}).\end{aligned}$$

Because $E(\boldsymbol{\epsilon})=\mathbf{0}$ and $\boldsymbol{\epsilon}$ and $\boldsymbol{Z}$ are independent, we have $E((\mathbf{Z}'\mathbf{Z})^{-1}\mathbf{Z}'\boldsymbol{\epsilon})=\mathbf{0}$. It is easy to prove that, by the Weak Law of Large Numbers, $Var((\mathbf{Z}'\mathbf{Z})^{-1}\mathbf{Z}'\boldsymbol{\epsilon})\rightarrow\mathbf{E}_n^{-1}\sigma^2$ in probability, $n\mathbf{E}_n^{-1}\xrightarrow{P}\mathbf{E}^{-1}$, so $\sqrt{n}(\mathbf{Z}'\mathbf{Z})^{-1}\mathbf{Z}'\boldsymbol{\epsilon}\xrightarrow{d}N(0,\mathbf{E}^{-1}\sigma^2)$ by the Central Limit Theorem.

Now we need to show that $\sqrt{n}(\mathbf{Z}'\mathbf{Z})^{-1}\mathbf{Z}'(\mathbf{g}-\hat{\mathbf{g}}) \xrightarrow{P} \mathbf{0}$. As stated above, $n(\mathbf{Z}'\mathbf{Z})^{-1} = n\mathbf{E}_n^{-1} \xrightarrow{P} \mathbf{E}^{-1} = O(1)$, it remains to show that $\frac{1}{\sqrt{n}}\mathbf{Z}'(\mathbf{g}-\hat{\mathbf{g}}) \xrightarrow{P} \mathbf{0}$. Towards this, we will show that

$$\sup_{\boldsymbol{\beta}^*} \| \frac{1}{\sqrt{n}} \sum_{j=1}^{n} Z_j(\hat{g}(\boldsymbol{X}_j'\boldsymbol{\beta}^*) - g(\boldsymbol{X}_j'\boldsymbol{\beta}))\| \xrightarrow{P} 0. \tag{15}$$

Using an argument as in Lemma 3, we will let $q = 1$ without loss of generality. We can thus write Z in place of $\boldsymbol{Z}$, and the norm becomes an absolute value.

Note that Z is independent of $\hat{g}(\boldsymbol{X}_j'\boldsymbol{\beta}^*) - g(\boldsymbol{X}_j'\boldsymbol{\beta})$. Using arguments similar to those found in the first three lemmas, we have

$$\begin{aligned} &P(\sup_{\boldsymbol{\beta}^*} |n^{-1/2} \sum_{j=1}^{n} Z_j(\hat{g}(\boldsymbol{X}_j'\boldsymbol{\beta}^*) - g(\boldsymbol{X}_j'\boldsymbol{\beta}))| > \epsilon) \\ \le& 4E\left[Dn^w \sup_{\boldsymbol{\beta}^*} \exp\left(\frac{-\epsilon^2/128}{n^{-1}\sum_{j=1}^n (Z_j(\hat{g}(\boldsymbol{X}_j'\boldsymbol{\beta}^*) - g(\boldsymbol{X}_j'\boldsymbol{\beta})))^2}\right)\right]. \end{aligned} \tag{16}$$

Lemma 5 implies that

$$\sup_{\mathbf{X}\in\Re^p} \sup_{\boldsymbol{\beta}^*:\|\boldsymbol{\beta}^*-\boldsymbol{\beta}\|=O(n^{-1/2})} |\hat{g}(\mathbf{x}'\boldsymbol{\beta}^*) - g(\mathbf{x}'\boldsymbol{\beta})| = O_P\left(\frac{1}{\sqrt{nh}} + h^2\right),$$

so $n^{-1}\sum_{j=1}^n |Z_j|^2 = O_P(1)$, and

$$n^{-1}\sum_{j=1}^{n} |Z_j(\hat{g}(\boldsymbol{X}_j'\boldsymbol{\beta}^*) - g(\boldsymbol{X}_j'\boldsymbol{\beta}))|^2 = O_P(((1/\sqrt{nh}) + h^2)^2).$$

Therefore, the probability in (16) goes to zero, which implies (15) . The proof of Theorem 2 is now completed. □

Proof of Theorem 3: Let $g(u|\boldsymbol{\beta}) = E(Y - \boldsymbol{Z}'\boldsymbol{\theta}|\boldsymbol{X}'\boldsymbol{\beta} = u)$. Here and below, $\boldsymbol{\beta}$ is always a unit p-vector and $g(\cdot)$ is estimated by a local polynomial smoother. Let $\mathbf{X} = (\mathbf{X_1}, \cdots, \mathbf{X_n})'$, $\boldsymbol{Y} = (Y_1, \cdots, Y_n)'$ and $\mathbf{Z} = (\mathbf{Z_1}, \cdots, \mathbf{Z_n})'$. The estimator is defined as

$$\hat{g}(\mathbf{X}\boldsymbol{\beta}|\boldsymbol{\beta}) = \mathbf{S}_{\mathbf{X}\boldsymbol{\beta}}(\boldsymbol{Y} - \mathbf{Z}\hat{\boldsymbol{\theta}}),$$

where $\mathbf{S}_{\mathbf{X}\boldsymbol{\beta}}$ is the smoothing matrix based on the variables $\boldsymbol{X}_1'\boldsymbol{\beta}, \cdots, \boldsymbol{X}_n'\boldsymbol{\beta}$ similar to the situation right before Lemma 5 . For the convenience of notations, we define $\tilde{\boldsymbol{Y}} = \boldsymbol{Y} - \mathbf{Z}\boldsymbol{\theta}$ and $\tilde{g}(u|\boldsymbol{\beta}) = \mathbf{S}_{\mathbf{X}\boldsymbol{\beta}}\tilde{\boldsymbol{Y}}$. Since $g(u|\boldsymbol{\beta}) =$

$g(\boldsymbol{X}'\boldsymbol{\beta})$ we may estimate $\boldsymbol{\beta}$ by selecting the orientation $\boldsymbol{\beta}^*$ which minimizes a measure of the distance $g(\cdot|\boldsymbol{\beta}^*) - g$. To this end, define

$$\begin{aligned}\hat{D}(\boldsymbol{\beta}^*, h) &= \sum_{i=1}^{n}[Y_i - \boldsymbol{Z}_i'\hat{\boldsymbol{\theta}} - \hat{g}(\boldsymbol{X}_i'\boldsymbol{\beta}^*|\boldsymbol{\beta}^*)]^2 \\ &= (\boldsymbol{Y} - \mathbf{Z}\hat{\boldsymbol{\theta}})'(\mathbf{I} - \mathbf{S}_{\mathbf{X}\boldsymbol{\beta}^*})'(\mathbf{I} - \mathbf{S}_{\mathbf{X}\boldsymbol{\beta}^*})(\boldsymbol{Y} - \mathbf{Z}\hat{\boldsymbol{\theta}}).\end{aligned}$$

Note that our initial estimator $\tilde{\boldsymbol{\beta}}$ of $\boldsymbol{\beta}$ is $\sqrt{n}$-consistent. Therefore, the minimization only needs to be taken over $\boldsymbol{\beta}^*$ such that $|\boldsymbol{\beta}^* - \tilde{\boldsymbol{\beta}}| = O(1/\sqrt{n})$, that is, $|\boldsymbol{\beta}^* - \boldsymbol{\beta}| = O(1/\sqrt{n})$. We then define the minimizer $\hat{\boldsymbol{\beta}}$ as the estimator of $\boldsymbol{\beta}$.

It is clear that

$$\begin{aligned}\hat{D}(\boldsymbol{\beta}^*, h) = &\tilde{\boldsymbol{Y}}'(I - \mathbf{S}_{\mathbf{X}\boldsymbol{\beta}^*})'(I - \mathbf{S}_{\mathbf{X}\boldsymbol{\beta}^*})\tilde{\boldsymbol{Y}} \\ &+ (\hat{\boldsymbol{\theta}} - \boldsymbol{\theta})'\mathbf{Z}(\mathbf{I} - \mathbf{S}_{\mathbf{X}\boldsymbol{\beta}^*})'(\mathbf{I} - \mathbf{S}_{\mathbf{X}\boldsymbol{\beta}^*})\mathbf{Z}(\hat{\boldsymbol{\theta}} - \boldsymbol{\theta}) \\ &- (\hat{\boldsymbol{\theta}} - \boldsymbol{\theta})'\mathbf{Z}(\mathbf{I} - \mathbf{S}_{\mathbf{X}\boldsymbol{\beta}^*})'(\mathbf{I} - \mathbf{S}_{\mathbf{X}\boldsymbol{\beta}^*})\tilde{\boldsymbol{Y}} \\ &- \tilde{\boldsymbol{Y}}'(I - \mathbf{S}_{\mathbf{X}\boldsymbol{\beta}^*})'(I - \mathbf{S}_{\mathbf{X}\boldsymbol{\beta}^*})\mathbf{Z}(\hat{\boldsymbol{\theta}} - \boldsymbol{\theta}) \\ =: &\tilde{D}(\boldsymbol{\beta}^*, h) + I_{n1}(\boldsymbol{\beta}^*, h) + I_{n2}(\boldsymbol{\beta}^*, h) + I_{n3}(\boldsymbol{\beta}^*, h).\end{aligned}$$

Invoking the arguments used to prove the Theorem of Härdle, Hall and Ichimura (1993), we have

$$\tilde{D}(\boldsymbol{\beta}^*, h) = \tilde{D}(\boldsymbol{\beta}^*) + T(h) + R_1(\boldsymbol{\beta}^*, h) + R_2(h),$$

where

$$\begin{aligned}\tilde{D}(\boldsymbol{\beta}^*) &= \sum(\tilde{Y}_i - g(\boldsymbol{Z}_i'\boldsymbol{\beta}^*|\boldsymbol{\beta}^*))^2 \\ T(h) &= \sum(\hat{g}(\boldsymbol{Z}_i'\boldsymbol{\beta}|\boldsymbol{\beta}) - g(\boldsymbol{Z}_i'\boldsymbol{\beta}))^2;\end{aligned}$$

and uniformly over $\boldsymbol{\beta}^*$ and h such that $\|\boldsymbol{\beta}^* - \boldsymbol{\beta}\| = O(1/\sqrt{n}), h = O(n^{-1/5})$,

$$\|R_1(\boldsymbol{\beta}^*, h)\| = o_p(n^{1/5}), \qquad \text{and} \qquad \|R_2(h)\| = o_p(1).$$

Furthermore, from their arguments, we have for some constants A_1 and A_2,

$$\begin{aligned}\tilde{D}(\boldsymbol{\beta}^*) &= n\Big[\mathbf{W}^{1/2}(\boldsymbol{\beta}^* - \boldsymbol{\beta}) - n^{-1/2}(\mathbf{W}^-)^{1/2}U_n\Big] \\ &\quad \times \Big[\mathbf{W}^{1/2}(\boldsymbol{\beta}^* - \boldsymbol{\beta}) - n^{-1/2}(\mathbf{W}^-)^{1/2}U_n\Big] + R_3 + R_4(\boldsymbol{\beta}^*), \\ T(h) &= A_1 h^{-1} + A_2 n h^4 + R_5(h),\end{aligned}$$

where

$$U_n = \sum[\boldsymbol{X}_i - E(\boldsymbol{X}_i|\boldsymbol{X}_i'\boldsymbol{\beta})]g'(\boldsymbol{X}_i'\boldsymbol{\beta})e_i,$$

$$\sup_{\|\boldsymbol{\beta}^* - \boldsymbol{\beta}\| = O(1/\sqrt{n})} \|R_4(\boldsymbol{\beta}^*)\| = o_p(1), \qquad \sup_{h = O(n^{-1/5})} \|R_5(h)\| = o_p(n^{1/5}),$$

g' is the derivative of g, and R_3 is a constant independent of $\boldsymbol{\beta}^*$ and h. Note that our initial estimator $\hat{\boldsymbol{\theta}}$ is $\sqrt{n}$-consistent for $\boldsymbol{\theta}$. By the independence between $\tilde{\boldsymbol{Y}}$ and $\boldsymbol{Z}$ and the $\sqrt{n}$-consistency of $\tilde{\boldsymbol{\beta}}$ for $\boldsymbol{\beta}$, we obtain easily that, and uniformly over $\boldsymbol{\beta}^*$ and h such that $\|\boldsymbol{\beta}^* - \boldsymbol{\beta}\| = O(1/\sqrt{n}), h = O(n^{-1/5})$,

$$\|I_{nl}(\boldsymbol{\beta}^*, h)\| = o_p(1), \quad l = 1, 2, 3;$$
$$\|\frac{1}{\sqrt{n}}\boldsymbol{Z}'(I - \mathbf{S}_{\mathbf{X}\boldsymbol{\beta}^*})'(I - \mathbf{S}_{\mathbf{X}\boldsymbol{\beta}^*})\tilde{\boldsymbol{Y}}\| = o_p(1).$$

Therefore, uniformly over $\boldsymbol{\beta}^*$ and h

$$\hat{D}(\boldsymbol{\beta}^*, h) = \tilde{D}(\boldsymbol{\beta}^*) + T(h) + o_p(n^{1/5}) + C_n,$$

where C_n is a constant independent of $\boldsymbol{\beta}^*$ and h. Hence the minimum of $\hat{D}(\boldsymbol{\beta}^*, h)$ within a radius $O(n^{-1/2})$ of $\boldsymbol{\beta}$ for the first variable and on a scale of $n^{-1/5}$ for the second variable satisfies, for any unit vector $u \neq \boldsymbol{\beta}$,

$$\begin{aligned} u'(\hat{\boldsymbol{\beta}}^* - \boldsymbol{\beta}) &= n^{-1/2}u'(\mathbf{W}^-)U_n + o_p(n^{-1/2}) \\ &= n^{-1/2}u'(\mathbf{W}^-)\sum[\boldsymbol{X}_i - E(\boldsymbol{X}_i|\boldsymbol{\beta}'\boldsymbol{X}_i)]g'(\boldsymbol{\beta}'\boldsymbol{X}_i)\varepsilon_i + o_p(n^{-1/2}). \end{aligned}$$

In other words,

$$n^{1/2}u'(\hat{\boldsymbol{\beta}}^* - \boldsymbol{\beta}) \Longrightarrow N(0, u'\sigma^2(\mathbf{W}^-)u).$$

This completes the proof. □

APPENDIX B

Linear Smoother: We first consider the simple case of one-dimensional smoothing to estimate the mean response, $m(x) = E\{Y|X = x\}$, based on data $\{(X_i, Y_i), i = 1, \cdots, n\}$. For a given scalar point x and bandwidth h, the local polynomial smoother (Fan and Gijbels (1996)) is based on a window, $(x - h, x + h)$, and a kernel weight function to fit locally a weighted polynomial regression, and then uses the fitted value at x as the estimate for $m(x)$. For instance, a locally linear smoother with a kernel K, using a linear polynomial to estimate the regression function via the least squares method, yields the following estimate:

$$\hat{m}(x) = \arg\min_a \min_b \sum_{i=1}^{n} [y_i - a - b(x_i - x)]^2 K_h(x_i - x),$$

where $K_h(u) = h^{-1}K(u/h)$. Define $Q_i = K_h(x_i - x)$ and $Q_j = K_h(x_j - x)$. The solution to the minimization equation $\hat{m}(x)$ equals

$$\frac{\sum_{i=1}^{n} Q_i[\sum_{j=1}^{n} Q_j(x_j - x)^2 - (x_i - x)\sum_{j=1}^{n} Q_j(x_j - x)]y_i}{\sum_{i=1}^{n} Q_i[\sum_{j=1}^{n} Q_j(x_j - x)^2 - (x_i - x)\sum_{j=1}^{n} Q_j(x_j - x)]}.$$

This smoother belong to a class called the class of linear smoothers, which is a linear combination of the observed responses. For a linear smoother, we may construct a matrix $\mathbf{S_x}$ such that the estimated mean response is $\hat{\boldsymbol{y}} = \mathbf{S_x}\boldsymbol{y}$, where the subscript $\mathbf{x}$ denotes the covariate variables, $\{x_1, \cdots, x_n\}$, on which the smoothing is based. We will call $\mathbf{S_x}$ the smoothing matrix. It depends on the type of smoother and kernel function K used, the observed values of the covariates $\mathbf{x}$, and the smoothing parameter h.

Suppose, for example, that $\mathbf{x} = (x_1, \ldots, x_n)$ is observed and that we are using a kernel K that has support $[-1, 1]$. Below, let $Q_{rs} = K_h(x_r - x_s)$ for two subscripts r and s.The matrix $\mathbf{S}$ corresponding to the locally linear smoother above will have elements

$$S_{ij} = \frac{Q_{ij}[\sum_{k=1}^{n} Q_{ki}(x_k - x_i)^2 - (x_j - x_i)\sum_{k=1}^{n} Q_{ki}(x_k - x_i)]}{\sum_{k=1}^{n} Q_{ki}[\sum_{l=1}^{n} Q_{li}(x_l - x_i)^2 - (x_k - x_i)\sum_{l=1}^{n} Q_{li}(x_l - x_i)]}.$$

Automatic bandwidth choices based on Generalized cross validation: The bandwidth h which minimizes

$$GCV(h) = \frac{\frac{1}{n}\sum_{i=1}^{n}(Y_i - \hat{g}_h(X_i))^2}{(\frac{1}{n}\mathrm{tr}(\mathbf{I} - \mathbf{S}_h))^2}, \tag{17}$$

is the generalized cross-validated bandwidth, where $\mathbf{S}_h$ is the smoothing matrix corresponding to a bandwidth of h and $\hat{g}_h(X_i)$ is the estimated regression function corresponding to a bandwidth of h, evaluated at X_i.

References

1. BHATTACHARYA, P. K. AND ZHAO, P.-L. (1997), Semiparametric inference in a partial linear model, *Ann. Statist.* **25**, 244-262.

2. CARROLL, R. J., FAN, J., GIJBELS, I. AND WAND, M. P. (1997), Generalized partially linear single-index models, *J. Amer. Statist. Assoc.* **92**, 477-489.

3. CHAUDHURI, P., DOKSUM, K. AND SAMAROV, A. (1997), On average derivative quantile regression, *Ann. Statist.* **25**, 715-744.

4. CHEN, C.-H. AND LI, K. C. (1998), Can SIR be as popular as multiple linear regression?, *Statistica Sinica* **8**, 289-316.

5. CHEN, H. (1988), Convergence rates for parametric components in a partly linear model, *Ann. Statist.* **16** 136-146.

6. CHEN, H. AND SHIAU, J.-J. H. (1994), Data-driven efficient estimators for a partially linear model, *Ann. Statist.* **22**, 211-237.

7. CHIOU, J. M. AND MÜLLER, H. G. (1998), Quasi-likelihood regression with unknown link and variance functions, *J. Amer. Statist. Assoc.* **93**, 1376-1387.

8. CHIOU, J. M. AND MÜLLER, H. G. (1999), Nonparametric quasi-likelihood, *Ann. Statist.* **27**, 36-64.

9. CHONG, Y. S. (1999), *Dimension reduction mehtods for discrete and continuos covariates*, Unpublished Ph.D. Dissertaiton of the University of California.

10. COOK, R. D. (1998), Principal Hessian directions revisited (with discussion), *J. Amer. Statist. Assoc.* **93**, 84-100.

11. COOK, R. D. AND WEISBERG, S. (1991), Discussion of "Sliced Inverse Regression," *J. Amer. Statist. Assoc.* **86**, 328-332.

12. CRAVEN, P. AND WAHBA, G. (1979), Smoothing and noisy data with spline functions: estimating the correct degree of smoothing by the method of generalized cross-validation, *Numer. Math.* **31**, 377-403.

13. DABROWSKA, D. AND DOKSUM, K. (1988a), Partial likelihood in transformaiton models with censored data, *Scand. J. Statist.***15**, 1-23.

14. DABROWSKA, D. AND DOKSUM, K. (1988b), Estimation and testing in a two sample generalized odds rate model, *J. Amer. Statist. Assoc.* **83**, 744-749.

15. DENBY, L. (1986), Smooth regression functions, *Statistical Research Report 26*, AT&T Bell Laboratories, Murray Hill.

16. DOKSUM, K. (1987), An extension of partial likelihood methods for proportional harzards model to general transformaiton models, *Ann. Statist.* **15**, 325-345.

17. DOKSUM, K. AND GASKO, M. (1990), On a correspondence between modles in binary regression analysis and in survival analysis, *Internat. Statist. Rev.* **58**, 243-252.

18. DOKSUM, K. AND SAMAROV, A. (1995), Nonparametric estimation of global functionals and a measure of the explanatory power of covariates in regression, *Ann. of Statist.* **23**, 1443-1473.

19. ENGLE, R., GRANGER, C., RICE, J. AND WEISS, A. (1986), Semiparametric estimates of the relation between weather and electricity sales, *J. Amer. Statist. Assoc.* **81**, 310-320.

20. FAN, J. (1993), Local linear regression smoothers and their minimax efficiency, *Ann. Statist.* **21**, 196-216.

21. FAN, J. AND GIJBELS, I. (1996), *Local Polynomial Modelling and Its Applications*, Chapman and Hall, London.

22. FRIEDMAN, J. H. AND STUETZLE, W. (1981), Projection pursuit regression, *J. Amer. Statist. Assoc.* **76**, 817-823.

23. HALL, P. (1989), On projection pursuit regression, *Ann. Statist.* **17**, 573-588.

24. HAMILTON, S. A. AND TRUONG, Y. K. (1997), Local linear estimation in partly linear models, *J. Multivariate Analysis* **60**, 1-19.

25. HÄRDLE, W., HALL, P. AND ICHIMURA, H. (1993), Optimal smoothing in single-index models, *Ann. Statist.* **21**, 157-178.

26. HÄRDLE, W. AND STOKER, T. M. (1989), Investigating smooth multiple regression by the method of average derivatives, *J. Amer. Statist. Assoc.* **84**, 986-995.

27. HECKMAN, N. E. (1986), Spline smoothing in a partly linear model, *J. Royal Statist. Soc. B* **48**, 244-248.

28. HRISTACHE, M. AND JUDITSKY, A. AND SPOKOINY, V. (2001), Direct estimation of the index coefficient in a single-index model, *Ann. Statist.* **29**, 595-623.

29. LI, K.-C. (1991), Sliced inverse regression for dimension reduction (with discussion), *J. Amer. Statist. Assoc.* **86**, 316-342.

30. LI, K.-C. (1992), On principal Hessian directions for data visualization and dimension reduction: another application of Stein's lemma, *J. Amer. Statist. Assoc.* **87**, 1025-1039.

31. POLLARD, D. (1984), *Convergence of Stochastic Processes*, Springer-Verlag, New York.

32. RICE, J. (1986), Convergence rates for partially splined models, *Statist. Prob. Lett.* **4**, 203-208.

33. SAMAROV, A. (1993), Exploring structure using nonparametric functional estimation, *J. Amer. Statist. Assoc.* **88**, 836-847.

34. SEVERINI, T. A. AND STANISWALIS, J. G. (1994), Quasi-likelihood estimation in semiparametric models, *J. Amer. Statist. Assoc.* **89**, 501-511.

35. SPECKMAN, P. (1988), Kernel smoothing in partial linear models, *J. Royal Statist. Soc. B* **50**, 413-436.

36. STOKER, T. M. (1986), Consistent estimation of scaled coefficient, *Economoetrics.* **54**, 1461-1481.

37. STUTE, W. AND ZHU, L.-X. (2005), Nonparametric checks for single-index models, *Annals of Statistics*, in press.

38. WAHBA, G.(1984), Partial spline models for the semiparametric estimation of functions of several variables, *Statistical Analyses for Time Series*, pp. 319-329, Institute of Statistical Mathematics, Tokyo.

39. XIA, Y.,TONG, H., LI, W. K. AND ZHU, L.-X. (2002), An adaptive estimation of optimal regression subspace, *Journal of Royal Statistical Society, Series B*, **64**, 363-410.

40. YU, Y. AND RURRERT, D. (2002). Penalized spline estomation for partial linear single-index models, *J. Amer. Statist. Assoc.* **97**, 1042-1054.

41. ZHU, L.-X. AND NG, K. W. (1995), Asymptotics of sliced inverse regression, *Statistica Sinica* **5**, 727-736.

42. ZHU, L.-X. AND FANG, K. T. (1996), Asymptotics for the kernel estimates of sliced inverse regression, *Ann. Statist.* **24**, 1053-1067.

Chapter 11

ON SINGLE INDEX REGRESSION MODELS FOR MULTIVARIATE SURVIVAL TIMES

Probal Chaudhuri

Theoretical Statistics and Mathematics Unit
Indian Statistical Institute, Kolkata, INDIA

E-mail: probal@isical.ac.in

An extension of rank based partial likelihood method of Cox (1975) for general transformation model was introduced by Doksum (1987), and Chaudhuri, Doksum and Samarov (1997) introduced average derivative quantile regression estimates of parameters in semiparametric single index regression models that generalize transformation models. An important requirement for rank and quantile based methods to be applicable to any such model is an intrinsic monotonicity property of the underlying link function. In this note, we explore certain extensions of such semiparametric single index models for multivariate life time data and the possibility of estimation of index coefficients by average derivative quantile regression techniques. Monotonicity properties of the link functions associated with such models are also investigated.

Key words: Frailty model, Log-concave density, Multivariate monotonicity, Quantile regression, Stochastic ordering.

1 Introduction: Semiparametric Regression Models in Survival Analysis

An intriguing connection between proportional hazard model (see Cox 1972) and general transformation model was pointed out by Doksum (1987), who extended rank based partial likelihood methods (see Cox 1975) to general transformation models. If T denotes the survival time and $\mathbf{X} = (X_1, \ldots, X_d)$ denotes a d-dimensional vector of covariates, Cox's proportional hazard model can be described by the equation $\lambda(t) = \lambda_0(t)\exp(\sum_{i=1}^{d} \beta_i X_i)$, where $\lambda(.)$ is the hazard function associated with the distribution of T, $\lambda_0(.)$ is the baseline hazard function associated with a baseline distribution function F_0, and the β_i's are the regression coefficients. Here $F_0(t)$ is an absolutely continuous distribution with a continuous and positive density for $t > 0$. An equivalent formulation of this proportional hazard model can be given in the form of a linear regression model with a transformed response given as $h(T) = -\sum_{i=1}^{d} \beta_i X_i + \epsilon$, where $h(t) = \ln[-\ln\{1 - F_0(t)\}]$, and ϵ has the distribution $F_\epsilon(s) = 1 - \exp\{-\exp(s)\}$. Since $h(.)$ is an unknown monotonically increasing transformation, this formulation makes the reason for using the rank based partial likelihood in the proportional hazard model quite transparent. As ranks remain invariant under strictly increasing transformation, partial likelihood does not depend on the unknown transformation $h(.)$. All these were amply exploited by Doksum (1987) to investigate extensions of partial likelihood method in more general linear regression models with transformed response variables.

Another very well known model used in the regression analysis of survival time data is the proportional odds rate model, and that too can be viewed as a linear regression model with transformed response : $h(T) = -\sum_{i=1}^{d} \beta_i X_i + \epsilon$. Here, $h(t) = \ln[F_0(t)/\{1 - F_0(t)\}]$, $F_0(t)$ is a continuous and strictly increasing distribution function for $t > 0$, and ϵ has the logistic distribution $F_\epsilon(s) = \{1 + \exp(-s)\}^{-1}$. Readers are referred to Doksum and Gasko (1990) for a discussion of this model. The accelerated failure time model, which is also fairly popular in survival analysis (see e.g., Kalbfleish and Prentice 2002), is another example of linear regression model with transformed response, where $h(t) = \ln(t)$, and the distribution of ϵ is unspecified.

Let us now consider the single index model

$$T = \psi\left(\sum_{i=1}^{d} \beta_i X_i, \epsilon\right) \tag{1}$$

where $\psi(.,.)$ is an unknown link function, which is monotonically increasing in its second argument, and ϵ is a continuously distributed unobserved random variable whose distribution is assumed not to depend on $\mathbf{X}$. Clearly, all of the survival analysis models mentioned in the preceding two paragraphs are special cases of this more general model with specific choices for the function $\psi(.,.)$. Further, various extensions and variations of those models available in the literature, like the proportional mean residual time model studied by Oakes and Dasu (1990), different semiparametric versions of generalized proportional hazard models considered by Bagdonavicius and Nikulin (1999) and transformation models investigated by Cai and Cheng (2004) can be obtained as special cases with appropriate forms of the function $\psi(.,.)$. An interesting feature of the model in (1) is that if the function

$\psi(.,.)$ is completely unspecified except for the fact that it is a monotonically increasing function of its second argument, one can assume ϵ to have uniform distribution on $[0,1]$ without any loss of generality. This is a consequence of the fact that any continuously distributed random variable can be viewed as a continuous and monotonically increasing function of a uniform random variable on $[0,1]$.

An important point to note at this stage is that in the general case with a completely unknown $\psi(.,.)$ in (1), rank based partial likelihood is no longer useful for estimating the index coefficients β_i's as the distributions of the ranks now depend on that unknown link function $\psi(.,.)$. However, as shown in Chaudhuri, Doksum and Samarov (1997), one can use local polynomial quantile regression (see, e.g., Chaudhuri 1991 and Chaudhuri and Loh 2002) to construct $n^{1/2}$-consistent and asymptotically normal estimates of properly normalized β_i's, which are identifiable only up to a scalar multiple. Quantiles are equivariant under monotonically increasing transformation, and consequently, for any $0 < \alpha < 1$, the conditional α-th quantile of T given $\mathbf{X}$ is $\psi\left(\sum_{i=1}^{d} \beta_i X_i, \alpha\right)$, assuming ϵ to have uniform distribution on $[0,1]$. This was one of the key ideas used by Chaudhuri, Doksum and Samarov (1997) in their average derivative quantile regression estimation of parameters in the single index model in (1). Quantile regression in Cox's proportional hazard model has been considered earlier by Dabrowska and Doksum (1987). The equivariance of quantiles under monotonically increasing transformations and a related use of quantile regression in survival analysis can also be found in Koenker and Geling (2001).

In this paper, we intend to investigate multivariate versions of semiparametric index models for survival analysis that retain the above mentioned monotonicity property, which is of fundamental importance in rank or quantile based regression analysis, in some natural sense, even when the response is multivariate in nature. In particular, we will try to investigate the monotonicity properties of the well-known frailty model of Vaupel, Manton and Stallard (1979) for general baseline and frailty distributions. In course of our investigation, we will also indicate possible extensions of quantile regression and related techniques to estimate the index coefficients when the response is multivariate in nature.

2 Single Index Regression Models for Multivariate Survival Time Data

Dependent multivariate life time data may arise in many situations, and they have been widely studied in the literature (see e.g., Hougard 2000). Well known examples of dependent multivariate failure time data include the time of onset of schizophrenia among family members who are genetically related (see Pulver and Liang 1991, Lin 1994), tumor occurrence in litter-matched tumorigenesis experiments involving animals (see Mantel, Bohidar and Ciminera 1977), occurrence times for multiple tumors in bladder cancer study (see Wei, Lin and Weissfeld 1989), etc.

Let $\mathbf{T} = (T_1, T_2, \ldots, T_p)$ denote a p-dimensional vector of survival times and $\mathbf{X}_j = (X_{j1}, X_{j2}, \ldots, X_{jd})$ be the d-dimensional vector of covariates corresponding to the j-th component of the multivariate survival time $\mathbf{T}$, where $1 \leq j \leq p$. We now state two theorems that lead to some interesting generalisations of the single index model in (1) for multivariate survival times.

Theorem 1. *Suppose that the conditional distribution of each of the marginal survival time T_j $(1 \leq j \leq p)$ given $\mathbf{X}_j$ is the same as the conditional distribution of T_j given the linear function $Z_j = \sum_{i=1}^{d} \beta_i X_{ji}$. Assume also that the conditional distribution of each of the T_j's given Z_j is continuous and strictly increasing on its support. Then there exists a functions $\psi_j(.,.)$ from R^2 into R, which is monotonically increasing in its second argument, and a random vector $\mathbf{U} = (U_1, U_2, \ldots, U_p)$ with each marginal U_j $(1 \leq j \leq p)$ having a uniform distribution on $[0,1]$ that does not depend on $\mathbf{X}_j$ such that*

$$T_j = \psi_j(Z_j, U_j) \tag{2}$$

Theorem 2. *Suppose that the conditional distribution of the vector $\mathbf{T}$ of survival times given the covariate vectors $\mathbf{X}_j$ for $1 \leq j \leq p$ is the same as the conditional distribution of $\mathbf{T}$ given the vector of linear functions $\mathbf{Z} = (Z_1, Z_2, \ldots, Z_p)$, where $Z_j = \sum_{i=1}^{d} \beta_i X_{ji}$, and the regressor vector $\mathbf{X}_j$ corresponds to the marginal survival time T_j. Assume further that the conditional distribution of T_j given $T_1, \ldots, T_{j-1}$ for each $1 < j \leq p$ as well as given $Z_1, Z_2, \ldots Z_p$ is continuous and strictly increasing on its support, and conditioned on $Z_1, Z_2, \ldots, Z_p$, the survival time T_j is stochastically increasing in $T_1, \ldots, T_{j-1}$. Then there exist : (i) a function $\psi_j(.,.)$ from $R^p \times R^p$ into R, which is monotonically increasing in each of its last p*

arguments, and (ii) a p-dimensional random vector $\mathbf{U}$ *such that*

$$T_j = \psi_j(\mathbf{Z}, \mathbf{U}) \tag{3}$$

where $\mathbf{U}$ *has uniform distribution on the unit p-dimensional hypercube* $[0,1]^p$ *(i.e., the marginals of* $\mathbf{U}$ *are i.i.d uniform random variables), and it is independent of the* $\mathbf{X}_j$*'s.*

The proofs of both of Theorems 1 and 2 are given at the end of the paper. For the definition and a thorough discussion of stochastic increasing properties of random vectors, readers are referred to Barlow and Proschan (1975). The most interesting aspect of these two theorems is that they make very little assumption on the type of the conditional distribution of $\mathbf{T}$ or its marginals given $\mathbf{Z}$. As a result, they lead to a very flexible and substantially distribution free regression modeling for multivariate failure time data, which includes many of the standard models proposed and studied in the literature with more restricted distributional assumptions as special cases.

Both the theorems in the preceding section lead to versions of single index regression models with some intrinsic monotonicity properties for the link functions ψ_j's $(1 \leq j \leq p)$ in a multi-dimensional set-up. In a sense, Theorem 1 attempts to model only the marginal life times and leaves the structure of dependence among these marginals completely unspecified. Average derivative quantile regression estimates of the index coefficients β_i's can be be obtained first for each marginal life time separately using the procedure proposed in Chaudhuri, Doksum and Samarov (1997), and then those estimates can be combined by suitable weighted averaging. The local polynomial estimation (see e.g., Chaudhuri 1991, Chaudhuri and Loh 2002) of the conditional quantile functions and its derivatives for different marginal life times conditioned on the covariates can be carried out leading to the average derivative estimates of the index coefficients. This approach is comparable to marginal modeling and estimation based on "quasi partial likelihood estimating equations" obtained using marginal ranks and "independence working assumption", which have been widely used in the literature (see, e.g., Wei, Lin and Weissfeld 1989, Lin 1994, Spiekerman and Lin 1998). In particular, this gives a flexible modeling with very weak assumptions, and this approach of modeling includes marginal proportional hazard models considered by those earlier authors as a special case. The $n^{1/2}$-consistency and the asymptotic normality of the average derivative quantile regression estimates of the index coefficients β_i's can be established in this case using the same asymptotic analysis as in Chaudhuri, Doksum and Samarov (1997) applied to each of the marginal life times.

3 Multivariate Monotonicity of the Link Function

Theorem 2 in the preceding section attempts to completely model the joint distribution of the vector of life times taking into consideration the nature of dependence among the marginal life times. However, unlike Theorem 1, this theorem requires an extra condition, namely a stochastic monotonicity property of the vector of survival times. In this section, we will discuss this property for some well known models in the literature for dependent multivariate life time data.

In the frailty model for multivariate failure time data (see e.g., Vaupel, Manton and Stallard 1979, Clayton and Cuzick 1985, Hougaard 2000, Oakes 1989, Nielsen, Gill, Andersen and Sorensen 1992), the dependency among marginal failure times is induced by a frailty variable ξ, which is common to all the marginals. Conditional on ξ and the $\mathbf{X}_j$'s, the marginal failure times T_j's are assumed to follow independent distributions such that the conditional hazards function of T_j is $\xi\lambda_j(t|Z_j)$, where $Z_j = \sum_{i=1}^{d} \beta_i X_{ji}$ as before. Let us further assume that each of the hazard functions $\lambda_j(.|.)$'s satisfies the proportional hazard model (see e.g., Nielsen et al. 1992). Then the conditional hazard function of the j-th marginal failure time T_j $(1 \leq j \leq p)$ given the regressor $\mathbf{X}_j$ and the frailty variable ξ is given as $\xi\lambda_j(t|Z_j)\exp(\sum_{i=1}^{d} \beta_i X_{ji}) = \xi\mu_j(t)\exp(Z_j)$. In this case, in view of our discussion in Section 1, it is straight-forward to verify that for $1 \leq j \leq p$, there exist monotonic transformations $h_j(.)$'s from R into R such that $h_j(T_j) = \alpha_0 + \sum_{i=1}^{d} \beta_i X_{ji} + \epsilon_j = \alpha_0 + Z_j + \epsilon_j$, where the ϵ_j's are i.i.d random variables having the common distribution $F_\epsilon(s) = 1 - \exp\{-\exp(s)\}$, as mentioned in Section 1, and $\alpha_0 = \ln\xi$. Here ξ and the ϵ_j's are all independent, and they do not depend on the regressor vectors $\mathbf{X}_j$'s. This gives a formulation of the frailty model with a proportional hazard type model for each marginal life time in terms of a multi-response linear regression model, where each co-ordinate of the response vector is transformed by a monotonic transformation. In this formulation, there is a common random intercept term in the linear regresession that induces the dependence among the marginal failure times, and conditioned on that common intercept term, the marginal survival times follow independent linear regression models with different transformations for different marginals but the same regression coefficients $\beta_1, \beta_2, \ldots, \beta_d$ and the same error distribution. We now state a theorem which provides further insights into intrinsic monotonicity of such general frailty type models for multivariate life time data.

Theorem 3. *Consider the multi-response linear regression model with* *transformed marginals and a common random intercept term given as* $h_j(T_j) = \alpha_0 + \sum_{i=1}^{d} \beta_i X_{ji} + \epsilon_j = \alpha_0 + Z_j + \epsilon_j$. *Here, for* $1 \leq j \leq p$, *the* $h_j(.)$*'s are monotonically increasing functions from* R *into* R, α_0 *is a common random intercept term having an absolutely continuous distribution for all of the marginal life times, and the* ϵ_j*'s are independent with a common smooth positive density on the real line that is log-concave in nature. Then, conditioned on* $\mathbf{Z} = (Z_1, Z_2, \ldots, Z_p)$, *the survival time* T_j *is stochastically increasing in* $T_1, T_2, \ldots, T_{j-1}$ *for all* $1 < j \leq p$. *In particular, when the standard proportional hazard model holds for the hazard function* $\lambda_j(.|.)$ *in the frailty model discussed above, the vector of life times* $\mathbf{T}$ *satisfies a single index model with link function having the monotonicity property described in Theorem 2 for any absolutely continuous frailty distribution.*

Let us next consider a situation where the marginal life times $T_1 < T_2 < \ldots < T_p$ are ordered, and they form the order statistics from a common absolutely continous distribution supported on an interval $(0, \gamma)$, where γ may be ∞. For example, when there are multiple occurrences of tumors in the same subject, and T_j denotes the appearance time of the j-th tumor, such a model might be appropriate. Here we assume that the vectors of covariates $\mathbf{X}_j$'s are the same for different j's (i.e., $\mathbf{X}_j = \mathbf{X}$) as for different j's, these are all associated with the same subject. In such a situation, given $\mathbf{X}$, the T_j's may be viewed as order statistics derived from p i.i.d random samples from the conditional distribution of the life time. In this case, one can show that the monotonicity property of the link function described in Theorem 2 will again hold because after we condition on $\mathbf{X}$, for any $1 < j \leq p$, T_j will be stochastically increasing in $T_1, T_2, \ldots, T_{j-1}$. This is a consequence of the fact the conditional distribution of the j-th order statistic given the 1st, the 2nd, ... , the $(j-1)$-th oder statistics satisfies the monotone likelihood ratio property with respect to each of the conditioning variables, and this can be verified in a straight-forward way from the joint density of the order statistics from an i.i.d sample. It is a well known result that monotone likelihood ratio property implies the stochastic increasing property in our set up (see e.g., Barlow and Proschan 1975).

4 Proofs of Theorems

Proof of Theorem 1 : For any $1 \leq j \leq p$, let $F_j(t_j) = Pr(T_j \leq t_j | Z_j)$, which is the conditional c.d.f. of T_j given Z_j. Set $\psi_j(Z_j, U_j)$ to be the U_j-th quantile of the distribution F_j for any $0 < U_j < 1$. Clearly, ψ_j is monotonically increasing in its second argument. Now, if we take $U_j =$

$F_j(T_j)$, it will be a uniformly distributed random variable on $[0,1]$ with its distribution independent of $\mathbf{X}_j$. This completes the proof. □

Proof of Theorem 2 : Let $F_1(t_1) = Pr(T_1 \le t_1|Z_1, Z_2, \ldots Z_p)$, which is the conditional c.d.f. of T_1 given the Z_j's for all $1 \le j \le p$. Next, define $\psi_1(\mathbf{Z}, \mathbf{U})$ to be the U_1-th quantile of the distribution F_1, where $0 < U_1 < 1$ is the first co-ordinate of the p-dimensional vector $\mathbf{U}$. Then, if we set $U_1 = F_1(T_1)$, it will be a uniformly distributed random variable on $[0,1]$ with a distribution that is independent of the $\mathbf{X}_j$'s for all $1 \le j \le p$. Clearly, ψ_1 depends only on the first co-ordinate U_1 of $\mathbf{U}$, and it is monotonically increasing in U_1. Now, for $2 \le j \le p$, we sequentially define $U_j = F_j(T_j)$, where $F_j(t_j) = Pr(T_j \le t_j|Z_1, Z_2, \ldots Z_p, T_1, T_2, \ldots, T_{j-1}) = Pr(T_j \le t_j|Z_1, Z_2, \ldots Z_p, U_1, U_2, \ldots, U_{j-1})$, which is the conditional c.d.f. of T_j given $Z_1, Z_2, \ldots, Z_p$ and $T_1, T_2, \ldots, T_{j-1}$. This ensures that $U_1, U_2, \ldots, U_p$ are i.i.d random variables uniformly distributed on $[0,1]$, and their joint distribution would be independent of $\mathbf{X}_1, \mathbf{X}_2, \ldots, \mathbf{X}_p$. Note also that U_j is a monotonically increasing function of T_j. Finally, we set $\psi_j(\mathbf{Z}, \mathbf{U})$ to be the U_j-th quantile of the distribution F_j. Then, ψ_j depends only on the first j co-ordinates $U_1, U_2, \ldots, U_j$ of $\mathbf{U}$, and it is clearly a monotonically increasing function of U_j. Further, ψ_j will be a monotonically increasing function of each of $U_1, U_2, \ldots, U_{j-1}$ if conditioned on $Z_1, Z_2, \ldots, Z_p$, the distribution of T_j is stochastically increasing in $T_1, T_2, \ldots, T_{j-1}$. This completes the proof of the theorem. □

Proof of Theorem 3 : It is easy to verify that for two independent random variables Q and S with absolutely continuous distributions and for $W = Q + S$, the conditional distribution of Q given W will have the monotone likelihood ratio property if the density of S is log-concave. One way to verify this is by considering the conditional density of Q given W. Since monotone likelihood ratio property implies stochastic increasing property (see e.g., Barlow and Proschan 1975), the distribution of Q will be stochastically increasing in W in this case. This result can be extended for independent $S_1, S_2, \ldots, S_p$ and $W_1 = Q + S_1, W_2 = Q + S_2, \ldots, W_p = Q + S_p$ in a straight-forward way to yield the stochastic increasing property of the conditional distribution of Q (and consequently that of the distribution of W_j) given $W_1, W_2, \ldots, W_{j-1}$ for any $2 \le j \le p$. This completes the proof of the first assertion in the theorem. The second assertion in the theorem is an immediate consequence of the fact that in the linear regression model with a monotonically transformed response, which is a reformulation the proportional hazard model of Cox, the residual term has an extreme value distribution that has a log-concave probability density function. □

References

1. BAGDONAVICIUS, V. AND NIKULIN, M. (1999). Generalized proportional hazards model based on modified partial likelihood. *Life Time Data Analysis*, **5**, 329-350.

2. BARLOW, R. E. AND PROSCHAN, F. (1975). *Statistical Theory or Reliability and Life Testing : Probability Models.* Hold, Rinehart and Winston, New York.

3. CAI, T. AND CHENG, S. (2004). Semiparametric regrssion analysis for doubly censored data. *Biometrika*, **91**, 277–290.

4. CHAUDHURI, P. (1991). Nonparametric estimates of regression quantiles and their local Bahadur representation. *The Annals of Statistics*, **19**, 760–777.

5. CHAUDHURI, P., DOKSUM, K. AND SAMAROV, A. (1997). On average derivative quantile regression. *The Annals of Statistics*, **25**, 715–744.

6. CHAUDHURI, P. AND LOH, W. Y.(2002) Nonparametric estimation of conditional quantiles using quantile regression trees. *Bernoulli*, **8**, 561–576.

7. CLAYTON, D. G. AND CUZICK, J. (1985). Multivariate generalizations of the proportinal hazards model (with discussion). *Journal of the Royal Statistical Society, Series B*, **148**, 82–117.

8. COX, D. R. (1972). Regression models and life tables (with discussion). *Journal of the Royal Statistical Society, Series B*, **34**, 187–202.

9. COX, D. R. (1975). Partial likelihood. *Biometrika*, **62**, 269–276.

10. DABROWSKA, D. AND DOKSUM, K. (1987). Estimates and confidence intervals for median and mean life in the proportional hazard model. *Biometrika*, **74**, 799–807.

11. DOKSUM, K. (1987). An extension of partial likelihood methods for proportional hazard models to general transformation models. *The Annals of Statistics*, **15**, 325–345.

12. DOKSUM, K. AND GASKO, M. (1990). On a correspondence between models in binary regression analysis and in survival analysis. *International Statistical Review*, **58**, 243–252.

13. HOUGAARD, P. (2000). *Analysis of Multivariate Survival Data.* Springer, New York

14. KALBFLEISCH, J. D. AND PRENTICE, R. L. (2002). *The Statistical Analysis of Failure Time Data.* Wiley, New York.

15. KOENKER, R. AND GELING, O. (2001). Reappraising medfly longevity : a quantile regression survival analysis. *Journal of the American Statistical Associations*, **96**.

16. LIN, D. Y. (1994). Cox regression analysis of multivariate failure time data : the marginal approach. *Statistics in Medicine*, **13**, 2233–2247.

17. MANTEL, N., BOHIDAR, N. R. AND CIMINERA, J. L. (1977). Mantel-Haenszel analysis of litter-matched time-to-response data, with modifications for recovery of interlitter information. *Cancer Research*, **37**, 3863–3868.

18. NIELSEN, G. G., GILL, R. D., ANDERSEN, P. K. AND SORENSEN, T. I. A. (1992). A counting process approach to maximum likelihood estimation in frailty models. *Scandinavian Journal of Statistics*, **19**, 25–43.

19. OAKES, D. (1989). Bivariate survival models induced by frailties. *Journal of the American Statistical Association*, **84**, 487–493.

20. OAKES, D. AND DASU, T. (1990). A note on residual life. *Biometrika*, **77**, 409–410.

21. PULVER, A. E. AND LIANG, K. Y. (1991). Estimating effects of proband characteristics and familial risk : II, the association between age at onset and familial risk in the Maryland schizophrenia sample. *Genetic Epidemiology*, **8**, 339–350.

22. SPIEKERMAN, C. F. AND LIN, D. Y. (1998). Marginal regression models for multivariate failure time data. *Journal of the American Statistical Association*, **93**, 1164–1175.

23. VAUPEL, J. W., MANTON, K. G. AND STALLARD, E. (1979). The impact of heterogeneity in individual frailty on the dynamics of mortality. *Demography*, **16**, 439–454.

24. WEI, L. J., LIN, D. Y. AND WEISSFELD, L. (1989). Regression analysis of multivariate incomplete failure time data by modeling marginal distributions. *Journal of the American Statistical Association*, **84**, 1065–1073.

Chapter 12

AGGREGATION OF DENSITY ESTIMATORS AND DIMENSION REDUCTION

Alexander Samarov and Alexandre Tsybakov

University of Massachusetts-Lowell and MIT, U.S.A.

Laboratoire de Probabilités et Modèles Aléatoires, Université Paris VI, FRANCE

E-mail: samarov@mit.edu & tsybakov@ccr.jussieu.fr

We consider the problem of model-selection-type aggregation of arbitrary density estimators using MISE risk. Given a collection of arbitrary density estimators, we propose a data-based selector of the best estimator in the collection and prove a general ready-to-use oracle inequality for the selected aggregate estimator. We then apply this inequality to the adaptive estimation of a multivariate density in a "multiple index" model. We show that the proposed aggregate estimator adapts to the unknown index space of unknown dimension in the sense that it allows us to estimate the density with the optimal rate attainable when the index space is known.

Key words: Nonparametric density estimation; Aggregation of estimators; Dimensionality reduction model.

1 Introduction

The problem of aggregation of M arbitrary estimators has been recently studied by many authors (see, e.g., Nemirovski (2000), Yang (2000), Devroye and Lugosi (2000), Catoni (2004), Wegkamp (2003), Tsybakov (2003), Birgé (2003), Bunea, Tsybakov and Wegkamp (2004), Rigollet and Tsybakov (2004) and the references cited therein). A motivating factor is that in frequently used statistical models (such as regression or density estimation) there exists a great variety of possible competing estimators, and it is often difficult to decide which estimator to choose. Assume that a Statistician is given a list of size M of such estimators: $p_1, \ldots, p_M$. A natural idea is then to look for a new, improved, estimator constructed by combining $p_1, \ldots, p_M$ in a suitable way. A combined "super-estimator" obtained from $p_1, \ldots, p_M$ is usually called *aggregate* and its construction is called aggregation.

One can distinguish between three main types of aggregation: model selection (MS) aggregation, convex (C) aggregation and linear (L) aggregation. The objective of (MS) is to select the optimal single estimator from the list; that of (C) is to select the optimal convex combination of the given estimators; and that of (L) is to select the optimal linear combination of the given estimators. The notion of optimality mentioned here is defined with respect to a given risk function, and it can be formalized in a minimax sense leading to the concept of optimal rates of aggregation [Tsybakov (2003)]. A standard approach to establishing this kind of optimality is to show that the aggregate satisfies a sufficiently precise oracle inequality.

Most of the currently available results on aggregation were obtained for the regression model (see a recent overview in Bunea, Tsybakov and Wegkamp (2004)). The literature on aggregation of density estimators is not as large: Catoni (2004) and Yang (2000) investigated the (MS) aggregation with the Kullback-Leibler divergence as a loss function; Devroye and Lugosi (2000) developed a method of (MS) aggregation of density estimators under the L_1 loss. Another approach to density aggregation under the L_1 loss was proposed by Birgé (2003). Finally, we mention the recent paper of Rigollet and Tsybakov (2004) on optimal convex (C) and linear (L) aggregation of density estimators under the L_2 loss, and the work of Juditsky, Nazin, Tsybakov and Vayatis (2005a, 2005b) where a recursive aggregation procedure is proposed for various statistical contexts, including density estimation, classification and regression.

In this paper we consider the (MS) aggregation of arbitrary density estimators under the L_2 loss (MISE). The main precursor of our study is the paper of Wegkamp (1999) who treated a more particular problem of bandwidth selection for kernel density estimation, but some of his results can be interpreted in general aggregation framework. For instance, some oracle inequalities can be deduced from Wegkamp's work, although he does not derive them explicitly. Our first aim is to obtain a ready-to-use oracle inequality for the L_2 (MS) aggregation using techniques that are somewhat different from those of Wegkamp (1999). Then we consider an example of application of this inequality, namely, to the adaptive estimation of a multivariate density in a *multiple index model*. We show that the proposed aggregate adapts to the unknown index matrix B in the sense that it allows to estimate the density with the optimal rate attainable when B is known.

2 A density aggregation theorem

Let $X_1, \ldots, X_n$ be i.i.d. random vectors with common probability density p on $\mathbf{R}^d$. Suppose that we are given M candidate estimators $p_1, \ldots, p_M$ of the density p based on the sample $X_1, \ldots, X_n$. Our goal here is the model selection (MS) aggregation, that is, we would like to choose $\tilde{N} \in \{1, \ldots, M\}$, a random index based on the data, such that the aggregate $p_{\tilde{N}}$ satisfies an oracle inequality of the form

$$\mathbf{E}\|p_{\tilde{N}} - p\|^2 \leq (1 + \delta_n) \min_{1 \leq N \leq M} \mathbf{E}\|p_N - p\|^2 + r_n, \tag{1}$$

where the value $\delta_n = \delta_{n,M} > 0$ and the remainder term $r_n = r_{n,M} > 0$ are small enough (they tend to 0, as $n \to \infty$; their dependence on M may be suppressed because M will be chosen to grow with n, see Assumption 1 below), and

$$\|p\| = \left(\int p^2 \right)^{1/2} = \left(\int_{\mathbf{R}^d} p^2(x) dx \right)^{1/2}.$$

We interpret the inequality (1) as the fact that the aggregate $p_{\tilde{N}}$ mimics asymptotically the best among the estimators $p_1, \ldots, p_M$ (in the sense of MISE), up to a small remainder term. Note that here $p_1, \ldots, p_M$ are arbitrary estimators, not necessarily belonging to a specific family of nonparametric estimators. In particular, some estimators in the list can be parametric and others can be nonparametric of different nature (kernel, spline, wavelet etc.). To apply the inequality (1) in the nonparametric density estimation context, it is usually sufficient that the remainder r_n were smaller in order than the standard nonparametric MISE rates, for example, $r_n = (\log n)^a / n$ for some $a > 0$. This will be the case in the result that we prove below.

In order to define a specific aggregation algorithm, we split the sample $X_1, \ldots, X_n$ into two parts: I_1, used for constructing "base" estimators p_N, and I_2, used for their aggregation. Let $n_1 = \mathrm{Card}(I_1)$, $n_2 = \mathrm{Card}(I_2)$, $n = n_1 + n_2$. We select $\tilde{N}$ using the rule:

$$\tilde{N} = \arg \min_{1 \leq N \leq M} J_N, \tag{2}$$

where

$$J_N = -\frac{2}{n_2} \sum_{I_2} p_N(X_i) + \int p_N^2. \tag{3}$$

Here and later we abbreviate $\sum_{X_i \in I_2} = \sum_{I_2}$. Note that, because subsamples I_1 and I_2 are independent,

$$\mathbf{E}\left(\frac{1}{n_2} \sum_{I_2} p_N(X_i) \right) = \mathbf{E}\left(\int p_N(x) p(x) dx \right). \tag{4}$$

Therefore, J_N is such that

$$\mathbf{E}(J_N) = \mathbf{E}\|p - p_N\|^2 - \|p\|^2, \quad N = 1, \ldots, M,$$

i.e. J_N is an unbiased estimator of the MISE of p_N, up to the summand $\|p\|^2$ free from N.

To state the aggregation theorem, we need the following assumptions.

Assumption 1. *There exist finite positive constants a_1, a_2, and C_1, C_2 such that*

$$\sum_{N=1}^{M} \mathbf{E}\|p_N - p\| \le C_1 n^{a_1} \tag{5}$$

with $M \ge 2$ satisfying

$$M \ge C_2 n^{a_2}. \tag{6}$$

Assumption 2. *There exists a finite constant C_3 and a constant $\gamma_0 \le 1/12$ such that*

$$\sum_{N=1}^{M} \mathbf{E}\left[\|p_N - p\|_\infty \exp\left(-\frac{\gamma_0 \log^{7/4} M}{\|p_N - p\|_\infty}\right)\right] \le C_3 \log^2 M, \tag{7}$$

where $\|f\|_\infty = \sup_{x \in D} |f(x)|$ and $D \in \mathbf{R}^d$ is the support of the density $p(\cdot)$.

Assumption 3. *The density p is uniformly bounded: there exists a constant $p_{max} < \infty$ such that $\|p\|_\infty \le p_{max}$.*

Remark 1. Assumptions 1 – 3 are not very restrictive. First of all, note that the (MS) aggregation has the largest oracle risk and the smallest order of the remainder term among the three types of aggregation mentioned in the introduction (Tsybakov (2003), see also Bunea, Tsybakov and Wegkamp (2004), where these issues are discussed for the regression model). Therefore, it is not crucial to use (MS) aggregation when the number M of base estimators is small, for example, when M grows as a power of $\log n$. In this case one can efficiently mimic more powerful convex or linear oracles (Rigollet and Tsybakov (2004)). However, if the number M of estimators to aggregate is polynomial in n or bigger, the remainder terms of convex and linear aggregation become too large as compared to the typical nonparametric MISE rates. This does not happen for the (MS) aggregation remainder term. Therefore, the (MS) aggregation is the type of aggregation which is especially important for polynomial M, explaining why assumption (6) is natural.

The assumption (5) is usually satisfied: it suffices to have the risks $\mathbf{E}\|p_N - p\|$ uniformly bounded and M bounded by a power of n. Typically p_N are consistent with rates, and we have even a stronger bound.

Finally, Assumption 2 looks rather technical, but it is also quite a mild one. For example, it is satisfied if

$$\max_{N=1,\dots,M} \mathbf{E}\left[\|p_N - p\|_\infty I(\|p_N - p\|_\infty > \gamma_0 \log^{3/4} M)\right] \le \frac{\log^2 M}{M}, \qquad (8)$$

where $I(\cdot)$ denotes the indicator function. Below we give examples showing that (8) is not a restrictive condition in density estimation. For instance, a sufficient condition for (8) is that the probability $\mathbf{P}(\|p_N - p\|_\infty > t)$ decreases exponentially in t, as $t \to \infty$ (an example is given in Section 3), but often it suffices to check a weaker and quite natural condition that the deviation of the stochastic part of the estimator $\mathbf{P}(\|p_N - \mathbf{E}p_N\|_\infty > t)$ is exponentially small (see the example below).

To show that (8) implies (7), define the event $W = \{\|p_N - p\|_\infty \le \gamma_0 \log^{3/4} M\}$ and write

$$\begin{aligned}
&\mathbf{E}\left[\|p_N - p\|_\infty \exp\left(-\frac{\gamma_0 \log^{7/4} M}{\|p_N - p\|_\infty}\right)\right] \\
&\quad \le \frac{\gamma_0}{M} \log^{3/4} M + \mathbf{E}\left[\|p_N - p\|_\infty \exp\left(-\frac{\gamma_0 \log^{7/4} M}{\|p_N - p\|_\infty}\right) I(W^c)\right] \\
&\quad \le \frac{\gamma_0}{M} \log^{3/4} M + \mathbf{E}\left[\|p_N - p\|_\infty I(\|p_N - p\|_\infty > \gamma_0 \log^{3/4} M)\right].
\end{aligned}$$

Consider a simple example illustrating that (8) is indeed a mild assumption: let p be supported on $[0,1]$ and let p_N be a kernel density estimator with bandwidth $h_N > 0$ and with a bounded Lipschitz continuous kernel $K \ge 0$ such that $\int K = 1$:

$$p_N(x) = \frac{1}{n_1 h_N} \sum_{i \subset I_1} K\left(\frac{X_i - x}{h_N}\right), \quad N = 1, \dots, M.$$

Then, clearly, $\|\mathbf{E}p_N\|_\infty \le p_{\max}$ and $\|p_N\|_\infty \le D_1/h_{\min}$ where $D_1 > 0$ is a constant and $h_{\min} = \min\{h_1, \dots, h_M\}$. Hence $\|p_N - p\|_\infty \le 2p_{\max} + \|p_N - \mathbf{E}p_N\|_\infty$ and $\|p_N - p\|_\infty \le D_1/h_{\min} + p_{\max}$, so that we get $\mathbf{E}\left[\|p_N - p\|_\infty I(\|p_N - p\|_\infty > \gamma_0 \log^{3/4} M)\right] \le (D_1/h_{\min} + p_{\max})\mathbf{P}(\|p_N - \mathbf{E}p_N\|_\infty > D_2 \log^{3/4} M)$ with some constant $D_2 > 0$. Now, using Bernstein's inequality, the Lipschitz condition on K and bounding $\|p_N - \mathbf{E}p_N\|_\infty$ by the maximum over a fine enough grid on $[0,1]$ with step $n_1^{-\alpha}$ for some large enough $\alpha > 0$ we get the bound on the probability $\mathbf{P}(\|p_N - \mathbf{E}p_N\|_\infty > D_2 \log^{3/4} M) \le$

$D_3 n_1^{\alpha} \exp(-D_4 n_1 h_N \log^{3/4} M) \le D_3 n_1^{\alpha} \exp(-D_4 n_1 h_{\min} \log^{3/4} M)$ with some constants $D_3, D_4 > 0$. Finally, if $M \asymp n_1^a$ with $a > 0$ and if the bandwidths are such that $h_{\min} \ge n_1^{-1} \log^{3/4} n_1$ we get the bound $\mathbf{E}\left[\|p_N - p\|_\infty I(\|p_N - p\|_\infty > \gamma_0 \log^{3/4} M)\right] \le D_5 M^{a'} \exp(-D_6 \log^{3/2} M)$ with some constants $D_5, D_6, a' > 0$, which implies (8) for n_1 large enough. Thus, Assumption 2 holds under quite standard conditions on the kernel K and on the bandwidths h_N.

Theorem 1. *If* $n_2 = \lfloor \frac{cn}{\log M} \rfloor$ *for some constant* $c > 0$ *such that* $1 \le n_2 < n$, *then, under Assumptions 1 – 3, we have*

$$\mathbf{E}\|p_{\tilde N} - p\|^2 \le \left(1 + \frac{C^*}{\log^{1/4} M}\right) \min_{1 \le N \le M} \mathbf{E}\|p_N - p\|^2 + C^* \frac{\log^3 M}{n}, \qquad (9)$$

where $C^* > 0$ *is a constant which depends only on* $p_{max}, a_1, a_2, C_1, C_2, C_3, c$.

Proof. Note first that, by definition, $J_{\tilde N} \le J_N$ for all $1 \le N \le M$. Using this and (4), we have

$$\begin{aligned}
&\mathbf{E}\|p_{\tilde N} - p\|^2 - \mathbf{E}\|p_N - p\|^2 \\
&\quad = \mathbf{E}\left(-2\int p p_{\tilde N} + \int p_{\tilde N}^2\right) - \mathbf{E}\left(-\frac{2}{n_2}\sum_{I_2} p_N(X_i) + \int p_N^2\right) \\
&\quad = \mathbf{E}(J_{\tilde N}) - \mathbf{E}(J_N) + \mathbf{E}\left(\frac{2}{n_2}\sum_{I_2} p_{\tilde N}(X_i) - 2\int p p_{\tilde N}\right) \\
&\quad \le 2\mathbf{E}\left(\frac{1}{n_2}\sum_{I_2} p_{\tilde N}(X_i) - \int p p_{\tilde N}\right) \\
&\quad = 2\mathbf{E}[Z_{\tilde N}], \qquad (10)
\end{aligned}$$

where

$$Z_N \triangleq \frac{1}{n_2}\sum_{I_2}(p_N(X_i) - p(X_i)) - \left(\int p p_N - \int p^2\right).$$

Set $W_N = \gamma(\|p_N - p\|^2 + r)$, where $r = (\log M)^2/n_2$ and $\gamma > 0$ will be chosen later. Denoting by $I(A)$ the indicator of a set A, we have

$$\begin{aligned}
\mathbf{E}(|Z_{\tilde N}|) &\le \mathbf{E}(|Z_{\tilde N}| I(|Z_{\tilde N}| < W_{\tilde N})) + \mathbf{E}(|Z_{\tilde N}| I(|Z_{\tilde N}| \ge W_{\tilde N})) \\
&\le \gamma \mathbf{E}[\|p_{\tilde N} - p\|^2 + r] + \mathbf{E}(|Z_{\tilde N}| I(|Z_{\tilde N}| \ge W_{\tilde N})) \\
&\le \gamma \mathbf{E}\|p_{\tilde N} - p\|^2 + \gamma r + \sum_{N=1}^{M} \mathbf{E}(|Z_N| I(|Z_N| \ge W_N)). \qquad (11)
\end{aligned}$$

Now,

$$\mathbf{E}(|Z_N| I(|Z_N| \ge W_N)) = \mathbf{E}\{\mathbf{E}[|Z_N| I(|Z_N| \ge W_N) | I_1]\}. \qquad (12)$$

Note that $Z_N = n_2^{-1}\sum_{I_2}[\zeta_{iN} - \mathbf{E}(\zeta_{iN}|I_1)]$ where, for fixed subsample I_1, the random variables $\zeta_{iN} = p_N(X_i) - p(X_i), X_i \in I_2$, are i.i.d., and

$$\mathbf{E}(\zeta_{iN}|I_1) = \int pp_N - \int p^2,$$

$$\mathbf{E}(\zeta_{iN}^2|I_1) = \int (p_N(x) - p(x))^2 p(x)dx \le p_{max}\|p_N - p\|^2,$$

by Assumption 3. To evaluate (12) we will use Bernstein's inequality (see, e.g., Serfling (1980)):

$$\mathbf{P}(|Z_N| \ge t|I_1) \le 2\rho(t) \quad \text{for all } t > 0,$$

where

$$\rho(t) = \exp\left(-\frac{n_2 t^2}{2p_{max}\|p_N - p\|^2 + 2t\|p_N - p\|_\infty/3}\right).$$

We have

$$\begin{aligned}\mathbf{E}[|Z_N|I(|Z_N| \ge W_N)|I_1] &= W_N\mathbf{P}(|Z_N| \ge W_N|I_1) + \int_{W_N}^{\infty}\mathbf{P}(|Z_N| \ge t|I_1)dt \\ &\le A_0 + A_1, \end{aligned} \tag{13}$$

where

$$A_0 = 2W_N\rho(W_N) \quad \text{and} \quad A_1 = 2\int_{W_N}^{\infty}\rho(t)dt.$$

We first bound from above the integral A_1. Consider the following two sets:

$$T_1 = \{t > 0 : \ t\|p_N - p\|_\infty \le 3p_{max}\|p_N - p\|^2\},$$
$$T_2 = \{t > 0 : \ t\|p_N - p\|_\infty > 3p_{max}\|p_N - p\|^2\}.$$

On T_1 we evaluate:

$$\rho(t) \le \exp\left(-\frac{n_2 t^2}{4p_{max}\|p_N - p\|^2}\right), \quad \text{for all } \ t \in T_1, \tag{14}$$

while on T_2:

$$\rho(t) \le \exp\left(-\frac{3n_2 t}{4\|p_N - p\|_\infty}\right), \quad \text{for all } \ t \in T_2. \tag{15}$$

Consider first the set T_1. Setting $u = t\sqrt{n_2}/(\sqrt{2p_{max}}\ \|p_N - p\|)$ and $W_N' = W_N\sqrt{n_2}/(\sqrt{2p_{max}}\ \|p_N - p\|)$, we get

$$\begin{aligned} A_{11} &\triangleq \int_{W_N}^{\infty}\exp\left(-\frac{n_2t^2}{4p_{max}\|p_N - p\|^2}\right)I(t \subset T_1)dt \\ &\le \frac{\sqrt{2p_{max}}\ \|p_N - p\|}{\sqrt{n_2}}\int_{W_N'}^{\infty} e^{-u^2/2}du \\ &\le C\frac{\sqrt{p_{max}}\ \|p_N - p\|}{\sqrt{n_2}}\exp(-(W_N')^2/2) \\ &= C\frac{\sqrt{p_{max}}\ \|p_N - p\|}{\sqrt{n_2}}\exp\left(-\frac{n_2W_N^2}{4p_{max}\|p_N - p\|^2}\right) \\ &\le C\sqrt{\frac{p_{max}}{n_2}}\ \|p_N - p\|\exp\left(-\frac{\gamma^2\log^2 M}{p_{max}}\right), \end{aligned}$$

where we have used $W_N \geq 2\gamma(\log M)\|p_N - p\|/\sqrt{n_2}$, and C, here and later, denotes a positive constant, not always the same.

Consider now the set T_2. Setting $W_N'' = 3n_2 W_N/(4\|p_N - p\|_\infty)$, we find

$$\begin{aligned}
A_{12} &\triangleq \int_{W_N}^{\infty} \exp\left(-\frac{3n_2 t}{4\|p_N - p\|_\infty}\right) I(t \in T_2) dt \\
&\leq \frac{4\|p_N - p\|_\infty}{3n_2} \int_{W_N''}^{\infty} e^{-u} du \\
&= \frac{4\|p_N - p\|_\infty}{3n_2} \exp\left(-\frac{3n_2 W_N}{4\|p_N - p\|_\infty}\right) \\
&\leq \frac{4\|p_N - p\|_\infty}{3n_2} \exp\left(-\frac{3\gamma \log^2 M}{4\|p_N - p\|_\infty}\right),
\end{aligned}$$

where we have used $W_N \geq \gamma(\log M)^2/n_2$. Therefore we have

$$\begin{aligned}
A_1 \leq 2(A_{11} + A_{12}) &\leq C\sqrt{\frac{p_{max}}{n_2}}\,\|p_N - p\| \exp\left(-\frac{\gamma^2 \log^2 M}{p_{max}}\right) \\
&\quad + \frac{8\|p_N - p\|_\infty}{3n_2} \exp\left(-\frac{3\gamma \log^2 M}{4\|p_N - p\|_\infty}\right). \qquad (16)
\end{aligned}$$

We turn now to the evaluation of A_0. The argument here is similar to that used above. If $W_N \in T_1$, then using (14) and the inequality $x\exp(-x^2) \leq \exp(-x^2/2)$, for all $x > 0$, we get

$$\begin{aligned}
A_0 &\leq 2W_N \exp\left(-\frac{n_2 W_N^2}{4p_{max}\|p_N - p\|^2}\right) \\
&\leq 4\sqrt{\frac{p_{max}}{n_2}}\,\|p_N - p\| \exp\left(-\frac{n_2 W_N^2}{8p_{max}\|p_N - p\|^2}\right) \\
&\leq 4\sqrt{\frac{p_{max}}{n_2}}\,\|p_N - p\| \exp\left(-\frac{\gamma^2 \log^2 M}{2p_{max}}\right). \qquad (17)
\end{aligned}$$

Similarly, if $W_N \in T_2$, then using (15) and the inequality $x\exp(-x) \leq \exp(-x/2)$, for all $x > 0$, we find

$$\begin{aligned}
A_0 &\leq 2W_N \exp\left(-\frac{3n_2 W_N}{4\|p_N - p\|_\infty}\right) \\
&\leq \frac{8\|p_N - p\|_\infty}{3n_2} \exp\left(-\frac{3n_2 W_N}{8\|p_N - p\|_\infty}\right) \\
&\leq \frac{8\|p_N - p\|_\infty}{3n_2} \exp\left(-\frac{3\gamma \log^2 M}{8\|p_N - p\|_\infty}\right). \qquad (18)
\end{aligned}$$

Returning now to (12) and (13) and using (16) – (18), we obtain

$$\begin{aligned}\mathbf{E}(|Z_N|I(|Z_N| \geq W_N)) &\leq \mathbf{E}(A_0) + \mathbf{E}(A_1)\\ &\leq \frac{C}{\sqrt{n_2}} \exp(-C^{-1}\gamma^2 \log^2 M)\mathbf{E}\|p_N - p\|\\ &+ \frac{C}{n_2}\mathbf{E}\left[\|p_N - p\|_\infty \exp\left(-\frac{3\gamma \log^2 M}{8\|p_N - p\|_\infty}\right)\right].\end{aligned}$$

This together with (11) gives

$$\begin{aligned}2\mathbf{E}(|Z_{\tilde N}|) &\leq 2\gamma\left(\mathbf{E}\|p_{\tilde N} - p\|^2 + \frac{\log^2 M}{n_2}\right)\\ &+ \frac{C}{\sqrt{n_2}} \exp(-C^{-1}\gamma^2 \log^2 M) \sum_{N=1}^{M} \mathbf{E}\|p_N - p\|\\ &+ \frac{C}{n_2}\sum_{N=1}^{M}\mathbf{E}\left[\|p_N - p\|_\infty \exp\left(-\frac{3\gamma \log^2 M}{8\|p_N - p\|_\infty}\right)\right]\\ &\triangleq 2\gamma\mathbf{E}\|p_{\tilde N} - p\|^2 + R. \qquad (19)\end{aligned}$$

From (19) and (10) we get

$$(1 - 2\gamma)\mathbf{E}\|p_{\tilde N} - p\|^2 \leq \mathbf{E}\|p_N - p\|^2 + R,$$

and, with $0 < \gamma < 1/4$,

$$\mathbf{E}\|p_{\tilde N} - p\|^2 \leq (1 + 4\gamma)\mathbf{E}\|p_N - p\|^2 + (1 + 4\gamma)R.$$

Set now $\gamma = (8\gamma_0/3)(\log M)^{-1/4}$ where $\gamma_0 \leq 1/12$ is the constant in Assumption 2. Then $0 < \gamma \leq 2(\log 2)^{-1/4}/9 < 1/4$ for all $M \geq 2$, and we have the following bound on the remainder term R defined in (19):

$$\begin{aligned}R \leq C\Bigg\{&\frac{\log^{7/4} M}{n_2} + \frac{1}{\sqrt{n_2}} \exp(-C^{-1}\log^{3/2} M) \sum_{N=1}^{M} \mathbf{E}\|p_N - p\|\\ &+ \frac{1}{n_2}\sum_{N=1}^{M}\mathbf{E}\left[\|p_N - p\|_\infty \exp\left(-\frac{\gamma_0 \log^{7/4} M}{\|p_N - p\|_\infty}\right)\right]\Bigg\}.\end{aligned}$$

The theorem follows from the last two displays by applying Assumptions 1 and 2.

Remark 2. Inspection of the proof shows that Assumption 2 can be slightly generalized and the remainder term $(\log M)^3/n$ in (9) can be reduced to $(\log M)^{1+\varepsilon}/n$ for an arbitrarily small $\varepsilon > 0$. To obtain this, it suffices to fix an arbitrarily small $\nu > 0$, to replace $\log^2 M$ by $(\log M)^{1+\nu}$ in the definition of r, and to take $\gamma \asymp (\log M)^{-\nu'}$ with $\nu' < \nu/2$, $n_2 = \lfloor cn/(\log M)^\nu \rfloor$. Then $\log^{7/4} M$ and $\log^2 M$ in (7) can be replaced by $(\log M)^{1+\nu-\nu'}$ and $(\log M)^{1+2\nu-\nu'}$, respectively. We did not include these extensions in Theorem 1, because they require more notation but seem not to be crucial for application of the result.

3 Application to a dimensionality reduction model

Let $X_1, \ldots, X_n$ be i.i.d. random vectors with common probability density p on $\mathbf{R}^d, d \geq 2$. We consider the problem of nonparametric estimation of the density p assuming that it has the form

$$p(x) \equiv f_B(x) \triangleq \phi_d(x) g(B^T x), \quad x \in \mathbf{R}^d, \tag{20}$$

where B is an unknown $d \times m$ matrix with orthonormal columns, $1 \leq m \leq d$, the function $g : \mathbf{R}^m \to [0, \infty)$ is unknown, and $\phi_d(\cdot)$ is the density of the standard d-variate normal distribution. Our goal is to show, using Theorem 1, that one can estimate the density (20), without knowing B and m, with the same rate as the optimal rate attainable when B and m are known.

Note that the representation (20) is not unique. In particular, if Q_m is an $m \times m$ orthogonal matrix, the density p in (20) can be rewritten as $p(x) = \phi_d(x) g_1(B_1^T x)$ with $g_1(y) = g(Q_m y)$ and $B_1 = BQ_m$. However, the linear subspace $\mathcal{M}$ spanned by the columns of B is uniquely defined by (20). By analogy with regression models, e.g. Li (1991), Hristache, et al. (2001), we will call $\mathcal{M}$ the *index space*. In particular, if the dimension of $\mathcal{M}$ is 1, (20) can be viewed as a density analog of the single index model in regression. In general, if the dimension of $\mathcal{M}$ is arbitrary, we call (20) the *multiple index model*. The directions where the density of projections of X_i is standard normal are interpreted as non-interesting ("pure noise" directions).

The model (20) can be viewed as a modification of the projection pursuit density estimation (PPDE) model, e.g. Huber (1985). A common PPDE model corresponds to the special case of (20) where the function g can be represented as a product of densities corresponding to one-dimensional projections. In this case, the density can be estimated with one-dimensional rate (Samarov and Tsybakov (2004)), and thus the dimension reduction principle is realized. Models similar to (20) also arise in biased, or weighted, sampling, where a direct sampling from a density f is, for some reason, impossible, and an observation $X = x$ from f may be available with a relative probability proportional to a so-called biasing function $w(x)$. The biased observations have the density $p(x) = f(x)w(x)/\int w(x)f(x)dx$, and a typical problem in biased estimation is: having observations from p, estimate f, when $w(\cdot)$ is known, e.g. Cox (1969), Patil and Rao (1977). In our setting, $f = \phi_d$ is known while the biasing function has the form $g(B^T x)$ and is unknown, and our goal is to estimate $p(\cdot)$.

When the dimension m and an index matrix B (i.e. any of the matrices, equivalent up to an orthogonal transformation, that define the index space

$\mathcal{M}$) are specified, the density (20) can be estimated using a kernel estimator

$$\hat{p}_{m,B}(x) = \frac{\phi_d(x)}{\phi_m(B^T x)} \frac{1}{nh^m} \sum_{i=1}^{n} K\left(\frac{B^T(X_i - x)}{h}\right), \tag{21}$$

with appropriately chosen bandwidth $h > 0$ and kernel $K : \mathbf{R}^m \to \mathbf{R}^1$. We will assume the following.

Assumption 4. *The function $g : \mathbf{R}^m \to [0, \infty)$ in (20) is bounded on $\mathbf{R}^m$ with its gradient ∇g and Hessian $\nabla^2 g$, so that $\max\{g(z), |\nabla g(z)|_m, \|\nabla^2 g(z)\|_2\} \leq L_g$, for all $z \in \mathbf{R}^m$, where L_g is a constant, $|\cdot|_m$ denotes the Euclidean norm in $\mathbf{R}^m$ and $\|A\|_2 = \mathrm{Tr}^{1/2}(AA^T)$ denotes the Frobenius norm of the matrix A.*

Assumption 5. *The kernel $K : \mathbf{R}^m \to \mathbf{R}^1$ is a bounded function supported on $[-1,1]^m$ and such that $\int_{\mathbf{R}^m} K(t)dt = 1$ and $\int_{\mathbf{R}^m} K(t)t_j dt = 0$, $j = 1, \ldots, m$, where t_j is the jth component of $t \in \mathbf{R}^m$.*

Kernels satisfying Assumption 5 can be easily constructed as products of m one-dimensional kernels.

We first suppose that the dimension m and an index matrix B are known and establish the rate of convergence of the estimator (21).

Proposition 1. *Let the density p be of the form (20) with g satisfying Assumption 4. Then, for the estimator (21) with kernel K satisfying Assumption 5, we have the following bounds on the L_2-bias and variance terms*

$$\|\mathbf{E}(\hat{p}_{m,B}) - p\|^2 \leq C_4 h^4, \tag{22}$$

$$\mathbf{E}\left(\|\mathbf{E}(\hat{p}_{m,B}) - \hat{p}_{m,B}\|^2\right) \leq \frac{C_5}{nh^m}. \tag{23}$$

Here $0 < h \leq h_0$ with some $h_0 < \infty$ and any integer $n \geq 1$ and C_4 and C_5 are constants depending only on d, L_g, h_0 and on $K_{\max} \triangleq \sup_{z \in \mathbf{R}^m} |K(z)|$.

Proof. For every $x \in \mathbf{R}^d$, the expectation of $\hat{p}_{m,B}(x)$ can be written as follows:

$$\begin{aligned}
&\mathbf{E}(\hat{p}_{m,B}(x)) \\
&= \frac{\phi_d(x)}{h^m \phi_m(B^T x)} \int_{\mathbf{R}^d} K\left(\frac{B^T(y-x)}{h}\right) \phi_d(y) g(B^T y) dy \\
&= \frac{\phi_d(x)}{h^m \phi_m(B^T x)} \int_{\mathbf{R}^{d-m}} \left[\int_{\mathbf{R}^m} K\left(\frac{u - B^T x}{h}\right) \phi_m(u) g(u) du\right] \phi_{d-m}(v) dv
\end{aligned} \tag{24}$$

with new variables $u = B^T y$ and $v = \tilde{B}^T y$, where $\tilde{B}$ is a $d \times (d-m)$ matrix with orthonormal columns such that $(B|\tilde{B})$ is a $d \times d$ orthogonal matrix.

Making in (24) the change of variables $t = (u - B^T x)/h$, we find that the bias of $\hat{p}_{m,B}(x)$ equals

$$\mathbf{E}(\hat{p}_{m,B}(x)) - p(x) = \frac{\phi_d(x)}{\phi_m(B^T x)} \int_{\mathbf{R}^m} K(t)\phi_m(B^T x + th) g(B^T x + th) dt - \phi_d(x) g(B^T x), \tag{25}$$

and, under the above assumptions about g and K, the standard Taylor expansion argument gives

$$\begin{aligned} \|\mathbf{E}(\hat{p}_{m,B}) - p\|^2 &= \int_{\mathbf{R}^d} (\mathbf{E}(\hat{p}_{m,B}(x)) - p(x))^2 dx \\ &= \int_{\mathbf{R}^d} \left(\frac{\phi_d(x)}{\phi_m(B^T x)} \int_{\mathbf{R}^m} K(t) \frac{h^2}{2} t^T D(B^T x + a^* t) t \, dt \right)^2 dx, \end{aligned} \tag{26}$$

where $0 \le a^* \le h$ and $D(z) = \nabla^2(\phi_m(z) g(z)) = [(zz^T - \mathbf{I}_m) g(z) - \nabla g(z) z^T - z \nabla^T g(z) + \nabla^2 g(z)] \phi_m(z)$. Here and in what follows $\mathbf{I}_m$ stands for the identity matrix of dimension m. Using Assumption 4 and the fact that $a^* \le h_0$, we get

$$\begin{aligned} t^T D(B^T x + a^* t) t &\le C L_g |t|_m^2 (1 + h_0^2 |t|_m^2 + |B^T x|_m^2) \phi_m(B^T x + a^* t) \\ &\le C L_g |t|_m^2 (1 + h_0^2 |t|_m^2 + |B^T x|_m^2) \exp(|B^T x|_m h_0 |t|_m) \\ &\quad \times \phi_m(B^T x), \end{aligned}$$

with some constant $C > 0$. Because $K(t)$ has bounded support, (22) follows from (26). For the variance term, we have

$$\begin{aligned} Var(\hat{p}_{m,B}(x)) &= \frac{\phi_d^2(x)}{n h^{2m} \phi_m^2(B^T x)} Var\left(K\left(\frac{B^T(X - x)}{h} \right) \right) \\ &\le \frac{1}{(2\pi)^{d-m} n h^{2m}} \exp(-x^T (\mathbf{I}_d - BB^T) x) \\ &\quad \times \int_{\mathbf{R}^d} K^2 \left(\frac{B^T(y - x)}{h} \right) \phi_d(y) g(B^T y) dy \\ &\le \frac{L_g}{(2\pi)^{d-m} n h^{2m}} \int_{\mathbf{R}^d} K^2 \left(\frac{B^T(y - x)}{h} \right) \phi_d(y) dy, \end{aligned}$$

and after making the same changes of variables as for the bias, we obtain

$$\mathbf{E}\left(\|\mathbf{E}(\hat{p}_{m,B}) - \hat{p}_{m,B}\|^2 \right) = \int_{\mathbf{R}^d} Var(\hat{p}_{m,B}(x)) dx = O(n^{-1} h^{-m}).$$

■

Consider the mean integrated mean squared error (MISE) of the estimator $\hat{p}_{m,B}$:

$$MISE(\hat{p}_{m,B}, p) \triangleq \mathbf{E}\|\hat{p}_{m,B} - p\|^2 \equiv \mathbf{E}\|\hat{p}_{m,B} - f_B\|^2. \tag{27}$$

Proposition 1 implies that, under Assumptions 4 and 5,

$$MISE(\hat{p}_{m,B}, p) = O(n^{-4/(m+4)}), \tag{28}$$

if the bandwidth h is chosen of the order $h \asymp n^{-1/(m+4)}$. Using the standard techniques of the minimax lower bounds (e.g. Tsybakov (2004)), it is easy to show that the rate $n^{-4/(m+4)}$ given in (28) is the optimal MISE rate for the model (20) on the class of densities p defined by Assumption 4, and thus the estimator $\hat{p}_{m,B}$ with $h \asymp n^{-1/(m+4)}$ has the optimal rate for this class of densities.

Consider now the case where the dimension m and the index matrix B are unknown. We will use the procedure of Section 2 to aggregate estimators of the type (20) corresponding to candidate pairs $(m, B) = (k, A)$ with $k = 1, \ldots, d$ and with A that runs over a finite net on the set of all admissible $d \times k$ index matrices. The latter is the set $\mathcal{B}_k$ of all $d \times k$ matrices A with orthonormal columns. This set is bounded in the Frobenius norm $\|A\|_2 = \mathrm{Tr}^{1/2}(AA^T)$. Consider an ϵ-net Q_k on $\mathcal{B}_k$ constructed using the Frobenius norm. Note that orthogonal transformations preserve the norm, so that both estimators (21) and the ϵ-net Q_k are invariant under orthogonal transformations, and thus are not affected by the non-uniqueness of representation (20). The set $\mathcal{B}_k$ is bounded and can be imbedded in $\mathbf{R}^s$ with $s = k(d-(k+1)/2)$, and therefore we can construct an ϵ-net Q_k with cardinality

$$\mathrm{Card}(Q_k) = O(\epsilon^{-k(d-(k+1)/2)}), \tag{29}$$

e.g. Wellner and van der Vaart (1996). Doing this for $k = 1, \ldots, d$, we obtain a collection $Q_1, \ldots, Q_k$ of ϵ-nets with the property (29) each, and in what follows we set $\epsilon = n^{-a}$ with $a > 2/5$ for all $k = 1, \ldots, d$.

We can now define the aggregate. As in Section 2, we split the sample $X_1, \ldots, X_n$ into two parts, I_1 and I_2 with $n_1 = \mathrm{Card}(I_1)$, $n_2 = \mathrm{Card}(I_2)$, $n = n_1 + n_2$. From the first subsample we construct estimators

$$\hat{p}_{k,A}(x) = \frac{\phi_d(x)}{\phi_k(A^T x)} \frac{1}{n_1 h_k^k} \sum_{I_1} K\left(\frac{A^T(X_i - x)}{h_k}\right), \quad k = 1, \ldots, d, \quad A \in Q_k, \tag{30}$$

where $h_k \asymp n^{-1/(k+4)}$. These estimators are of the form (21), but here we plug in k and A that are not necessarily equal to the true unknown values m and B and we use only the first subsample I_1. Nevertheless, we preserve the same notation as in (21) since this will not cause ambiguity.

Let now $p_{\tilde{N}}$ be the aggregate defined as in (2) and (3) using as $\{p_1, \ldots, p_M\}$ the collection of estimators $\{\hat{p}_{k,A}, k = 1, \ldots, d, \ A \in Q_k\}$ of the form (30) with bandwidths $h_k \asymp n^{-\frac{1}{k+4}}$ and ϵ-nets Q_k such that $\epsilon = n^{-a}$, $a > 2/5$. In view of (29), the cardinality M of this set of estimators is

$$M \asymp \sum_{k=1}^{d} n^{ak(d-(k+1)/2)} \asymp n^{ad(d-1)/2}. \tag{31}$$

In this case, the aggregate $p_{\tilde{N}}$ of (2) and (3) can be written in the form $\hat{p}_{\tilde{k},\tilde{A}}$ where $(\tilde{k},\tilde{A})$ are given by

$$(\tilde{k},\tilde{A}) = \arg \min_{k=1,\dots,d,A\in Q_k} \left(-\frac{2}{n_2} \sum_{I_2} \hat{p}_{k,A}(X_i) + \int \hat{p}^2_{k,A} \right). \tag{32}$$

We can now state the main result of this section.

Theorem 2. *Let Assumptions 4 and 5 hold and let* $n_2 = \lfloor \frac{cn}{\log n} \rfloor$ *for some constant* $c > 0$ *such that* $1 \le n_2 < n$. *Assume in addition that the kernel* $K(\cdot)$ *is Lipschitz continuous. Then for the aggregate* $\hat{p}_{\tilde{k},\tilde{A}}$ *we have*

$$\mathbf{E}\|\hat{p}_{\tilde{k},\tilde{A}} - p\|^2 = O(n^{-4/(m+4)}), \tag{33}$$

as $n \to \infty$, *so that* $\hat{p}_{\tilde{k},\tilde{A}}$ *estimates* p *with the best rate attainable when dimension* m *and matrix* B *are known.*

Proof. We first verify the assumptions of Theorem 1. Clearly, Assumption 4 implies Assumption 3. With a bounded kernel K, $\|\hat{p}_{k,A} - p\| \le Ch_k^{-k} = O(n^{k/(k+4)})$, so that Assumption 1 holds with $a_1 = d/(d+4) + ad(d-1)/2$ and $a_2 = ad(d-1)/2$.

In order to verify Assumption 2, we will show that (8) holds for estimators $p_N = \hat{p}_{k,A}$ with $k = 1,\dots,d$ and $A \in Q_k$.

Proof of (8). For any estimator $\hat{p}_{k,A}$ we have $\|\hat{p}_{k,A} - p\|_\infty \le \|\hat{p}_{k,A} - \mathbf{E}(\hat{p}_{k,A})\|_\infty + \|\mathbf{E}(\hat{p}_{k,A}) - p\|_\infty$, so that (25), written with $\hat{p}_{k,A}$ instead of $\hat{p}_{m,B}$, implies that $\|\hat{p}_{k,A} - p\|_\infty \le \|\hat{p}_{k,A} - \mathbf{E}(\hat{p}_{k,A})\|_\infty + C$ for some constant $C > 0$ which depends on g but not on k and A. Therefore we have

$$\begin{aligned} &\mathbf{E}[\|\hat{p}_{k,A} - p\|_\infty I(\|\hat{p}_{k,A} - p\|_\infty > \gamma_0 \log^{3/4} M)] \\ &\quad \le \mathbf{E}[(\|\hat{p}_{k,A} - \mathbf{E}(\hat{p}_{k,A})\|_\infty + C) I(\|\hat{p}_{k,A} - \mathbf{E}(\hat{p}_{k,A})\|_\infty > \gamma_0 \log^{3/4} M - C)]. \end{aligned} \tag{34}$$

Note that, for any $x \in \mathbf{R}^d$,

$$\hat{p}_{k,A}(x) - \mathbf{E}(\hat{p}_{k,A}(x)) = (2\pi)^{-(d-k)/2} \exp(-x^T(\mathbf{I}_d - AA^T)x/2) \sum_{I_1} \zeta_{i,n}(z), \tag{35}$$

where $z = A^T x$ and

$$\zeta_{i,n}(z) = \zeta'_{i,n}(z) - \mathbf{E}(\zeta'_{i,n}(z)), \qquad \zeta'_{i,n}(z) = \frac{1}{n_1 h_k^k} K\left(\frac{A^T X_i - z}{h_k}\right).$$

Introduce the truncated variables

$$\xi_{i,n}(z) = \xi'_{i,n}(z) - \mathbf{E}(\xi'_{i,n}(z)),$$

$$\xi'_{i,n}(z) = \frac{1}{n_1 h_k^k} K\left(\frac{A^T X_i - z}{h_k}\right) I(|X_i|_d \le \log n),$$

and note that

$$\mathbf{P}(|X_1|_d > \log n) \le C(\log n)^d \exp\left(-\frac{\log^2 n}{2}\right), \tag{36}$$

where the constant C depends only on $g_{\max}$ and d. This follows from the relations

$$\begin{aligned}\mathbf{P}(|X_1|_d > \log n) &= \int_{\mathbf{R}^d} I(|x|_d > \log n)\phi_d(x)g(B^T x)dx \\ &\le g_{\max}\int_{|x|_d>\log n} \phi_d(x)dx,\end{aligned}$$

followed by evaluation of the tail of d-dimensional standard normal distribution. Consider the random event $\mathcal{A} = \{|X_i|_d \le \log n,\ i = 1,\dots,n\}$. In view of (36), the probability of the complementary event satisfies

$$\mathbf{P}(\mathcal{A}^c) \le Cn(\log n)^d \exp\left(-\frac{\log^2 n}{2}\right). \tag{37}$$

Using (35) and the fact that $\mathbf{I}_d - AA^T \ge 0$ for all matrices $A \in \mathcal{B}_k$, we get

$$\|\hat{p}_{k,A} - \mathbf{E}(\hat{p}_{k,A})\|_\infty \le (2\pi)^{-(d-k)/2} \sup_{z\in E^k}\Big|\sum_{I_1}\zeta_{i,n}(z)\Big|,$$

where E^k is the linear subspace of $\mathbf{R}^d$ spanned by the columns of A. Now, (36) implies that for any $D > 0$ there exists a constant C depending only on $g_{\max}$, $K_{\max}$ and d, such that $\mathbf{E}|\zeta'_{i,n}(z) - \xi'_{i,n}(z)| \le Cn^{-D}$. Therefore,

$$\|\hat{p}_{k,A} - \mathbf{E}(\hat{p}_{k,A})\|_\infty \le (2\pi)^{-(d-k)/2} \sup_{z\in E^k}\Big|\sum_{I_1}\Big[\zeta'_{i,n}(z) - \mathbf{E}(\xi'_{i,n}(z))\Big]\Big| + Cn^{-D}. \tag{38}$$

Setting

$$\eta \triangleq \sup_{z\in E^k}\Big|\sum_{I_1}\Big[\zeta'_{i,n}(z) - \mathbf{E}(\xi'_{i,n}(z))\Big]\Big|,$$

we note that, in view of the inequalities (31), (34) and (38), to prove (8) it is enough to show that

$$\mathbf{P}(\eta > C\log^{3/4} n) + \mathbf{E}[\eta I(\eta > C\log^{3/4} n)] \le \frac{\log^{?} M}{M}. \tag{39}$$

We will in fact prove a stronger result, namely, that the left-hand side of (39) decreases faster than any power of n. Since on the event $\mathcal{A}$ it holds that $\zeta'_{i,n}(z) = \xi'_{i,n}(z)$ for all $z \in \mathbf{R}^k$, we obtain

$$\begin{aligned}\mathbf{P}(\eta > s) &\le \mathbf{P}(\mathcal{A}^c) + \mathbf{P}\left(\sup_{z\in E^k}\Big|\sum_{I_1}\xi_{i,n}(z)\Big| > s\right) \\ &= \mathbf{P}(\mathcal{A}^c) + \mathbf{P}\left(\sup_{z\in S\cap E^k}\Big|\sum_{I_1}\xi_{i,n}(z)\Big| > s\right),\end{aligned} \tag{40}$$

where the last equality is due to the fact that $\xi'_{i,n}(z) = 0$ for all $z \notin S$, with $S = \{x \in \mathbf{R}^d : |x|_d \le 1 + \log n\}$.

As kernel K is Lipschitz continuous, we get

$$\Big|\sum_{I_1}(\xi_{i,n}(z) - \xi_{i,n}(y))\Big| \le C_L h_k^{-(k+1)}|z - y|_d, \quad \forall\, z, y \in E^k, \qquad (41)$$

where C_L is a constant. Next, fix some $\delta > 0$, and let $z_1, ..., z_L$ be a δ-net in Euclidean metric on the bounded set $S \cap E^k$ such that $L \le C(\frac{\log n}{\delta})^d$. Clearly, a δ-net of cardinality L satisfying the latter inequality exists, since the cardinality of the minimal δ-net on the larger set S is of the order $(\frac{\log n}{\delta})^d$. In view of (41), we have, for $s > 2C_L\delta h_k^{-(k+1)}$,

$$\begin{aligned}\mathbf{P}\left(\sup_{z\in S\cap E^k}\Big|\sum_{I_1}\xi_{i,n}(z)\Big| > s\right) &\le \mathbf{P}\left(\max_{1\le j\le L}\Big|\sum_{I_1}\xi_{i,n}(z_j)\Big| > s/2\right)\\ &\le L\sup_{z\in S\cap E^k}\mathbf{P}\left(\Big|\sum_{I_1}\xi_{i,n}(z)\Big| > s/2\right). \quad (42)\end{aligned}$$

We have $\mathbf{E}(\xi_{i,n}(z)) = 0$ and $\sup_{z\in S\cap E^k}|\xi_{i,n}(z)| \le c_1 n_1^{-1}h_k^{-k}$, for some constant $c_1 > 0$. Also, using (20) and Assumption 4, we find

$$\begin{aligned}Var(\xi_{i,n}(z)) \le \mathbf{E}\zeta'^{\,2}_{i,n}(z) &= \frac{1}{n_1^2 h_k^{2k}}\int_{\mathbf{R}^d} K^2\left(\frac{A^T y - z}{h_k}\right) f_B(y)dy\\ &\le \frac{L_g}{n_1^2 h_k^{2k}}\int_{\mathbf{R}^d} K^2\left(\frac{A^T y - z}{h_k}\right)\phi_d(y)dy\\ &= \frac{L_g}{n_1^2 h_k^{k}}\int_{\mathbf{R}^k} K^2(t)\phi_k(th_k + z)\left[\int_{\mathbf{R}^{d-k}}\phi_{d-k}(u)du\right]dt\end{aligned}$$

with new variables $t = (A^T y - z)/h_k$ and $u = \tilde{A}^T y$, where $\tilde{A}$ is a $d \times (d-k)$ matrix with orthonormal columns such that $(A|\tilde{A})$ is a $d \times d$ orthogonal matrix. Therefore we have $\sup_{z\in S\cap E^k} Var(\xi_{i,n}(z)) \le c_2 n_1^{-2}h_k^{-k}$, for some constant $c_2 > 0$.

Choosing now $\delta = h_k^{k+1}$, applying in (42) the Bernstein inequality and recalling that $h_k \asymp n^{-\frac{1}{k+4}}$, $n_1 = n - n_2 = n(1+o(1))$ we get, for $s > 2C_L$,

$$\begin{aligned}
\mathbf{P}\left(\sup_{z\in S\cap E^k}\left|\sum_{I_1}\xi_{i,n}(z)\right| > s\right)
&\leq 2L\exp\left(-\frac{(s/2)^2}{2c_2 n_1^{-1}h_k^{-k} + c_1 n_1^{-1}h_k^{-k}s/3}\right)\\
&\leq C\left(\frac{\log n}{\delta}\right)^d \exp\left(-\frac{s^2 n_1 h_k^k}{8c_2 + 2c_1 s}\right)\\
&\leq C(\log n)^d n^{\frac{d(k+1)}{k+4}}\exp\left(-\frac{s^2 n^{4/(k+4)}(1+o(1))}{8c_2+2c_1 s}\right)\\
&\leq C(\log n)^d n^{\frac{d(d+1)}{d+4}}\exp\left(-\frac{s^2 n^{4/(d+4)}}{C(1+s)}\right), \qquad (43)
\end{aligned}$$

where the last inequality is valid for n large enough. From (40), (37) and (43) we deduce that, for n large enough,

$$\begin{aligned}
\mathbf{P}(\eta > C\log^{3/4} n) \leq C(\log n)^d &\left[n\exp\left(-\frac{\log^2 n}{2}\right)\right.\\
&\left. + \; n^{\frac{d(d+1)}{d+4}}\exp\left(-\frac{n^{4/(d+4)}\log^{3/4} n}{C}\right)\right]. \qquad (44)
\end{aligned}$$

On the other hand, $\eta \leq 2K_{\max}h_k^{-k} = O(n^{k/(k+4)}) = O(n^{d/(d+4)})$, and therefore $\mathbf{E}[\eta I(\eta > C\log^{3/4} n)] \leq O(n^{d/(d+4)})\mathbf{P}(\eta > C\log^{3/4} n)$. This inequality and (44) combined with (31) prove that (39) holds for n large enough. The proof of (8) is thus complete.

All the assumptions of Theorem 1 are therefore satisfied. Applying Theorem 1 we get the oracle inequality

$$\mathbf{E}\|\hat{p}_{\tilde{k},\tilde{A}} - p\|^2 \leq \left(1 + \frac{C^*}{\log^{1/4} n}\right)\min_{k=1,\dots,d}\min_{A\in Q_k} MISE(\hat{p}_{k,A}, p) + C^*\frac{\log^3 n}{n}.$$

To complete the proof of Theorem 2, we now show that

$$\min_{k=1,\dots,d}\min_{A\in Q_k} MISE(\hat{p}_{k,A}, p) = O(n^{-4/(m+4)}). \qquad (45)$$

In fact,

$$\min_{k=1,\dots,d}\min_{A\in Q_k} MISE(\hat{p}_{k,A}, p) \leq MISE(\hat{p}_{m,B^*}, p), \qquad (46)$$

where B^* is a matrix in Q_m closest to B in the Frobenius norm, and thus satisfying $\|B^* - B\|_2 \leq \epsilon$. We have (recall that $p \equiv f_B$)

$$\begin{aligned}
\|\hat{p}_{m,B^*} - p\|^2 &\leq 2(\|\hat{p}_{m,B^*} - f_{B^*}\|^2 + \|f_{B^*} - p\|^2)\\
&= 2(\|\hat{p}_{m,B^*} - f_{B^*}\|^2 + \|f_{B^*} - f_B\|^2). \qquad (47)
\end{aligned}$$

It follows from (27) and (28) that

$$\mathbf{E}\|\hat{p}_{m,B^*} - f_{B^*}\|^2 = O(n^{-4/(m+4)}). \tag{48}$$

(Note that we proved (28) for the estimator (21), while here the estimator $\hat{p}_{m,B^*}$ is defined by (30) and based on the sample of size n_1; nevertheless the result remains valid, since $n_1 = n(1 + o(1))$.) Using (48) and applying Assumption 4 to bound from above the last summand in (47), we obtain

$$MISE(\hat{p}_{m,B^*}, p) \leq b_1 n^{-4/(m+4)} + b_2\epsilon^2$$

with some constants b_1, b_2. Since $\epsilon = n^{-a}$ with $a > 2/5 \geq 2/(m+4)$ we get $MISE(\hat{p}_{m,B^*}, p) = O(n^{-4/(m+4)})$. Together with (46) this implies (45). ∎

Remark 3. The aggregate estimator for model (20) suggested here automatically accomplishes dimension reduction. In fact, if the unknown true dimension m is small, it achieves the rate $O(n^{-4/(m+4)})$ that can be much faster than the best attainable rate $O(n^{-4/(d+4)})$ for a model of full dimension. The aggregate can be interpreted as an adaptive estimator, but in contrast to adaptation to unknown smoothness usually considered in nonparametrics, here we deal with adaptation to unknown dimension m and to the index space $\mathcal{M}$ determined by a matrix B. The procedure provides explicit estimates $(\tilde{k}, \tilde{A})$ of (m, B) that are optimal in the sense of Theorem 2. The tools of this paper do not allow us, however, to evaluate how close is $(\tilde{k}, \tilde{A})$ to (m, B) (or, equivalently, how close is the estimated index space to the true one $\mathcal{M}$).

References

1. Birgé, L. (2003). Model selection via testing: an alternative to (penalized) maximum likelihood estimators. Preprint n.862, Laboratoire de Probabilités et Modèles Aléatoires, Universités Paris 6 and Paris 7 (available at http://www.proba.jussieu.fr/mathdoc/preprints).
2. Bunea, F., Tsybakov, A. and Wegkamp, M. (2004). Aggregation for regression learning. Preprint n.948, Laboratoire de Probabilités et Modèles Aléatoires, Universités Paris 6 and Paris 7 (available at arXiv:math.ST/0410214, 8 Oct. 2004.)
3. Catoni, O. (2004). *Statistical Learning Theory and Stochastic Optimization. Ecole d'Eté de Probabilités de Saint-Flour XXXI - 2001.* Lecture Notes in Mathematics, vol.1851, Springer, New York.
4. Cox, D. R. (1969). Some Sampling Problems in Technology. In: Johnson, N. and Smith, H. (eds.), *New Developments in Survey Sampling*, Wiley-Interscience, New York, pp.506-529.

5. DEVROYE, L. AND LUGOSI, G. (2000). *Combinatorial Methods in Density Estimation.* Springer, New-York.

6. HRISTACHE, M., JUDITSKY, A., POLZEHL J. AND SPOKOINY, V. (2001). Structure Adaptive Approach for Dimension Reduction. *Ann. Statist.*, **29**, 1537-1566.

7. HUBER, P. (1985). Projection Pursuit. *Ann. Statist.*, **13**, 435-475.

8. JUDITSKY, A. B., NAZIN, A. V., TSYBAKOV, A. B. AND VAYATIS, N. (2005a) Generalization error bounds for aggregation by mirror descent. *Proceedings of NIPS-2005* (to appear).

9. JUDITSKY, A. B., NAZIN, A. V., TSYBAKOV, A. B. AND VAYATIS, N. (2005b) Recursive aggregation of estimators by a mirror descent method with averaging. *Problems of Information Transmission*, **41**, n.4 (to appear).

10. LI, K-C.(1991). Sliced inverse regression for dimension reduction. *J. Amer. Statist. Assoc.*, **86**, 316-342.

11. NEMIROVSKI, A. (2000). Topics in Non-parametric Statistics. In: *Ecole d'Eté de Probabilités de Saint-Flour XXVIII - 1998*, Lecture Notes in Mathematics, vol. 1738, Springer, New York.

12. PATIL, G. AND RAO, C. R. (1977). The Weighted Distributions: the Survey of Their Applications. In: P.R. Krishnaiah (ed.), *Applications of Statistics*, Amsterdam, North Holland, pp. 383-405.

13. RIGOLLET, PH. AND TSYBAKOV, A. B. (2004) Linear and convex aggregation of density estimators. Submitted.

14. SAMAROV, A. AND TSYBAKOV, A. (2004) Nonparametric Independent Component Analysis. *Bernoulli*, **10**, 565-582.

15. SERFLING, R. (1980) *Approximation Theorems of Mathematical Statistics*, J. Wiley, New York.

16. TSYBAKOV, A. (2003). Optimal rates of aggregation. In: *Computational Learning Theory and Kernel Machines*, (B.Schölkopf and M.Warmuth, eds.), Lecture Notes in Artificial Intelligence, v.2777. Springer, Heidelberg, 303-313.

17. TSYBAKOV, A. (2004). *Introduction à l'estimation non-paramétrique.* Springer, Berlin-Heidelberg.

18. WEGKAMP, M. H. (1999). Quasi-universal bandwidth selection for kernel density estimators. *Canad. J. Statist.*, **27**, 409-420.

19. WEGKAMP, M.H. (2003). Model selection in nonparametric regression. *Ann. Statist.*, **31**, 252 - 273.

20. WELLNER, J. AND VAN DER VAART, A. (1996). *Weak convergence and empirical processes.* Springer, New York.

21. YANG, Y. (2000). Mixing strategies for density estimation. *Ann. Statist.*, **28**, 75-87.

PART 4

Transformation Models

Chapter 13

A STURM-LIOUVILLE PROBLEM IN SEMIPARAMETRIC TRANSFORMATION MODELS

Chris A.J. Klaassen

Korteweg-de Vries Institute for Mathematics
University of Amsterdam, Amsterdam, THE NETHERLANDS

E-mail: chrisk@science.uva.nl

A general class of semiparametric transformation models is considered. A second order differential equation of Sturm-Liouville type is derived that determines the semiparametric information on the Euclidean parameter involved. Under quite general conditions properties are proved of the solution of the resulting boundary value problem. A frailty model of Clayton and Cuzick for survival data is studied in some detail.

Key words: Clayton-Cuzick frailty model; Semiparametric model; Transformation model; Sturm-Liouville equation.

1 Introduction

We will discuss a quite general class of semiparametric transformation models, which includes the famous Cox proportional hazards model. To describe this class of models we start out with a model for the random vector (T, Z), the so-called core model. The random variable T lives on the closed interval $[a, b] \subset \mathbb{R}$ with $-\infty \leq a < b \leq \infty$. Let Z have distribution Q on some measurable space $\mathcal{Z}$. The parameter space Θ will be an open subset of $\mathbb{R}^d$. Denote the conditional distribution function of T at t given $Z = z$ by

$$F_0(t \mid z, \theta), \quad t \in [a, b], z \in \mathcal{Z}, \theta \in \Theta. \tag{1}$$

This conditional distribution function is assumed to be continuous in its argument t. If the distribution Q of Z is unknown and belongs to some parametric or nonparametric class $\mathcal{Q}$ of distributions, then the core model is called parametric or semiparametric, respectively.

We observe i.i.d. copies of (Y, Z) with $\psi(Y) = T$, where $\psi : [a, b] \to [a, b]$ is a nondecreasing transformation that is onto. Consequently, the conditional distribution function $F(y\,|\,z,\theta,\psi)$ of Y given $Z=z$ at y equals

$$F(y\,|\,z,\theta,\psi) = F_0(\psi(y)\,|\,z,\theta)\,, \quad y \in [a,b]\,, z \in \mathcal{Z}\,, \theta \in \Theta\,. \tag{2}$$

Both the Euclidean parameter θ and the transformation ψ are unknown to the statistician. With ψ varying over a nonparametric function class, we have a semiparametric transformation model here, even if the core model is parametric. In order to avoid identifiability problems, we have to and do assume that Z is nondegenerate. Indeed, if Z would be degenerate, the only observable containing information about the parameters would be Y. However, the probability integral transform yields a transformation that turns Y into a uniformly distributed random variable, whatever θ. This argument shows that θ and ψ would be confounded for Z degenerate and hence that θ would not be identifiable.

Note that, for

$$F_0(t\,|\,z,\theta) = 1 - \exp\left\{-e^{\theta^T z} t\right\}, \quad t \in [0,\infty]\,, z \in \mathbb{R}^d\,, \theta \in \mathbb{R}^d\,, \tag{3}$$

our model leads to

$$F(y\,|\,z,\theta,\psi) = 1 - \exp\left\{-e^{\theta^T z}\psi(y)\right\}, \quad y \in [0,\infty]\,, z \in \mathbb{R}^d\,, \theta \in \mathbb{R}^d\,,$$

with $\psi(0) = 0, \psi(\infty) = \infty$, and ψ nondecreasing. This is the conditional distribution function of the famous Cox proportional hazards model with regression parameter θ and baseline cumulative hazard function ψ. In the corresponding core model, T has an exponential distribution with parameter $\exp\{\theta^T z\}$, given $Z=z$. In order to explain population heterogeneity, one might introduce an unobservable "frailty" parameter η into the core model (3), thus T being exponentially distributed with parameter $\eta \exp\{\theta^T z\}$, given $Z=z$. Clayton and Cuzick (1985) suggested to take η random with a gamma distribution with mean 1 and variance $c \geq 0$. Of course, $c=0$ leads back to (3), but in general the conditional distribution function of T given $Z=z$ is Pareto, to wit

$$F_0(t\,|\,z,\theta) = 1 - \left\{1 + c e^{\theta^T z} t\right\}^{-1/c}, \quad t \in [0,\infty]\,, z \in \mathbb{R}^d\,, \theta \in \mathbb{R}^d\,. \tag{4}$$

Another important model that fits into this framework is a semiparametric generalization of the Box-Cox model $F_0(t|z,\theta) = \Phi(t - \theta^T z)$, where Φ is the standard normal distribution function. So, if ϵ has a standard normal distribution we observe i.i.d. copies of (Y,Z), where $\psi(Y) = \theta^T Z + \epsilon$ holds with θ and ψ unknown.

A path-breaking analysis of the general transformation model (2) has been given by Bickel (1986), who showed the connection to a Sturm-Liouville problem within a hypothesis testing frame work. We shall demonstrate this connection again within estimation and via a different route. In

Section 2 we will describe the semiparametric Fisher information and its natural relation to projection in the generic semiparametric model. Section 3 shows that the projection of Section 2 for our transformation models translates into the second order differential equation that gives rise to the Sturm-Liouville problem of Bickel (1986). Properties of the solution to this Sturm-Liouville problem that are relevant to semiparametric inference, are proved in Section 4. These properties are derived under weaker conditions than in Bickel (1986) and than in Sections 4.7 and 7.6 of Bickel, Klaassen, Ritov and Wellner (1993). This is the main result of our paper, which thus redeems the promises of the last sentences of Example 4.7.3, continuation 1, page 171, and of Section 7.6, page 382, of Bickel et al. (1993). The final Section 5 applies these results to the Clayton-Cuzick frailty model introduced above.

Interpretation of an estimate of the Euclidean parameter in our transformation model has to depend on an estimate of the unknown transformation. In an unusually heated debate about the *parametric* Box-Cox model, it has become clear that there is ample room for disagreement here. See Doksum (1984) for a lucid exposition. Both the Box-Cox model and the Cox proportional hazards model triggered research on semiparametric transformation models resulting in a long series of papers; we just mention Bickel (1986), Doksum (1987), Dabrowska and Doksum (1988a,b), Murphy (1994, 1995), Murphy, Rossini, and van der Vaart (1997), Bickel and Ritov (1997), Lenstra (1998), Dabrowska (2002), and Gørgens (2003).

2 Semiparametrics

In the preceding section we have introduced a semiparametric class of transformation models. A basic issue in any semiparametric model is minimization of Fisher information. We will discuss this issue here by an approach that slightly differs from the one in Bickel et al. (1993). Consider a general semiparametric model $\mathcal{P}$ parametrized by (θ, G), where $\theta \in \Theta \subset \mathbb{R}^d$ is the Euclidean parameter and G the Banach parameter, which varies over some large set $\mathcal{G}$. Let $\mathbf{X}$ be a random quantity governed by some distribution $P_{(\theta,G)} \in \mathcal{P}$. By $\mathbf{l}(\mathbf{X} \mid \theta, G)$ we denote the loglikelihood of $\mathbf{X}$ at $\mathbf{X}$ under (θ, G) with respect to some dominating measure and we assume that we have a so-called regular parametric model in θ for each fixed $G \in \mathcal{G}$. This implies that there exists a score function $\dot{\mathbf{l}}_\theta(\mathbf{X} \mid \theta, G)$, which is a d-vector of one-dimensional score functions, the derivatives with respect to the components of θ of the loglikelihood; more precisely,

$$\dot{\mathbf{l}}_\theta(\mathbf{X} \mid \theta, G) = \left(\frac{\partial}{\partial \theta_i} \mathbf{l}(\mathbf{X} \mid \theta, G) \right)_{i=1}^{d} . \tag{5}$$

Regularity of the parametric submodel implies also that Θ is open and that the Fisher information matrix

$$I(\theta \mid G) = E_\theta \dot{\mathbf{l}}_\theta(\mathbf{X} \mid \theta, G) \dot{\mathbf{l}}_\theta^T(\mathbf{X} \mid \theta, G) \tag{6}$$

exists, is nonsingular, and is continuous in θ. Let $\mathcal{G}' = \{G_\eta \in \mathcal{G} \mid \eta \in \Theta\}$ be a surface in $\mathcal{G}$ such that for fixed θ the distributions of $\mathbf{X}$ with parameters (θ, G_η) constitute a regular parametric model in η with score function

$$\dot{\mathbf{l}}_\eta(\mathbf{X} \mid \theta, \eta\,; \mathcal{G}') = \left(\frac{\partial}{\partial \eta_i} \mathbf{l}(\mathbf{X} \mid \theta, G_\eta) \right)_{i=1}^{d} . \tag{7}$$

Under additional regularity conditions

$$\mathbf{l}(\mathbf{X} \mid \theta\,; \mathcal{G}') = \mathbf{l}(\mathbf{X} \mid \theta, G_\theta) , \quad \theta \in \Theta ,$$

are the loglikelihoods of a d-dimensional regular parametric family $\mathcal{P}_{\mathcal{G}'}$ with score functions

$$\dot{\mathbf{l}}_\theta(\mathbf{X} \mid \theta\,; \mathcal{G}') = \dot{\mathbf{l}}_\theta(\mathbf{X} \mid \theta, G_\theta) + \dot{\mathbf{l}}_\eta(\mathbf{X} \mid \theta, \eta\,; \mathcal{G}')\Big|_{\eta=\theta} , \quad \theta \in \Theta . \tag{8}$$

Within $\mathcal{P}_{\mathcal{G}'} \subset \mathcal{P}$ the degree of difficulty for estimation based on $\mathbf{X}$ of θ, may be measured by the Fisher information matrix

$$I(\theta \mid \mathcal{G}') = E_\theta \dot{\mathbf{l}}_\theta(\mathbf{X} \mid \theta\,; \mathcal{G}') \dot{\mathbf{l}}_\theta^T(\mathbf{X} \mid \theta\,; \mathcal{G}') . \tag{9}$$

Fix θ. A d-dimensional submodel $\mathcal{P}_{\mathcal{G}(\theta)}$ for which the trace of this Fisher information matrix is minimal at this θ under all d-dimensional submodels $\mathcal{P}_{\mathcal{G}'}$ of the above type that we wish to consider, is called *least favorable* at θ, but need not exist. If the class of submodels $\mathcal{P}_{\mathcal{G}'}$ is sufficiently rich, minimization of this trace boils down to minimization in $\mathcal{G}'$ for each j, $j = 1, \ldots, d$, of

$$E_\theta \dot{\mathbf{l}}_{\theta j}^2(\mathbf{X} \mid \theta\,; \mathcal{G}') = E_\theta \left(\dot{\mathbf{l}}_{\theta j}(\mathbf{X} \mid \theta, G_\theta) + \dot{\mathbf{l}}_{\eta j}(\mathbf{X} \mid \theta, \eta\,; \mathcal{G}')\Big|_{\eta=\theta} \right)^2 . \tag{10}$$

Note that the terminology 'least favorable' is natural for $\mathcal{P}_{\mathcal{G}(\theta)}$, since for each component of the Euclidean parameter unbiased estimation at θ is most difficult within $\mathcal{P}_{\mathcal{G}(\theta)}$, at least in principle, in view of the Cramér-Rao inequality; a recent reference for this inequality is Lenstra (2005).

Within the Hilbert space $\mathcal{L}_2^0(P_{(\theta, G_\theta)})$ of random variables with mean zero and finite variance under (θ, G_θ), minimization of (10) is related to projection as follows. Consider the closed linear span $\dot{\mathcal{P}}_{\mathcal{G}}$ within $\mathcal{L}_2^0(P_{\theta, G_\theta})$ of the components of the d-dimensional score functions $\dot{\mathbf{l}}_\eta(\mathbf{X} \mid \theta, \eta\,; \mathcal{G}') \mid_{\eta=\theta}$ that correspond to the d-dimensional submodels $\mathcal{P}_{\mathcal{G}'}$ that we wish to consider. Projection of $\dot{\mathbf{l}}_{\theta j}(X \mid \theta, G_\theta)$ on $\dot{\mathcal{P}}_{\mathcal{G}}$ yields

$$\Pi \left(\dot{\mathbf{l}}_{\theta j}(\mathbf{X} \mid \theta, G_\theta) \,\Big|\, \dot{\mathcal{P}}_{\mathcal{G}} \right)$$

and consequently the semiparametric Fisher information matrix for θ satisfies

$$\min_{\mathcal{P}_{\mathcal{G}'}} E_\theta \mathbf{i}^2_{\theta j}(\mathbf{X} \mid \theta ; \mathcal{G}') \geq E_\theta \left(\dot{\mathbf{l}}_{\theta j}(\mathbf{X} \mid \theta, G_\theta) - \Pi \left(\dot{\mathbf{l}}_{\theta j}(\mathbf{X} \mid \theta, G_\theta) \middle| \dot{\mathcal{P}}_{\mathcal{G}} \right) \right)^2 . \tag{11}$$

Quite often equality holds here for $j = 1, \ldots, d$ and the minimal value of the trace of the Fisher information matrix may be determined via projection on $\dot{\mathcal{P}}_{\mathcal{G}}$ then.

In the i.i.d.-case we consider $\mathbf{X} = (X_1, \ldots, X_n)$ with $X_1, \ldots, X_n$ i.i.d. random variables. In view of

$$\mathbf{l}(\mathbf{X} \mid \theta, G) = \sum_{i=1}^n \ell(X_i \mid \theta, G)$$

it makes sense to restrict attention to one random variable X that has the same distribution as each of the X_i. It has loglikelihood $\ell(X \mid \theta, G)$. Proceeding as above and adapting the notation in that the model $\mathcal{P}$ and a submodel $\mathcal{P}_{\mathcal{G}'}$ denote collections now of distributions of X in stead of $\mathbf{X}$, we see that the minimal value of the trace of the Fisher information matrix might still be determined via projection on $\dot{\mathcal{P}}_{\mathcal{G}}$. One calls the d-vector of functions

$$\ell^*(X \mid \theta, G_\theta) = \dot{\ell}_\theta(X \mid \theta, G_\theta) - \Pi \left(\dot{\ell}_\theta(X \mid \theta, G_\theta) \middle| \dot{\mathcal{P}}_{\mathcal{G}} \right) , \tag{12}$$

where the projection is componentwise, the semiparametrically efficient score function for θ. The terminology of 'least favorable' and 'efficient' is justified asymptotically as $n \to \infty$ by asymptotic versions of the Cramér-Rao inequality, like the convolution theorem; see Bickel et al. (1993) for a comprehensive exposition.

3 Efficient scores in transformation models

In this section the semiparametric transformation model (2) with core model (1) is studied with T taking values in the interval $[a, b]$, $-\infty \leq a < b \leq \infty$. We denote the class of transformations ψ by

$$\Psi = \{\psi : [a, b] \to [a, b] \mid \psi \text{ onto, absolutely continuous, derivative } \psi' \geq 0\} . \tag{13}$$

We assume that the conditional distribution function F_0 has conditional density p_0 with respect to some dominating measure and that the loglikelihood

$$\ell(t \mid z, \theta) = \log p_0(t \mid z, \theta) \tag{14}$$

is differentiable in $\theta \in \Theta$, an open subset of $\mathbb{R}^d$, and continuously differentiable in $t \in (a,b)$, and we denote the derivatives by $\dot{\ell}(t \mid z, \theta)$ (a d-dimensional column vector of functions) and $\ell'(t \mid z, \theta)$, respectively. Furthermore, we assume that

$$E_\theta |\dot{\ell}(T \mid Z, \theta)|^2 < \infty \,,\; E_\theta \left(\ell'(T \mid Z, \theta)\right)^2 < \infty \,, \tag{15}$$

where $|\cdot|$ is the Euclidean norm in $\mathbb{R}^d$. Fix the Euclidean parameter at $\theta_0 \in \Theta$, the transformation at $\psi_0 \in \Psi$, and the distribution of the covariate Z at $Q_0 \in \mathcal{Q}$, where $\mathcal{Q}$ denotes a class of nondegenerate distributions Q of Z. The score function for θ at (θ_0, ψ_0, Q_0) equals

$$\frac{\partial}{\partial\theta} \log p(y \mid z, \theta, \psi_0, Q_0)\Big|_{\theta=\theta_0} = \dot{\ell}\left(\psi_0(y) \mid z, \theta_0\right) . \tag{16}$$

Varying in the model (2) $\psi \in \Psi$ and $Q \in \mathcal{Q}$ in a smooth way, one can obtain 'score functions' for these infinite dimensional parameters also. To this end we choose a smoothing function $\chi : \mathbb{R} \to (0,2)$ that is differentiable with derivative χ' satisfying $\chi(0) = \chi'(0) = 1$, $0 \le \chi' \le 1$, and with χ'/χ bounded. Such functions exist. Take $\chi(x) = 2/(1 + e^{-2x})$, for example. Let $\alpha : \mathbb{R} \to \mathbb{R}$ be absolutely continuous with derivative α' satisfying

$$E_0(\ell'(T \mid Z, \theta_0)\alpha(T))^2 < \infty ,\; E_0(\alpha'(T))^2 < \infty \tag{17}$$

and let $g : \mathcal{Z} \to \mathbb{R}$ be a measurable function with $E_0 g^2(Z) < \infty$. Here E_0 denotes expectation under P_0, the distribution corresponding to (θ_0, ψ_0, Q_0) or $(\theta_0, \text{identity}, Q_0)$. We denote the set of possible α's and g's by A and $\mathcal{G}$, respectively. In an approach slightly different from the one in Section 2, we construct a regular parametric submodel

$$\mathcal{P}_{\alpha,g} = \{p(y \mid z, \theta, \psi_\eta) dQ_\zeta/dQ_0(z) \,,\, y \in [a,b] \,, z \in \mathcal{Z} \mid \theta \in \Theta \,,\, \eta, \zeta \in \mathbb{R}\}$$

of (2) by defining

$$dQ_\zeta(z) = dQ_0(z)\chi(\zeta g(z)) \Big/ \int \chi(\zeta g(\tilde{z})) dQ_0(\tilde{z}) \tag{18}$$

and by defining for the case $-\infty < a < b < \infty$

$$\psi_\eta(y) = a + (b-a)\left\{\int_a^b \chi(\eta\alpha'(s))ds\right\}^{-1} \int_a^{\psi_0(y)} \chi(\eta\alpha'(t))dt \,, \tag{19}$$

for the case $-\infty < a < b = \infty$

$$\psi_\eta(y) = a + \tfrac{1}{2}\int_a^{\psi_0(y)} \{1 + \chi(2\eta\alpha'(t))\}dt \,, \tag{20}$$

for the case $-\infty = a < b < \infty$

$$\psi_\eta(y) = b + \tfrac{1}{2}\int_b^{\psi_0(y)} \{1 + \chi(2\eta\alpha'(t))\}dt \,, \tag{21}$$

and for the case $-\infty = a < b = \infty$

$$\psi_\eta(y) = c + \tfrac{1}{2}\int_c^{\psi_0(y)} \{1 + \chi(2\eta\alpha'(t))\}dt\,, \tag{22}$$

where the constant c is chosen such that α vanishes at c. Furthermore, in each case the function α has to satisfy $\alpha(a) = \alpha(b) = 0$. Indeed, Q_ζ is a distribution on $\mathcal{Z}$, ψ_η belongs to Ψ and the score functions for ζ and η at (θ_0, ψ_0, Q_0) equal

$$\begin{aligned}\frac{\partial}{\partial\zeta}\log\frac{dQ_\zeta}{dQ_0}(z)\bigg|_{\zeta=0} &= g(z) - E_0 g(Z)\,,\\ \frac{\partial}{\partial\eta}\log p(y \mid z, \theta_0, \psi_\eta)\bigg|_{\eta=0} &= \ell'(\psi_0(y) \mid z, \theta_0)\alpha(\psi_0(y)) + \alpha'(\psi_0(y))\,.\end{aligned} \tag{23}$$

Next we define, for $j = 1, \ldots, d$,

$$\begin{aligned} I_j(\theta_0) &= \inf_{\alpha\in A, g\in\mathcal{G}} E_0\Big(\dot\ell_j(\psi_0(Y) \mid Z, \theta_0) - \ell'(\psi_0(Y) \mid Z, \theta_0)\alpha(\psi_0(Y))\\ &\qquad - \alpha'(\psi_0(Y)) - g(Z) + E_0 g(Z)\Big)^2 \qquad\qquad (24)\\ &= \inf_{\alpha\in A, g\in\mathcal{G}} E_0\Big(\dot\ell_j(T \mid Z, \theta_0) - \ell'(T \mid Z, \theta_0)\alpha(T) - \alpha'(T)\\ &\qquad -g(Z) + E_0 g(Z)\Big)^2. \end{aligned}$$

Note that this infimum is independent of ψ_0 and is in fact the minimum distance, in an $\mathcal{L}_2^0(P_0)$-sense, of $\dot\ell_j(T \mid Z, \theta_0)$ to the set

$$\{\ell'(T \mid Z, \theta_0)\alpha(T) + \alpha'(T) + g(Z) - E_0 g(Z) \mid \alpha \in A\,,\ g \in \mathcal{G}\}\,.$$

Note also that this approach differs from the one in Section 2, in that for each j separately, $j = 1, \ldots, d$, we minimize the Fisher information in (24) by considering appropriate parametric submodels $\mathcal{P}_{\alpha,g}$, whereas in Section 2 we do this for all j simultaneously by minimizing the trace of the Fisher information matrix and by considering parametric submodels that in this case would have dimension $3d$; cf. (7).

Denoting as before differentiation with respect to θ and t by $\dot{}$ and $'$ respectively, we note that under regularity conditions and appropriate assumptions on α

$$E_0\left(\dot\ell_j(T \mid Z, \theta_0) \mid Z\right) = \int \frac{\partial}{\partial\theta_j} p_0(t \mid Z, \theta_0)dt = 0 \quad \text{a.s.},$$

$$\begin{aligned} E_0\left(\ell'(T \mid Z, \theta_0)\alpha(T) + \alpha'(T) \mid Z\right) &= \int \{\alpha(t)p_0'(t \mid Z, \theta_0) + \alpha'(t)p_0(t \mid Z, \theta_0)\}\, dt\\ &= \alpha(t)p_0(t \mid Z, \theta_0)\Big]_a^b = 0 \quad \text{a.s.} \end{aligned}$$

Therefore, we shall assume that

$$E_0\left(\dot{\ell}_j(T\,|\,Z,\theta_0)\,|\,Z\right) = 0 \quad \text{a.s.}, \quad j = 1,\dots,d, \tag{25}$$

$$E_0\left(\ell'(T\,|\,Z,\theta_0)\alpha(T) + \alpha'(T)\,|\,Z\right) = 0 \quad \text{a.s.}, \quad \alpha \in A. \tag{26}$$

Consequently, $\dot{\ell}_j(T\,|\,Z,\theta_0)\,, j = 1.\dots,d$, and $\ell'(T\,|\,Z,\theta_0)\alpha(T) + \alpha'(T)$ are orthogonal to $\{g(Z) - E_0 g(Z)\,|\,g \in G\}$ in the $\mathcal{L}_2^0(P_0)$-sense and it suffices to project the $\dot{\ell}_j(T\,|\,Z,\theta_0)$ onto the linear space $B = \{\ell'(T\,|\,Z,\theta_0)\alpha(T) + \alpha'(T)\,|\,\alpha \in A\}$. Therefore, we will look for $\alpha_j \in A$ such that for all $\alpha \in A$

$$E_0\Big(\{\dot{\ell}_j(T\,|\,Z,\theta_0) - \ell'(T\,|\,Z,\theta_0)\alpha_j(T) - \alpha_j'(T)\}$$
$$\{\ell'(T\,|\,Z,\theta_0)\alpha(T) + \alpha'(T)\}\Big) = 0\,. \tag{27}$$

We assume that with

$$\chi_j(t) = \int_{\mathcal{Z}} \left\{\dot{\ell}_j(t\,|\,z,\theta_0) - \ell'(t\,|\,z,\theta_0)\alpha_j(t) - \alpha_j'(t)\right\} p_0(t\,|\,z,\theta_0)dQ_0(z)$$

we have

$$\chi_j(t) - \chi_j(a) = \int_a^t \int_{\mathcal{Z}} \Big\{\dot{p}_{0j}'(s\,|\,z,\theta_0) - p_0''(s\,|\,z,\theta_0)\alpha_j(s)$$
$$-2p_0'(s\,|\,z,\theta_0)\alpha_j'(s) - p_0(s\,|\,z,\theta_0)\alpha_j''(s)\Big\}dQ_0(z)ds, \quad a \le t \le b, \tag{28}$$

and

$$\int_a^b \int_a^t \left|\alpha'(t) \int_{\mathcal{Z}} \Big\{\dot{p}_{0j}'(s\,|\,z,\theta_0) - p_0''(s\,|\,z,\theta_0)\alpha_j(s) - 2p_0'(s\,|\,z,\theta_0)\alpha_j'(s)\right.$$
$$\left. -p_0(s\,|\,z,\theta_0)\alpha_j''(s)\Big\}dQ_0(z)\right|dsdt < \infty, \quad \alpha \in \tilde{A}\,, \tag{29}$$

where $\tilde{A}$, containing α_j, is a subset of A (defined after (17)), such that $\alpha\chi_j$ vanishes at a and b for all $\alpha \in \tilde{A}$, $j = 1,\dots,d$. Then (28), (29), and Fubini's theorem (or, more specifically, partial integration) yield

$$E_0\Big(\alpha'(T)\{\dot{\ell}_j(T\,|\,Z,\theta_0) - \ell'(T\,|\,Z,\theta_0)\alpha_j(T) - \alpha_j'(T)\}\Big)$$
$$= \int_a^b \alpha(s) \int_{\mathcal{Z}} \Big\{ - \dot{p}_{0j}'(s\,|\,z,\theta_0) + p_0''(s\,|\,z,\theta_0)\alpha_j(s) + 2p_0'(s\,|\,z,\theta_0)\alpha_j'(s)$$
$$+p_0(s\,|\,z,\theta_0)\alpha_j''(s)\Big\}dQ_0(z)ds$$

and consequently (27) holds for all $\alpha \in \tilde{A}$, if $\alpha_j \in \tilde{A}$ satisfies the Sturm-Liouville equation

$$\int_{\mathcal{Z}} \Big\{ p_0(t \mid z, \theta_0)\alpha_j''(t) + p_0'(t \mid z, \theta_0)\alpha_j'(t) + [p_0'' - (p_0')^2 p_0^{-1}](t \mid z, \theta_0)\alpha_j(t)$$
$$+ [\dot{p}_{0j} p_0' p_0^{-1} - \dot{p}_{0j}'](t \mid z, \theta_0) \Big\} dQ_0(z) = 0\,, \quad a \le t \le b\,. \tag{30}$$

We have shown

Proposition 1. *Consider the model (2) with (13) through (15), and with $\tilde{A}$ as in (17) and as in the line after (29). If the conditions (25) and (26) are satisfied and if $\alpha_j \in \tilde{A}$ satisfies (28) through (30), then the infimum $I_j(\theta_0)$ of (24) with $\tilde{A}$ replacing A is attained by α_j.*

Assume now, for every $\theta \in \Theta$, that $\alpha_{\theta j}$ is as in this proposition yielding $I_j(\theta) > 0\,, j = 1, \ldots, d$, and define $\alpha_{\theta,Q} = (\alpha_{\theta 1}, \ldots, \alpha_{\theta d})^T$, $\alpha_{\theta,Q}' = (\alpha_{\theta 1}', \ldots, \alpha_{\theta d}')^T$,

$$S(t \mid z, \theta, Q) = \dot{\ell}(t \mid z, \theta) - \ell'(t \mid z, \theta)\alpha_{\theta,Q}(t) - \alpha_{\theta,Q}'(t)\,, \tag{31}$$

$$I(\theta, Q) = E_{\theta,Q} S(T \mid Z, \theta, Q) S^T(T \mid Z, \theta, Q)\,, \tag{32}$$

$$J(y, z \mid \theta, \psi, Q) = I^{-1}(\theta, Q) S(\psi(y) \mid z, \theta, Q). \tag{33}$$

Consider an estimator sequence $\{T_n\}$ for θ, which is (locally) regular at (θ_0, ψ_0, Q_0) for a parametric model containing the 'directions' given by $\alpha_{\theta_0 1}, \ldots, \alpha_{\theta_0 k}$; cf. (19)–(22). By the convolution theorem its limit distribution is the convolution of a normal distribution with covariance matrix $I^{-1}(\theta_0, Q_0)$ and another distribution. The latter distribution is degenerate at 0 iff $\{T_n\}$ is locally asymptotically linear at (θ_0, ψ_0, Q_0) with influence function $J(y, z \mid \theta_0, \psi_0, Q_0)$. Note that $\dot{\ell}, \ell', \alpha_{\theta,Q}$ and hence $I(\theta, Q)$ do not depend on ψ, but they do depend on θ and Q.

In view of this we will call $\{T_n\}$ an efficient estimator sequence for θ at (θ_0, ψ_0, Q_0) in the semiparametric model (2), (13), if for every sequence $\{\theta_n\}$ with $\theta_n = \theta_0 + \mathcal{O}(n^{-1/2})$, every sequence $\{\psi_n\}$ with $\psi_n \in \Psi, \psi_n(t) \to \psi_0(t)\,, t \in \mathbb{R}$, and every sequence of distributions $\{Q_n\}$ converging weakly to Q_0 in such a way that $\{Q_n^n\}$ and $\{Q_0^n\}$ are contiguous,

$$\sqrt{n}\left(T_n - \theta_n - n^{-1} \sum_{i=1}^n J(Y_i, Z_i \mid \theta_n, \psi_n, Q_n) \right) = o_P(1) \tag{34}$$

under (θ_n, ψ_n, Q_n). Note that we ask for more regularity here than one usually does. Note also that if we take $Q_n = Q_0\,, \psi_n = \psi_{\eta_n}\,, \eta_n = \mathcal{O}\left(n^{-1/2}\right)$, with ψ_η as in (19)–(22) with α any linear combination of the $\alpha_{\theta_0 j}$, then in this parametric model (34) implies (local) regularity of $\{T_n\}$ at (θ_0, ψ_0, Q_0) under smoothness conditions on J, and the convolution theorem states that we cannot do better than (34) with $\theta_n = \theta_0, \psi_n = \psi_0$, even if we knew that we were in this parametric submodel. Since $I^{-1}(\theta_0, Q_0)$ is by (24) a supremum in a certain sense over all $\alpha \in \tilde{A}$, the functions $\alpha_{\theta_0 j}$ represent least favorable directions and it is not a priori impossible that efficient estimators in the sense of (34) exist. In fact, they do exist in the Cox and Clayton-Cuzick models. Much more on the theory of efficient estimation in semiparametric models is presented in Bickel et al. (1993).

In the Cox model (3), (2), the second order differential equation (30) is trivial, since $p_0'' - (p_0')^2 p_0^{-1}$ vanishes and the equation is in fact of first order. If we assume existence of a finite C with

$$P(|Z| \leq C) = 1, \tag{35}$$

straightforward computation shows that

$$\begin{aligned}\alpha_{\theta_0}(t) = & \int_0^t \left\{ \int_{\mathcal{Z}} z e^{\theta_0 z} \exp\left(-e^{\theta_0 z} s\right) h_0(z) d\mu(z) \right\} \\ & \qquad \left\{ \int_{\mathcal{Z}} e^{\theta_0 z} \exp(-e^{\theta_0 z} s) h_0(z) d\mu(z) \right\}^{-1} ds \\ = & \int_0^t E_0(Z \mid T = s) ds \end{aligned} \tag{36}$$

satisfies the conditions of Proposition 1 with information (cf. (24))

$$I(\theta_0) = E_0 \mathrm{Var}_0(Z \mid T). \tag{37}$$

For the two sample Cox model with another parametrization such a bound has been derived by Begun and Wellner (1983), and the well-known Cox estimator has been proved efficient by Begun (1987). Efficiency in the sense of (34) of the Cox estimator for this model with $Q = Q_0$ known has been shown by Klaassen (1989), which also contains a less general version of the present section, namely with Q fixed and known and $d = 1$.

In the Clayton-Cuzick model (4), (2) we do not have explicit solutions α_j for the differential equations (30). However, we shall study the existence and some properties of the solutions for these equations in Section 4 and roughly indicate construction of an estimator satisfying (34) in Section 5.

4 The differential equation

In this section we will study the Sturm-Liouville equation (30) in more detail. We suppress the subscripts 0 and j and denote the marginal density of T by

$$f_\theta(t) = f_{\theta,Q}(t) = \int_{\mathcal{Z}} p(t \mid z, \theta) dQ(z) \,. \tag{38}$$

Under regularity conditions on p, (30) reads as follows

$$\begin{aligned} f_\theta \alpha'' + f_\theta' \alpha' + & \left[f_\theta'' - \int_{\mathcal{Z}} \frac{(p')^2}{p} (\cdot \mid z, \theta) dQ(z) \right] \alpha \\ & + \int_{\mathcal{Z}} \left[\frac{\dot{p} p'}{p} - \dot{p}' \right] (\cdot \mid z, \theta) dQ(z) = 0 \,. \end{aligned} \tag{39}$$

Let F_θ^{-1} be the quantile function of the distribution function F_θ corresponding to f_θ. By the transformation

$$\begin{aligned} &\Delta(u) = f_\theta \left(F_\theta^{-1}(u) \right) \alpha \left(F_\theta^{-1}(u) \right) \,, \quad 0 \le u \le 1 \,, \\ &F_\theta^{-1}(0) = a \,, \quad F_\theta^{-1}(1) = b \,, \end{aligned} \tag{40}$$

the differential equation (39) becomes

$$\Delta''(u) - \beta(u)\Delta(u) - \gamma(u) = 0 \,, \quad 0 \le u \le 1 \,, \tag{41}$$

with

$$\begin{aligned} \beta(u) &= f_\theta^{-3} \left(F_\theta^{-1}(u) \right) \left[\int_{\mathcal{Z}} \frac{(p')^2}{p} \left(F_\theta^{-1}(u) \mid z, \theta \right) dQ(z) - \frac{(f_\theta')^2}{f_\theta} \left(F_\theta^{-1}(u) \right) \right] \\ &= f_\theta^{-2} \left(F_\theta^{-1}(u) \right) \operatorname{Var} \left(\frac{p'}{p} (T \mid Z, \theta) \,\middle|\, T = F_\theta^{-1}(u) \right) \,, \end{aligned} \tag{42}$$

$$\gamma(u) = f_\theta^{-2} \left(F_\theta^{-1}(u) \right) \int_{\mathcal{Z}} \left[\dot{p}' - \frac{\dot{p} p'}{p} \right] \left(F_\theta^{-1}(u) \mid z, \theta \right) dQ(z) \,. \tag{43}$$

It makes sense to choose the set $\tilde{A}$ in such a way that $\Delta(0) = \Delta(1) = 0$; see also the sentences after (22) and (29). We have been led to the boundary value problem

$$\Delta''(u) - \beta(u)\Delta(u) - \gamma(u) = 0, \quad \Delta(0) = \Delta(1) = 0 \,, \tag{44}$$

with

$$\beta(u) \ge 0 \,. \tag{45}$$

For the testing problem $\theta = \theta_0$ versus $\theta > \theta_0$ in model (2), with Z taking only two distinct values, Bickel (1986) has suggested a 'quadratic rank statistic', which he proved to be locally asymptotically most powerful, conditionally on $Z_1, \ldots, Z_n$. This rank statistic is defined in terms of the solution of the Sturm-Liouville problem (44) with essentially the same function β, but with a different function γ; compare (42), (43), (44), and (51) below with Bickel's (1986) formulas (1.14) and (1.16). We extend Bickel's treatment of the Sturm-Liouville problem (44) to obtain (see also Section 4.7 of Bickel et al. (1993)) the following bounds.

Lemma 1. *If for the Sturm-Liouville problem (44) there exist positive constants β, B, γ and Γ with $\beta + 2\gamma > 1$, such that*

$$0 \le \beta(u) \le B(1-u)^{\beta-2}, \quad |\gamma(u)| \le \Gamma(1-u)^{\gamma-2}, \tag{46}$$

then the unique solution Δ of (44) satisfies

$$|\Delta(u)| \le D\Big[(1-u)^{(\beta\wedge 1+1)/2}\left\{1-\mathbf{1}_{[\beta=1]}\log(1-u)\right\}^{1/2} + (1-u)^{\gamma}\left\{1-\mathbf{1}_{[\gamma=1]}\log(1-u)\right\}\Big], \quad 0\le u\le 1, \tag{47}$$

and

$$|\Delta'(u)| \le D\Big[1+(1-u)^{(3\beta-1)/2}+(1-u)^{\gamma-1} - \left(\mathbf{1}_{[\beta+\gamma=1]}+\mathbf{1}_{[\gamma=1]}\right)\log(1-u)\Big], \quad 0\le u\le 1, \tag{48}$$

for some $D>0$. Moreover Δ depends continuously on $\beta(\cdot)$ and $\gamma(\cdot)$, where $\beta(\cdot),\gamma(\cdot)\in\mathcal{L}_1([0,1],(1-t)dt)$ and $\Delta\in\ell^\infty([0,1])$.

Proof. If Δ_1 and Δ_2 are solutions of (44), then $\Delta_0 = \Delta_1 - \Delta_2$ satisfies

$$\Delta''(u) = \beta(u)\Delta(u), \quad \Delta(0)=\Delta(1)=0. \tag{49}$$

Since Δ_0 is continuous, the set of points where it does not vanish is open and hence is a countable union of open intervals. Let (x_0, x_1) be such an interval with $0 \le x_0 < x_1 \le 1$. If Δ_0 were positive on (x_0,x_1) with $\Delta_0(x_0)=\Delta_0(x_1)=0$, then (45) would imply that Δ_0 is convex on $[x_0,x_1]$. However, a convex positive function on an interval cannot vanish at the endpoints of this interval and consequently such intervals (x_0,x_1) do not exist. Similarly there do not exist intervals where Δ_0 is negative and hence concave. Consequently $\Delta_0(u)=0$ holds on $[0,1]$ and (44) has at most one solution.

We use the approach of Lemma 2.2 of Bickel (1986). For $K(s,t) = s\wedge t - st$ (the autocovariance function of the Brownian bridge), we have

$$K(s,t) \le t(1-t), \quad 0\le s,t\le 1. \tag{50}$$

From (44) it follows that Δ is continuous and hence bounded on [0,1]. Hence (44) and (46) imply $\Delta''(u) = \mathcal{O}\left((1-u)^{\beta\wedge\gamma-2}\right)$. Together with (46) and (50) this implies that multiplication of (44) with $K(\cdot,\cdot)$ and integration over

(0, 1) are possible and yield the Fredholm integral equation of the second kind

$$\Delta(u)+\int_0^1 K(u,s)\Delta(s)\beta(s)ds+\int_0^1 K(u,s)\gamma(s)ds=0\,,\quad 0\le u\le 1\,,\tag{51}$$

where we note that, by Fubini,

$$\int_0^1 K(u,s)\Delta''(s)ds=\int_0^1\int_s^1\left[u-\mathbf{1}_{[0,u]}(t)\right]\Delta''(s)dtds$$
$$=\int_0^1\left[u-\mathbf{1}_{[0,u]}(t)\right]\int_0^t\Delta''(s)dsdt=\int_0^1\left[u-\mathbf{1}_{[0,u]}(t)\right]\left[\Delta'(t)-\Delta'(0)\right]dt$$
$$=-\Delta(u).$$

On the other hand, writing (51) as

$$\Delta(u)+u\int_0^1(1-s)\{\Delta(s)\beta(s)+\gamma(s)\}ds+\int_0^u(s-u)\{\Delta(s)\beta(s)+\gamma(s)\}ds=0$$

and noting

$$\lim_{s\to u}(s-u)\{\Delta(s)\beta(s)+\gamma(s)\}=0,\quad 0<u<1\,,$$

we see that Δ is differentiable with

$$\Delta'(u)=-\int_0^1(1-s)\{\Delta(s)\beta(s)+\gamma(s)\}ds+\int_0^u\{\Delta(s)\beta(s)+\gamma(s)\}ds\,.$$

Consequently, $\Delta'(\cdot)$ is absolutely continuous with derivative $\Delta(\cdot)\beta(\cdot)+\gamma(\cdot)$, which is just the meaning of (44).

If the Green's function $\Delta(\cdot,v)$ solves

$$\Delta(u,v)+\int_0^1 K(u,s)\Delta(s,v)\beta(s)ds+K(u,v)=0\,,\tag{52}$$

then we have

$$\Delta(u)=\int_0^1\Delta(u,v)\gamma(v)dv\tag{53}$$

as the unique solution of (51) or equivalently (44). With the notation

$$\phi(u,v)=\sqrt{\beta(u)}\Delta(u,v)\,,\tag{54}$$

(52) can be rewritten as

$$L\left(\phi(\,\cdot\,,v)\right)(u)=-\sqrt{\beta(u)}K(u,v)\,,\tag{55}$$

where $L:\mathcal{L}_2(0,1)\to\mathcal{L}_2(0,1)$ is the operator given by $L=I+K$. Here I is the identity and

$$K(\chi)(u)=\int_0^1\sqrt{\beta(s)}K(s,u)\sqrt{\beta(u)}\chi(s)\,ds\,.\tag{56}$$

In view of (46) we have

$$\int_0^1 K^2(s,u)\beta(s)ds \le B\left\{\int_0^u (1-u)^2(1-s)^{\beta-2}ds + \int_u^1 (1-s)^\beta ds\right\}$$
$$= \begin{cases} \mathcal{O}((1-u)^{\beta+1}), & 0<\beta<1, \\ \mathcal{O}((1-u)^2\{1-\log(1-u)\}), & 0<\beta=1, \\ \mathcal{O}((1-u)^2), & 1<\beta. \end{cases} \quad (57)$$

The first part of (57) also yields

$$\|K(\chi)\|^2 = \int_0^1 \{K(\chi)(u)\}^2 du \le \int_0^1 \|\chi\|^2 \int_0^1 K^2(s,u)\beta(s)\beta(u)dsdu$$
$$\le B^2\|\chi\|^2 \int_0^1 (1-u)^{\beta-2}\left\{\int_0^u (1-u)^2(1-s)^{\beta-2}ds + \int_u^1 (1-s)^\beta ds\right\} du$$
$$= B^2\|\chi\|^2 \left\{\int_0^1\int_s^1 (1-u)^\beta du(1-s)^{\beta-2}ds + (\beta+1)^{-1}\int_0^1 (1-u)^{2\beta-1}du\right\}$$
$$= B^2\beta^{-1}(\beta+1)^{-1}\|\chi\|^2 . \quad (58)$$

We see that the operator K is bounded and nonnegative definite since the Brownian bridge covariance function is. Consequently (Theorem 12.32 of Rudin (1973)) the spectrum of K is contained in $[0,\infty)$ and hence that of L in $[1,\infty)$. Therefore L is invertible, the spectrum of L^{-1} is contained in $(0,1]$, and the Green's function $\Delta(\cdot,v)$ exists by (52), (54) and (55). Since K is self-adjoint, L and hence L^{-1} are. Consequently, L^{-1} is normal and Theorem 11.28(b) of Rudin (1973) implies that $\|L^{-1}\|$ equals the spectral radius of L^{-1} which is bounded by 1; a different argument for this is given immediately after formula (4.7.34) of Bickel et al. (1993). By (55) this yields

$$\int_0^1 \phi^2(u,v)du = \|L^{-1}\left(-\sqrt{\beta(\cdot)}K(,v)\right)\|^2 \le \int_0^1 K^2(u,v)\beta(u)du . \quad (59)$$

Combining (52), (54) and (59) we obtain

$$|\Delta(u,v)| \le K(u,v) + \left|\int_0^1 K(u,s)\Delta(s,v)\beta(s)ds\right|$$
$$\le K(u,v) + \left\{\int_0^1 K^2(u,s)\beta(s)ds \int_0^1 \phi^2(s,v)ds\right\}^{1/2}$$
$$\le K(u,v) + \left\{\int_0^1 K^2(u,s)\beta(s)ds \int_0^1 K^2(v,s)\beta(s)ds\right\}^{1/2} \quad (60)$$
$$= K(u,v) + \begin{cases} \mathcal{O}(\{(1-u)(1-v)\}^{(\beta\wedge 1+1)/2}), & 0<\beta\neq 1, \\ \mathcal{O}((1-u)(1-v)\{1-\log(1-u)\}^{1/2} \\ \qquad \{1-\log(1-v)\}^{1/2}), & 0<\beta=1 . \end{cases}$$

Together with (53) and the condition $\beta+2\gamma>1$ this implies (47) and with the help of (52) also

$$
\begin{aligned}
|\Delta'(u)| &= \left| \frac{\partial}{\partial u} \int_0^1 \left\{ K(u,v) + \int_0^1 K(u,s)\Delta(s,v)\beta(s)ds \right\} \gamma(v)dv \right| \\
&= \left| \frac{\partial}{\partial u} \left\{ u \int_0^1 \int_0^1 \left[1 - v + (1-s)\Delta(s,v)\beta(s)\right] \gamma(v)dvds \right.\right. \\
&\qquad \left.\left. + \int_0^u (s-u) \left[\gamma(s) + \int_0^1 \Delta(s,v)\gamma(v)dv\beta(s) \right] ds \right\} \right| \\
&\le \left| \int_0^1 \left[\mathbf{1}_{(u,1)}(v) - v \right] \gamma(v)dv \right| \\
&\qquad + \left| \int_0^1 \int_0^1 \left[\mathbf{1}_{(u,1)}(s) - s \right] \Delta(s,v)\beta(s)\gamma(v)dvds \right| \qquad (61) \\
&= \mathcal{O}\left(1 - \mathbf{1}_{[\gamma=1]}\log(1-u) + (1-u)^{\gamma-1}\right) \\
&\quad + \mathcal{O}\Bigg(\int_0^1 \left| \mathbf{1}_{(u,1)}(s) - s \right| [(1-s)^{((3\beta)\wedge 1-3)/2} + (1-s)^{\beta+\gamma-2} \\
&\qquad\qquad - \mathbf{1}_{[\gamma=1]}(1-s)^{\beta-1}\log(1-s)]ds \Bigg) \\
&= \mathcal{O}\Big(1 + (1-u)^{((3\beta)\wedge 1-1)/2} + (1-u)^{\gamma-1} \\
&\qquad - \mathbf{1}_{[\beta+\gamma=1]}\log(1-u) - \mathbf{1}_{[\gamma=1]}\log(1-u) \Big) \ .
\end{aligned}
$$

Finally, let $\beta_n(\cdot)$ and $\gamma_n(\cdot)$ be functions satisfying (46) and converging to $\beta(\cdot)$ and $\gamma(\cdot)$ Lebesgue almost everywhere. Let $\Delta_n(\cdot)$ be the solution of (51) with $\beta_n(\cdot)$ and $\gamma_n(\cdot)$ replacing $\beta(\cdot)$ and $\gamma(\cdot)$, respectively. In view of (48) $\Delta_n(\cdot)$ can be written uniquely as the difference $\Delta_n(\cdot) = \Delta_n^+(\cdot) - \Delta_n^-(\cdot)$ of two nondecreasing bounded functions on $[0,1]$ with $\Delta_n^+(0) = \Delta_n^-(0) = 0$ and $\Delta_n^+(1) = \Delta_n^-(1)$ minimal. By e.g. Helly's selection theorem there exists for every subsequence of $\{\Delta_n(\cdot)\}, \Delta_n(\cdot) = \Delta_n^+(\cdot) - \Delta_n^-(\cdot)$, a further subsequence $\{\Delta_{\tilde{n}}(\cdot)\}, \Delta_{\tilde{n}}(\cdot) = \Delta_{\tilde{n}}^+(\cdot) - \Delta_{\tilde{n}}^-(\cdot)$, converging pointwise and hence in the supnorm to $\tilde{\Delta}(\cdot)$, say. The boundedness of $K(\cdot,\cdot)$ and of $\Delta_n(\cdot)$ uniformly in n, the dominated convergence theorem and the unicity of a solution of (51) show that $\tilde{\Delta}(\cdot)$ is the solution of (51). We have shown that $\Delta(\cdot)$ depends on $\beta(\cdot)$ and $\gamma(\cdot)$ continuously in the given norms. □

It is easy to derive from (47) and (48) the inequalities

$$|\alpha(t)| \le D\Big[(1-F_\theta(t))^{(\beta\wedge 1+1)/2}\left\{1-\mathbf{1}_{[\beta=1]}\log(1-F_\theta(t))\right\}^{1/2}$$
$$+(1-F_\theta(t))^{\gamma}\left\{1-\mathbf{1}_{[\gamma=1]}\log(1-F_\theta(t))\right\}\Big]/f_\theta(t), \qquad (62)$$
$$|\alpha'(t)| \le |\alpha(t)f_\theta'(t)/f_\theta(t)| + D\Big[1+(1-F_\theta(t))^{(3\beta-1)/2}$$
$$+(1-F_\theta(t))^{\gamma-1}-(\mathbf{1}_{[\beta+\gamma=1]}+\mathbf{1}_{[\gamma=1]})\log(1-F_\theta(t))\Big]. \qquad (63)$$

Together with (39) these inequalities yield bounds for

$$S(t\,|\,z,\theta,Q)=\dot{\ell}(t\,|\,z,\theta)-\ell'(t\,|\,z,\theta)\alpha(t)-\alpha'(t) \qquad (64)$$

and its derivatives S' and S''. As we will indicate in the next section these bounds might be relevant when constructing estimators and proving their semiparametric efficiency (see also (31) and the rest of Section 3).

5 The frailty model of Clayton and Cuzick

In the frailty model (4) for survival data suggested by Clayton and Cuzick (1985) we take the regression parameter 1-dimensional for notational simplicity, we call it ν in stead of θ, and we are interested in semiparametrically efficient estimation of the 2-dimensional Euclidean parameter $\theta=(\nu,c)$ where the Pareto shape parameter c is assumed to be positive. Note that at $\nu=0$ the Pareto shape parameter c and the transformation ψ, a nuisance parameter, are confounded. In order to avoid the ensuing identifiability problems it is assumed that ν does not vanish. Suppressing again the subscript 0 we note that the conditional core density equals

$$p(t\,|\,z,\theta)=e^{\nu z}(1+ce^{\nu z}t)^{-1-1/c}\mathbf{1}_{(0,\infty)}(t), \quad \theta=(\nu,c).$$

Straightforward verification under (35) shows that the assumptions leading to (44) are fulfilled with $a=0$ and $b=\infty$ and that,

$$c^{-1}e^{-C|\nu|}\left((1-u)^{-c}-1\right) \le F_\theta^{-1}(u) \le c^{-1}e^{C|\nu|}\left((1-u)^{-c}-1\right). \qquad (65)$$

By the expansion, as $t\to\infty$,

$$\left[r(1+crt)^{-1}\right]^j=\left[(ct)^{-1}\left(1+(crt)^{-1}\right)^{-1}\right]^j$$
$$=(ct)^{-j}\left(1-j(crt)^{-1}+\mathcal{O}(t^{-2})\right)$$

and with the notation

$$I_j=\int_{\mathcal{Z}} e^{-j\nu z}\left(1+ce^{\nu z}t\right)^{-1/c}dQ(z)$$

we have, uniformly in $c \in [c_0, c_1]$, $0 < c_0 \le c_1 < \infty$,

$$\beta(F_\theta(t)) = (1+c)^2 f_\theta^{-4}(t)(ct)^{-4}\Big\{ \left[I_0 - 3(ct)^{-1}I_1\right]\left[I_0 - (ct)^{-1}I_1\right] - \left[I_0 - 2(ct)^{-1}I_1\right]^2 + \mathcal{O}\left(t^{-2-2/c}\right)\Big\}$$
$$= \mathcal{O}\left(t^{-2+2/c}\right)$$

and, together with (65),

$$\beta(u) = \mathcal{O}\left((1-u)^{2c-2}\right). \tag{66}$$

Furthermore, for the projection of the score function for the regression parameter ν we are led to

$$\gamma(F_\theta(t)) = -f_\theta^{-2}(t)(1+c)\int_{\mathcal{Z}} z e^{\nu z}(1+ce^{\nu z}t)^{-2} p(t\,|\,z,\theta)\,dQ(z)$$
$$= \mathcal{O}\left((1+t)^{-1+1/c}\right),$$

or

$$\gamma(u) = \mathcal{O}\left((1-u)^{c-1}\right). \tag{67}$$

From (66) and (67) it follows that Lemma 1 may be applied with $\beta = 2c$ and $\gamma = c+1$. In this case, some computation and (44), (62) through (64) yield

$$\frac{\partial^j}{\partial t^j}\alpha_1(t) = \mathcal{O}\left((1+t)^{(2c)^{-1}\vee 1-j}\left\{1 + \mathbf{1}_{[c=1/2]}\log(1+t)\right\}^{1/2}\right) \tag{68}$$

for $j = 0, 1, 2, 3$ and hence

$$\frac{\partial^j}{\partial t^j} S_1(t\,|\,z,\theta,Q)$$
$$= \mathcal{O}\left((1+t)^{(2c)^{-1}\vee 1-j-1}\left\{1 + \mathbf{1}_{[c=1/2]}\log(1+t)\right\}^{1/2}\right) \tag{69}$$

for $j = 0, 1, 2$ or, more crudely,

$$\frac{\partial^j}{\partial t^j} S_1(t\,|\,z,\theta,Q) = \mathcal{O}\left((1+t)^{(2c)^{-1}\vee 1-j-1+\epsilon}\right), \quad \varepsilon > 0, \tag{70}$$

for $j = 0, 1, 2$.

For projecting the score function for c we are led to

$$\gamma(F_\theta(t)) = -f_\theta^{-2}(t)\int_{\mathcal{Z}} e^{\nu z}(1-e^{\nu z}t)(1+ce^{\nu z}t)^{-2} p(t\,|\,z,\theta)\,dQ(z)$$
$$= \mathcal{O}\left((1+t)^{1/c}\right),$$

or

$$\gamma(u) = \mathcal{O}\left((1-u)^{-1}\right) . \tag{71}$$

From (66) and (71) it follows that Lemma 1 may be applied with $\beta = 2c$ and $\gamma = 1$. Again, by (44) and (62) through (64) we arrive at

$$\frac{\partial^j}{\partial t^j}\alpha_2(t) = \mathcal{O}\Big((1+t)^{(2c)^{-1}\vee 1-j}\left\{1+\mathbf{1}_{[c=1/2]}\log(1+t)\right\}^{1/2} + (1+t)^{1-j}\left\{1+\log(1+t)\right\}\Big) \tag{72}$$

for $j = 0, 1, 2, 3$ and hence

$$\frac{\partial^j}{\partial t^j} S_2(t \mid z, \theta, Q) = \mathcal{O}\Big((1+t)^{(2c)^{-1}\vee 1-j-1}\left\{1+\mathbf{1}_{[c=1/2]}\log(1+t)\right\}^{1/2} + (1+t)^{-j}\left\{1+\log(1+t)\right\}\Big) \tag{73}$$

for $j = 0, 1, 2$ or, more crudely

$$\frac{\partial^j}{\partial t^j} S_2(t \mid z, \theta, Q) = \mathcal{O}\left((1+t)^{(2c)^{-1}\vee 1-j-1+\epsilon}\right), \quad \epsilon > 0\,, j = 0, 1, 2\,. \tag{74}$$

Note that both (70) and (74) are uniform in $c \in [c_0, c_1]\,, 0 < c_0 \le c_1 < \infty$.

We note that it is not hard to check that the above solutions α_j of (44) make it possible to construct a subset $\tilde{A}$ of A such that $\tilde{A}$ contains α_j and the conditions of Proposition 1 are satisfied for this frailty model of Clayton and Cuzick. This means that we have found a linear subspace (corresponding to $\tilde{A}$) of the nuisance parameter score functions onto which we have been able to project the score functions of the parameters of interest. As described after Proposition 1 and in Section 2 this gives us a lower bound for the covariance matrix of limiting distributions of regular estimators.

To show that this lower bound is optimal we have to construct an estimator attaining it. This might be done along the lines of the general sample splitting procedure of Klaassen (1987) or of Theorem 7.8.1 of Bickel et al. (1993). This procedure is based on a $\sqrt{n}$-consistent estimator of θ and a consistent, $\sqrt{n}$-unbiased estimator of the efficient influence function, given θ.

Here we present a simple way to construct a $\sqrt{n}$-consistent estimator of θ, assuming without loss of generality that n is even. We define

$$\Delta_i = \mathbf{1}_{[Y_{2i-1}\le Y_{2i}]}, \quad i = 1, \ldots, n/2\,, \tag{75}$$

we note that $(\Delta_i, Z_{2i-1}, Z_{2i}), i = 1, \ldots, n/2$, are i.i.d. and we compute

$$P_\theta(\Delta_1 = 1 \mid Z_1, Z_2) = P_\theta(T_1 \le T_2 \mid Z_1, Z_2) = \int_0^1 (1 + e^{\nu(Z_2 - Z_1)}[u^{-c} - 1])^{-1/c} du\,, \quad \text{a.s.} \tag{76}$$

In this way we are back in the classical i.i.d. parametric case and a maximum likelihood procedure should yield a $\sqrt{n}$-consistent estimator. Note that the resulting preliminary estimator has a distribution independent of ψ.

Studying (31), (32), and (33), we see that in order to construct a consistent, $\sqrt{n}$-unbiased estimator of the efficient influence function, given θ, we need an estimator of the marginal distribution of the covariates and an estimator of the transformation ψ. The empirical $\hat{Q}_n$ of $Z_1, \dots, Z_n$ will do as an estimator of the distribution of the covariates. Let now $\hat{F}_n$ be the empirical distribution function of $Y_1, \dots, Y_n$ and let $F_{\theta,Q}$ be the marginal distribution function in the core model (1) of the survival time T under parameters θ and Q. In view of

$$E_{\theta,\psi,Q}\hat{F}_n(y) = E_Q F_0(\psi(y) \mid Z, \theta) = E_Q F_{\theta,\hat{Q}_n}(\psi(y)), \quad 0 < y, \qquad (77)$$

it is natural to estimate, given θ, the transformation ψ by

$$\hat{\psi}_n(y) = F^{-1}_{\theta,\hat{Q}_n}\left(\hat{F}_n(y) \wedge (1 - n^{-1})\right), \quad 0 < y. \qquad (78)$$

It may be shown that these estimators are sufficiently accurate for their goal, namely to construct a consistent, $\sqrt{n}$-unbiased estimator of the efficient influence function, given θ. However, the technical details will not be pursued here. If for fixed known θ this estimator of ψ or another estimator is efficient, then Klaassen and Putter (2005) gives a construction to transform it into an efficient estimator in the full semiparametric model with θ unknown, using an efficient estimator of θ itself. See e.g. Dabrowska (2002) and Gørgens (2003) for alternative estimators of the transformation, the efficiency of which is not discussed there either.

Acknowledgement. I would like to thank Peter Bickel, Dorota Dabrowska, Kjell Doksum, Andries Lenstra, and Jon Wellner for illuminating discussions on the topic of this paper, most of those discussions many years ago.

References

1. BEGUN, J. M. (1987), Estimates of relative risk. *Metrika* **34**, 65–82.
2. BEGUN, J. M. AND WELLNER, J. A. (1983), Asymptotic efficiency of relative risk estimates, *Contributions to statistics*, P.K. Sen (ed.), North-Holland, Amsterdam, 47–62.
3. BICKEL, P. J. (1986), Efficient testing in a class of transformation models, *Papers on Semiparametric Models at the ISI Centenary Session, Amsterdam*, Report MS-R8614, Centrum voor Wiskunde en Informatica, Amsterdam, 63–81.

4. Bickel, P. J., Klaassen, C. A. J., Ritov, Y. and Wellner, J. A. (1993), *Efficient and Adaptive Estimation in Semiparametric Models*, Johns Hopkins Univ. Press, Baltimore; reprint (1998), Springer, New York.

5. Bickel, P. J. and Ritov, Y. (1997), Local asymptotic normality of ranks and covariates in transformation models, *Festschrift for Lucien Le Cam*, D. Pollard, E. Torgerson, and G.L. Yang (eds.), Springer, New York, 43–54.

6. Clayton, D. G. and Cuzick, J. (1985), The semi-parametric Pareto model for regression analysis of survival times. *Bull. Int. Statist. Inst.* **51**, 23.3.175–23.3.180.

7. Cox, D. R. (1972), Regression models and life-tables, *J. Roy. Statist. Soc. Ser. B* **34**, 187–220.

8. Dabrowska, D. M. (2002), Averaged non-parametric regression in analysis of transformation models, *Limit theorems in probability and statistics (Balatonlelle, 1999)* János Bolyai Math. Soc., Budapest, **I**, 479–494.

9. Dabrowska, D. M. and Doksum, K. A. (1988a), Estimation and testing in a two-sample generalized odds-rate model, *J. Amer. Statist. Assoc.* **83**, 744–749.

10. Dabrowska, D. M. and Doksum, K. A. (1988b), Partial likelihood in transformation models with censored data, *Scand. J. Statist.* **15**, 1–23.

11. Doksum, K. A. (1984), The Analysis of Transformed Data: Comment, *J. Amer. Statist. Assoc.* **79**, 316–319.

12. Doksum, K. A. (1987), An extension of partial likelihood methods for proportional hazard models to general transformation models. *Ann. Statist.* **15**, 325–345.

13. Elbers, C. and Ridder, G. (1982), True and spurious duration dependence: the identifiability of the proportional hazard model, *Rev. Econom. Stud.*, **XLIX**, 403–409.

14. Gørgens, T. (2003), Semiparametric estimation of censored transformation models, *J. Nonparametr. Stat.* **15**, 377–393.

15. Ibragimov, I. A. and Has'minskiĭ, R. Z. (1981), *Statistical Estimation*, Springer, New York.

16. Klaassen, C. A. J. (1987), Consistent estimation of the influence function of locally asymptotically linear estimators, *Ann. Statist.* **15**, 1548–1562.

17. Klaassen, C. A. J. (1989), Efficient estimation in the Cox model for survival data, *The proceedings of the Fourth Prague Symposium on Asymptotic Statistics, 29 August–2 September 1988*, P. Mandl and M. Hušková (eds.), Charles University, Prague, 313–319.

18. Klaassen, C. A. J. and Putter, H. (2005), Efficient estimation of Banach parameters in semiparametric models. *Ann. Statist.* **33**, 307–346.

19. Lenstra, A. J. (1998), *Analyses of the nonparametric mixed proportional hazards model*, Ph D thesis, Universiteit van Amsterdam, Amsterdam.

20. LENSTRA, A. J. (2005), Cramér-Rao revisited, *Bernoulli* **11**, 263-282.

21. MASON, D. M. (1983), The asymptotic distribution of weighted empirical distribution functions, *Stochastic Process. Appl.* **15**, 99–109.

22. MURPHY, S. A. (1994), Consistency in a proportional hazards model incorporating a random effect, *Ann. Statist.* **22**, 712–731.

23. MURPHY, S. A. (1995), Asymptotic theory for the frailty model, *Ann. Statist.* **23**, 182–198.

24. MURPHY, S. A., ROSSINI, A. J. AND VAN DER VAART, A. W. (1997), Maximum likelihood estimation in the proportional odds model, *J. Amer. Statist. Assoc.* **92**, 968–976.

25. RUDIN, W. (1973), *Functional Analysis*, McGraw-Hill, New York.

26. SHORACK, G. R. AND WELLNER, J. A. (1986), *Empirical Processes with Applications to Statistics*, Wiley, New York.

Chapter 14

ESTIMATION FOR THE SEMIPARAMETRIC TRANSFORMATION MODEL FOR SURVIVAL DATA AND EXTENSIONS

Thomas H. Scheike

Department of Biostatistics
University of Copenhagen, Copenhagen, DENMARK

Email: ts@biostat.ku.dk

The chapter considers the semiparametric transformation model and compare the finite sample properties of the modified partial likelihood estimator with a simple un-weighted estimating equation estimator. For the semiparametric transformation model, resampling methods may be used to provide uniform confidence bands for the nonparametric baseline function and the survival function. It is also shows how a score process (defined by the estimating equations) may be used to validate the assumption about constant proportional odds. Sometimes the transformation model will not be sufficiently flexible to deal with for example time-varying effects, and an extension of the transformation model is suggested. The extended model specifies a time-varying regression structure for the transformation, and this may be thought of as a first-order Taylor series expansion of a completely non-parametric covariate dependent baseline. Special cases include a stratified version of the usual semiparametric transformation model. The method is illustrated by a simulation study. The added flexibility increases the practical use of the model considerably.

Key words: Counting process; Estimating equation; Modified partial likelihood; Resampling inference; Survival data; Timevarying effect; Transformation model; Stratified transformation model.

1 Introduction

The semiparametric transformation model has recently received considerable attention. The model extends the Cox regression model (Cox, 1972) as well as the proportional odds model (Bennett, 1983; Murphy, Rossini and Van der Vaart 1997). The model may be of interest in its own right but, but it is not clear how to interpret the regression coefficients in other than these cases. Some authors have dealt with the model quite generally by inverse probability weighting techniques Cheng, Wei and Ying (1995, 1997), Fine, Ying and Wei (1998) and Cai, Wei and Wilcox (2000), these approaches has the drawback that one is forced to model the censoring distribution. Further in the standard set-up of inverse probability weighting the censoring distribution is not allowed to depend on covariates, but this assumption may be relaxed although this lead to a more complicated analysis. Alternatively one may consider an estimating equations approach that avoids direct modeling of the censoring distribution as in Bagdonavicius and Nikulin (1999,2001) and Chen, Jin and Ying (2002).

Given the satisfactory apparatus that has been developed to deal with the Cox model the transformation model is primarily of interest as a method for dealing with the proportional odds model where there only recently has been developed estimation procedures that are easy to use. Murphy, Rossini, and Van der Vaart (1997)non-parametric maximum likelihood estimator but from a practical point of view the procedure is difficult to work with primarily because of the difficulty in getting quick and reliable standard errors to go along with estimates. Another approach is to use a partial likelihood estimator Dabrowska and Doksum (1988). In recent work Slud and Vonta (2004) considers a non-parametric maximum-likelihood estimation (NPMLE) procedure for the transformation model. I here review the modified partial likelihood estimator Bagdonavicius and Nikulin (1999, 2001), and a similar estimating equation approach Chen, Jin and Ying (2002) and compare their finite sample properties.

To focus ideas, let T be a survival time and Z a covariate vector that do not depend on time. The transformation model now assumes that

$$\log(H(T)) = -Z^T\beta + \epsilon \tag{1}$$

where H is an un-specified monotone transformation and the error ϵ has a known distribution. The two special cases of the Cox and proportional odds model are obtained when ϵ has an extreme value distribution and when ϵ is a standard logistic distribution, respectively.

Denoting the survival distribution given covariates as $S_Z(t)$. It follows in the Cox case that

$$\log(-\log((S_Z(t)))) = \log(H(t)) + Z^T\beta$$

with cumulative baseline $H(t)$. Similarly for the proportional odds model

$$\text{logit}(1 - S_Z(t)) = \log(H(t)) + Z^T\beta$$

where $H(t)$ is the cumulative baseline odds that corresponds to a baseline survival for a subject with covariates equal to zero. For this model it is crucial that the covariates lead to constant proportional odds. To examine this hypothesis I suggest a simple goodness-of-fit test for checking this assumption. The practical implementation and computation of p-values is based on resampling techniques. I also suggest a resampling based approach for approximating confidence bands for the survival function.

The proportional odds model can be written alternatively as

$$S_Z(t) = \frac{\exp(-Z^T\beta)}{\exp(-Z^T\beta) + H(t)}.$$

In the case of stratified sampling the baseline, $H(t)$, may need to be stratified, and more generally the baseline may depend on various covariates. Also when the effects of covariates are not well described as being constant, as is illustrated in a worked example in the following, it may be necessary to extend the model to deal with this. One way of formally extending the model is to consider

$$\text{logit}(1 - S_{X,Z}(t)) = \log(H(t)) + X^T\beta(t) + Z^T\gamma$$

where some effects lead to constant proportional effects, modeled by Z, and some do not, modeled by X. This leads to a survival function on the form

$$S_{X,Z}(t) = \frac{\exp(-Z^T\beta)}{\exp(-Z^T\gamma) + \exp(X^T\beta(t))H(t)}.$$

The term $\exp(X^T\beta(t))H(t)$ gives a covariate dependent baseline. This model is impossible to identify without smoothness assumptions because $H(0) = 0$ and then $\beta(t)$ can not be identified close to 0.

More generally one may consider a fully unstructured covariate dependent baseline, $H(t|X)$, and such a model may be fitted along the lines of Dabrowska (1997) assumptions are made and X is continuously varying. When the dimension of X is large and the baseline is completely unstructured the model will be difficult to identify and it will be difficult to summarize the effect of the covariates X.

One practical compromise between bias and variance is to consider a linear first order approximation of the baseline. We therefore consider a flexible time-varying regression model where the covariate dependent baseline is given by

$$H(t|X) = X^T A(s) \tag{2}$$

where $A(t) = \int_0^t \alpha(s)ds$. One problem when fitting the model is that $H(t|X)$ must be increasing. This is ignored in the approach taken here just as for the additive hazards model suggested by Aalen (1989) If the model provides a good approximation of the true underlying survival functions it will tend to lead increasing baselines and then one can avoid serious bias on the effects of Z from not correctly modeling the effects of X. The model is considered in further details in Scheike (2006). One simple important model that is contained in this framework is when X gives a simple stratification.

The paper is organized as follows. Section 2 reviews some approaches for the transformation model and extends these ideas with resampling techniques and robust variance estimators. Section 3 outlines the extended model and presents a simulation study and a worked example. Finally, Section 4 contains some closing remarks.

2 Estimation

The intensity of T can be written as

$$\lambda(t)dt = Y(t)\exp(Z^T\beta)h(t)\lambda_0(\exp(Z^T\beta)H(t-)), \tag{3}$$

where $Y(t)$ is the at-risk indicator, $\lambda_0(t)$ is the hazard associated with $\exp(\epsilon)$, Z is p-dimensional bounded covariate and H is an unknown strictly increasing function. We assume that the derivative of $H(t)$ exists and is denoted as $h(t) = \frac{\partial}{\partial t}H(t)$.

Assume that i.i.d. triplets $(N_i, Z_i, Y_i())$ for $i = 1, ...n$, representing survival times, covariates and independent at risk processes, are being observed subject to this generic hazard model over the time-period $[0, \tau]$. We here consider the specific assumption that the censoring distribution and survival time of interest are independent given the covariate and that the support of the censoring distribution given the covariate does not depend on the covariate. Define $N(t) = (N_1(t), .., N_n(t))^T$ the n-dimensional counting process of all subjects with intensity $\lambda(t) = (\lambda_1(t), ..., \lambda_n(t))^T$. Based on this we can define the basic martingales $M_i(t) = N_i(t) - \int_0^t \lambda_i(s)ds$ $i = 1, .., n$. We organize the covariates into a matrix of dimension $n \times p$: $Z(t) = (Y_1(t)Z_1, ..., Y_n(t)Z_n)^T$ where $Y_i(t)$ $i = 1, ..., n$ are the at-risk indicators and $Y(t) = (Y_1(t), ..., Y_n(t))^T$. Let $dN.(t) = \sum_{i=1}^n dN_i(t)$.

2.1 *Estimating equation approach*

We now review the work of Chen, Jin and Ying (2002) where additional details and asymptotic results can be found. Some new robust variance estimators as well as resampling techniques for checking the goodness-of-fit of the model and for describing the the variability of the cumulative baseline estimator are also suggested.

To estimate β and $H(t)$ one may consider the following estimating equations

$$\int Z^T(t)W(t)(dN(t) - \lambda(t)dt) = 0, \tag{4}$$

$$Y^T(t)V(t)(dN(t) - \lambda(t)dt) = 0, \tag{5}$$

where $W(t)$ and $V(t)$ are known diagonal weight matrices. The efficient choice of both these matrices are not simple, and we here consider the case with $W = V = I$ as in Chen, Jin and Ying (2002).

For known β (5) is solved recursively for

$$d\tilde{H}(t,\beta) = \frac{1}{S_0(t,H,\beta)} dN.(t) \tag{6}$$

where

$$S_j(t,H,\beta) = \sum_{i=1}^{n} Z_i^j Y_i(t)\exp(Z_i^T\beta)\lambda_0(\exp(Z_i^T\beta)H(t-))$$

for $j = 0, 1$. Thus leading to

$$\tilde{H}(t) = \tilde{H}(t,\beta) = \int_0^t \frac{1}{S_0(t,\tilde{H},\beta)} dN.(t), \tag{7}$$

a recursive structure for computing $\tilde{H}$.

Now, with the increment estimator of $dH(t)$ the equation for β reads

$$\tilde{U}(\tau,\beta) = \sum_{i=1}^{n} \int_0^{\tau} (Z_i - \frac{S_1(t,\tilde{H},\beta)}{S_0(t,\tilde{H},\beta)}) dN_i(t) = 0. \tag{8}$$

Let the solution to this estimating equation be denoted $\tilde{\beta}$ and based on this estimate $H(t)$ by

$$\tilde{H}(t,\tilde{\beta}).$$

Chen, Jin and Ying (2002) show that the solution is consistent and asymptotically normal and provide estimators of the asymptotic variance for a reparameterized version of the problem that considers β and $\log(H)$. We here just point out that the estimating function can be written

$$n^{-1/2}\tilde{U}(t,\beta_0) = n^{-1/2}\sum_i \int_0^t q_i(s,\beta_0)dM_i(s) + o_p(1),$$

where q_i $i = 1,..,n$ are i.i.d. processes (the explicit expression is given in Scheike (2006) and follows as in Bagdonavicius and Nikulin (1999) or Chen, Jin and Ying (2002)). Since it is a sum of i.i.d. terms (or a martingale) and therefore converges to a normal distribution with variance that is estimated by the robust estimator

$$\hat{\Psi} = n^{-1}\sum_i \left\{ \int_0^{\tau} Y_i(t)\hat{q}_i(t,\tilde{\beta})d\hat{M}_i(t) \right\}^{\otimes 2}, \tag{9}$$

where $\hat{q}_i$ and $\hat{M}_i$ are estimators of q_i and M_i obtained by using the estimates and empirical versions of all covariances. An alternative estimator of the variance of the estimating function process is given by the (estimated) optional variation process

$$n^{-1}\sum_i \int_0^\tau Y_i(t)\left\{\hat{q}_i(t,\tilde{\beta})\right\}^{\otimes 2} dN_i(t).$$

This suggest that the variance of $\tilde{\beta}-\beta$ is estimated by

$$\mathcal{I}^{-1}(\tau,\tilde{\beta})\hat{\Psi}\mathcal{I}^{-1}(\tau,\tilde{\beta})$$

where $\mathcal{I}(t,\tilde{\beta})$ is the derivative of the estimating function $\tilde{U}(t,\beta)$ evaluated at $\tilde{\beta}$. The derivative is given for an extended model in the next section. Let $I(t,\beta)$ denote the limit of $n^{-1}\mathcal{I}(t,\beta)$.

A similar i.i.d. decomposition may be established for $\tilde{H}(t,\tilde{\beta})$, such that

$$\sqrt{n}(\tilde{H}(t,\tilde{\beta})-H(t)) = n^{-1/2}\sum_i H_i(t,\beta_0)+o_p(1)$$

where

$$H_i(t,\beta) = P(t,\beta_0)I^{-1}(\tau,\beta_0)\int_0^\tau q_i(t,\beta_0)dM_i(t)+\int_0^t \frac{1}{s_o(t,\beta_0)}dM_i(s),$$

where s_o is the limit of S_0 and where $P(t,\beta)$ the limit of

$$\tilde{P}(t,\tilde{\beta}) = n^{-1}\int_0^t -\frac{D_\beta\{S_0(t,\tilde{H}(\tilde{\beta},t-),\tilde{\beta})\}}{S_0^2(t,\tilde{H},\tilde{\beta})}dN.(t),$$

that must be computed recursively, and with the numerator being the derivative with respect to β, $D_\beta(S_0(t,H(\beta,t-),\beta))$, evaluated in $\tilde{\beta}$. Therefore a robust variance estimator for $\tilde{H}(t,\tilde{\beta})$ is given by

$$\hat{\Sigma}(t) = \sum_i \hat{H}_i^2(t), \tag{10}$$

where $\hat{H}_i(t)$ is the obvious estimator of H_i. To construct confidence bands one may resample the residuals

$$\Delta(t) = \sum_i G_i\hat{H}_i(t)$$

where $G_1,...,G_n$ are independent standard normals independent of the counting processes and their covariates Lin, Wei, and Ying (1993)

This may be used to construct a confidence band for the survival function $S_Z(t)$. Consider the situation (without loss of generality) where $Z=0$

then $S_Z(t)$ has a confidence band that is expressed directly from the confidence band of $\tilde{H}(t)$. To construct a uniform confidence band for $H(t)$ first compute the 95 % percentile, C_{95}, of

$$\sup_{t\in[0,\tau]} \frac{|\Delta_k(t)|}{\hat{\Sigma}^{1/2}(t)}$$

among the resampling processes $\Delta_1(t), ..., \Delta_K(t)$, and then an approximate 95 % confidence band is given as

$$\tilde{H}(t) \pm C_{95}\hat{\Sigma}^{1/2}(t) = [H_l(t), H_u(t)].$$

One consequence of this is that the survival function has an approximate 95 % confidence interval given by

$$[\frac{1}{1+H_l(t)}, \frac{1}{1+H_u(t)}].$$

To evaluate the constant proportional odds assumption for the covariates consider the estimating function, $\tilde{U}$, as a function of time. It has already been established that $\tilde{U}(t)$ could be written as a sum of i.i.d. terms. When evaluated at $\tilde{\beta}$ one gets

$$\begin{aligned}\tilde{U}(t,\tilde{\beta}) &= \tilde{U}(t,\beta_0) + \mathcal{I}(t,\beta_o)(\tilde{\beta}-\beta_0) + o_p(n^{-1/2})\\ &= \tilde{U}(t,\beta_0) + \mathcal{I}(t,\beta_o)\mathcal{I}^{-1}(\tau,\beta_o)\tilde{U}(\tau,\beta_0) + o_p(n^{-1/2}).\end{aligned}$$

With $\hat{U}_i(t) = \int_0^t \hat{q}_i(s,\tilde{\beta})d\hat{M}_i(s)$ the distribution of $\tilde{U}(t,\tilde{\beta})$ can be approximated by the resampling processes

$$\sum_i G_i\left\{\hat{U}_i(t) + \mathcal{I}(t,\tilde{\beta})\mathcal{I}^{-1}(\tau,\tilde{\beta})\hat{U}_i(\tau)\right\}$$

where $G_1, ..., G_n$ are independent standard normals.

2.2 *Modified partial likelihood approach*

The approach of Bagdonavicius and Nikulin (1999) is based on the partial likelihood. They substitute H with its estimator (7) and use $dH(t)$ to replace $h(t)$ to get a **modified partial likelihood**, a pseudo profile-likelihood, for β on the form

$$\mathcal{PL}(\beta) = \left(\prod_{i=1}^{n}\prod_{t\geq 0}\left\{Y_i(t)\exp(Z_i^T\beta)d\tilde{H}(t)\lambda_0(\exp(Z_i^T\beta)\tilde{H}(t-))\right\}^{\Delta N_i(t)}\right).$$

The derivative of the log pseudo profile-likelihood is given as

$$\tilde{U}_m(\beta) = \sum_i \int \left\{\frac{\dot{w}_i(t,\beta,\tilde{H})}{w_i(t,\beta,\tilde{H})} - \frac{\frac{\partial}{\partial\beta}S_0(t,\beta,\tilde{H})}{S_0(t,\beta,\tilde{H})}\right\}dN_i(t), \qquad (11)$$

where $w_i(t,\beta,\tilde{H}) = \exp(Z_i^T\beta)\lambda_0(\exp(Z_i^T\beta)\tilde{H}(t-,\beta))$ and $\dot{w}_i(t,\beta,\tilde{H}) = D_\beta(w_i(t,\beta,\tilde{H}))$. Denote the second derivative of the modified partial likelihood as $\mathcal{I}_m(\beta)$.

It can be shown that there exist a consistent solution with probability tending to one. With $\hat{\beta}$ the proper root of (11) and with additional regularity conditions it follows that $\sqrt{n}(\hat{\beta}-\beta_0)$ is asymptotically zero-mean normal with covariance matrix that is estimated consistently by

$$\mathcal{I}_m^{-1}(\hat{\beta})\hat{\Psi}_m\mathcal{I}_m^{-1}(\hat{\beta})$$

where $\hat{\Psi}_m$ may be estimated using optional variation or robust estimators as in the previous section.

Further, to estimate $H_0(t)$ use the estimator $\hat{H}(t) = \tilde{H}(\hat{\beta},t)$. It can be shown that $\sqrt{n}(\hat{H}(t)-H_0(t))$ converges to a Gaussian process with a variance that can be estimated by expressions similarly to those given in the previous section. Further, to get uniform bands one may resample the residuals and one may also resample the estimating-function $\tilde{U}_m(\hat{\beta},t)$ as a function of time to evaluate the goodness-of-fit with respect to the constant proportional odds assumption.

2.3 *Simulations and data example*

In this section compare the performance of the modified partial likelihood and the simple estimating equation are compared under the semiparametric proportional odds model.

In this case the Chen, Jin and Ying (2002) estimating equations becomes $(\lambda_0(t) = 1/(1+t))$

$$\sum_{i=1}^{n}\int (Z_i - \frac{\sum_{i=1}^n Z_iY_i(t)/[\exp(-Z_i^T\beta)+\tilde{H}(t-)]}{\sum_{i=1}^n Y_i(t)/[\exp(-Z_i^T\beta)+\tilde{H}(t-)]})dN_i(t) = 0,$$

and with a baseline odds of failure by time t

$$\tilde{H}(t) = \int_0^t \frac{1}{\sum_{i=1}^n Y_i(t)/[\exp(-Z_i^T\tilde{\beta})+\tilde{H}(t-)]}dN.(t).$$

The modified partial likelihood estimation equation becomes

$$\sum_{i=1}^{n}\int\left\{\frac{Z_i\exp(-Z_i^T\beta)-\frac{\partial}{\partial\beta}\tilde{H}(t-)}{\exp(-Z_i^T\beta)+\tilde{H}(t-)} - \frac{\frac{\partial}{\partial\beta}S_0(t,\tilde{H},\beta)}{S_0(t,\tilde{H},\beta)}\right\}dN_i(t) = 0.$$

We consider a 4-dimensional covariate Z that consist of independent standard normals and has constant proportional odds $0.1, -0.1, 0.5, -0.5$ and let $H(t) = t$. All survival times were censored at time 50 to avoid some numerical problems towards the end of the time-period. This lead to approximately 5 % censorings. The modified partial likelihood estimator converged considerably more quickly than the estimating equation estimator.

We only report the findings for $\beta = 0.1$ based on the estimating equation approach and the modified partial likelihood approach for 3 different sample sizes and with 1000 repetitions.

We computed the mean of the estimates, the standard error of the estimates (SE), the mean of the estimated standard errors based on the optional variation estimator (mSE), and the mean of the estimated standard error based on the robust estimator (mRSE).

Table 1 Independent censoring. Mean of 1000 replications for first covariate (see text), SE of estimates, mean of SE (mSE) and mean of estimated robust standard errors (mRSE). Based on the two methods and different sample sizes.

Method	n	β	SE	mSE	mRSE
EE	50	0.10	0.29	0.31	0.29
EE	100	0.10	0.20	0.20	0.20
EE	200	0.10	0.14	0.14	0.15
MPL	50	0.10	0.27	0.26	0.27
MPL	100	0.10	0.18	0.18	0.18
MPL	200	0.10	0.13	0.12	0.13

We see that both approaches lead to essentially unbiased estimates and that the the standard error was well estimated by both estimators of the variation. The two variance estimators performed quite similarly. It is also evident that the modified partial likelihood estimator was more efficient, with a gain about 7 % for all sample sizes when considering the standard error.

The baseline showed a similar behavior. That is, the estimators in both situations where essentially unbiased and its variance where well estimated by the robust estimator. The resampling based confidence bands also lead to a coverage close to the true level (95 %). The modified partial likelihood estimator was superior to the estimating equation approach.

We now consider a more complex situation. We consider a 4-dimensional covariate Z with the first two components independent standard normals and the second two components standard log-normals. The aim is to examine the influence of skewness on the performance. We simulate from a constant proportional odds with regression coefficients $0.1, -0.5, 0.1, -0.5$ and let $H(t) = t^d$ where $d = 0.5, 1, 2$. The right-censoring times were drawn from either an independent proportional odds model with the same baseline as the survival time and without covariates or a proportional odds model with the same 4 covariates and same regression coefficients as the survival data. All simulations were calibrated by multiplying the censoring times with a constant to give 40% censorings.

Table 2 contains the results for the dependent censoring and Table 3 the results for the independent censoring. Table 2 contains the mean of the estimates, the standard error (SE) and the mean of the robust estimated standard errors (mRSE) based on 1000 replications. Comparing with Table 3 it appears that results are quite similar. I also computed coverage probabilities and these lead to a level close to the nominal level. Standard errors are well estimated by the robust standard errors. The robust standard errors were slightly better than those based on martingales (the optional variation form, not shown). Within the two tables the MPL appears to lead to slightly smaller standard errors than the EE approach, but the effect is only marginal compared to the previous simpler simulation study.

Within each of the approaches the shape of the baseline does not appear to have much influence on the performance of the two non-skew covariates. For the skew covariates, however, it appears that the timing of events later rather than early gives smaller SE's and this is consistent across methods and does not depend on the type of censoring.

2.4 *Veterans data*

The methodology is illustrated on the Veterans' Administration lung cancer trial consisting of the 97 patients that did not receive any prior treatment. This data was also considered by Murphy, Rossini and Van der Vaart (1997) and Chen, Jin and Ying (2002) and is available in the R package.

The 97 survival times contains 37 ties; these ties were broken by adding a little random noise. The estimates revealed some dependence on how the ties were resolved.

The considered covariates were the kanofsky score (karno) and celltype (squamous, small, adeno, large).

The estimating equation function converged after 33 iterations. The model was also fitted by the modified partial likelihood, that converged after only 4 iterations.

Table 2 Dependent censoring. Mean of 1000 repetitions for 4 covariates (see text), SE of estimates and mean of estimated robust standard errors (mRSE). Comparison of estimating equations (EE) and modified partial likelihood (MPL).

Method	H(t)	n	0.1	−0.5	0.1	−0.5
EE-mean	$\sqrt{t}$	100	0.12	−0.49	0.09	−0.53
EE-mean	$\sqrt{t}$	200	0.11	−0.53	0.09	−0.52
EE-mean	t	100	0.10	−0.51	0.08	−0.51
EE-mean	t	200	0.10	−0.51	0.10	−0.51
EE-mean	t^2	100	0.11	−0.51	0.09	−0.51
EE-mean	t^2	200	0.11	−0.51	0.09	−0.51
EE-SE	$\sqrt{t}$	100	0.22	0.22	0.16	0.17
EE-SE	$\sqrt{t}$	200	0.15	0.15	0.11	0.12
EE-SE	t	100	0.22	0.23	0.17	0.15
EE-SE	t	200	0.15	0.16	0.12	0.11
EE-SE	t^2	100	0.21	0.22	0.16	0.14
EE-SE	t^2	200	0.15	0.15	0.11	0.10
EE-mRSE	$\sqrt{t}$	100	0.21	0.21	0.15	0.17
EE-mRSE	$\sqrt{t}$	200	0.15	0.15	0.11	0.12
EE-mRSE	t	100	0.21	0.22	0.16	0.15
EE-mRSE	t	200	0.15	0.15	0.11	0.10
EE-mRSE	t^2	100	0.21	0.21	0.15	0.14
EE-mRSE	t^2	200	0.14	0.15	0.10	0.10
MPL	$\sqrt{t}$	100	0.10	−0.50	0.08	−0.53
MPL	$\sqrt{t}$	200	0.11	−0.51	0.09	−0.53
MPL	t	100	0.11	−0.51	0.10	−0.52
MPL	t	200	0.10	−0.50	0.09	−0.51
MPL	t^2	100	0.08	−0.51	0.08	−0.52
MPL	t^2	200	0.09	−0.51	0.08	−0.50
MPL-SE	$\sqrt{t}$	100	0.20	0.21	0.15	0.17
MPL-SE	$\sqrt{t}$	200	0.14	0.15	0.10	0.11
MPL-SE	t	100	0.21	0.22	0.16	0.15
MPL-SE	t	200	0.14	0.15	0.11	0.10
MPL-SE	t^2	100	0.20	0.21	0.15	0.14
MPL-SE	t^2	200	0.14	0.14	0.10	0.10
MPL-mRSE	$\sqrt{t}$	100	0.20	0.21	0.15	0.17
MPL-mRSE	$\sqrt{t}$	200	0.14	0.15	0.10	0.11
MPL-mRSE	t	100	0.21	0.21	0.16	0.15
MPL-mRSE	t	200	0.14	0.15	0.11	0.10
MPL-mRSE	t^2	100	0.20	0.21	0.15	0.14
MPL-mRSE	t^2	200	0.14	0.14	0.10	0.09

Both estimators lead to similar estimates. To validate the model I also show the estimating function process for estimating β with 50 resampled processes under the null. This reveals that the assumption of constant proportional odds appears somewhat unreasonable for the karnofsky score but

Table 3 Independent censoring. Mean of 1000 repetitions for 4 covariates (see text), SE of estimates and mean of estimated robust standard errors (mRSE). Comparison of estimating equations (EE) and modified partial likelihood (MPL).

Method	H(t)	n	0.1	−0.5	0.1	−0.5
EE-mean	$\sqrt{t}$	100	0.12	−0.53	0.10	−0.54
EE-mean	$\sqrt{t}$	200	0.10	−0.51	0.10	−0.53
EE-mean	t	100	0.10	−0.51	0.09	−0.53
EE-mean	t	200	0.10	−0.50	0.11	−0.51
EE-mean	t^2	100	0.09	−0.53	0.12	−0.51
EE-mean	t^2	200	0.10	−0.50	0.10	−0.51
EE-SE	$\sqrt{t}$	100	0.21	0.22	0.11	0.18
EE-SE	$\sqrt{t}$	200	0.15	0.15	0.07	0.12
EE-SE	t	100	0.21	0.22	0.11	0.17
EE-SE	t	200	0.15	0.15	0.07	0.12
EE-SE	t^2	100	0.21	0.22	0.11	0.17
EE-SE	t^2	200	0.15	0.15	0.07	0.12
EE-mRSE	$\sqrt{t}$	100	0.20	0.21	0.10	0.17
EE-mRSE	$\sqrt{t}$	200	0.14	0.15	0.07	0.12
EE-mRSE	t	100	0.21	0.21	0.10	0.17
EE-mRSE	t	200	0.15	0.15	0.07	0.12
EE-mRSE	t^2	100	0.21	0.22	0.10	0.17
EE-mRSE	t^2	200	0.15	0.15	0.07	0.12
MPL-mean	$\sqrt{t}$	100	0.10	−0.51	0.10	−0.52
MPL-mean	$\sqrt{t}$	200	0.10	−0.51	0.10	−0.52
MPL-mean	t	100	0.11	−0.51	0.10	−0.53
MPL-mean	t	200	0.10	−0.50	0.10	−0.52
MPL-mean	t^2	100	0.11	−0.50	0.10	−0.52
MPL-mean	t^2	200	0.09	−0.51	0.11	−0.51
MPL-SE	$\sqrt{t}$	100	0.20	0.21	0.10	0.17
MPL-SE	$\sqrt{t}$	200	0.14	0.14	0.07	0.12
MPL-SE	t	100	0.20	0.21	0.10	0.16
MPL-SE	t	200	0.14	0.14	0.07	0.11
MPL-SE	t^2	100	0.20	0.21	0.10	0.16
MPL-SE	t^2	200	0.14	0.14	0.07	0.11
MPL-mRSE	$\sqrt{t}$	100	0.20	0.21	0.10	0.17
MPL-mRSE	$\sqrt{t}$	200	0.14	0.14	0.07	0.12
MPL-mRSE	t	100	0.20	0.20	0.10	0.16
MPL-mRSE	t	200	0.14	0.14	0.06	0.11
MPL-mRSE	t^2	100	0.20	0.20	0.10	0.16
MPL-mRSE	t^2	200	0.14	0.14	0.06	0.11

acceptable for the celltypes. Karnofsky score lead to a p-value at around 5 % when using the supremum as a test-statistic and comparing to the resample distribution with 1000 repetitions. The plot is based on the estimating equations and a similar plot was obtained when using the modified partial

Table 4 Veterans data. Regression coefficients based on the two different methods.

	Estimating Equation			Modified Partial Likelihood		
	β	SE	Robust SE	β.	SE	Robust SE
karno	-0.044	0.007	0.006	-0.053	0.009	0.010
celltype squamous	-0.449	0.460	0.502	-0.178	0.592	0.598
celltype smallcell	1.230	0.500	0.543	1.401	0.492	0.464
celltype adeno	1.504	0.568	0.447	1.331	0.498	0.431

likelihood estimator.

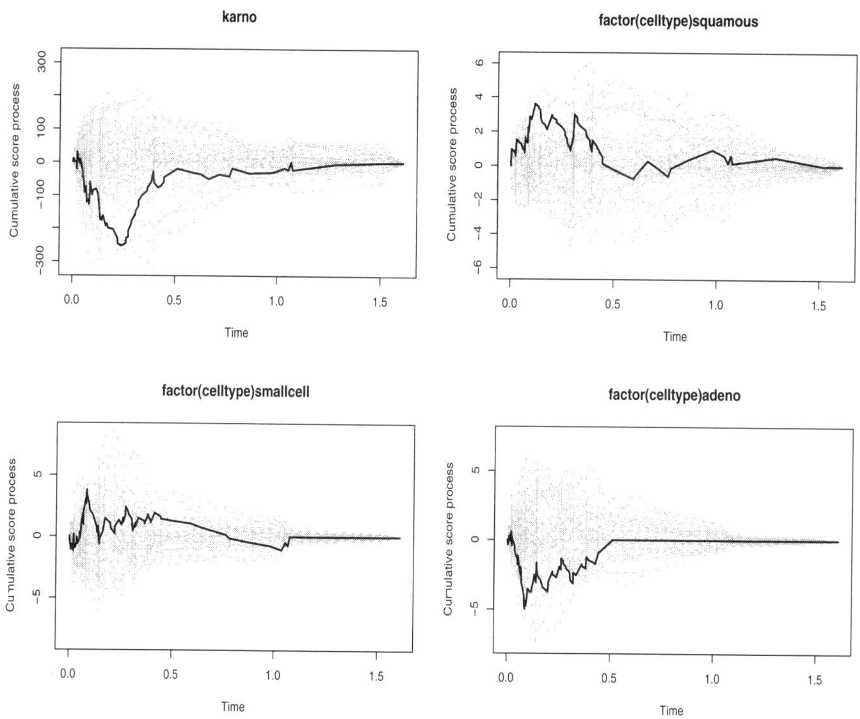

Figure 1 Goodness-of-fit plot for Veterans data with 50 resampled processes under model.

We conclude that there appears to be a problem with karnofsky score and how this affects the other estimates is unclear. The next section gives an extended model.

3 A flexible semiparametric transformation model

One way of extending the model to become much more flexible is to use a flexible time-varying regression model to model the baseline given covariates such that

$$H(t|X) = X^T A(s) \tag{12}$$

where $A(t) = \int_0^t \alpha(s)ds$.

Now, the intensity of T can be written as

$$\lambda(t) = Y(t)\exp(Z^T\beta)(X^T\alpha(t))\lambda_0(\exp(Z^T\beta)H(t-|X)), \tag{13}$$

where $\lambda_0(t)$ is the hazard associated with $\exp(\epsilon)$. When ϵ has the extreme value distribution then $\exp(\epsilon)$ is exponentially distributed ($\lambda_0(t) = 1$), and then the model is a Cox-Aalen regression model Scheike and Zhang (2002)

Assume that i.i.d. observations $(N_i(), Y_i(), X_i, Z_i)$ are are being observed over a time interval $[0, \tau]$ subject to this generic hazard model. Censoring is assumed to be independent given the covariates as in the previous Section. We organize the covariates into an $n \times q$ matrix $Z(t) = (Y_1(t)Z_1, ..., Y_n(t)Z_n)^T$ and an $n \times p$ matrix $X(t) = (Y_1(t)X_1, ..., Y_n(t)X_n)^T$ where $Y_i(t)$ $i = 1, ..., n$ are the at-risk indicators and the $n \times q$ matrix $\dot{\boldsymbol{H}}_\beta(t-|X) = (Y_1(t)\dot{H}_\beta(t-|X_1), ..., Y_n(t)\dot{H}_\beta(t-|X_n))^T$ where $\dot{H}_\beta(t-|X_i)$ is the $q \times 1$ vector of derivatives of the estimator $H_\beta(t|X_i)$ with respect to β. Define also diagonal matrices $D(\beta, A) = \text{diag}(\exp(Z_i^T\beta)\lambda_0\{H(t-|X_i)\exp(Z_i^T\beta)\})$ and $D^*(\beta, A) = \text{diag}(\exp(Z_i^T\beta)\dot{\lambda}_0(H(t-|X_i)\exp(Z_i^T\beta)))$ with $\dot{\lambda}_0(t) = \frac{\partial}{\partial t}\lambda_0(t)$. Also, $S_0(\beta, A) = X(t)D(t, \beta, A)X(t)$ and $X^-(t, \beta, A) = S_0^{-1}(\beta_0, A)X(t)^T$.

3.1 *Estimating equations*

We now consider estimating equations for β and $H(t)$ based on the counting processes

$$\int Z^T(t)(dN(t) - \lambda(t)dt) = 0, \tag{14}$$

$$X^T(t)(dN(t) - \lambda(t)dt) = 0. \tag{15}$$

For known β the increments of $A(t)$ based on (15) is solved for the recursive formulae

$$d\tilde{A}(t) = X^-(t, \beta_0, \tilde{A})dN(t),$$

thus leading to

$$\tilde{A}(t) = \int_0^t X^-(s, \beta_0, \tilde{A})dN(s), \tag{16}$$

a recursive formulae for $\tilde{A}$. To simplify the notation let all definitions depending on of both β and A and evaluated in β and $\tilde{A}(t)$ (that is a function of β be written as a function of only β, such that for example $X^-(t,\beta_0) = X^-(t,\beta_0,\tilde{A})$.

Now, with the increment estimator of $dA(t)$ the equation for β reads

$$\tilde{U}_e(\beta) = \int \{Z^T - Z^T D(\beta) X S_0^{-1}(\beta) X^T\} \, dN. \tag{17}$$

The estimating equation has a strong similarity with the estimating equation for the related Cox-Aalen survival model Scheike and Zhang (2002, 2003), that it reduces to in the Cox case. For the proportional odds model the estimating equation considered earlier is achieved.

The derivative of the estimating function is

$$\begin{aligned}\mathcal{I}_e(\beta) = &- \int Z^T D^*(\beta) d\hat{\Lambda} \left\{\text{diag}(\exp(Z_i^T\beta))Z + \dot{\boldsymbol{H}}_\beta(t-|X)\right\} \\ &+ \int Z^T D(\beta) X S_0^{-1}(\beta) X D^*(\beta) d\hat{\Lambda} \left\{\text{diag}(\exp(Z_i^T\beta))Z + \dot{\boldsymbol{H}}_\beta(t-|X)\right\} \\ &- \int Z^T D(\beta) d\hat{\Lambda} Z + \int Z^T D(\beta) X S_0^{-1}(\beta) X D(\beta) d\hat{\Lambda} Z,\end{aligned} \tag{18}$$

with $d\hat{\Lambda}(t) = \text{diag}(X(t)S_0^{-1}(t,\beta)X(t)^T dN(t))$.

Similarly, to the one-dimensional case and by following the proofs in this case it follows that with $\hat{\beta}$ the proper root of (17) and with additional regularity conditions then a consistent solution exist with probability tending to one. Further, $\sqrt{n}(\hat{\beta}-\beta_0)$ is asymptotically zero-mean normal with covariance matrix that is estimated consistently by

$$n\mathcal{I}_e^{-1}(\hat{\beta})\hat{\Psi}_e\mathcal{I}_e^{-1}(\hat{\beta})$$

where an expression for $\hat{\Psi}_e$ is on a similar form the one given in the one-dimensional case.

Further, to estimate $A_0(t)$ use the estimator $\hat{A}(t) = \tilde{A}(\hat{\beta},t)$. It follows that $\sqrt{n}(\hat{A}(t) - A_0(t))$ converges to a Gaussian process with a variance that can be estimated by a formula similar the one given in the one-dimensional case. Further, to get uniform bands one may resample the residuals.

To evaluate the the goodness-of-fit of the model consider the observed estimating function process $U_e(\hat{\beta},t)$, and this process can be approximated by

$$U_e(t,\beta_0) + \mathcal{I}_e(t,\beta_0)\mathcal{I}_e^{-1}(\tau,\beta_0)U_e(\tau,\beta_0)$$

based on which resampling is possible.

3.2 *Simulations*

Consider a two-dimensional covariate Z, with Z_1 Bernoulli with $p = 0.5$ and Z_2 standard normal with mean Z_1 and variance Z_1+1 , now based on these X_1 is Bernoulli with $p = 0.3+0.2*Z_1$ and X_2 is log-normal with mean Z_2* 0.3 and standard deviation 0.4. This implies a positive correlation between X and Z, and positive correlation within the pairs as well. Now assume that data is generated from the flexible proportional odds model where the survival function is given as $S(t|X, Z) = \exp(-Z^T\beta)/(\exp(-Z^T\beta) + H(t|X))$ where $H(t|X) = 0.1 * t + \sqrt{t}X_1 + 0 * X_2$. The proportional odds effects are $\beta = (0.1, -0.1)$. Subjects are censored at time 10 and this leads to approximately 11 % censorings.

First I censored all observations at time 8 and this lead to results similar to those presented below for the estimates of β but with a small bias and a small bias for the estimates of the three nonparametric effects. The censoring was approximately 40 %. The results that I will present are based on censoring at time 5. This lead to approximately 50 % censorings. The simulations were done based on a sample size of 200 and 400 respectively and were repeated 1000 times.

Table 5 Flexible proportional odds model. Mean of 1000 repetitions for proportional odds regression, SE of estimates and mean of estimated robust standard errors (mRSE).

n	β	SE	mRSE
200	0.08	0.30	0.30
200	-0.11	0.14	0.13
400	0.10	0.21	0.21
400	-0.10	0.09	0.09

For sample size 200 the estimates were slightly biased but their variability were well estimated, and the bias was removed when the sample size increased to 400.

To illustrate the performance of the baseline components I plot the estimates in 50 simulations. Figure 2 shows that the estimates were almost unbiased with the mean of all 1000 estimates thick broken lines) being almost equivalent to the true functions (thick full line). Some slight bias remained for the first and third component, and this disappeared when the sample increased further. The variability were reasonably well described by the suggested estimator, but further analysis will take place before these findings are reported. It is evident that the estimates show an erratic performance in some of the simulations.

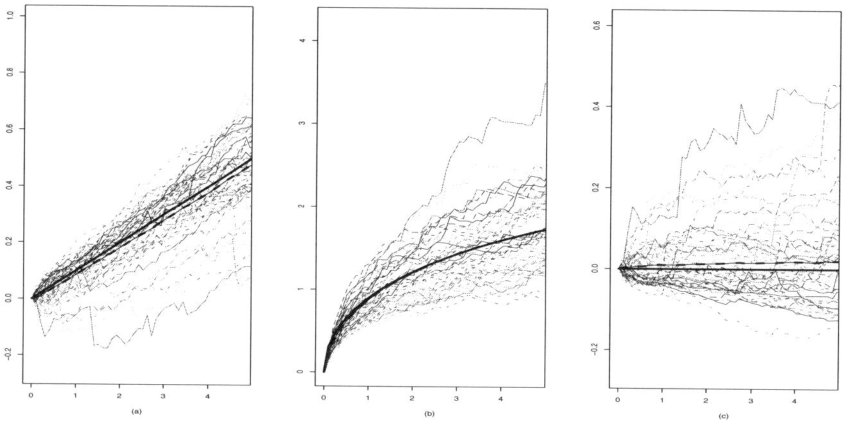

Figure 2 50 randomly chosen estimates of the non-parametric components of the model and the average of 1000 repetitions (thick line) and true functions (thick broken line).

3.3 *PBC data*

I also consider the PBC data of Fleming and Harrington (1991) a Mayo Clinic trial in primary biliary cirrhosis (PBC) where the proportional odds model does not provide a good fit. The PBC data comprises of 418 patients that are followed until death or censoring and is also available in the R survival package. Again ties were broken by adding a little random noise. The considered covariates were age, edema, albumin, and protime. This data set has been analyzed for various models and it is known that the Cox model does not fit it well. For simplicity I grouped protime in four groups based on the quartiles. Edema and protime are known to have strongly time-varying effects.

I first fitted a normal semiparametric proportional odds model with a simple baseline. The goodness of fit plots revealed that there were a serious lack of fit of the model. Figure 3 shows the goodness-of-fit processes based on the estimating equation function.

The p-values and the plots for edema and grouped protime indicated that the constant proportional odds model did not describe the data well. Age and albumin, however, indicated no lacking fit. Thus indicating severe problems with the fit for the model, and suggesting that edema and protime is the cause of these problems. The edema component shows a estimating function that rises quickly, thus indicating more deaths than described by the model initially and less later in time.

To deal with the strongly time-dependent effects of edema and protime I included edema and protime in the baseline. This lead to a model that

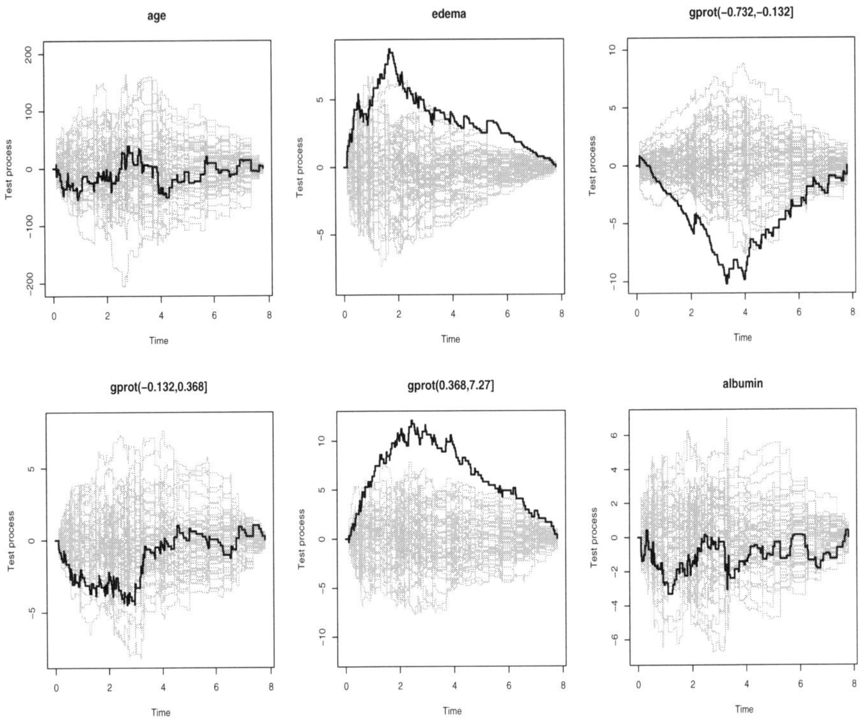

Figure 3 Goodness-of-fit plot for PBC data with 50 resampled processes under model.

indicated no problems with the fit for the two components age and albumin. The baseline components of this model further summarizes how the fit is lacking for edema and the protime quartiles.

In Figure 4 I have plotted the baseline components of the model with flexible modeling of edema and grouped protime. Edema, for example, shows an increase in the odds-ratio initially that is diminishing over time and is followed by a drop after about 5 years of study. Each baseline component is shown with 95 % pointwise confidence intervals (solid lines) and a 95 % confidence band (broken lines).

To illustrate that the added flexibility has important consequences for the models fit and ability to predict important quantities such as survival probabilities I consider the survival predictions for the flexible model and the standard semiparametric proportional odds model. In Figure 5 I have plotted the survival functions for a subject with mean albumin and mean age and depending on edema and protime for the two models. The pro-

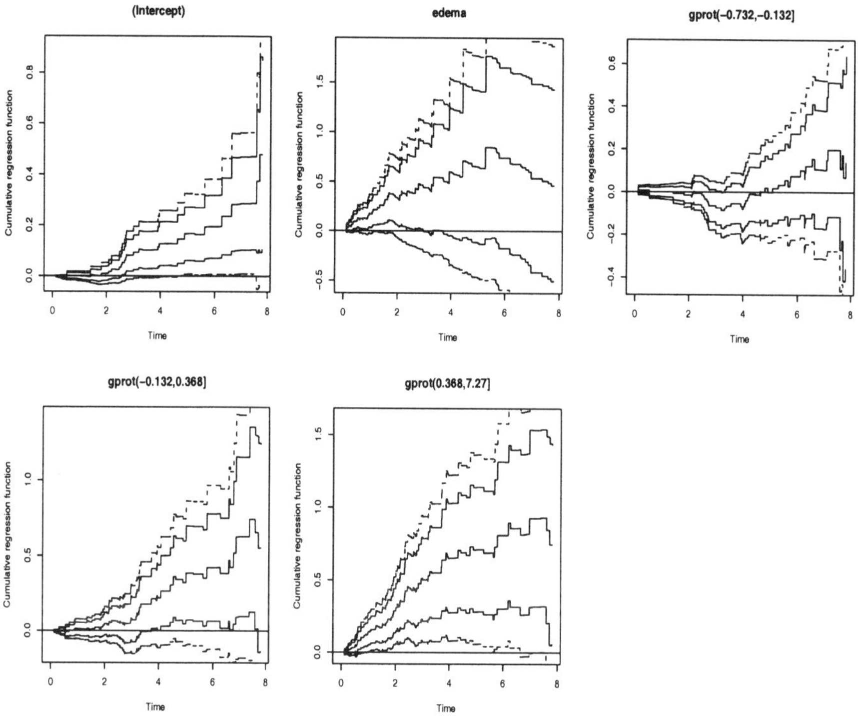

Figure 4 Odd-ratio baseline components for PBC data with 95 % pointwise confidence intervals (solid lines) and resampling based confidence bands (broken lines).

portional odds model corrects for the effects of age, albumin, edema and protime in quartiles (2., 3. and 4.) log-proportional odds parameters (sd) at 0.030, (0.0106), -1.570 (0.2890), 1.070 (0.334), 0.190 (0.296), 0.916 (0.304) and 1.450 (0.2990), respectively. For the extended model the estimated log-proportional odds effect of age and albumin were 0.0284 i(0.0106) and $-1.510(0.3210)$, respectively.

This leads to estimates of the survival function for the 8 groups depending on edema and protime. Figure 5a shows the survival function estimates for the proportional odds model. Similarly, Figure 5b gives the estimated survival based on the flexible proportional odds model that gave a much improved fit. The added flexibility is noticeable in the Figure 5b, where the presence of edema leads to survival curves with an initial steep slope that flattens considerably out towards the end of the considered time-period. Similarly, the different quartiles of protime leads to a markedly different behavior over time. The 4. quartile also has an initial strong effect reduc-

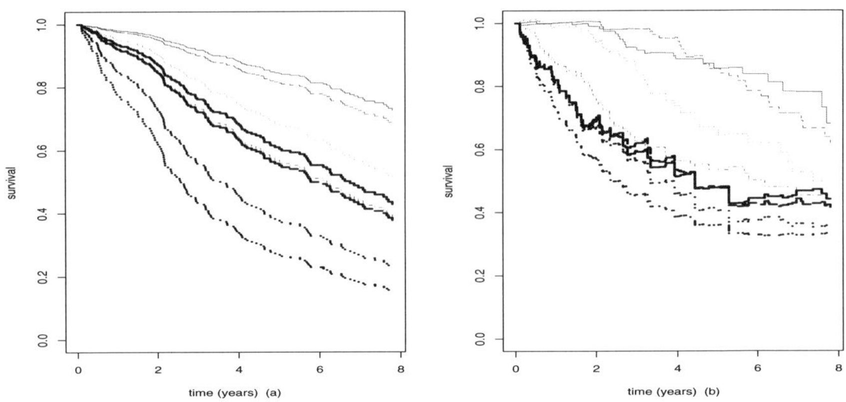

Figure 5 Estimated survival for proportional odds model (a) and for extended model (b) for for subjects with mean level of albumin, no edema (edema=0) and protime in quartiles (thin lines, full=1. quartile, dashed=2. quartile, dotted=3. quartile,dotdash=4. quartile), and for subjects with edema and protime in the quartiles (thick lines).

ing survival and then flattens out, in contrast to this the 2. and 3. quartiles on the contrary starts out with almost no effect and then increases towards the end of the time-period.

4 Discussion

I have reviewed two recent approaches for estimation in semiparametric transformation models and compared their finite sample properties in a simulation study. The modified partial likelihood estimator appears to be slightly more efficient than the one based on estimating equations. This has to do with the choice of the weight function. An alternative is to use the maximum-likelihood score with respect to β with $d\tilde{H}(t)$ instead of $h(t)$. The resulting pseudo-score equation is quite similar to the modified partial likelihood estimator. The only differ because the modified partial likelihood leads to a score for β that also includes the derivative of $\tilde{H}$ with respect to β. For known H the two approaches would be equivalent.

A simple goodness-of-fit test was suggested and was easy to implement by the use of resampling techniques. One practical limitation of the the models are that all covariate effects must lead to constant regression effects and to deal with this an extension of the model was suggested. In the simple case with just one baseline the estimate will automatically be increasing over time, but in the general regression situation the odds-ratio baseline may show some non-monotone behavior thus leading to negative hazards, this is similar to what happens in the additive hazard model and one may remedy the problem by similar techniques. Additional work is needed to fully understand the asymptotic properties and performance of the estimators of the extended partly proportional odds model.

All the methods are implemented in an R package available at the authors homepage (`http://staff.pubhealth.ku.dk/~ts/`).

Acknowledgments

I am grateful to the referees and Professor Dorota Dabrowska for constructive comments that have resulted in a much improved paper. Part of the work was done while I visited Center for Advanced Study in Oslo.

References

1. Aalen, O. O. (1989). A linear regression model for the analysis of life times. *Statist. Med.* **8**, 907–925.
2. Bagdonavicius, V. and Nikulin, M. (1999). Generalised proportional hazards model based on modified parital likelihood. *Lifetime Data Anal.* **5**, 329–350.
3. Bagdonavicius, V. and Nikulin, M. (2001). *Accelerated life models: Modelling and statistical analysis.* Chapman & Hall, London.
4. Bennett, S. (1983). Analysis of survival data by the proportional odds model. *Statist. Med.* **2**, 273–7.
5. Cai, T., Wei, L. J. and Wilcox, M. (2000). Semiparametric regression analysis for clustered failure time data. *Biometrika* **87**, 867–878.
6. Chen, K., Jin, Z. and Ying, Z. (2002). Semiparametric analysis of transformation models with censored data. *Biometrika* **89**, 659–668.
7. Cheng, S. C., Wei, L. J. and Ying, Z. (1995). Analysis of transformation models with censored data. *Biometrika* **82**, 835–845.
8. Cheng, S. C., Wei, L. J. and Ying, Z. (1997). Prediction of survival probabilities with semi-parametric transformation models. *J. Amer. Statist. Assoc.* **92**, 227–235.
9. Cox, D. R. (1972). Regression models and life tables (with discussion). *J. Roy. Statist. Soc. Ser. B* **34**, 187–220.
10. Dabrowska, D. M. (1997). Smoothed Cox regression. *Ann. Statist.* **25**, 1510–1540.
11. Dabrowska, D. M. and Doksum, K. A. (1988). Partial likelihood in transformation models with censored data. *Scand. J. Statist.* **15**, 1–24.

12. Fine, J., Ying, Z. and Wei, L. J. (1998). On the linear transformation model with censored data. *Biometrika* **85**, 980–986.
13. Fleming, T. R. and Harrington, D. P. (1991). *Counting Processes and Survival Analysis.* Wiley, New York.
14. Lin, D. Y., Wei, L. J. and Ying, Z. (1993). Checking the Cox model with cumulative sums of martingale-based residuals. *Biometrika* **80**, 557–572.
15. Murphy, S., Rossini, A. and van der Vaart, A. (1997). Maximum likelihood estimation in the proportional odds model. *J. Amer. Statist. Assoc.* **92**, 968–976.
16. Scheike, T. H. (2006). A flexible semiparametric transformation model for survival data. *Lifetime Data Analysis* (to appear).
17. Scheike, T. H. and Zhang, M.-J. (2002). An additive-multiplicative Cox-Aalen model. *Scand. J. Statist.* **28**, 75–88.
18. Scheike, T. H. and Zhang, M.-J. (2003). Extensions and applications of the Cox-Aalen survival model. *Biometrics* **59**, 1033–1045.
19. Slud, E. and Vonta, F. (2004). Consistency of the NPML estimator in the right-censored transformation model. *Scand. J. Statist.* **31**, 21–41.

Chapter 15

STATISTICAL ISSUES INVOLVED WITH EXTENDING STANDARD MODELS

Jeremy M. G. Taylor and Ning Liu

Department of Biostatistics
University of Michigan, Ann Arbor, MI 48109, USA

E-mails: jmgt@umich.edu & liuning@umich.edu

In this paper we discuss, via some specific examples, some of the issues associated with embedding a standard model in a larger family of models, indexed by an additional parameter. The examples considered are the Box-Cox transformation family, a family of models for binary responses that includes the logit and complementary log log as special cases, and a new family that includes two formulations of cure models as special cases. We discuss parameter interpretations, inflation in variance due to the addition of the extra parameter, predictions on an observable scale, ratios of parameters and score tests. We review the literature on these topics for the Box-Cox and binary response models and provide more details for the cure model.

Keywords: Cure models; Transformation; Variance inflation.

1 Introduction

There are a number of situations in statistics where standard models can be generalized by embedding the model in a family of models indexed by an additional scalar parameter. A well known example of this is the Box-Cox transformation family in which standard linear regression $Y = X\beta + e$, $e \sim N(0, \sigma^2)$, is generalized to $Y^{(\lambda)} = X\beta + e$ where $Y^{(\lambda)} = (Y^\lambda - 1)/\lambda$ for $\lambda \neq 0$ and $Y^{(\lambda)} = log(Y)$ for $\lambda = 0$.

The usual reason for considering families of models is because of the potential that the model may fit the data better. However such model extension raises a number of interesting statistical issues. Some of the issues are specific to the particular family, whereas others have a common theme in many of the families. Some of these common themes are concerned with interpretation of parameters, ratios of parameters, how to estimate variances, predictions on the original scale, the inflation in variance due to the addition of an extra parameter and score tests.

Having a model that fit the data substantially better would nearly always trump other concerns that may arise when an extra parameter is added to a model. However, these other concerns may be non-trivial. For example, in the above Box-Cox example the interpretation of β depends on the value of λ. But not all extensions of standard models lead to such difficulties. For example, for the standard regression model $Y = X\beta + e$, $e \sim N(0, \sigma^2)$, a different one parameter extension is to assume that e has a T distribution with ν degrees of freedom (Lange et al, 1989). Such an extension does not alter the interpretation of β, and estimates of ν and β are asymptotically orthogonal.

The set of possible examples of extending standard models is obviously very large. We will limit ourselves to nice fully parametric models in a regression setting with independent observations, for which the parameters can be estimated at a $\sqrt{n}$ rate.

2 Power transformation family

One of the earliest papers in which the idea of transforming data was formulated as a statistical model was Box and Cox (1964). They described estimation methods for the model $Y_i^{(\lambda)} = X_i\beta + e_i$, where $e_i \sim N(0, \sigma^2)$. They formulated both maximum likelihood and Bayesian estimation techniques for the parameters (β, σ, λ). A key point is that the log-likelihood

$$logL = -(n/2)log(2\pi\sigma^2) - (1/2\sigma^2)\Sigma(Y_i^{(\lambda)} - X_i\beta)^2 + (\lambda - 1)\Sigma log(Y_i)$$

includes the Jacobian. Thus standard least squares estimation methods to find the maximum likelihood estimate can't be applied.

There are different philosophies on the role of the Box-Cox model and the associated likelihood. One is that the likelihood is a method to find a transformation of the data, then standard linear model techniques are applied to these transformed data. A different philosophy is that the addition of λ is simply a means of making the model more flexible. Rewriting the model in the non-linear form $Y_i = 1 + \lambda(X_i\beta + e_i)^{1/\lambda}$ makes this philosophy more obvious.

An important aspect of this model is that, except for $\beta = 0$, interpretation of β depends on the value of λ. For this reason some have suggested that one only considers a small number of convenient values for λ, such as -1, 0, 1/3, 1/2, 1 and 2. The calculation of the variance of $\hat{\beta}$ has been the most controversial. Box and Cox(1964) and others (Box and Cox(1982),

Hinkley and Runger(1984)) took the position that the model is a way to determine how the data should be transformed, but after it is transformed the estimate of β is obtained conditional of that value of λ as if it were known beforehand. With this conditional view the estimate of the variance of $\hat{\beta}$ can be obtained from a Hessian that does not include a row and column for the parameter λ. This closely mimics what is done in practice. Bickel and Doksum (1982) pointed out that this method does not incorporate the uncertainty associated with the estimation of the parameter λ. When one calculates the variance of $\hat{\beta}$ using a Hessian that does include a column and row for λ, then the variance of $\hat{\beta}$ can be an order of magnitude larger. This inflation in variance was quite controversial and there is not a satisfactory solution to this day. The problem is somewhat similar to that of inference after variable selection or model selection (Faraway 1992). Some variables are selected to include in a model, whereas others are removed, then the final inference is based on the selected model. It is well recognized that this tends to give fits that are too optimistic and standard errors that are too small, yet it remains the common practice.

A different way to approach this issue is in a prediction framework, by retransforming back to the original scale of the observations (Carroll and Ruppert 1981, Taylor 1986). Consider $\hat{Y} = 1 + \hat{\lambda}(\mathbf{X}_0\hat{\beta})^{1/\hat{\lambda}}$, when $\hat{\lambda} \neq 0$ or $\exp(\mathbf{X}_0\hat{\beta})$ when $\hat{\lambda} = 0$, which is the predicted median of the distribution of Y given $\mathbf{X}_0$. This has an interpretation irrespective of the value of λ. It has been shown that the $Var(\hat{Y})$, calculated considering λ as a parameter, tends to be larger than but not substantially larger than $Var_{\hat{\lambda}}(\hat{Y})$ calculated treating λ as if it were fixed and known to equal $\hat{\lambda}$.

While the individual parameters β_1 and β_2 in the model $Y^{(\lambda)} = \beta_0 + \beta_1 X_1 + \beta_2 X_2 + e$ can be hard to interpret, it is interesting to note that the ratio of two parameters, such as β_1/β_2, does have an interpretation as the substitutability of one variable for another, irrespective of the value of λ. In particular, if X_1 is increased by one unit, then β_1/β_2 is the amount by which X_2 would need to be decreased to give the same response. This property that the ratio of parameters can be interpreted and more robustly estimated is a general result in statistics (Brillinger 1983, Li and Duan 1989). It is also the case that the inflation in variance of this ratio due to the estimation of λ is close to 1 (Taylor 1989).

A more recent article on Box-Cox transformations is provided in Chen et al (2002). They and the discussants address many issues concerned with Box-Cox transformations. They argue for reparametrization in terms of parameters β/σ and $\lambda\sigma/(1 + \lambda\mu)$, whose estimates are more stable with respect to λ. This parametrization is useful for asymptotics and related to the orthogonality developed by Cox and Reid (1987). A more demanding discussion of parametrization is given by McCullagh (2002) who argues that the source of the problem is that β and σ are not identifiable parameters.

The value of $\beta = 0$ is a special case, in that the interpretation is the same irrespective of the value of λ. One consequence of this is that tests of the null hypothesis $\beta = 0$, do not suffer from the same inference problems as those associated with estimation of β (Doksum and Wong 1983).

Score tests can be applied as a general strategy to assess whether a standard model needs to be extended. A score test of $\lambda = 1$, would be a simple way to assess whether transformations might be needed. An approach to obtain a confidence interval for λ is via profile likelihood. It is not uncommon for small sample sizes, for this profile likelihood to be far from quadratic, suggesting that Wald tests for λ may be unreliable.

3 Binary response regression models

Aranda-Ordaz (1981) suggested a family of regression models for binary outcomes, which included the logit and complementary log log link functions as special cases. The model for $p_i = P(Y_i = 1)$ takes the form

$\log(((1 - p_i)^{-\lambda} - 1)/\lambda) = X_i\beta$, for $\lambda \neq 0$,

$\log(-\log(1 - p_i)) = X_i\beta$, for $\lambda = 0$.

The well known special cases are the logit link for λ=1 and the complementary log log link for λ=0.

A graphical way to think about this model is that the probability of response follows a sigmoid shape as a function of the covariates. With λ=1, this sigmoid curve is symmetric, and other values of the parameter λ index the amount of left or right asymmetry of the sigmoid curve. With binary data the ability to distinguish between symmetric and asymmetric link functions will obviously be difficult unless the sample size is large.

A number of the statistical issues that arise for the Box-Cox model, also arise for this model. Writing the model as $p = 1 - (1 + \lambda \exp(X\beta))^{-1/\lambda}$ makes it clear that the model can simply be viewed as a more flexible way to describe the relationship between a binary response and covariates.

The variance of $\hat{\beta}$, calculated from the information matrix, will be much larger if the uncertainty in λ is incorporated, compared to assuming λ is known.

Again for this model the coefficients β have interpretations that depend on the value of λ, for $\beta \neq 0$; but $\beta = 0$ has the same interpretation for all values of λ. Also the ratios of two β's have a substitutability interpretation independent of λ.

The model can be used for estimating a predicted probability for a fixed X_0 using the equation

$\hat{p} = 1 - (1 + \hat{\lambda}\exp(X_0\hat{\beta}))^{-1/\hat{\lambda}}$ for $\hat{\lambda} \neq 0$ and
$\hat{p} = 1 - \exp(-\exp(X_0\hat{\beta}))$ for $\hat{\lambda} = 0$.

It has been shown (Taylor 1988) that the average inflation in variance of this predicted probability due to estimation of λ, when summarized in a specific way has a nice algebraic result. In particular

$$\frac{\Sigma_{i=1}^{n} w_i Var(\hat{p}_i)}{\Sigma_{i=1}^{n} w_i Var_{\hat{\lambda}}(\hat{p}_i)} = 1 + 1/q$$

where $w_i = (p_i(1 - p_i))^{-1}$.

In this expression the variance in the numerator is based on the full Hessian, while the Hessian used in the denominator ignores the row and column corresponding to λ, and q is the dimension of X. This result is intuitively appealing because it says the inflation in variance is on average proportional to the number of parameters. It has also been shown that this result generalizes to broader families of models (Taylor, Siqueira and Weiss, 1996).

Another use of this family of models is as a basis of a score test to assess goodness-of-fit of a particular model. For example, to assess if the logit link is appropriate, one could fit the model assuming a logit link, calculate $\partial log(L)/\partial\lambda$ and apply the score test. The advantage of this approach is that it doesn't require estimation of λ, which could be computationally cumbersome.

4 Cure models

A common situation, particularly in cancer research, is that the event of interest will never happen, even if the person were to be followed for a long time. For example, if the person is treated for cancer and cured by the treatment, and if the event of interest is recurrence of the disease, then this event will never occur. A good example of this is in head and neck cancer. The typical treatment for localized disease is either radiation therapy or surgery. Both of these are effective in eradicating the tumor cells, however if any tumor cells do remain they will tend to divide and grow quickly, such that within three years they will be clinically detectable. Thus if a person is followed for more than three years after treatment without detectable recurrence of the disease, there is a high likelihood that they are actually cured of the cancer. Models to analyze data in this situation have been called cure models.

One formulation of a model for such a situation is as a mixture model (Farewell 1982) or equivalently as a special case of a frailty model. In such a model a person is in the cured group with probability p and in the non-cured or susceptible group with probability $1-p$, and conditional of being in the susceptible group the survival distribution is given by S_0. Observations for which the event has occured are in the susceptible group, but observations that are censored could be in either group.

The overall, non-proper, survival distribution for this model is given by $S(t) = p + (1-p)S_0(t)$. Covariates X, could be allowed to affect both p and $S_0(t)$. The model has a number of nice features (i) it has a sensible interpretation in many applications, (ii) the probability p can depend on covariates and these can be interpreted as being associated with whether the event will occur, (iii) the model for $S_0(t)$ can depend on different covariates, which might be important factors in determining when the event occurs, given that it is not cured. A logistic model is frequently assumed for p, although below we will be using a log-log link function, and Weibull (Farewell 1977), accelerated failure time (Yamaguchi 1992, Li and Taylor 2002), non-parametric (Taylor 1995) and semi-parametric proportional hazards models (Sy and Taylor 2000, Peng and Dear 2000) have been suggested for S_0. There are some well known potential identifiability problems with this model (Fare! well 1986, Li et al 2001), due to the improper survival distribution. This can lead to estimates of intercept parameters in p being highly collinear with estimates of shape parameters in S_0. Thus care is needed in estimation and interpretation.

A different class of cure models has been suggested (Yakovlev and Tsodikov 1996, Chen, Ibrahim and Sinha 1999) and recently reviewed (Tsodikov et al 2003). The easiest representation of this class of cure models is via the cumulative hazard, $H(t)$, which is a non-decreasing function. For a cure model, $H(t)$ must be bounded as t becomes large, thus $H(t)$ can be written as $H(t) = \theta F_0(t)$, where $F_0(t)$ has the form of a distribution function of a positive random variable. But we note that $F_0(t)$ is not the distribution function of T. The survival distribution of T can be written as $S(t) = exp(-\theta F_0(t))$. Note that the probability of eventual cure is given by $p = exp(-\theta)$, thus $S(t)$ can be written as $S(t) = exp(log(p)F_0(t))$. Covariates can be included in the model in both $\theta(X)$ and $F_0(t, X)$. Note that if F_0 does not depend on covariates, then a proportional hazards assumption is satisfied.

An alternative derivation for this cure model comes from the consideration of recurrence after cancer therapy. Suppose the treatment leaves N independent clonogenic cells, and that N has a Poisson(θ) distribution. Each cell grows independently and becomes large enough to be detected at a time that has a distribution $F_0(t)$. The recurrence is recorded when the first of these N clonogens becomes detectable, ie $T = min(T_1, T_2, ..., T_N)$. Thus $P(T > t) = P(N = 0) + \Sigma_{n=1}^{\infty} P(N = n)(1 - F_0(t))^n$. After some algebra it can be shown that this simplifies to $P(T > t) = \exp(-\theta F_0(t))$. This provides a nice motivation, although in most cases the biological assumption implied by $T = min(T_1, T_2, ..., T_N)$ is not realistic. Never-the-less this algebraic motivation has proved useful in suggesting an algorithm for Bayesian estimation that has been utilized by some authors (Chen, Ibrahim and Sinha 1999).

The following more general model has both the above two cure models as special cases. Let $S(t)$ be the survival probability of all subjects. The model can be written as

$$\frac{S(t)^\lambda - 1}{\lambda} = \frac{p^\lambda - 1}{\lambda} F_0(t) \tag{1}$$

Or equivalently,

$$S(t) = [1 + (p^\lambda - 1)F_0(t)]^{\frac{1}{\lambda}}. \tag{2}$$

The extra parameter λ in this model has no real interpretation, its role is to provide a more flexible model to apply to real data. We note that a similar generalization has recently been proposed by Yin and Ibrahim (2006), although they only apply the power transformation to the left hand side of equation 1.

From equation 2 it is clear that $S(t) \to p$ as $t \to \infty$, thus p is the probability of eventual cure irrespective of the value of λ.

We can see that

$$S(t) = e^{(log(p))F_0(t)}$$

when $\lambda = 0$, and

$$S(t) = p + (1-p)S_0(t)$$

when $\lambda = 1$, where $F_0(t) = 1 - S_0(t)$.

We can allow covariates to affect both p and F_0. For example, using a log-log link with two covariates we could assume

$$prob(cure) = p = exp(-exp(\alpha_0 + \alpha_1 X_1 + \alpha_2 X_2)).$$

Assuming a Weibull form for $S_0(t)$ gives

$$S_0(t) = exp(-\tau t^\gamma exp(\beta_1 X_1 + \beta_2 X_2)).$$

Assuming n independent observations, let T_i be the event time and C_i be the censoring time for each subject. Suppose we observe (t_i, δ_i) where $t_i = min(T_i, C_i)$ and $\delta_i = I(t_i = T_i)$. Under standard assumptions about independent censoring, the likelihood function is

$$L(\alpha, \beta, \tau, \gamma, \lambda) = \prod_{i=1}^{n} [(1 - p_i^\lambda) f_0(t_i)]^{\delta_i} \lambda^{-\delta_i} S(t_i)^{1-\lambda\delta_i}$$

Table 1 Estimates and standard errors for simulated datasets.

Parameters (True values)	Estimate (SE) assuming $\lambda = 0$	Estimate (SE) assuming $\lambda = 1$	Estimate with λ estimated	SE assuming λ known	SE assuming λ unknown
		Dataset 1, True $\lambda = 0$			
α_0	0.05	0.02	0.05		
(0.0)	(0.13)	(0.12)		(0.13)	(0.13)
α_1	-1.34	-1.30	-1.37		
(-1.5)	(0.23)	(0.24)		(0.24)	(0.25)
α_2	-0.92	-0.83	-0.94		
(-1.0)	(0.24)	(0.23)		(0.24)	(0.24)
α_1/α_2	1.46	1.57	1.46		
(1.5)	(0.40)	(0.47)		(0.39)	(0.40)
β_1	0.76	0.12	0.67		
(1.0)	(0.27)	(0.21)		(0.25)	(0.33)
β_2	1.23	0.64	1.14		
(1.5)	(0.29)	(0.20)		(0.27)	(0.35)
β_2/β_1	1.62	5.33	1.70		
(1.5)	(0.70)	(12.14)		(0.78)	(0.81)
λ	0	1	0.12		
(0)				—	(0.26)
log-likelihood	-234.07	-235.61	-233.95	—	—
		Dataset 2, True $\lambda = 1$			
α_0	0.03	0.08	0.07		
(0.0)	(0.10)	(0.14)		(0.14)	(0.14)
α_1	-0.98	-1.11	-1.10		
(-1.5)	(0.20)	(0.25)		(0.25)	(0.25)
α_2	-0.52	-0.77	-0.77		
(-1.0)	(0.21)	(0.29)		(0.28)	(0.28)
α_1/α_2	1.88	1.43	1.42		
(1.5)	(0.89)	(0.49)		(0.48)	(0.48)
β_1	1.43	0.96	0.88		
(1.0)	(0.24)	(0.22)		(0.21)	(0.36)
β_2	1.98	1.70	1.62		
(1.5)	(0.23)	(0.23)		(0.22)	(0.35)
β_2/β_1	1.38	1.76	1.84		
(1.5)	(0.32)	(0.45)		(0.49)	(0.57)
λ	0	1	1.23		
(0.0)				—	(0.87)
log-likelihood	-224.49	-222.80	-197.51	—	—

where $f_0(t)$ is the density corresponding to distribution function $F_0(t)$.

For all values of λ in these cure models the probability of eventual cure is given by p. Since we assume $log(-log(p)) = \alpha_0 + \alpha_1 X_1 + \alpha_2 X_2$, the interpretation of the α's is the same in all models, and this is independent of λ. The parameter λ determines the shape of the distribution of time to event, amongst those who are not cured. This distribution is associated with covariates X through the linear combination $\beta_1 X_1 + \beta_2 X_2$, thus the

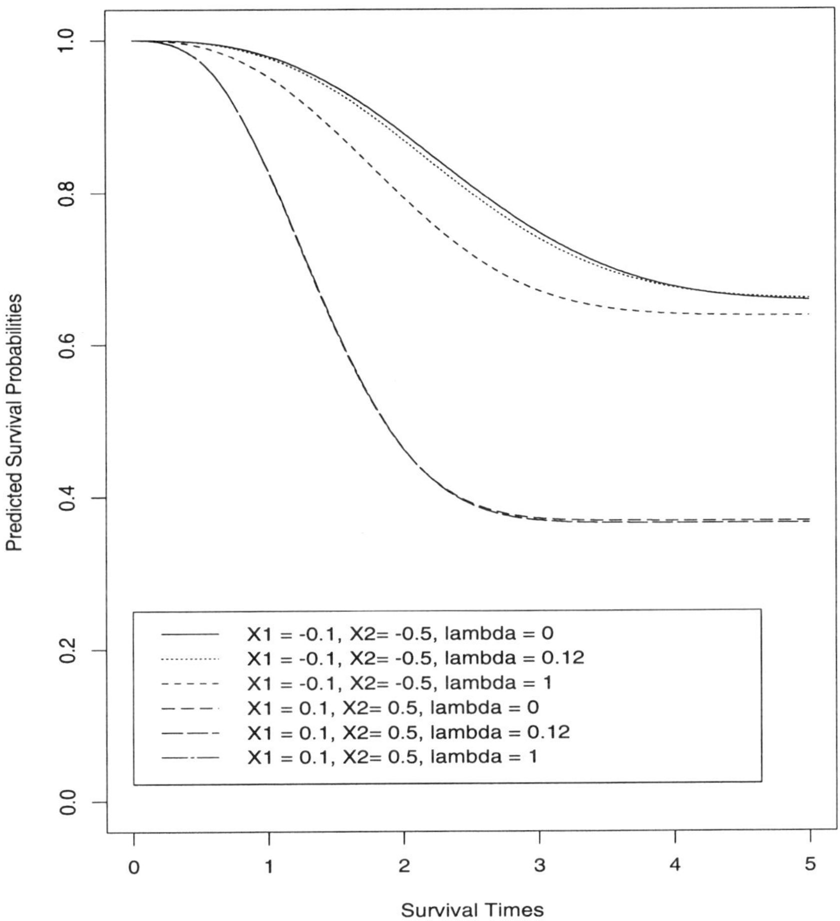

Figure 1 Predicted survival distribution for two values of (X_1, X_2) and the three values of λ (0, 1 and $\hat{\lambda}$), for dataset 1.

interpretation of β_1 and β_2 will depend on λ. The ratio of parameter estimates α_1/α_2 and β_1/β_2 again have interpretation as the substitutability of one covariate for another, however the fact that there are two ratios makes this a less useful concept.

Many of the statistical issues that arose for the Box-Cox model and the Aranda-Ordaz model, also arise for this more general cure model. As discussed above, interpretation of the regression coefficients β depend on the value of λ, score tests can be used for testing $\lambda = 0$ and $\lambda = 1$, the

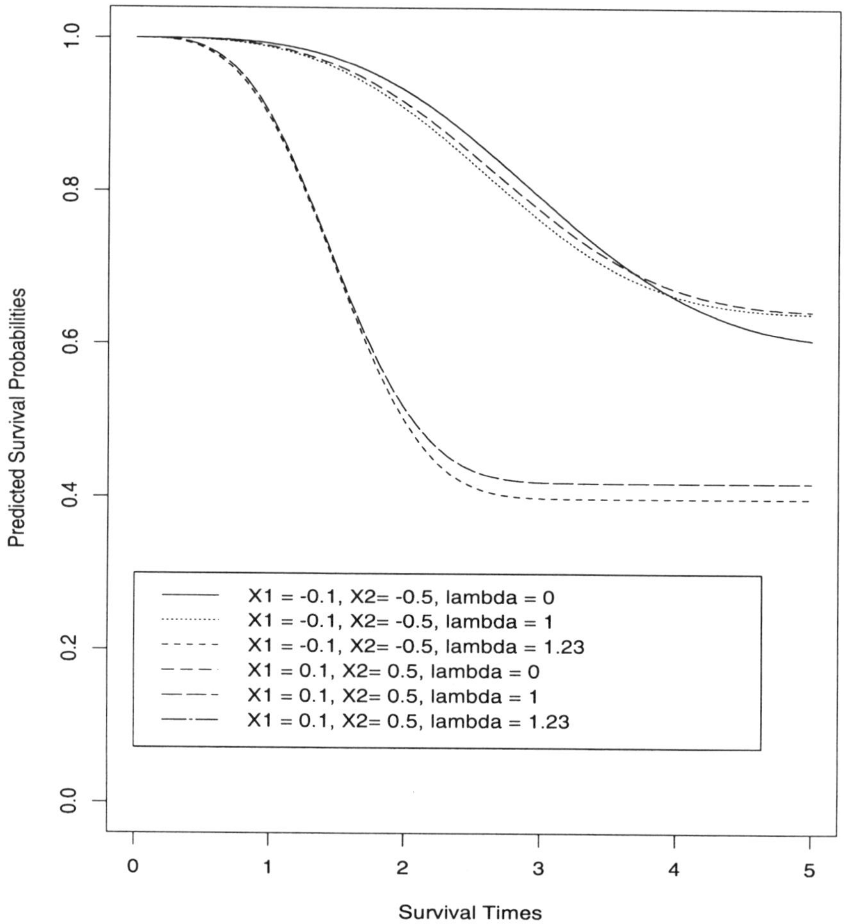

Figure 2 Predicted survival distribution for two values of (X_1, X_2) and the three values of λ (0, 1 and $\hat{\lambda}$), for dataset 2.

inferences about the regression parameters can differ depending on whether one regards λ as fixed or as an unknown parameter, and the model allows predictions back to the original scale of the observations. To address these issues, we conducted a small simulation experiment that investigated the properties of estimates from this model. We simulated two datasets each of size n=300, one with $\lambda = 0$ and one with $\lambda = 1$. There were two covariates and uniform censoring was included. The distribution of the covariates, X_1 and X_2, are both Uniform(-1,1). The values of τ and γ are 0.1 and

3.0 respectively. For each dataset we estimated the parameters under three scenarios, one with λ fixed at 0, one with λ fixed at 1 and one with λ estimated. When λ is estimated we consider two different standard errors, one based on the Hessian without a row and column for λ (labelled as SE assuming λ known) and one based on the full Hessian (labelled as SE assuming λ unknown).

The true values of the parameters along with estimates, standard errors and log-likelihoods are given in Table 1. The results show that the estimate of α are not sensitive to the value of λ, but the estimates of β are sensitive to the choice of λ. There appears to be almost no inflation in variance of $\hat{\alpha}$ due to estimation of λ, and as expected, the inflation in variance of $\hat{\beta}$ due to estimation of λ is more substantial. The results for the ratios of parameters α_1/α_2 and β_2/β_1 are also presented. There is a slight suggestion that these ratios are less dependent on λ than the parameters themselves, but a more thorough evaluation is required.

Profile likelihood confidence intervals and score tests were calculated for each dataset. The 95% confidence intervals for λ are (-0.5,1.2) and (-0.1,4.2) for the two datasets respectively. Both these are quite wide, suggesting it will be hard to obtain precise estimates of λ unless the sample size is large. The p-values from the score tests of $\lambda = 0$ and $\lambda = 1$ are 0.79 and 0.009 for dataset 1 respectively, and 0.0001 and 0.94 for dataset 2 respectively. These are broadly compatible with the log-likelihood differences as seen in Table 1, but in general we found that there could be differences between score tests and tests based on likelihood ratios at the sample sizes we considered.

The predicted survival distribution is given by equation 2. These predictions are of the marginal distribution of T given X. They are on an observable scale and have the same interpretation, irrespective of the value of λ. We calculated this survival distribution at two values of (X_1, X_2) for each of the datasets. The results are presented graphically in Figures 1 and 2. There are 6 lines on each figure, corresponding to the two values of (X_1, X_2) and the three values of λ (0, 1 and $\hat{\lambda}$). The three lines for each value of λ are close to each other, suggesting that one would get similar interpretations from the various models.

The results from this specific small simulation study, suggest that the distinction between the two cure models formulations are not so great. An interesting research question is what type of designs and sample sizes would be needed to see a practical difference between the two formulations.

Overall we see that adding an extra parameter to a standard model can lead to difficulties in parameter interpretation and inference. However, many of these difficulties are less important if the results of fitting the model are presented on the original scale of the observations.

References

1. ARANDA-ORDAZ, F. J. (1981), "On Two Families of Transformations to Additivity for Binary Response Data", *Biometrika*, **68**, 357-364.
2. BICKEL, P. J. AND DOKSUM, K. A. (1981), "An analysis of transformations revisited", *Journal of the American Statistical Association*, **76**, 296 - 311.
3. BOX, G. E. P. AND COX D. R. (1964) "An analysis of transformations (with discussion)". *Journal of the Royal Statistical Society, Series B*, **26**, 211-252.
4. BOX, G. E. P. AND COX D. R. (1982) "An analysis of transformations revisited, rebutted". *Journal of the American Statistical Association*, **77**, 209-210.
5. BRILLINGER D. R. (1983). "A generalised linear model with "Gaussian" regressor variables". *Festschrift for Erich Lehmann* (P.J. Bickel, K.A. Doksum and J.L. Hodges, Jr eds.), Wadworth, Belmont, CA, pp97-114.
6. CARROLL, R. J. AND RUPPERT, D. (1981), "On prediction and the power transformation family", *Biometrika*, **68**, 609 - 615.
7. CHEN, M. H., IBRAHIM, J. G. AND SINHA, D. (1999)."A new Bayesian model for survival data with a survival fraction." *Journal of the American Statistical Association*, **94**, 909-919.
8. CHEN, G., LOCKHART R. A. AND STEPHENS M. A. (2002)."Box-Cox transformations in linear models: Large sample theory and tests of normality." *The Canadian Journal of Statistics*, **30**, 177-234.
9. COX, D. R. AND REID N. M. (1987). "Parameter orthoganality and approximate conditional inference." *Journal of the Royal Statistical Society, Series B*, **49**, 1-39.
10. DOKSUM, K. A. AND WONG C-W. (1983). "Statistical tests based on transformed data." *Journal of the American Statistical Association*, **78**, 411-417.
11. FAREWELL, V. T. (1977). "A model for a binary variable with time censored observations." *Biometrika*, **64**, 43-46.
12. FAREWELL, V. T. (1982). "The use of mixture models for the analysis of survival data with long-term survivors." *Biometrics*, **38**, 1041-1046.
13. FAREWELL, V. T. (1986). "Mixture models in survival analysis: are they worth the risk?" *The Canadian Journal of Statistics*, **14**, 257-262.
14. FARAWAY J. J. (1992) "On the cost of data analysis". *J. Computational and Graphical Statist.*, **1**, 215-231.
15. HINKLEY, D. V. AND RUNGER, G. (1984) "The analysis of transformed data (with discussion)". *Journal of the American Statistical Association*, **79**, 302-309.
16. LANGE, K., LITTLE, R. J. A. AND TAYLOR, J. M. G. (1989) "Robust estimation using the t distribution". *Journal of the American Statistical Association*, **84**, 881-896.

17. Li, K-C. and Duan N. (1989) "Regression analysis under link violation." *The Annals of Statistics*, **17**, 1009-1052.

18. Li, C. S. and Taylor, J. M. G. (2002) "A semi-parametric accelerated failure time cure model." *Statistics in Medicine*, **21**, 3235-3247.

19. Li, C. S., Taylor, J. M. G. and Sy, J. P. (2001). "Identifiability of cure models." *Statistics & Probability Letters*, **54**, 389-395.

20. McCullagh, P. (2002) "What is a statistical model (with discussion)." *The Annals of Statistics*, **30**, 1225-1267.

21. Peng, Y. and Dear, K. B. G. (2000). A nonparametric mixture model for cure rate estimation. *Biometrics*, **56**, 237-243.

22. Sy, J. P. and Taylor, J. M. G. (2000). Estimation in a Cox proportional hazards cure model. *Biometrics*, **56**, 227-236.

23. Taylor, J. M. G. (1986), "The retransformed mean after a fitted power transformation," *Journal of the American Statistical Association*, **81**, 114-118.

24. Taylor, J. M. G. (1988), "The Cost of Generalizing Logistic Regression", *Journal of the American Statistical Association*, **87**, 1078-1083.

25. Taylor, J. M. G. (1989), "A note on the cost of estimating the ratio of regression parameters after fitting a power transformation ," *Journal of Statistical Planning and Inference*, **21**, 223-230.

26. Taylor, J. M. G. (1995). "Semi-parametric estimation in failure time mixture models." *Biometrics*, **51**, 899-907.

27. Taylor, J. M. G., Siqueira, A. L. and Weiss, R. E. (1996), "The Cost of Adding Parameters to a Model", *Journal of the Royal Statistical Society, Series B*, **58**, 593-608.

28. Tsodikov, A., Ibrahim, J. G. and Yakovlev A. Y. (2003). "Estimating cure rates from survival data: An alternative to two-component mixture models." *Journal of the American Statistical Association*, **98**, 1063-1078.

29. Yamaguchi, K. (1992). "Accelerated failure time regression models with a regression time model of surviving fraction: An application to the analysis of 'permanent employment'." *Journal of the American Statistical Association*, **87**, 284-292.

30. Yakovlev, A. Y. and Tsodikov, A. D. (1996). Stochastic models of tumor latency and their biostatistical applications. River Edge, New Jersey: World Scientific.

31. Yin, G and Ibrahim, J. G. (2006). "Cure rate models: a unified approach." *The Canadian Journal of Statistics*, **33**, 559-570.

PART 5
Nonparametric Regression

Chapter 16

LINEARLY UNBIASED ESTIMATION OF CONDITIONAL MOMENT AND CORRELATION FUNCTIONS

Hans-Georg Müller

Department of Statistics
University of California, Davis, CA, U.S.A

E-mail: mueller@wald.ucdavis.edu

We consider a random-design regression model with vector-valued observations and develop nonparametric estimation of smooth conditional moment functions in the predictor variable. This includes estimation of higher order mixed moments and also functionals of the moments, such as conditional covariance, correlation, variance, and skewness functions. Our asymptotic analysis targets the limit distributions. We find that some seemingly reasonable procedures do not reproduce the identity or other linear functions without undesirable bias components, i.e., they are *linearly biased.* Alternative *linearly unbiased* estimators are developed which remedy this bias problem without increasing the variance. A general linearly unbiased estimation scheme is introduced for arbitrary smooth functionals of moment functions.

Key words: Covariance function; Moment functional; Identity reproducing estimation; Local linear fitting; Mean squared errors; Nonparametric regression; Skewness; Smoothing; Variance function.

1 Introduction

We consider the situation of a nonparametric regression model with a random predictor X and a vector of dependent variables $Y \in \Re^p$, $p \geq 1$. It is assumed that one observes a sample of n pairs (X_i, Y_i), $i = 1, ..., n$, of independent and identically distributed (i.i.d.) bivariate data, drawn from a joint distribution $F(u, v)$. Extending the basic problems of estimating the mean regression function $E(Y|X = x)$ for univariate responses Y [Fan and Gijbels (1996), Wand and Jones (1995)] or of estimating a variance function $\text{var}(Y|X = x)$ [Müller and Stadtmüller (1993)], we consider estimation of mixed conditional moment functions of the type

$$\mu_\alpha(x) = E(Y^\alpha|X = x) = E(Y_1^{\beta_1}Y_2^{\beta_2}...Y_p^{\beta_p}|X = x), \quad (1)$$

where $\alpha = (\beta_1, ..., \beta_p)$ is a multi-index of nonnegative integers and $Y = (Y_1, ..., Y_p)^T$. Given a number of $k \geq 1$ such conditional moments $\mu_{\alpha_1}, ..., \mu_{\alpha_k}$, and a smooth mapping G from $\Re^k$ to $\Re$, our main object of interest is the functional

$$g(x) = G\left(\mu_{\alpha_1}(x), ..., \mu_{\alpha_k}(x)\right). \quad (2)$$

Interest in estimating the function $g(\cdot)$ is motivated by the following examples. (Whenever $p = 1$, we write Y for Y_1.)

Example 1. *Conditional Moment Function.*
For $k = 1$, $p = 1$, $\alpha_1 = 1$, and $G(x) = x$, one has the *conditional moment function*

$$\mu_\ell(x) = E(Y^\ell|X = x), \quad (3)$$

which includes the classical regression function for $\ell = 1$.

Example 2. *Conditional Variance Function.*
For $k = 2$, $p = 1$, $\alpha_1 = 2$, $\alpha_2 = 1$ and $G(x, y) = x - y^2$, we obtain the *conditional variance function*

$$\begin{aligned} g(x) = v(x) = \text{var}(Y|X = x) &= E(Y^2|X = x) - \{E(Y|X = x)\}^2 \\ &= \mu_2(x) - (\mu_1(x))^2. \end{aligned} \quad (4)$$

Example 3. *Conditional Skewness Function.*
The choices $k = 3$, $p = 1$, $\alpha_1 = 1$, $\alpha_2 = 2$, $\alpha_3 = 3$ and $G(x_1, x_2, x_3) = (x_3 - 3x_1x_2 + 2x_1^3)/((x_2 - x_1^2)^{3/2})$ lead to the *conditional skewness function*

$$g(x) = s(x) = E\{(Y - \mu(x))^3|X = x\} = \left\{\frac{\mu_3 - 3\mu_2\mu_1 + 2\mu_1^3}{(\mu_2 - \mu_1^2)^{3/2}}\right\}(x). \quad (5)$$

Example 4. *Conditional Covariance Function.*
For $k = 3$, $p = 2$, $\alpha_1 = (1, 1)$, $\alpha_2 = (1, 0)$, $\alpha_3 = (0, 1)$ and $G(x_1, x_2, x_3) = x_1 - x_2x_3$, we arrive at the *conditional covariance function*

$$g(x) = v_{12}(x) = E(Y_1Y_2|X = x) - E(Y_1|X = x)E(Y_2|X = x). \quad (6)$$

Example 5. *Conditional Correlation Function.*
For $k = 5$, $p = 2$, $\alpha_1 = (1,1)$, $\alpha_2 = (1,0)$, $\alpha_3 = (0,1)$, $\alpha_4 = (2,0)$, $\alpha_5 = (0,2)$ and $G(x_1, x_2, x_3, x_4, x_5) = (x_1 - x_2 x_3)/((x_4 - x_2^2)(x_5 - x_3^2))^{1/2}$, we obtain the *conditional correlation function*

$$g(x) = \rho_{12}(x) = \frac{v_{12}(x)}{\{v_{11}(x)v_{22}(x)\}^{1/2}}, \tag{7}$$

where $v_{k\ell}(x) = E(Y_k Y_\ell | X = x) - E(Y_k | X = x)E(Y_\ell | X = x)$.

Conditional moment functions, especially for the first moment, are naturally of interest in applications of nonparametric regression. The variance function has long been recognized as a valuable tool in applied statistics. Applications include the construction of confidence regions, adjustments of least squares estimators in parametric regression models to heteroscedasticity, local bandwidth selection, and volatility modeling [see, e.g., Dette and Munk (1998), Eubank and Thomas (1993), Fan and Yao (1998), Müller and Stadtmüller (1987) and Picard and Tribouley (2000)].

For example, the estimation of a conditional skewness function will be of particular interest in cases where the skewness of responses Y changes sign for varying predictors x, in which case an overall skewness estimate is less meaningful. In an obvious manner, conditional curtosis functions, conditional cumulant functions and conditional moment generating functions can be defined along the same lines. Conditional covariance and correlation functions are of particular interest. These functions are relevant whenever the response is multivariate, and when the relation between any two response variables changes as a predictor variable varies.

Another type of conditional correlation function has been introduced by Bjerve and Doksum (1993) and was further analyzed in Doksum et al. (1994) and Doksum and Samarov (1995). The topic of these papers is measuring the local strength of a relation between a univariate dependent variable Y and a univariate predictor variable X, which is modeled as varying with predictor value. In contrast, our focus here is on the dependency of the correlation between two response variables, conditional on the level of a predictor variable. This dependency could be characterized as a conditional partial correlation function, which we will model nonparametrically in the following.

A criterion which has been shown to be of practical as well as theoretical interest for discriminating between various possible function estimates focuses on whether an estimator can reproduce a linear function with a zero leading (first order) bias term. This is a desirable feature, as then bias is controlled irrespective of the locations and design of the predictors. Different aspects of this property with regard to the comparison of specific smoothers have been pointed out by various authors, among them Chu and Marron (1991), Jennen-Steinmetz and Gasser (1987), Jones and Park (1994), and Müller (1997). Specifically, the notion of an *identity reproducing function estimator* was introduced in Müller and Song (1993), where it was shown how given function estimators can be modified by incorporating an identity reproducing transformation in such a way that they become identity reproducing. Additional results along these lines were obtained by Mammen and Marron (1997) and Park, Kim and Jones (1997). A defining feature is that scatterplot data (X_i, X_i), fed into an identity-reproducing estimator of $E(Y|X = x)$, will return the identity function as function estimate, which is of course the true underlying function in this situation.

We formalize here the notion of vanishing leading bias terms in the following way: We call a function estimator $\hat{g}(x)$ of $g(x)$ *linearly unbiased*, if the leading term of its asymptotic bias is proportional to the second derivative $g^{(2)}(x)$ and does not involve any further terms depending on $g(\cdot)$ or the joint distribution $F(\cdot,\cdot)$ of X and Y. This notion is motivated by several appealing properties of linearly unbiased estimators: not only do they reproduce the identity function, but they also are associated with a bias structure that is predictable from the curve estimate itself, since bias depends only on the second derivative of the function g that is to be estimated; in particular the bias does not depend on properties of the underlying design such as the density of the design points. For example, for Nadaraya-Watson quotient type kernel estimators of the mean regression function it is well known that these estimators are linearly biased, while convolution type kernel estimators and local polynomial smoothers are linearly unbiased [Bhattacharya and Müller (1993)].

The paper is organized as follows. In Section 2, we provide further details on linearly unbiased curve estimators. We then state the main results and obtain a general construction for linearly unbiased estimates of conditional moment functionals in Section 3. In Section 4, we provide examples of linearly unbiased estimators, including estimators for skewness, covariance and correlation functions. A simulation example is included in Section 5, while proofs, auxiliary results and assumptions can be found in Section 6.

2 Preliminaries on linearly unbiased curve estimators

A generalized version of the concept of linear unbiasedness is *r-th order polynomial unbiasedness*, which would imply that the leading term of the asymptotic bias is proportional to $g^{(r+1)}(x)$. Examples of such estimators are provided by local polynomial fitting of polynomials of degree $(r+1)$, applied for estimating functions that are $(r+1)$ times continuously differentiable, and also by convolution kernel estimators using kernels of order $(r+1)$ [see Gasser, Müller and Mammitzsch (1985)]. When targeting shapes other than linear or polynomial, other notions of target unbiasedness might be of interest, extending the concept of polynomial unbiasedness to other functional shapes which one desires to estimate without leading bias terms. For the sake of simplicity, we consider here only the case $r = 1$, corresponding to linear unbiasedness.

Note that linear or first order unbiasedness implies that the bias, when estimating linear functions $g(\cdot)$, corresponds to a relatively small remainder term. This is of course a highly desirable requirement in nonparametric curve estimation, since parametric estimates based on the assumption of a linear underlying function will be unbiased when the underlying function to be estimated is indeed linear. If this is not the case, however, then such parametric estimates will be inconsistent. The idea is that the price one pays in terms of bias in nonparametric estimation, which is much more flexible than parametric modeling and yields consistent estimates as long as the underlying function is smooth, should be kept reasonably small when estimating functions with common parametric shapes.

If linear unbiasedness is not satisfied, curve estimates will show unpredictable systematic deviations that are often dependent on the underlying design, whereas under linear unbiasedness, the leading bias term depends only on the local curvature of the function to be estimated. This bias can then be at least roughly assessed from the estimated function. For the special case of estimating a variance function, the need for a bias correction of this sort has been recognized in Ruppert et al. (1997) and accordingly included in their estimation procedure by dividing smoothed squared residuals by a constant.

The analysis in Ruppert et al. (1997) also provides one of many examples of the commonly adopted conditional approach, where one focuses on the behavior of the conditional mean squared errors $E\{(\hat{g}(X) - g(X))^2|X_1, ..., X_n\}$. While such conditional measures of performance are valuable in their own right, they can only provide partial reassurance to a user who encounters new data with different designs. For instance, it is well known, and indeed corresponds to a practical problem, that unconditional mean squared error does not even exist for local polynomial smoothers, including the Nadaraya-Watson kernel estimator, while it does exist for convolution type kernel estimators (Seifert and Gasser (1996) discuss these issues in great detail). Unconditional asymptotics for local polynomial fitting and related estimation methods can still be achieved by discarding moment based criteria such as mean squared error and instead focussing on asymptotic bias and variance, defined as the bias and variance obtained from an asymptotic limiting distribution; this is the approach we adopt here. In fact, a prime example where the large sample limit of conditional bias and variance differs from asymptotic bias and variance defined in this sense is provided by local polynomial fitting.

Assume a sample of i.i.d. random vectors $(X_i, Y_i) \in \Re^{p+1}$, $i = 1, ..., n$, is given, and consider the problem of estimating the conditional moment function $\mu_\alpha(x) = E(Y^\alpha | X = x)$ for a fixed x in the domain of g. Given a sequence of bandwidths $b > 0$ and a kernel or weight function $K \geq 0$, we define kernel weights

$$w_i(x) = (nb)^{-1} K\{b^{-1}(x - X_i)\}, \tag{8}$$

usually assuming that for the sequence of bandwidths $b \to 0$, $nb \to \infty$ as $n \to \infty$, and that the kernel K is a square integrable probability density function with finite variance that is centered around 0.

The most common estimators are linear smoothers of the form

$$\hat{\mu}_\alpha(x) = \sum_{i=1}^{n} W_i(x) Y_i^\alpha, \tag{9}$$

i.e., weighted averages of the responses, where the $W_i(\cdot)$ are weight functions which characterize a particular smoothing method. Linear smoothers include splines and kernel estimators. For *Nadaraya-Watson kernel estimators* which are of quotient type, the smoothing weights W_i are explicitly given by

$$W_{i,NW}(x) = w_i(x) / \sum_{j=1}^{n} w_j(x), \tag{10}$$

leading to estimates $\hat{\mu}_{\alpha,NW}$.

A second form of kernel estimation is local linear fitting by weighted least squares. Here the smoothing weight functions W_i are obtained by solving the weighted least squares problem (compare Fan and Gijbels (1996))

$$\hat{\mu}_{\alpha,LS}(x) = \arg\min_{a_0} \left[\min_{a_1} \left\{ \sum_{i=1}^{n} w_i(x) \left[Y_i^\alpha - (a_0 + a_1 (X_i - x)) \right]^2 \right\} \right], \tag{11}$$

for which the explicit smoothing weights are found to be

$$W_{i,LS}(x) = \frac{w_i(x)}{\sum_{j=1}^n w_j(x)} - \frac{\sum_{j=1}^n w_j(x)(X_j - x)}{\sum_{j=1}^n w_j(x)} \times$$
$$\left[\frac{w_i(x)(X_i - x)\sum_{j=1}^n w_j(x) - w_i(x)\sum_{j=1}^n w_j(x)(X_j - x)}{\sum_{j=1}^n w_j(x)\sum_{j=1}^n w_j(x)(X_j - x)^2 - (\sum_{j=1}^n w_j(x)(X_j - x))^2} \right]. \tag{12}$$

One obtains the asymptotic distributions of both of these estimates under suitable regularity conditions (compare Bhattacharya and Müller (1993)) as

$$(nb)^{1/2} \left[\hat{\mu}_\alpha(x) - \mu_\alpha(x) \right] \to \mathcal{N}(B, V) \tag{13}$$

in distribution. Here the expression for the asymptotic variance is the same for both estimators $\hat{\mu}_{\alpha,NW}$ and $\hat{\mu}_{\alpha,LS}$, and is given by

$$V = f_X^{-1}(x)\,[\mu_{2\alpha}(x) - (\mu_\alpha(x))^2] \int K^2(u)\,du, \tag{14}$$

where $f_X(\cdot)$ denotes the marginal density of X. However, the asymptotic bias terms B differ between Nadaraya-Watson and local least squares estimators. Assuming $nb^5 \to d^2$ for a constant $d \geq 0$, we find for the asymptotic bias term B_{NW} of Nadaraya-Watson kernel estimators and for the bias term B_{LS} for local linear fitting that

$$B_{NW} = \frac{d}{2} \frac{\mu_\alpha^{(2)}(x) f_X(x) + 2\mu_\alpha^{(1)}(x) f_X^{(1)}(x)}{f_X(x)} \int K(u)u^2\,du \tag{15}$$

and

$$B_{LS} = \frac{d}{2}\mu_\alpha^{(2)}(x) \int K(u)u^2\,du. \tag{16}$$

We conclude that Nadaraya-Watson estimators $\hat{\mu}_{\alpha,NW}$ are linearly biased for μ_α, while local linear estimators $\hat{\mu}_{\alpha,LS}$ are linearly unbiased. This is not entirely surprising, given that Nadaraya-Watson kernel estimators can be derived as the local weighted least squares solutions of fitting local constants, which naturally leads to less flexible biases as compared to fitting local least squares lines.

3 Main results on linearly unbiased estimation

For the following, we assume that regularity conditions (C1)-(C8), listed in section 6, are in force. Relevant for the formulation of the following result on the estimation of functionals of moment functions is the condition $nb^5 \to d^2$, as $n \to \infty$ for a $d > 0$. We consider functionals (2), $g(x) - G(\mu_{\alpha_1}(x), ..., \mu_{\alpha_k}(x))$, as described in the Introduction. In order to estimate $g(x)$, a natural approach are the plug-in estimators

$$\hat{g}(x) = G\{\hat{\mu}_{\alpha_1,LS}(x), ..., \hat{\mu}_{\alpha_k,LS}(x)\}. \tag{17}$$

As the following result demonstrates, these estimators generally do not possess the desirable property of linear unbiasedness. Specifically, writing $\mu = (\mu_{\alpha_1}, ..., \mu_{\alpha_k})$, and using the abbreviations $c_B = \frac{1}{2}d \int K(u)u^2 du$ and $c_V = \int K^2(u)du$, we have the following.

Theorem 1. *Under regularity conditions (C1)-(C8),*

$$(nb)^{1/2}\{\hat{g}(x) - g(x)\} \to \mathcal{N}(\tilde{B}, \tilde{V})$$

in distribution, where

$$\tilde{B}=c_B \sum_{m=1}^{k} \left\{ \frac{dG}{dx_m}|_\mu \frac{d^2}{d^2x}\mu_{\alpha_m}(x) \right\},$$

$$\tilde{V}=\frac{c_V}{f_X(x)} \left(\sum_{l,m=1}^{k} \frac{dG}{dx_l}|_\mu \frac{dG}{dx_m}|_\mu \right) \{\mu_{\alpha_l+\alpha_m}(x) - \mu_{\alpha_l}(x)\mu_{\alpha_m}(x)\}.$$

The proof of this and the next theorem can be found in Section 6. As a simple illustration of this result, consider the conditional covariance function (Example 4, eq.(6))

$$v_{12}(x) = E(Y_1Y_2|X = x) - E(Y_1|X = x)E(Y_2|X = x).$$

The plug-in estimator is $\hat{v}_{12}(x) = \hat{\mu}_{11}(x) - \hat{\mu}_{10}(x)\hat{\mu}_{01}(x)$ and Theorem 1 leads to

$$(nb)^{1/2}\{\hat{v}_{12}(x) - v_{12}(x)\} \to \mathcal{N}(\tilde{B}_{12}, \tilde{V}_{12})$$

in distribution, where

$$\begin{aligned}\tilde{B}_{12}=&c_B(\mu_{11}^{(2)} - \mu_{01}\mu_{10}^{(2)} - \mu_{01}^{(2)}\mu_{10})(x),\\ \tilde{V}_{12}=&\frac{c_V}{f_X(x)}\{\mu_{22} - \mu_{11}^2 + \mu_{01}^2(\mu_{02} - \mu_{01}^2) + \mu_{10}^2(\mu_{20} - \mu_{10}^2)\\ &-2\mu_{01}(\mu_{21} - \mu_{11}\mu_{10}) - 2\mu_{10}(\mu_{12} - \mu_{11}\mu_{01})\\ &+2\mu_{10}\mu_{01}(\mu_{11} - \mu_{10}\mu_{01})\}(x).\end{aligned}$$

As

$$v_{12}^{(2)}(x) = \{\mu_{11}^{(2)} - (\mu_{10}^{(2)}\mu_{01} + 2\mu_{10}^{(1)}\mu_{01}^{(1)} + \mu_{10}\mu_{01}^{(2)})\}(x),$$

this estimator is found to be linearly biased. The problem is that in the asymptotic bias $\tilde{B}_{12}$ the term $-2\mu_{10}^{(1)}\mu_{01}^{(1)}(x)$ is missing.

In order to achieve linear unbiasedness for the general case, one needs to target the asymptotic bias term

$$\begin{aligned}B^*=&c_B\, g^{(2)}(x)\\ =&c_B \sum_{m=1}^{k} \left\{ \frac{dG}{dx_m}|_\mu \frac{d^2}{d^2x}\mu_{\alpha_m}(x) \right\}\\ &+c_B \sum_{l,m=1}^{k} \frac{d^2G}{dx_l dx_m}|_\mu \frac{d}{dx}\mu_{\alpha_l}(x)\frac{d}{dx}\mu_{\alpha_m}(x).\end{aligned} \quad (18)$$

Therefore the problem of linear bias in estimates $\hat{g}$ arises because the second summand in the desirable asymptotic bias term B^* is missing from the actual bias term $\tilde{B}$.

To remedy this problem, we introduce the bias correction term,

$$\hat{\Delta}(x) = \frac{1}{2} s_x^2 \sum_{l,m=1}^{k} \left\{ \frac{d^2 G}{dx_l dx_m} |_{\hat{\mu}} \ \hat{\delta}_{\alpha_l}(x) \hat{\delta}_{\alpha_m}(x) \right\}, \tag{19}$$

where

$$\hat{\delta}_\alpha(x) = \tag{20}$$
$$\frac{\sum_{i=1}^n w_i(x)(X_i - x)Y_i^\alpha \sum_{i=1}^n w_i(x) - \sum_{i=1}^n w_i(x)(X_i - x) \sum_{i=1}^n w_i(x) Y_i^\alpha}{\sum_{i=1}^n w_i(x)(X_i - x)^2 \sum_{i=1}^n w_i(x) - \left[\sum_{i=1}^n w_i(x)(X_i - x)\right]^2}$$

and

$$s_x^2 = \frac{\sum_{i=1}^n w_i (X_i - x)^2}{\sum_{i=1}^n w_i} - \left\{ \frac{\sum_{i=1}^n w_i (X_i - x)}{\sum_{i=1}^n w_i} \right\}^2. \tag{21}$$

We then propose the bias corrected estimators

$$\hat{g}^*(x) = \hat{g}(x) + \hat{\Delta}(x) = G(\hat{\mu}_{\alpha_1}, \ldots, \hat{\mu}_{\alpha_k}) + \hat{\Delta}(x). \tag{22}$$

Interestingly, this bias correction has no effect on the variance, and the modified estimator $\hat{g}^*$ is justified by the following result.

Theorem 2. *Under regularity conditions (C1)-(C8),*

$$(nb)^{1/2}\{\hat{g}^*(x) - g(x)\} \to \mathcal{N}(B^*, V^*)$$

in distribution, where

$$B^* = c_B g^{(2)}(x), \quad V^* = \tilde{V}.$$

Continuing the example of the covariance function, the bias-corrected estimator is given by

$$\hat{v}_{12}^*(x) = \hat{\mu}_{12}(x) - \hat{\mu}_{10}(x)\hat{\mu}_{01}(x) - s_x^2 \hat{\delta}_{01}(x)\hat{\delta}_{10}(x).$$

According to Theorem 2, this estimator has the desirable asymptotic bias term $B_{12}^* = c_B v_{12}^{(2)}(x)$ and therefore is linearly unbiased.

4 Variance, skewness and correlation function estimation

We resume the discussion of the examples given in the Introduction in the light of Theorem 2. Multi-indices are replaced by single indices if only one coordinate of the response vector Y needs to be considered; for simplicity the first coordinate is chosen by default. Regarding the simple moment functions (3), $\mu_\ell(x) = E(Y_1^\ell | X = x)$, we find that $\hat{\delta} \equiv 0$, so that according to Theorem 2, the plug-in estimators $\hat{\mu}_{\alpha,LS}$ defined in (11), (12) are linearly unbiased. This can also be seen from the fact that G is the identity function in this case. Indeed, applying Theorem 1 yields an asymptotic normal distribution with

$$B_\ell = c_B \mu_\ell^{(2)}(x), \qquad V_\ell = \frac{c_V}{f_X(x)}[\mu_{2\ell}(x) - \{\mu_\ell(x)\}^2]. \tag{23}$$

We note that in the general case, the asymptotic quantities B^*, V^* or $\tilde{B}$, $\tilde{V}$ are useful for the construction of approximate asymptotic confidence intervals for estimates $\hat{g}^*(x)$ (22) or $\hat{g}(x)$ (17). There exist various possibilities for the actual construction of such asymptotic pointwise confidence intervals. Common approaches include to simply ignore the asymptotic bias, centering the asymptotic confidence intervals symmetrically around the curve estimates, or to use undersmoothing for the construction of confidence intervals, so as to justify that bias becomes asymptotically negligible and can be safely ignored. Other approaches are to approximate the asymptotic variance in the limiting normal distribution by the square root of mean squared error, while centering the intervals around the curve estimates, in an effort to make the intervals wider to account for the bias, or bootstrapping (see, e.g., Claeskens and van Keilegom (2003), Eubank and Speckman (1993), Hall (1992) and Picard and Tribouley (2000)).

Continuing Example 2 concerning the nonparametric estimation of variance functions $v(x) = \mathrm{var}(Y_1|X = x)$ (4), the plug-in estimator is

$$\hat{v}(x) = \hat{\mu}_{2,LS}(x) - \{\hat{\mu}_{1,LS}(x)\}^2. \tag{24}$$

With the local weighted least squares smoothing weight functions $W_{i,LS}(x)$ (12) we may write

$$\begin{aligned}\hat{v}(x) &= \sum_{i=1}^n W_{i,LS}(x) Y_{1i}^2 - \left\{\sum_{i=1}^n W_{i,LS}(x) Y_{1i}\right\}^2 \\ &= \sum_{i=1}^n W_{i,LS}(x)\{Y_{1i} - \hat{\mu}(x)\}^2, \end{aligned} \tag{25}$$

so that this estimator is seen to be equivalent to smoothing squared residuals obtained from an initial local linear fit.

Another proposal for variance function estimation that is closely related to the work of Doksum et al. (1994) and Doksum and Samarov (1995) is to estimate the variance function by using the error mean square as in classical regression, but now formed from properly weighted residuals within the local smoothing window. The starting point is the well-known classical formula for the mean square due to error, $MSE = (\tilde{s}_y^2 - \hat{b}_1^2 \tilde{s}_x^2)(n-1)/(n-2)$ in simple linear regression, where $\tilde{s}_y^2 = (n-1)^{-1}\sum_{i=1}^n (Y_i - \bar{Y})^2$, $\tilde{s}_x^2 = (n-1)^{-1}\sum_{i=1}^n (X_i - \bar{X})^2$ and $\hat{b}_1$ is the least squares estimate of the slope parameter of the regression line. Since $E(MSE) = \sigma^2 = \mathrm{var}(Y_i)$, this relationship can be exploited for estimating a local variance function. Localizing and introducing weights, this motivates the estimate

$$\hat{v}_{MSE}(x) = \frac{\sum_{i=1}^n w_i(x)Y_i^2}{\sum_{i=1}^n w_i(x)} - \left\{\frac{\sum_{i=1}^n w_i(x)Y_i}{\sum_{i=1}^n w_i(x)}\right\}^2 - \hat{\delta}_1(x)^2 s_x^2, \qquad (26)$$

with $w_i, \hat{\delta}_1$ and s_x^2 as defined in (8), (20) and (21).

Finally, the bias-corrected version of estimator (25) is obtained as

$$\hat{v}^*(x) = \hat{v}(x) + \hat{\Delta}(x) = \hat{\mu}_2(x) - \{\hat{\mu}_1(x)\}^2 - s_x^2\{\hat{\delta}_1(x)\}^2. \qquad (27)$$

We note that for all three estimators $\hat{v}$, $\hat{v}_{MSE}$ and $\hat{v}^*$, we obtain the same asymptotic variance term

$$V_V = f_X^{-1}(x)c_V\{\mu_4 - \mu_2^2 + 4(\mu_2 - \mu_1^2)\mu_1^2 - 4\mu_1(\mu_3 - \mu_1\mu_2)\}(x). \qquad (28)$$

For $\hat{v}$ and $\hat{v}^*$ this follows from Theorems 1 and 2, and for $\hat{v}_{MSE}$ it is shown in Section 6. For these estimators, a more interesting comparison concerns the leading asymptotic bias terms. According to Theorems 1, 2, we find for $\hat{v}$ and $\hat{v}^*$,

$$\tilde{B} = c_B\{v^{(2)} + 2\mu_1^{(1)2}\}(x), \qquad B^* = c_B v^{(2)}, \qquad (29)$$

the latter being the desired expression for linearly unbiased estimation. As seen in Section 6, the asymptotic bias term for $\hat{v}_{MSE}$ is

$$B_{MSE} = c_B\{v^{(2)} + 2v^{(1)}\frac{f_X^{(1)}}{f_X}\}(x). \qquad (30)$$

The estimator $\hat{v}_{MSE}$ thus is seen to suffer from the drawback that the bias depends on the marginal density f_X which means artificial curvature in the curve estimates may be introduced by just replacing a uniform marginal distribution of the predictor variable X by a normal marginal distribution.

Continuing now the discussion of the conditional skewness function (5) in Example 3, the straightforward "plug-in" estimate is

$$\hat{s}(x) = \left\{\frac{\hat{\mu}_{3,LS} - 3\hat{\mu}_{2,LS}\hat{\mu}_{1,LS} + 2\hat{\mu}_{1,LS}^3}{\left(\hat{\mu}_{2,LS} - \hat{\mu}_{1,LS}^2\right)^{3/2}}\right\}(x).$$

This estimate is linearly biased. According to Theorem 2, the linearly unbiased skewness function estimate is obtained by introducing an additional bias correction,

$$\hat{s}^*(x) = \hat{s}(x) + \frac{(3/2)s_x^2}{\left(\hat{\mu}_2 - \hat{\mu}_1^2\right)^{7/2}}\left[\left(\hat{\mu}_2\hat{\mu}_3 + 4\hat{\mu}_1^2\hat{\mu}_3 - 5\hat{\mu}_2^2\hat{\mu}_1\right)\hat{\delta}_1^2\right]$$
$$+\left(\hat{\mu}_2^2 + 4\hat{\mu}_2\hat{\mu}_1^2 - 5\hat{\mu}_1\hat{\mu}_3\right)\hat{\delta}_1\hat{\delta}_2 - \frac{1}{4}\left(3\hat{\mu}_1\hat{\mu}_2 + 2\hat{\mu}_1^3 - 5\hat{\mu}_3\right)\hat{\delta}_2^2$$
$$+2\left[\left(\hat{\mu}_1\hat{\mu}_2 - \hat{\mu}_1^3\right)\hat{\delta}_1\hat{\delta}_3 - \left(\hat{\mu}_2 - \hat{\mu}_1^2\right)\hat{\delta}_2\hat{\delta}_3\right],$$

where the argument x has been omitted in terms on the r.h.s., and all $\hat{\mu}_\ell = \hat{\mu}_{\ell,LS}(x)$. Analogously one can construct linearly unbiased estimators for kurtosis functions, etc.

Of some interest for applications is Example 5, the estimation of the conditional correlation function (7). The plug-in estimate is

$$\hat{\rho}_{12}(x) = \frac{\hat{v}_{12}(x)}{\{\hat{v}_{11}(x)\hat{v}_{22}(x)\}^{1/2}} = \frac{\hat{\mu}_{11} - \hat{\mu}_{10}\hat{\mu}_{01}}{[\{\hat{\mu}_{20} - \hat{\mu}_{10}^2\}\{\hat{\mu}_{02} - \hat{\mu}_{01}^2\}]^{1/2}},$$

where $\hat{\mu}_{lm} = \hat{\mu}_{lm,LS}(x)$. This estimate is linearly biased. The construction of the bias correction requires estimation of the mixed partial derivatives $\frac{\partial G}{\partial x_l \partial x_m}$, $1 \le l, m \le 5$ of $G(x_1, x_2, x_3, x_4, x_5) = (x_1 - x_2 x_3)/\{(x_4 - x_1^2)(x_5 - x_3^2)\}^{1/2}$. These derivatives are easiest calculated by using a package that includes symbolic calculus. According to Theorem 2, the linearly unbiased estimator is found to be, setting $A = \{\hat{\mu}_{(20),LS}(x) - \hat{\mu}^2_{(10),LS}(x)\}^{-1/2}$, $B = \{\hat{\mu}_{(02),LS}(x) - \hat{\mu}^2_{(01),LS}(x)\}^{-1/2}$ and omitting arguments x on the r.h.s.,

$$\begin{aligned}
\hat{\rho}^*_{12}(x) =& \hat{\rho}_{12}(x) \\
&+ \frac{1}{2} s_x^2 [2A^3 B \hat{\mu}_{(10)} \hat{\delta}_{(11)} \hat{\delta}_{(10)} + 2AB^3 \hat{\mu}_{(01)} \hat{\delta}_{(11)} \hat{\delta}_{(01)} - A^3 B \hat{\delta}_{(11)} \hat{\delta}_{(20)} \\
&- A^3 B \hat{\delta}_{(11)} \hat{\delta}_{(02)} + A^5 B\{-3\hat{\mu}_{(10)}\hat{\mu}_{(01)}\hat{\mu}_{(20)} + \hat{\mu}_{(11)}(2\hat{\mu}^2_{(10)} + \hat{\mu}_{(20)})\}\hat{\delta}^2_{(10)} \\
&+ 2A^3 B^3 (\hat{\mu}_{(11)}\hat{\mu}_{(10)}\hat{\mu}_{(01)} - \hat{\mu}_{(02)}\hat{\mu}_{(20)}) \hat{\delta}_{(10)} \hat{\delta}_{(01)} \\
&+ A^5 B\{-3\hat{\mu}_{(11)}\hat{\mu}_{(10)} + \hat{\mu}_{(01)}(2\hat{\mu}^2_{(10)} + \hat{\mu}_{(20)})\}\hat{\delta}_{(10)})\hat{\delta}_{(20)} \\
&+ A^3 B^3 (-\hat{\mu}_{(11)}\hat{\mu}_{(10)} + \hat{\mu}_{(01)}\hat{\mu}_{(20)}) \hat{\delta}_{(10)} \hat{\delta}_{(02)} \\
&+ A^5 B\{-3\hat{\mu}_{(10)}\hat{\mu}_{(01)}\hat{\mu}_{(02)} + \hat{\mu}_{(11)}(2\hat{\mu}^2_{(01)} + \hat{\mu}_{(02)})\}\hat{\delta}^2_{(01)} \\
&+ A^3 B^3 (-\hat{\mu}_{(11)}\hat{\mu}_{(01)} + \hat{\mu}_{(10)}\hat{\mu}_{(02)}) \hat{\delta}_{(01)} \hat{\delta}_{(20)} \\
&+ A^5 B\{-3\hat{\mu}_{(11)}\hat{\mu}_{(01)} + \hat{\mu}_{(10)}(2\hat{\mu}^2_{(01)} + \hat{\mu}_{(02)})\}\hat{\delta}_{(01)} \hat{\delta}_{(02)} \\
&+ \frac{3}{4} A^5 B (\hat{\mu}_{(11)} - \hat{\mu}_{(10)}\hat{\mu}_{(01)}) \hat{\delta}^2_{(20)} + \frac{1}{2} A^3 B^3 (\hat{\mu}_{(11)} - \hat{\mu}_{(10)}\hat{\mu}_{(01)}) \hat{\delta}_{(20)} \hat{\delta}_{(02)} \\
&+ \frac{3}{4} AB^5 (\hat{\mu}_{(11)} - \hat{\mu}_{(10)}\hat{\mu}_{(01)}) \hat{\delta}^2_{(02)}].
\end{aligned}$$

5 A simulation example

The finite sample behavior of estimators $\hat{v}$ (24), $\hat{v}_{MSE}$ (26) and $\hat{v}^*$ (27) was investigated in a small scale simulation study. Since according to (28), these estimators behave identically with respect to asymptotic variance, and as the bias behavior is the focus of this paper, only the bias was investigated.

This was done graphically by averaging estimated variance functions for the various methods over 200 Monte Carlo runs.

According to (29) and (30), asymptotic biases for $\hat{v}, \hat{v}_{MSE}$ and $\hat{v}^*$ are determined by the behavior of $v^{(2)} + 2\mu_1^{(1)^2}$, $v^{(2)} + 2v^{(1)}f^{(1)}/f$, respectively. The following example was used to investigate the influence of these terms and Monte Carlo runs were made assuming that $n = 50, 250, 1250$ observations were available.

Pseudo-random pairs (X_i, Y_i), $i = 1, \ldots, n$ were generated according to

$$X_i = \sqrt{0.25}Z_i, \qquad Y_i = (X_i + 2)^2 + Z_i\sigma_i,$$

with

$$\sigma_i = 0 \text{ for } X_i \leq -0.5, \text{ and } \sigma_i = (X_i + 0.5)^{1/2} \text{ for } X_i > -0.5,$$

where Z_i are independent standard normal pseudo random numbers. The relevant functions are seen to be

$$\mu_1(x) = (x+2)^2, \quad \mu_1^{(1)}(x) = 2(x+2),$$
$$v(x) = (x + 0.5), \quad v^{(1)}f^{(1)}/f = -8x, \; x \geq -0.5.$$

Several bandwidths and also other variance functions were chosen, and the results were qualitatively the same for a broad range of cases. We report the results for the bandwidth $b = 0.4$ and sample size $n = 250$.

A typical data sample is shown in Figure 1, while Figure 2 displays the average curve estimates from 200 Monte Carlo runs for the three estimators considered along with the target variance function which is linear in this example.

As expected, the asymptotic bias term $2\mu_1^{(1)^2}$ is so large that as a consequence the estimators $\hat{v}$ are unacceptable. The differences between $\hat{v}_{MSE}$ and $\hat{v}^*$ are more subtle. It is obvious that $\hat{v}_{MSE}$ exhibits an upward bias for small predictor levels below 0, and a downward bias for larger predictor levels above 0. This behavior is expected from the asymptotic bias expression (30). The linearly unbiased estimator $\hat{v}^*$ has a very small, more or less constant bias. It emerges as the clearly preferred estimator, as predicted by theory.

6 Technical details and proofs

6.1 *Assumptions and auxiliary results*

We first list the necessary regularity conditions. For a given x in the domain of the predictor variable X, let $N(x)$ be a neighborhood of x. We denote convergence in distribution by $\xrightarrow{\mathcal{L}}$ and convergence in probability by $\xrightarrow{p}$, as $n \to \infty$.

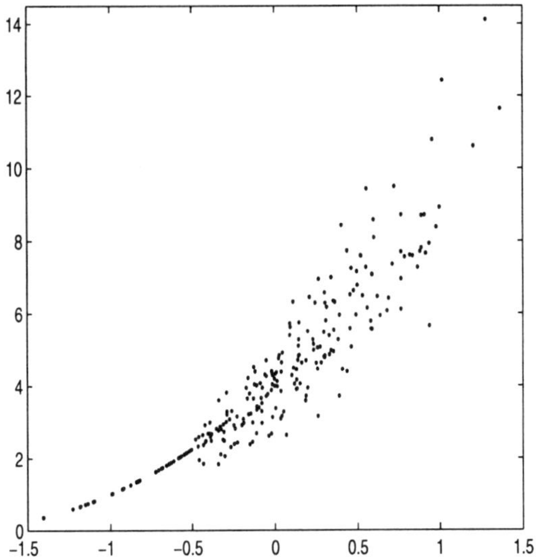

Figure 1 Sample points (X_i, Y_i) for $n = 250$.

(C1) The joint distribution $F(\cdot,\cdot)$ of (X, Y) has a density $f(u, v)$, which is continuous on $N(x) \times \Re^p$.

(C2) $f(u, v)$ is twice continuously differentiable in the first argument on $N(x) \times \Re^p$.

(C3) The marginal density f_X of X is twice continuously differentiable on $N(x)$ and satisfies $f_X(x) > 0$.

(C4) The function $g(x) = G\{\mu_{\alpha_1}(x), ..., \mu_{\alpha_k}(x)\}$ is twice continuously differentiable on $N(x)$; this implies corresponding differentiability conditions for G and moment functions $\mu_{\alpha_m}(x)$.

(C5) The bandwidth sequence satisfies

$$b \to 0, \quad nb \to \infty \quad \text{and} \quad nb^5 \to d^2 \quad \text{as} \quad n \to \infty \quad \text{for} \quad \text{a} \quad d \geq 0.$$

(C6) The kernel function K satisfies

$$K \geq 0, \quad \int K = 1, \quad \int Ku = 0, \quad \int Ku^2 < \infty, \quad \int K^2 < \infty.$$

Consider now weighted averages

$$\Psi_{\lambda n} = (nb)^{-1} \sum_{i=1}^{n} \psi_\lambda (X_i, Y_i) K \left(b^{-1} (x - X_i)\right), \quad \lambda = 1, ..., m,$$

where the real valued functions ψ_λ satisfy:

(C7) All ψ_λ are bounded and continuous on $\{x\} \times \Re^p$.

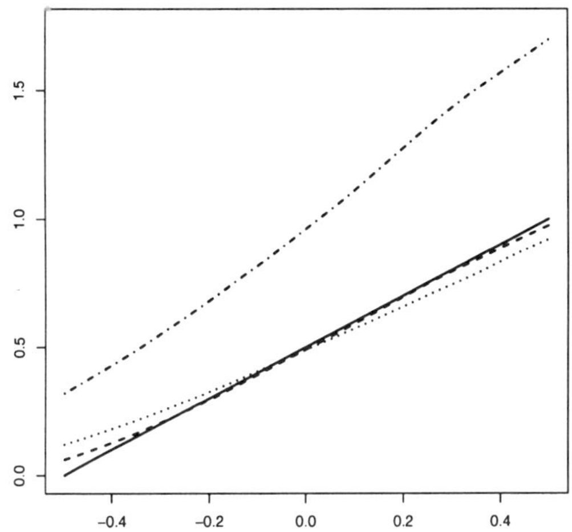

Figure 2 Variance function estimators averaged over 200 Monte Carlo runs for $n = 250$ data pairs: Target variance function (solid), linearly biased estimators $\hat{v}$ (24) (dash-dotted), $\hat{v}_{MSE}$ (26) (dotted) and linearly unbiased estimator $\hat{v}^*$ (27) (dashed).

(C8) The second derivatives with respect to the first argument exist for all ψ_λ and are continuous on $\{x\} \times \Re^p$.

Define

$$\zeta_\lambda(x) = \int \psi_\lambda(x,v) f(x,v)\, dv \tag{31}$$

$$\sigma_{\kappa\lambda}(x) = \int \psi_\kappa(x,v)\psi_\lambda(x,v) f(x,v) dv - \int \psi_\kappa(x,v) f(x,v) dv \int \psi_\lambda(x,v) f(x,v)\, dv, \tag{32}$$

and let $H : \Re^q \to \Re$ be a function with continuous second derivatives and $\zeta = (\zeta_1, ..., \zeta_q)$. The following auxiliary result is a direct consequence of Theorem 4.1 in Bhattacharya and Müller (1993); the proof is omitted.

Lemma 1. *Under (C1)-(C8),*

$$(nb)^{1/2} \left[H(\Psi_{1n}, ..., \Psi_{qn}) - H(\zeta_1, ..., \zeta_q) \right] \xrightarrow{\mathcal{L}} \mathcal{N}(B, V),$$

where

$$B = c_B \sum_{\lambda=1}^{q} \frac{dH}{dx_\lambda}|_\zeta \frac{d^2}{d^2x}\zeta_\lambda(x),\ V = c_V \sum_{\kappa,\lambda=1}^{q} \frac{dH}{dx_\kappa}|_\zeta \frac{dH}{dx_\lambda}|_\zeta\, \sigma_{\kappa\lambda}(x).$$

Note for the following that in the case where

$$H(x_1, ..., x_q) = H(zx_1, ..., zx_q)$$

for all $z \neq 0$, a simple chain rule argument shows that $\sigma_{\kappa\lambda}$ in (33) can be replaced by

$$\tilde{\sigma}_{\kappa\lambda} = \int \psi_\kappa(x, v)\psi_\lambda(x, v) f(x, v) dv.$$

6.2 *Proof of Theorem 1*

For $\hat{\delta}$ as defined in (20), it follows from Corollary 4.3 in Bhattacharya and Müller (1993) that

$$(nb)^{1/2} \left\{ \hat{\delta}_{e_1}(x) - \mu_{e_1}^{(1)}(x) \right\} \xrightarrow{p} 0,$$

where $e_1 = (1, 0, ..., 0)$. This generalizes easily to

$$(nb)^{1/2} \left\{ \hat{\delta}_\alpha(x) - \mu_\alpha^{(1)}(x) \right\} \xrightarrow{p} 0 \qquad (33)$$

for any multi-index α. Define the function

$$H_1(x_1, x_2, x_3) = \frac{x_1 - x_2 \mu_\alpha^{(1)}(x)}{x_3}.$$

Choosing $x_q = \Psi_{qn}$, $q = 1, 2, 3$, with $\psi_1(u, v) = v^\alpha$, $\psi_2(u, v) = u - x$, $\psi_3(u, v) = 1$, we find that (33), Slutsky's Theorem and Lemma 1 imply that

$$(nb)^{1/2} \{\hat{\mu}_\alpha(x) - \mu_\alpha(x)\} \xrightarrow{\mathcal{L}} \mathcal{N}(B_\alpha, V_\alpha), \qquad (34)$$

with $B_\alpha = c_B \, \mu_\alpha^{(2)}(x)$, $V_\alpha = c_V \{\mu_{2\alpha}(x) - \mu_\alpha^2(x)\}/f_X(x)$.

Furthermore, for any given constants $c_1, c_2 \neq 0$, and any multi-indices α_1, α_2 with $|\alpha_1| > 0, |\alpha_2| > 0$, generalizing (34), one obtains

$$(nb)^{1/2}\{c_1(\hat{\mu}_{\alpha_1}(x) - \mu_{\alpha_1}(x)) + c_2(\hat{\mu}_{\alpha_2}(x) - \mu_{\alpha_2}(x))\} \qquad (35)$$
$$\to \mathcal{N}(B_{\alpha_1,\alpha_2}, V_{\alpha_1,\alpha_2})$$

where

$$B_{\alpha_1,\alpha_2} = c_B\{c_1\mu_{\alpha_1}^{(2)}(x) + c_2\mu_{\alpha_2}^{(2)}(x)\},$$
$$V_{\alpha_1,\alpha_2} = c_V\{c_1^2\mu_{2\alpha_1}(x) + c_2^2\mu_{2\alpha_2}(x) + 2c_1c_2\mu_{\alpha_1+\alpha_2}(x) - (c_1\mu_{\alpha_1}(x) + c_2\mu_{2\alpha_2}(x))^2\}/f_X(x).$$

This follows from Lemma 1, choosing

$$H_2(x_1, x_2, x_3, x_4) = c_1 \frac{x_1 - x_2\mu_{\alpha_1}^{(1)}(x)}{x_3} + c_2 \frac{x_4 - x_2\mu_{\alpha_2}^{(1)}(x)}{x_3},$$

with $x_q = \Psi_{qn}$, $q = 1, ..., 4$, $\psi_1(u, v) = v^{\alpha_1}, \psi_2(u, v) = u - x$, $\psi_3(u, v) = 1$, and $\psi_4(u, v) = v^{\alpha_2}$. Extending (35) to a linear combination of more than two estimators and a Taylor expansion

$$\hat{g}(x) - g(x) = \sum_{m=1}^{k} \frac{dG}{dx_m}|_\mu \{\hat{\mu}_{\alpha_m}(x) - \mu_{\alpha_m}(x)\} + o\left[\sum_{m=1}^{k} \{\hat{\mu}_{\alpha_m}(x) - \mu_{\alpha_m}(x)\}\right]$$

conclude the proof.

6.3 *Proof of (28) and (30) for estimators* $\hat{v}_{MSE}$

Let $p = 1$ and set

$$H_3(x_1, x_2, x_3) = (x_1/x_2) - (x_3^2/x_2^2),\ x_q = \Psi_{qn},\ q = 1, 2, 3$$

with

$$\psi_1(u, v) = v^2,\ \psi_2(u, v) \equiv 1,\ \psi_3(u, v) = v.$$

Applying Lemma 1, we obtain

$$(nb)^{1/2}\left\{\frac{\sum_{i=1}^n w_i(x)Y_i^2}{\sum_{i=1}^n w_i(x)} - \left(\frac{\sum_{i=1}^n w_i(x)Y_i}{\sum_{i=1}^n w_i(x)}\right)^2\right\} \xrightarrow{\mathcal{L}} \mathcal{N}(\bar{B}, \bar{V}), \qquad (36)$$

with

$$\bar{B} = c_B(v_{(2)} + 2\mu_1^{(1)^2} + 2v^{(1)}\frac{f_X^{(1)}}{f_X}(x), \quad \bar{V} = V_V$$

as in (34).

We note in passing that (36) provides a result for plug-in variance function estimation based on the Nadaraya-Watson quotient type kernel estimator. Furthermore, by (33),

$$\hat{\delta}_1(x) \xrightarrow{P} \mu_1^{(1)}(x). \qquad (37)$$

Note that s_x^2 (21) can be equivalently written as

$$s_x^2 = \frac{\sum_{i=1}^n w_i(x)X_i^2}{\sum_{i=1}^n w_i(x)} - \left(\frac{\sum_{i=1}^n w_i(x)X_i}{\sum_{i=1}^n w_i(x)}\right)^2.$$

Therefore, (36) applies, with Y_i replaced by X_i, and analogous changes in the definition of $v(\cdot)$ and $\mu_\ell(\cdot)$. This leads to $\bar{B} = 2c_B$ and $\bar{V} = 0$, so that

$$(nb)^{1/2}s_x^2 \xrightarrow{P} 2c_B. \qquad (38)$$

The result now follows by combining (37)-(38) and applying Slutsky's theorem.

6.4 *Proof of Theorem 2*

Using analogous arguments as in (37) and (38), we find

$$(nb)^{1/2}\hat{\Delta}(x) \xrightarrow{P} c_B\left(\sum_{l,m=1}^{k}\frac{d^2G}{dx_l dx_m}|_\mu\right)\delta_{\alpha_l}(x)\delta_{\alpha_m}(x).$$

The result follows from Theorem 1 and Slutsky's theorem.

Acknowledgments

Part of this research was carried out while I was visiting the Statistical Institute of Academia Sinica at Taipei, and I wish to express my sincere gratitude for the hospitality received there. Special thanks are due to Philip Cheng and Zhen Zhang for helpful feedback on a previous version and to two anonymous referees for their comments and suggestions. This research was supported in part by NSF grants DMS03-54448 and DMS05-05537.

References

1. Bhattacharya, P.K. and Müller, H.G. (1993). Asymptotics for nonparametric regression. *Sankhyā* A 55, 420-441.
2. Bjerve, O. and Doksum, K. (1993). Correlation curves as a measure of dependence conditional on a covariate. *Ann. Statist.* **21**, 890-902.
3. Chu, C.K. and Marron, J.S. (1991). Choosing a kernel regression estimator (with discussion). *Statist. Science* **6**, 404-436.
4. Claeskens, G. and van Keilegom, I. (2003). Bootstrap confidence bands for regression curves and their derivatives. *Ann. Statist.* **31**, 1852-1884.
5. Dette, H. and Munk, A. (1998). Testing heteroscedasticity in nonparametric regression. *J. R. Statist. Soc. B* **60**, 693-708.
6. Doksum, K., Blyth, S., Bradlow, E., Meng, X.L. and Zhao, H.Y. (1994). Correlation curves as local measures of variance explained by regression. *J. Am. Statist. Assoc.* **89**, 571-582.
7. Doksum, K. and Samarov, A. (1995). Nonparametric estimation of global functionals and a measure of the explanatory power of covariates in regression. *Ann. Statist.* **23**, 1443-1473.
8. Eubank, R.L. and Speckman, P.L. (1993). Confidence bands in nonparametric regression. *J. Am. Statist. Assoc.* **88**, 1287-1301.
9. Eubank, R. and Thomas, W. (1993). Detecting heteroscedasticity in nonparametric regression. *J. R. Statist. Soc. B* **55**, 145-155.
10. Fan, J. and Gijbels, I. (1996). *Local Polynomial Modelling.* Chapman and Hall, London.
11. Fan, J. and Yao, Q.W. (1998). Efficient estimation of conditional variance functions in stochastic regression. *Biometrika* **85**, 645-660.
12. Gasser, T., Müller, H.G. and Mammitzsch, V. (1985). Kernels for nonparametric curve estimation. *J. R. Statist. Soc. B* **47**, 238-252.
13. Hall, P. (1992). Effect of bias estimation on coverage accuracy of bootstrap confidence intervals for a probability density. *Ann. Statist.* **20**, 675-694.

14. JENNEN-STEINMETZ, C. AND GASSER, T. (1987). A unifying approach to nonparametric regression estimation. *J. Am. Statist. Assoc.* **83**, 1084-1089.

15. JONES, M.C. AND PARK, B.U. (1994). Versions of kernel-type regression estimators. *J. Am. Statist. Assoc.* **89**, 825-832.

16. MAMMEN, E. AND MARRON, J.S. (1997). Mass recentred kernel smoothers. *Biometrika* **84**, 765-777.

17. MÜLLER, H.G. (1997). Density adjusted kernel estimators for random design nonparametric regression. *Statist. Prob. Letters* **36**, 161-172.

18. MÜLLER, H.G. AND SONG, K.S. (1993). Identity reproducing multivariate nonparametric regression. *J. Multiv. Anal.* **46**, 237-253.

19. MÜLLER, H.G. AND STADTMÜLLER, U. (1987). Estimation of heteroscedasticity in regression analysis. *Ann. Statist.* **15**, 610-625.

20. MÜLLER, H.G. AND STADTMÜLLER, U. (1993). On variance function estimation with quadratic forms. *J. Statist. Plann. Inf.* **35**, 213-231.

21. PARK, B.U., KIM, W.C. AND JONES, M.C. (1997). On identity reproducing nonparametric regression estimators. *Statist. Prob. Letters* **32**, 279-290.

22. PICARD, D. AND TRIBOULEY, K. (2000). Adaptive confidence interval for pointwise curve estimation. *Ann. Statist.* **28**, 298-335.

23. RUPPERT, D., WAND, M.P., HOLST, U. AND HÖSSJER, O. (1997). Local polynomial variance function estimation. *Technometrics* **39**, 262-273.

24. SEIFERT, B. AND GASSER, T. (1996). Finite-sample variance of local polynomials – analysis and solutions. *J. Am. Statist. Assoc.* **91**, 267-275.

25. WAND, M.P. AND JONES, M.C. (1995). *Kernel Smoothing.* Chapman and Hall, London.

Chapter 17

SERIAL AUTOREGRESSION AND REGRESSION RANK SCORES STATISTICS

Marc Hallin, Jana Jurečková, and Hira L. Koul

Université Libre de Bruxelles, Brussels, BELGIUM

Charles University, Prague, CZECH REPUBLIC

Michigan State University, East Lansing, Michigan, U.S.A.

E-mails: koul@stt.msu.edu, mhallin@ulb.ac.be & jurecko@karlin.mff.cuni.cz

This paper establishes an asymptotic representation for regression and autoregression rank score statistics of the serial type. Applications include rank-based versions of the Durbin-Watson test, tests of AR(p) against AR($p+1$) dependence, or detection of the presence of random components in AR processes.

Key words: Time series; Robustness; Serial autoregression rank score; Rank test.

1 Introduction

1.1 *Rank tests*

Rank tests are known to be robust, distribution-free yet powerful alternative to Gaussian testing methods under a broad set of model assumptions. These models include a class of semiparametric models under which the distribution $P_{n;\theta;f}$ of the observation vector $\boldsymbol{X}_n := (X_{n1}, \cdots, X_{nn})'$ belongs to a family $\mathcal{P}_n := \{P_{n;\theta;f};\ \theta \in \Theta,\ f \in \mathcal{F}\}$, where $\theta \in \Theta \subseteq \mathbb{R}^k$ is some parameter of interest, and $\mathcal{F}$ is a class of densities f on $\mathbb{R}$. More specifically, rank tests can be constructed whenever

(A) for all n, there exists a $(\theta, \boldsymbol{X}_n)$-measurable *residual function*

$$(\theta, \boldsymbol{X}_n) \mapsto \boldsymbol{\varepsilon}_n(\theta, \boldsymbol{X}_n) := (\varepsilon_{n,1}(\theta, \boldsymbol{X}_n), \cdots, \varepsilon_{n,n}(\theta, \boldsymbol{X}_n))'$$

such that the distribution of $\boldsymbol{X}_n$ is $P_{n;\theta;f}$ iff the components of the vector $\varepsilon_n(\theta, \boldsymbol{X}_n)$ are i.i.d. with common density f. Henceforth, we shall write $\varepsilon_{n,t}(\theta)$ instead of $\varepsilon_{n,t}(\theta, \boldsymbol{X}_n)$, $t = 1, \cdots, n$.

Let $R_{n,t}(\theta)$ denote the rank of the residual $\varepsilon_{n,t}(\theta)$ and $\boldsymbol{R}_n(\theta) := (R_{n,1}(\theta), \cdots, R_{n,n}(\theta))'$. The rank tests for the simple null hypothesis $\mathcal{H}_0 : \theta = \theta_0$, where θ_0 is a given value of θ, are based $\boldsymbol{R}_n(\theta_0)$.

The use of the rank tests can be justified by several main arguments.

(a) The vector of ranks is a maximal invariant with respect to the group of *order-preserving transformations* of residuals for a broad class of densities over $\mathbb{R}$. In such invariant situations, every invariant statistic and test depend only on the maximal invariant, and hence are distribution-free under the null hypothesis.
(b) The rank tests are more robust with respect to some outliers than their parametric counterparts.
(c) In linear regression or ARMA models where semiparametric and parametric efficiencies coincide, asymptotically most powerful tests against contiguous alternatives at a given f can be found among rank tests, cf. Chernoff and Savage (1958), Hajék and Sidák (1967), Hallin (1994), Paindaveine (2004, 2005), among others.

A general result by Hallin and Werker (2003) shows that under Le Cam LAN formalism with *central sequences* $\Delta_{n;f}(\theta)$ and under some conditions on $\mathcal{F}$, a semiparametrically efficient inference about θ, at given (θ_0, f), can be based on the rank-based efficient central sequence obtained by conditioning $\Delta_{n;f}(\theta_0)$ on the vector $\boldsymbol{R}_n(\theta_0)$, under $\mathcal{H}_0$. In linear regression models where for some known non-random $p \times 1$ design vector $\{c_{n,t};\ 1 \le t \le n\}$, $\varepsilon_{n,t}(\theta) = X_{n,t} - c'_{n,t}\theta$ are i.i.d., rank-based efficient central sequences take the form of *linear rank statistics* vectors

$$S_{n,\varphi}(\theta) := \sum_{t=1}^{n} \varphi\Big(\frac{R_{n,t}(\theta)}{n+1}\Big) c_{n,t},$$

where φ is a *score-generating function* from (0, 1) to $\mathbb{R}$. Under some general conditions and under $P_{n,\theta,f}$, one obtains

$$S_{n,\varphi}(\theta) = \sum_{t=1}^{n} \varphi\big(F(\varepsilon_t(\theta))\big) c_{n,t} + o_P(n^{1/2}), \quad n \to \infty, \quad \forall\, \theta \in \Theta,$$

where F is the distribution function associated with f.

In ARMA models, rank-based central sequences can be expressed (Hallin, Ingenbleek and Puri 1985; Hallin and Puri 1988; Bentarzi and Hallin 1996) as linear combinations of *serial linear rank statistics*

$$S_{n,\varphi_1\varphi_2}(\theta) := \sum_{t=i+1}^{n} \varphi_1\Big(\frac{R_{n,t}(\theta)}{n+1}\Big)\varphi_2\Big(\frac{R_{n,t-i}(\theta)}{n+1}\Big), \quad i = 1, 2, \cdots, \tag{1}$$

where φ_1 and φ_2 are adequately centered and scaled score generating functions. In more general problems– detection of random coefficients (Akharif and Hallin 2003), detection of nonlinearities (Benghabrit and Hallin 1992; Allal and El-Melhaoui 2005, among others)– rank-based efficient central sequences involve more complex serial linear rank statistics of the form

$$S_{n,\varphi_1\cdots\varphi_m}(\theta) := \sum_{t=i_{m-1}+1}^{n} \varphi_1\big(\frac{R_{n,t}(\theta)}{n+1}\big)\cdots\varphi_m\big(\frac{R_{n,t-i_{m-1}}(\theta)}{n+1}\big), \qquad (2)$$

$1 \leq i_1 \leq \cdots \leq i_{m-1}$, where $\varphi_1, \varphi_2 \cdots, \varphi_m$ are m score functions. Hallin et al. (1985) show under some general conditions that $\forall\,\theta \in \Theta$, as $n \to \infty$,

$$S_{n,\varphi_1\cdots\varphi_m}(\theta) = \sum_{t=i_{m-1}+1}^{n} \varphi_1\big(F(\varepsilon_t(\theta))\big)\cdots\varphi_m\big(F(\varepsilon_{t-i_{m-1}}(\theta))\big) + o_P(n^{1/2}).$$

In most problems of practical interest, however, one is interested in testing the composite null hypothesis $\tilde{\mathcal{H}}_0 : \theta \in \Theta_0$, where Θ_0 is a subset of Θ. It is then natural to first obtain an estimate $\hat{\theta}$ of θ under $\tilde{\mathcal{H}}_0$ and use the aligned ranks test statistics $S_{n,\varphi}(\hat{\theta})$ or $S_{n,\varphi_1\cdots\varphi_m}(\hat{\theta})$, cf., e.g., Koul (1970) and Jurečková (1971) for linear regression models; Hallin and Puri (1994) for the ARMA models. These statistics are not asymptotically distribution-free (ADF), and thus are unsuitable for testing purposes.

1.2 *Autoregression and regression rank scores*

The lack of robustness of aligned rank statistics motivated Gutenbrunner and Jurečkovà (1992) to introduce *regression rank scores* in the context of linear regression models with independent observations, as an alternative to the aligned ranks. The regression rank scores are n functions $\hat{a}_n(u) = (\hat{a}_{n,1}(u), \cdots, a_{n,n}(u))'$ with $\hat{a}_{n,t} : [0,1] \mapsto [0,1]$, $t = 1, \cdots, n$, obtained from the observations as the solution of a linear programming problem itself depending on $\tilde{\mathcal{H}}_0$; see Section 2.1 below for details. The regression rank score statistic (RSS) corresponding to a function φ is defined as

$$\tilde{S}_{n,\varphi} := -\sum_{t=1}^{n}\int_0^1 \varphi(u)d\hat{a}_{n,t}(u)\ c_{n,t}.$$

Note that this is like $S_{n,\varphi}(\theta)$ but where $\varphi(R_{n,t}(\theta)/(n+1))$ are replaced by $-\int_0^1 \varphi(u)d\hat{a}_{n,t}(u)$. These scores remedy the lack of invariance of aligned ranks. If not exactly (for fixed n) distribution-free, $\tilde{S}_{n,\varphi}$, indeed, contrary to $S_{n,\varphi}(\hat{\theta})$, is asymptotically equivalent to $S_{n,\varphi}(\theta)$ in probability under $P_{n,\theta,f}$, for each θ, hence asymptotically invariant with respect to the group of order-preserving transformations acting on residuals and, therefore, ADF. Being moreover *regression-invariant* over Θ_0, it is robust against the influence of possible outliers—if not against the possible leverage effect of certain regression constants. And, the asymptotic performance of tests based on $\tilde{S}_{n,\varphi}$ is matching that of the tests based on $S_{n,\varphi}(\theta)$, for all $\theta \in \Theta_0$; see Gutenbrunner et al. (1993).

Koul and Saleh (1995) and Hallin and Jurečková (1999) developed similar ideas for linear autoregressive models where $\theta' = (\rho_0, \rho_1, \cdots, \rho_p)$, and $\varepsilon_t(\theta) = X_i - \rho_0 - \rho_1 X_{t-1} - \cdots - \rho_p X_{t-p}$ are i.i.d. innovations with mean zero. The autoregression rank score statistics are of the form

$$\tilde{S}^*_{n,\varphi_1} := -\sum_{t=i+1}^{n} \int_0^1 \varphi_1(u) d\hat{a}_{n,t}(u) X_{t-i},$$

where $\hat{a}_{n,t}(\cdot)$ are the autoregression rank scores defined in Section 2.1 below and φ_1 is a function like φ.

Unlike the linear regression models, in autoregressive models the outliers in the errors affect the leverage points X_{t-i} also. This fact renders the statistics $\tilde{S}^*_{n,\varphi_1}$ non-robust against outliers in the errors. Genuine autoregression rank scores statistics are the serial autoregression rank score statistics obtained from $\tilde{S}^*_{n,\varphi_1}$ after replacing X_{t-i} by $-\int_0^1 \varphi_2(v) d\hat{a}_{n,t-i}(v)$, yielding

$$\tilde{S}_{n,\varphi_1\varphi_2} := \sum_{t=i+1}^{n} \int_0^1 \int_0^1 \varphi_1(u)\varphi_2(v) da_{n,t}(u)\ d\hat{a}_{n,t-i}(v), \tag{3}$$

(when the lag is to be emphasized, we write $\tilde{S}_{n,\varphi_1\varphi_2;i}$) or, more generally,

$$\tilde{S}_{n,\varphi_1\cdots\varphi_m} \tag{4}$$
$$=(-1)^m \sum_{t=i_{m-1}+1}^{n} \int_0^1 \cdots \int_0^1 \varphi_1(u_1)\ldots\varphi_m(u_m) d\hat{a}_{n,t}(u_1)\ldots d\hat{a}_{n,t-i_{m-1}}(u_m),$$

analogous to the serial rank statistics (1) and (2). Here $\varphi_j; 1 \leq j \leq m$ are m functions from $(0,1)$ to $\mathbb{R}$.

The main objective of this paper is to obtain an asymptotic representation of these *serial* regression or autoregression rank score statistics for possibly unbounded functions φ_j's, which so far have not been considered in the literature. This result is useful in obtaining their limiting distributions and in showing that the tests based on them are ADF.

This paper is organized as follows. Section 2 provides the precise conditions under which serial regression or autoregression rank score statistics can be used in hypothesis testing. Section 2.2 describes three potential applications: a version of the classical Durbin-Watson test based on regression rank scores, a test of AR(p) against AR($p+1$) dependence based on autoregression rank scores, and a test, based on serial autoregression rank scores, detecting the presence of a random component in the autoregressive coefficient of an AR(1) model. Technical assumptions are collected in Section 2.3. The main result of this paper is Proposition 1 giving an asymptotic representation result for a class of serial autoregression rank score statistics.

2 Notation and basic assumptions

2.1 *Autoregression and regression quantiles and rank scores*

We shall now recall the definition of autoregression and regression quantiles and rank scores. First consider the stationary linear autoregressive time series model, where starting with an observable p-vector $X_{1-p}, \cdots, X_0$, one observes the process

$$X_t = \rho_0 + \sum_{j=1}^{p} \rho_j X_{t-j} + \varepsilon_t, \quad (\rho_0, \rho_1, \cdots, \rho_p)' \in \mathbb{R}^{1+p}. \tag{5}$$

The errors ε_t are assumed to be i.i.d. with zero mean and variance σ^2. The parameters $\rho^* := (\rho_1, \cdots, \rho_p)'$ are such that all solutions of the equation $1-\sum_{t=1}^{p} \rho_i z^i = 0$ lie outside the unit sphere and for each t, ε_t is independent of the vector $y_{t-1}^* := (X_{t-1}, \cdots, X_{t-p})'$. Note that this model satisfies the assumption (A) with $k = 1+p$, $\theta' = (\rho_0, \rho^{*\prime})$, $\boldsymbol{X}_n' = (y_0^{*\prime}, X_1, X_2, \cdots, X_n)$, and $\varepsilon_{n,t}(\theta) = X_t - \rho_0 - \rho' y_{t-1}^*$. Now, let $y_{t-1}' := (1, y_{t-1}^{*\prime})$, and

$$h_\alpha(z) := |z|\big(\alpha I[z > 0] + (1-\alpha) I[z \le 0]\big), \quad z \in \mathbb{R}, \ \alpha \in (0, 1).$$

Then αth autoregression quantiles $\rho_n(\alpha)' = (\rho_{n0}(\alpha), \rho_n^*(\alpha)')$, for an $0 < \alpha < 1$, are defined as an $\mathrm{argmin}_{r_0 \in \mathbb{R}, r \in \mathbb{R}^p} \sum_{t=1}^{n} h_\alpha\big(X_t - r_0 - y_{t-1}^{*\prime} r\big)$. The corresponding autoregression rank scores are defined to be an n-vector $\hat{a}_n(\alpha) := (\hat{a}_{n,1}(\alpha), \cdots, \hat{a}_{n,n}(\alpha))'$ in $[0,1]^n$ maximizing $\sum_{t=1}^{n} X_t a_t$ with respect to vectors $a \in [0,1]^n$, subject to the conditions

$$Y_n'(a - (1-\alpha)1_n) = 0, \tag{6}$$

where Y_n is the $n \times (1+p)$ matrix whose tth row is y_{t-1}', $t = 1, \cdots, n$, $1_n := (1, \cdots, 1)'$, an $n \times 1$ vector of 1's, and 0 in the right hand side is the $(1+p) \times 1$ vector of zeros.

These autoregression quantiles and rank scores are the analogs of their counterparts in linear regression model $X_{n,t} = \beta_0 + c_{n,t}'\beta + \varepsilon_t$, as defined in Koenker and Bassett (1978) and Gutenbrunner and Jurečková (1992). Let C_n denote the $n \times (1+p)$ matrix whose tth row consists of $(1, c_{n,t}')$,

$1 \le t \le n$. An αth *regression quantile* vector $\hat{\theta}_n(\alpha) := (\hat{\beta}_{0n}(\alpha), \hat{\beta}_n(\alpha)')$, for an $\alpha \in (0,1)$, is defined as a minimizer vector of $\sum_{t=1}^n h_\alpha(X_{n,t} - b_0 - c'_{n,t}b)$, w.r.t. $b_0 \in \mathbb{R}$, $b \in \mathbb{R}^p$. The corresponding regression rank scores are defined to be an n-vector $\hat{a}_n(\alpha) := (\hat{a}_{n,1}(\alpha), \cdots, \hat{a}_{n,n}(\alpha))'$ in $[0,1]^n$ maximizing $\sum_{t=1}^n X_t a_t$ w.r.t. vectors $a \in [0,1]^n$, such that $C'_n(a - (1-\alpha)1_n) = 0$.

2.2 *Examples*

2.2.1 *The Durbin-Watson problem*

The objective of the classical Durbin-Watson test is the detection of first-order autocorrelation in the noise of a traditional regression model; its extension to higher-order dependencies is straightforward.

The general overarching model is a linear regression with AR(1) errors

$$X_t = \beta_0 + c'_{n,t}\beta + e_t, \qquad e_t = \rho e_{t-1} + \varepsilon_t, \qquad t = 1, \cdots n,$$

where $\rho \in [0,1)$, $\beta_0 \in \mathbb{R}$, $\beta' := (\beta_1, \cdots, \beta_p) \in \mathbb{R}^p$, and $\varepsilon_1, \cdots, \varepsilon_n$ are i.i.d. with density f. The null hypothesis of interest here is $\mathcal{H}_0 : \rho = 0$, against the alternatives of the form $\mathcal{H}_1 : \rho > 0$. Thus, here $\Theta = \mathbb{R}^{1+p} \times [0,1)$ and $\Theta_0 = R^{1+p} \times \{0\}$. The regression parameters β_0, β play the role of nuisance parameters. Let $\hat{\theta}' := (\hat{\beta}_0, \hat{\beta}')$ be the least square estimators of (β_0, β') under the above null hypothesis, and let $\hat{\varepsilon}_t := \varepsilon_t(\hat{\beta}_0, \hat{\beta}) = X_t - \hat{\beta}_0 - c'_{n,t}\hat{\beta}$. The traditional Durbin-Watson test is based on the first-order residual autocorrelation $\hat{r}_{n1} := \sum_{t=2}^n \hat{\varepsilon}_t \hat{\varepsilon}_{t-1} / \sum_{t=1}^n \hat{\varepsilon}_t^2$. When $F(x) \equiv \Phi(x/\sigma)$, $n\hat{r}_{n1}$ coincides with

$$\frac{\sigma^2 \sum_{t=2}^n \Phi^{-1}(F(\hat{\varepsilon}_t))\Phi^{-1}(F(\hat{\varepsilon}_{t-1}))}{n^{-1}\sum_{t=1}^n \hat{\varepsilon}_t^2} = \sum_{t=2}^n \varphi_1(F(\varepsilon_t))\varphi_2(F(\varepsilon_{t-1})) + o_P(\frac{1}{\sqrt{n}}).$$

with $\varphi_1 = \varphi_2 = \Phi^{-1}$, provided $n^{-1}\sum_{t=1}^n c_{n,t}c'_{n,t} = O(1)$, and $\max_{1 \le t \le n} \frac{1}{\sqrt{n}}\|c_{n,t}\| = o(1)$. The aligned rank based version of $n\hat{r}_{n1}$ is the serial statistic $S_{n,\varphi_1\varphi_2}(\hat{\theta})$ defined in (1), with $i = 1$ and the van der Waerden scores $\varphi_1 = \varphi_2 = \Phi^{-1}$; an asymptotic representation result of Hallin, Ingenbleek and Puri (1985) establishes the equivalence $S_{n,\varphi_1\varphi_2}(\hat{\theta}) = T_{n,\varphi_1\varphi_2} + o_P(n^{1/2})$, where

$$T_{n,\varphi_1\varphi_2} := \sum_{t=2}^n \varphi_1(F(\varepsilon_t))\varphi_2(F(\varepsilon_{t-1})).$$

By Proposition 1 below it follows that the autoregression rank score statistic $\tilde{S}_{n,\varphi_1\varphi_2}$ of (3) is also asymptotically equivalent in probability to $T_{n,\varphi_1\varphi_2}$, under the above $\mathcal{H}_0$. An advantage of using $\tilde{S}_{n,\varphi_1\varphi_2}$ is that one does not need any preliminary estimates of the nuisance parameters.

In the case of non-Gaussian errors one uses the above serial autoregression rank score statistics with $\varphi_2(v) = F^{-1}(v)$ and $\varphi_1(u) = -\dot{f}(F^{-1}(u))/f(F^{-1}(u))$, to perform an asymptotically optimal test of $\tilde{\mathcal{H}}_0$, see, e.g., Hallin and Werker (1998).

2.2.2 *AR order identification*

The objective here is to test AR(p) against AR($p+1$) dependence. The overarching model is thus the AR($p+1$) model, where

$$X_t = \rho_0 + \sum_{i=1}^{p+1} \rho_i X_{t-i} + \varepsilon_t,$$

with $\rho_1, \cdots, \rho_{p+1}$ being such that the corresponding characteristic polynomial has all its roots outside the unit disc, $\rho_p \neq 0$, and $\varepsilon_1, \cdots, \varepsilon_n$ are i.i.d. with density f. The null hypothesis of interest here is $\mathcal{H}_0 : \rho_{p+1} = 0$, against the alternatives of the form $\mathcal{H}_1 : \rho_{p+1} \neq 0$. The autoregressive parameters $\rho_1, \cdots, \rho_p$ play the role of nuisance parameters.

The classical Gaussian test for this problem is based on a Lagrange multiplier type statistic

$$\hat{r}_{ni} := \sum_{t=i+1}^{n} \hat{\varepsilon}_t \hat{\varepsilon}_{t-i} \Big/ \sum_{t=1}^{n} \hat{\varepsilon}_t^2, \quad i = 1, 2, \cdots,$$

where the estimated residuals $\hat{\varepsilon}_t$ are computed from fitting an AR(p) model to the data: see Garel and Hallin (1999) for details. Arguing as in the previous example, a rank-based version of this test statistic is obtained by substituting the aligned serial rank statistics $(n-i)^{-1} S_{n,\varphi_1\varphi_2;i}(\hat{\theta})$'s of (1) for the residual autocorrelations $\hat{r}_{ni}$ into the quadratic test statistic. But such tests are not ADF, while by Proposition 1, the tests based on the analogous quadratic form using serial autoregression rank score statistic $\tilde{S}_{n,\varphi_1\varphi_2;i}$ will be ADF. Here again, asymptotically optimal tests at non-Gaussian errors case can be handled by an adequate choice of φ_1 and φ_2.

Contrary to the previous case, $S_{n,\varphi_1\varphi_2}(\hat{\theta})$ and $\tilde{S}_{n,\varphi_1\varphi_2}$ are no longer asymptotically equivalent: $S_{n,\varphi_1\varphi_2}(\hat{\theta})$ suffers from an *alignment effect* (which is not distribution-free), whereas $\tilde{S}_{n,\varphi_1\varphi_2}$ remains unaffected. Hallin and Jurečková (1999) constructed ADF tests of $\mathcal{H}_0$ against $\mathcal{H}_1$ based on non-serial autoregression rank score statistics of the type $\tilde{S}^*_{n,\varphi}(\theta)$'s. A simulation study of these tests can be found in Hallin et al. (1997) and an application to meteorological data in Kalvová et al. (2000).

2.2.3 Detection of random coefficients in AR models

The general overarching model is the autoregressive model (for simplicity, a first-order one) with random coefficients, of the form

$$X_t = (\rho + \tau u_t)X_{t-1} + \varepsilon_t,$$

where $\rho \in (0,1)$, $\tau \geq 0$, $u_1, \cdots, u_n$ are i.i.d. standardized r.v.'s with density g, and $\varepsilon_1, \cdots, \varepsilon_n$ are i.i.d. with density f, independent of the u_t's. The null hypothesis of interest here is $\mathcal{H}_0 : \tau = 0$ (ordinary AR(1) dependence), against the alternatives of the form $\mathcal{H}_1 : \tau > 0$. The autoregression parameter ρ and the densities g and f are nuisance parameters. Here $\Theta = \{\theta = (\rho, \tau)' \in (0,1) \times [0,1);\, \rho^2 + \tau^2 < 1\}$, $\Theta_0 = \{\theta = (\rho, 0)'; 0 < \rho < 1\}$, and $\varepsilon(\theta) = X_t - \rho X_{t-1}$, for a $\theta \in \Theta_0$.

Ramanathan and Rajarshi (1994) provide aligned rank tests for this problem. In the more general AR(p) case, Akharif and Hallin (2003) have studied this problem from a pseudo-Gaussian point of view. The locally asymptotically optimal Gaussian test statistic for this problem is the combination

$$\sum_{k=1}^{n-1} \hat{\rho}^{2(k-1)}(n-k)^{-1/2} \sum_{t=k+1}^{n} \Big(1 - \frac{\hat{\varepsilon}_t^2}{\hat{\sigma}^2}\Big)\Big(\frac{\hat{\varepsilon}_{t-k}}{\hat{\sigma}}\Big)^2$$
$$+2 \sum\sum_{1\leq k<\ell\leq n-1} \hat{\rho}^{k-1}\hat{\rho}^{\ell-1}(n-\ell)^{-1/2} \sum_{t=\ell+1}^{n} \Big(1 - \frac{\hat{\varepsilon}_t^2}{\hat{\sigma}^2}\Big)\Big(\frac{\hat{\varepsilon}_{t-k}}{\hat{\sigma}}\Big)\Big(\frac{\hat{\varepsilon}_{t-\ell}}{\hat{\sigma}}\Big) \quad (7)$$

of the statistics of the form

$$\frac{1}{\sqrt{n-k}} \sum_{t=k+1}^{n} \Big(1 - \frac{\hat{\varepsilon}_t^2}{\hat{\sigma}^2}\Big)\Big(\frac{\hat{\varepsilon}_{t-k}}{\hat{\sigma}}\Big)^2, \quad \frac{1}{\sqrt{n-\ell}} \sum_{t=\ell+1}^{n} \Big(1 - \frac{\hat{\varepsilon}_t^2}{\hat{\sigma}^2}\Big)\Big(\frac{\hat{\varepsilon}_{t-k}}{\hat{\sigma}}\Big)\Big(\frac{\hat{\varepsilon}_{t-\ell}}{\hat{\sigma}}\Big),$$

where $\hat{\rho}$ is an arbitrary root-n consistent (under $\mathcal{H}_0$) of ρ, $\hat{\varepsilon}_t := X_t - \hat{\rho}X_{t-1}$, and $\hat{\sigma}^2 := n^{-1}\sum_{t=1}^{n} \hat{\varepsilon}_t^2$. Just as in the Durbin-Watson case, the diagonality of the information matrix (relative to ρ and σ^2) implies that the impact of estimating ρ in (7) is $o_P(1)$, under $\mathcal{H}_0$. In the case $F(x) = \Phi(x/\sigma)$, these statistics coincide, respectively, up to $o_P(1)$ terms, with

$$\frac{1}{\sqrt{n-k}} T_{\varphi_1, \varphi_{k;1}} := \frac{1}{\sqrt{n-k}} \sum_{t=k+1}^{n} \big(1 - (\Phi^{-1}(F(\varepsilon_t)))^2\big)\big(\Phi^{-1}(F(\varepsilon_{t-k}))\big)^2,$$

$$\frac{1}{\sqrt{n-\ell}} T_{n,\varphi_1,\varphi_{k;2},\varphi_\ell}$$
$$:= \frac{1}{\sqrt{n-\ell}} \sum_{t=\ell+1}^{n} \big(1 - (\Phi^{-1}(F(\varepsilon_t)))^2\big)\Phi^{-1}(F(\varepsilon_{t-k}))\Phi^{-1}(F(\varepsilon_{t-\ell})),$$

respectively, with $\varphi_1(u) := 1 - (\Phi^{-1}(u))^2$, $\varphi_{k;1}(u) := (\Phi^{-1}(u))^2$, and $\varphi_{k;2}(u) = \varphi_\ell(u) := \Phi^{-1}(u)$. Asymptotic representation results for serial aligned rank statistics again imply the asymptotic equivalence, up to $o_P(n^{1/2})$, of $T_{n,\varphi_1,\varphi_{k;1}}$ and $T_{n,\varphi_1,\varphi_{k;2},\varphi_\ell}$ with the serial rank statistics

$$S_{n,\varphi_1,\varphi_{k;1}}(\rho) := \sum_{t=k+1}^{n} \Big[\Big(1 - \big(\Phi^{-1}\big(\frac{R_{n,t}(\rho)}{n+1}\big)\big)^2\Big)\Big(\Phi^{-1}\big(\frac{R_{n,t-k}(\rho)}{n+1}\big)\Big)^2\Big],$$

$$S_{n,\varphi_1,\varphi_{k;2},\varphi_\ell}(\rho) := \sum_{t=\ell+1}^{n} \Big[\Big(1 - \big(\Phi^{-1}\big(\frac{R_{n,t}(\rho)}{n+1}\big)\big)^2\Big)\Phi^{-1}\big(\frac{R_{n,t-k}(\rho)}{n+1}\big) \times \\ \times \Phi^{-1}\big(\frac{R_{n,t-\ell}(\rho)}{n+1}\big)\Big],$$

respectively, where $R_t(\rho)$ is the rank of $\varepsilon_t(\rho) = X_t - \rho X_{t-1}$. These statistics, based on exact residual ranks, cannot be computed from the observations. However, in view of Proposition 1 below, $S_{n\varphi_1,\varphi_{k;1}}$ and $S_{n,\varphi_1,\varphi_{k;2},\varphi_\ell}$ in turn are asymptotically equivalent to their autoregression rank score counterparts $\tilde{S}_{n,\varphi_1,\varphi_{k;1}}$ and $\tilde{S}_{n,\varphi_1,\varphi_{k;2},\varphi_\ell}$, which are measurable with respect to the observations.

Perhaps it should be emphasized that serial autoregression rank scores based tests for a given choice of φ_j's can be always implemented regardless of the knowledge of the error density.

2.3 *Assumptions on f and the score functions*

We shall now state additional assumptions needed for obtaining the asymptotic representation result for serial autoregression rank scores. Besides the structural assumption (A), we also need some technical assumptions on the density f and the score functions $\varphi_1, \cdots, \varphi_m$. As usual, these assumptions cannot be separated: stronger assumptions on φ's allow for weaker assumptions on the densities, and vice-versa. Therefore, we formulate two sets of assumptions, (F1)-(F4), (φ-1) and (F1), (F5) and (φ-2), that can be used equivalently. We assume that all densities f in the class $\mathcal{F}$ are such that

(F1) $\int_{-\infty}^{\infty} x dF(x) = 0, \qquad 0 < \int_{-\infty}^{\infty} x^2 dF(x) = \sigma^2 < \infty$;

(F2) The density f is positive on $\mathbb{R}$ and absolutely continuous, with a.e. derivative $\dot{f}$, satisfying $\mathcal{I}_f := \int_{-\infty}^{\infty} (\dot{f}(x)/f(x))^2 f(x)dx < \infty$.

(F3) There exists a constant $K = K_f \geq 0$ such that, for $|x| \geq K$, f has two bounded derivatives, f' and f'', respectively.

(F4) As $x \longrightarrow \pm\infty$, $f(x)$ is monotonically decreasing to 0 and,

$$\lim_{x \longrightarrow -\infty} \frac{-\log F(x)}{b|x|^r} = 1 = \lim_{x \longrightarrow \infty} \frac{-\log(1 - F(x))}{b|x|^r}$$

for some $b = b_f > 0$ and $r = r_f \geq 1$.

As for the functions $\varphi_1, \cdots, \varphi_m$, we assume the following:

(φ-1) The functions $\varphi_1, \cdots, \varphi_m$ from $(0,1)$ to $\mathbb{R}$ are square integrable, nondecreasing, differentiable, with respective derivatives $\dot{\varphi}_1, \cdots, \dot{\varphi}_m$, and satisfy $\int_0^1 \varphi_j(u)du = 0$, for at least one $j = 1, \cdots, m$,

$$|\dot{\varphi}_j| \leq C(u(1-u))^{-1-\delta}, \quad \forall j = 1, \cdots, m,$$

for some $0 < C < \infty$ and $0 < \delta < 1/4$.

The second set of assumptions consists of (F1), (F5) and (φ-2), where

(F5) f is uniformly continuous and a.e. positive on $\mathbb{R}$.

(φ-2) The functions $\varphi_1, \cdots, \varphi_m$ from $(0,1)$ to $\mathbb{R}$ are nondecreasing bounded and $\int_0^1 \varphi_j(u)du = 0$, for some $j = 1, \cdots, m$.

3 Asymptotic representation

The following main result of the paper gives the asymptotic representation of the serial autoregression rank score statistics. It enables one to construct the asymptotic rejection regions of the pertaining tests and their asymptotic powers against the Pitman alternatives. A similar result holds for serial regression rank scores of linear regression models with bounded designs.

Proposition 1. *Suppose the linear AR(p) model (5) holds. Suppose additionally either (F1)-(F4) and (φ-1) or (F1), (F5) and (φ-2) hold. Then, under $P_{n,\theta;f}$, as $n \to \infty$,*

$$\tilde{S}_{n,\varphi_1\cdots\varphi_m} = T_{n,\varphi_1\cdots\varphi_m} + o_P(n^{1/2}) = S_{n,\varphi_1\cdots\varphi_m}(\theta) + o_P(n^{1/2}), \qquad (8)$$

where $\tilde{S}_{n,\varphi_1\cdots\varphi_m}$ is the serial autoregression rank score statistic (4), $S_{n,\varphi_1\cdots\varphi_m}(\theta)$ the serial rank statistic (2), and

$$T_{n,\varphi_1\cdots\varphi_m} := \sum_{t=i_{m-1}+1}^{n} \varphi_1(F(\varepsilon_{n,t}))\cdots\varphi_m(F(\varepsilon_{n,t-i_{m-1}})).$$

Proof: Without loss of generality, we restrict the proof to the autoregressive case for $m = 2$ and $i_1 = 1$, hence to statistics of the form $\tilde{S}_{n,\varphi_1\varphi_2}$, $T_{n,\varphi_1\varphi_2}$ and $S_{n,\varphi_1\varphi_2}$. Additionally, we shall assume the first set of conditions (F1)-(F4) and (φ-1) with the proviso that $\int \varphi_j = 0$, for both $j = 1, 2$. See Remark 1 below for the case when one of the φ's is not centered. The proof is much simpler under the second set of assumptions. We systematically drop subscripts n in the proof.

Let $0 < \alpha_0 < 1/2$ be a fixed number, $\alpha_n := n^{-1}(\log n)^2(\log\log n)^2$, and take n large enough so that $\alpha_n < \alpha_0$. Define, for a $0 < u < 1$,

$$\rho(u) := \rho + F^{-1}(u)e_1, \quad \text{with} \quad e_1' := (1, 0, \cdots, 0),$$

where $\rho := (1, \rho_1, \cdots, \rho_p)'$ of the model (5). For all $0 < u < 1$, put

$$\begin{aligned} D_t(u) &:= I(X_t \le y_{t-1}'\hat{\rho}(u)) - I(X_t \le y_{t-1}'\rho(u)), \\ \tilde{a}_t(u) &:= I(\varepsilon_t > F^{-1}(u)) - (1-u), \\ \hat{a}_t(u) - (1-u) &:= \tilde{a}_t(u) - D_t(u) + \hat{a}_t(u) I(X_t = y_{t-1}'\hat{\rho}(u)). \end{aligned} \tag{9}$$

Also, for $j = 1, 2$, let

$$\hat{b}_{j,t} := -\int_0^1 \varphi_j(u) d\hat{a}_t(u), \quad \hat{b}_{n;j,t} := \int_{\alpha_n}^{1-\alpha_n} [\hat{a}_t(u) - (1-u)] d\varphi_j(u).$$

Note that $\int_0^1 \varphi_j(u)du = 0$ and integration by parts yield that, for all t,

$$\hat{b}_{j,t} = -\int_0^1 \varphi_j(u) d[\hat{a}_t(u) - (1-u)] = \int_0^1 [\hat{a}_t(u) - (1-u)] d\varphi_j(u).$$

Decomposing this further gives, with $\bar{a}_t(u) := \hat{a}_t(u) - (1-u)$,

$$\begin{aligned} \hat{b}_{j,t} &= \int_0^{\alpha_n} \bar{a}_t(u) d\varphi_j(u) + \hat{b}_{n;j,t} + \int_{1-\alpha_n}^1 \bar{a}_t(u) d\varphi_j(u), \\ \hat{b}_{n;j,t} &= \int_{\alpha_n}^{\alpha_0} \bar{a}_t(u) d\varphi_j(u) + \int_{\alpha_0}^{1-\alpha_0} \bar{a}_t(u) d\varphi_j(u) + \int_{1-\alpha_0}^{1-\alpha_n} \bar{a}_t(u) d\varphi_j(u) \\ &= \hat{c}_{n1;j,t} + \hat{c}_{n2;j,t} + \hat{c}_{n3;j,t}, \qquad \text{say.} \end{aligned}$$

We start with analyzing the sum $n^{-1/2} \sum_{t=2}^n \hat{c}_{n1;1,t} \hat{c}_{n1;2,t-1}$. The analysis of the similar sum involving $\hat{c}_{n3;j,t}$'s is exactly similar, while the similar sum corresponding to the $\hat{c}_{n2;j,t}$ terms can be analyzed using the results for bounded scores. The analysis of the cross product sums is also similar and relatively less involved. For the ease of writing, let $\varphi_{jn}(u) := \varphi_j(u) I(\alpha_n < u \le \alpha_0)$. Using (9), rewrite

$$\hat{c}_{n1;j,t} = \int \tilde{a}_t d\varphi_{jn} - \int D_t d\varphi_{jn} + \int \hat{a}_t(u) I(X_t = y_{t-1}'\hat{\rho}(u)) d\varphi_{jn}(u).$$

Letting $A_{j,t} := \int \tilde{a}_t d\varphi_{jn}$, we thus obtain

$$
\begin{aligned}
&\frac{1}{\sqrt{n}}\sum_{t=2}^{n} \hat{c}_{n1;j,t}\hat{c}_{n1;j,t-1} \qquad (10)\\
=&\frac{1}{\sqrt{n}}\sum_{t=2}^{n} \Big[A_{1,t}A_{2,t} - \int D_t d\varphi_{1n} A_{2,t}\\
&\quad + \int \hat{a}_t(u) I\big(X_t = y_{t-1}'\hat{\rho}(u)\big) d\varphi_{1n}(u) A_{2,t} - A_{1,t}\int D_{t-1} d\varphi_{2n}\\
&\quad + \int D_t d\varphi_{1n}\int D_{t-1}d\varphi_{2n} - \Big\{\int \hat{a}_t(u) I(X_t = y_{t-1}'\hat{\rho}(u)) d\varphi_{1n}(u)\\
&\qquad \times \int D_{t-1} d\varphi_{2n}\Big\} + A_{1,t}\int \hat{a}_{t-1}(u) I\big(X_{t-1} = y_{t-2}'\hat{\rho}(u)\big) d\varphi_{2n}(u)\\
&\quad - \int D_t d\varphi_{1n}\int \hat{a}_{t-1}(u) I\big(X_{t-1} = y_{t-2}'\hat{\rho}(u)\big) d\varphi_{2n}(u)\\
&\quad + \Big\{\int \hat{a}_t(u) I\big(X_t = y_{t-1}'\hat{\rho}(u)\big) d\varphi_{1n}(u)\\
&\qquad \times \int \hat{a}_{t-1}(u) I\big(X_{t-1} = y_{t-2}'\hat{\rho}(u)\big) d\varphi_{2n}(u)\Big\}\Big]\\
=&C_1 - C_2 + C_3 - C_4 + C_5 - C_6 + C_7 - C_8 + C_9, \qquad \text{say.}
\end{aligned}
$$

In order to show that the first term $C_1 := \frac{1}{\sqrt{n}}\sum_{t=2}^{n} A_{1,t}A_{2,t}$ provides the approximating terms to the left hand side above, we shall verify that all of the remaining terms tend to zero in probability. Let $d_{jn} := \varphi_j(\alpha_0) - \varphi_j(\alpha_n)$, $j = 1, 2$, and $d_n := \max(d_{1n}, d_{2n})$. From the linear programming definition of $\hat{a}_t(u)$'s, we obtain that for all $0 < u < 1$,

$$\sum_{t=2}^{n} \hat{a}_t(u) I\big(X_t = y_{t-1}'\hat{\rho}(u)\big) \le (p+1), \quad \text{a.s.} \tag{11}$$

This in turn implies

$$\frac{1}{\sqrt{n}}\sum_{t=2}^{n}\int \hat{a}_t(u) I\big(X_t = y_{t-1}'\hat{\rho}(u)\big) d\varphi_{1n}(u) \le \frac{1}{\sqrt{n}}(p+1)d_n, \tag{12}$$

$$\frac{1}{\sqrt{n}}\sum_{t=2}^{n}\int \hat{a}_{t-1}(u) I\big(X_{t-1} = y_{t-2}'\hat{\rho}(u)\big) d\varphi_{2n}(u) \le \frac{1}{\sqrt{n}}(p+1)d_n, \quad \text{a.s.}$$

Now, consider the term C_9. The fact that $\hat{a}_t \le 1$ and (12) imply

$$C_9 \le d_n^2 \frac{1}{\sqrt{n}}(p+1), \quad \text{a.s.} \tag{13}$$

Similarly using $|D_t| \le 1$, $|A_{j,t}| \le d_n$, for all t and (12), we obtain

$$\max\{|C_3|,\ |C_6|,\ |C_7|, |C_8|\} \le d_n^2 \frac{1}{\sqrt{n}}(p+1), \qquad \text{a.s.} \tag{14}$$

The assumption (φ-1) on φ_1, φ_2 and the definition of α_n imply that

$$\begin{aligned} d_n \le & \int_{\alpha_n}^{\alpha_0} (u(1-u))^{-1-\delta} du \le \int_{\alpha_n}^{\alpha_0} u^{-1-\delta} du \\ \le & \frac{C}{\delta}\left(\alpha_n^{-\delta} - \alpha_0^{-\delta}\right) = O(n^{\delta}(\log n)^{-2\delta}(\log \log n)^{-2\delta}), \end{aligned}$$

so that, because $0 < \delta < 1/4$, $d_n^2\, n^{-1/2} = o(1)$. Hence,

$$\max\{|C_3|,\ |C_6|,\ |C_7|, |C_8|, |C_9|\} = o(1), \qquad \text{a.s.} \tag{15}$$

Note also that, because of the $n^{-1/2}$ factor, the same conclusions hold if φ_j, $j = 1, 2$, is replaced by $\tilde{\varphi}_{jn} := \varphi_j I\big[[\alpha_n, \alpha_n(1+\epsilon)]\big]$.

To deal with C_4, we need to center the factor involving D_{t-1} properly. For this the r.v.'s involved in the indicators need to be suitably standardized. This standardization is done differently for the u-quantiles in the tail and in the middle, because in the tail the consistency rate of $\hat{\rho}(u)$ is different from $\frac{1}{\sqrt{n}}$ and also depends on u, as was shown in Hallin and Jurečková (1999). We need to use these facts in the following analysis. Accordingly, let $q(u) := f(F^{-1}(u))$,

$$\begin{aligned} \sigma_u :=& (u(1-u))^{1/2}/q(u), \quad \Delta(u) := \sigma_u^{-1} n^{1/2}(\hat{\rho}(u) - \rho(u)), \\ \mu_t(u) :=& F\big(F^{-1}(u) + \sigma_u \frac{1}{\sqrt{n}} y'_{t-1} \Delta(u)\big) - F\big(F^{-1}(u)\big), \\ \nu_t(u) :=& \mu_t(u) - \sigma_u \frac{1}{\sqrt{n}} y'_{t-1} \Delta(u)\, q(u). \end{aligned}$$

Rewrite

$$D_t(u) = I(\varepsilon_t \le F^{-1}(u) + \frac{1}{\sqrt{n}} \sigma_u y'_{t-1} \Delta(u)) - I(\varepsilon_t \le F^{-1}(u)).$$

Then,

$$\begin{aligned} C_4 :=& \frac{1}{\sqrt{n}} \sum_{t=2}^{n} A_{1,t} \int D_{t-1} d\varphi_{2n} \\ =& \frac{1}{\sqrt{n}} \sum_{t=2}^{n} A_{1,t} \int [D_{t-1} - \mu_{t-1}] d\varphi_{2n} + \frac{1}{\sqrt{n}} \sum_{t=2}^{n} A_{1,t} \int \nu_{t-1} d\varphi_{2n} \\ & + n^{-1} \sum_{t=2}^{n} A_{1,t} y'_{t-2} \int \Delta(u)\, \sigma_u q(u) d\varphi_{2n}(u) \\ =& C_{41} + C_{42} + C_{43}, \qquad \text{say.} \end{aligned} \tag{16}$$

But, because $\sigma_u q(u) = (u(1-u))^{1/2}$,

$$|C_{43}| \leq \|n^{-1}\sum_{t=2}^{n} A_{1,t}y_{t-2}\| \, \|\int \Delta(u)\sigma_u q(u)d\varphi_{2n}(u)\|$$
$$= O_P(n^{-1/2}) \sup_{\alpha_n < u \leq \alpha_0} \|\Delta(u)\| \int (u(1-u))^{1/2} d\varphi_{2n}(u).$$

The first factor of $O_P(n^{-1/2})$ comes from the fact that $\sum_{t=2}^{n} A_{1,t}y_{t-2}$ is a vector of zero mean martingales, and hence

$$E\|n^{-1}\sum_{t=2}^{n} A_{1,t}y_{t-2}\|^2 = n^{-2}\sum_{t=2}^{n} Ey'_{t-2}y_{t-2}\, EA_{1,t}^2 = O(n^{-1}).$$

Also, by (φ-1), $\int (u(1-u))^{1/2} d\varphi_{jn}(u) \leq \int_0^{\alpha_0} u^{-1/2-\delta} du < \infty$, $j = 1, 2$. Next, recall from Hallin & Jurečková (1999) (H-J) that under (F1)-(F4),

$$\sup_{\alpha_n \leq u \leq 1-\alpha_n} \|\Delta(u)\| = O_P(\log\log n)^{1/2}. \tag{17}$$

Upon combining these observations we obtain

$$|C_{43}| = O_P(n^{-1/2}(\log\log n)^{1/2}) = o_P(1). \tag{18}$$

Next consider C_{42}. Let $\delta_{n,t,u} := n^{-1/2}\sigma_u\, y'_{t-2}\Delta(u)$, $\epsilon_n := C\,(\log n)^{2/r-2} \times (\log\log n)^{-1/4}$, $r \geq 1$, and $K_n := C\,(\log\log n)^{1/2}$. We need the following results from H-J obtained under (F1)-(F4). By (A.5) and (A.9) in there, for any $r \geq 1$,

$$\max_{1\leq t\leq n, \alpha_n \leq u \leq 1-\alpha_n} |\delta_{n,t,u}| = O_P(\epsilon_n), \tag{19}$$

$$\frac{|\dot{f}(x)|}{f(x)}|x|^{1-r} = O(1), \quad \text{as } x \to \pm\infty. \tag{20}$$

Let $x_n = F^{-1}(\alpha_n)$, $x_0 = F^{-1}(\alpha_0)$, $\tau_y = \sigma_{F(y)}$, $\tilde{\Delta}_y = \Delta(F(y))$, $\tilde{\delta}_{n,t,y} = \delta_{n,t,F(y)}$, and $dL(y) = F^{-1/2-\delta}(y)dF(y)$. Since $\alpha_n < \alpha_0 < 1/2$, we have $x_n < x_0 < 0$. Also, in the left tail $|d\varphi_j(F)| \leq C\,F^{-1-\delta}dF$. Let $\mathcal{A}_n := \{\sup_{\alpha_n \leq u \leq 1-\alpha_n} \|\Delta(u)\| \leq K_n\}$. For $x_n \leq y \leq x_0 < 0$, on the event $\mathcal{A}_n$, $|\tilde{\delta}_{n,t,y}| \leq \frac{1}{\sqrt{n}}(F^{1/2}(y)/f(y))\|y_{t-2}\|K_n$, and

$$|C_{42}| = |\frac{1}{\sqrt{n}}\sum_{t=2}^{n} A_{1,t}\int \nu_{t-1}(u)d\varphi_{2n}(u)|$$
$$\leq \frac{1}{\sqrt{n}}\sum_{t=2}^{n}|A_{1,t}|$$
$$\times |\int [F(y + \frac{\tau_y y'_{t-2}\tilde{\Delta}_y}{\sqrt{n}}) - F(y) - \frac{\tau_y y'_{t-2}\tilde{\Delta}_y}{\sqrt{n}} f(y)]d\varphi_{2n}(F(y))|$$
$$\leq n^{-1}\sum_{t=2}^{n}|A_{1,t}|\|y_{t-2}\|K_n \int_{x_n}^{x_0}\int_0^1 \frac{|f(y + v\,\tilde{\delta}_{n,t,y}) - f(y)|}{f(y)}\, dv dL(y).$$

But, on the event $\max_{1\le t\le n, x_n\le y\le x_0}|\tilde{\delta}_{n,t,y}|\le C\epsilon_n$, see (19), the double integral in this bound is further bounded above by $C\epsilon_n$ times the integral

$$\begin{aligned}
&\max_{1\le t\le n}\int_{x_n}^{x_0}\int_0^1\int_0^1 v\frac{|\dot{f}(y+\tilde{\delta}_{n,t,y}vz)|}{f(y)}\,dvdzdL(y)\\
&\quad\le \max_{1\le t\le n}\int_0^1 v\int_0^1\int_{x_n}^{x_0}\frac{|\dot{f}(y+\tilde{\delta}_{n,t,y}vz)|}{f(y)}\,dL(y)\,dzdv\\
&\quad\longrightarrow_P (1/2)\int_{-\infty}^{x_0}\frac{|\dot{f}(y)|}{f(y)}\,dL(y)\le C\int_{-\infty}^{x_0}|y|^{r-1}dL(y)\qquad\text{(by (20))}\\
&\quad\le C\Big(\int|y|^{p(r-1)}dF(y)\Big)^{1/p}\Big(\int F^{-q(\frac{1}{2}+\delta)}(y)dF(y)\Big)^{1/q},\qquad (21)
\end{aligned}$$

where p, q are positive integers, $1/p+1/q=1$, and such that $1>q(1/2+\delta)$ so that the second integral in the above bound is finite. Such a q always exists since $\delta<1$. Also note that $(2/r)-2<0$, because $r>1$. Hence

$$K_n\epsilon_n=C(\log n)^{\frac{2}{r}-2}(\log\log n)^{1/4}=o(1),\quad \sum_{t=2}^{n}|A_{1,t}|\|y_{t-2}\|=O_P(n),$$

and, in view of (19) and (20),

$$|C_{42}|=O_P(K_n\epsilon_n)=o_P(1).\qquad (22)$$

Next, we treat the C_{41} term. Let $a_n=\frac{1}{\sqrt{n}}K_n$. Define, for y, $a\in\mathbb{R}$ and $s\in\mathbb{R}^{m+1}$ such that $\|s\|\le 1$,

$$\begin{aligned}
C_{41}^{\pm}(y,s,a):=&\frac{1}{\sqrt{n}}\sum_{t=2}^{n}A_{1,t}^{\pm}[I(\varepsilon_{t-1}\le y+a_n\tau_y(y'_{t-2}s+\|y_{t-2}\|a))\\
&-I(\varepsilon_{t-1}\le y)-m_{t-1}(y,s,a)],
\end{aligned}$$

$$m_{t-1}(y,s,a):=[F(y+a_n\tau_y(y'_{t-2}s+\|y_{t-2}\|a))-F(y)],$$

where $A_t^{\pm}$ stand for the positive and negative parts of A_t. Write $C_{41}^{\pm}(y,s)$ for $C_{41}^{\pm}(y,s,0)$ and let

$$\begin{aligned}
C_{41}(y,s)&:=C_{41}^{+}(y,s)-C_{41}^{-}(y,s)\\
&=\frac{1}{\sqrt{n}}\sum_{t=2}^{n}A_{1,t}[I(\varepsilon_{t-1}\le y+a_n\tau_y y'_{t-2}s)-I(\varepsilon_{t-1}\le y)-\mu_{t-1}(F(y),s)].
\end{aligned}$$

Note that on the event $\mathcal{A}_n$,

$$|C_{41}|\le\int\sup_{\|s\|\le 1}|C_{41}^{+}(y,s)|d\psi_{2n}(y)+\int\sup_{\|s\|\le 1}|C_{41}^{-}(y,s)|d\psi_{2n}(y),$$

with $\psi_{jn}(y) := \varphi_{jn}(F(y))$. We shall show that

$$\int \sup_{\|s\|\le 1} |C_{41}^{\pm}(y,s)| d\psi_{2n}(y) = o_P(1), \tag{23}$$

which obviously will imply

$$|C_{41}| = o_P(1). \tag{24}$$

For an $s \in \mathbb{R}^{m+1}$, a, $y \in \mathbb{R}$, let

$$\ell_n(y,s,a) := \int_0^1 \frac{E\{\|y_0\| f(y + z\,\tau_y a_n(y_0's + \|y_0\| a))\}}{f(y)} dz,$$

$$\gamma_n(y,s,a) := \int_0^1 \frac{E\{\|y_0\| f(y + \tau_y a_n(y_0's - 2\|y_0\| a\,z))\}}{f(y)} dz.$$

Arguing as above and conditionally, we have for all $y \in \mathbb{R}$, with $b = EA_{1,t}^2$,

$$\begin{aligned}
&E|C_{41}^{\pm}(y,s,a)|^2 \\
&\quad = E(A_{1,t}^{\pm})^2 E\big[I(\varepsilon_1 \le y + a_n\tau_y(y_0's + \|y_0\|a)) - I(\varepsilon_1 \le y) \\
&\qquad\qquad - F(y + a_n\tau_y(y_0's + \|y_0\|a)) + F(y)\big]^2 \\
&\quad \le b\, E|F(y + a_n\tau_y(y_0's + \|y_0\|a)) - F(y)| \\
&\quad \le b\, a_n\tau_y(\|s\| + a) \int_0^1 E\{\|y_0\| f(y + z\,\tau_y a_n(y_0's + \|y_0\|a))\} dz \\
&\quad = b\, \frac{1}{\sqrt{n}} K_n\, (\|s\| + |a|)\, [F(y)(1 - F(y))]^{1/2}\, \ell_n(y,s,a).
\end{aligned} \tag{25}$$

Similarly, for any s, $t \in \mathbb{R}^p$, and y, $a \in \mathbb{R}$,

$$\begin{aligned}
&E|C_{41}^{\pm}(y,t,a) - C_{41}^{\pm}(y,s,a)|^2 \\
&\qquad \le b\, \|t - s\|\, \frac{1}{\sqrt{n}} K_n\, [F(y)(1 - F(y))]^{1/2}\, \ell_n(y,s,a), \\
&E|C_{41}^{\pm}(y,t,a) - C_{41}^{\pm}(y,t,0)|^2 \\
&\qquad \le b\, |a|\, \frac{1}{\sqrt{n}} K_n\, [F(y)(1 - F(y))]^{1/2}\, \gamma_n(y,t,a).
\end{aligned} \tag{26}$$

Since the unit ball is compact, there is an $\eta > 0$ and a finite number k of points $s_1, \cdots, s_k$ in the unit ball such that for any $\|s\| \le 1$, there is an s_j in the unit ball with $\|s - s_j\| \le \eta$. We will need to choose η to depend on n and hence so also k. Now,

$$\begin{aligned}
\sup_{\|s\|\le 1} |C_{41}^{\pm}(y,s)| \le & \max_{1\le j\le k} \sup_{\|s - s_j\|\le \eta} |C_{41}^{\pm}(y,s) - C_{41}^{\pm}(y,s_j)| \\
& + \max_{1\le j\le k} |C_{41}^{\pm}(y,s_j)|.
\end{aligned} \tag{27}$$

But, $\|s - s_j\| \le \eta$ implies that for all $y \in \mathbb{R}, 1 \le j \le k,\ n \ge 1,\ 1 \le t \le n$,

$$a_n \tau_y (y'_{t-2} s_j - \|y_{t-2}\| \eta) \le a_n \tau_y y'_{t-2} s \le a_n \tau_y (y'_{t-2} s_j + \|y_{t-2}\| \eta).$$

This, the monotonicity of the indicators and nonnegativity of $A^{\pm}_{1,t}$'s, imply

$$\begin{aligned} &|C^{\pm}_{41}(y,s) - C^{\pm}_{41}(y,s_j)| \\ &\quad \le |C^{\pm}_{41}(y,s_j,\eta) - C^{\pm}_{41}(y,s_j,0)| + |C^{\pm}_{41}(y,s_j,-\eta) - C^{\pm}_{41}(y,s_j,0)| \\ &\quad +2\frac{1}{\sqrt{n}} \sum_{t=2}^{n} A^{\pm}_{1,t} [m_{t-1}(y,s_j,\eta) - m_{t-1}(y,s_j,-\eta)] \end{aligned} \tag{28}$$

Moreover, by (26), and the Cauchy-Schwarz inequality,

$$\begin{aligned} &E(\max_{1 \le j \le k} |C^{\pm}_{41}(y,s_j,\pm\eta) - C^{\pm}_{41}(y,s_j,0)|) \\ &\qquad \le k\,\{2b\,\eta\,\frac{1}{\sqrt{n}} K_n\,[F(y)(1-F(y))]^{1/2} \max_{1 \le j \le k} \gamma_n(y,s_j,\eta)\}^{1/2}. \end{aligned}$$

Let $g_n(y,\eta) := \max_{1\le j \le k} \gamma_n(y,s_j,\eta)$ and $\nu(y) := \int_{-\infty}^{y} F^{-1/4} dL(y)$. Note that for $\delta < 1/4$ this is a finite measure. Also note that $g_n(y,\eta) \to E\|y_0\| < \infty$. Arguing as for (21), we thus obtain for all η,

$$\begin{aligned} B_n :=& \int [F(y)(1-F(y))]^{1/4}\, g_n(y,\eta) d\psi_n(y) \\ \le& \int_{x_n}^{x_0} F(y)^{-3/4-\delta} g_n^{1/2}(y,\eta) dF(y) \le \int g_n^{1/2}(y,\eta) d\nu(y) = O(1). \end{aligned}$$

Hence,

$$\int E \max_{1\le j\le k} |C^{\pm}_{41}(y,s_j,\pm\eta) - C^{\pm}_{41}(y,s_j,0)| d\psi_{2n}(y) \le B_n k (\frac{\eta K_n}{\sqrt{n}})^{1/2}. \tag{29}$$

Next, let $d_{n,t,y,s} := a_n \tau_y y'_{t-1} s$. The third term in the upper bound of (28) is bounded above by

$$\begin{aligned} &\frac{1}{\sqrt{n}} \sum_{t=2}^{n} A^{\pm}_{1,t} \{F(y + d_{n,t-1,y,s_j} + a_n \tau_y \|y_{t-2}\| \eta) - F(y) \\ &\qquad\qquad -[d_{n,t-1,y,s_j} + a_n \tau_y \|y_{t-2}\| \eta] f(y)\} \\ &\quad -\frac{1}{\sqrt{n}} \sum_{t=2}^{n} A^{\pm}_{1,t} \{F(y + d_{n,t-1,y,s_j} - a_n \tau_y \|y_{t-2}\| \eta) - F(y) \\ &\quad -[d_{n,t-1,y,s_j} - a_n \tau_y \|y_{t-2}\| \eta] f(y)\} + 2\eta\,\frac{1}{\sqrt{n}} a_n \tau_y \sum_{t=2}^{n} A^{\pm}_{1,t} \|y_{t-2}\| f(y) \\ &= M_{1,j}(y) - M_{2,j}(y) + 2\eta K_n [F(y(1-F(y))]^{1/2} n^{-1} \sum_{t=2}^{n} A^{\pm}_{1,t} \|y_{t-2}\|. \end{aligned}$$

Note that for all $\|s_j\| \le 1$, and all t, and all y with $F(y)$ close to α_n, on the event $\{\|y_{t-2}\| \le C(\log n)^{1/r}(\log\log n)^{1/4}\}$,

$$[d_{n,t-1,y,s_j} + a_n\tau_y\|y_{t-2}\|\eta] \le a_n\tau_y\|y_{t-2}\|(1+\eta) \le \epsilon_n(1+\eta).$$

Hence, by arguing as for (21),

$$\begin{aligned}
&\int \max_{1\le j\le k} |M_{1,j}(y)| d\psi_{2n}(y) \\
&\le \int \frac{1}{\sqrt{n}} \sum_{t=2}^{n} |A_{1,t}| [d_{n,t-1,y,s_j} + a_n\tau_y\|y_{t-2}\|\eta] \\
&\qquad \times \max_{1\le j\le k} \int_0^1 |f(y + [d_{n,t-1,y,s_j} + a_n\tau_y\|y_{t-2}\|\eta]z) - f(y)| dz\, d\psi_{2n}(y) \\
&\le K_n n^{-1} \sum_{t=2}^{n} |A_{1,t}| \|y_{t-2}\|(1+\eta) \int \max_{1\le j\le k} [d_{n,t-1,y,s_j} + a_n\tau_y\|y_{t-2}\|\eta] \\
&\quad \times \int_0^1 z \int_0^1 \frac{|\dot f(y + [d_{n,t-1,y,s_j} + a_n\tau_y\|y_{t-2}\|\eta]z\, v)| dv dz}{f(y)} dL(y) \\
&\le K_n\epsilon_n\, (1+\eta)^2\, n^{-1} \sum_{t=2}^{n} |A_{1,t}| \|y_{t-2}\|^2 \\
&\quad \times \int \frac{\max_{1\le j\le k} \int_0^1 z \int_0^1 |\dot f(y + [d_{n,t-1,y,s_j} + a_n\tau_y\|y_{t-2}\|\eta]z\, v)| dv dz}{f(y)} dL(y) \\
&= O_P(K_n\epsilon_n) = o_P(1), \quad \forall \eta > 0.
\end{aligned}$$

Similarly,

$$\int \max_{1\le j\le k} |M_{2,j}(y)| d\psi_{2n}(y) = O_P(K_n\epsilon_n) = o_P(1), \quad \forall \eta > 0.$$

Thus we obtain that the integral of the maximum over $1 \le j \le k$ of the third term in the bound (28) is bounded above by

$$\begin{aligned}
&O_P(K_n\epsilon_n(1+\eta)^2) + 2K_n\eta n^{-1} \sum_{t=2}^{n} |A_{1,t}| \|y_{t-2}\| \int dL(y) \qquad (30) \\
&\quad = O_P(\eta\, K_n) + O_P(K_n\epsilon_n).
\end{aligned}$$

Next, let $L_n(y) := \max_{1\le j\le k} \ell_n(y, s_j, 0)$. By (25) applied with $a = 0$,

$$\begin{aligned}
&E\big(\max_{1\le j\le k} |C_{41}^{\pm}(y, s_j)|\big) \\
&\quad \le k\, \Big\{b\, \frac{1}{\sqrt{n}} K_n\, [F(y)(1-F(y))]^{1/2} \sum_{j=1}^{k} \|s_j\| \ell_n(y, s_j, 0)\Big\}^{1/2} \\
&\quad \le k\, \Big\{b\, \frac{1}{\sqrt{n}} K_n\, [F(y)(1-F(y))]^{1/2}\, L_n(y)\Big\}^{1/2},
\end{aligned}$$

so that

$$\int E\big(\max_{1\le j\le k}|C_{41}^{\pm}(y,s_j)|\big)d\psi_{2n}(y)\le C\,k\,n^{-1/4}C_n^{1/2}\int L_n(y)\frac{dF(y)}{F^{3/4+\delta}(y)}.$$

All these bounds together with (27), (28) and (29), yield

$$\begin{aligned}&\int\sup_{\|s\|\le 1}|C_{41}^{\pm}(y,s)|d\psi_{2n}(y)\\ =&O_P(k\eta^{1/2}n^{-1/4}K_n^{1/2})+O_P(\eta\,K_n)+O_P(\epsilon_n K_n),\\ =&O_P(\eta^{1/2-p}n^{-1/4}K_n^{1/2})+O_P(\eta\,K_n).\end{aligned}$$

This in turn implies (23), by choosing η suitably. For example $\eta=K_n^{-a}$, $a>1$, will suffice. The results (24), (22) and (18) together imply

$$C_4=o_P(1). \tag{31}$$

Similarly one can prove

$$C_2=o_P(1). \tag{32}$$

Next, consider

$$\begin{aligned}|C_5|:=&\frac{1}{\sqrt{n}}\Big|\sum_{t=2}^{n}\int D_t d\varphi_{1n}\int D_{t-1}d\varphi_{2n}\Big|\\ \le&\int\int\sup_{\|s\|\le 1}\frac{1}{\sqrt{n}}\sum_{t=2}^{n}\big\{|I(\varepsilon_t\le x+\tau_x a_n y_{t-1}'s)-I(\varepsilon_t\le x)|\\ &\times|I(\varepsilon_{t-1}\le y+\tau_y a_n y_{t-2}'s)-I(\varepsilon_{t-1}\le y)|\big\}d\psi_{1n}(x)d\psi_{2n}(y)\\ \le&\int\int\frac{1}{\sqrt{n}}\sum_{t=2}^{n}\big\{I(x-\tau_x a_n\|y_{t-1}\|<\varepsilon_t\le x+\tau_x a_n\|y_{t-1}\|)\\ &\times I(y-\tau_y a_n\|y_{t-2}\|<\varepsilon_{t-1}\le y+\tau_y a_n\|y_{t-2}\|)\big\}d\psi_{1n}(x)d\psi_{2n}(y),\end{aligned}$$

so that, by a conditioning argument,

$$
\begin{aligned}
E|C_5| \leq & \int\int n^{1/2} E\{I(y - \tau_y a_n \|y_0\| < \varepsilon_1 \leq y + \tau_y a_n \|y_0\|) \\
& \quad \times |F(x + \tau_x a_n \|Y_1\|) - F(x - \tau_x a_n \|Y_1\|)|\} d\psi_{1n}(x) d\psi_{2n}(y) \\
\leq & K_n \int \int_{y=x_n}^{y=x_0} E\{I(y - \tau_y a_n \|y_0\| < \varepsilon_1 \leq y + \tau_y a_n \|y_0\|) \\
& \quad \times \|Y_1\| \frac{\int_{-1}^{1} f(x + \tau_x a_n \|Y_1\| \, v) dv}{f(x)}\} \, dL(x) d\psi_{2n}(y) \\
\leq & K_n \int \int_{y=x_n}^{y=x_0} E^{1/2} |F(y + \tau_y a_n \|y_0\|) - F(y - \tau_y a_n \|y_0\|)| \\
& \quad \times E^{1/2} \{\|Y_1\| \frac{\int_{-1}^{1} f(x + \tau_x a_n \|Y_1\| \, v) dv}{f(x)}\}^2 dL(x) d\psi_{2n}(y) \\
\leq & K_n \int \int_{y=x_n}^{y=x_0} E^{1/2} \tau_y a_n \|y_0\| \int_{-1}^{1} f(y + \tau_y a_n \|y_0\| v) dv \\
& \quad \times E^{1/2} \{\|Y_1\| \frac{\int_{-1}^{1} f(x + \tau_x a_n \|Y_1\| \, v) dv}{f(x)}\}^2 dL(x) d\psi_{2n}(y) \\
\leq & n^{-1/4} C_n^2 \int_{y=x_n}^{y=x_0} E^{1/2} \{\|y_0\| \frac{\int_{-1}^{1} f(y + \tau_y a_n \|y_0\| v) dv}{f(y)}\} dL(y) \\
& \quad \times \int_{x_n}^{x_0} E^{1/2} \{\|Y_1\| \frac{\int_{-1}^{1} f(x + \tau_x a_n \|Y_1\| \, v) dv}{f(x)}\}^2 dL(x) \\
= & o(1).
\end{aligned}
$$

The above bounds clearly prove the following

Lemma 1. *Under the conditions of Proposition 1, we have*

$$
\begin{aligned}
& \frac{1}{\sqrt{n}} \sum_{t=2}^{n} \hat{c}_{n1;1,t} \hat{c}_{n1;2,t-1} \\
& \quad = \frac{1}{\sqrt{n}} \sum_{t=2}^{n} \int \left[I(\varepsilon_t > F^{-1}(u)) - (1-u) \right] d\varphi_{n1}(u) \\
& \qquad\qquad \times \int \left[I(\varepsilon_{t-1} > F^{-1}(v)) - (1-v) \right] d\varphi_{n2}(v) + o_P(1).
\end{aligned}
$$

Next, consider the sum $n^{-1/2} \sum_{t=2}^{n} \hat{c}_{n2;1,t} \hat{c}_{n2;2,t-1}$. This is similar to the above sum with φ_{jn} replaced with $\varphi_j^0 := \varphi_j(u) I(\alpha_0 \leq u \leq 1 - \alpha_0)$. Thus several calculations are similar to those in the case considered in the proof of Lemma 1. To begin with, the decomposition (10) remains valid with φ_{jn} replaced by φ_j^0. Denote these terms by C_i^0, $i = 1, \cdots, 9$. That is, C_i^0 is the C_i of (10) with φ_j^0 substituted for φ_{jn}. The bounds (13) and (14) now hold with d_{jn} replaced by $d_j^0 := \varphi_j(1-\alpha_0) - \varphi_j(\alpha_0)$, so that the analog of (15) clearly holds here. The places where one uses a different argument is in the handling of the remaining terms C_1^0, C_2^0, C_4^0 and C_5^0.

Consider C_4^0. As mentioned earlier, here one needs to standardize the random variables involved in the indicators of D_t differently. Accordingly, now let $\gamma(u) := n^{1/2}(\hat{\rho}(u) - \rho(u))$, and rewrite

$$\mu_t(u) := F\big(F^{-1}(u) + \frac{1}{\sqrt{n}} y_{t-1}'\gamma(u)\big) - F\big(F^{-1}(u)\big),$$
$$\nu_t(u) := \mu_t(u) - \frac{1}{\sqrt{n}} y_{t-1}'\gamma(u)\, q(u).$$

Then,

$$\begin{aligned} C_4^0 =& \frac{1}{\sqrt{n}} \sum_{t=2}^n A_{1,t}\Big[\int [D_{t-1} - \mu_{t-1}] d\varphi_2^0 + \int \nu_{t-1} d\varphi_2^0 \\ &+ \int \frac{1}{\sqrt{n}} y_{t-2}'\gamma(u)\, d\varphi_2^0(u)\Big] \\ =& C_{41}^0 + C_{42}^0 + C_{43}^0, \quad \text{say.} \end{aligned}$$

Now recall from Koul and Saleh (1995) that

$$\sup_{\alpha_0 \le u \le 1-\alpha_0} \|\gamma(u)\| = O_P(1). \tag{33}$$

Using this, the fact φ_2^0 is bounded and arguing as for (18), we have

$$|C_{43}^0| \le \|n^{-1} \sum_{t=2}^n A_{1,t} y_{t-2}\| \sup_{\alpha_0 \le u \le 1-\alpha_0} \|\gamma(u)\| = O_P(\frac{1}{\sqrt{n}}) = o_P(1).$$

Next, consider C_{42}^0. Let $\psi_j^0(y) := \varphi_j^0(F(y))$, $a_0 = F^{-1}(\alpha_0)$, $a_1 = F^{-1}(1-\alpha_0)$, $\tilde{\gamma}_y = \gamma(F(y))$, $y \in \mathbb{R}$, $\zeta_n := \sup_{1 \le t \le n, a_0 \le y \le a_1} \frac{1}{\sqrt{n}} |y_{t-2}'\tilde{\gamma}_y|$. The stationarity of the time series, $E\|y_0\|^2 < \infty$ and (33) imply that $\zeta_n = o_P(1)$, and $n^{-1}\sum_{t=2}^n |A_{1,t}|\|y_{t-2}\| = O_P(1)$. Hence,

$$\begin{aligned} |C_{42}^0| \le& \frac{1}{\sqrt{n}} \sum_{t=2}^n |A_{1,t}| \\ &\times \int \big|F(y + \frac{1}{\sqrt{n}} y_{t-2}'\tilde{\gamma}_y) - F(y) - \frac{1}{\sqrt{n}} y_{t-2}'\tilde{\gamma}_y f(y)\big| d\psi_2^0(y) \\ \le& n^{-1} \sum_{t=2}^n |A_{1,t}|\|y_{t-2}\| \sup_{|y-x| \le \zeta_n} |f(y) - f(x)| = o_P(1). \end{aligned}$$

Next, we treat the C_{41}^0 term. Define, for $s \in \mathbb{R}^{p+1}$, y, $a \in \mathbb{R}$,

$$\begin{aligned}
m_{t-1}(y,s,a) &:= [F(y + \frac{1}{\sqrt{n}}(y'_{t-2}s + \|y_{t-2}\|a)) - F(y)] \\
C_{41}^{0\pm}(y,s,a) &:= \frac{1}{\sqrt{n}} \sum_{t=2}^{n} A_{1,t}^{\pm}[I(\varepsilon_{t-1} \le y + \frac{1}{\sqrt{n}}(y'_{t-2}s + \|y_{t-2}\|a)) \\
&\qquad - I(\varepsilon_{t-1} \le y) - m_{t-1}(y,s,a)], \\
C_{41}^0(y,s) &:= C_{41}^{0,+}(y,s) - C_{41}^{0,-}(y,s) \\
&:= \frac{1}{\sqrt{n}} \sum_{t=2}^{n} A_{1,t}[I(\varepsilon_{t-1} \le y + \frac{1}{\sqrt{n}} y'_{t-2}s) \\
&\qquad - I(\varepsilon_{t-1} \le y) - \mu_{t-1}(F(y),s)],
\end{aligned}$$

where $C_{41}^{0,\pm}(y,s) = C_{41}^{0,\pm}(y,s,0)$. On the event $\sup_{\alpha_0 \le u \le 1-\alpha_0} \|\gamma(u)\| \le b$,

$$|C_{41}^0| \le \int \sup_{\|s\| \le b} |C_{41}^{0,+}(y,s)| d\psi_2^0(y) + \int \sup_{\|s\| \le b} |C_{41}^{0,-}(y,s)| d\psi_2^0(y).$$

We shall show that for every $0 < b < \infty$,

$$\int \sup_{\|s\| \le b} |C_{41}^{\pm}(y,s)| d\psi_2^0(y) = o_P(1) \tag{34}$$

which together with (33) will imply $|C_{41}^0| = o_P(1)$.

Let $c = EA_{1,t}^2$. Arguing as before and because f is bounded, for all $y \in \mathbb{R}$, $0 \le b < \infty$, $s \in \mathbb{R}^{p+1}$ with $\|s\| \le b$ and for all $a \in \mathbb{R}$, we obtain

$$\begin{aligned}
E|C_{41}^{0\pm}(y,s,a)|^2 &= E(A_{1,2}^{\pm})^2 E\big[I(\varepsilon_1 \le y + \frac{1}{\sqrt{n}}(y'_0 s + \|y_0\|a)) - I(\varepsilon_1 \le y) \\
&\qquad - F(y + \frac{1}{\sqrt{n}}(y'_0 s + \|y_0\|a)) + F(y)\big]^2 \\
&\le c\, E|F(y + \frac{1}{\sqrt{n}}(y'_0 s + \|y_0\|a)) - F(y)| \\
&\le C\, \frac{1}{\sqrt{n}}(b+a)E\|y_0\|.
\end{aligned} \tag{35}$$

Similarly, and for any s, $t \in \mathbb{R}^{p+1}$, and y, $a \in \mathbb{R}$,

$$\begin{aligned}
E|C_{41}^{0\pm}(y,t,a) - C_{41}^{0\pm}(y,s,a)|^2 &\le c\,\|t-s\|\, \frac{1}{\sqrt{n}}\, E\|y_0\|, \\
E|C_{41}^{0\pm}(y,t,a) - C_{41}^{0\pm}(y,t,0)|^2 &\le c\,|a|\, \frac{1}{\sqrt{n}}\, E\|y_0\|.
\end{aligned} \tag{36}$$

Since the ball $\{\|s\| \le b\|\}$ is compact, there is an $\eta > 0$ and a finite number of points $s_1, \cdots, s_k$ in the unit ball such that for any $\|s\| \le b$, there is an s_j with $\|s_j\| \leq b$, $\|s - s_j\| \leq \eta$. Now, because,

$$\sup_{\|s\|\leq b} |C_{41}^{0\pm}(y,s)| \tag{37}$$
$$\leq \max_{1\leq j\leq k} \sup_{\|s-s_j\|\leq\eta} |C_{41}^{0\pm}(y,s) - C_{41}^{0\pm}(y,s_j)| + \max_{1\leq j\leq k} |C_{41}^{0\pm}(y,s_j)|;$$

$\|s - s_j\| \leq \eta$ implies that for all $y \in \mathbb{R}, 1 \leq j \leq k$, $n \geq 1$, $1 \leq t \leq n$,

$$y_{t-2}'s_j - \|y_{t-2}\|\eta \leq y_{t-2}'s \leq y_{t-2}'s_j + \|y_{t-2}\|\eta.$$

This, the monotonicity of the indicators, and nonnegativity of $A_{1,t}^{\pm}$'s, imply

$$|C_{41}^{0\pm}(y,s) - C_{41}^{0\pm}(y,s_j)|$$
$$\leq |C_{41}^{0\pm}(y,s_j,\eta) - C_{41}^{0\pm}(y,s_j,0)| + |C_{41}^{0\pm}(y,s_j,-\eta) - C_{41}^{0\pm}(y,s_j,0)|$$
$$+2\frac{1}{\sqrt{n}}\{\sum_{t=2}^{n} A_{1,t}^{\pm}[m_{t-1}(y,s_j,\eta) - m_{t-1}(y,s_j,-\eta)]. \tag{38}$$

Moreover, by (36), and the Cauchy-Schwarz inequality,

$$E\big(\max_{1\leq j\leq k} |C_{41}^{0\pm}(y,s_j,\pm\eta) - C_{41}^{0\pm}(y,s_j,0)|\big) \leq C\,k\,\eta^{1/2}\,n^{-1/4}.$$

Because f is bounded, the third term in the upper bound of (38) is bounded above by

$$\max_{1\leq j\leq k} \frac{1}{\sqrt{n}} \sum_{t=2}^{n} |A_{1,t}| \big| [F(y + \frac{1}{\sqrt{n}} y_{t-2}'s_j + \frac{1}{\sqrt{n}}\|y_{t-2}\|\eta)$$
$$-F(y + \frac{1}{\sqrt{n}} y_{t-2}'s_j - \frac{1}{\sqrt{n}}\|y_{t-2}\|\eta)] \big|$$
$$\leq C\eta\, n^{-1} \sum_{t=2}^{n} |A_{1,t}| \|y_{t-2}\| = O_P(\eta).$$

Finally, by (35) applied with $a = 0$,

$$\int E\big(\max_{1\leq j\leq k} |C_{41}^{0\pm}(y,s_j)|\big) d\psi_2^0(y) \leq d_2^0 \sup_y E\big(\max_{1\leq j\leq k} |C_{41}^{0\pm}(y,s_j)|\big)$$
$$\leq d_2^0 \sum_{j=1}^{k} E\big|C_{41}^{0\pm}(y,s_j)\big| \leq C\,k\,n^{-1/4},$$

All these bounds together with (35), (37) and (38), yield

$$\int \sup_{\|s\|\leq b} |C_{41}^{0\pm}(y,s)| d\psi_2^0(y) = O_P(k\,\eta^{1/2}\,n^{-1/4}) + O_P(k\,n^{-1/4}) + O_P(\eta)$$
$$= O_P(\eta^{-p}\,n^{-1/4}) \quad \forall\, \eta > 0.$$

Letting $\eta \to 0$ at a suitable rate (such as, for instance, $\eta = n^{-a}$, with $ap < 1/4$), this in turn implies (34), and completes the proof of $C_4^0 = o_P(1)$. Similarly one can prove $C_2^0 = o_P(1)$.

Next, note that for $\|s\| \le b$, $-\|y_{t-2}\|b \le y'_{t-2}s \le \|y_{t-2}\|b$. Therefore,

$$\begin{aligned}
|C_5^0| &:= \Big|\frac{1}{\sqrt{n}}\sum_{t=2}^{n}\int D_t d\varphi_1^0 \int D_{t-1} d\varphi_2^0\Big| \\
&\le \int\int \sup_{\|s\|\le b}\frac{1}{\sqrt{n}}\sum_{t=2}^{n}\{|I(\varepsilon_t \le x + \frac{1}{\sqrt{n}}y'_{t-1}s) - I(\varepsilon_t \le x)| \\
&\quad \times |I(\varepsilon_{t-1} \le y + \frac{1}{\sqrt{n}}y'_{t-2}s) - I(\varepsilon_{t-1} \le y)|\}d\psi_1^0(x)d\psi_2^0(y) \\
&\le \int\int \frac{1}{\sqrt{n}}\sum_{t=2}^{n}\{I(x - \frac{1}{\sqrt{n}}\|y_{t-1}\|b < \varepsilon_t \le x + \frac{1}{\sqrt{n}}\|y_{t-1}\|b) \\
&\quad \times I(y - \frac{1}{\sqrt{n}}\|y_{t-2}\|b < \varepsilon_{t-1} \le y + \frac{1}{\sqrt{n}}\|y_{t-2}\|b)\}d\psi_1^0(x)d\psi_2^0(y).
\end{aligned}$$

Therefore, a conditioning argument, $\|f\|_\infty < \infty$ and that ψ_j^0, $j = 1, 2$ are finite measures on $\mathbb{R}$, imply

$$\begin{aligned}
E|C_5^0| &\le \int\int n^{1/2}E\{|F(x + \frac{1}{\sqrt{n}}\|Y_1\|b) - F(x - \frac{1}{\sqrt{n}}\|Y_1\|b)| \\
&\quad \times I(y - \frac{1}{\sqrt{n}}\|y_0\|b < \varepsilon_1 \le y + \frac{1}{\sqrt{n}}\|y_0\|b)\}d\psi_1^0(x)d\psi_2^0(y) \\
&\le C\int E\{I(y - \frac{1}{\sqrt{n}}\|y_0\|b < \varepsilon_1 \le y + \frac{1}{\sqrt{n}}\|y_0\|b)\|Y_1\|\}d\psi_2^0(y) \\
&\le C\int E^{1/2}|F(y + \frac{1}{\sqrt{n}}\|y_0\|b) - F(y - \frac{1}{\sqrt{n}}\|y_0\|b)|d\psi_2^0(y) \\
&\le C\, n^{-1/4} = o(1).
\end{aligned}$$

To summarize, we have proved the following

$$\begin{aligned}
\frac{1}{\sqrt{n}}\sum_{t=2}^{n}\hat{c}_{n2;1,t}\hat{c}_{n2;2,t-1} &= \frac{1}{\sqrt{n}}\sum_{t=2}^{n}\{\int[I(\varepsilon_t > F^{-1}(u) - (1-u)]d\varphi_1^0(u) \\
&\quad \times \int[I(\varepsilon_{t-1} > F^{-1}(v) - (1-v)]d\varphi_2^0(v)\} + o_P(1).
\end{aligned}$$

Along the same lines as on page 1412 of Hallin and Jurečková (1999), one can show that the remaining cross-product terms are negligible. For example consider

$$T_n := \frac{1}{\sqrt{n}}\sum_{t=2}^{n}\int_0^{\alpha_n}\bar{a}_t d\varphi_1 \int_0^{\alpha_n}\bar{a}_{t-1}d\varphi_2.$$

The Cauchy-Schwarz inequality, the facts that $|u-(1-\hat{a}_t(u)|^2 \leq u+(1-\hat{a}_t(u))$, and $\sum_{t=1}^n (1-\hat{a}_t(u)) = nu$, $\forall\, 0 \leq u \leq 1$, imply

$$\begin{aligned}
&\frac{1}{\sqrt{n}}\sum_{t=2}^{n}\big|[\hat{a}_t(u)-(1-u)][\hat{a}_{t-1}(v)-(1-v)]\big| \\
&\leq \frac{1}{\sqrt{n}}\Big\{\sum_{t=2}^{n}[u-(1-\hat{a}_t(u)]^2\Big\}^{1/2}\Big\{\sum_{t=2}^{n}[v-(1-\hat{a}_{t-1}(v)]^2\Big\}^{1/2} \\
&\leq \frac{1}{\sqrt{n}}\{2n\,u\}^{1/2}\{2n\,v\}^{1/2} = 2n^{1/2}u^{1/2}v^{1/2}.
\end{aligned}$$

Thus,

$$\begin{aligned}
|T_n| &\leq \int_0^{\alpha_n}\int_0^{\alpha_n}\frac{1}{\sqrt{n}}\sum_{t=2}^{n}\big|[\hat{a}_t(u)-(1-u)][\hat{a}_{t-1}(v)-(1-v)]\big|d\varphi_1(u)d\varphi_2(v) \\
&\leq 2n^{1/2}\Big(\int_0^{\alpha_n}u^{-1/2-\delta}du\Big)^2 = O(n^{-1/2+\delta}) = o(1).
\end{aligned}$$

The next one is even easier because $|\int \bar{a}_t d\varphi_j^0| \leq \varphi_j(1-\alpha_0)) - \varphi_j(\alpha_0)$, $j=1,2$, so that

$$\begin{aligned}
\Big|\frac{1}{\sqrt{n}}\sum_{t=2}^{n}\int \bar{a}_t d\varphi_1^0\int_0^{\alpha_n}\bar{a}_{t-1}d\varphi_2^0\Big| &\leq \int_0^{\alpha_n}\frac{1}{\sqrt{n}}\sum_{t=2}^{n}|u-(1-\hat{a}_{t-1}(u)|d\varphi_2 \\
&\leq 2n^{1/2}\int_0^{\alpha_n}u^{-\delta}du = O(n^{-1/2+\delta}).
\end{aligned}$$

Exactly similar arguments can be used to show that the cross-product terms involving the right tail integrals are also negligible. Putting all these conclusions together implies

$$\begin{aligned}
\frac{1}{\sqrt{n}}\sum_{t=2}^{n}\hat{b}_t\hat{b}_{t-1} &= \frac{1}{\sqrt{n}}\sum_{t=2}^{n}\int[I(\varepsilon_t > F^{-1}(u)-(1-u)]d\varphi_1(u) \\
&\quad\times\int[I(\varepsilon_{t-1} > F^{-1}(v)-(1-v)]d\varphi_2(v) + o_P(1) \\
&= \frac{1}{\sqrt{n}}\sum_{t=2}^{n}\varphi_1(F(\varepsilon_t))\,\varphi_2(F(\varepsilon_{t-1})) + o_P(1).
\end{aligned}$$

This proves the left hand side of the claim (8) while the right hand side of the claim follows from Hallin and Puri (1988) and Hallin and Werker (1998). □

Remark 1. Here we indicate the proof when not all φ's are centered at the origin. For example consider the case $m=2$. Assume that $\mu_1 :=$

$\int \varphi_1(u)du = 0$ and $\int \varphi_2(u)du =: \mu_2 \neq 0$. Put $\varphi_2^0 = \varphi_2 - \mu_2$. Let $\bar{a}_t(u) := \hat{a}_t(u) - (1-u)$ and note that integration by parts and $\mu_1 = 0$ yields that

$$\int \varphi_1(u)d\hat{a}_t(u) \int \varphi_2^0(v)d\hat{a}_{t-1}(v) = \int \varphi_1(u)d\bar{a}_t(u) \int \varphi_2^0(v)d\bar{a}_{t-1}(v)$$
$$= \int \bar{a}_t(u)d\varphi_1(u) \int \bar{a}_{t-1}(v)d\varphi_2(v), \quad \forall\, t.$$

Hence $\tilde{S}_{n,\varphi_1\varphi_2^0} = \tilde{S}_{n,\varphi_1\varphi_2}$, and $\tilde{S}_{n,\varphi_1\varphi_2} = \tilde{S}_{n,\varphi_1\varphi_2^0} = S_{n,\varphi_1\varphi_2^0}(\theta) + o_P(n^{1/2})$, by Proposition 1. But,

$$\frac{1}{\sqrt{n}} S_{n,\varphi_1\varphi_2^0}(\theta) = \frac{1}{\sqrt{n}} \sum_{t=1}^{n} \varphi_1\Big(\frac{R_t(\theta)}{n+1}\Big)\Big[\varphi_2\Big(\frac{R_{t-1}(\theta)}{n+1}\Big) - m_2\Big]$$
$$= \frac{1}{\sqrt{n}} S_{\varphi_1\varphi_2}(\theta) - n^{1/2} m_2 [n^{-1} \sum_{i=1}^{n} \varphi_1(i/(n+1))].$$

Since φ_1 is square integrable, the difference between the Rieman sum $n^{-1}\sum_{i=1}^{n} \varphi_1(i/(n+1)$ and its limit $\int \varphi_1(u)du = 0$ is $o(n^{-1/2})$, so that centering φ_2 has asymptotically negligible influence on $n^{-1/2}S_{n\varphi_1\varphi_2}$, and the conclusion of Proposition 1 continues to hold.

Acknowledgements

This research was supported in part by a P.A.I. contract of the Belgian Federal Government and an Action de Recherche Concertée of the Communauté française de Belgique, the Czech Republic Grant 201/05/2340, and the FNRS of the Communauté française de Belgique.

References

1. Akharif, A. and M. Hallin (2003). Efficient detection of random coefficients in AR(p) models. *Ann. of Statist.* **31**, 675-704.

2. Allal, J. and S. El Melhaoui (2005). Optimal detection of exponential components in autoregressive models, submitted.

3. Benghabrit, Y. and M. Hallin (1992). Optimal rank-based tests against first-order super diagonal bilinear dependence. *Journal of Statistical Planning and Inference* **32**, 45-61.

4. Bentarzi, M. and M. Hallin (1996). Locally optimal tests against periodic autoregression : parametric and nonparametric approaches. *Econometric Theory* **12**, 88-112.

5. CHERNOFF, H. AND SAVAGE, I. R. (1958). Asymptotic normality and efficiency of certain nonparametric tests. *Ann. Math. Statist.* **29**, 972-994.

6. GAREL, B. AND M. HALLIN (1999). Rank-based AR order identification. *Journal of the American Statistical Association* **94**,1357-1371.

7. GUTENBRUNNER, C. AND J. JUREČKOVÁ (1992). Regression rank scores and regression quantiles. *Ann. Statist.* **20**,305-330.

8. GUTENBRUNNER, C., J. JUREČKOVÁ, R. KOENKER AND S. PORTNOY (1993). Tests of linear hypotheses based on regression rank scores. *J. Nonparametric Statist.* **2**, 307-331.

9. HALLIN, M. (1994). On the Pitman non-admissibility of correlogram based methods. *J. Time Ser. Anal.* **15**, 607-612.

10. Hallin, M., J.-Fr. Ingenbleek and M.L. Puri. (1985). Linear serial rank tests for randomness against *ARMA* alternatives. *Ann. Statist.* **13**, 1156-1181.

11. HALLIN, M. AND J. JUREČKOVÁ (1999). Optimal tests for autoregressive models based on autoregression rank scores. *Ann. of Statist.* **27**, 1385-1414.

12. HALLIN, M., J. JUREČKOVÁ, J. KALVOVÁ, J. PICEK AND T. ZAHAF (1997). Non-parametric tests in AR-models, with applications to climatic data. *Environmetrics* **8**, 651–660.

13. HALLIN, M. AND M. L. PURI (1988). Optimal rank-based procedures for time-series analysis: testing an *ARMA* model against other *ARMA* models. *Ann. Statist.* **16**, 402-432.

14. HALLIN, M. AND M. L. PURI (1994). Aligned rank tests for linear models with autocorrelated errors. *J. Multiv. Anal.* **50**, 175-237.

15. HALLIN, M. AND B. WERKER. (1998). Optimal testing for semiparametric *AR* models: from Gaussian Lagrange multipliers to autoregression rank scores and adaptive tests. In S. Ghosh, Ed.: *Asymptotics, Nonparametrics and Time Series*, M. Dekker, New York, 295-350.

16. HALLIN, M. AND B. WERKER. (2003). Semiparametric efficiency, distribution-freeness, and invariance. *Bernoulli* **9**, 137-165.

17. JUREČKOVÁ, J. (1971). Asymptotic independence of rank test statistic for testing symmetry on regression. *Sankhyā Ser. A* **33**, 1–18.

18. KALVOVÁ, J., J. JUREČKOVÁ, I. NEMEŠOVÁ AND J. PICEK. (2000). On the order of autoregression (AR) models in temperature series. *Meteorologický časopis* **3**, 19–23.

19. KOENKER, R. AND G. BASSETT (1978). Regression quantiles. *Econometrica* **46**, 33-50.

20. KOUL, H. L. (1970). A class of ADF tests for sub-hypothesis in the multiple linear regression. *Ann. Math. Statist.* **41**, 1273–1281.

21. KOUL, H. L. (2002). *Weighted Empirical Processes in Dynamic Nonlinear Models, Second Edition*, Springer Lecture Notes in Statistics, **166**, New York.

22. Koul, H. L. and A. K. Md. E. Saleh (1995). Autoregression quantiles and related rank scores processes. *Ann. Statist.* **23**, 670-689.

23. Paindaveine, D. (2004). A unified and elementary proof of serial and nonserial, univariate and multivariate, Chernoff-Savage results. *Statistical Methodology* **1**, 81-91.

24. Paindaveine, D. (2005). A Chernoff-Savage result for shape: on the non-admissibility of pseudo-Gaussian methods. To appear in *J. Multiv. Anal.*

Chapter 18

A NEW CONVOLUTION ESTIMATOR FOR NONPARAMETRIC REGRESSION

Baard Støve and Dag Tjøstheim

Department of Mathematics
University of Bergen, Bergen, NORWAY

E-mails: baards@mi.uib.no & dagt@mi.uib.no

We present a convolution smoother for nonparametric regression. Its asymptotic behavior is examined, and its asymptotic total squared error is found to be smaller than standard kernel estimators, such as Nadaraya-Watson and local linear regression. Results based on some simulation studies are given, including a comparison with a fourth order kernel. Asymptotic normality for the proposed estimator is proved.

Key words: Convolution; Kernel function; Mean squared errors; Nonparametric estimation.

1 Introduction

There are many nonparametric estimators of the conditional mean $\mathrm{E}(Y|X=x)$ for independent identically distributed observations (X_i, Y_i), $i = 1, ..., n$, with a joint density $f(\cdot,\cdot)$, and a marginal density $f(\cdot)$ of X_i. The three most common are the local polynomial estimator; see Stone (1977), Cleveland (1979), Müller (1987), and Fan (1992) the Nadaraya-Watson estimator; see Nadaraya (1964) and Watson (1964) and the Gasser-Müller estimator; see Gasser and Müller (1979). In the case

$$Y_i = m(X_i) + \epsilon_i, \tag{1}$$

where $\{X_i\}$ and $\{\epsilon_i\}$ consist of i.i.d. zero-mean random variables with $\{\epsilon_i\}$ independent of $\{X_i\}$, $\mathrm{E}(Y|X=x) = m(x)$. Neither of the three above-mentioned estimators require a regression relationship like (1) to work, and one might think that if one was able to construct an estimator of $m(x)$ making explicit use of the extra information extra information contained in (1), then possibly one could improve on the standard nonparametric regression estimators. This is the basic idea of this paper, and it leads to what we have called "the convolution estimator". We will show that this new estimator generally has a smaller asymptotic total squared error and also that in a number of finite sample experiments it gives better results, although in many cases these improvements are not dramatic.

Before we define the new estimator, let us briefly mention that several authors have proposed adjustments and improvements of the kernel estimators. Both the Gasser-Müller and Nadaraya-Watson estimators have a large order bias when estimating a curve in its boundary region. Thus the idea of boundary kernels, which are weight functions that are used only within the boundary region, were introduced and studied by Gasser and Müller (1979) and Gasser et al. (1985). Another approach has been the reflection method; see Schuster (1985) and Hall and Werly (1991). In the papers Hjort and Glad (1995), Efron and Tibshirani (1996), and Glad (1998), the possibility of parametrically guided nonparametric density and regression estimation are examined. Several authors have studied the use of higher order kernels to improve the asymptotic bias; see e.g. Marron and Wand (1992) for a quantification of the practical gain, in density estimation.

A brief summary of the paper is as follows: In Section 2 the estimator is introduced, and its asymptotic behavior is examined and discussed in Sections 3, 4 and 5. In Section 6 some simulation results are given. Section 7 introduces a variant of the new estimator, and finally, Section 8 gives some concluding remarks.

2 The estimator

The regression function of interest is

$$m(x) = \mathrm{E}(Y|X = x) = \int y f(y|x) \mathrm{d}y. \tag{2}$$

Under the assumption that the equation (1) holds, $f(y|x)$ can be written

$$f(y|x) = f_\epsilon(y - m(x)),$$

where f_ϵ is the density of ϵ. Inserting this into (2) gives the convolution integral equation

$$m(x) = \int y f_\epsilon(y - m(x)) \mathrm{d}y, \tag{3}$$

where both $m(x)$ and f_ϵ are unknown. However, $m(x)$ may be replaced by a standard estimator $\tilde{m}(x)$, e.g. the local linear estimator, and f_ϵ can be estimated using a kernel estimate of $f_{\hat{\epsilon}}$ with $\hat{\epsilon}_i = Y_i - \tilde{m}(X_i)$. Based on the relation (3), the proposed estimator is

$$\hat{m}(x)=\int y\hat{f}_{\hat{\epsilon}}\big(y-\tilde{m}(x)\big)\mathrm{d}y$$
$$=n^{-1}\sum_{i=1}^{n}\int yK_{h_D}\big(y-\tilde{m}(x)-Y_i+\tilde{m}(X_i)\big)\mathrm{d}y, \qquad (4)$$

where $\tilde{m}(x)$ is, as mentioned, a nonparametric regression estimator and $K_{h_D}(\cdot)=K(\cdot/h_D)/h_D$, where K is a kernel function. We have chosen the local linear estimator, that is, the local polynomial estimator of degree 1, with bandwidth h_R and kernel function $K^L(\cdot)$,

$$\tilde{m}(x)=n^{-1}\sum_{i=1}^{n}\frac{\{\hat{s}_2(x)-\hat{s}_1(x)(x-X_i)\}K_{h_R}^L(x-X_i)Y_i}{\hat{s}_2(x)\hat{s}_0(x)-\hat{s}_1(x)^2},$$

where

$$\hat{s}_r(x)=n^{-1}\sum_{i=1}^{n}(x-X_i)^rK_{h_R}^L(x-X_i),\quad r=0,1,2,$$

and $K_{h_R}^L(\cdot)=K^L(\cdot/h_R)/h_R$, see e.g. Fan and Gijbels (1996). The expression $\hat{f}_{\hat{\epsilon}}\big(y-\tilde{m}(x)\big)$ in (4) is the kernel density estimator with bandwidth equal to h_D,

$$\hat{f}_{\hat{\epsilon}}\big(y-\tilde{m}(x)\big)=\frac{1}{nh_D}\sum_{i=1}^{n}K\Big(\frac{y-\tilde{m}(x)-\hat{\epsilon}_i}{h_D}\Big)$$

with $\hat{\epsilon}_i=Y_i-\tilde{m}(X_i)$, see e.g. Wand and Jones (1995) page 11. Observe that the new estimator is computationally more demanding than standard methods.

It is also possible to iterate the estimator (4) using a previous estimate of $m(x)$ as input for the next iteration. Set $\tilde{m}_0$ equal to the local linear estimator for the regression curve. Then the convolution estimator is

$$\hat{m}_1(x)=\int y\hat{f}_{\hat{\epsilon}^0}\big(y-\tilde{m}_0(x)\big)\mathrm{d}y,$$

where $\hat{\epsilon}_i^0=Y_i-\tilde{m}_0(X_i)$. Iterating further gives, for $j=1,2,...$

$$\hat{m}_{j+1}(x)=\int y\hat{f}_{\hat{\epsilon}^j}\big(y-\hat{m}_j(x)\big)\mathrm{d}y, \qquad (5)$$

where again $\hat{\epsilon}_i^j=Y_i-\hat{m}_j(X_i)$. Note that in this estimator x can only be equal to the observed x_i for $i=1,...,n$, because we update $\hat{\epsilon}_i^j$ at each iteration. One would perhaps believe that this "iterated convolution estimator" will give better results than the convolution estimator. However, this is not the case, unless one uses a special type of kernel function in the estimation of $\hat{f}_{\hat{\epsilon}^j}$. This special kernel is introduced, and some simulation results are given in Section 7.

3 Intuitive discussion of bias reduction

Asymptotic analysis of nonparametric estimators is usually based on the asymptotic bias and variance of the estimator. It is well known that the Gasser-Müller estimator has an asymptotic variance 1.5 times that of the Nadaraya-Watson estimator, but its asymptotic bias is superior; see Mack and Müller (1989) and Chu and Marron (1991). The local polynomial estimator of order one and higher has been examined by several authors and has been found to have better properties than the above mentioned estimators. In particular, it provides automatic boundary bias correction; see Fan (1994), Fan and Gijbels (1992), Fan (1993), and Hastie and Loader (1993). For a more complete discussion of the different estimators and comparisons, see the books Wand and Jones (1995), Simonoff (1996), Fan and Gijbels (1996), Fan and Yao (2003) and references therein.

We now discuss the asymptotic properties of the estimator (4). In the sequel, the bandwidth h refers to both h_D and h_R, since most of the time we assume that these two bandwidths are equal. Standard conditions on the kernels, the random variables and the regression function are assumed to be fulfilled, see e.g. Wand and Jones (1995) page 120.

The relation (4) can be written as

$$\hat{m}(x) = \sum_{i=1}^{n} \int y \frac{1}{nh_D} K\left(\frac{y - \tilde{m}(x) - \hat{\epsilon}_i}{h_D}\right) dy. \tag{6}$$

By a simple substitution and using assumptions on $K(\cdot)$, (6) gives

$$\hat{m}(x) = \tilde{m}(x) + \frac{1}{n} \sum_{i=1}^{n} \hat{\epsilon}_i. \tag{7}$$

Further,

$$\begin{aligned} \hat{\epsilon}_i = Y_i - \tilde{m}(X_i) &= Y_i - m(X_i) + m(X_i) - \tilde{m}(X_i) \\ &= \epsilon_i + m(X_i) - \tilde{m}(X_i). \end{aligned} \tag{8}$$

Substituting (8) in (7), we obtain

$$\hat{m}(x) - m(x) = \tilde{m}(x) - m(x) - \frac{1}{n} \sum_{i=1}^{n} [\tilde{m}(X_i) - m(X_i)] + \frac{1}{n} \sum_{i=1}^{n} \epsilon_i. \tag{9}$$

From this relation it is possible to find the asymptotic bias of $\hat{m}(x)$.

Let us recall the asymptotic bias formula for the local linear estimator (e.g. Wand and Jones (1995) page 124). With a slight abuse of notation,

$$\text{As.Bias}\big(\tilde{m}(x)\big) = \mathrm{E}\big(\tilde{m}(x) - m(x)\big) \sim \frac{h^2}{2} m''(x) \int z^2 K(z) \mathrm{d}z. \tag{10}$$

Since

$$\begin{aligned}\mathrm{E}\big(\tilde{m}(X_i)-m(X_i)\big) &= \mathrm{E}[\mathrm{E}\big(\tilde{m}(X_i)\big)-m(X_i)|X_i] \\ &\sim \mathrm{E}\big(\frac{h^2}{2}m''(X_i)\int z^2K(z)\mathrm{d}z\big),\end{aligned}$$

we obtain

$$\mathrm{E}\big(\frac{1}{n}\sum_{i=1}^{n}[\tilde{m}(X_i)-m(X_i)]\big) \sim \frac{h^2}{2}\int z^2K(z)\mathrm{d}z\int m''(y)f(y)\mathrm{d}y.$$

Further,

$$\mathrm{E}\big(\frac{1}{n}\sum_{i=1}^{n}\epsilon_i\big)=0.$$

Hence the asymptotic bias of $\hat{m}(x)$ is

$$\begin{aligned}\text{As.Bias}\big(\hat{m}(x)\big)&\sim\text{As.Bias}\big(\tilde{m}(x)\big)-\frac{h^2}{2}\int z^2K(z)\mathrm{d}z\int m''(y)f(y)\mathrm{d}y \\ &=\frac{h^2}{2}\int z^2K(z)\mathrm{d}z\Big[\int\big(m''(x)-m''(y)f(y)\big)\mathrm{d}y\Big]. \qquad (11)\end{aligned}$$

Let us consider the following special cases:

(1) $m''(x)$ = constant (i.e. $m(x)=a+bx+cx^2$). The bias of $\hat{m}(x)$ is of higher order and improvement of the bias can be expected.
(2) $m''(x)=0$ (linear case). The bias is of higher order, both for $\hat{m}(x)$ and $\tilde{m}(x)$, and it is uncertain whether improvement is obtained.
(3) x close to maximum and minimum values (peaks and valleys). If $\hat{m}(x)-m(x)$ has one maximum or minimum, even though the bias correction in (11) is x-independent, visually the reduction will be larger at this point (cf. Figure 2, which is explained in more detail in Section 6).

Observe, however, that if one has a curve with several peaks and valleys it may be difficult to gain any bias reduction. This is because the integral $\int m''(y)f(y)\mathrm{d}y$ can be equal to zero in this case.

As mentioned before, performing iterations of the estimator in equation (5) will not give any improved bias effect. In this case, the equation (7) will be

$$\hat{m}_{j+1}(x)=\hat{m}_j(x)+\frac{1}{n}\sum_{i=1}^{n}\hat{\epsilon}_i^j.$$

For $j=1$, the same argument as above gives

$$\text{As.Bias}\big(\hat{m}_2(x)\big)=\text{As.Bias}\big(\hat{m}_1(x)\big),$$

and further

$$\text{As.Bias}\big(\hat{m}_{j+1}(x)\big) = \text{As.Bias}\big(\hat{m}_j(x)\big).$$

However, we introduce a special kernel in Section 7, such that a bias reduction may occur at each iteration.

Another possible improvement, suggested by a referee, is that instead of (7) one could think about localized corrections where the average over all residuals is replaced by a locally weighted average of residuals in the neighborhood of x only. This could alleviate some of the disadvantages that are associated with the current global adjustment, and should be a part of further research.

4 Distributional properties

By (9)

$$\hat{m}(x) - \mathrm{E}\big(\hat{m}(x)\big) = \tilde{m}(x) - \mathrm{E}\big(\tilde{m}(x)\big) - \Big[\frac{1}{n}\sum_{i=1}^{n}\big(\tilde{m}(X_i) - m(X_i)\big) - \mathrm{E}\Big[\sum_{i=1}^{n}\big(\tilde{m}(X_i) - m(X_i)\big)\Big]\Big] + \frac{1}{n}\sum_{i=1}^{n}\epsilon_i. \tag{12}$$

We have

$$\tilde{m}(x) - \mathrm{E}\big(\tilde{m}(x)\big) = O_p\Big(\frac{1}{\sqrt{nh}}\Big)$$

and

$$\frac{1}{n}\sum_{i=1}^{n}\epsilon_i = O_p\Big(\frac{1}{\sqrt{n}}\Big).$$

If we can show that the convergence in probability of the remaining terms in (12) is of higher order, then the asymptotic distribution of $\hat{m}(x) - \mathrm{E}(\hat{m}(x))$ is the same as $\tilde{m}(x) - \mathrm{E}(\tilde{m}(x))$, but with a different asymptotic bias.

Although this is not necessary, we do our formal calculations as if all expectations exist. Let us consider again

$$\mathrm{E}\Big[\frac{1}{n}\sum_{i=1}^{n}\big(\tilde{m}(X_i) - m(X_i)\big)\Big]^2 = \mathrm{E}\Big[\frac{1}{n^2}\sum_{i=1}^{n}\big(\tilde{m}(X_i) - m(X_i)\big)^2\Big] + \mathrm{E}\Big[\frac{1}{n^2}\sum_{\substack{i,j=1\\ i\neq j}}^{n}\Big\{\big(\tilde{m}(X_i) - m(X_i)\big) \times \big(\tilde{m}(X_j) - m(X_j)\big)\Big\}\Big]. \tag{13}$$

Further

$$\mathrm{E}\big(\tilde{m}(X_i) - m(X_i)\big)^2 = \mathrm{E}\big[\mathrm{E}\big(\tilde{m}(X_i) - m(X_i)|X_i\big)^2\big] = o(1).$$

Thus

$$\mathrm{E}\Big[\frac{1}{n^2}\sum_{i=1}^{n}\big(\tilde{m}(X_i)-m(X_i)\big)^2\Big]=o(n^{-1}).$$

Let us examine the second term in (13). By independence,

$$\begin{aligned}
&\mathrm{E}\big[\big(\tilde{m}(X_i)-m(X_i)\big)\big(\tilde{m}(X_j)-m(X_j)\big)\big]\\
&\quad=\mathrm{E}\Big[\mathrm{E}\big[\big(\tilde{m}(X_i)-m(X_i)\big)\big(\tilde{m}(X_j)-m(X_j)\big)|X_i,X_j\big]\Big]\\
&\quad\sim\int\mathrm{E}\big[\big(\tilde{m}(x)-m(x)\big)\big(\tilde{m}(y)-m(y)\big)\big]f(x)f(y)\mathrm{d}x\mathrm{d}y. \qquad (14)
\end{aligned}$$

The expectation in the above integral satisfies

$$\begin{aligned}
&\mathrm{E}\big[\big(\tilde{m}(x)-m(x)\big)\big(\tilde{m}(y)-m(y)\big)\big]\\
&\quad=\mathrm{E}\big[\tilde{m}(x)\tilde{m}(y)\big]-m(x)\mathrm{E}\big[\tilde{m}(y)-m(y)\big]\\
&\quad\quad-m(y)\mathrm{E}\big[\tilde{m}(x)-m(x)\big]-m(x)m(y). \qquad (15)
\end{aligned}$$

Let us examine the term $\mathrm{E}[\tilde{m}(x)\tilde{m}(y)]$. By a conditioning argument and by independence, we obtain

$$\begin{aligned}
\mathrm{E}[\tilde{m}(x)\tilde{m}(y)]=&\mathrm{E}\Big[n^{-2}\sum_{i=1}^{n}\frac{\{\hat{s}_2(x)-\hat{s}_1(x)(x-X_i)\}K^L_{h_R}(x-X_i)Y_i}{\hat{s}_2(x)\hat{s}_0(x)-\hat{s}_1(x)^2}\\
&\times\sum_{j=1}^{n}\frac{\{\hat{s}_2(y)-\hat{s}_1(y)(y-X_j)\}K^L_{h_R}(y-X_j)Y_j}{\hat{s}_2(y)\hat{s}_0(y)-\hat{s}_1(y)^2}\Big]\\
\sim&\frac{n-1}{n^2}\mathrm{E}\big(\tilde{m}(x)\big)\mathrm{E}\big(\tilde{m}(y)\big)\\
&+\mathrm{E}\Big[n^{-2}\sum_{i=1}^{n}\frac{Y_i^2\{\hat{s}_2(x)-\hat{s}_1(x)(x-X_i)\}K^L_{h_R}(x-X_i)}{[\hat{s}_2(x)\hat{s}_0(x)-\hat{s}_1(x)^2][\hat{s}_2(y)\hat{s}_0(y)-\hat{s}_1(y)^2]}\\
&\times K^L_{h_R}(y-X_i)\{\hat{s}_2(y)-\hat{s}_1(y)(y-X_i)\}\Big]. \qquad (16)
\end{aligned}$$

From Wand and Jones (1995) p. 123, asymptotically

$$\hat{s}_l(x)\sim\begin{cases}h^l\int z^lK^L(z)\mathrm{d}zf(x)+o_P(h^l) & l\text{ even,}\\ h^{l+1}\int z^{l+1}K^L(z)\mathrm{d}zf'(x)+o_P(h^{l+1}) & l\text{ odd.}\end{cases}$$

Therefore the order of the denominator in (16) is

$$h^4\Big[\int z^2K^L(z)\mathrm{d}z\Big]^2f^2(x)f^2(y)+o_P(h^4). \qquad (17)$$

The only term contributing to the last term of the numerator in (16) is

$$\frac{1}{n^2h^2}\hat{s}_2(x)\hat{s}_2(y)\mathrm{E}\Big[\sum_{i=1}^{n}K^L\big(\frac{x-X_i}{h}\big)K^L\big(\frac{y-X_i}{h}\big)Y_i^2\Big],$$

since all the other terms are of higher order.

Further, by a simple substitution, Taylor expansion, using the equation (17) and since

$$\hat{s}_2(x)\hat{s}_2(y) = h^4\Big[\int z^2K^L(z)\mathrm{d}z\Big]^2 f(x)f(y) + o_P(h^4),$$

the last term in (16) becomes asymptotically,

$$\begin{aligned}
&\frac{1}{n^2h^2f(x)f(y)}\mathrm{E}\Big[\sum_{i=1}^{n}K^L\Big(\frac{x-X_i}{h}\Big)K^L\Big(\frac{y-X_i}{h}\Big)Y_i^2\Big]\\
&\quad= \frac{1}{nhf(x)f(y)}\int v^2K^L(z)K^L\Big(\frac{zh+y-x}{h}\Big)f_{X,Y}(x-zh,v)\mathrm{d}z\mathrm{d}v\\
&\quad= \frac{1}{nhf(y)}K_2^L\Big(\frac{y-x}{h}\Big)\mathrm{E}(Y^2|X=x),
\end{aligned}$$

where $K_2^L(w) = \int K(z)K(z+w)\mathrm{d}z$.

Writing

$$\begin{aligned}
\mathrm{E}\big(\tilde{m}(x)\big)\mathrm{E}\big(\tilde{m}(y)\big) =& m(x)m(y) + m(x)\mathrm{E}[\tilde{m}(y) - m(y)]\\
&+m(y)\mathrm{E}[\tilde{m}(x) - m(x)] + \Big\{\mathrm{E}[\tilde{m}(y) - m(y)]\\
&\times \mathrm{E}[\tilde{m}(x) - m(x)]\Big\},
\end{aligned}$$

the equation (15) becomes

$$\begin{aligned}
&\mathrm{E}\big[\big(\tilde{m}(x) - m(x)\big)\big(\tilde{m}(y) - m(y)\big)\big]\\
&\qquad\sim \Big(1-\frac{1}{n}\Big)\mathrm{E}\big[\tilde{m}(y) - m(y)\big]\mathrm{E}\big[\tilde{m}(x) - m(x)\big]\\
&\qquad\quad+\frac{1}{nhf(y)}K_2^L\Big(\frac{y-x}{h}\Big)\mathrm{E}(Y^2|X=x).
\end{aligned}$$

Inserting this in (14) gives

$$\begin{aligned}
&\mathrm{E}\big[\big(\tilde{m}(X_i) - m(X_i)\big)\big(\tilde{m}(X_j) - m(X_j)\big)\big]\\
&\quad\sim \Big(1-\frac{1}{n}\Big)\int \mathrm{E}[\tilde{m}(y) - m(y)]\mathrm{E}[\tilde{m}(x) - m(x)]f(x)f(y)\mathrm{d}x\mathrm{d}y\\
&\qquad+\frac{1}{nh}\int K_2^L\Big(\frac{y-x}{h}\Big)\mathrm{E}(Y^2|X=x)f(x)\mathrm{d}x\mathrm{d}y\\
&\quad= \Big(1-\frac{1}{n}\Big)\Big(\frac{h^2}{2}\Big)^2\Big[\int z^2K_2^L(z)\mathrm{d}z\Big]^2\Big[\int m''(x)f(x)\mathrm{d}x\Big]^2\\
&\qquad+\frac{1}{n}\int K_2^L(z)\mathrm{d}z\int \mathrm{E}(Y^2|X=x)f(x)\mathrm{d}x. \qquad (18)
\end{aligned}$$

Clearly the first term in (18) can be identified with the squared expectation in the decomposition

$$\mathrm{E}\Big[\frac{1}{n}\sum_{i=1}^{n}\big(\tilde{m}(X_i)-m(X_i)\big)\Big]^2 = \mathrm{Var}\Big[\frac{1}{n}\sum_{i=1}^{n}\big(\tilde{m}(X_i)-m(X_i)\big)\Big] + \Big[\mathrm{E}\Big(\frac{1}{n}\sum_{i=1}^{n}[\tilde{m}(X_i)-m(X_i)]\Big)\Big]^2. \quad (19)$$

The term

$$\frac{1}{n}\int K_2^L(z)\mathrm{d}z\int \mathrm{E}(Y^2|X=x)f(x)\mathrm{d}x = \frac{1}{n}\int K_2^L(z)\mathrm{d}z[\sigma_\epsilon^2+\int m^2(x)f(x)\mathrm{d}x] \quad (20)$$

can be identified with the variance. We have

$$\frac{1}{n}\sum_{i=1}^{n}[\tilde{m}(X_i)-m(X_i)]-\mathrm{E}\Big(\frac{1}{n}\sum_{i=1}^{n}\tilde{m}(X_i)-m(X_i)\Big)\sim o_P\Big(\frac{1}{\sqrt{n}}\Big).$$

From the equation (12), it follows that $\hat{m}(x)-\mathrm{E}(\hat{m}(x))$ has the same asymptotic normal distribution as $\tilde{m}(x)-\mathrm{E}(\tilde{m}(x))$, i.e., the asymptotic variance is the same for the estimators $\hat{m}(x)$ and $\tilde{m}(x)$, but

$$\text{As.Bias}\big(\hat{m}(x)\big)=\text{As.Bias}\big(\tilde{m}(x)\big)-\frac{h^2}{2}\int z^2K(z)\mathrm{d}z\int m''(y)f(y)\mathrm{d}y.$$

5 Total squared error

We would like to compare the asymptotic total squared error, i.e.

$$\mathrm{E}\Big[\sum_{i=1}^{n}\big(\hat{m}(X_i)-m(X_i)\big)^2\Big] \text{ against } \mathrm{E}\Big[\sum_{i=1}^{n}\big(\tilde{m}(X_i)-m(X_i)\big)^2\Big].$$

From equation (9)

$$\hat{m}(X_i)-m(X_i)=\tilde{m}(X_i)-m(X_i)-\frac{1}{n}\sum_{j=1}^{n}[\tilde{m}(X_j)-m(X_j)]+\frac{1}{n}\sum_{j=1}^{n}\epsilon_j. \quad (21)$$

Further,

$$
\begin{aligned}
&[\hat{m}(X_i) - m(X_i)]^2 \\
&\quad = [\tilde{m}(X_i) - m(X_i)]^2 - \frac{2}{n}[\tilde{m}(X_i) - m(X_i)]\sum_{j=1}^{n}[\tilde{m}(X_j) - m(X_j)] \\
&\quad + \frac{1}{n^2}\Big[\sum_{j=1}^{n}\big(\tilde{m}(X_j) - m(X_j)\big)\Big]^2 + \frac{2}{n}[\tilde{m}(X_i) - m(X_i)]\sum_{j=1}^{n}\epsilon_j \\
&\quad - \frac{2}{n^2}\sum_{j=1}^{n}[\tilde{m}(X_j) - m(X_j)]\sum_{k=1}^{n}\epsilon_k + \frac{1}{n^2}\sum_{j=1}^{n}\sum_{k=1}^{n}\epsilon_j\epsilon_k. \qquad (22)
\end{aligned}
$$

This implies

$$
\begin{aligned}
\frac{1}{n}\sum_{i=1}^{n}[\hat{m}(X_i) - m(X_i)]^2 = &\frac{1}{n}\sum_{i=1}^{n}[\tilde{m}(X_i) - m(X_i)]^2 \qquad (23) \\
&- \frac{1}{n^2}\Big[\sum_{i=1}^{n}\big(\tilde{m}(X_i) - m(X_i)\big)\Big]^2 + \frac{1}{n^2}\sum_{j=1}^{n}\sum_{k=1}^{n}\epsilon_j\epsilon_k.
\end{aligned}
$$

Taking expectation in (23) gives us the total squared error of $\hat{m}(x)$. Thus the order of the different terms is

$$
\begin{aligned}
\mathrm{E}\Big[\frac{1}{n}\sum_{i=1}^{n}[\tilde{m}(X_i) - m(X_i)]^2\Big] \sim &\int \Big[\mathrm{Var}\big(\tilde{m}(x)\big) + \mathrm{Bias}^2\big(\tilde{m}(x)\big)\Big] f(x)\mathrm{d}x \\
= &O\Big(\frac{1}{nh} + h^4\Big),
\end{aligned}
$$

from the decomposition (19), and the calculated bias and variance

$$
\mathrm{E}\Big[\frac{1}{n^2}\Big[\sum_{i=1}^{n}\big(\tilde{m}(X_i) - m(X_i)\big)\Big]^2\Big] = O(h^4)
$$

and finally

$$
\mathrm{E}\Big[\frac{1}{n^2}\sum_{j=1}^{n}\sum_{k=1}^{n}\epsilon_j\epsilon_k\Big] = \frac{\sigma_\epsilon^2}{n}.
$$

This means that if $\mathrm{E}\big[\frac{1}{n}\sum_{i=1}^{n}\big(\tilde{m}(X_i) - m(X_i)\big)\big] \neq 0$ asymptotically, then

$$
\mathrm{E}\Big[\sum_{i=1}^{n}\big(\hat{m}(X_i) - m(X_i)\big)^2\Big] < \mathrm{E}\Big[\sum_{i=1}^{n}\big(\tilde{m}(X_i) - m(X_i)\big)^2\Big]
$$

i.e., the total asymptotic squared error of $\hat{m}(x)$ is smaller than $\tilde{m}(x)$.

6 Simulation study

We compare the estimator (4) with the local linear estimator in several situations. The comparisons are based on the mean squared error (MSE) of the estimators. For $\hat{m}(x)$ the MSE, if it exists, is

$$\text{MSE}(\hat{m}(x)) = \text{E}\big[\{\hat{m}(x) - m(x)\}^2\big].$$

In the simulation study, we use the empirical mean squared error

$$\hat{\text{MSE}}(\hat{m}) = \frac{1}{n}\sum_{i=1}^{n}\{\hat{m}(X_i) - m(X_i)\}^2, \tag{24}$$

with similar expressions for the local linear estimator.

We simulated between 100 and 500 realizations of sample size 100 to 10 000 of (X_i, Y_i) and calculated the empirical MSE (24). The bandwidth choice in the kernel density estimation for f_ϵ is the Solve-the-Equation Plug-in approach proposed in Sheater and Jones (1991), while the bandwidth used in the local linear estimator is the Direct Plug-In methodology described in Ruppert et al. (1995) , if nothing else is stated. Gaussian kernels are always used, both in the regression estimation of $m(x)$ and the density estimation of f_ϵ. The integration in the estimator (4) is calculated using the trapezoidal rule between $[2\min(Y_i), 2\max(Y_i)]$, when $\min(Y_i) < 0$ and $\max(Y_i) > 0$. We consider the case where observations of (X_i, Y_i) are independent.

Our first simulation experiment was based on the model $Y_i = X_i^2 + \epsilon_i$, where ϵ_i are i.i.d normal with expectation 0 and variance 0.1 and X_i is uniformly distributed on $[-2, 2]$. A hundred realizations each with sample size 500 were simulated, and the convolution and local linear estimators were used to estimate the regression curves. In this case, the estimated MSE for the local linear estimator is: $2.301 \cdot 10^{-3}$ and for the convolution estimator: $1.986 \cdot 10^{-3}$. Thus we obtain an improvement of 13.690%. Figure 1 shows the estimated variance and bias of the two estimators. The Figure 1(a) 1(a) displays the estimated variance; here there is no difference between the two estimators. Figure 1(b) shows the estimated bias. The dashed line is the estimated bias for the convolution estimator. This is clearly smaller than for the local linear estimator; thus our predictions that the improvement occurs in the bias is supported. The results were similar for the other simulation experiments. See Table 1 with sample size equal to 100, 1000 and 5000. The improvement is also

The next simulation was based on the same model, except that the interval is now $[-0.5, 0.5]$. Only one realization with 500 sample points was simulated, and the estimated lines are given in Figure 2. The solid line is the true function, the non-filled points are the local linear estimates and the

Table 1 The estimated MSE for a parabola using the convolution estimator

Sample size	Local linear	Convolution type	Improvement in %
100	$9.086 \cdot 10^{-3}$	$7.997 \cdot 10^{-3}$	11.985
500	$2.301 \cdot 10^{-3}$	$1.986 \cdot 10^{-3}$	13.690
1000	$1.321 \cdot 10^{-3}$	$1.120 \cdot 10^{-3}$	15.216
5000	$3.489 \cdot 10^{-4}$	$2.957 \cdot 10^{-4}$	15.248

Table 2 The estimated MSE for a parabola using the convolution estimator and the Nadaraya-Watson estimator with a fourth order kernel

Sample size	Nadaraya-Watson	Convolution type	Improvement in %
100	$2.448 \cdot 10^{-2}$	$1.002 \cdot 10^{-2}$	59.069
500	$5.839 \cdot 10^{-3}$	$2.175 \cdot 10^{-3}$	62.750
1000	$3.396 \cdot 10^{-3}$	$1.243 \cdot 10^{-3}$	63.398
5000	$1.020 \cdot 10^{-3}$	$2.978 \cdot 10^{-4}$	70.804

black points are the convolution estimates. These results clearly indicate that the asymptotic bias formula in equation (11) is reasonable. For each estimated local linear point, the estimated convolution point is below by a fixed amount, but the visual impression is that the convolution estimator does much better at the bottom points of the parabola.

As mentioned in the introduction, a common bias reduction technique in nonparametric estimation is the use of higher order kernels [see e.g. Wand and Jones (1995), page 32]. Thus we have included a comparison between the Nadaraya-Watson estimator with a fourth-order kernel and the proposed convolution estimator. The fourth-order kernel used is $K_4(x) = 0.5(3 - x^2)\phi(x)$, where $\phi(x)$ is the standard normal distribution. The bandwidth used in the Nadaraya-Watson estimator is the same as for the above local linear estimator. It is thus not optimal in this situation, but other choices of the bandwidth were examined without a large impact on the results. Again we performed simulations for the parabola model on the interval $[-2, 2]$. The results from the simulations are given in Table 2. The convolution estimator clearly outperforms the fourth order kernel method when comparing the MSE. Figure 3 may explain these results. Here the bias and variance of both estimator are plotted for the simulations with sample size 500. The bias of both estimators seems to be reasonably equal. But the variance is much larger for the fourth order kernel method , the solid line, than the variance for the convolution estimator, the dashed line. This behavior of the fourth-order kernel method is not unexpected [see e.g. Simonoff (1996) page 60].

The last simulation experiment in this section was based on a straight line regression $Y_i = a + bX_i + \epsilon_i$, with $a = 1$, $b = 1$, ϵ as before, and

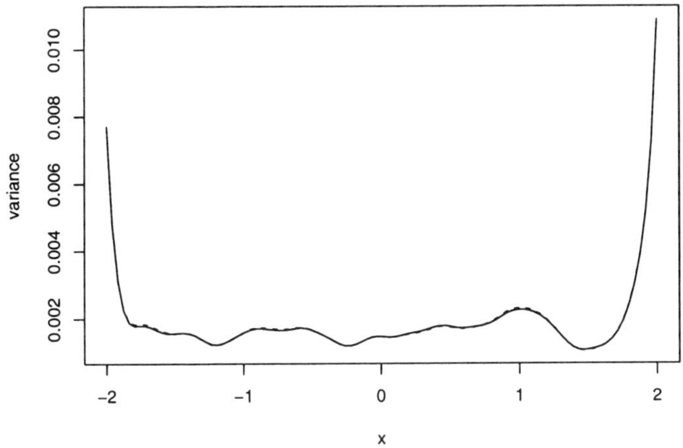

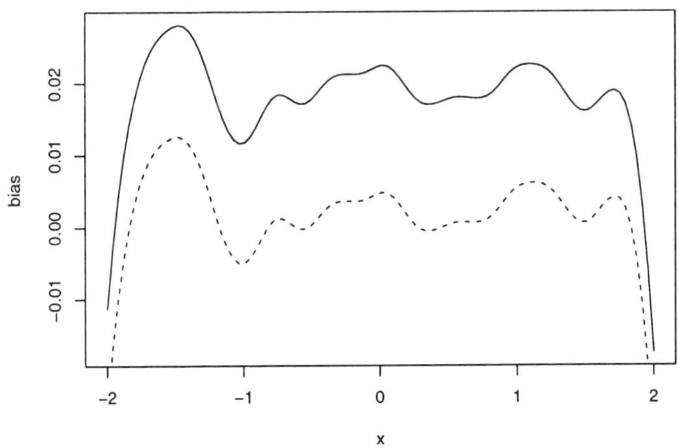

Figure 1 The estimated variance (top – (a)) and bias (below – (b)) for the parabola experiment (dashed line - convolution estimator, solid line - local linear estimator)

X_i uniformly distributed on $[0, 2]$. From this model, 100 realizations of sample size 100 to 5000 were simulated. The integration in the estimator was now performed on the interval $[-2\max(Y_i), 2\max(Y_i)]$. The results, given in Table 3 for the convolution estimator, indicate that the convolution estimator is almost as good as the local linear estimator. We cannot expect the convolution estimator to do better here, since $m''(x)$ is zero if $m(x)$ is

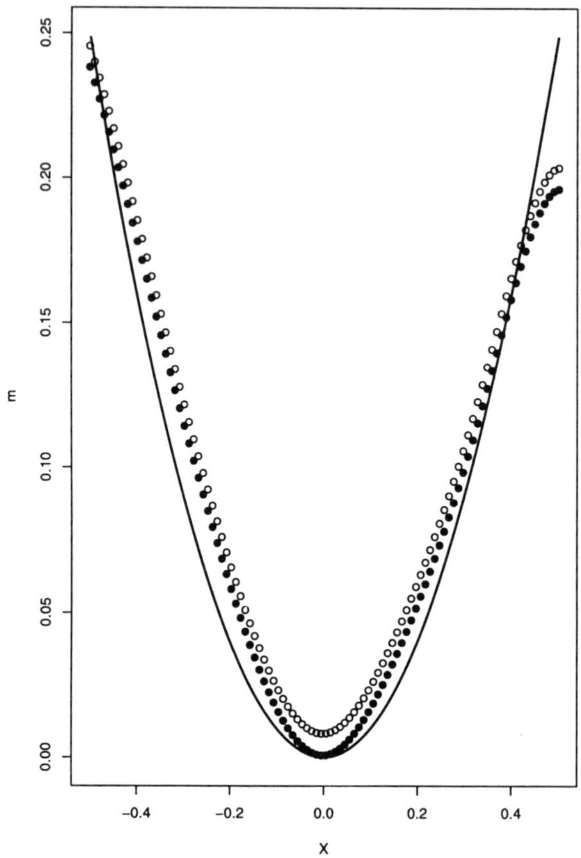

Figure 2 The solid true line, the estimated local linear points (non-filled) and the estimated convolution points (black) from one realization for the parabola model

a straight line, thus no bias improvement occurs in the formula (11).

7 A special kernel variant

Let us consider equation (4) again. After a substitution, we obtain

$$\hat{m}(x) = \int y \hat{f}_{\hat{\epsilon}}(y - \tilde{m}(x))\mathrm{d}y = \int z \hat{f}_{\hat{\epsilon}}(z)\mathrm{d}z + \tilde{m}(x) \int \hat{f}_{\hat{\epsilon}}(z)\mathrm{d}z.$$

If $\int \hat{f}_{\hat{\epsilon}}(z)\mathrm{d}z > 1$, the estimator $\hat{m}(x)$ will clearly be closer to the true function than $\tilde{m}(x)$ in locations where the function has a large curvature

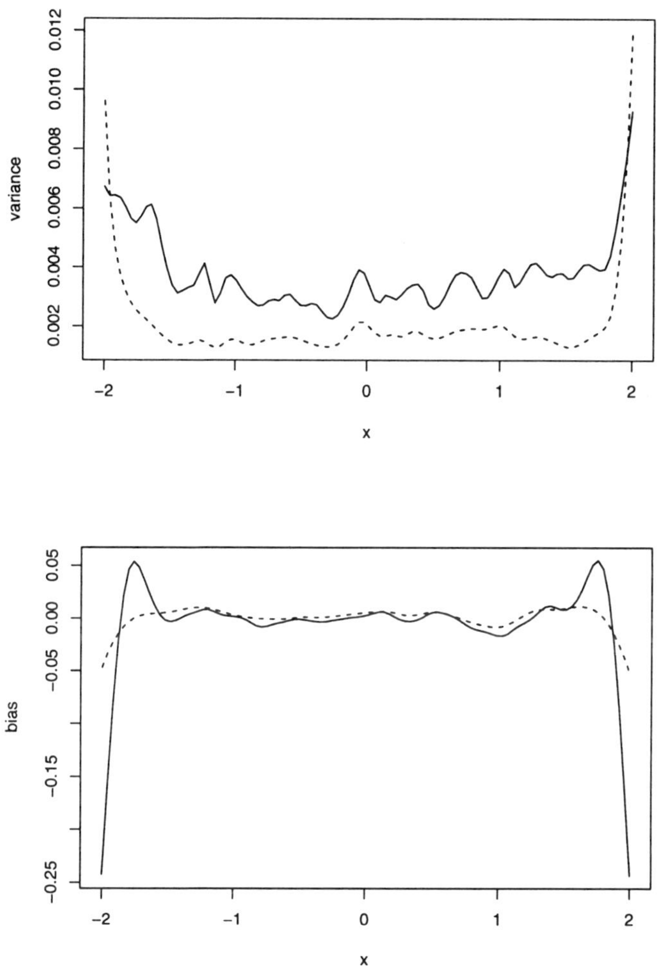

Figure 3 The estimated variance (top) and bias for the parabola experiment (dashed line - convolution estimator, solid line - Nadaraya-Watson estimator with fourth order kernel)

due to the bias formula (10). Of course one could adjust with an estimate $\tilde{m}''(x)$ of $m''(x)$, but this increases the variance. Instead we have chosen to introduce a kernel function with the property

$$\int K(z)\mathrm{d}z > 1. \tag{25}$$

Table 3 The estimated MSE for a straight line using the convolution estimator

Sample size	Local linear	Convolution type	Improvement in %
100	$5.675 \cdot 10^{-3}$	$5.653 \cdot 10^{-3}$	0.388
500	$1.240 \cdot 10^{-3}$	$1.246 \cdot 10^{-3}$	-0.484
1000	$5.627 \cdot 10^{-4}$	$5.630 \cdot 10^{-4}$	-0.053
5000	$1.009 \cdot 10^{-4}$	$1.098 \cdot 10^{-4}$	-8.821

Table 4 The estimated MSE for $m_1(x)$

Sample size	Local linear	Convolution type	Improvement in %
100	$1.442 \cdot 10^{-2}$	$1.445 \cdot 10^{-2}$	-0.208
500	$4.071 \cdot 10^{-3}$	$4.049 \cdot 10^{-3}$	0.540
1000	$2.382 \cdot 10^{-3}$	$2.374 \cdot 10^{-3}$	0.336
5000	$6.766 \cdot 10^{-4}$	$6.723 \cdot 10^{-4}$	0.636
10 000	$4.028 \cdot 10^{-4}$	$3.988 \cdot 10^{-4}$	0.993

This could be considered as an alternative to allowing the kernel to be negative, which is a known device for reducing bias, as seen in section 6 for the higher order kernel used there.

In this case, it may pay to perform iterations, as equation 5 suggests. However, $\hat{m}(x)$ will also be larger than $\tilde{m}(x)$ in absolute value in locations where $m''(x) \approx 0$, and this is not desirable. The following simulation experiments should therefore be considered just as a part of a preliminary investigation where at least some promising results are obtained, but where more work is needed to find a more optimal procedure. In these experiments we have chosen the kernel K such that $\int \hat{f}_{\hat{\epsilon}}(z)\mathrm{d}z = 1.001$, a very modest overestimation indeed, and clearly other choices can be examined.

Two regression functions have been studied: from Härdle (1990) chapter 5, $m_1(x) = \sin^3(2\pi x^3)$ and from Prewitt (2003) $m_2(x) = \sin(2x) + 2\exp(-16x^2)$. See Figure 4 for these curves. The observations were generated by simulating X_i as uniformly distributed on an interval $[0,1]$ for $m_1(x)$ and $[-2,2]$ for $m_2(x)$. The response observations have been generated through $Y_i = m(X_i) + \epsilon_i$ where ϵ_i are i.i.d normal with expectation 0 and variance 0.1. Here 100 realizations of sample size 100 to 10 000 of (X_i, Y_i) have been simulated. Since both functions have at least one peak and one valley, we may not expect to get much bias reduction. The estimated MSE, using the convolution estimator with the adjusted kernel the adjusted kernel (25), is shown in Table 4 expected, the results show only a very modest improvement. The results for m_2 were similar.

In Table 5 and 6, the iterated convolution estimator (5) with the adjustment (25) has been used with $i = 10$ iterations, for the regression of $m_1(x)$ and $m_2(x)$. Again 100 realizations were different sample size, and the MSE has been estimated. Here the normal reference bandwidth selector (see e.g.

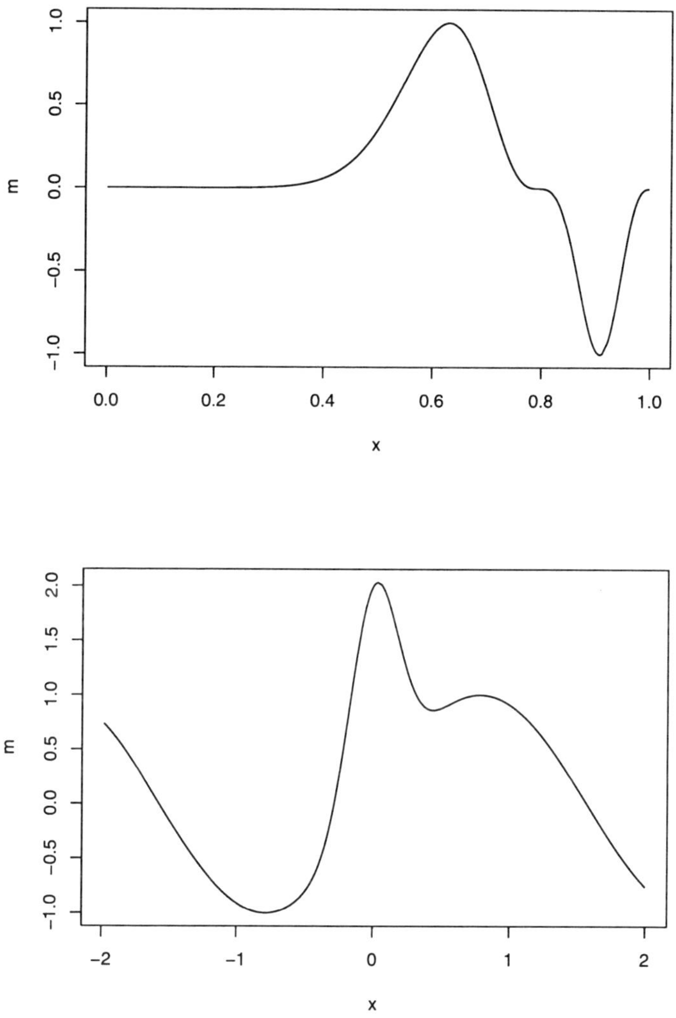

Figure 4 The curve $m_1(x)$ at the top, $m_2(x)$ at the bottom

Härdle (1990), page 91) has been used for the selection of the bandwidth in the density estimation.

Both tables show that the iterated convolution type estimator performes better than the local linear estimator. The results also indicate that performing iterations improves the first order convolution estimator $\hat{m}_1(x)$. Clearly, one cannot improve the estimates indefinitely. The improvement

Table 5 The estimated MSE for $m_1(x)$, using the iteration estimator $\hat{m}_i(x)$ with $i = 10$

Sample size	Local linear	Iterated convolution type	Improvement in %
100	$1.438 \cdot 10^{-2}$	$1.434 \cdot 10^{-2}$	0.278
500	$4.292 \cdot 10^{-3}$	$4.235 \cdot 10^{-3}$	1.328
1000	$2.464 \cdot 10^{-3}$	$2.387 \cdot 10^{-3}$	3.125
5000	$6.648 \cdot 10^{-4}$	$6.319 \cdot 10^{-4}$	4.949

Table 6 The estimated MSE for $m_2(x)$, using the iteration estimator $\hat{m}_i(x)$ with $i = 10$

Sample size	Local linear	Iterated convolution type	Improvement in %
100	$1.773 \cdot 10^{-2}$	$1.741 \cdot 10^{-2}$	1.801
500	$4.669 \cdot 10^{-3}$	$4.531 \cdot 10^{-3}$	2.956
1000	$2.696 \cdot 10^{-3}$	$2.576 \cdot 10^{-3}$	4.451
5000	$7.245 \cdot 10^{-4}$	$6.887 \cdot 10^{-4}$	4.941

Table 7 The estimated MSE for $m_1(x)$ on the interval $[0.55, 0.7]$ using the iterated convolution estimator with $i = 10$

Sample size	Local linear	Iterated convolution type	Improvement in %
100	$1.373 \cdot 10^{-2}$	$1.373 \cdot 10^{-2}$	0
500	$3.273 \cdot 10^{-3}$	$3.127 \cdot 10^{-3}$	4.461
1000	$2.241 \cdot 10^{-3}$	$2.058 \cdot 10^{-3}$	8.166
5000	$4.766 \cdot 10^{-4}$	$4.205 \cdot 10^{-4}$	11.771

will be smaller and smaller, and at the same time there will be more and more higher order terms which may lead to trouble unless n is increased. It is important to carry out more detailed calculations and to find a good stopping criterion. Possibly some cross-validation type criterion can be used, but this is left for future research.

The point of the adjustment (25) is to improve results in peaks and valleys. To check this, 100 simulations with sample sizes from 100 to 5000 were performed for $m_1(x)$ and $m_2(x)$. However, the MSE has been calculated only for specific intervals which contain the peaks and valleys of the curves.

Tables 7 and 8 show the results from simulations of $m_1(x)$ using the iterated convolution estimator, with $i = 10$ and adjusted with (25). The intervals considered are $[0.55, 0.7]$, where the curve has a peak, and $[0.85, 0.95]$, which is a valley. The curve does not have a very large curvature on the interval $[0.55, 0.7]$, thus the results in the 7 show a modest improvement. However, Table 8 8 shows that the new estimator is much better when parts of the function with high curvature. Similar results were obtained for $m_2(x)$.

The above results show that the proposed estimator is better for parts of functions with high curvature, and in some cases this improvement is substantial. This is in contrast to the modest improvement we obtained

Table 8 The estimated MSE for $m_1(x)$ on the interval $[0.85, 0.95]$ using the iterated convolution estimator with $i = 10$

Sample size	Local linear	Iterated convolution type	Improvement in %
100	$3.621 \cdot 10^{-2}$	$3.414 \cdot 10^{-2}$	5.717
500	$1.005 \cdot 10^{-2}$	$8.802 \cdot 10^{-3}$	12.418
1000	$5.740 \cdot 10^{-3}$	$4.849 \cdot 10^{-3}$	15.523
5000	$1.653 \cdot 10^{-3}$	$1.229 \cdot 10^{-3}$	25.650

when we compared the MSE of the whole curve.

A real data set with peaks and valleys was also examined for completeness. We used the motorcycle data set from Härdle (1989 page 70) where the X-values represent time after a simulated impact with motorcycles and the response variable Y is the head acceleration of a post mortem human test object. Figure 5 shows the results, together with the data points. The upper graphs show the result from the local linear estimator (solid estimator (solid line) and the estimator in equation (4) (dashed line) with the adjusted kernel (25). These two approximately the same result. However, the lower graphs are different. Here the dashed line is the iteration estimator from (5), using 50 iterations and adjusted kernel, the solid line is as above. Again the convolution estimator is lower in valleys and higher in peaks compared to the local linear estimator.

8 Concluding remarks

This paper introduced a convolution estimator for nonparametric regression. Its asymptotic total squared error was proved to be smaller than standard kernel methods. The bias reduction will be large in cases where the function of interest has only one maximum (i.e. a peak) or one minimum (i.e. a valley).

Since the convolution estimator has two bandwidths, their choice is important. This has not been studied in this paper, and one might believe that the bias reduction can be larger if one is able to choose more optimal bandwidths.

An adjusted kernel has also been introduced and simulation results indicate that by using this kernel, even more bias reduction can be achieved. However, more theoretical analysis is needed here.

Acknowledgments

This work is supported by Grant 147231/432 from the Norwegian Research Council.

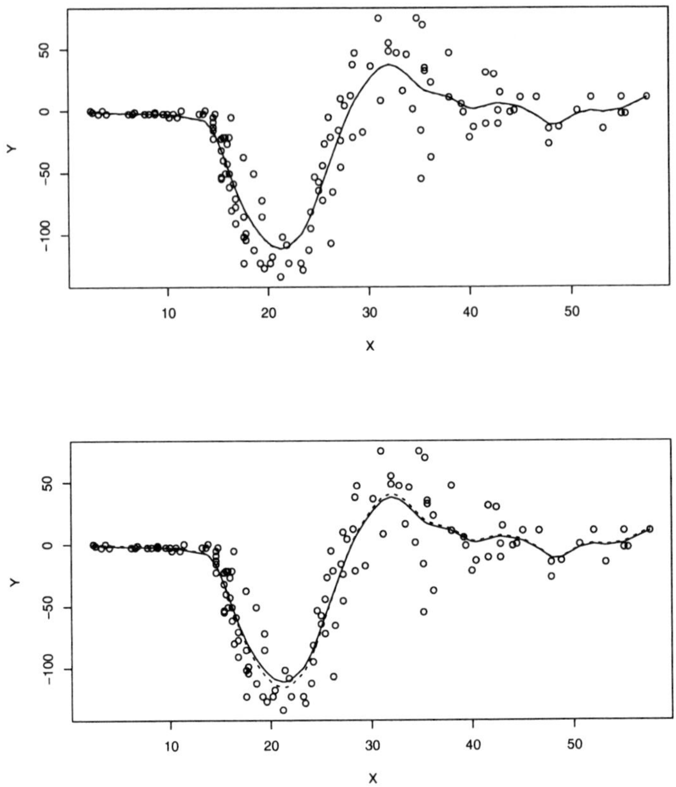

Figure 5 Nonparametric regression of the motorcycle data set: The data is given as points in both plots. Upper plot: solid line - local linear estimate, dashed line - convolution estimator with special kernel. Lower plot: solid line - local linear estimate, dashed line - iterated convolution estimator with special kernel (50 iterations)

References

1. Chu, C. K. and Marron, J. S (1991). Choosing a kernel regression estimator (with discussion). *Statist. Science* **6** 404-436.

2. Cleveland, W. S. (1979). Robust locally weighted regression and smoothing scatterplots. *J. Amer. Statist. Assoc.* **74** 829-836.

3. Efron, B. and Tibshirani, R. (1996). Using specially designed exponential families for density estimation. *Ann. Statist.* **24** 2431-2461.

4. Fan, J. (1992). Design-adaptive nonparametric regression. *J. Amer. Statist. Assoc.* **87** 998-1004.

5. FAN, J. (1993). Local linear regression smoothers and their minimax efficiency. *Ann. Statist.* **21** 196-216.
6. FAN, J. AND GIJBELS, I. (1992). Variable bandwidth and local linear regression smoothers. *Ann. Statist.* **20** 2008-2036.
7. FAN, J. AND GIJBELS, I. (1996). *Local Polynomial Modelling and Its Applications.* Chapman & Hall.
8. FAN, J. AND YAO, Q. (2003). *Nonlinear Time Series.* Springer-Verlag.
9. GASSER, T. AND MÜLLER, H. G. (1979). Kernel Estimation of Regression Functions. In *Smoothing Techniques for Curve Estimation* (T. Gasser and M. Rosenblatt, eds.) Springer-Verlag, Heidelberg, 23-68.
10. GASSER, T., MÜLLER, H.-G. AND MAMMITZSCH, V. (1985). Kernels for nonparametric curve estimation. *J. Roy. Statist. Soc. Series B* **47** 238-252.
11. GLAD, I. K. (1998). Parametrically guided non-parametric regression. *Scand. J. Statist.* **25** 649-668.
12. HALL, P. AND WEHRLY, T. E. (1991). A geometrical method for removing edge effects from kernel-type nonparametric regression estimators. *J. Amer. Statist. Assoc.* **86** 665-672.
13. HÄRDLE, W. (1989). *Applied Nonparametric Regression.* Cambridge University Press.
14. HÄRDLE, W. (1990). *Smoothing Techniques: With Implementation in S.* Springer-Verlag.
15. HASTIE, T. AND LOADER, C. (1993). Local regression: Automatic kernel carpentry (with comments). *Statist. Science* **8** 120-143.
16. HJORT, N. L. AND GLAD, I. K. (1995). Nonparametric density estimation with a parametric start. *Ann. Statist.* **23** 882-904.
17. MACK, Y. P. AND MÜLLER, H. G. (1989). Convolution type estimators for nonparametric regression. *Statist. Probab. Lett.* **7** 229-239.
18. MARRON, J. S. AND WAND, M. P. (1992). Exact mean integrated squared error. *Ann. Statist.* **20** 712-736.
19. MÜLLER, H. G. (1987). Weigthed local regression and kernel methods for nonparametric curve fitting. *J. Amer. Statist. Assoc.* **82** 231-238.
20. NADARAYA, E. A. (1964). On estimating regression. *Theo. Probab. and Applic.* **9** 141-142.
21. PREWITT, K. A. (2003). Efficient bandwidth selection in non-parametric regression. *Scand. J. of Statist.* **30** 75-92.
22. RUPPERT, D., SHEATHER, S. J. AND WAND, M. P. (1995). An effective bandwidth selector for local least squares regression. *J. Amer. Statist. Assoc.* **90** 1257-1270.
23. SCHUSTER, E. F. (1985). Incorporating support constraints into nonparametric estimates of densities. *Comm. Statist. - Theo. Meth.* **14** 1123-1126.

24. Sheather, S. J. and Jones, M. C. (1991). A reliable data-based bandwidth selection method for kernel density estimation. *J. Roy. Statist. Soc. Series B* **53** 683-690.

25. Simonoff, J. S. (1996). *Smoothing Methods in Statistics.* Springer-Verlag.

26. Stone, C. J. (1977). Consistent Nonparametric Regression. *Ann. Statist.* **5** 596-620.

27. Wand, M. P. and Jones, M. C. (1995). *Kernel Smoothing.* Chapman & Hall.

28. Watson, G. S. (1964). Smooth Regression Analysis. *Sankhyā Ser. A* **26** 359-372.

PART 6
Clustering and Mixture Models

Chapter 19

SEMIPARAMETRIC AND NONPARAMETRIC GENE MAPPING

Fei Zou, Brian S. Yandell and Jason P. Fine

Department of Biostatistics
The University of North Carolina, Chapel Hill, NC, U.S.A

Departments of Statistics and Horticulture
University of Wisconsin, Madison, WI, U.S.A

Departments of Statistics and Biostatistics & Medical Informatics
University of Wisconsin, Madison, WI, U.S.A

E-mails: fzou@bios.unc.edu, byandell@wisc.edu & fine@stat.wisc.edu

We review gene mapping, or inference for quantitative trait loci, in the context of recent research in semi-parametric and non-parametric inference for mixture models. Gene mapping studies the relationship between a phenotypic trait and inherited genotype. Semi-parametric gene mapping using the exponential tilt covers most standard exponential families and improves estimation of genetic effects. Non-parametric gene mapping, including a generalized Hodges-Lehmann shift estimator and Kaplan-Meier survival curve, provide a general framework for model selection for the influence of genotype on phenotype. Examples and summaries of reported simulations show the power of these methods when data are far from normal.

Keywords: Statistical genetics; Empirical process; Exponential tilt; Mixture model.

1 Introduction

Gene mapping concerns the statistical relationship between a phenotype, or measured response known as a trait, and the genotype, or heritable information measured at genetic markers scattered across the genome. Genetic information is incomplete, requiring consideration of mixture models across unknown genotypes. While gene mapping was initially developed for normally distribution of traits, the framework extends readily to both semi-parametric and non-parametric models.

Commonly, individuals in a gene mapping study are sampled from an experimental cross such as a backcross or intercross. First, two inbred lines (A and B, say) are crossed to create the F1, which is heterogeneous everywhere. That is, at any selected genetic marker, the inbred parents are AA and BB, respectively, while the F1 is always AB. An F1 back-crossed to an inbred line, say A, produces backcross offspring that are either homozygous (AA) or heterozygous (AB) at every marker, with equal likelihood. The intercross, or F2, results from brother-sister mating of F1 children, yielding marker genotypes AA:AB:BB in an idealized 1:2:1 ratio. The backcross or intercross individuals are genetic mosaics of their inbred grandparents, due to meioses in the F1 parent(s). Other inbred experimental crosses are possible but are not considered further here (see Kao and Zeng 1997).

Each individual in a sample from an experimental cross is genetically unique. The different genetic patterns scored at markers spread across the genome allow us to associate the phenotype with genomic regions, or quantitative trait loci (QTL), where differences in genotype are inferred to affect the phenotype. QTL have great importance in revealing the genetic basis of phenotypic differences (Belknap *et al.*, 1997; Haston et al., 2002; Wang *et al.*, 2003). In plant and laboratory animals, backcross or F2 individuals are widely used for mapping quantitative traits (see Lynch and Walsh 1998).

The basic model selection problems for QTL mapping are: (i) detecting the presence of one or more QTLs, (ii) estimating QTL map position(s), and (iii) estimating the genetic effects of the QTLs. This model selection process is often referred to as inferring the genetic architecture (Mackay 2001). Complications arise due to lack of genotype data between genetic markers, leading to a likelihood based on a mixture of distributions across the possible QTL genotypes. Initially, Weller (1986), and later Lander and Botstein (1989), assumed the phenotype distribution given the genotype is normal. A general framework was sketched by Jansen (1992) and others.

The basic problem involves relating observed genetic marker information, m, to observed phenotypic trait measurements, y through two coupled models,

$$\mathrm{pr}(y|m,\lambda) = \sum_q \mathrm{pr}(y|q)\mathrm{pr}(q|m,\lambda),$$

with the sum over all possible genotypes, q, at the putative QTL(s), λ. In this paper, we allow the phenotype model, $\mathrm{pr}(y|q)$, to be semi-parametric (exponential tilt, including many generalized linear models) or fully non-parametric. The recombination model, $\mathrm{pr}(q|m,\lambda)$, can be directly calculated using the binomial based on markers, m, that flank the QTL, λ, and plays the role here of mixture weight (Kao and Zeng 1997).

The first gene mapping study involved single marker t-tests (Sax 1923), which was essentially the standard until the introduction of interval mapping (Lander and Botstein 1989; Haley and Knott 1992; Kruglyak and Lander 1995; see Doerge et al. 1997). The normal mixture, with a normal phenotype distribution, is the default in the widely used software Mapmaker/QTL (Lander et al. 1987), QTL/Cartographer (Basten et al. 1995) and R/qtl (Broman et al. 2003).

Nettleton (Nettleton and Praestgaard 1998; Nettleton 1999; Nettleton 2002) considered hypothesis testing for QTL against ordered alternatives, assuming an underlying normal model. Several investigators studied other, non-normal, parametric phenotype models, including binomial and threshold models (Visscher et al. 1996; Xu and Atchley 1996; McIntyre et al. 2000; Rao and Li 2000; Yi and Xu 2000; Broman 2003), Poisson (Mackay and Fry 1996; Shepel *et al.* 1998), negative binomial (Lan et al. 2001). Hackett and Weller (1995) considered ordinal threshold models. Broman (2003) proposed a two-part parametric model for phenotype with a spike at one value, including structural zeroes and type I censoring. Parametric Cox proportional hazard model with a specified baseline function was examined by Diao et al. (2004).

Inference on the QTL map position(s) is fairly robust to normality. However, model misspecification may lead to reduced power to detect genes affecting a trait or to biased estimates of the genetic architecture (Hackett 1997; Wright and Kong 1997). Further, genetic differences may involve more than a mean shift, as modeled for normal data. Perhaps the phenotype has a different shaped distribution for individuals with different genotypes, as opposed to a difference in the means or center of location? While these issues have been widely studied with single QTL models, there has been little work on more complex multigene models. One might expect that naively using normal models with highly non-normal data might cause greater difficulty in this set-up, where inferences about subtle gene-gene interactions may be misleading. Therefore it is useful to consider semi-parametric and non-parametric generalizations for QTL, providing more robust inference about the genetic architecture, including insight about possible parametric models for the phenotype given the genotype.

Semi-parametric QTL were first considered by Zou and coauthors (Zou et al. 2000; Zou and Fine 2002; Jin et al. 2003) using the exponential tilt. Lange and Whittaker (2001) investigated QTL using generalized estimating equations; however, GEE may be biased for the mixture model necessary for QTL. Symons et al. (2002) and Epstein et al. (2003) considered a semi-parametric Cox proportional hazards model and a Tobit model, respectively, for gene mapping with censored survival data.

Kruglyak and Lander (1995) proposed model-free tests using Wilcoxon rank statistics for a backcross, where there are two genotypes. Broman (2003) considered an omnibus generalization of the Wilcoxon test for the intercross. Poole and Drinkwater (1996) used the Jonckheere-Terpstra generalization of the Wilcoxon test to ordered alternatives for the intercross. Hoff et al. (2002) considered stochastic ordering with respect to genotype as an alternative to no QTL. Zou et al. (2003) and Fine et al. (2004) provided non-parametric estimators that generalize the Hodges-Lehman shift and the Kaplan-Meier survival curve to mixture models.

In this chapter, we present semi-parametric models for QTL in Section 2, and non-parametric inference applied to QTL model selection problems in Section 3. An example on tumour counts of rats is used to illustrate both semi-parametric and non-parametric inference for QTL.

2 Semi-parametric Models

It is well known that statistical methods work best when they use all available information, and in particular here, knowledge about the exact form of the phenotype model. In the best cases, this arises from extensive knowledge from previous studies and an understanding of the underlying mechanism. This ideally focuses attention on a few key parameters, such as the center (mean) and spread (variance) in a population of individuals with identical genotype. However, in many cases, a suitable parametric form is not known. We consider here semi-parametric models that encompass most common parametric models, allowing us to separate the question of model form from detection of QTL.

In the best situation, a researcher believes from previous research that a particular parametric model, such as binomial, is suitable. For instance, Poisson is often appropriate for counts of instances of some event, such as the number of offspring, while binomial is pertinent for proportions, such as germination success or disease resistance. Concentrations often follow a log-normal distribution. Generalizations that allow dispersion may be appropriate in other situations. Caution is in order if a model choice is made on the basis of raw phenotype data, as part of the histogram shape may be due to genetic variation in the sample.

When considering a model, there are three primary options: (1) just use the normal and hope it is satisfactory; (2) build a method streamlined to the 'correct' phenotype model; (3) find a transformation that makes the normal model more tenable. Instead, we propose using semi-parametric models, leaving validation of parametric form to a later investigation by the researcher once the genetics is better understood.

2.1 *Exponential tilt models*

A natural choice for the phenotype model is a common shape that is slightly modified by genotype through an 'exponential tilt':

$$\mathrm{pr}(Y = y|q, \theta) = f(y)\gamma(y|q, \beta)$$

with $\theta = (\beta, f)$, $\log(\gamma)$ a low-order polynomial tilt function that is usually linear or quadratic in y, β a vector with unknown polynomial coefficients and f an unknown density. Note that $\mathrm{pr}(y|q, \theta)$ must be a density for every genotype q, which places some technical constraints on β. If we estimate f with 'point mass' at the observed phenotypes for a sample of n individuals, these constraints become

$$\sum_{i=1}^{n} f(y_i)\gamma(y_i|q, \beta) = 1$$

regardless of the genotype q.

A test for QTL with this semi-parametric phenotype model is simply a test that $\beta = 0$ while leaving the shape of f unspecified. Many parametric models are special cases of this semi-parametric model, including normal, Poisson and binomial (Anderson 1979). Thus this approach can be used to aide in selection of a parametric model. Interestingly, we can even approximate parametric models that do not fit this form, such as negative binomial.

We draw on empirical likelihoods, which use distributions that have point mass at the observed phenotypes. Recent work (see Owen 2001) shows how we can use much of the standard likelihood machinery for point mass empirical distributions with only slight modification. Thus we can use already developed QTL interval mapping for normal data once we can evaluate the likelihood, which is

$$\begin{aligned}\mathrm{pr}(y|m, \theta, \lambda) &= \prod_{i=1}^{n} \sum_q \mathrm{pr}(q|m_i, \lambda) f(y_i)\gamma(y_i|q, \beta) \\ &= \prod_{i=1}^{n} w(y_i|m_i, \beta, \lambda) f(y_i)\end{aligned}$$

with weights $w(y_i|m_i, \beta, \lambda) = \sum_q \mathrm{pr}(q|m_i, \lambda)\gamma(y_i|q, \beta)$ that rely only on the phenotype and on flanking markers around the QTL. Ideally, we profile the likelihood across loci λ in the genome. Unfortunately, the profile empirical likelihood may not exist for all β in a small compact neighborhood of the null value. That is, there may be no β that make $f(y)\gamma(y|q, \beta)$ a density for all possible q.

Zou et al. (2002) proposed a partial empirical likelihood, treating markers m as fixed, by noting that the profile empirical log-likelihood can be factored as

$$\log(\mathrm{pr}(y|m, \theta, \lambda)) = \ell_1(\beta, \alpha(\beta)) + \ell_2(\beta) - n \log n.$$

The first term involves a nuisance parameter to enforce the density constraints. It uses a clever trick concerning the Lagrange multiplier α for the constraints on β, leading to point mass estimates

$$\hat{f}(y_i|m,\beta,\lambda) = \left[\sum_q \gamma(y_i|q,\beta)\sum_{i=1}^n \mathrm{pr}(q|m_i,\lambda)\right]^{-1}.$$

The second term is the partial empirical likelihood,

$$\ell_2 = \sum_{i=1}^n \log(w(y_i|m_i,\beta,\lambda)) - \sum_{i=1}^n \log\left(\sum_q w(y_i|m_i,\beta,\lambda)\rho(m_i)\right),$$

with $\rho(m_i)$ estimated as the empirical proportion of flanking markers with the genotype agreeing with m_i (for a backcross, there are four possible flanking marker genotype combinations). Notice that the partial empirical likelihood ℓ_2 does not depend on the shape of the density f.

Zou and Fine (2002) justified this partial empirical likelihood using a conditioning argument. They assumed that the marker genotypes are random, as in breeding experiments, and that the flanking marker probabilities $\rho(m_i)$ may be determined directly by the breeding design, the map function and the marker map, which are typically known. They then demonstrated that one may construct a conditional likelihood based on distribution of flanking marker genotype given phenotype not involving the baseline density f. The partial empirical likelihood is this conditional likelihood with $\rho(m_i)$ replaced by estimates. Zou and Fine (2002) and Jin et al. (2003) showed that ℓ_2 gives valid inferences regardless whether or not m_i are treated as fixed or random.

Thus we profile ℓ_2 with respect to λ, maximizing β for each possible locus. This semi-parametric profiling yields the same formal behavior as the normal-based profile likelilhood the maximum profile likelihood (see Discussion). This semi-parametric approach can be used to examine the robustness of normal or other parametric phenotype models. First, does the estimated QTL, at the maximum LOD, agree between normal and semi-parametric approaches? Second, are the data consistent with a particular parametric model, using the cumulative distributions conditional on QTL genotype in a graphical goodness-of-fit test?

Mammary Tumors in Rats

Study has shown that female rats from the Wistar-Kyoto (WKy) strain resistant to carcinogenesis were crossed with male rats from the Wistar-Furth (WF) strain (Lan et al. 2000). To identify carcinogenesis resistant genes, 383 female BC rats were generated by mating F1 progeny to WF animals. These backcross rats were scored for number of mammary carcinomas and were genotyped at 58 markers on chromosome 5. Using Mapmaker/QTL,

Lan et al. (2000) found that marker D5Rat22 was strongly associated with lower tumor counts. The mean numbers of counts estimated from the normal mixture are 2.68 and 5.43 for the WKy/WF and WF/WF genotypes, respectively at the putative QTL identified.

Zou et al. (2002) applied the semiparametric method to this rat data and the results are summarized in Figures 1 and 2. In Figure 1, the partial likelihood ratio statistic is shown as a function of location on chromosome 5. The LOD score calculated from the partial likelihood ratio statistic is also given. For comparison, the profile from a normal mixture using MapMaker/QTL is displayed. Both curves are very similar with peaks near D5Rat22. The estimated distribution functions for Wky/WF and WF/WF genotypes were computed at the locus giving the maximum LOD score under the semiparametric and normal mixtures. These are displayed in Figure 2 along with 0.95 pointwise confidence intervals. The plots exhibit that WF/WF rats have higher tumor counts. The estimated means for carcinomas in WKy/WF and WF/WF rats are 2.69 and 5.45, respectively. The estimated distributions from the normal mixture are rather different from the semiparametric estimates and may lie outside the confidence intervals. Other estimates (not shown) from a negative binomial model (Drinkwater and Klotz 1981) fall entirely within the 0.95 limits.

2.2 *Measuring the shift of center*

Another way to generalize the normal model is to suppose that QT genotypes can shift the center but not otherwise change the shape of the model. That is,

$$\text{pr}(y|q,\theta) = F(y + q\beta)$$

with $\theta = (\beta, F)$, β consisting of a few parameters and F a completely unspecified distribution. This semi-parametric shift model has a natural estimator of shift suggested by Hodges and Lehmann. All one has to do is divide the phenotypes into groups based on QT genotype q and find β that shifts the medians of all groups to coincide.

Suppose we knew the shift, say β, and we knew the genotypes q. Then the shifted values $y_i(\beta) = y_i + (q_i - \bar{q}_.)\beta$ would all have the same distribution F. Consider the linear rank statistic

$$T(b|y,q) = \sum_{i=1}^{n} (q_i - \bar{q}_.) \frac{\text{rank}(y_i(\beta))}{n+1} ,$$

which depends on the phenotypes only through the ranks of their shifted values. In the next section, we develop this into a formal test for $\beta = 0$, but here we are interested in estimating the shift. If we knew q, then we

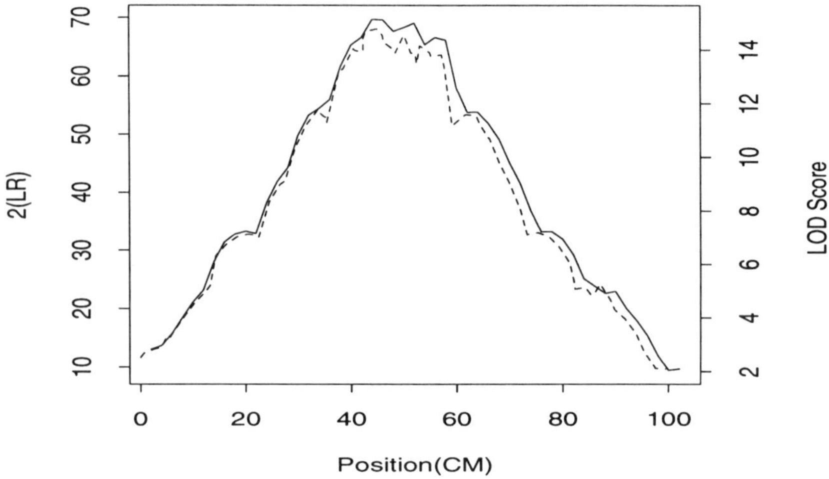

Figure 1 Likelihood ratio statistics and LOD score on chromosome 5. Solid line is the semiparametric mixture and the dashed is the normal mixture. (From Zou et al. 2002.)

could use the Hodges-Lehmann estimator $\hat{\beta} = \text{median}\{b|T(b) \approx 0\}$. Note that the linear rank statistic may not reach zero, so in practice we take the closest values on either side and average them.

This seems rather difficult to do in practice since q are unknown. However, Haley-Knott regression provides a decent approximation. In other words, we can substitute unknown q with its expectation when estimating β:

$$\text{pr}(y|q,\theta) = F(y + E(q)\beta),$$

with $E(q)$ the expectation of q given flanking markers to the loci λ (Haley and Knott 1992). Haley-Knott least squares estimators are consisten, but may be inefficient, while modified Hodges-Lehmann (HL) estimators may have bias, since they are nonlinear in q, depending on the median. Nevertheless, our HL estimators perform well in simulations. Our investigation for a single QTL shows that (Zou et al. 2003) the approximation works well for linkage maps that are relatively dense (when the average marker distance is no larger than 20 cM) which is true for most of the modern QTL mapping studies. The proposed estimator of β is more efficient than its traditional estimator based on the normality assumption when the data

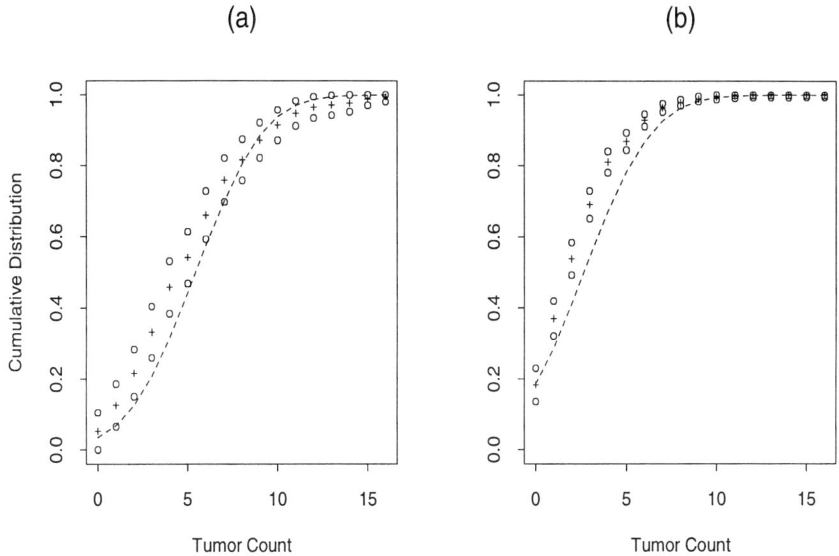

Figure 2 Point estimates (+) and 0.95 pointwise confidence limits (0) for cumulative distributions at location of maximum partial likelihood ratio statis tic. Dashed lines are point estimates from the normal mixture model. (a) WF/WF; (b) WKy/WF. (From Zou et al. 2002.)

is not normally distributed. Further, Haley-Knott (1992) regression gives valid estimates and testing when data are not normal.

Listeria Monocytogene Time-to-Death in Mice

Our second example relates to the date on the time-to-death following infection with Listeria monocytogenes of 116 F2 mice from an intercross between the BALB/cByJ and C57BL/6ByJ strains (Boyartchuk et al 2001). The histograms of the log time-to-death of the non-survivors are given in Figure 3. 31 mice which is roughly 30% of mice survive beyond 264 days. From the histogram it is hard to justify that the log time-to-death of the non-survivors is normally distributed. Broman (2003) applied four different methods, including both the standard interval mapping and non-parametric interval mapping, to this data set and showed that the locus on chromosome 1 appears to have effect only on the average time-to-death among the non-survivors. For this reason, our analysis will be restricted on chromosome 1 for those non-survivors.

The additive and dominance estimators from standard interval mapping are 0.262, 0.059, respectively while they are 0.257, 0.038, respectively

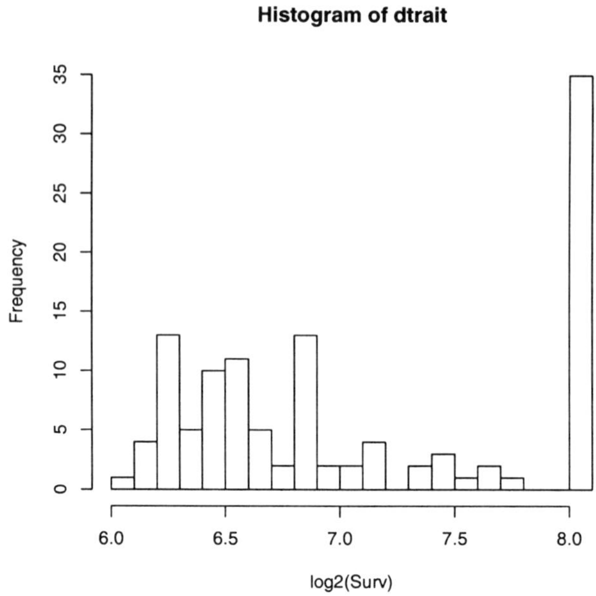

Figure 3 Histogram of $log2$(survival time), following infection with *Lis teria Monocytogenes*. 31 mice recovered from the infection and survived to the end of experiment $264hr$ ($log2(264) = 8$).

based on the rank based method. Therefore, the non-parametric rank based analysis confirms the results by Broman (2003).

3 Non-parametric Models

The semi-parametric models are quite useful, but they still rely on some common shape in some sense. What if we want to allow completely arbitrary shaped distributions with different QTL genotypes?

Here we examine non-parametric methods that make no assumptions about the shape of the distribution, that is we focus on cumulative distributions conditional only on the QT genotype

$$\text{pr}(Y \leq y|q) = F_q(y) \; .$$

This approach is more robust to heavy-tailed phenotype distributions and to occasional outliers.

Estimates of shift discussed in the previous section could be useful here, but they are actually semi-parametric. We wish to estimate the conditional

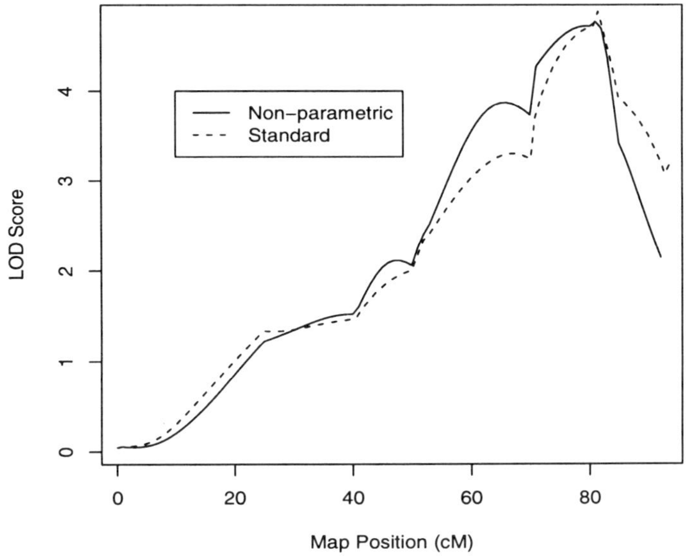

Figure 4 LOD score curves from standard interval mapping (dashed line) and nonparametric interval mapping (solid line).

distributions F_q without any assumptions of shape. Here is the basic idea. We estimate the cumulative distributions given flanking markers, $\text{pr}(Y \leq y|m, \lambda)$, by dividing phenotypes into groups based on flanking markers and summing up the corresponding histograms (details below). Now notice that the phenotype distributions conditional on QTL are mixtures of these flanking-marker distributions:

$$\text{pr}(Y \leq y|m, \lambda) = \sum_q \text{pr}(q|m, \lambda) F_q(y) \ .$$

Given QTL λ, we can calculate $\text{pr}(q|m, \lambda)$. If there are m QTL, then in a backcross there are 2^{2m} possible flanking marker values but only $L = 2^m$ possible QT genotypes. Thus we have fewer unknowns (F_q) than knowns in a set of linear equations, and we can estimate. This argument can be extended to handle missing marker genotypes and other types of experimental crosses.

To be specific, consider the cumulative distributions

$$H_i(y) = \text{pr}(y_i \leq y|m_i, \lambda) \ .$$

Here is a way to get the estimator of H_i. Let $N_i(y) = I(y_i \leq y)$, being 1 if $y_i \leq y$ or 0 if $y_i > y$. Divide experimental units up into sets based on the value of their flanking markers around the loci λ. Let s be one such set. For each unit i in this set s, average the indicators across this set:

$$\hat{H}_i(y) = \sum_{k \in s} N_k(y)/n_s$$

with n_s being the size of the set s. This gives an empirical estimator of $H_i(y)$ which increases from 0 to 1 as y increases, taking steps of size $1/n_s$. All individuals in set s have this same estimator. Thus,

$$\sum_{i \in s} \hat{H}_i(y) = \sum_{i \in s} N_i(y) \ .$$

Let $H = (H_1, \cdots, H_n)^{\mathrm{T}}$ be the cumulative phenotype distributions conditioned on flanking markers, and F be a column vector across the QT genotypes of F_q. Combine the segregation model into an $n \times 2^m$ matrix R with $R_{iq} = \mathrm{pr}(q|m_i, \lambda)$. Thus

$$H(y) = RF(y) \ .$$

In the case of fully informative flanking markers, the 'best' (least squares) estimator of $F_q(y)$ given QTL λ is

$$\hat{F}(y|\lambda) = (R^{\mathrm{T}}R)^{-1}R^{\mathrm{T}}\hat{H}(y) = W\hat{H}(y) = WN(y)$$

with $N = (N_1, \cdots, N_n)^{\mathrm{T}}$. The last equality holds since we are effectively summing first over individuals with the same flanking markers. This makes sense, since we can think of the problem as having the cumulative distribution as the phenotype of interest, with data being $N_i(y) = I(y_i \leq y)$. The least squares estimator of $F_q(y)$ minimizes the following sum of squares:

$$\sum_{i=1}^{n} \left[I(y_i \leq y) - \sum_{q} \mathrm{pr}(q|m_i, \lambda) F_q(y) \right]^2 \ .$$

That is, we find the best fit to the cumulative distribution of phenotypes y based on the segretation model and on phenotype model given QTL at λ.

The covariances of the phenotype cumulative distribution arise directly from the binomial model, since we are estimating a probability. For $y \leq y\prime$,

$$\mathrm{cov}\left(\hat{F}(y|\lambda), \hat{F}(y\prime|\lambda)\right) = WH(y)(I - H(y\prime))W^{\mathrm{T}} \ ,$$

which can be estimated by $WN(y)(I - N(y\prime))W^{\mathrm{T}}$.

The linear rank test provides a formal non-parametric testing framework to infer QTL, assuming common shape. Following this localization, the above estimators can provide graphical assessment of the shape of the distribution for each genotype.

The rank tests of Kruglyak and Lander (1995) may have low power to detect differences between the phenotypic distributions. A test for homogeneity of the components may also be conducted using the proposed nonparametric estimators. For given y, the null hypothesis is $H_0 : A\hat{F}(t) = 0$, where A is an $(L-1) \times L$ matrix containing $(L-1)$ linearly independent contrasts of $F_Q(y)$s corresponding all possible QTL genotypes q. Under H_0, the statistic

$$\mathcal{L}(y) = \{A\hat{F}(y)\}\{A\hat{\Sigma}(y,y)A^T\}^{-1}\{A\hat{F}(y)\}^T$$

has a chi-squared distribution with $L-1$ degrees of freedom. Evaluating the distribution of $\mathcal{L}$ as a process in $y \in [0,\tau]$ (τ is the maximum y value observed would) enable omnibus testing procedures which are sensitive to differences amongst the component distributions at all time points. For example, using $\sup_y \mathcal{L}(y)$ would provide a statistic which is sensitive to all alternatives, unlike the test of Kruglyak and Lander (1995). The theoretical developments of $\sup_y \mathcal{L}(y)$ appear to be rather challenging and deserves further investigation. In practice, one might consider using the bootstrap to approximate the distribution of the sup test under H_0 across the genome.

Again, the proposed method has been applied to the mammary tumor rat data (Fine et al. 2004). We compute nonparametric estimates of the carcinoma distributions for the WKy/WF and WF/WF genotypes at the estimated QTL and the estimated distributions are displayed in Figure 5 along with 0.95 pointwise confidence intervals. The plots exhibit that WF/WF rats have higher tumor counts. Further, the estimated distribution $\hat{F}(y)$ provides another goodness of fit method of the traditional parametric QTL mapping. The estimated means in the WKy/WF and WF/WF groups are 2.64 and 5.46, respectively, which agrees with Mapmaker/QTL. However, the estimated distributions from the normal mixture are rather different from the nonparametric estimates; these are not shown. Instead, the estimated components from a model with $F_{WKy/WF}$ and $F_{WF/WF}$ assumed to be negative binomial, which was fitted by Lan *et al.* (2001), are displayed in Figure 5. These fall entirely within the 0.95 limits, indicating that this model matches the data well.

4 Discussion

The Wilcoxon rank-sum test was extended to interval mapping by Kruglyak and Lander (1995). For related sum of scores tests that might be used as alternatives, see Puri and Sen (1985) or other texts on non-parametric statistics.

Technical details for the QTL exponential tilt can be found in Zou,

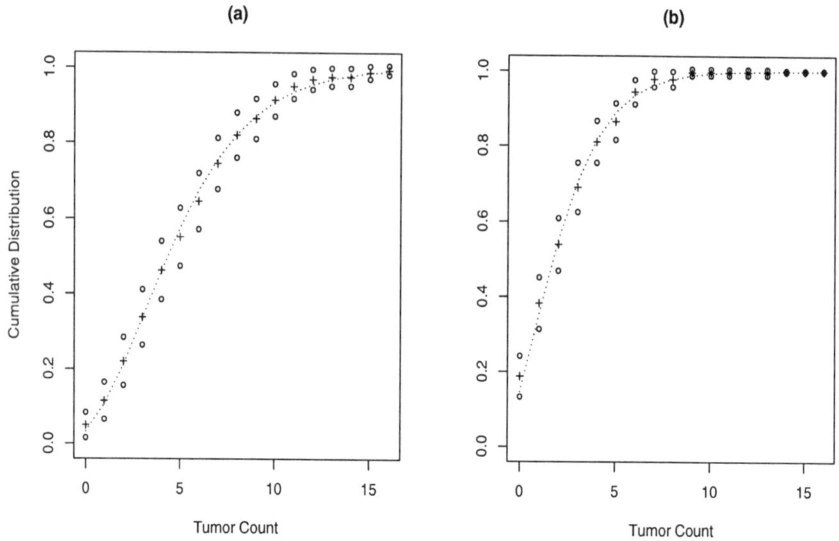

Figure 5 Point estimates (+) and 0.95 pointwise confidence limits (0) for cumulative distributions at location of maximum partial likelihood ratio statis tic. Dashed lines are point estimates from the negative binomial mixture model. (a) WF/WF; (b) WKy/WF. (From Fine et al. 2004.)

Fine and Yandell (2002), based on empirical likelihood work of Qin (Qin & Lawless 1994; Qin 1999). See Owen (2001) for a comprehensive treatment of empirical likelihoods. Zou and Fine (2002) showed how the partial empirical likelihood is closely related to the conditional likelihood. This connection raises interesting robustness issues with respect to selective genotyping and selective phenotyping that are discussed in Jin et al. (2003).

Fine, Zou and Yandell (2001) developed non-parametric cumulative distributions for QTL phenotypes for uncensored and censored data. Speed (pers. comm.) developed a QTL version of the Cox proportional hazards. Recent research has touched on time series and repeated measures analysis in the QTL context.

Calculating thresholds and power are important practical issues in the design and analysis of any QTL study. However, the usual point-wise significance level based on chi-square approximation is inadequate because the entire genome is tested for the presence of a QTL. Theoretical approximations based on the Ornstein-Uhlenbeck diffusion process have been developed to determine threshold and power (Lander and Botstein 1989; Dupuis and Siegmund 1999; Rebai et al. 1994, 1995; Zou et al. 2001, 2002) in some simple experimental crosses. However, permutation procedure is time consuming and may not be applicable under some conditions. The theoretical approximation is not readily available for any study designs and hard to obtain for complicated models. Empirical permutation procedures to estimate genome-wide threshold values for traditional interval mapping proposed by Churchill and Doerge (1994) and widely used for normal data can be readily applied to the semiparametric and nonparametric methods reviewed here. Recently, Zou et al. (2004) proposed a new resampling procedure to assess the significance of genome-wide QTL mapping that is computationally much less intensive than Churchill and Doerge (1994). Further, it is applicable to complicated QTL mapping models that the permutation and theoretical methods cannot handle.

Acknowledgements

This work was supported in part by United States Department of Agriculture Cooperative State Research, Education, and Extension Service and by National Institutes of Health/NIDDK, 5803701 and 66369-01.

References

1. ANDERSON J. A(1979). Multivariate logistic compounds. *Biometrika* **66**, 17-26.
2. BASTEN CJ, WEIR BS, ZENG ZB (1995) QTL Cartographer: A Reference Manual and Tutorial for QTL Mapping. Center for Quantitative Genetics, NC State Univeristy.
3. BELKNAP JK, RICHARDS SP, O'TOOLE LA, HELMS ML, AND PHILLIPS TJ (1997). Short-term selective breeding as a tool for QTL mapping: ethanol preference drinking in mice. *Behavior Genetics* **27**, 55-66.
4. BROMAN KW (2003). Quantitative trait locus mapping in the case of a spike in the phenotype distribution. *Genetics* **163**, 1169-1175.
5. BROMAN KW, WU H, SEN S, CHURCHILL GA (2003) R/qtl: QTL mapping in experimental crosses. *Bioinformatics* **19**, 889-890.
6. CHURCHILL GA AND DOERGE RW(1994). Empirical threshold values for quantitative trait mapping. *Genetics* **138**, 963-971.
7. DIAO G, LIN DY AND ZOU F (2004). Mapping quantitative trait loci with censored observations. *Genetics* **168**, 1689-1698.
8. DOERGE RW, ZENG ZB, WEIR BS (1997). Statistical issues in the search for genes affecting quantitative traits in experimental populations. *Statist Sci* **12**, 195-219.

9. Dupuis J, Siegmund D. (1999) Statistical methods for mapping quantitative trait loci from a dense set of markers. *Genetics* **151**: 373-386.

10. Epstein MP, Lin X, Boehnke M (2003) A Tobit variance-component method for linkage analysis of censored trait data. *Amer J Hum Genet* **72**, 611-620.

11. Fine JP, Zou F and Yandell BS (2004). Nonparametric estimation of mixture models, with application to quantitative trait loci. *Biostatistics* **5**, 501-513.

12. Hackett CA(1997). Model diagnostics for fitting QTL models to trait and marker data by interval mapping. *Heredity* **79**, 319-328.

13. Hackett CA and Weller JI (1995) Genetic mapping of quantitative trait loci for traits with ordinal distributions. *Biometrics* **51**, 1254-1263.

14. Haley C. and Knott S. (1992) A simple regression method for mapping quantitative trait loci of linked factors. *J Genetics* **8**, 299-309.

15. Haston CK, Zhou X, Gumbiner-Russo L, Irani R, Dejournett R, Gu X, Weil M, Amos CI and Travis EL(2002). Universal and radiation-specific loci influence murine susceptibility to radiation-induced pulmonary fibrosis. *Cancer Research* **62**, 3782-3788.

16. Hoff PD, Halberg RB, Shedlovsky A, Dove WF, Newton MA (2002) Identifying carriers of a genetic modifier using nonparametric Bayes methods. in *Case Studies in Bayesian Statistics* **5**, Springer-Verlag, 327-342.

17. Jansen RC (1992) A general mixture model for mapping quantitative trait loci by using molecular markers. *Theor Appl Genet* **85**, 252-260.

18. Jin C, Fine J, Yandell B (2003) A unified semiparametric framework for QTL analyses, with application to spike phenotypes. *J Amer Statist Assoc* (in review).

19. Kao CH and Zeng ZB (1997) General formulae for obtaining the MLEs and the asymptotic variance-covariance matrix in mapping quantitative trait loci when using the EM algorithm. *Biometrics* **53**, 653-665.

20. Kruglyak L and Lander ES (1995). A nonparametric approach for mapping quantitative trait loci. *Genetics* **139**, 1421-1428.

21. Lan H, Kendziorski CM, Haag JD, Shepel LA, Newton MA and Gould MN (2001). Genetic loci controlling breast cancer susceptibility in the Wistar-Kyoto rat. *Genetics* **157**, 331-339.

22. Lander E, Green P, Abrahamson J, Barlow A, Daley M, Lincoln S, and Newburg L (1987). MAPMAKER: An interactive computer package for constructing primary genetic linkage maps of experimental and natural populations. *Genomics* **1**, 174-181.

23. Lander ES, Botstein D (1989). Mapping Mendelian factors underlying quantitative traits using RFLP linkage maps. *Genetics* **121**, 185-199.

24. LANGE C, WHITTAKER JC (2001) Mapping quantitative trait loci using generalized estimating equations. *Genetics* **159**, 1325-1337.

25. LYNCH M AND WALSH B (1998). *Genetics and analysis of quantitative traits.* Sunderland, Mass., Sinauer.

26. MACKAY TFC (2001) The genetic architecture of quantitative traits. *Ann Rev Genet* **35**, 303-339.

27. MACKAY TF AND FRY JD (1996) Polygenic mutation in Drosophila melanogaster: Genetic interactions between selection lines and candidate quantitative trait loci. *Genetics* **144**, 671-688.

28. MCINTYRE LM AND COFFMAN C Doerge RW (2000) Detection and location of a single binary trait locus in experimental populations. *Genet Res* **78**, 79-92.

29. NETTLETON D (1999) Order restricted hypothesis testing in a variation of the normal mixture model. *Can J Statist* **27**, 383-394.

30. NETTLETON D (2002) Testing for ordered means in a variation of the normal mixture model. *J Statist Plan Infer* **107**, 143-153.

31. NETTLETON D AND DOERGE RW (2000) Accounting for variability in the use of permutation testing to detect quantitative trait loci. *Biometrics* **56**, 52-58.

32. NETTLETON D AND PRAESTGAARD J (1998) Interval mapping of quantitative trait loci through order-restricted inference. *Biometrics* **54**, 74-87.

33. OWEN AB (2001) *Empirical Likelihood.* Monographs on Statistics and Applied Probability, v. 92.

34. POOLE TM AND DRINKWATER NR (1996) Two genes abrogate the inhibition of murine hepatocarcinogenesis by ovarian hormones. *Proc Nat Acad Sci USA* **93**: 5848-5853.

35. PURI ML AND SEN PK (1985). *Nonparametric methods in general linear models.* John Wiley & Sons.

36. RAO SQ AND LI X (2000) Strategies for genetic mapping of categorical traits. *Genetica* **109**, 183-197.

37. REBAI A, GOFFINET B AND MANGIN B (1994) Approximate thresholds of interval mapping tests for QTL detection. *Genetics* **138**: 235-240.

38. REBAI A, GOFFINET B AND MANGIN B (1995) Comparing power of different methods of QTL detection. *Biometrics* **51**: 87-99.

39. SHEPEL LA, LAN H, HAAG JD, BRASIC GM, GHEEN ME, SIMON JS, HOFF P, MA NEWTON AND GOULD MN (1998) Genetic identification of multiple loci that control breast cancer susceptibility in the rat. *Genetics* **149**, 289-299.

40. QIN, J (1999). Empirical likelihood ratio based confidence intervals for mixture proportions. *The Annals of Statistics* **27**, 1368-84.

41. Qin J and Lawless JF (1994). Empirical likelihood and general estimating equations. *The Annals of Statistics* **22**, 300-25.

42. Sax K (1923). The association of size differences with seed-coat pattern and pigmentation in Phaseoulus vulgaris. *Genetics* **8**, 552-560.

43. Symons RCA, Daly MJ, Fridlyand J, Speed TP, Cook WD, Gerondakis S, Harris AW and Foote SJ (2002). Multiple genetic loci modify susceptibility to plasmacytoma-related morbidity in E5-v-abl transgenic mice. *Proc Natl Acad Sci USA* **99**, 11299-11304.

44. Visscher PM, Haley CS and Knott SA (1996) Mapping QTLs for binary traits in backcross and F-2 populations. *Genet Research* **68**, 55-63.

45. Wang X, Le Roy I, Nicodeme E, Li R, Wagner R, Petros C, Churchill GA, Harris S, Darvasi A, Kirilovsky J, Roubertoux PL, and Paige B (2003). Using Advanced Intercross Lines for High-Resolution Mapping of HDL Cholesterol Quantitative Trait Loci. *Genome Research* **13**, 1654-1664.

46. Weller JI (1986) Maximum likelihood techniques for the mapping and analysis of quantitative trait loci with the aid of genetic markers. *Biometrics* **42**, 627-640.

47. Wright FA and Kong A (1997). Linkage mapping in experimental crosses: the robustness of single-gene models. *Genetics* **146**, 417-425.

48. Xu S and Atchley WR (1995). A random model approach to interval mapping of quantitative genes. *Genetics* **141**, 1189-1197.

49. Yi N and Xu S (2000) Bayesian mapping of quantitative trait loci for complex binary traits. *Genetics* **155**, 1391-1403.

50. Zou F and Fine JP (2002) Note on a partial empirical likelihood. *Biometrika* **89**, 958-961.

51. Zou F, Fine JP, Hu J and Lin DY (2004). An efficient resampling method for assessing genome-wide statistical significance in mapping quantitative trait loci. *Genetics* **168**, 2307-2316.

52. Zou F, Fine JP and Yandell BS (2002). On empirical likelihood for a semiparametric mixture model. *Biometrika* **89**, 61-75.

53. Zou F, Yandell BS and Fine JP (2003). Rank-based statistical methodologies for quantitative trait locus mapping. *Genetics* **165**, 1599-1605.

54. Zou F, Yandell BS and Fine JP (2001). Statistical issues in the analysis of quantitative traits in combined crosses. *Genetics* **158**, 1339-1346.

Chapter 20

MODEL-BASED CLUSTERING AND WEIGHTED CHINESE RESTAURANT PROCESSES

John W. Lau and Albert Y. Lo

Department of Information and Systems Management
Hong Kong University of Science and Technology
Clear Water Bay, HONG KONG

E-mails: john.w.lau@googlemail.com & imaylo@ust.hk

Work in the last two decades on Bayesian nonparametric methods for mixture models finds that a posterior distribution is a double mixture. One first selects a partition of the objects based on a distribution on the partitions, and then performs a traditional parametric posterior analysis on the data corresponding to each cluster of the given partition. It is known that such a partition distribution favors partitions for which the clustering process is defined by predictive quantities such as predictive densities or weights. If a posterior distribution is a statistical guide to the unknown, this partition distribution could be used as a basis for a statistical model for clustering in which the partition is a parameter. The corresponding maximum likelihood estimator or posterior mode is used as an estimator of the partition. We also discuss methods to approximate these estimators based on a weighted Chinese restaurant process. A numerical example on a leukemia data set is given.

Key words: Cluster model, likelihood function, distribution on partitions, weighted Chinese restaurant processes, predictive density

1 Introduction

Clustering is the grouping of similar objects. Recent texts on clustering identify three key questions to be asked: (i) How many clusters? (ii) How do we define similarity between objects? (iii) How do we know that our results are good? See for example Gordon (1999), Duda, Hart and Stork (2001), and Amaratunga and Cabrera (2004), and Kuncheva (2004). Traditional methods of clustering (K-means clustering, single linkage and nearest-neighbor) select the number of clusters (i), and then work on (ii) by defining a similarity measure, usually through a distance, between two objects. Two objects belong to the same cluster if the distance between their corresponding measurements is relatively small. To evaluate the cluster method, the cluster obtained is usually empirically compared with the known cluster structure of the objects. These existing numerical methods partition the objects deterministically, and they are foreign to statistical inference methods. A model-based method for statistical clustering differs from deterministic clustering in that the numerical measurements of the n objects are assumed to have a joint model density. Once a model density is assumed, either Bayesian or frequentist methods can be applied. Scott and Symons (1971) assume a fixed number of clusters and a multivariate Normal components model, and discuss the maximum likelihood estimator for a 'classification likelihood;' see also Friedman and Rubin (1967). Banfield and Raftery (1993) cover the case that the covariance matrices of the Normal components varies and propose an eigenvalue decomposition to facilitate the analysis. Another approach for clustering is based on a finite mixture method [Wolfe (1970), Symons (1981); see McLachlan and Basford (1988) for references.] Fraley and Raftery (2002) suggest the use of the EM algorithm to approximate the MLEs in the finite mixture likelihood and give reviews and references.

Analytic work in Bayesian nonparametric mixture models suggests another possibility. Using an updating technique developed for Gamma-type processes, Lo (1984) and Lo and Weng (1989) eliminate the finite mixture restriction and obtain posterior quantities as averages over partitions; the averaging is defined by a distribution over the collection of all partitions of the objects. Lo, Brunner and Chan (1996) recognize a key property of this partition distribution in that it gives more weight to partitions that cluster the data in terms of Bayesian prediction. They propose a sequential seating algorithm, called the weighted Chinese restaurant process (WCR), which generates random partitions that also enjoy this predictive property. This algorithm and its Gibbs relative can serve as a basis of a Monte Carlo approximation to the posterior quantities. For further development along this line, see Ishwaran and James (2003a, 2003b), Lau and So (2004), and James (2002, 2005). The latter author extends the updating technique to prior processes that can be represented as an integral of a planar Poisson process, and finds that the partition distribution structure continues to prevail. [An example of such an integral is given by the Levy-Ito representation for an increasing process; see for example Section 1.11 of Ito (2004) or page 227 of Loeve (1977).] Brunner and Lo (1999) suggest the use of a partition distribution as a model for clustering; see also Lo (1999, Chapter 6). The unknown parameter is a partition, and the posterior mode is used to estimate the "true" partition. The traditional maximum likelihood estimator is but a posterior mode with respect to a uniform prior distribution on the space of partitions. However, the problem of searching for a modal partition for the partition likelihood is of a combinatorial nature and is known to be difficult [See Section 5.1 in Fraley and Raftery (2002).] Brunner and Lo (1999) suggest approximations to these modes based on a run of the Gibbs version of the WCR [Lo (2005)]. The WCR clustering is computational intensive. To cut down the computation time, Cabrera, Lau and Lo (2005) develop a WCR algorithm based on randomly sampling blocks of data during the Gibbs cycles, which reduces the computation to a manageable level. Other possible approaches to clustering based on Dirichlet mixture models emphasize the simulation of missing values rather than partitions; see Basu and Chib (2003) for references.

Recent work by Quintana and Iglesias (2003) point out the relationship between the partition distributions appearing in the Bayesian solution to the mixture density problem [Lo (1984)] and the 'product partition models' discussed by Hartigan (1990), Barry and Hartigan (1992), and Crowley (1997). In both cases, the products are over component densities. On the other hand, related studies in Bayesian mixture hazard rate models [Lo and Weng (1989) and James (2002, 2005, Section 4)] suggest that the partition distributions could be products over component hazard rates that are not subject to the normalization restriction of a density. It remains to be seen if the partition distribution is a natural way to describe a clustering phenomenon, or if it is but a consequence of the tools, i.e., planar Poisson representations, used. It seems that, despite a mounting number of clustering references, still more work needs to be done to better-understand the underlying structure of the clustering problem, and to present a cohesive and unified treatment.

Section 2 describes the partition likelihood of the clustering model and proposes the use of Monte Carlo methods to approximate the posterior modes. Section 3 covers the sequential seating WCR which is based on iid sampling. Its Markov chain relative, the Gibbs WCR, is discussed in Section 4. Section 5 spells out the prediction property of the seating probability of the two WCRs. The method is illustrated by an example in 't-density clustering' using a leukemia data set in Section 6. This example shows that the random-block WCR provides a better separation than traditional deterministic clustering algorithms.

2 A statistical model for clustering

Given a partition $\mathbf{p} = \{C_1, \ldots, C_{n(\mathbf{p})}\}$ of n objects $\{1, \ldots, n\}$, the measurements of the objects are modeled by a likelihood given $\mathbf{p}$ as

$$f(\mathbf{x}\,|\mathbf{p}) = \prod_{i=1}^{n(\mathbf{p})} k(x_j, j \in C_i), \tag{1}$$

where $k(x_j, j \in \{1, \ldots, n\})$ is a joint density of the data $\mathbf{x} = \{x_1, \ldots, x_n\}$. The joint density yields a marginal density of $\{x_q, q \in C\}$, $k(x_q, q \in C)$, for each subset C of $\{1, \ldots, n\}$. For $j \notin C$, with an abuse of notation, $k(x_j\,|x_q, q \in C)$ denotes the predictive density of x_j given $\{x_q, q \in C\}$ evaluated at x_j. For independent and identically distributed measurements x_js, $k(x_j, j \in \{1, \ldots, n\})$ is a symmetric function in x_js; see the following Remark 4. Note that this method is applicable to regression and time series models in which the existence of symmetric kernels is implicit; see for example Lau and So (2004).

One can proceed with either a frequentist analysis or a Bayesian analysis of the model likelihood function $f(\mathbf{x}\,|\mathbf{p})$. For example, a frequentist locates the maximum likelihood estimator $\hat{\mathbf{p}}$ so that

$$f(\mathbf{x}\,|\hat{\mathbf{p}}) = \max_{\mathbf{p}} \prod_{i=1}^{n(\mathbf{p})} k(x_j, j \in C_i);$$

the max is over all partitions $\mathbf{p}$ of $\{1, \ldots, n\}$. Alternatively, a Bayesian starts by assuming a prior density on $\mathbf{p}$s, i.e., $\pi(\mathbf{p})$ such that $\sum_{\mathbf{p}} \pi(\mathbf{p}) = 1$, and computes a posterior distribution on partitions given data as

$$\pi(\mathbf{p}\,|\text{data}) \propto f(\mathbf{x}|\mathbf{p}) \times \pi(\mathbf{p})$$

with $\sum_{\mathbf{p}} \pi(\mathbf{p}\,|\text{data}) = 1$. Here the posterior mode $\mathbf{p}^*$, $\pi(\mathbf{p}^*\,|\text{data}) = \max_{\mathbf{p}} \pi(\mathbf{p}\,|\text{data})$, is conveniently used as a Bayesian estimator of the true partition. To facilitate the construction of a posterior density, a conjugate prior could be used. A conjugate prior density of $\mathbf{p}$ should be proportional to the likelihood $f(\mathbf{x}\,|\mathbf{p})$ when the latter is viewed as a function of the "parameter" $\mathbf{p}$. This suggests that a conjugate prior density for $\mathbf{p}$ is of the product form

$$\pi(\mathbf{p}\,|g) \propto \prod_{i=1}^{n(\mathbf{p})} g(C_i), \tag{2}$$

where the parameter $g(\cdot)$ is a function defined on the space of subsets of $\{1, \ldots, n\}$. Note that C_0 denotes an empty table and $g(C_0)$ is defined to be

1. In this case, the posterior distribution $\pi(\mathbf{p}\,|\text{data})$ is also of the product form (2) in which g is updated to

$$g^*(C_i) = g(C_i) \times k(x_j, j \in C_i). \tag{3}$$

That is,

$$\pi(\mathbf{p}\,|\text{data}) = \pi(\mathbf{p}\,|g^*) \propto \prod_{i=1}^{n(\mathbf{p})} g^*(C_i). \tag{4}$$

An inspection of (2) and (4) suggests that the search of the maximum likelihood estimator $\hat{\mathbf{p}}$ and that of the posterior mode $\mathbf{p}^*$ are identical problems: Both look for the maximum of a distribution on the space of partitions, which have the product form (2). The resulting modal partition, $\hat{\mathbf{p}}$ or $\mathbf{p}^*$, answers (i) and (ii) simultaneously. It should be noted that while this method answers questions (i) and (ii), the idea of model checking, i.e., answer to question (iii), has not been entirely clear to us yet.

It suffices to describe Monte Carlo methods to locate a posterior model $\mathbf{p}^*$. The data $\mathbf{x}$ is fixed and given. One convenient way to locate $\mathbf{p}^*$ is through sampling $\pi(\mathbf{p}\,|\text{data})$: If $\mathbf{p}_1, \ldots, \mathbf{p}_M$ are iid from $\pi(\mathbf{p}\,|\text{data})$, a $\mathbf{p}_k$ with a larger $\pi(\mathbf{p}_k\,|\text{data})$ will have a higher probability to be sampled. Compute $\pi(\mathbf{p}_k\,|\text{data})$; $k = 1, \ldots, M$; the $\mathbf{p}_k$ that gives the largest $\pi(\mathbf{p}_k\,|\text{data})$ is an approximation to $\mathbf{p}^*$. However, the exact (perfect) sampling [Propp and Wilson (1996)] from $\pi(\mathbf{p}\,|\text{data})$ does not seem to be available. The sequential seating WCR [Lo Brunner and Chan (1996)] that samples a partition from a distribution close to $\pi(\mathbf{p}\,|\text{data})$ is a possibility. A Gibbs sampler [Geman and Geman (1984)] based on the sequential WCR that has a stationary distribution $\pi(\mathbf{p}\,|\text{data})$ is another reasonable alternative. The sequential seating and the Gibbs WCRs are close relatives. Due to the former's elementary nature, we shall discuss it first.

Remark 1. The structure of the posterior distribution for a Bayesian mixture model with respect to Gamma-type priors [Lo (1984) and Lo and Weng (1989)] suggests that the class of partition distributions (2) could be key. It turns out that the situation is not that different if one uses the more general Levy-type process priors as the class (2) appears again as a key component of the posterior distributions; see Section 4 in James (2002) and in James (2005). Models defined by (1) and (2) are also called product partition models [Hartigan (1990), Crowley (1997) and Quintana and Iglesias (2003).]

3 The sequential seating weighted Chinese restaurant process

The sequential seating weighted Chinese restaurant process is developed to approximate posterior quantities of the form $\xi = \sum_{\mathbf{p}} h(\mathbf{p})\, \pi(\mathbf{p}\,|g^*)$ where the function $h(\cdot)$ is known. The seating algorithm achieves this by sequentially generating a random partition $\mathbf{p}$ that has a distribution close to the posterior distribution $\pi(\mathbf{p}\,|g^*)$. Sampling $\mathbf{p}$ repeatedly and independently results in a Monte Carlo weighted average that approximates ξ. [The Monte Carlo weighted average is biased. That however does not seem to affect the approximation too much; see numerical Examples in Lo, Brunner and Chan (1996).]

We consider the $\pi(\mathbf{p}\,|g)$ case first. To initiate the sequential seating, customer 1 is seated at an empty C_0. Suppose $j-1$ customers are seated, resulting in occupied tables denoted by $C_1, \ldots, C_{n(\mathbf{p})}$. Customer j will be seated at an empty table with probability proportional to $g(\{j\})$; otherwise, he/she is seated at an occupied table C_i with probability proportional to the predictive ratios

$$g(\{j\}\,|C_i) \equiv \frac{g(\{j\} \cup C_i)}{g(C_i)}, \quad i = 0, \ldots, n(\mathbf{p}); \tag{5}$$

the normalization constant of this conditional probability at seating j is

$$\lambda(j-1) = g(\{j\}) + \sum_{i=1}^{n(\mathbf{p})} g(\{j\}\,|C_i).$$

After n customers are seated, the product rule gives the joint density of the resulting random partition $\mathbf{p}$ as [Lemma 1.1 in Lo, Brunner and Chan (1996)]

$$q(\mathbf{p}\,|g) = \frac{\prod_{i=1}^{n(\mathbf{p})} g(C_i)}{\prod_{j=1}^{n} \lambda(j-1)}.$$

The numerator agrees with that of $\pi(\mathbf{p}\,|g)$, and in this sense $q(\mathbf{p}\,|g)$ is close to $\pi(\mathbf{p}\,|g)$. On the other hand, the denominator $\prod_{j=1}^{n} \lambda(j-1)$ depends on the seating order; this dependence seems to retard the closeness of $q(\mathbf{p}\,|g)$ to $\pi(\mathbf{p}\,|g)$.

Letting g^* play the role of g in the seating algorithm, the resulting sequential WCR distribution is given by $q(\mathbf{p}\,|g^*)$, which is close to the posterior distribution $\pi(\mathbf{p}\,|g^*)$. The seating probability is proportional to

$$k(x_j|x_q, q \in C_i) \times \frac{g(\{j\} \cup C_i)}{g(C_i)}, \quad i = 0, \ldots, n(\mathbf{p}). \tag{6}$$

Replacing g by g^* in $\lambda(j-1)$ results in the normalization constant $\lambda^*(j-1)$. [For an empty table C_0, $k(x_j|x_q, q \in C_0)$ is defined to be $k(x_j)$.]

Remark 2. The seating probability (6) contains two predictive quantities: The prior part $g(\{j\} \cup C_i)/g(C_i)$ is an extension of the seating probability of a (unweighted) Chinese restaurant process; see the following Example 1. The word 'weighted' in WCR addresses the contribution of $k(x_j|x_q, q \in C_i)$.

4 The Gibbs weighted Chinese restaurant process

A Gibbs sampler version of the WCR has the posterior distribution $\pi(\mathbf{p}|g^*)$ as the stationary distribution, and the Gibbs average can also be used to approximate ξ. It suffices to consider the $\pi(\mathbf{p}|g)$ case. The general theory of Gibbs sampler states that the Markov transition consists of a Gibbs cycle which is based on the idea of predicting a new variable given the rest of the variables and then rotating (cycling) the new variable among the existing ones for predictions. [See Geman and Geman (1984).] In the present situation, the prediction part can be read off from the last seating step of the sequential WCR, and the cycling part amounts to a rotation of reseating j. The Gibbs WCR cycle of moving the current $\mathbf{p}$ to a new partition is completed by cycling through the following two steps for $j = 1, 2, \ldots, n$.

Step 1. The current partition of $\{1, \ldots, n\}$ is denoted by $\mathbf{q}$. Delete integer j from (the table containing j in) $\mathbf{q}$ and denote the resulting partition by $\mathbf{p} = \{C_0, C_1, \ldots, C_{n(\mathbf{p})}\}$.

Step 2. Reseat integer j to the cluster C_i, with a reseating probability proportional to

$$g(\{j\}|C_i) \equiv \frac{g(\{j\} \cup C_i)}{g(C_i)}, \quad i = 0, 1, \ldots, n(\mathbf{p}). \tag{7}$$

It follows then the Gibbs sampler having the posterior $\pi(\mathbf{p}|g^*)$ as a stationary distribution has a reseating probability proportional to

$$k(x_j|x_q, q \in C_i) \times \frac{g(\{j\} \cup C_i)}{g(C_i)}, \quad i = 0, 1, \ldots, n(\mathbf{p}) \tag{8}$$

Remark 3. A Gibbs sampler on partitions is discussed by MacEachern (1994) for a location mixture of Normals model. Crowley (1997) considers essentially similar problems from the product partition models viewpoint. Quintana and Iglesias (2003) cover extensions and give more references. See also the following Example 3.

5 A predictive property of WCR clustering

The use of a seating probability (6) or (8) to generate a random partition is of interest and deserves a careful examination. For the rest of the Section, we shall discuss the case of the Gibbs WCR, i.e., (8), as the sequential seating case is rather similar. The way a new customer, say j, is assigned to an occupied table C_i, reveals a clustering process for a set of data $\{x_j, j = 1, \ldots, n\}$ by means of predictive properties between measurements rather than the traditional one based on a distance function defined between (groups of) data. The 'next' customer j is assigned to table C_i with seating probability proportional to (8). Identify the observation x_j with integer j, $j = 1, \ldots, n$, and regard C_i as a cluster of data. The predictive weight $k\,(x_j\,|x_q, q \in C_i\,)$ is really the value of a predictive density, conditional on $\{x_q, q \in C_i\}$, and evaluated at a new observation x_j . Note that the predictive density $k\,(x_j\,|x_q, q \in C_i\,)$ is large if j is close to C_i (that is, if x_j is close to $\{x_q, q \in C_i\}$); otherwise $k\,(x_j\,|x_q, q \in C_i\,)$ is small. Hence if j is close to C_i, the seating probability that it will be grouped into C_i is also large. The second ratio $g\,(\{j\} \cup C_i)\,/g\,(C_i)$ is a similar 'predictive' quantity based on the prior (2). The product of the predictive density and the prior predictive ratio provides a rather natural balancing effect to the random seating, and it also defines the closeness of the clustering process. The following example illustrates the balancing effect of the seating probabilities based on various prior $\pi(\mathbf{p}\,|g)$. For more examples of $\pi(\mathbf{p}\,|g)$s and the corresponding $\pi(\mathbf{p}\,|g^*)$s, see Section 4 in James (2002, 2005).

Example 1 A Chinese restaurant process (CR) with parameter $e_0 > 0$ is a distribution on the space of partitions that has a density

$$\pi(\mathbf{p}\,|g) \propto \prod_{i=1}^{n(\mathbf{p})} [e_0 \times (e_i - 1)!]\,.$$

This is (2) with $g\,(C_i) = e_0 \times (e_i - 1)!$ for $i = 0, \ldots, n\,(\mathbf{p})$. [Recall $g\,(C_0) = 1$.] In this case, the predictive ratio $g\,(\{j\} \cup C_i)\,/g\,(C_i)$ reduces to e_i for $i = 0, \ldots, n\,(\mathbf{p})$. The seating probability for the posterior (8) becomes

$$k\,(x_j\,|x_q, q \in C_i\,) \times e_i\ , \quad i = 0, \ldots, n\,(\mathbf{p})$$

Here a short distance between x_j and $\{x_q, q \in C_i\}$ could be balanced out by a large table size e_i, yielding a seating probability of moderate size.

Example 2 The case of a uniform prior on $\mathbf{p}$ gives another perspective of the seating probability. Let $g\,(\cdot)$ be the constant 1, $\pi(\mathbf{p}\,|g)$ becomes a uniform prior on the space of partitions. The prior predictive ratio at (8) vanishes, and the Gibbs cycle is entirely data-driven. The resulting Gibbs sampler yields an approximation to a maximum likelihood estimator $\hat{\mathbf{p}}$.

Example 3 Model (1) can be fine-tuned to accommodate additional regression-type parameters ϕ. The general situation can be summarized by $\phi \sim \beta(d\phi) = \beta'(\phi)\, d\phi$, $\mathbf{p}\,|\phi$ has a density proportional to $\prod_{i=1}^{n(\mathbf{p})} g(C_i, \phi)$, and $\mathbf{x}\,|(\mathbf{p}, \phi)$ has a model density $\prod_{i=1}^{n(\mathbf{p})} k(x_j, j \in C_i\,|\phi)$. Here the model is

$$f(\mathbf{x}\,|\mathbf{p}, \phi) = \prod_{i=1}^{n(\mathbf{p})} k(x_j, j \in C_i\,|\phi),$$

and the prior is

$$\pi(\mathbf{p}, \phi\,|g) \propto \beta'(\phi) \prod_{i=1}^{n(\mathbf{p})} g(C_i, \phi).$$

It follows then given $\mathbf{x}$, the joint posterior density of $(\mathbf{p}, \phi)\,|\mathbf{x}$ is

$$\pi(\mathbf{p}, \phi\,|\mathbf{x}) \propto \beta'(\phi) \prod_{i=1}^{n(\mathbf{p})} g^*(C_i, \phi), \tag{9}$$

where $g^*(C_i, \phi)$ denotes $k(x_j, j \in C_i\,|\phi) \times g(C_i, \phi)$. That is, $\pi(\mathbf{p}, \phi\,|g)$ is a conjugate family of priors for this model. A Gibbs chain that samples the posterior (9) is obtained by alternately sampling $\mathbf{p}$ and ϕ:

(i) Given ϕ, the density of $\mathbf{p}$ is proportional to $\prod_{i=1}^{n(\mathbf{p})} g^*(C_i, \phi)$; move $\mathbf{p}$ to the next $\mathbf{p}'$ by cycle through Steps 1 and 2 for $j = 1, \ldots, n$.
(ii) Given $\mathbf{p}$, sample a next ϕ' from (9).

If sampling $\phi'\,|\mathbf{p}$ in (ii) is difficult (such as the case of a regression parameter ϕ with an error density being a scale mixture of Normals) one nests a Metropolis-Hastings rejection step [Hastings (1970)] to move ϕ to the next value ϕ'. This procedure replaces Step (ii) by a randomization step

(ii') Given $\mathbf{p}$, and ϕ from step (i), sample a ϕ^* from $\beta'(\psi)$. Let $\phi' = \phi^*$ with probability $\min\left\{\prod_{i=1}^{n(\mathbf{p})} \dfrac{g^*(C_i, \phi^*)}{g^*(C_i, \phi)}, 1\right\}$; otherwise $\phi' = \phi$.

The case of hierarchical Bayesian models is another example. Let $\phi = (u, v)$, and suppose $k(x_j, j \in C_i\,|\phi)$ depends on ϕ through u, and $g(C_i, \phi)$ depends on ϕ only through v, where u is a regression parameter and v is a mixing prior parameter. The posterior distribution (9) specializes to

$$\pi(\mathbf{p}, u, v, |\mathbf{x}) \propto \beta'(u, v) \prod_{i=1}^{n(\mathbf{p})} \left[k(x_j, j \in C_i, |u)\, g(C_i, v)\right]. \tag{10}$$

A Gibbs sampler samples $\mathbf{p}$, u, and v alternatively (given the other two), and a Metropolis-Hastings step may be needed to move u and v given $\mathbf{p}$.

Remark 4. The foregoing discussion reveals the importance of predictive properties between different objects, or their corresponding measurements. This presents no difficulty for a Bayesian. In the case that $\{x_1, \ldots, x_n\}$ are iid measurements from a mixture model, a Bayesian constructs a symmetric $k(x_j, j \in \{1, \ldots, n\})$ by averaging out a "nuisance parameter" (de Finetti's theorem)

$$k(x_j, j \in \{1, \ldots, n\}) = E\left[\prod_{j=1}^{n} \tau K(\tau(x_j - \mu))\right], \qquad (11)$$

where $K(\cdot)$ is a prescribed density on the line, and E is the expectation with respect to a distribution of $(\tau, \mu), \tau > 0$, which is independent of the data $\{x_1, \ldots, x_n\}$. Extensions to multivariate observations amount to a change of notation. The next Section discusses the case of a standard Normal $K(\cdot)$, which benefits from an integral-free expectation (11). A symmetric $k(x_j, j \in \{1, \ldots, n\})$ can also be constructed by conditioning on some sufficient statistic. This is a frequentist way to eliminate a nuisance parameter. For example, if $K(\cdot)$ is a standard Normal density, let $S =$ (Sample variance, Sample average), the conditional density of $\{x_1, \ldots, x_n\}$ given S can be used as $k(x_j, j \in \{1, \ldots, n\})$. The latter conditional density is independent of (τ, μ) by sufficiency; it is symmetric as it is a function of S.

6 The multivariate t-density clustering

There are a variety of kernels $k(\ldots)$s that generate a clustering method and Cabrera, Lau and Lo (2005) propose clustering n objects with multivariate measurements using a multivariate t-density kernel $k(\ldots)$ [i.e., the following (15)]. Some of the clustering algorithms existing in the literature use Normal component densities; see Remark 3. From the Bayesian mixture model viewpoint, a Normal component density appears only in location mixtures of Normals models, and methods described in Example 3 can be applied to cluster data. However, a location mixture family consists of densities that are convolutions with a fixed density, and they are necessarily bounded by the sup-norm of the fixed density (which is a Normal density in this case.) This restriction limits their ability to cluster data from an arbitrary density. The t-density kernel arises naturally in the Bayesian estimation of an arbitrary density using the mixture method [Section 3 in Lo

(1984)] and the t-density clustering does not suffer from this restriction; an added twist is that it yields integral-free seating probabilities.

To illustrate the idea of the t-density clustering, it suffices to consider the case that measurements corresponding to the n objects are univariate. Work in Bayesian nonparametric mixture models cited before and Remark 4 yield a model for clustering (1) determined by

$$k\left(x_j, j \in C_i\right) = E\left[\prod_{j \in C_i} \sqrt{\tau}\phi\left(\sqrt{\tau}\left(x_j - \mu\right)\right)\right]. \tag{12}$$

where $\phi(\cdot)$ is a standard $N(0,1)$ density. The expectation is with respect to the distribution of (τ, μ): τ is Gamma with mean α/β and variance α/β^2, and $\mu\,|\tau$ is Normal with mean m and precision $t\tau$. See Section 9.6 in De Groot (1970). In this case, $k\left(x_j, j \in \{1, \ldots, n\}\right)$ is a symmetric density with t-marginals.

The maximization of the criterion function $\log\left[\pi(\mathbf{p}\,|\text{data})\right]$ given by

$$J\left(\mathbf{p}\,|g^*\right) = \sum_{i=1}^{n(\mathbf{p})} \log\left[k(x_q, q \in C_i)\right] + \sum_{i=1}^{n(\mathbf{p})} \log\left[g\left(C_i\right)\right] \tag{13}$$

over all possible partition $\mathbf{p}$s is a criterion for clustering. We shall use the Chinese restaurant process prior with parameter e_0 (Example 1) for illustration. In this case the criterion function becomes

$$J\left(\mathbf{p}\right) = \sum_{i=1}^{n(\mathbf{p})} \log\left[k(x_q, q \in C_i)\right] + \sum_{i=1}^{n(\mathbf{p})} \log\left[(e_i - 1)!\right] + n(\mathbf{p})\log\left(e_0\right). \tag{14}$$

The resulting $\mathbf{p}^*$ obtained is then a posterior mode for model (1) with a Chinese restaurant process prior with parameter e_0.

The clustering of n objects with D dimensional multivariate measurements $\{\mathbf{x}_1, \ldots, \mathbf{x}_n\}$ is identical except for a change of notation: $\phi(\cdot)$ is a multivariate standard normal, $\boldsymbol{\tau}$ is a $D \times D$ Wishart matrix with α degrees of freedom and precision matrix $\boldsymbol{\beta}$, $\boldsymbol{\mu}\,|\boldsymbol{\tau}$ is multivariate Normal with mean vector $\mathbf{m}$ and precision matrix $t\boldsymbol{\tau}$. See Section 9.10 in De Groot (1970). Using this vector notation, the kernel $k\left(\mathbf{x}_j, j \in C_i\right)$ is given by (12), and the expectation there reduces to

$$k\left(\mathbf{x}_j, j \in C_i\right) = \frac{\prod_{r=1}^{D} \Gamma\left(\frac{\alpha + e_i + 1 - r}{2}\right)}{\prod_{r=1}^{D} \Gamma\left(\frac{\alpha + 1 - r}{2}\right)} \frac{t^{\frac{D}{2}}}{\pi^{\frac{De_i}{2}}\left(t + e_i\right)^{\frac{D}{2}}} \frac{|\boldsymbol{\beta}|^{\frac{\alpha}{2}}}{|\boldsymbol{\beta}^*|^{\frac{\alpha + e_i}{2}}}, \tag{15}$$

where

$$\boldsymbol{\beta}^* = \boldsymbol{\beta} + \sum_{j \in C_i}\left(\mathbf{x}_j - \bar{\mathbf{x}}_i\right)\left(\mathbf{x}_j - \bar{\mathbf{x}}_i\right)' + \frac{te_i}{t + e_i}\left(\bar{\mathbf{x}}_i - \mathbf{m}\right)\left(\bar{\mathbf{x}}_i - \mathbf{m}\right)'$$

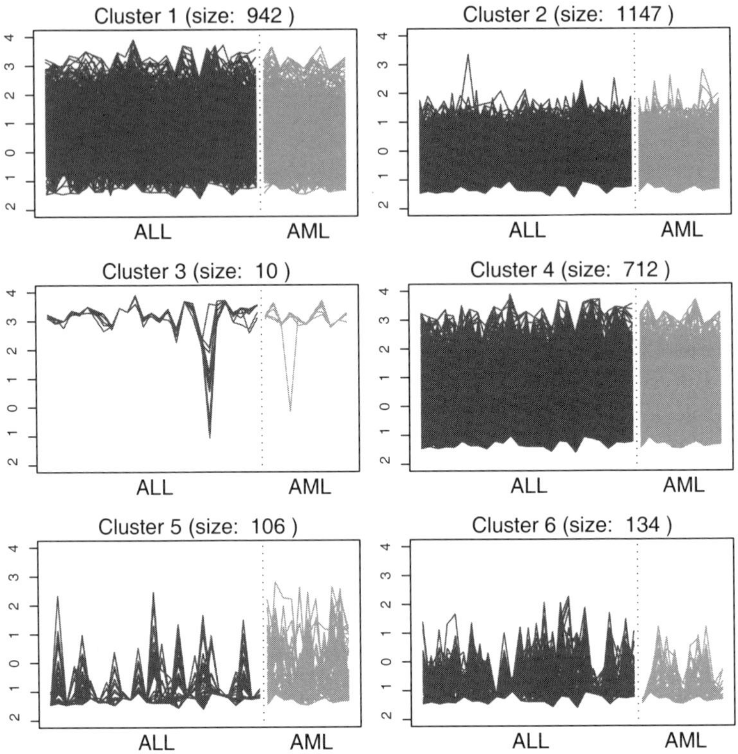

Figure 1 Profile plots of the gene expression clustered by a partition estimate from the t-density clustering model with Gibbs WCR

and $\bar{\mathbf{x}}_i = \frac{1}{e_i}\sum_{j\in C_i}\mathbf{x}_j$.

The following example illustrates the methodology. A leukemia data set [Golub et al. (1999)] contains 3051 gene expression levels of 38 patients, where 27 of them have acute lymphoblastic leukemia (ALL) and the rest of them have acute myeloid leukemia (AML). The data set is available in both MULTTEST and PLSGENOMICS packages of the statistical software R. Here the 3051 gene expression levels are to be clustered, and $D = 38$ is regarded as the dimension of a 'measurement'. See Amaratunga and Cabrera (2004) for the interexchangeable roles played by the 'objects/sample size' and 'variables/dimension'. The prior parameters $(\alpha, \boldsymbol{\beta}, \mathbf{m}, t)$ are chosen to be $(40, 0.1\mathbf{I}_{38\times 38}, \mathbf{0}_{38}, 0.1)$ where $\mathbf{I}_{38\times 38}$ is a 38×38 identity matrix and $\mathbf{0}_{38}$ is a 38 dimensional column vector with all entries 0. The parameter e_0 of the Chinese restaurant process is set to be 1. The Gibbs WCR based on randomly selected blocks of data, called random-block WCR [Cabrera,

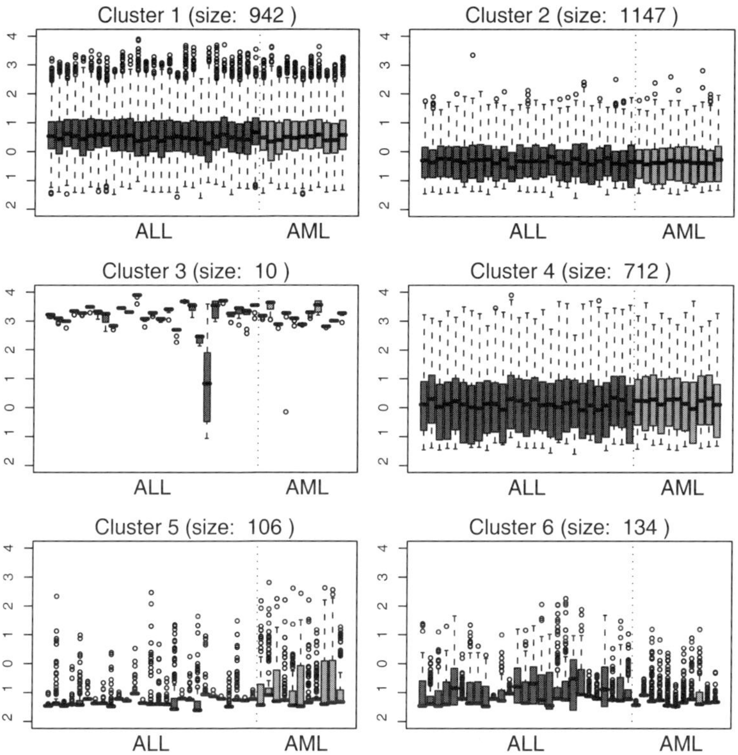

Figure 2 Box plots of the gene expression clustered by a partition estimate from the t-density clustering model with Gibbs WCR

Lau and Lo (2005)] is employed to reduce the search time. This procedure is a dimensional reduction technique. A block of 20 patients (out of 38) is randomly selected, and a Gibbs WCR reseating cycle is completed to obtain a new partition of the 3051 gene expression levels. For the next Gibbs cycle, another randomly sampled block of 20 patients is used. The reduced prior parameters are $(40, 0.1\mathbf{I}_{20\times 20}, \mathbf{0}_{20}, 0.1)$. The Gibbs sampler is initiated with an initial partition with $n = 3051$ singleton clusters. Among these iterations, the partition that maximizes the criterion function $J(\mathbf{p})$ is obtained. The partition has 6 clusters. Figure 1 shows the profile plots of the gene expression levels separated by the 6 clusters (The cluster sizes are located next to the cluster numbers.) Figure 2 shows the boxplots of the gene expression levels for each patient separated by the 6 clusters. The boxplots reveal the average and the variation differences of the gene expression levels between clusters. The gene expression levels vary constantly for

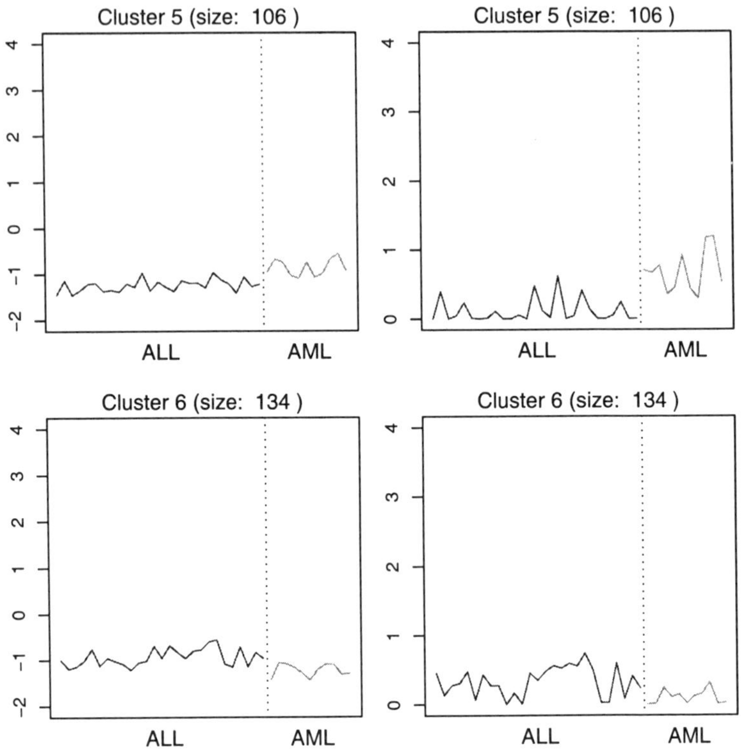

Figure 3 Means and Variances of Cluster 5 and 6 across patients from the t-density clustering model with Gibbs WCR. Left Column: Means across the patients of Cluster 5 and 6; Right Column: Variances across the patients of Cluster 5 and 6

patients in clusters 1, 2, and 4, which is not the case for patients in clusters 3, 5, and 6.

The Gibbs WCR successfully separates AML patients and ALL patients in terms of gene expression levels in both cluster 5 and cluster 6. For this particular gene expression data, Golub et al.(1999) defines "idealized expression pattern;" that is, a class of patients have uniformly high gene expression levels and the other class have uniformly low gene expression levels. The patterns are revealed in clusters 5 and 6, AML patients and ALL patients perform oppositely in both clusters. Compared with ALL patients, AML patients have uniformly higher averages and variations in Clusters 5, and yet uniformly lower averages and variations in Cluster 6 [Figure 3 plots the sample means and sample variances of the gene expression levels across patients of both clusters.] Thus, gene expression levels of AML and ALL

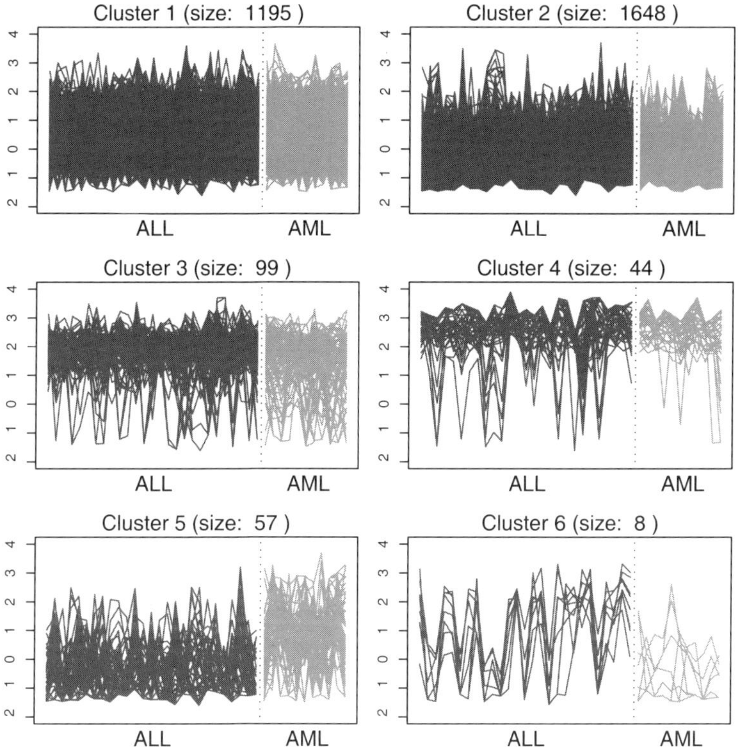

Figure 4 Profile plots of the gene expression clustered by a partition estimate from the complete linkage method

patients are well separated in clusters 5 and 6.

Gene clustering looks for both known and unknown patterns across samples, e.g. Golub et al. (1999)'s "idealized expression pattern". The multivariate t-density clustering model is an appropriate model to discover the patterns as the multivariate t-density clustering model allows clusters to have different covariances and different means. In each cluster, each patient has his/her own mean, own variance and covariance with other patients. This flexible feature prompts the success of the t-density clustering model. Upon setting the number of clusters at 6, several hierarchical agglomerative methods, single linkage, average linkage, median linkage, centroid, ward [Chapter 4, Gordon (1989)], and the K-mean method are implemented and the performances are compared. The hierarchical agglomerative methods use Euclidean distance as the similarity measure. The t-density model performs very well, and has highest $J(\mathbf{p})$ among other selected methods,

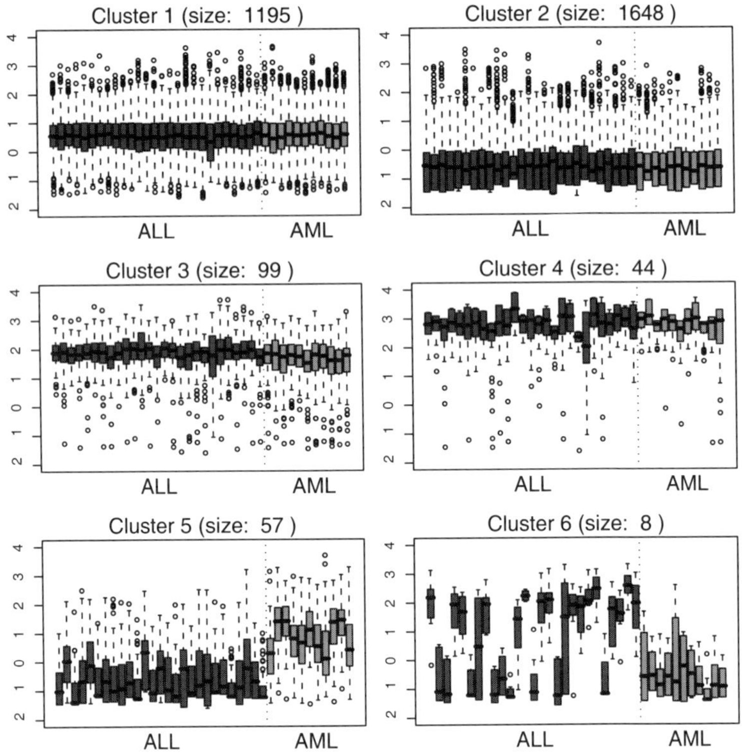

Figure 5 Box plots of the gene expression clustered by a partition estimate from the complete linkage method

Table 1 The criterion quantities (13) under the partitions generated by the selected clustering methods. The parameters are $(40, 0.1\mathbf{I}_{38\times 38}, \mathbf{0}_{38}, 0.1)$ and $e_0 = 1$

Methods	$J(\mathbf{p})$
Multivariate t clustering model with Gibbs WCR	-69678.83
Median linkage	-72240.54
Single linkage	-72395.95
Centroid	-73546.03
Average linkage	-73880.66
Complete linkage	-78216.41
Ward	-84084.50
K-mean	-84782.51

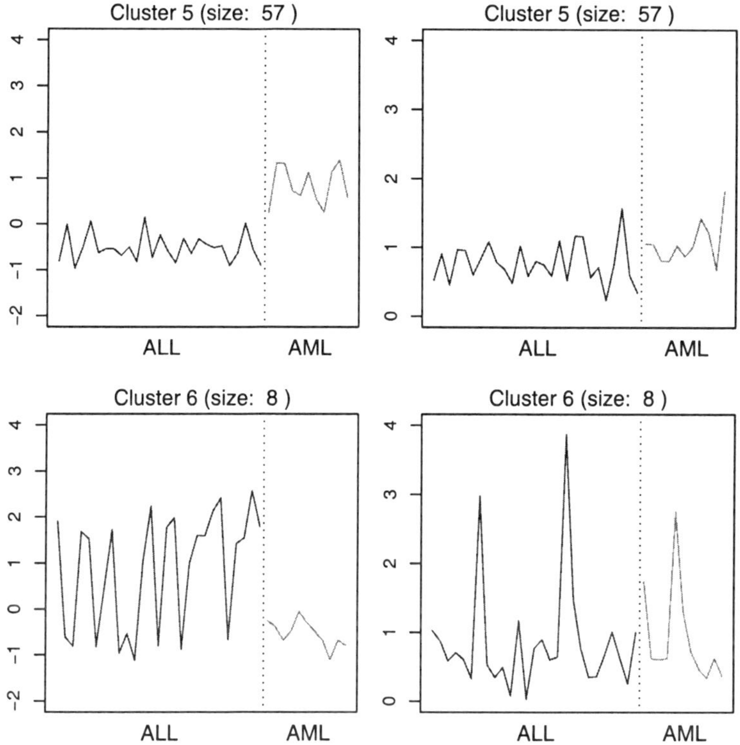

Figure 6 Means and Variances of Cluster 5 and 6 across patients from the complete linkage method. Left Column: Means across the patients of Cluster 5 and 6; Right Column: Variances across the patients of Cluster 5 and 6

see Table 1. With respect to Golub et al. (1999)'s "idealized expression pattern," the complete linkage method provides clearly better separation than the other selected methods. Figures 4 and 5 show the profile plots and the boxplots of the gene expression levels in the six clusters obtained by the complete linkage method. Figure 6 shows the sample means and sample variances across the patients of clusters 5 and 6. Compared with ALL patients, AML patients have higher averages and variations in Clusters 5. On the other hand, AML patients have lower averages and higher variations in Cluster 6; furthermore, the averages and variances do not exhibit a uniform pattern.

Acknowledgements

The paper is written on the occasion of Kjell A. Doksum's 65th birthday. The authors are grateful to Jerry Brunner and Javier Cabrera for enlightening discussions on WCR clustering, and to an anonymous referee for pointing out some key references.

References

1. Amaratunga, D. and Cabrera, J. (2004). *Exploration and analysis of DNA microarray and protein array data.* John Wiley and Sons.
2. Banfield, J. D. and Raftery, A. E. (1993). Model-based Gaussian and Non-Gaussian Clustering. *Biometrics*, **49**, 803–821
3. Barry, D. and Hartigan, J. A. (1992). Product partition models for change point problems. *The Annals of Statistics*, **20**, 260–279
4. Basu, S. and Chib, S. (2003). Marginal likelihood and Bayes factors for Dirichlet process mixture models. *Journal of the American Statistical Association*, **98**, 224–235
5. Brunner, L. J. and Lo, A. Y. (1999). Bayesian Classifications. *Preprint.* University of Toronto, Canada. Available at http://www.erin.utoronto.ca/~jbrunner/papers/BayesClass.pdf
6. Cabrera, J., Lau, J. W. and Lo, A. Y. (2005). Unsupervised Learning and Cluster Analysis. *Research Report*, ISMT Department, The University of Science and Technology, Hong Kong.
7. Crowley, E. M. (1997). Product partition models for normal means. *Journal of the American Statistical Association*, **92**, 192–198
8. De Groot, M. H. (1970). *Optimal statistical decisions.* McGraw-Hill.
9. Duda, R. O., Hart P. E. and Stork, D. G. (2001). *Pattern Classification.* John Wiley & Sons.
10. Fraley, C. and Raftery, A. E. (2002). Model-based clustering, Discriminant analysis, and density estimation. *Journal of the American Statistical Association*, **97**, 611–631.
11. Friedman, H. P. and Rubin, J. (1967). On Some Invariant Criteria for Grouping Data. *Journal of the American Statistical Association*, **62**, 1159–1178.
12. Geman, A. and Geman, D. (1984). Stochastic relaxation, Gibbs distributions, and the Bayesian restoration of images. *IEEE Transactions on Pattern Analysis and Machine Intelligence*, **6**, 721-741.
13. Golub, T. R., Slonim, D. K., Tamayo, P., Huard, C., Gaasenbeek, M., Mesirov, J. P., Coller, H., Loh, M. L., Downing, J. R.,

CALIGIURI, M. A., BLOOMFIELD, C. D. AND LANDER, E. S. (1999). Molecular classification of cancer: class discovery and class prediction by gene expression monitoring. *Science*, **286**, 531–537.

14. GORDON, A. D. (1999). *Classification*, 2nd edition. Chapman & Hall
15. HARTIGAN, J. A. (1990). Partition models. *Communications in Statistics, Part A - Theory and Methods*, **19**, 2745–2756
16. HASTINGS, W. K. (1970). Monte Carlo sampling methods using Markov chains and their applications. *Biometrika*, **57**, 97–109.
17. ISHWARAN, H. AND JAMES, L. F. (2003a). Generalized weighted Chinese restaurant processes for species sampling models. *Statisica Sinica*, **13**, 1211–1235.
18. ISHWARAN, H. AND JAMES, L. F. (2003b). Some further developments for stick-breaking priors: finite and infinite clustering and classification. *Sankhya Series A*, **65**, 577–592.
19. ITO, K. (2004). *Stochastic Processes: Lectures given at Aarhus University*, Springer-Verlag, Berlin-Heidelberg.
20. JAMES, L. F. (2002). Poisson process partition calculus with applications to exchangeable models and Bayesian Nonparametrics. Available at http://arXiv.org/abs/math/0205093
21. JAMES, L. F. (2005). Bayesian Poisson process partition calculus with an application to Bayesian Levy moving averages. *The Annals of Statistics*, **33**, 1771–1799.
22. KUNCHEVA, L. I. (2004). *Combining pattern classifiers: methods and algorithms.* John Wiley & Sons.
23. LAU, J. W. AND SO, M. K. P. (2004). Bayesian Mixture of Autoregressive Models. *Research Report.* ISMT Department, Hong Kong University of Science and Technology, Hong Kong.
24. LO, A. Y. (1984). On a class of Bayesian nonparametric estimates: I. Density estimates. *The Annals of Statistics*, **12**, 351–357.
25. LO, A. Y. (1999). *Notes for Bayesian nonparametric statistical methods and related topics. (Unpublished lecture notes.)* ISMT Department, Hong Kong University of Science and Technology, Hong Kong.
26. LO, A. Y. (2005). Weighted Chinese restaurant processes. *COSMOS*, **1**, 59–63. World Scientific Publishing Co..
27. LO, A. Y., BRUNNER, L. J. AND CHAN, A. T. (1996). Weighted Chinese restaurant processes and Bayesian mixture models. *Research Report*, ISMT Department, The University of Science and Technology, Hong Kong. Available at http://www.erin.utoronto.ca/~jbrunner/papers/wcr96.pdf
28. LO, A. Y. AND WENG, C. S. (1989). On a class of Bayesian nonparametric estimates. II. Hazard rate estimates. *Annals of the Institute of Statistical Mathematics series A*, **41**, 227–245.

29. LOEVE, M. (1977). *Probability Theory. Vol. II*, 4th Ed., Springer-Verlag, New York-Berlin-Heidelberg.

30. MACEACHERN, S. N. (1994). Estimating normal means with a conjugate style Dirichlet Process Prior. *Communications in statistics – Simulation and computation*, **23**, 727–741.

31. MCLACHLAN, G. AND BASFORD, K. (1988). *Mixture models: Inference and Applications to Clustering*, Marcel Dekker, New York.

32. PROPP, J. AND WILSON, D. (1996). Exact sampling with coupled Markov chains and applications to statistical mechanics. *Random structure and algorithm*, **9**, 223-252.

33. QUINTANA, F. A. AND IGLESIAS, P. L. (2003). Bayesian clustering and product partition models. *Journal of the Royal Statistical Society, Series B*, **65**, 557–574

34. SCOTT, A. J. AND SYMONS, M. J. (1971). Clustering methods based on likelihood ratio criterteria. *Biometrics*, **27**, 387–397

35. SYMONS, M. J. (1981). Clustering criteria and multivariate normal mixtures. *Biometrics*, **37**, 35–43

36. WOLFE, J. H. (1970). Pattern clustering by multivariate mixture analysis. *Multivariate Behavioral Research*, **5**, 329–350

Chapter 21

NEUTRAL-TO-THE-RIGHT SPECIES SAMPLING MIXTURE MODELS

Lancelot F. James

Department of Information and Systems Management
Clear Water Bay, HONG KONG

E-mail: lancelot@ust.hk

This paper describes briefly how one may utilize a class of species sampling mixture models derived from Doksum's (1974) neutral to the right processes. For practical implementation we describe an ordered/ranked variant of the generalized weighted Chinese restaurant process.

Key words: Chinese Restaurant process, Dirichlet process, Lévy process, Neutral to the right process, Species sampling model.

1 Introduction

The field of Bayesian nonparametric statistics involves the idea of assigning prior and posterior distributions over spaces of probability measures or more general measures. That is, similar to the classical parametric Bayesian idea of assigning priors to an unknown parameter, say θ, which lies in a Euclidean space, one views, for instance, an unknown cumulative distribution function, say $F(t)$, as being a stochastic process. More generally for an unknown probability measure P, a Bayesian views it as a random probability measure. This is currently a well-developed and active area of research that has links to a variety of areas where Lévy and more general random processes are commonly used. However, as discussed in Doksum and James (2004), in the late 1960's, noting the high activity and advance in nonparametric statistics, David Blackwell and others wondered how one could assign priors which were both flexible and tractable. Arising from these questions were two viable answers which till this day remain at the cornerstone of Bayesian nonparametric statistics.

Ferguson (1973, 1974) proposed the use of a Dirichlet process prior [see also Freedman (1963)]. For this prior if P is a probability on some space $\mathscr{X}$, and $(B_1, \ldots, B_k)$ is a measurable partition of $\mathscr{X}$, then $P(B_1), \ldots, P(B_k)$ has a Dirichlet distribution. Moreover, the posterior distribution of P given a sample $\mathbf{X} = (X_1, \ldots, X_n)$ is also a Dirichlet process. For a specified probability measure H and a scalar $\theta > 0$, one can say that $P :\stackrel{d}{=} P_{\theta H}$ is a Dirichlet process with shape parameter θH, if the Dirichlet distributions discussed above have parameters given by $\mathbb{E}[P(A_i)] = \theta H(A_i)$.

Following this, Doksum (1974) introduced the class of Neutral to the Right (NTR) random probability measures on the real line. In these models, if P is a distribution on the real line, then for each partition $B_1, \ldots, B_k$, with $B_j = (s_{j-1}, s_j]$, $j = 1, \ldots, k$, $s_0 = -\infty, s_k = \infty$, $s_i < s_j$ for $i < j$; $P(B_1), \ldots, P(B_k)$ is such that $P(B_i)$ has the same distribution as $V_i \prod_{j=1}^{i-1}(1 - V_j)$, where $V_1, \ldots, V_2, \ldots$ is a collection of independent non-negative random variables. This represents a remarkably rich choice of models defined by specifying different distributions for the V_i. Notably if V_i is chosen to be beta random variable with parameters (α_i, β_i) and $\beta_i = \sum_{j=1}^{k-1} \alpha_j$, then this gives the Dirichlet process as described in Doksum (1974). Doksum (1974) shows that if P is a NTR distribution then the posterior distribution of P give a sample $X_1, \ldots, X_n$ is also an NTR. Subsequently, Ferguson and Phadia (1979) showed that this type of conjugacy property extends to the case of right censored survival models. This last fact coupled with the subsequent related works of Hjort (1990), Kim (1999), Lo (1993) and Walker and Muliere (1997) have popularized the usage of NTR processes in models related to survival and event history analysis.

Despite these attractive points, the usage of NTR processes in more complex statistical models, such as mixture models, has been notably absent. This is in contrast to the Dirichlet process which, coupled with the advances in MCMC and other computational procedures, is regularly used in nonparametric or semi-parametric statistical models. The theoretical framework for Dirichlet process mixture models can be traced back to Lo (1984) who proposed to model a density as a convolution mixture model of a known kernel density $K(y|x)$ and a Dirichlet process P as,

$$f(y|P) = \int_{\mathscr{X}} K(y|x)P(dx). \tag{1}$$

This may be equivalently expressed in terms of a missing data model where for a sample $\mathbf{Y} = (Y_1, \ldots, Y_n)$ based on (1), one has $Y_1, \ldots, Y_n|\mathbf{X}, P$ are such that Y_i are independent with distributions $K(\cdot|X_i)$, $X_i|P$ are iid P and P is a Dirichlet process. It is clear that the description of the posterior distribution of P and related quantities is much more complex than in the setting discussed in Ferguson (1973). However, Lo (1984) shows that its description is facilitated by the descriptions of the posterior distribution of $P|\mathbf{X}$ given by Ferguson (1973) and the exchangeable marginal distribution of $\mathbf{X}$ discussed in Blackwell and MacQueen (1973). Blackwell and MacQueen describe the distribution via what is known as the Blackwell-MacQueen Pólya urn scheme where $\mathbb{P}(X_1 \in A) = H(A)$ and for $n > 1$

$$\mathbb{P}(X_n \in \cdot|X_1, \dots, X_{n-1}) = \frac{\theta}{\theta + n - 1} H(\cdot) + \frac{1}{\theta + n - 1} \sum_{j=1}^{n-1} \delta_{X_i}(\cdot). \quad (2)$$

Note that (2) clearly indicates that there can be ties among $(X_1, \dots, X_n)$ and that the $n(\mathbf{p}) \le n$ unique values, say $X_1^*, \dots, X_{n(\mathbf{p})}^*$ are iid with common distribution H. Letting $\mathbf{p} = \{C_1, \dots, C_{n(\mathbf{p})}\}$ denote a partition of the integers $\{1, \dots, n\}$, where one can write $C_j = \{i : X_i = X_j^*\}$, with size $n_j = |C_j|$ for $j = 1, \dots, n(\mathbf{p})$. This leads to the following important description of the distribution of $\mathbf{X}$,

$$\pi(d\mathbf{X}|\theta H) = \mathrm{PD}(\mathbf{p}|\theta) \prod_{j=1}^{n(\mathbf{p})} H(dX_j^*)$$

where

$$PD(\mathbf{p}|\theta) = \frac{\theta^{n(\mathbf{p})}\Gamma(\theta)}{\Gamma(\theta + n)} \prod_{j=1}^{n(\mathbf{p})} (n_j - 1)! := p_\theta(n_1, \dots, n_{n(\mathbf{p})})$$

is a variant of Ewens sampling formula [see Ewens (1972) and Antoniak (1974)], often called the Chinese restaurant process. It can be interpreted as $\mathbb{P}(C_1, \dots, C_{n(\mathbf{p})}) = p_\theta(n_1, \dots, n_{n(\mathbf{p})})$ where p_θ, being symmetric in its arguments, is the most notable example of an *exchangeable partition probability function*(EPPF) [see Pitman (1996)]. It is easily seen that a Dirichlet Process with shape θH is characterized by the pair (p_θ, H). Letting $p(n_1, \dots, n_k)$, for $n(\mathbf{p}) = k$, denote an arbitrary EPPF, Pitman (1996) shows that the class of random probability measures whose distribution is completely determined by the pair (p, H) must correspond to the class of *species sampling random probability measures.* General species sampling random probability measures constitute all random probability measures that can be represented as

$$P(\cdot) = \sum_{i=1}^{\infty} P_i \delta_{Z_i}(\cdot) + (1 - \sum_{k=1}^{\infty} P_k) H(\cdot) \quad (3)$$

where $0 \le P_i < 1$ are random weights such that $0 < \sum_{i=1}^{\infty} P_i \le 1$, independent of the Z_i which are iid with some non-atomic distribution H. Furthermore the law of the (P_i) is determined by the EPPF p. Noting these points, Ishwaran and James (2003) described the class of species sampling mixture models by replacing a Dirichlet process in (1) with P specified by (3). See also Müller and Quintana (2004).

Except for the special case of the Dirichlet process, NTR processes are not species sampling models and this is one of the factors which makes analysis a bit difficult. Nonetheless, James (2003, 2006) was able to extend the definition of NTR processes to a class of random probability measures on more general spaces, which he called Spatial NTR processes. Additionally a tractable description of the marginal distribution of this class of models was obtained. These two ingredients then allow for the implementation of NTR mixture models. Our goal here is not to describe the mechanisms for a full-blown NTR mixture model, as this requires much more overhead, but rather mixture models based on species sampling models which are derived from NTR processes. James (2003, 2006) introduced these *NTR species sampling models.* Quite specifically, though the NTR processes are not species sampling models they produce EPPF's p that, along with the specification of H, are uniquely associated with an NTR species sampling model. This produces a very rich and flexible class of random priors that are a bit simpler analytically than NTR processes. An interesting fact is that this class contains the two-parameter (α, θ) Poisson-Dirichlet random probability measures for parameters $0 \leq \alpha < 1$ and $\theta > 0$. That is the Dirichlet process and a class of random probabilities defined by normalizing a stable law process and further power tempering the stable law distribution, which are discussed in Pitman (1996) and Pitman and Yor (1997). Implementations of these latter models, being quite special, may be treated by computational procedures involving random partitions discussed in Ishwaran and James (2003) or by the methods in Ishwaran and James (2001). Here we will discuss a ranked weighted Chinese restaurant procedure which applies more generally.

2 NTR and related processes

2.1 *NTR processes*

Let $F(t)$ denote a cumulative distribution function on the positive real line. Additionally, let $S(t) = 1 - F(t)$ denote a survival function. Doksum (1974, Theorem 3.1) shows that F is an NTR process if and only if it can be represented as

$$F(t) = 1 - e^{-Y(t)} \tag{4}$$

where $Y(t)$ is an independent increment process which is non-decreasing and right continuous almost surely and furthermore $\lim_{t\to\infty} Y(t) = \infty$ and

$\lim_{t\to-\infty} Y(t) = 0$ almost surely. In other words Y belongs to the class of positive Lévy processes.

We shall suppose hereafter that T is a positive random variable such that, conditional on F, its distribution function is F where F is an NTR process. Then T has an interpretation as a survival time with "conditional" survival distribution $S(t) = 1 - F(t) := P(T > t|F)$. It is evident from (4) that the distribution of F is completely determined by the law of Y which is determined by its Laplace transform

$$E\left[\mathrm{e}^{-\omega Y(t)}\right] = \mathrm{e}^{-\int_0^t \phi(\omega|s)\Lambda_0(ds)} := E[(S(t))^{\omega}]$$

where $\phi(\omega|s)$ is equal to

$$\begin{aligned}\int_0^{\infty}(1-\mathrm{e}^{-v\omega})\tau(dv|s) &= \int_0^1 (1-(1-u)^{\omega})\rho(du|s)\\ &= \int_0^1 \omega(1-u)^{\omega-1}\left[\int_u^1 \rho(dv|s)\right]du, \qquad (5)\end{aligned}$$

τ and ρ are Lévy densities on $[0,\infty]$ and $[0,1]$ respectively which are in correspondence via the mapping $y \to 1 - e^{-y}$. Without loss of generality we shall assume that $\int_0^1 u\rho(du|s) = 1$ for each fixed s, which implies that $\phi(\omega|s) = 1$. Hence we have that

$$E[S(t)] = \mathrm{e}^{-\Lambda_0(t)} = 1 - F_0(t)$$

where F_0 represents one's prior belief about the true distribution and $\Lambda_0(dt) = F_0(dt)/S_0(t-)$ is its corresponding cumulative hazard with $S_0(t-) = 1 - F_0(t-) = \mathbb{P}(T \geq t)$.

Note that for each fixed s, $\phi(\omega|s)$ corresponds to the log Laplace transform of an infinitely-divisible random variable. It follows that different specifications for τ or equivalently ρ lead to different NTR processes. When τ and ρ do not depend on s, then F, Y, and all relevant functionals are said to be *homogeneous*. We also apply this name to τ and ρ. Additionally $\phi(\omega|s)$ specializes to

$$\begin{aligned}\phi(\omega) := \int_0^{\infty}(1-\mathrm{e}^{-v\omega})\tau(dv) &= \int_0^1 (1-(1-u)^{\omega})\rho(du)\\ &= \int_0^1 \omega(1-u)^{\omega-1}\left[\int_u^1 \rho(dv)\right]du.\end{aligned}$$

Consider now the cumulative hazard process of F, say Λ, defined by $\Lambda(dt) = F(dt)/S(t-)$. The idea of Hjort (1990) was to work directly with Λ rather than F. He showed importantly that if one specified Λ to be a positive completely random measure on $[0,1]$, whose law is specified by the Laplace transform

$$\mathbb{E}[\mathrm{e}^{-\omega\Lambda(t)}] = \mathrm{e}^{-\int_0^t \psi(\omega|s)\Lambda_0(ds)}$$

where $\psi(\omega|s) := \int_0^1 (1-e^{-u\omega})\rho(du|s)$, then F and S must be NTR processes specified by (5). James (2003, 2006) shows that one can extend the definition of an NTR process to a spatial NTR process on $[0,\infty] \times \mathscr{X}$ by working with the concept of a random hazard measure, say $\Lambda_H(dt, dx)$. Λ_H is a natural extension of Λ in the sense that $\Lambda_H(dt, \mathscr{X}) = \Lambda(dt)$ and is otherwise specified by replacing the intensity $\rho(du|s)\Lambda_0(ds)$ by $\rho(du|s)\Lambda_0(ds, dx)$, where,

$$\Lambda_0(ds, dx) = H(dx|s)\Lambda_0(ds)$$

is a hazard measure and $H(\cdot|s)$ may be interpreted as the conditional distribution of $X|T = s$. A spatial NTR process (SPNTR) is then defined as

$$P_S(dt, dx) = S(t-)\Lambda_H(dt, dx) \tag{6}$$

The SPNTR in (6) has marginals such that $P_S(dt, d\mathscr{X}) = F(dt)$ is an NTR and

$$P_S([0,\infty), dx) = \int_0^\infty S(t-)\Lambda_H(ds, dx), \tag{7}$$

represents an entirely new class of random probability measures.

2.2 *NTR species sampling models*

NTR species sampling models arise as a special case of (7) by setting $H(dx|s) := H(dx)$. Here we will work only with the class of homogeneous processes and hence we will additionally choose $\rho(du|s) = \rho(du)$. Thus an NTR species sampling model is of the form

$$P_{\rho,H}(dx) = \int_0^\infty S(s-)\Lambda_H(ds, dx) = \sum_{k=1}^\infty P_k \delta_{Z_k}(dx).$$

Furthermore, if $P \stackrel{d}{=} P_{\rho,H}$, we denote its law as $\mathscr{P}(\cdot|\rho, H)$. It follows that for practical usage in mixture models one needs a tractable description of the corresponding EPPF, say p_ρ. However, before we do that we will need to introduce additional notation which connects p_ρ with the NTR process. If we suppose that $X_1, \ldots, X_n|P_{\rho,H}$ are iid with distribution $P_{\rho,H}$, then these points come from a description of the n conditionally independent pairs $(T_1, X_1), \ldots, (T_n, X_n)|P_S$ where (T_i, X_i) are iid P_S, such that T_i are iid F, where F is an NTR, and X_i are iid $P_{\rho,H}$. Here P_S must be specified by the intensity $\rho(du)\Lambda_0(ds)H(dx)$. Now if one denotes the $n(\mathbf{p})$ unique pairs as (T_j^*, X_j^*) for j=1,..., n($\mathbf{p}$), then one may simply set each $C_j = \{i : T_i = T_j^*\}$. Furthermore we define $T_{(1:n)} > T_{(2:n)} > \ldots > T_{(n(\mathbf{p}):n)} > 0$ to be the ordered values of the unique values $(T_j^*)_{j \leq n(\mathbf{p})}$. Hence we can define $\mathbf{p}$ by setting $C_j := \{i : T_i = T_j^*\}$, and define $\mathbf{m} = \{D_1, \ldots, D_{n(\mathbf{p})}\}$ with cells $D_j = \{i : T_i = T_{(j:n)}\}$ with cardinality $d_j = |D_j|$. It is evident that given a partition $\mathbf{p} = \{C_1, \ldots, C_{n(\mathbf{p})}\}$, $\mathbf{m}$ takes its values over the symmetric group, say $\mathscr{S}_{n(\mathbf{p})}$, of all $n(\mathbf{p})!$ permutations of $\mathbf{p}$. Let $R_{j-1} = \bigcup_{k=1}^{j-1} D_k := \{i : T_i > T_{(j:n)}\}$ with cardinality $r_{j-1} = \sum_{k=1}^{j-1} d_k$. Then, in terms of survival analysis, the quantities d_j and $r_j = d_j + r_{j-1}$ have the interpretation as the number of deaths at time $T_{(j:n)}$, and the number at risk at time $T_{(j:n)}$, respectively. See James (2006) for further elaboration. Now from James (2003, 2006) it follows that

$$\pi_\rho(\mathbf{p}) = p_\rho(n_1, \ldots, n(\mathbf{p})) = \sum_{\mathbf{m} \in \mathscr{S}_{n(\mathbf{p})}} \frac{\prod_{j=1}^{n(\mathbf{p})} \kappa_{d_j, r_{j-1}}(\rho)}{\prod_{j=1}^{n(\mathbf{p})} \phi(r_j)} \tag{8}$$

where

$$\kappa_{d_j, r_{j-1}}(\rho) = \int_0^1 u^{d_j}(1-u)^{r_{j-1}} \rho(du).$$

The form of the EPPF is in general not tractable. However by augmentation one sees that the distribution of $\mathbf{m}$ is given by

$$\pi_\rho(\mathbf{m}) = \frac{\prod_{j=1}^{n(\mathbf{p})} \kappa_{d_j, r_{j-1}}(\rho)}{\prod_{j=1}^{n(\mathbf{p})} \phi(r_j)} \tag{9}$$

and has a nice product form. This suggests that one can work with a joint distribution of $(\mathbf{X}, \mathbf{m})$ given by

$$\pi_\rho(\mathbf{m}) \prod_{j=1}^{n(\mathbf{p})} H(dX_j^*).$$

Related to this, James (2006) shows that a prediction rule of $X_{n+1}|\mathbf{X}, \mathbf{m}$ is given by

$$\mathbb{P}(X_{n+1} \in dx | \mathbf{X}, \mathbf{m}) = (1 - \sum_{j=1}^{n(\mathbf{p})} p_{j:n}) P_0(dx) + \sum_{j=1}^{n(\mathbf{p})} p_{j:n} \delta_{X_j^*}(dx),$$

with $(1 - \sum_{j=1}^{n(\mathbf{p})} p_{j:n}) = \sum_{j=1}^{n(\mathbf{p})+1} q_{j:n}$, and where

$$p_{j:n} = \frac{\kappa_{d_j+1, r_{j-1}}(\rho) \prod_{l=j+1}^{n(\mathbf{p})} \kappa_{d_l, r_{l-1}+1}(\rho)}{\kappa_{d_j, r_{j-1}}(\rho) \prod_{l=j+1}^{n(\mathbf{p})} \kappa_{d_l, r_{l-1}}(\rho)} \prod_{l=j}^{n(\mathbf{p})} \frac{\phi(r_l)}{\phi(r_l+1)},$$

and

$$q_{j:n} = \frac{\kappa_{1, r_{j-1}}(\rho)}{\phi(r_{j-1}+1)} \frac{\prod_{l=j}^{n(\mathbf{p})} \kappa_{d_l, r_{l-1}+1}(\rho)}{\prod_{l=j}^{n(\mathbf{p})} \kappa_{d_l, r_{l-1}}(\rho)} \prod_{l=j}^{n(\mathbf{p})} \frac{\phi(r_l)}{\phi(r_l+1)},$$

with $q_{n(\mathbf{p})+1:n} = \kappa_{1,n}(\rho)/\phi(n+1)$, are transition probabilities derived from $\pi_\rho(\mathbf{m})$. Note that in the calculation of $\kappa_{1,r_{j-1}}(\rho)$, $r_{j-1}+1$ is to be used rather than $r_j = r_{j-1} + m_j$. As an example, consider the choice of a homogeneous beta process [Hjort (1990), see also Ferguson (1974), Ferguson and Phadia (1979) and Gnedin (2004)] defined by

$$\rho(du) = \theta u^{-1}(1-u)^{\theta-1}du.$$

Then it is easily seen that $\phi(r_j) = \sum_{l=1}^{r_j} \theta/(\theta+l-1)$, and it follows that in this case

$$p_{j:n} = \frac{d_j}{n+\theta} \prod_{l=j}^{n(\mathbf{p})} \frac{\phi(r_l)}{\phi(r_l+1)}, \quad \text{and}$$

$$q_{j:n} = \frac{1}{n+\theta} \frac{1}{\sum_{i=1}^{r_{j-1}+1} 1/(\theta+i-1)} \prod_{l=j}^{n(\mathbf{p})} \frac{\phi(r_l)}{\phi(r_l+1)}.$$

Remark 1. Gnedin and Pitman (2005a) also obtained the expressions (8) and (9) independent of James (2003, 2006) in a different context. See James (2006) for more details.

Remark 2. Related to this, Gnedin and Pitman (2005a) [see additionally Gnedin and Pitman (2005b)] showed that the EPPF in (8) corresponds to that of the two-parameter (α, θ) Poisson-Dirichlet process with parameters $0 \leq \alpha < 1$ and $\theta > 0$ if $\rho := \rho_{\alpha,\theta}$ is chosen such that

$$\int_u^1 \rho_{\alpha,\theta}(dv) = \frac{\Gamma(\theta+2-\alpha)}{\Gamma(1-\alpha)\Gamma(1+\theta)} u^{-\alpha}(1-u)^\theta.$$

From this, James (2006) deduced that $P_{\rho_{\alpha,\theta},H} = \sum_{k=1}^\infty W_k \prod_{i=1}^{k-1}(1 - W_i)\delta_{Z_k}$ where (W_k) are independent beta $(1-\alpha, \theta+k\alpha)$ random variables independent of the (Z_k) which are iid H. That is a two-parameter (α, θ) Poisson-Dirichlet process, for $0 \leq \alpha < 1$ and $\theta > 0$ can be represented as the marginal probability measure of a spatial NTR process, as described above. See Pitman and Yor (1997) and Ishwaran and James (2001) for more on the stick-breaking representation of the two parameter Poisson-Dirichlet process.

3 NTR species sampling mixture models

3.1 *General mixture models*

Now setting $P = P_{\rho,H}$ in (1) yields a special case of the species sampling models described in Ishwaran and James (2003). That is

$$\int_{\mathscr{X}} K(y|x)P_{\rho,H}(dx) = \int_{\mathscr{X}}\int_0^\infty K(y|x)S(s-)\Lambda_H(ds,dx) \qquad (10)$$

is called an NTR species sampling models. We look at the situation where $Y_1, \ldots, Y_n | P_{\rho,H}$ are iid with density or *pmf* (10). This translates into the hierarchical model

$$\begin{aligned} Y_i | X_i, P &\overset{ind}{\sim} K(Y_i | X_i) \text{ for } i = 1, \ldots, n \\ X_i | P &\overset{iid}{\sim} P \\ P &\sim \mathscr{P}(\cdot | \rho, H). \end{aligned} \tag{11}$$

Since we have a description of the EPPF, in principle the theoretical results and computational procedures described in Ishwaran and James (2003) apply. However, as we have noted, in general $\pi_\rho(\mathbf{p})$ is not as simple to work with as $\pi_\rho(\mathbf{m})$. So we develop here results that allows us to sample from a posterior distribution of $\mathbf{m}$ rather than partitions. We summarize these results in the next proposition.

Proposition 1. *Suppose that one has the model specified in (11). Then the following results holds*

(i) The distribution of $X_1, \ldots, X_n | \mathbf{Y}, \mathbf{m}$ is such that the unique values X_j^* for $j = 1, \ldots, n(\mathbf{p})$ are conditionally independent with distributions

$$\pi(dX_j^* | D_j) \propto H(dX_j^*) \prod_{i \in D_j} K(Y_i | X_j^*).$$

(ii) The posterior distribution of $\mathbf{m}|\mathbf{Y}$ is

$$\pi_\rho(\mathbf{m}|\mathbf{Y}) \propto \pi_\rho(\mathbf{m}) \prod_{j=1}^{n(\mathbf{p})} \int_{\mathscr{X}} \prod_{i \in D_j} K(Y_i | x) H(dx).$$

(iii) The posterior distribution of $\mathbf{p}|\mathbf{Y}$ is

$$\sum_{\mathbf{m} \in \mathscr{S}_{n(\mathbf{p})}} \pi_\rho(\mathbf{m}|\mathbf{Y}) \propto \pi_\rho(\mathbf{p}) \prod_{j-1}^{n(\mathbf{p})} \int_{\mathscr{X}} \prod_{i \in C_j} K(Y_i | x) H(dx).$$

From this result one can compute a Bayesian predictive density of $Y_{n+1} | \mathbf{m}, \mathbf{Y}$ as,

$$\begin{aligned} l(n) = f(Y_{n+1} | \mathbf{m}, \mathbf{Y}) = &\left[\sum_{j=1}^{n(\mathbf{p})+1} q_{j:n} \right] \int_{\mathscr{X}} K(Y_{n+1} | x) H(dx) \\ &+ \sum_{j=1}^{n(\mathbf{p})} p_{j:n} \int_{\mathscr{X}} K(Y_{n+1} | x) \pi(dx | D_j). \end{aligned}$$

A Bayesian density estimate analogous to Lo (1984) is then obtained by summing this expression relative to the distribution of $\mathbf{m}|\mathbf{Y}$.

Corollary 1. *For the model in Proposition 1, a Bayesian predictive density estimator of $Y_{n+1}|\mathbf{Y}$ is given by*

$$\mathbb{E}[f(Y_{n+1}|P)|\mathbf{Y}] = \sum_{\mathbf{p}} \sum_{\mathbf{m}\in S_{n(\mathbf{p})}} f(Y_{n+1}|\mathbf{m}, \mathbf{Y})\pi_\rho(\mathbf{m}|\mathbf{Y}).$$

3.2 *Ordered/Ranked generalized weighted Chinese restaurant processes*

The significance of the expression for the predictive density is that we can use $l(n)$ in precisely the same manner as the predictive densities given $\mathbf{p}, \mathbf{Y}$, used in Ishwaran and James (2003) [see also Lo, Brunner and Chan (1996)] to construct computational procedures for approximating posterior quantities. In fact, all the major computational procedures for Dirichlet process mixture models [see for instance Escobar (1994) and Escobar and West (1995)] utilize some type of predictive density. Here, in analogy to the gWCR algorithms in Lo, Brunner and Chan (1996) and Ishwaran and James (2003), we define a weighted version of the *Ordered/Ranked generalized Chinese restaurant process* developed in James (2003, 2006), to approximate a draw from $\pi_\rho(\mathbf{m}|\mathbf{Y})$ as follows. For each $n \geq 1$, let $\{D_{1:n}, \ldots, D_{n(\mathbf{p}):n}\}$, denote a seating configuration of the first n customers, where $D_{j:n}$ denotes the set of the n customers seated at a table with common rank j.

(i) Given this configuration, the next customer $n+1$ is seated at an occupied table $D_{j:n}$, denoting that customer $n+1$ is equivalent to the jth largest seated customers, with probability

$$\frac{p_{j:n}}{l(n)} \int_{\mathscr{X}} K(Y_{n+1}|x)\pi(dx|D_{(j:n)}) \tag{12}$$

for $j = 1, \ldots, n(\mathbf{p})$.

(ii) Otherwise, the probability that customer $n+1$ is new and is the jth largest among $n(\mathbf{p})+1$ possible ranks is,

$$\frac{q_{j:n}}{l(n)} \int_{\mathscr{X}} K(Y_{n+1}|x)H(dx) \tag{13}$$

for $j = 1, \ldots, n(\mathbf{p})+1$.

Similar to the gWCR SIS algorithms [see Ishwaran and James (2003, Lemma 2)], by appealing to the product rule of probability, repeating this procedure for customers $\{1, \ldots, n\}$, produces a draw of $\mathbf{m}$ from a density of $\mathbf{m}$ depending on $\mathbf{Y}$, say $q(\mathbf{m})$, that satisfies the relationship

$$L(\mathbf{m})q(\mathbf{m}) = \pi_\rho(\mathbf{m}) \prod_{j=1}^{n(\mathbf{p})} \int_{\mathscr{X}} \prod_{i \in D_j} k(Y_i|x)H(dx)$$

where $L(\mathbf{m}) = \prod_{i=1}^{n} l(i-1)$. Hence for any functional $h(\mathbf{m})$, it follows that

$$\sum_{\mathbf{p}} \sum_{\mathbf{m} \in S_{n(\mathbf{p})}} h(\mathbf{m})\pi_\rho(\mathbf{m}|\mathbf{Y}) = \frac{\sum_{\mathbf{p}} \sum_{\mathbf{m} \in S_{n(\mathbf{p})}} h(\mathbf{m})L(\mathbf{m})q(\mathbf{m})}{\sum_{\mathbf{p}} \sum_{\mathbf{m} \in S_{n(\mathbf{p})}} L(\mathbf{m})q(\mathbf{m})}. \tag{14}$$

If the functional $h(\mathbf{m})$ has a closed form, such as the predictive density $\mathbb{E}[f(y|P)|\mathbf{m}, \mathbf{Y}] = f(y|\mathbf{m}, \mathbf{Y})$, then one approximates (14) by using the rules in (12) and (13) to draw $\mathbf{m}$. Repeating this procedure say B times, results in iid realizations, say $(\mathbf{m}_{(b)})$ for $b = 1, \ldots, B$ and one can approximate (14) by

$$\frac{\sum_{b=1}^{B} h(\mathbf{m}_{(b)})L(\mathbf{m}_{(b)})}{\sum_{b=1}^{B} L(\mathbf{m}_{(b)})}.$$

When the kernels K are set to 1, this procedure reduces to that described in James (2003, 2006) producing an exact draw from $\pi_\rho(\mathbf{m})$. For more intricate models one can incorporate a draw from the unique values $X_1^*, \ldots, X_{n(\mathbf{p})}^*$ which has the same distribution that arises for the Dirichlet process. One can also incorporate draws from the posterior distribution of $P_{\rho,H}(dx)$ which is described in James (2006). Otherwise it is a simple matter to modify all the computational procedures discussed in Ishwaran and James (2003, Section 4).

3.3 *Normal mixture example*

One of the most studied and utilized Bayesian mixture models is the Normal mixture model. It is specified by the choice of

$$f_\sigma(y|P) = \int_{-\infty}^{\infty} \phi_\sigma(y - x)P(dx) \tag{15}$$

where

$$\phi_\sigma(z) = \frac{1}{\sqrt{2\pi}\sigma} \exp\left(-\frac{z^2}{2\sigma^2}\right)$$

is a Normal density and a natural candidate for density estimation. In the case of the Dirichlet process, this model was introduced by Lo (1984) and popularized by the development of feasible computational algorithms in Escobar (1994) and Escobar and West (1995). Suppose that $\{Y_i\}$ are iid with true density f_0, a recent result of Lijoi, Prünster and Walker (2005) shows that $f_\sigma(\cdot|P)$ in (15) based on very general random probability measures, and a suitable prior distribution for σ, have posterior distributions that are strongly consistent in terms of estimating the unknown density f_0 under rather mild conditions. In particular their result validates the use of rather arbitrary NTR species sampling models in this context with the classical choice of H set to be a Normal distribution with mean 0 and variance A. Here, setting $\sigma = \sqrt{\theta}$, one has

$$K(Y_i|X_i) = \frac{1}{\sqrt{2\pi\theta}} \exp\left(\frac{1}{2\theta}(Y_i - X_i)^2\right).$$

Using these specifications we present the details of the proposed algorithm:

(i) Customer $n+1$ is seated to a new table and assigned rank j among $n(\mathbf{p})+1$ possible ranks with probability

$$\frac{q_{j:n}}{\lambda_\theta(n+1)} \frac{1}{\sqrt{2\pi(\theta+A)}} \exp\left(-\frac{Y_{n+1}^2}{2(\theta+A)}\right)$$

(ii) Customer $n+1$ is seated to an existing table and is assigned rank j with probability

$$\frac{p_{j:n}}{\lambda_\theta(n+1)} \sqrt{\frac{\theta + Ad_j}{2\pi\theta[\theta + A(d_j+1)]}} \exp\left[-\frac{1}{2\theta}\left(Y_{n+1}^2 - \frac{A\sum_{i\in D_j} Y_i + Y_{n+1}}{\theta + A(d_j+1)} + \frac{A\sum_{i\in D_j} Y_i}{\theta + Ad_j}\right)\right]$$

(iii) Additionally each $X_j^*|\mathbf{Y}, \mathbf{m}, \theta$ is normally distributed with parameters

$$\frac{1}{\sigma_j} = \frac{d_j}{\theta} + \frac{1}{A} \text{ and } \mu_j = \frac{\sigma_j}{\theta} \sum_{i\in D_j} Y_i.$$

$\lambda_\theta(n+1)$ is the appropriate normalizing constant which is a special case of $l(n)$.

Remark 3. For comparison, the setup and notation we use is similar to that used in Ishwaran and James (2003, 6.1) which is based on weighted Chinese restaurant sampling of partitions $\mathbf{p}$.

4 Concluding remarks

We have given a brief account of how one can use Doksum's NTR models to create a new class of species sampling random probability measures which can be applied to complex mixture models. These models exhibit many features of the NTR models, in terms of clustering behavior, but as we have shown are simpler to use. Ideally one would like to describe parallel schemes for the more complex spatial NTR models. However, this requires a considerably more involved study which we shall report elsewhere. More details can be found in James (2003, 2006) where explicit examples can be easily constructed.

The representation in 4 is important as it connects NTR processes to a large body of work on exponential functionals of Lévy processes which have applications in many fields including physics and finance. For a recent survey, see Bertoin and Yor (2005). Some recent papers which exploit this representation and are directly linked to NTR processes are Epifani, Lijoi and Prünster (2003) and James (2003, 2006). Additionally, outside of a Bayesian context, there is a notable body of recent work which has some overlaps with James (2003, 2006) and hence NTR processes by Gnedin and Pitman (2005a) and subsequent papers Gnedin and Pitman (2005b), Gnedin and Pitman and Yor (2005) and Gnedin, Pitman and Yor (2006). Although outside of a specific Bayesian context these papers contain results which are relevant to statistical analysis such as results related to the behavior of the number of ties $n(\mathbf{p})$. The fact that these models arise from different considerations and different points of emphasis attests to their rich nature. It will be interesting to see what future connections will be made.

References

1. Antoniak, C. E. (1974) Mixtures of Dirichlet processes with applications to Bayesian nonparametric problems. *Ann. Statist.* **2**, 1152-1174.

2. Bertoin, J. and Yor, M. (2005) Exponential functionals of Lévy processes. *Probab. Surv.* **2**, 191-212.

3. Blackwell, D. and MacQueen, J. B. (1973) Ferguson distributions via Pólya urn schemes. *Ann. Statist.* **1**, 353-355.

4. Doksum, K. A. (1974) Tailfree and neutral random probabilities and their posterior distributions. *Ann. Probab.* **2**, 183-201.

5. Doksum, K. A. and James, L. F. (2004) On spatial neutral to the right processes and their posterior distributions. In *Mathematical Reliability: An Expository Perspective* (Editors: Mazzuchi, Singpurwalla and Soyer). International Series in Operations Research and Management Science. Kluwer Academic Publishers

6. Epifani, I., Lijoi, A. and Pruenster, I. (2003) Exponential functionals and means of neutral to the right priors. *Biometrika*, **90**, 791-808.

7. Escobar, M. D. (1994) Estimating normal means with the Dirichlet process prior. *J. Amer. Stat. Assoc.* **89**, 268-277

8. Escobar, M. D. and West, M. (1995) Bayesian density estimation and inference using mixtures. *J. Amer. Stat. Assoc.* **90**, 577-588.

9. Ewens, W. J. (1972) The sampling theory of selectively neutral alleles, *Theor. Popul. Biol.* **3**, 87-112

10. Ferguson, T. S. (1973) A Bayesian analysis of some nonparametric problems, *Ann. Statist.*, **1**, 209-230

11. Ferguson, T. S. (1974) Prior distributions on spaces of probability measures, *Ann. Statist.* **2**,615-629

12. Ferguson, T. S. and Phadia, E. (1979) Bayesian nonparametric estimation based on censored data, *Ann. Statist.* **7**, 163-186

13. Freedman, D. A. (1963) On the asymptotic behavior of Bayes estimates in the discrete case. *Ann. Math. Statist.* **34**, 1386-1403

14. Gnedin, A. V. (2004) Three sampling formulas. *Combin. Probab. Comput.* **13**, 185-193

15. Gnedin, A. V. and Pitman, J. (2005a) Regenerative composition structures, *Ann. Probab.* **33**, 445-479

16. Gnedin, A. V. and Pitman, J. (2005b) Self-similar and Markov composition structures. In *Representation Theory, Dynamical Systems, Combinatorial and Algorithmic Methods. Part 13* (Editor: A. A. Lodkin). Zapiski Nauchnyh Seminarov POMI, Vol. 326, PDMI, 59-84

17. Gnedin, A. V., Pitman, J. and Yor, M. (2005) Asymptotic laws for regenerative compositions: gamma subordinators and the like, *Probab. Th. and Rel. Fields.* Published online November 2005

18. Gnedin, A. V., Pitman, J. and Yor, M. (2006) Asymptotic laws for compositions derived from transformed subordinators, *Ann. Probab.* **34**, 468-492

19. Hjort, N. L. (1990) Nonparametric Bayes estimators based on Beta processes in models for life history data *Ann. Statist.* **18**,1259-1294

20. Ishwaran, H. and James, L. F. (2001) Gibbs sampling methods for stick-breaking priors, *J. Amer. Stat. Assoc.* **96**, 161-173

21. Ishwaran, H. and James, L. F. (2003) Generalized weighted Chinese restaurant processes for species sampling mixture models *Statistica Sinica* **13**, 1211-1235

22. James, L. F. (2003) Poisson calculus for spatial neutral to the right processes. Available at http://arxiv.org/abs/math.PR/0305053

23. James, L. F. (2006) Poisson calculus for spatial neutral to the right processes. *Ann. Statist.* **34**, 416-440

24. Kim, Y. (1999) Nonparametric Bayesian estimators for counting processes. *Ann. Statist.* **27**, 562-588

25. LIJOI, A., PRÜNSTER, I. AND WALKER, S. G. (2005) On consistency of nonparametric normal mixtures for Bayesian density estimation. *J. Amer. Stat. Assoc.* **100**, 1292-1296

26. LO, A. Y. (1993) A Bayesian bootstrap for censored data¿ *Ann. Statist.* **21**, 100-123

27. LO, A. Y. (1984) On a class of Bayesian nonparametric estimates: I. density estimates. *Ann. Statist.* **12**, 351-357

28. LO, A. Y., BRUNNER, L. J. AND CHAN, A. T. (1996) Weighted Chinese restaurant processes and Bayesian mixture model. Research Report Hong Kong University of Science and Technology

29. MÜLLER, P. AND QUINTANA, F. A. (2004) Nonparametric Bayesian data analysis. *Statist. Sci.* **19**, 95-110

30. PITMAN, J. (1996) Some developments of the Blackwell-MacQueen urn scheme. In *Statistics, Probability and Game Theory* (Editors: T.S. Ferguson, L.S. Shapley and J.B. Macqueen). IMS Lecture Notes-Monograph series, Vol 30, 245-267

31. PITMAN, J. AND YOR, M. (1997) The two-parameter Poisson-Dirichlet distribution derived from a stable subordinator. *Ann. Probab.* **25**, 855-900

32. WALKER, S. AND MULIERE, P. (1997) Beta-Stacy processes and a generalization of the Pólya-urn scheme. *Ann. Statist.* **25**, 1762-1780

PART 7
Bayesian Nonparametric Inference

Chapter 22

NONPARAMETRIC BAYESIAN INFERENCE ABOUT PERCENTILES

Richard A. Johnson and Songyong Sim

Department of Statistics
University of Wisconsin-Madison, Madison, WI, U.S.A.

Department of Statistics
Hallym University, Chuncheon, Kangwon-do, S. KOREA

E-mails: rich@stat.wisc.edu & sysim@hallym.ac.kr

We investigate the posterior distribution of a percentile and several percentiles In the Dirichlet process nonparametric setting. Our main result is an asymptotic expansion for the posterior distribution of a percentile that has a leading normal term. We also introduce a procedure for sampling from the posterior distribution.

Keywords: Empirical distribution; Posterior distribution.

1 Introduction

This research is motivated by the first author's study of statistical procedures for making inferences about the ratio of percentiles (Johnson and Hwang 2003). Here, we consider nonparametric Bayesian procedures for making inferences about one or more percentiles from the same distribution.

Let $\boldsymbol{X} = (X_1, X_2, \ldots, X_n)$ be a random sample from a distribution $G(\cdot|\xi)$ and let

$$\xi_p = F^{-1}(p) = \inf\{x|F(x) \geq p\}$$

be the 100p-th percentile of a $G(\cdot)$ for $0 < p < 1$. Without priors, the properties of percentiles are summarized in Serfling (1980).

We consider the Bayesian setting with a Dirichlet process, with prior measure α_0, over the possible cdf's F. Assume $0 < p_1 < p_2 < \cdots < p_k < 1$ and $t_1 < t_2 < \cdots < t_k$. Then given the observations $X_1, X_2, \ldots, X_n$,

$$\alpha(t_i) = \alpha(-\infty, t_i] = \#(X_j \leq t_i) + \alpha_0(t_i)$$

for $1 \le i \le k$, so

$$\begin{aligned}\alpha(t_i, t_{i+1}) &= \alpha(-\infty, t_{i+1}] - \alpha(-\infty, t_i] \\ &= \#(t_i < X_j \le t_{i+1}) + \alpha_0(t_i, t_{i+1})\end{aligned}$$

where $\alpha_0(t_i) = \alpha_0(-\infty, t_i]$ and $\alpha_0(t_i, t_{i+1}) = \alpha_0(-\infty, t_{i+1}] - \alpha_0(-\infty, t_i]$. For simplicity, we denote $\alpha_R = \alpha_0(\boldsymbol{R})$. By Ferguson (1973), the posterior distribution of F is Dirichlet process with the measure α and

$$\begin{aligned}(F(t_1), &F(t_2) - F(t_1), \ldots, F(t_k) - F(t_{k-1}), 1 - F(t_k)) \\ &\sim \text{Dirichlet}(\alpha(t_1), \alpha(t_1, t_2), \ldots, n + \alpha_R - \alpha(t_k)),\end{aligned}$$

for $t_1 < t_2 < \cdots < t_k$.

Then, for $i < j$, it is well known that the marginal and conditional posterior distributions of $F(t_i)$, $F(t_j) - F(t_i)$ and $(F(t_j) - F(t_i))/(1 - F(t_i))|_{F(t_i)}$ are

$$F(t_i) \sim \text{Beta}(\alpha(t_i), n + \alpha_R - \alpha(t_i)),$$
$$F(t_j) - F(t_i) \sim \text{Beta}(\alpha(t_i, t_j), n + \alpha_R - \alpha(t_i, t_j))$$

and

$$\left.\frac{F(t_j) - F(t_i)}{1 - F(t_i)}\right|_{F(t_i)} \sim \text{Beta}(\alpha(t_i, t_j), n + \alpha_R - \alpha(t_i) - \alpha(t_i, t_j)). \quad (1)$$

(see Johnson, Kotz and Balakrishnan 2002 and Ferguson 1973)

In this chapter, we consider the joint posterior distribution of the percentiles. In Section 2, we propose a method to generate random quantities from the posterior distribution. Examples are given in Section 3. In Section 4 we establish asymptotic joint normality for the posterior distribution of several percentiles. Then, we employ these results to obtain an asymptotic expansion of the posterior distribution of a percentile in Section 5. Numerical comparisons are given in Section 6.

2 Random Number Generation from the Posterior

Suppose we wish to explore the joint posterior distribution of two percentiles (ξ_{p_1}, ξ_{p_2}) with $p_1 < p_2$. To generate random values for (ξ_{p_1}, ξ_{p_2}) from the posterior

(1) Generate a value from $\text{Beta}(\alpha(t_1), n + \alpha_R - \alpha(t_1))$ and call it ξ_{p_1}. Since the marginal posterior distribution of $F(t_1)$ is $\text{Beta}(\alpha(t_1), n + \alpha_R - \alpha(t_1))$, we have (Ferguson 1973)

$$\Pr[\xi_{p_1} \le t_1] = \Pr[F^{-1}(p_1) \le t_1] = \Pr[p_1 \le F(t_1)]$$

Hence we generate ξ_{p_1} as follows;

(a) Generate a value u_1 from $U(0,1)$.

(b) Find t_1 such that

$$u_1 = \int_{p_1}^{1} \frac{\Gamma(n+\alpha_R)}{\Gamma(\alpha(t_1))\Gamma(n+\alpha_R-\alpha(t_1))} \times z^{\alpha(t_1)-1}(1-z)^{n+\alpha_R-\alpha(t_1)-1}dz \tag{2}$$

Take t_1 as ξ_{p_1}.

(2) Given the value ξ_{p_1} generated in the previous step, we next generate a value from Beta$(\alpha(t_1,t_2), n+\alpha_R-\alpha(t_1)-\alpha(t_1,t_2))$ and call it η. Take $\xi_{p_1}+\eta$ as ξ_{p_2}. Let $\eta = t_2 - t_1$. To implement this procedure, we note that the conditional posterior probability

$$\begin{aligned}
&\Pr[\xi_{p_2} - \xi_{p_1} \leq \eta | F(t_1) = p_1] \\
=&\Pr[\xi_{p_2} \leq t_1 + \eta | F(t_1) = p_1] \\
=&\Pr[p_2 \leq F(t_1+\eta) | F(t_1) = p_1] \\
=&\Pr\Big[\frac{p_2-p_1}{1-p_1} \leq \frac{F(t_1+\eta)-p_1}{1-p_1}\Big| F(t_1) = p_1\Big] \\
=&\Pr\Big[\frac{p_2-p_1}{1-p_1} \leq \frac{F(t_1+\eta)-F(t_1)}{1-F(t_1)}\Big| F(t_1) = p_1\Big]
\end{aligned}$$

and by equation (1), the a value of ξ_{p_2} given $\xi_{p_1} = F^{-1}(p_1) = t_1$ can be generated as follows ;

(a) Generate $U_2 = u_2$ from $U(0,1)$.

(b) Find η such that

$$\begin{aligned}
u_2 =& \int_{\frac{(p_2-p_1)}{(1-p_1)}}^{1} \frac{\Gamma(n+\alpha_R-\alpha(t_1))}{\Gamma(\alpha(t_1,t_1+\eta))\Gamma(n+\alpha_R-\alpha(t_1)-\alpha(t_1,t_1+\eta))} \\
&\times z^{\alpha(t_1,t_1+\eta)-1}(1-z)^{n+\alpha_R-\alpha(t_1)-\alpha(t_1,t_1+\eta)-1}dz \\
=& \int_{(p_2-p_1)/(1-p_1)}^{1} \frac{\Gamma(n+\alpha_R-\alpha(t_1))}{\Gamma(\alpha(t_1,t_1+\eta))\Gamma(n+\alpha_R-\alpha(t_1+\eta))} \\
&\times z^{\alpha(t_1,t_1+\eta)-1}(1-z)^{n+\alpha_R-\alpha(t_1+\eta)-1}dz
\end{aligned}$$

We note that $\alpha(t_1, t_1 + \eta) = \alpha(t_2) - \alpha(t_1)$.

(c) Take $t_1+\eta$ as ξ_{p_2}.

Finally, get the paired value (t_1,t_2) as a value of (ξ_{p_1},ξ_{p_2}).

For a higher dimensional quantities of $(t_1,t_2,\ldots,t_k)'$, we extend above arguments as follows; Let $-\infty = t_0 < t_1 < \cdots < t_k < t_{k+1} = \infty$. Given $0 < p_1 < p_2 < \ldots < p_k < 1$, generating a k-dimensional vector $(\xi_{p_1},\xi_{p_2},\ldots,\xi_{p_k})$ by expressing the posterior joint distribution

$$(F(t_1), F(t_2)-F(t_1),\ldots,1-F(t_k)) \sim \text{Dirichlet}(\alpha(t_1),\alpha(t_2)-\alpha(t_1),\ldots,n+\alpha_R-\alpha(t_k))$$

Table 1 107 observations(sorted) of MOR(lb/in^2 of 2 × 4 inches, Grade 2, Green(30% moisture content).

1228.4	1420.3	1931.2	2105.5	2327.6	2350.2	2444.5	2524.8
2566.1	2642.7	2748.2	2764.8	2822.7	3015.2	3022.1	3095.2
3168.1	3207.7	3340.9	3396.4	3440.7	3508.1	3556.0	3677.3
3747.0	3770.9	3854.5	3879.0	3917.7	3938.0	4044.5	4050.5
4161.0	4187.7	4230.9	4274.4	4281.8	4392.0	4414.5	4420.0
4429.6	4432.9	4451.7	4461.2	4533.1	4533.1	4607.6	4616.1
4658.4	4690.8	4760.3	4787.2	4795.9	4818.4	4872.4	4892.6
4896.3	4994.6	5078.2	5128.6	5143.3	5161.1	5231.6	5268.4
5270.6	5278.7	5289.3	5325.4	5325.4	5418.6	5465.8	5511.1
5530.9	5531.8	5651.6	5677.2	5681.0	5691.8	5817.3	5818.8
5851.4	5883.6	5894.0	5976.8	5988.6	6051.4	6082.7	6136.7
6245.9	6307.9	6351.2	6357.9	6455.7	6617.3	6674.1	6767.2
6843.5	6964.0	6997.6	7011.3	7061.8	7311.8	7529.2	7643.6
8145.4	8153.4	9213.0					

(1) Generate t_1 using the Beta$(\alpha(t_1), n+\alpha_R-\alpha(t_1))$ distribution as in (2) and take t_1 as ξ_{p_1}.

(2) For $i = 1, 2, \ldots, k$, use the posterior conditional distributions

$$\frac{F(t_i)-F(t_{i-1})}{1-F(t_{i-1})}\Big|_{F(t_{i-1}),\ldots,F(t_1)} \sim \text{Beta}(\alpha(t_i)-\alpha(t_{i-1}), n+\alpha_R-\alpha(t_{i-1})).$$

For values ξ_{p_i} given $\xi_{p_{i-1}}$, we generate u from $U(0,1)$ and solve

$$u = \int_{\frac{(p_i-p_{i-1})}{(1-p_{i-1})}}^{1} \frac{\Gamma(n+\alpha_R-\alpha(t_{i-1}))}{\Gamma(\alpha(t_{i-1},t_{i-1}+\eta))\Gamma(n+\alpha_R-\alpha(t_{i-1}+\eta))} \times z^{\alpha(t_{i-1}+\eta)-\alpha(t_{i-1})-1}(1-z)^{n+\alpha_R-\alpha(t_{i-1}+\eta)-1}dz$$

for η.

(3) Take $t_{i-1}+\eta$ as ξ_{p_i}.

3 Examples

We apply our algorithm to data on the modulus of rupture(MOR) of Douglas Fir specimens.

From the data in Table 1, we generated 5,000 separate random vectors for each of for cases $(\xi_{0.05}, \xi_{0.10})$, $(\xi_{0.05}, \xi_{0.20})$, $(\xi_{0.05}, \xi_{0.50})$, $(\xi_{0.05}, \xi_{0.95})$. The means and Pearson correlation coefficients are calculated in Table 2.

The priors measure α_0 used in our simulation study is the uniform on $(0, 9500)$ and we consider $p_1 = 0.05$ and $p_2 = 0.1, 0.2, 0.5$ and 0.95. We note that the estimates for $\xi_{0.05}$, $\xi_{0.10}$, $\xi_{0.20}$, $\xi_{0.50}$ and $\xi_{0.95}$ based on the linear

Table 2 Basic statistics from 5,000(each) vectors of percentiles.

	$(\xi_{0.05}, \xi_{0.10})$	$(\xi_{0.05}, \xi_{0.20})$
Mean vector	(2327.839, 2749.342)	(2322.128, 3490.037)
sd	(263.1465, 225.2865)	(265.8184, 267.9980)
Correlation	0.61588595	0.41460567
	$(\xi_{0.05}, \xi_{0.50})$	$(\xi_{0.05}, \xi_{0.95})$
Mean vector	(2317.516, 4810.936)	(2317.570 7105.646)
sd	(267.7407, 189.4894)	(268.1235, 248.2899)
Correlation	0.19598357	0.03640863

interpolation

$$\hat{\xi}_p = x_{[np]} + (x_{[np+1]} - x_{[np]})(np - [np]) \tag{3}$$

are 2335.51, 2716.55, 3467.66, 4807.15 and 7224.30, respectively, where $[a]$ is integer part of $a \in \boldsymbol{R}$. For comparisons, we consider the asymptotic properties of (ξ_{p_1}, ξ_{p_2}). For $p_1 < p_2$, we note that $(\hat{\xi}_{p_1}, \hat{\xi}_{p_2})$ is asymptotically normal with mean (ξ_{p_1}, ξ_{p_2}) and covariance σ_{12}/n where

$$\sigma_{12} = \frac{p_1(1-p_2)}{f(\xi_{p_1})f(\xi_{p_2})} \tag{4}$$

and $\sigma_{21} = \sigma_{12}$. Here, f is the probability density function of $X_1, \ldots, X_n$ and f is positive and continuous at ξ_{p_1} and ξ_{p_2} (Serfling 1980). Based on the data in Table 1, $\bar{x} = 4840.325$ and $s_x^2 = 2354470$. If we assume a normal distribution for the data,

$$\sigma_{12} = \sigma_{21} = \frac{0.05(1-0.1)}{0.00006721438 * 0.0001143738} = 5853610$$

$$\sigma_{11} = \frac{0.05(1-0.05)}{0.00006721438 * 0.00006721438} = 10514030$$

$$\sigma_{22} = \frac{0.1(1-0.1)}{0.0001143738 * 0.0001143738} = 6880015$$

so that the correlation is 0.6882473 for $(\xi_{0.05}, \xi_{0.1})$. We note that the correlation coefficients reduce to

$$\sqrt{\frac{p_1(1-p_2)}{p_2(1-p_1)}}$$

and the correlation coefficients for $(\xi_{0.05}, \xi_{0.2})$, $(\xi_{0.05}, \xi_{0.5})$ and $(\xi_{0.05}, \xi_{0.95})$ are 0.4588315, 0.2294157 and 0.05263158, respectively.

The standard deviations of $\xi_{0.05}$, $\xi_{0.1}$, $\xi_{0.2}$, $\xi_{0.5}$, $\xi_{0.95}$ based on (4) under normal assumption are (313.4676 253.5730 211.9414 185.9151 313.4676)

The standard deviations with kernel density estimator of f obtained by `density` function in R-language, with bandwidth in equation (3.31) in

Silverman (1986) and Gaussian kernel, are (242.8418, 256.4729, 250.108, 188.9272, 267.2380).

For a higher dimensional example, we consider the same data given in Table 1, we generate 5,000 tuples of (Q_1, Q_2, Q_3) where Q_i is the i-th quartile. Note that, using the original order statistics, (3) gives the estimates $(3833.60, 4807.15, 5826.95)$. The mean vector of the 5,000 tuples is $(3825.409, 4823.007, 5787.796)$ and correlation matrix of generated vectors and based on the asymptotics in (4) are

$$\begin{bmatrix} 1.000000.575630.34629 \\ 0.575631.000000.57409 \\ 0.346290.574091.00000 \end{bmatrix} \text{ and } \begin{bmatrix} 1.000000.577350.33333 \\ 0.577351.000000.57735 \\ 0.333330.577351.00000 \end{bmatrix}$$

respectively. The standard deviations of (Q_1, Q_2, Q_3) based on the kernel density estimation of f from R-language function `library` with the bandwidth given in equation (3.31) of Silverman (1986) and Gaussian kernel is $(224.5768, 188.9272,, 197.9982)$ and based on normal assumption is $(202.1312, 185.9151, 202.1312)$ and from the 5,000 generated tuples is $(243.6205, 195.7481, 180.7857)$.

4 Asymptotic Normality of Joint Posterior Distribution

After establishing two preliminary lemmas, we give our direct proof of asymptotic normality. At the time we obtained Theorem 1 and presented it at ISI 2003 Berlin, Hjort (2003) obtained an even stronger result. He established the weak convergence of the posterior percentile process but by an indirect proof. Further results on convergence, and examples similar to ours for single percentiles, appear in Hjort and Petrone (2006) in this volume. The results of our direct proof here are needed in the development of asymptotic expansions in the next section. Our results do not require conditions on the tail behavior of the underlying distribution, but we also do not get weak convergence of the posterior process.

Lemma 1. *Let $X_1, X_2, \ldots, X_n$ be a random sample from a distribution G. Assume that G has its derivative g and g is uniformly continuous on* $\boldsymbol{R}$. *Let*

$$G_n(t) = \frac{\#(X_j \le t) + \alpha_0(t)}{n + \alpha_R} \tag{5}$$

and $t_n = \inf\{t | p \le G_n(t)\}$. Then for $w > 0$, as $n \to \infty$,

$$\frac{\#(t_n < X_j \le t_n + w/\sqrt{n})}{n \cdot w/\sqrt{n}} = g(\eta_p) + o(1) \tag{6}$$

almost surely.

Proof. Denote

$$g_n(t) = \frac{\#(t < X_j \leq t + w/\sqrt{n})}{n \cdot w/\sqrt{n}} = \frac{1}{n \cdot w/\sqrt{n}} \sum_{j=1}^{n} K\Big(\frac{X_j - t}{w/\sqrt{n}}\Big) \tag{7}$$

where

$$K(u) = \begin{cases} 1 \text{if } 0 < u \leq 1 \\ 0 \text{otherwise} \end{cases}$$

Since K is bounded with finite variation and $w/\sqrt{n} \to 0$, $n \cdot w/\sqrt{n} \cdot (\log n)^{-1} \to \infty$ as $n \to \infty$, we have

$$\sup_x |g_n(x) - g(x)| \to 0$$

almost surely (p71–72 in Silverman 1986, Bertrand-Retali 1978). Because $t_n \to \eta_p$ almost surely, g is continuous and

$$\begin{aligned} |g_n(t_n) - g(\eta_p)| &\leq |g_n(t_n) - g(t_n)| + |g(t_n) - g(\eta_p)| \\ &\leq \sup_x |g_n(x) - g(x)| + |g(t_n) - g(\eta_p)|, \end{aligned}$$

the result follows. □

Corollary 1. *Under the same conditions in Lemma 1, (6) holds for $w < 0$.*

Proof: Let $w < 0$ and $c = |w|$. With $K(u) = I_{[-1/2<u<1/2]}$, we have

$$\frac{\#(t - c/\sqrt{n} < X_j \leq t + c/\sqrt{n})}{2n \cdot c/\sqrt{n}} = \frac{1}{n \cdot 2c/\sqrt{n}} \sum_{j=1}^{n} K\Big(\frac{X_j - t}{2c/\sqrt{n}}\Big)$$

$$\to g(\eta_p)$$

by a similar argument in Lemma 1. Since

$$\begin{aligned} &\frac{\#(t + w/\sqrt{n} < X_j \leq t)}{n \cdot c/\sqrt{n}} \\ &= 2\frac{\#(t - c/\sqrt{n} < X_j \leq t + c/\sqrt{n})}{2n \cdot c/\sqrt{n}} - \frac{\#(t < X_j \leq t + c/\sqrt{n})}{n \cdot c/\sqrt{n}}, \end{aligned}$$

same result holds for $w < 0$.

Remark 1. By Lemma 1, Corollary 1 and $G_n(t_n) = p + O(n^{-1})$, as $n \to \infty$,

$$\frac{G_n(t_n + w/\sqrt{n}) - p}{w/\sqrt{n}} \to g(\eta_p) \tag{8}$$

almost surely, for all w.

Lemma 2. *Let* $0 = p_0 < p_1 < p_2 < \cdots < p_k < p_{k+1} = 1$ *and* $t_{in} = \inf\{t | p_i \le G_n(t)\}$ *with* $G_n(t)$ *in (5). Let* $Z_1, Z_2, \ldots$ *be a random sample of size* $[n + \alpha_R] + 1$ *from a gamma distribution with parameters* $(1,1)$*. Let, with* $G_n(t_{0n}) = 0$ *and* $G_n(t_{k+1,n}) = 1$ *,*

$$\alpha_{in} = \alpha_{in}(t_{in} + w_i/\sqrt{n}) = (n + \alpha_R)G_n(t_{in} + w_i/\sqrt{n})$$

so that $\alpha_{in}/n \to p_i$ *as* $n \to \infty$*. Denote* $T_1 = Z_1 + Z_2 + \cdots Z_{[\alpha_{1n}]} + Z_{10}$ $T_i = Z_{[\alpha_{i-1,n}]+1} + Z_{[\alpha_{i-1,n}]+2} + \cdots + Z_{[\alpha_{i,n}]} + Z_{i0}$ *where* $Z_{i0} = Z_{\alpha_{in} - [\alpha_{in}]} \sim Gamma(\alpha_{in} - [\alpha_{in}], 1)$*, for* $i = 1, 2, \ldots, (k+1)$*. Then*

$$\left(\frac{T_1}{\sum_{i=1}^{k+1} T_i}, \frac{T_1 + T_2}{\sum_{i=1}^{k+1} T_i}, \ldots, \frac{T_1 + T_2 + \cdots + T_k}{\sum_{i=1}^{k+1} T_i}\right)' \to N_k(\boldsymbol{p}, \boldsymbol{V})$$

where $\boldsymbol{p} = (p_1, p_2, \ldots, p_k)'$ *and* $\boldsymbol{V} = (v_{ij})_{i,j=1,2,\ldots k} = p_i(1 - p_j)$ *for* $i \le j$

Proof. Since $Z_{i0}/\sqrt{n} \xrightarrow{p} 0$ and by CLT, we note that

$$\begin{aligned} &\sqrt{\alpha_{in} - \alpha_{i-1,n}}\left(\frac{T_i}{\alpha_{in} - \alpha_{i-1,n}} - 1\right) \\ =&\sqrt{n}\left(\frac{T_i}{n}\frac{\sqrt{n}}{\sqrt{\alpha_{in} - \alpha_{i-1,n}}} - \frac{\sqrt{\alpha_{in} - \alpha_{i-1,n}}}{\sqrt{n}}\right) \xrightarrow{d} N(0,1) \end{aligned}$$

so that

$$\sqrt{n}(T_i/n - (p_i - p_{i-1})) \xrightarrow{d} N(0, p_i - p_{i-1}) \tag{9}$$

Let $\boldsymbol{t} = (t_1, t_2, \ldots, t_{k+1})'$ and

$$\begin{aligned} h(\boldsymbol{t}) =& (h_1(\boldsymbol{t}), h_2(\boldsymbol{t}), \ldots, h_k(\boldsymbol{t}))' \\ =& \left(\frac{t_1}{\sum_{i=1}^{k+1} t_i}, \frac{t_1 + t_2}{\sum_{i=1}^{k+1} t_i}, \ldots, \frac{t_1 + t_2 + \cdots + t_k}{\sum_{i=1}^{k+1} t_i}\right)' \end{aligned}$$

and $\boldsymbol{p}_d = (p_1, p_2 - p_1, \ldots, p_k - p_{k-1}, 1 - p_k)'$. Then we have

$$h(\boldsymbol{p}_d) = (p_1, p_2 - p_1, \ldots, p_k - p_{k-1})$$

and, for $j = 1, 2, \ldots, k$ and $l = 1, 2, \ldots, k+1$,

$$\begin{aligned} \underset{k \times (k+1)}{\nabla h(\boldsymbol{t})} =& \left[\frac{\partial h_l(\boldsymbol{t})}{\partial t_j}\right]_{jl} \\ =& \begin{cases} (\sum t_i) - (t_1 + t_2 + \cdots + t_j)/(\sum t_i)^2 & \text{if } j \ge l \\ -(t_1 + t_2 + \cdots + t_j)/(\sum t_i)^2 & \text{if } j < l \end{cases} \end{aligned}$$

Then

$$\begin{aligned} [\nabla h(\boldsymbol{p}_d)]_{jl} =& \begin{cases} 1 - p_j & \text{if } j \ge l \\ -p_j & \text{if } j < l \end{cases} \text{for } j = 1, 2, \ldots, k \text{ and } l = 1, 2, \ldots, k+1. \\ =& \begin{bmatrix} 1 - p_1 & -p_1 & -p_1 & \cdots & -p_1 \\ 1 - p_2 & 1 - p_2 & -p_2 & \cdots & -p_2 \\ \vdots & \vdots & \vdots & & \vdots \\ 1 - p_k & 1 - p_k & 1 - p_k & \cdots & 1 - p_k \end{bmatrix} \end{aligned}$$

Since the covariance matrix $\mathbf{\Sigma}_{(k+1)\times(k+1)}$ of (9) is diagonal matrix of $(p_1, p_2 - p_1, \ldots, p_k - p_{k-1}, 1 - p_k)$, we have

$$\nabla h(\boldsymbol{p}_d)\mathbf{\Sigma} = \begin{cases} (1-p_j)(p_l - p_{l-1}) & \text{if } j \geq l \\ -p_j(p_l - p_{l-1}) & \text{if } j < l \end{cases},$$

which can be written as

$$\begin{bmatrix} p_1(1-p_1) - p_1(p_2-p_1) & \cdots - p_1(1-p_k) \\ p_1(1-p_2)(1-p_2)(p_2-p_1) & \cdots - p_2(1-p_k) \\ p_1(1-p_3)(1-p_3)(p_2-p_1) & \cdots - p_3(1-p_k) \\ \vdots \quad\quad \vdots & \vdots \quad \vdots \\ p_1(1-p_k)(1-p_k)(p_2-p_1) & \cdots (1-p_k)(1-p_k) \end{bmatrix}$$

Hence the diagonal elements of $\boldsymbol{V}_{k\times k} = \nabla h(\boldsymbol{p}_d)\mathbf{\Sigma}(\nabla h(\boldsymbol{p}_d))'$ is

$$v_{jj} = \sum_{m=1}^{j}(1-p_j)^2(p_m - p_{m-1}) + \sum_{m=j+1}^{k+1} p_j^2(p_m - p_{m-1}) = p_j(1-p_j)$$

for $j = 1, 2, \ldots, k$ and lower diagonal elements $(j > l)$ are

$$v_{jl} = \sum_{m=1}^{l}(1-p_j)(p_m - p_{m-1})(1-p_l) - \sum_{m=l+1}^{j} p_l(1-p_j)(p_m - p_{m-1})$$
$$+ \sum_{m=j+1}^{k+1} p_j p_l(p_m - p_{m-1}) = p_l(1-p_j).$$

Note that he upper diagonal elements are the same as those of lower diagonal by symmetry of $\boldsymbol{V}$. The final matrix can be written as

$$\begin{bmatrix} p_1(1-p_1) p_1(1-p_2) p_1(1-p_3) \cdots p_1(1-p_k) \\ p_1(1-p_2) p_2(1-p_2) p_2(1-p_3) \cdots p_2(1-p_k) \\ \vdots \quad\quad \vdots \quad\quad \vdots \quad\quad \vdots \quad \vdots \\ p_1(1-p_k) p_2(1-p_k) p_2(1-p_k) \cdots p_k(1-p_k) \end{bmatrix}$$

This completes proof. □

Theorem 1. *For $i = 1, 2, \ldots, k$, assume that G is continuous at η_i with $p_i = G(\eta_i)$ and has its derivative g. Assume the conditions on t_{in}, g, $\mathbf{p}$ and $X_1, X_2, \ldots, X_k$ in Lemma 1 and Lemma 2 hold. Then, as $n \to \infty$,*

$$\Pr[\sqrt{n}(\xi_{p_1} - t_{1n}) \leq w_1, \ldots, \sqrt{n}(\xi_{p_k} - t_{kn}) \leq w_k] \to N_k(\mathbf{0}, N_k(\mathbf{0}, \boldsymbol{V}_g)$$

where

$$\boldsymbol{V}_g = \begin{cases} p_i(1-p_j)/[g(\eta_i)g(\eta_j)] i \leq j \\ p_i(1-p_j)/[g(\eta_i)g(\eta_j)] i > j \end{cases}$$

Proof. Since the distribution of $F(t_{1n} + w_i/\sqrt{n}), \ldots, F(t_{k+1,n} + w_{k+1}/\sqrt{n}) - F(t_{k,n} + w_{k+1}/\sqrt{n}))$ is Dirichlet$(\alpha_{1n}, \ldots, \alpha_{k+1,n} - \alpha_{k,n})$, we have

$$(F(t_{in} + w_1/\sqrt{n}), \ldots, F(t_{kn} + w_k/\sqrt{n})) \stackrel{d}{=} \left(\frac{T_1}{\sum T_m}, \ldots, \frac{\sum^k T_m}{\sum T_m}\right)$$
$$\stackrel{d}{\rightarrow} N_k(\boldsymbol{p}, \boldsymbol{V})$$

By Lemma 2, asymptotically, we have $(j \geq i)$

$$\begin{aligned}
&\Pr[\sqrt{n}(\xi_{p_1} - t_{1n}) \leq w_1, \ldots, \sqrt{n}(\xi_{p_k} - t_{kn}) \leq w_k] \\
=&\Pr[p_1 \leq F(t_{1n} + w_1/\sqrt{n}), \ldots, p_k \leq F(t_{kn} + w_k/\sqrt{n})] \\
=&\Pr\left[p_1 \leq \frac{\sum_{m=1}^{1} T_m}{\sum_{m=1}^{k+1} T_m}, \ldots, p_k \leq \frac{\sum_{m=1}^{k} T_m}{\sum_{m=1}^{k+1} T_m}\right] \\
=&\Pr\left[p_1 - \frac{\alpha_{1n}}{n+\alpha_R} \leq \frac{\sum_{m=1}^{1} T_m}{\sum_{m=1}^{k+1} T_m} - \frac{\alpha_{1n}}{n+\alpha_R}, \ldots, \right. \\
&\qquad \left. p_k - \frac{\alpha_{kn}}{n+\alpha_R} \leq \frac{\sum_{m=1}^{k} T_m}{\sum_{m=1}^{k+1} T_m} - \frac{\alpha_{jn}}{n+\alpha_R}\right]
\end{aligned}$$

$$\begin{aligned}
=&\Pr\left[p_1 - G_n(t_{1n} + w_1/\sqrt{n}) \leq \frac{\sum_{m=1}^{1} T_m}{\sum_{m=1}^{k+1} T_m} - G_n(t_{1n} + w_1/\sqrt{n}), \ldots, \right. \\
&\qquad \left. p_k - G_n(t_{kn} + w_k/\sqrt{n}) \leq \frac{\sum_{m=1}^{k} T_m}{\sum_{m=1}^{k+1} T_m} - G_n(t_{kn} + w_k/\sqrt{n})\right] \\
=&\int_{-\infty}^{\sqrt{n}(G_n(t_{1n}+w_1/\sqrt{n})-p_1)} \cdots \int_{-\infty}^{\sqrt{n}(G_n(t_{kn}+w_k/\sqrt{n})-p_k)} \frac{1}{(2\pi)^{k/2}|\boldsymbol{V}|^{1/2}} \\
&\times \exp\left\{-\frac{1}{2}\boldsymbol{y}'\boldsymbol{V}^{-1}\boldsymbol{y}\right\} dy_1 \cdots dy_k \qquad (10)
\end{aligned}$$

Let $x_i = g_n(t_{in})y_i$ for $i = 1, 2, \ldots, k$ with g_n in (7). Hence, by Lemma 1, (10) is

$$\int_{-\infty}^{w_1} \cdots \int_{-\infty}^{w_k} \frac{1}{(2\pi)^{k/2}|\boldsymbol{V}_g|^{1/2}} \exp\left\{-\frac{1}{2}\boldsymbol{x}'\boldsymbol{V}_g^{-1}\boldsymbol{x}\right\} dx_1 \cdots dx_k$$

as $n \rightarrow \infty$. □

Note that the covariance is $p_i(1-p_j)/[g(\eta_i)g(\eta_j)]$.

5 An Asymptotic Expansion for the Posterior Distribution

In this section, we establish some higher order correction terms to the normal limit distribution. Correction terms, to the normal limit for the joint posterior distribution, are established in Johnson and Sim (2006).

We first state the main result as Theorem 2.

Theorem 2. *Let Φ and ϕ be the standard normal cdf and pdf, respectively, and write $G_n = G_n(t_n + w/\sqrt{n})$. Then, as $n \to \infty$,*

$$\frac{\Gamma(n+\alpha_R)}{\Gamma((n+\alpha_R)G_n)\Gamma((n+\alpha_R)(1-G_n))} \times \int_p^1 z^{(n+\alpha_R)G_n-1}(1-z)^{(n+\alpha_R)(1-G_n)-1}dz \tag{11}$$

$$=1-\Phi(k_n)+\frac{1}{\sqrt{n+\alpha_R}}\frac{(1-2G_n)(k_n^2-1)}{3\sqrt{G_n(1-G_n)}}\phi(k_n) + \frac{d_n}{n+\alpha_R}\phi(k_n)+O(n^{-3/2}) \tag{12}$$

where

$$k_n = \sqrt{\frac{n+\alpha_R}{G_n(1-G_n)}}(p-G_n), \tag{13}$$

and the d_n are given in (20). Here, k_n and d_n each converge to finite constants.

Proof. By Stirling's formula,

$$\ln\Gamma(a) = (a-\frac{1}{2})\ln a - a + \frac{1}{2}\ln(2\pi) + \frac{1}{12a} - \frac{1}{360a^3} + \cdots$$

we have

$$
\begin{aligned}
&\frac{1}{B((n+\alpha_R)G_n,(n+\alpha_R)(1-G_n))}\\
&=\frac{\Gamma(n+\alpha_R)}{\Gamma((n+\alpha_R)G_n)\Gamma((n+\alpha_R)(1-G_n))}\\
&=\exp\{\ln\Gamma(n+\alpha_R)-\ln\Gamma((n+\alpha_R)G_n)-\ln\Gamma((n+\alpha_R)(1-G_n))\}\\
&=\exp\Big\{(n+\alpha_R-\frac{1}{2})\ln(n+\alpha_R)-(n+\alpha_R)+\frac{1}{2}\ln 2\pi+\frac{1}{12(n+\alpha_R)}\\
&+O(n^{-3})-\Big[((n+\alpha_R)G_n-\frac{1}{2})\ln((n+\alpha_R)G_n)-(n+\alpha_R)G_n\\
&+\frac{1}{2}\ln 2\pi+\frac{1}{12(n+\alpha_R)G_n}+O(n^{-3})\Big]-\Big[((n+\alpha_R)(1-G_n)-\frac{1}{2})\times\\
&\ln((n+\alpha_R)(1-G_n))-(n+\alpha_R)(1-G_n)+\frac{1}{2}\ln 2\pi\\
&+\frac{1}{12(n+\alpha_R)(1-G_n)}+O(n^{-3})\Big]\Big\}\\
&=\exp\Big\{\frac{1}{2}\ln(n+\alpha_R)-\Big[(n+\alpha_R)G_n-\frac{1}{2}\Big]\ln G_n\\
&-\Big[(n+\alpha_R)(1-G_n)-\frac{1}{2}\Big]\ln(1-G_n)-\frac{1}{2}\ln(2\pi)\\
&+\frac{G_n(1-G_n)-(1-G_n)-G_n}{12(n+\alpha_R)G_n(1-G_n)}+O(n^{-3})\Big\}\\
&=\frac{(n+\alpha_R)^{1/2}}{\sqrt{2\pi}G_n^{(n+\alpha_R)G_n-1/2}(1-G_n)^{(n+\alpha_R)(1-G_n)-1/2}}\\
&\qquad\qquad\times\exp\Big\{\frac{-1+G_n-G_n^2}{12(n+\alpha_R)G_n(1-G_n)}+O(n^{-3})\Big\}\\
&=\frac{(n+\alpha_R)^{1/2}}{\sqrt{2\pi}G_n^{(n+\alpha_R)G_n-1/2}(1-G_n)^{(n+\alpha_R)(1-G_n)-1/2}}\\
&\qquad\qquad\times\Big[1+\frac{-1+G_n-G_n^2}{12(n+\alpha_R)G_n(1-G_n)}+O(n^{-2})\Big] \quad (14)
\end{aligned}
$$

Now, write the integrand in (11) as

$$z^{(n+\alpha_R)G_n-1}(1-z)^{(n+\alpha_R)(1-G_n)-1}=\frac{1}{z(1-z)}e^{(n+\alpha_R)[G_n\ln z+(1-G_n)\ln(1-z)]}$$

and let

$$h_n(z)=G_n\ln z+(1-G_n)\ln(1-z).$$

Then

$$h_n'(z)=\frac{G_n}{z}-\frac{1-G_n}{1-z}$$

so that the value $z_{n0} = z_0 = G_n$ giving the maximum of $h_n(z)$ satisfies $h'_n(z) = 0$. Also

$$e^{(n+\alpha_R)h_n(z_0)} = z_0^{(n+\alpha_R)z_0}(1-z_0)^{(n+\alpha_R)(1-z_0)} = [z_0^{z_0}(1-z_0)^{1-z_0}]^{n+\alpha_R} \quad (15)$$

and $h''_n(z_0) = -1/(z_0(1-z_0))$. Note that, for $l \geq 2$,

$$h_n^{(l)}(z_0) = (-1)^{l-1}\frac{(l-1)!G_n}{z^l} - \frac{(l-1)!(1-G_n)}{(1-z)^l}\bigg|_{z_0}$$
$$=(-1)^{l-1}\frac{(l-1)!}{z_0^{l-1}} - \frac{(l-1)!}{(1-z_0)^{l-1}}$$

Using a Taylor expansion for h_n at z_0, the integrand in (11) can be written

$$\frac{1}{z(1-z)}e^{(n+\alpha_R)h_n(z_0)}e^{(n+\alpha_R)h''_n(z_0)(z-z_0)^2/2}e^{(n+\alpha_R)(z-z_0)^3\psi(z,G_n)}$$

where

$$\psi(z, G_n) = \frac{h'''_n(z_0)}{3!} + \frac{h_n^{(4)}(z_0)(z-z_0)}{4!} + \cdots + \frac{h_n^{(l)}(z-z_0)^{l-3}}{l!} + \cdots$$

Let $v = (n+\alpha_R)(z-z_0)^3$ and set

$$P_n(v, z, G_n) = \frac{1}{z(1-z)}\exp\{v\psi(z, G_n)\}$$

Then P_n has a two variable expansion (see Johnson and Ladalla 1978 for similar expansions.)

$$P_n(v, z, G_n) = \sum_{l=0}^{\infty}\sum_{i+j=l} c_{ij}v^i(z-z_0)^j \quad (16)$$
$$= c_{00} + c_{10}v + c_{01}(z-z_0) + c_{20}v^2 + c_{11}v(z-z_0) + c_{02}(z-z_0)^2 + \cdots$$

where $c_{ij} = \frac{\partial^n P_n}{\partial v^i \partial z^j}\frac{1}{i!j!}\Big|_{(0,z_0)}$. The exact expressions for c_{ij}'s, up to order 2, are

$$c_{00} = P_n(v, z, G_n)\Big|_{(0,z_0)} = \frac{1}{z_0(1-z_0)},$$

and

$$c_{01} = \frac{\partial P_n}{\partial z}\Big|_{(0,z_0)} = \frac{2z_0 - 1}{z_0^2(1-z_0)^2} = \frac{1}{z_0(1-z_0)}c'_{01}$$

$$c_{10} = \frac{\partial P_n}{\partial v}\Big|_{(0,z_0)} = \frac{1}{z_0(1-z_0)}\frac{h_n^{(3)}(z_0)}{3!} = \frac{1}{z_0(1-z_0)}\frac{(1-2z_0)}{3z_0^2(1-z_0)^2}$$
$$=\frac{1}{z_0(1-z_0)}c'_{10}$$

$$
\begin{aligned}
c_{11} &= \frac{1}{1!}\frac{1}{1!}\frac{\partial^2 P_n}{\partial v \partial z}\bigg|_{(0,z_0)} = \frac{2z_0-1}{z_0^2(1-z_0)^2}\frac{h_n^{(3)}(z_0)}{3!} + \frac{1}{z_0(1-z_0)}\frac{h_n^{(4)}(z_0)}{4!} \\
&= -\frac{1}{z_0(1-z_0)}\left[\frac{(1-2z_0)^2}{3z_0^3(1-z_0)^3} + \frac{1-3z_0+3z_0^2}{4z_0^3(1-z_0)^3}\right] \\
&= \frac{1}{z_0(1-z_0)}\frac{-7+25z_0-25z_0^2}{12z_0^3(1-z_0)^3} = \frac{1}{z_0(1-z_0)}c'_{11}
\end{aligned}
$$

$$
c_{20} = \frac{1}{2!}\frac{\partial^2 P_n}{\partial v^2}\bigg|_{(0,z_0)} = \frac{1}{z_0(1-z_0)}\frac{(1-2z_0)^2}{18z_0^4(1-z_0)^4} = \frac{1}{z_0(1-z_0)}c'_{20}
$$

$$
\begin{aligned}
c_{02} &= \frac{1}{2!}\frac{\partial^2 P_n}{\partial z^2}\bigg|_{(0,z_0)} = \frac{1-3z_0+3z_0^2}{z_0^3(1-z_0)^3} = \frac{1}{z_0(1-z_0)}\frac{1-3z_0+3z_0^2}{z_0^2(1-z_0)^2} \\
&= \frac{1}{z_0(1-z_0)}c'_{02}
\end{aligned}
$$

Hence the integral in (11), with v being replaced by $(n+\alpha_R)(z-z_0)^3$, is

$$
\begin{aligned}
& e^{(n+\alpha_R)h_n(z_0)}\int_p^1 \frac{1}{z(1-z)}e^{(n+\alpha_R)h''_n(z_0)(z-z_0)^2/2}e^{(n+\alpha_R)(z-z_0)^3\psi(z,G_n)}dz \\
&= e^{(n+\alpha_R)h_n(z_0)}\int_p^1 e^{(n+\alpha_R)h''_n(z_0)(z-z_0)^2/2}P_n(n(z-z_0)^3, z, G_n)dz \\
&= \frac{e^{(n+\alpha_R)h_n(z_0)}}{z_0(1-z_0)}\int_p^1 e^{(n+\alpha_R)h''_n(z_0)(z-z_0)^2/2}\Big[1 + c'_{10}(n+\alpha_R)(z-z_0)^3 \\
&\quad + c'_{01}(z-z_0) + c'_{11}(n+\alpha_R)(z-z_0)^4 + c'_{20}(n+\alpha_R)^2(z-z_0)^6 \\
&\quad + c'_{02}(z-z_0)^2 + \cdots\Big]dz
\end{aligned}
\tag{17}
$$

We make a change of variable

$$
y = \sqrt{-(n+\alpha_R)h''_n(z_0)}(z-z_0) = \sqrt{\frac{n+\alpha_R}{z_0(1-z_0)}}(z-z_0) \tag{18}
$$

where $\dfrac{dz}{dy} = \sqrt{\dfrac{z_0(1-z_0)}{n+\alpha_R}}$ and the range of y is

$$
k_n = \sqrt{\frac{(n+\alpha_R)}{z_0(1-z_0)}}(p-z_0) < y < \sqrt{\frac{(n+\alpha_R)}{z_0(1-z_0)}}(1-z_0) = \sqrt{\frac{(n+\alpha_R)(1-z_0)}{z_0}}.
$$

Next, we turn our attention to the sequence of constants, which is the product of four terms: $1/(z_0(1-z_0))$, $e^{(n+\alpha_R)h_n(z_0)}$ in (15), $1/B((n+\alpha_R)G_n,$

$(n+\alpha_R)(1-G_n))$ in (14) and the Jacobian. By writing z_0 for G_n

$$\frac{e^{(n+\alpha_R)h_n(z_0)}}{z_0(1-z_0)}\frac{1}{B((n+\alpha_R)z_0,(n+\alpha_R)(1-z_0))}\sqrt{\frac{z_0(1-z_0)}{n+\alpha_R}}$$

$$=\frac{z_0^{(n+\alpha_R)z_0}(1-z_0)^{(n+\alpha_R)(1-z_0)}}{z_0(1-z_0)}\times$$

$$\frac{(n+\alpha_R)^{1/2}}{\sqrt{2\pi}z_0^{(n+\alpha_R)z_0-1/2}(1-z_0)^{(n+\alpha_R)(1-z_0)-1/2}}\times$$

$$\left[1+\frac{-1+z_0-z_0^2}{12(n+\alpha_R)z_0(1-z_0)}+O(n^{-2})\right]\sqrt{\frac{z_0(1-z_0)}{n+\alpha_R}}$$

$$=\frac{1}{\sqrt{(n+\alpha_R)z_0(1-z_0)}}\left[\frac{\sqrt{(n+\alpha_R)z_0(1-z_0)}}{\sqrt{2\pi}}\right.$$
$$\left.+\frac{-1+z_0-z_0^2}{12\sqrt{2\pi(n+\alpha_R)z_0(1-z_0)}}+O(n^{-3/2})\right]$$
$$=\frac{1}{\sqrt{2\pi}}+\frac{1}{\sqrt{2\pi}(n+\alpha_R)}a_n+O(n^{-2}) \tag{19}$$

where $a_n=\dfrac{-1+z_0-z_0^2}{12z_0(1-z_0)}$.

Returning to the integral (11), under the change of variable (18), by (17) and (19), we have

$$\frac{1}{\sqrt{2\pi}}\int_{k_n}^{\sqrt{(n+\alpha_R)(1-z_0)/z_0}}e^{-y^2/2}\left[1+\frac{1}{\sqrt{n+\alpha_R}}\left\{c'_{10}z_0^{3/2}(1-z_0)^{3/2}y^3\right.\right.$$
$$\left.+c'_{01}\sqrt{z_0(1-z_0)}y\right\}+\frac{1}{n+\alpha_R}\left\{c'_{11}z_0^2(1-z_0)^2y^4+c'_{20}z_0^3(1-z_0)^3y^6\right.$$
$$\left.\left.+c'_{02}z_0(1-z_0)y^2\right\}\right]dy+\frac{a_n}{n+\alpha_R}\frac{1}{\sqrt{2\pi}}\int_{k_n}^{\sqrt{(n+\alpha_R)(1-z_0)/z_0}}e^{-y^2/2}dy$$
$$+O(n^{-3/2})$$
$$=I_0+\frac{1}{\sqrt{n+\alpha_R}}\left\{c''_{10}I_3+c''_{01}I_1\right\}$$
$$+\frac{1}{n+\alpha_R}\left\{c''_{11}I_4+c''_{20}I_6+c''_{02}I_2+a_nI_0\right\}+O(n^{-3/2})$$

where

$$I_l=\frac{1}{\sqrt{2\pi}}\int_{k_n}^{\sqrt{(n+\alpha_R)(1-z_0)/z_0}}y^le^{-y^2/2}dy,\quad l=0,1,2,\ldots$$

and

$$c''_{10} = \frac{1-2z_0}{3\sqrt{z_0(1-z_0)}}, \quad c''_{20} = \frac{(1-2z_0)^2}{18z_0(1-z_0)}, \quad c''_{02} = \frac{1-3z_0+3z_0^2}{z_0(1-z_0)},$$

$$c''_{01} = -\frac{1-2z_0}{\sqrt{z_0(1-z_0)}} = -3c''_{10}, \quad c''_{11} = \frac{-7+25z_0-25z_0^2}{12z_0(1-z_0)}.$$

Note that

$$\begin{aligned} I_0 =& \frac{1}{\sqrt{2\pi}} \int_{k_n}^{\sqrt{(n+\alpha_R)(1-z_0)/z_0}} e^{-y^2/2} dy, \\ I_1 =& \frac{1}{\sqrt{2\pi}} \left[e^{-k_n^2/2} - e^{-[(n+\alpha_R)(1-z_0)/z_0]/2} \right] \\ I_l =& -\frac{1}{\sqrt{2\pi}} y^{l-1} e^{-y^2/2} \Big|_{k_n}^{\sqrt{(n+\alpha_R)(1-z_0)/z_0}} + (l-1) I_{l-2} \\ =& J_{l-1} + (l-1) I_{l-2}, \end{aligned}$$

for $l \geq 2$, where

$$J_l = -\frac{1}{\sqrt{2\pi}} y^l e^{-y^2/2} \Big|_{k_n}^{\sqrt{(n+\alpha_R)(1-z_0)/z_0}}.$$

Note that the upper limit of J_l is $O(n^{-m})$ for all $m > 0$ so that

$$J_l = \frac{1}{\sqrt{2\pi}} k_n^l e^{-k_n^2/2} + O(n^{-m}).$$

Further, $I_1 = J_0$, $I_3 = J_2 + 2I_1$ and $c''_{01} = -3c''_{10}$, so we have

$$\begin{aligned} & c''_{10} I_3 + c''_{01} I_1 \\ =& c''_{10}(J_2 + 2I_1) - 3c''_{10} I_1 = \frac{1-2z_0}{3\sqrt{z_0(1-z_0)}} (J_2 - J_0) \\ =& \frac{1-2z_0}{3\sqrt{z_0(1-z_0)}} (k_n^2 - 1)\phi(k_n) + O(n^{-m}) \text{ for any } m > 0. \end{aligned}$$

and similarly we have

$$\begin{aligned} & c''_{11} I_4 + c''_{20} I_6 + c''_{02} I_2 + a_n I_0 \\ =& c''_{11}(J_3 + 3J_1 + 3I_0) + c''_{20}(J_5 + 5J_3 + 15J_1 + 15I_0) \\ & + c''_{02}(J_1 + I_0) + a_n I_0 \\ =& c''_{20} J_5 + (c''_{11} + 5c''_{20}) J_3 + (3c''_{11} + 15c''_{20} + c''_{02}) J_1 \\ & + (3c''_{11} + 15c''_{20} + c''_{02} + a_n) I_0 \\ =& \frac{(1-2z_0)^2}{18z_0(1-z_0)} J_5 + \frac{-11+35z_0-35z_0^2}{36z_0(1-z_0)} J_3 + \frac{1-z_0+z_0^2}{12z_0(1-z_0)} J_1 \\ =& \left[\frac{(1-2z_0)^2}{18z_0(1-z_0)} k_n^5 + \frac{-11+35z_0-35z_0^2}{36z_0(1-z_0)} k_n^3 + \frac{1-z_0+z_0^2}{12z_0(1-z_0)} k_n \right] \phi(k_n) \\ & + O(n^{-m}) \\ =& d_n \phi(k_n) + O(n^{-m}) \text{ for any } m > 0. \end{aligned}$$

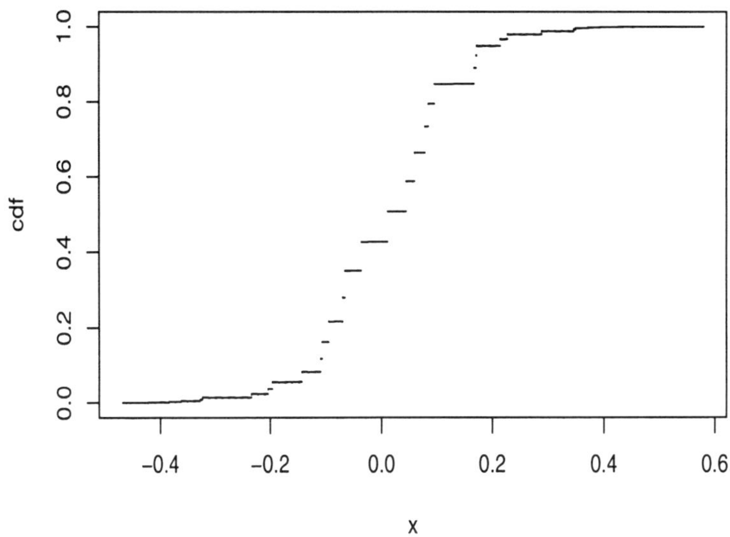

Figure 1 The posterior cdf and approximate posterior cdf overlaid for the 50-th percentile. Based on a random sample of 100 standard normal variables.

Here, d_n can also be written as

$$\frac{(1-2G_n)^2}{18G_n(1-G_n)}k_n^5+\frac{-11+35G_n-35G_n^2}{36G_n(1-G_n)}k_n^3+\frac{1-G_n+G_n^2}{12G_n(1-G_n)}k_n. \qquad (20)$$

Finally, since $I_0 = 1-\Phi(k_n)+O(n^{-m})$ for any $m>0$ (see p166 Feller 1960), we obtain the leading term in (12) and the result follows. □

Remark 2. By Lemma 1 and Corollary 1 or equation (8),

$$k_n=\sqrt{(n+\alpha_R)/z_0(1-z_0)}(p-z_0)\to -g(\eta_p)w/\sqrt{p(1-p)}.$$

6 Numerical Comparisons

We conclude with a brief numerical comparison of the exact posterior distribution and the approximate result using the first order correction term.

Random samples of standard normal variables were generated and the exact posterior distribution was evaluated using the equation (11) in Theorem 2 and the expansion using equation (12) without the error term $O^{-3/2}$. In our simulation study we chose a uniform prior whose support contained the generated random sample.

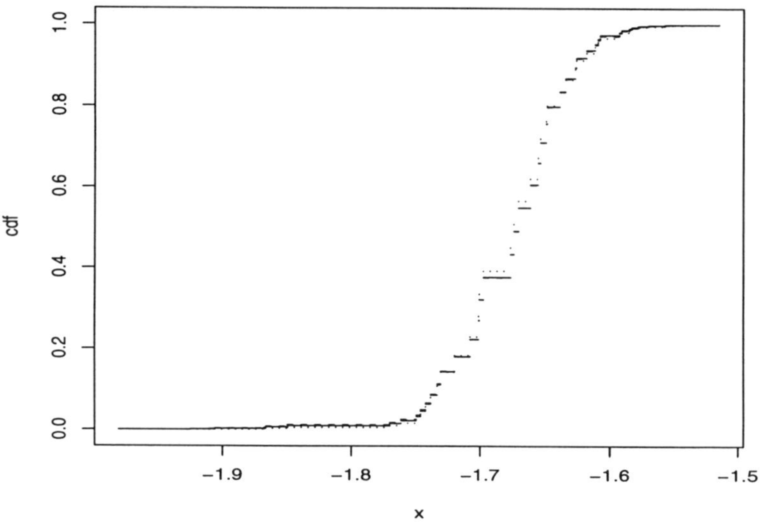

Figure 2 The posterior cdf and approximate posterior cdf overlaid for the 5-th percentile. Based on a random sample of 100 standard normal variables.

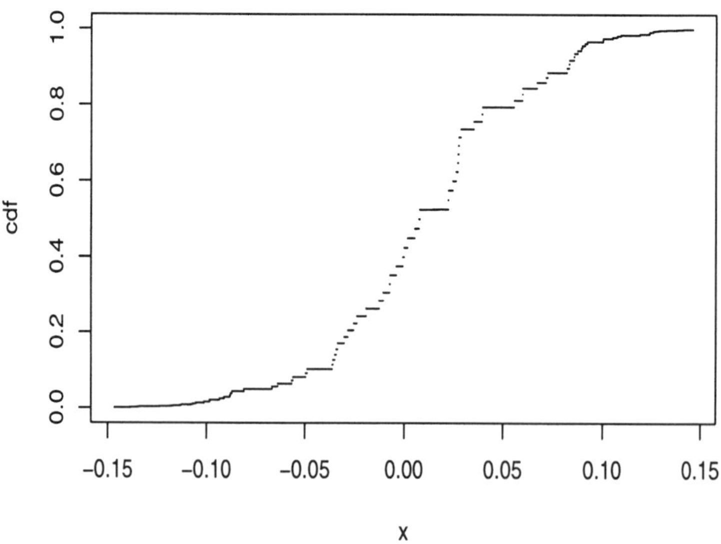

Figure 3 The posterior cdf and approximate posterior cdf overlaid for 50-th percentile. Based on a random sample of 1,000 standard normal variables.

Figure 1 concerns the cumulative distribution function of the 50-th percentile of the posterior distribution for a case where the sample size is 100. The exact and approximate distributions are almost identical. Figure 2 presents the exact and approximate distribution of the 5-th percentile for a case where the sample size is 100. In this figure, small differences between the exact and approximate distribution are apparent. The approximation is not as good as for the 50-th percentile.

Figure 3 presents the exact and approximate distribution of the 5-th percentile for a case where the sample size is increased to 1000. The graph becomes much smoother and the exact and approximate distributions are almost the same.

References

1. BERTRAND-RETALI, M. (1978), Convergence uniforme d'um estimateur la densité pa la méthode de noyau, *Rev. Roumaine Math. Pures. Appl.*, **23**, 361–385.

2. FELLER, W. (1960) *An Introduction to Probability Theory and Its Applications*, 2nd edition, John Wiley & Sons, New York.

3. FERGUSON, T. S. (1973), A Bayesian Analysis of Some Nonparametric Problems, *The Annals of Statistics*, **1**, No. 2, pp209–230.

4. HJORT, N. (2003), Topics in Nonparametric Statistics, *Highly Structured Stochastic Systems*, eds. Green, P. , Hjort, N. and Richardson, S., Oxford,

5. HJORT, N. AND PETRONE, S. (2007), Nonparametric Quantile Inference Using Dirichlet Processes, Chapter in *Advances in Statistical Modeling and Inference*, ed. by V. Nair.

6. JOHNSON, R. A. AND HWANG, L. (2003) Some Exact and Approximate Confidence Region for the Ratio of Percentiles from Two Different Distributions, *Statistical Methods in Reliability*, eds. Lindqvist, B. H. and Doksum, K. A., World Scientific, New Jersey.

7. JOHNSON, R. A. AND LADALLA, J. N. (1979) The Large Sample Behavior of Posterior Distributions When Sampling from Multiparameter Exponential Family Models, and Allied Results, *Sankhyā, Ser. B* **41**, 196–215.

8. JOHNSON, N. L., KOTZ, S. AND BALAKRISHNAN, N. (2002) *Continuous Multivariate Distributions, Volume 1, Models and Applications*, John Wiley and Sons, New York, NY.

9. JOHNSON, R. A. AND SIM, S. (2006) Asymptotic Expansions for the Joint Posterior Distribution of Percentiles in a Nonparametric Bayesian Setting. Technical Report. University of Wisconsin Department of Statistics.

10. SERFLING, R. J. (1980) *Approximation Theorems of Mathematical Statistics*, Academic Press. (1986), *Density Estimation for Statistics and Data Analysis*, Chapman and Hall, London.

Chapter 23

NONPARAMETRIC QUANTILE INFERENCE USING DIRICHLET PROCESSES

Nils Lid Hjort and Sonia Petrone

Department of Mathematics
University of Oslo, NORWAY

IMQ, Bocconi University
Milano, ITALY

E-mails: nils@math.uio.no & sonia.petrone@unibocconi.it

This chapter deals with nonparametric inference for quantiles from a Bayesian perspective, using the Dirichlet process. The posterior distribution for quantiles is characterised, enabling also explicit formulae for posterior mean and variance. Unlike the Bayes estimator for the distribution function, our Bayes estimator for the quantile function is a smooth curve. A Bernshteĭn–von Mises type theorem is given, exhibiting the limiting posterior distribution of the quantile process. Links to kernel-smoothed quantile estimators are provided. As a side product we develop an automatic nonparametric density estimator, free of smoothing parameters, with support exactly matching that of the data range. Nonparametric Bayes estimators are also provided for other quantile-related quantities, including the Lorenz curve and the Gini index, for Doksum's shift curve and for Parzen's comparison distribution in two-sample situations, and finally for the quantile regression function in situations with covariates.

Keywords: Bayesian bootstrap; Bayesian quantile regression; Bernshteĭn–von Mises theorem; Comparison distribution; Dirichlet process; Doksum's shift function; Lorenz curve; Nonparametric Bayes, Quantile inference.

1 Introduction and summary

Assume data $X_1, \ldots, X_n$ come from some unknown distribution F, and that interest focusses on one or more quantiles, say $Q(y) = F^{-1}(y)$. This chapter develops and discusses methods for carrying out nonparametric Bayesian inference for Q, based on a Dirichlet process prior for F. The methods also extend to various other quantile-related quantities in other contexts, notably to various functions and plots for comparing two samples, like Doksum's shift function (see Doksum, 1974a and Doksum and Sievers, 1976) and Parzen's (1979, 1982) comparison distribution, and to quantile regression. A guide-map of our chapter is as follows.

We start in Section 2 with setting the framework and by characterising the prior and posterior distributions of one or more quantiles. This makes it possible to derive explicit formulae for the posterior mean, variance and co-variance in Section 3. A noteworthy feature here is that the posterior mean function is a smooth curve $\widehat{Q}(y)$, unlike the traditional Bayes estimator $\widetilde{F}_n$ for F, which has jumps at the data points. Of particular interest is the non-informative limit of the Bayes estimator $\widehat{Q}_0$ when the strength parameter of the Dirichlet prior is sent to zero. It is seen to be a Bernshteĭn-type smoothed quantile method.

In Section 4 we consider Bayes estimators of the quantile density $q = Q'$ and of the probability density $f = F'$, formed by the appropriate operations on $\widehat{Q}$. A particular construction of interest is the density estimator $\widehat{f}_0$, computed by inversion and differentiation of $\widehat{Q}_0$. This estimator is nonparametric and automatic, requires no smoothing parameters, and is supported on the exact data range, say $[x_{(1)}, x_{(n)}]$. In Section 5 we discuss applications to the Lorenz curve and the Gini index, which are frequently used in econometric contexts. We obtain nonparametric Bayes estimators of these quantities. Then Section 6 provides Bayesian sister versions of two important nonparametric plotting strategies for comparing two populations: Doksum's shift curve $D(x)$ and Parzen's comparison distribution $\pi(y)$. Recipes for computing Bayesian credibility bands are also given. In Section 7 we study large-sample properties of our estimators, and reach Bernshteĭn–von Mises type theorems for the limits of the posterior processes $\sqrt{n}(Q - \widehat{Q})$, $\sqrt{n}(D - \widehat{D})$, $\sqrt{n}(\pi - \widehat{\pi})$. This can be used to form certain approximate credibility intervals for the quantile function, for the shift function, and for the comparison distribution. Then in Section 8 results are generalised to a semiparametric regression framework, where the regression parameters are given a prior independent of the quantile process of the error distribution. Our chapter ends with a list of concluding comments, some pointing to further research problems of interest.

2 The quantile process of a Dirichlet

This section derives the basic distributional results about the distribution of random quantiles for Dirichlet priors, pre and post data. Our point of departure is a Dirichlet process F with parameter measure $\alpha(\cdot) = aF_0(\cdot)$, written $F \sim \mathrm{Dir}(aF_0)$, splitting into constant $a = \alpha(\mathbb{R})$ and probability distribution $F_0 = \alpha/a$; for definitions and basic results one may consult Ferguson (1973, 1974). For a review of general Bayesian nonparametrics, see Hjort (2003).

2.1 *Prior distributions of quantiles*

For the random F, consider its accompanying quantile process

$$Q(y) = F^{-1}(y) = \inf\{t\colon F(t) \geq y\}.$$

For this left-continuous inverse of the right-continuous F it holds generally that $Q(y) \leq x$ if and only if $y \leq F(x)$, even for cases when F, like here, has jumps. It follows, by the basic Beta distribution property of marginals of Dirichlet processes, that the distribution of $Q(y)$ can be written

$$\begin{aligned} H_{0,a}(x) &= \Pr\{Q(y) \leq x\} \\ &= 1 - \mathrm{Be}(y; aF_0(x), a\bar{F}_0(x)) = \mathrm{Be}(1-y; a\bar{F}_0(x), aF_0(x)). \end{aligned} \quad (1)$$

Here and below we let $\mathrm{Be}(\cdot; b, c)$ and $\mathrm{be}(\cdot; b, c)$ denote respectively the distribution function and the density of a Beta variable with parameters (b, c), and $\bar{F}_0$ is the survival function $1 - F_0$. We allow Beta variables with parameters $(b, 0)$ and $(0, c)$; these are with probability one equal to respectively 1 and 0. Thus $\mathrm{Be}(y; b, 0) = 0$ and $\mathrm{Be}(y; 0, c) = 1$ for $y \in [0, 1]$.

Note that $H_{0,a}(x) = J_a(F_0(x))$, where $J_a(x) = \mathrm{Be}(1-y; a(1-x), ax)$ is the distribution of a random y-quantile for the special case of F_0 being uniform on $(0, 1)$, say $Q_{\mathrm{uni}}(y)$. This means that the distribution of $Q(y)$ in the general case is the same as the distribution of $F_0^{-1}(Q_{\mathrm{uni}}(y))$. If F_0 has a density f_0, this also implies that the prior density of $Q(y)$ is $h_0(x) = j_a(F_0(x))f_0(x)$, where

$$j_a(x) = \frac{\partial}{\partial x}\int_0^{1-y} \frac{\Gamma(a)}{\Gamma(a-ax)\Gamma(ax)} u^{a-ax-1}(1-u)^{ax-1}\,du \quad (2)$$

is the density of $Q_{\mathrm{uni}}(y)$. The point is that the prior densities can be computed and displayed via numerical integration and derivation; see Figure 1.

2.2 *Several quantiles simultaneously*

Consider now the joint distribution of two or more Q-values. For $y_1 < \cdots < y_k$, we have

$$\begin{aligned}\Pr\{Q(y_1) \le t_1, \ldots, Q(y_k) \le t_k\} &= \Pr\{y_1 \le F(t_1), \ldots, y_k \le F(t_k)\} \\ &= \Pr\{V_1 \ge y_1, \ldots, V_1 + \cdots + V_k \ge y_k\},\end{aligned}$$

in terms of a Dirichlet vector $(V_1, \ldots, V_k, V_{k+1})$ with parameters $(c_0, \ldots, c_k, c_{k+1})$, where $c_j = aF_0(t_{j-1}, t_j]$; here $F_0(A)$ is the probability assigned to the set A by the F_0 distribution, and $t_0 = -\infty$, $t_{k+1} = \infty$. This in principle determines all aspects of the simultaneous distribution of the vector of random quantiles.

To give somewhat more qualitative insights into the joint distribution of the random quantiles, we start recalling an important and convenient property of the Dirichlet process. When it is 'chopped up' into smaller pieces, conditioned to have certain total probabilities on certain sets, the individual daughter processes become independent and are indeed still Dirichlet. In detail, if F is Dirichlet aF_0, and one conditions on the event $F(B_1) = z_1, \ldots, F(B_m) = z_m$, where the B_is form a partition and the z_is sum to 1, then this creates m new and independent Dirichlet processes on $B_1, \ldots, B_m$. Specifically, $F(.)/z_i$ is Dirichlet on its 'local sample space' B_i with parameter aF_0, that is,

$$F(.)/z_i \sim \mathrm{Dir}(aF_0) = \mathrm{Dir}(aF_0(B_i)\, F_0(.)/F_0(B_i)).$$

See Hjort (1986, 1996) for this fact about pinned down Dirichlets and some of its consequences. Note the rescaling of the Dirichlet parameter, as a new prior strength parameter $aF_0(B_i)$ times the rescaled distribution $F_0(.)/F_0(B_i)$ on set B_i.

Consider two quantiles $Q(y_1)$ and $Q(y_2)$, where $y_1 < y_2$, for the prior process. Conditional on $y_2 = F(t_2)$, our F splits into two independent Dirichlet processes on $(-\infty, t_2]$ and (t_2, ∞). By the general result just described, and arguing as with equation (1), one finds for $t_1 \le t_2$ that

$$\begin{aligned}\Pr\{Q(y_1) \le t_1 \,|\, y_2 = F(t_2)\} &= \Pr\{y_1 \le y_2 F^*(t_1)\} \\ &= \mathrm{Be}(1 - y_1/y_2; aF_0(t_1, t_2], aF_0(-\infty, t_1]),\end{aligned}$$

where F^* is Dirichlet (aF_0) on $(-\infty, t_2]$. This argument may be extended to the case of three or more random quantiles, also suitable for simulation purposes.

2.3 *Posterior distributions of quantiles*

Conditionally on the randomly selected F, let $X_1, \ldots, X_n$ be independently drawn from F. Since F given data is an updated Dirichlet with parameter

$aF_0 + nF_n$, where F_n is the empirical distribution of the data points, the posterior distribution of $Q(y)$ may be written as in (1), with $aF_0 + nF_n$ replacing aF_0 there. Assume for simplicity that the data points are distinct, order them $x_{(1)} < \cdots < x_{(n)}$, and write $x_{(0)} = -\infty$ and $x_{(n+1)} = \infty$. Then

$$\begin{aligned} H_{n,a}(x) &= \Pr\{Q(y) \le x \,|\, \text{data}\} \\ &= 1 - \text{Be}(y; (aF_0 + nF_n)(x), (a\bar{F}_0 + n\bar{F}_n)(x)), \end{aligned} \tag{3}$$

in terms of $\bar{F}_0 = 1 - F_0$ and $\bar{F}_n = 1 - F_n$. For $x_{(i)} \le x < x_{(i+1)}$, this is equal to $\text{Be}(1 - y; a\bar{F}_0(x) + n - i, aF_0(x) + i)$. Thus $Q(y)$ has a density of the form

$$h_{n,a}(x) = (\partial/\partial x)\,\text{Be}(1 - y; a\bar{F}_0(x) + n - i, aF_0(x) + i) \quad \text{inside } (x_{(i)}, x_{(i+1)}),$$

cf. the calculations leading to (2), and posterior point mass

$$\begin{aligned} \Delta H_{n,a}(x_{(i)}) &= \text{Be}(y; aF_0(x_{(i)}-) + i - 1, a\bar{F}_0(x_{(i)}-) + n - i + 1) \\ &\qquad - \text{Be}(y; aF_0(x_{(i)}) + i, a\bar{F}_0(x_{(i)}) + n - i) \\ &= (n + a)^{-1}\text{be}(y; aF_0(x_{(i)}) + i, a\bar{F}_0(x_{(i)}) + n - i + 1) \end{aligned} \tag{4}$$

at point $x_{(i)}$. The partial integration formula (A1) of the Appendix is used here, and assumes continuity of F_0 at $x_{(i)}$.

If a is sent to zero here there is no posterior probability mass left between data points; the distribution concentrates on the data points with probabilities

$$\begin{aligned} p_n(x_{(i)}) &= \text{Be}(y; i - 1, n - i + 1) - \text{Be}(y; i, n - i) \\ &= \binom{n-1}{i-1} y^{i-1}(1 - y)^{n-i}. \end{aligned} \tag{5}$$

These binomial weights concentrate around y for moderate to large n. We also have the following result, proved in our Appendix, which says that even if a is large, the combined posterior probability that $Q(y)$ has of landing outside the data points goes to zero as n increases. In other words, the distribution function $H_{n,a}(x)$ becomes closer and closer to being concentrated in only the n sample points.

Proposition 1. *For fixed positive a, the sum of the posterior point masses $\Delta H_{n,a}(x_{(i)})$ that $Q(y)$ has at the data points goes to 1 as $n \to \infty$.*

The prior to posterior mechanism is illustrated in Figure 1 for the case of the upper quartile $Q(0.75)$, with prior guess $F_0 = \text{N}(0, 1)$, with $n = 100$ data points really coming from $\text{N}(1, 1)$. The right panel shows only the posterior probabilities (5) corresponding to $a = 0$; even for $a = 10$ the (4) probabilities are quite close to those of (5).

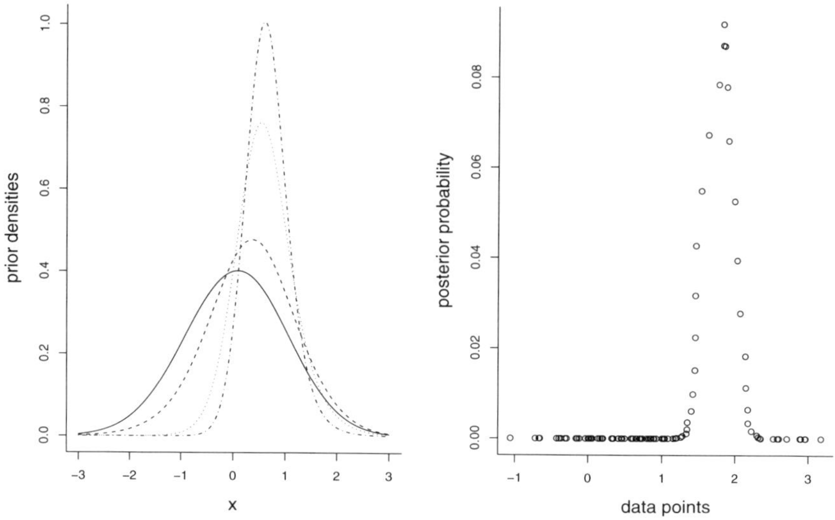

Figure 1 Prior to posterior for a given quantile: The left panel shows the prior densities $j_a(F_0(x))f_0(x)$ at quantile $y = 0.75$, for values $a = 0.1, 1, 5, 10$, for F_0 the standard normal, with smaller values of a closer to the f_0 and larger values of a tighter around $Q_0(y) = 0.675$. The right panel shows the posterior probabilities (5) after having observed $n = 100$ data points from the distribution N(1, 1), with true quartile 1.675. The posterior probability mass outside the data points equals 0.0002, 0.0017, 0.0085, 0.0181 for the four values of a, respectively.

Next consider random quantiles at positions $y_1 < \cdots < y_k$. Then the event $Q(y_1) \le t_1, \ldots, Q(y_k) \le t_k$, where $t_1 \le \cdots \le t_k$, is equivalent to

$$y_1 \le V_1,\ y_2 \le V_1 + V_2, \ldots,\ y_k \le V_1 + \cdots + V_k,$$

writing now $V_j = F(t_j) - F(t_{j-1})$ for $j = 1, \ldots, k+1$, where $t_0 = -\infty$ and $t_{k+1} = \infty$. The vector $(V_1, \ldots, V_k, V_{k+1})$ has the appropriate Dirichlet distribution with parameters $(c_1, \ldots, c_k, c_{k+1})$, where $c_j = (aF_0 + nF_n)(t_{j-1}, t_j]$. This fully defines $\Pr\{Q(y_1) \le t_1, \ldots, Q(y_k) \le t_k \,|\, \text{data}\}$. Its limit as $a \to 0$ is discussed below.

2.4 *The objective posterior quantile process*

For the non-informative prior case of $a = 0$ we have seen that $Q(y)$ concentrates on the observed data points with binomial probabilities given in (5). When considering two quantiles, we find that $\Pr\{Q(y_1) =$

$x_{(i)} \mid \text{data}, Q(y_2) = x_{(j)}\}$ becomes

$$\begin{aligned}\text{Be}(1 - y_1/y_2; j - i, i) \; - \; &\text{Be}(1 - y_1/y_2; j - i + 1, i - 1) \\ &= (1/j)\text{be}(1 - y_1/y_2; j - i + 1, i),\end{aligned}$$

using (A1) again. Combining this with (5) one finds that $(Q(y_1), Q(y_2))$ selects the pair $(x_{(i)}, x_{(j)})$ with probability $p_n(x_{(i)}, x_{(j)})$ equal to

$$\begin{aligned}&\frac{(n-1)!}{(j-1)!(n-j)!}\, y_2^{j-1}(1-y_2)^{n-j}\frac{(1/j)\,j!}{(j-i)!(i-1)!}\Big(\frac{y_2 - y_1}{y_2}\Big)^{j-i}\Big(\frac{y_1}{y_2}\Big)^{i-1} \\ &\qquad = \binom{n-1}{i-1, j-i, n-j} y_1^{i-1}(y_2 - y_1)^{j-i}(1-y_2)^{n-j} \qquad (6)\end{aligned}$$

for $1 \le i \le j \le n$. This trinomial structure generalises to a suitable multinomial one for more than two quantiles at a time.

In fact, the non-informative case corresponds to a random F which is concentrated at the data points $x_{(1)} < \cdots < x_{(n)}$ with probabilities $D_1, \ldots, D_n$ following a Dirichlet distribution with parameters $(1, \ldots, 1)$. This in turn means that

$$Q(y) = x_{(i)} \quad \text{if } D_1 + \cdots + D_i \le y < D_1 + \cdots + D_{i+1}.$$

In yet other words, $Q(y) = x_{(N(y))}$, where $N(y)$ is the smallest i at which the cumulative sum $S_i = D_1 + \cdots + D_i$ exceeds y. One may re-prove (5) from this, as well as the trinomial result (6) for

$$p_n(x_{(i)}, x_{(j)}) = \Pr\{S_{i-1} < y_1 \le S_i \le S_{j-1} < y_2 \le S_j\},$$

via integrations in the distribution for $(S_{i-1}, S_i - S_{i-1}, S_{j-1} - S_{i-1}, S_j - S_{j-1}, 1 - S_j)$, which is Dirichlet with parameters $(i-1, 1, j-1-i, 1, n-j)$. The easiest argument uses that $S_1, \ldots, S_{n-1}$ forms an ordered sample of size $n-1$ from the uniform distribution on the unit interval. For the general case of m quantiles one finds that $\Pr\{Q(y_1) = x_{(i_1)}, \ldots, Q(y_m) = x_{(i_m)}\}$ is equal to

$$\binom{n-1}{i_1 - 1, 1, \ldots, i_m - i_{m-1}, 1, n - i_m} y_1^{i_1 - 1}(y_2 - y_1)^{i_2 - i_1} \cdots (1 - y_m)^{n - i_m},$$

valid for $y_1 < \cdots < y_m$ and $i_1 \le \cdots \le i_m$. This 'multinomial structure' hints at connections to Brownian bridges; such are indeed studied in Section 7.

3 Bayesian quantile inference

To carry out Bayesian inference for $Q(y)$, for specific quantiles or for the full quantile function, several options are available.

One possibility is to repeatedly simulate full Q functions by numerically inverting simulated paths of F, these being drawn according to the $\mathrm{Dir}(aF_0 + nF_n)$ distribution. Another is to work directly with the explicit posterior distribution $H_{n,a}$ of (3) for $Q(y)$, or if necessary with the generalisations to several quantiles discussed in Section 2.3. One attractive estimator is

$$Q_n^*(y) = \mathrm{median}\{Q(y)\,|\,\mathrm{data}\} = H_{n,a}^{-1}(\tfrac{1}{2}),$$

which is the Bayes estimator under loss functions of the type $\int_0^1 w(y)|\widehat{Q}(y) - Q(y)|\,\mathrm{d}y$. It is not difficult to implement a programme that for each y finds the posterior median, from the formula for $H_{n,a}(x)$. For the special case of $y = \frac{1}{2}$, the posterior median of the random median is the median of the posterior expectation $\widetilde{F}_n = (aF_0 + nF_n)/(a+n)$. This may also naturally be supplemented with posterior credibility bands of the type $[H_{n,a}^{-1}(0.05), H_{n,a}^{-1}(0.95)]$. It follows from theory developed below that such a band is secured limiting 90% pointwise coverage probability, also in the frequentist sense. Here, however, we focus on directly computable Bayes estimators and on posterior variances.

We first set out to compute the posterior mean function of $Q(y)$, which is the Bayes estimator under quadratic loss. The informative case $a > 0$ is more cumbersome mathematically than the $a \to 0$ case, and is considered first. Ferguson (1973, p. 224) pointed out that the posterior expectation "is difficult to compute, and may, in fact, not even exist". Here we give both precise finiteness conditions and a formula; such have apparently not been given earlier in the literature. From our results in Section 2 it is clear that when the integrals exist, a formula for the posterior mean takes the form

$$\widehat{Q}_a(y) = \sum_{i=1}^n \Delta H_{n,a}(x_{(i)})x_{(i)} + \sum_{i=0}^n \int_{(x_{(i)},x_{(i+1)})} x h_{n,a}(x)\,\mathrm{d}x, \qquad (7)$$

with $H_{n,a}$ and $h_{n,a}$ as given in Section 2.3. Existence requires finiteness of the first and the last integrals here, over respectively $(-\infty, x_{(1)})$ and $(x_{(n)}, \infty)$. The following is proved in our Appendix.

Proposition 2. *Let $Q = F^{-1}$ have the prior process induced by a Dirichlet process prior with parameter aF_0 for F, where a is positive. Then the posterior mean $\widehat{Q}_a(y)$ of the quantile function $Q(y)$ is well-defined and finite if and only if the prior mean $\mathrm{E}_0|X| = \int |x|\,\mathrm{d}F_0(x)$ is finite. This result is independent of the sample size n and of the value of y, and is also valid for the prior situation.*

For implementation purposes, formula (7) is a little awkward. A simpler equivalent formula is

$$
\begin{aligned}
\widehat{Q}_a(y) &= \int_0^\infty \Pr\{Q(y) \geq x \,|\, \text{data}\}\,\mathrm{d}x - \int_{-\infty}^0 \Pr\{Q(y) \leq x \mid \text{data}\}\,\mathrm{d}x \\
&= \int_0^\infty \mathrm{Be}(y; aF_0(x) + nF_n(x), a\bar{F}_0(x) + n\bar{F}_n(x))\,\mathrm{d}x \qquad (8) \\
&\qquad - \int_{-\infty}^0 \mathrm{Be}(1-y; a\bar{F}_0(x) + n\bar{F}_n(x), aF_0(x) + nF_n(x))\,\mathrm{d}x.
\end{aligned}
$$

For large a dominating n in size, this estimator is close to the prior guess function $F_0^{-1}(y)$. Even a moderate or large a will however be 'washed out' by the data as n grows, as is apparent from Proposition 1 and made clearer in Section 7.

Particularly interesting is the nonparametric quantile estimator emerging by letting a tend to zero, since the posterior then concentrates on the data points alone. By (5), the result is

$$\widehat{Q}_0(y) = \sum_{i=1}^n \binom{n-1}{i-1} y^{i-1}(1-y)^{n-i} x_{(i)}. \qquad (9)$$

This is a $(n-1)$-degree polynomial function that smoothly climbs from $\widehat{Q}_0(0) = x_{(1)}$ to $\widehat{Q}_0(1) = x_{(n)}$. It may of course be used also outside the present Bayesian framework. Its frequentist properties have been studied, to various extents, in Hjort (1986), Sheather and Marron (1990), and Cheng (1995), and we learn more in Section 7 below. Interestingly, it can also be expressed as $n^{-1}\sum_{i=1}^n \mathrm{be}(y; i, n-i+1)\, x_{(i)}$, an even mixture of beta densities.

The posterior variance $\widehat{V}_a(y)$ may also be computed explicitly, via $\mathrm{E}\{Q(y)^2 \,|\, \text{data}\} = \int_0^\infty \Pr\{|Q(y)| \geq x^{1/2} \,|\, \text{data}\}\,\mathrm{d}x$, which as with other calculations above with some efforts also may be expressed in terms of finite sums of explicit terms. One may show as with Proposition 2 that the posterior variance is finite if and only if the prior variance is finite; this statement is valid for each n. In the $a \to 0$ case the variance simplifies to

$$\widehat{V}_0(y) = \sum_{i=1}^n \binom{n-1}{i-1} y^{i-1}(1-y)^{n-i}\,\{x_{(i)} - \widehat{Q}_0(y)\}^2. \qquad (10)$$

The posterior covariance between two quantiles can similarly be estimated explicitly, via (6). With the limiting normality results of Section 7 this implies for example that $\widehat{Q}_0(y) \pm 1.96\,\widehat{V}_0(y)^{1/2}$ becomes an asymptotic pointwise 95% confidence band in the frequentist sense, as well as an asymptotic pointwise 95% credibility band in the Bayesian posterior sense.

Remark 1. Note first that $X_{([nt])}$ is distributed as $F_{\mathrm{tr}}^{-1}(U_{([nt])})$, in terms of an ordered sample $U_{(1)}, \ldots, U_{(n)}$ from the uniform distribution on the unit interval, in terms of the true distribution F_{tr} for the X_is. Hence $X_{([nt])}$ is close to $F^{-1}(t)$ for moderate to large n. A kernel type estimator based on the order statistics would be of the form

$$\widetilde{Q}(y) = \int K_h(t-y)X_{([nt])}\,\mathrm{d}t \doteq n^{-1}\sum_{i=1}^{n} K_h(i/n-y)x_{(i)},$$

in terms of a scaled kernel function $K_h(u) = h^{-1}K(h^{-1}u)$ and its smoothing parameter h. One may now show, via approximate normality of the binomial weights used in (9), that $\widehat{Q}_0(y)$ is asymptotically identical to such a kernel estimator, with K the standard normal kernel, and $h = \{y(1-y)/n\}^{1/2}$; proving this is related to the classic de Moivre–Laplace result. This means under-smoothing if compared to the theoretically optimal bandwidths, which are of size $O(n^{-1/3})$ for moderate to large n. See Sheather and Marron (1990). ■

4 Quantile density and probability density estimators

Assume that the true $F = F_{\mathrm{tr}}$ governing data has a smooth density f_{tr}, positive on its support. The quantile function $Q_{\mathrm{tr}}(y) = F_{\mathrm{tr}}^{-1}(y)$ has derivative $q_{\mathrm{tr}}(y) = 1/f_{\mathrm{tr}}(Q_{\mathrm{tr}}(y))$, sometimes called the quantile density function. In this section we look at the relatives $\widehat{q}_a$ and $\widehat{f}_a$ following from $\widehat{Q}_a$ of the previous section, with $a = 0$ leading to particularly interesting estimators.

First consider the quantile density. The Bayes estimator with the Dirichlet process prior under squared error loss is, via results of Section 3, after an exchange of derivative and mean operations,

$$\begin{aligned}\widehat{q}_a(y) &= \int_0^\infty \mathrm{be}(y; aF_0(x)+nF_n(x), a\bar{F}_0(x)+n\bar{F}_n(x))\,\mathrm{d}x \\ &\quad + \int_{-\infty}^0 \mathrm{be}(1-y; a\bar{F}_0(x)+n\bar{F}_n(x), aF_0(x)+nF_n(x))\,\mathrm{d}x.\end{aligned}$$

The limiting non-informative case $\widehat{q}_0 = \widehat{Q}_0'$ can be written in several revealing ways, from (9) or as a limit of the above;

$$\begin{aligned}\widehat{q}_0(y) &= \sum_{i=1}^{n}\binom{n-1}{i-1}y^{i-1}(1-y)^{n-i}\Big(\frac{i-1}{y}-\frac{n-i}{1-y}\Big)x_{(i)} \\ &= \int_{x_{(1)}}^{x_{(n)}} \mathrm{be}(y, nF_n(x), n\bar{F}_n(x))\,\mathrm{d}x = \sum_{i=1}^{n-1}(x_{(i+1)}-x_{(i)})\mathrm{be}(y,i,n-i).\end{aligned}$$

Note that there is no smoothing parameter in this construction; the inherent smoothing comes 'for free' through the limiting Dirichlet process prior argument. The level of this inherent smoothing is about $\{y(1-y)/n\}^{1/2}$, as per Remark 1 above.

We have devised Bayesian ways of estimating $Q = F^{-1}$, and are free to invert back to the F scale, finding in effect new estimators of the distribution function. Thus let $\widehat{F}_a(x)$ be the solution to $x = \widehat{Q}_a(y)$. It can be computed from (8). This is not the same as the posterior mean or posterior median, but is a Bayes estimator in its own right, with loss function of the form $L(F, \widehat{F}) = \int_0^1 w(\widehat{Q} - Q)^2 \, dy$. It is noteworthy that $\widehat{F}_a$ is smooth and differentiable in x, unlike the posterior mean function $\{aF_0(x)+nF_n(x)\}/(a+n)$, which has jumps at each data point. When a dominates n, $\widehat{F}_a$ is close to F_0. The case $a = 0$ is again of particular interest, with $\widehat{F}_0$ climbing smoothly from zero at $x_{(1)}$ to one at $x_{(n)}$, with an everywhere positive density over this data range. The $\widehat{F}_0$ may be considered a smoother default alternative to the empirical distribution function F_n, for e.g. display purposes. It follows from theory of Section 7 that $\sqrt{n}(\widehat{F}_0 - F_n) \to_p 0$, so the two estimators are close.

It is well known that distribution functions chosen from the Dirichlet prior are discrete with probability one. Thus the random posterior quantile process is also discrete. That the posterior mean of $Q(y)$ happens to be a smooth function of y is not a contradiction, however. We have somehow 'gained smoothness' by passing from F to Q and back to F again. This should perhaps be viewed as mathematical happenstance; neither F nor Q is smooth, but the posterior mean function of Q is.

Our efforts also lead to new nonparametric Bayesian density estimators. We solved $\widehat{Q}_a(y) = x$ to reach the estimator $\widehat{F}_a(x)$, and its derivative $\widehat{f}_a(x)$ is a Bayes estimator of the underlying data density $f_{\rm tr}$. The result is a continuous bridge in a, from the prior guess f_0 for a large to something genuinely nonparametric and prior-independent for $a = 0$. One may contemplate devising methods for choosing a from data, smoothing between prior and data, perhaps in empirical Bayesian fashions, or via a hyperprior. Here we focus on the automatic density estimator $\widehat{f}_0$, corresponding to the non-informative prior.

From $\widehat{f}_0(x) = (\widehat{Q}_0^{-1})'(x)$ we may write

$$\widehat{f}_0(x) = \Big[\sum_{i=1}^{n-1} (x_{(i+1)} - x_{(i)}) \mathrm{be}(\widehat{F}_0(x); i, n-i)\Big]^{-1}, \tag{11}$$

where, for each x, the equation $\widehat{Q}_0(y) = x$ is numerically solved for y to get $\widehat{F}_0(x)$, for example using a Newton–Raphson method. From smoothness properties of $\widehat{F}_0$ noted above, one sees that $\widehat{f}_0(x)$ is strictly positive on the exact data range $[x_{(1)}, x_{(n)}]$, with unit integral.

The formula above for $\widehat{f}_0(x)$ is directly valid inside $(x_{(1)}, x_{(n)})$. At the end points some details reveal that

$$\widehat{f}_0(x_{(1)}) = 1/\widehat{q}_0(0) = \{(n-1)(x_{(2)} - x_{(1)})\}^{-1},$$
$$\widehat{f}_0(x_{(n)}) = 1/\widehat{q}_0(1) = \{(n-1)(x_{(n)} - x_{(n-1)})\}^{-1}.$$

It is interesting and perhaps surprising that this nonparametric Bayesian approach leads to such explicit advice about the behaviour of f near and at the endpoints; estimation of densities in the tails is in general a difficult problem with no clear favourite among frequentist proposals.

It is perhaps too adventurous to struggle for the abolition of all histograms, replacing them instead with the automatic Bayesian non-informative density estimator $\widehat{f}_0$ of (11). But as Figure 2 illustrates, it can be a successful data descriptor, with better smoothness properties than the histogram, and without the need for selecting smoothing parameters. It also has the pleasant property that $\int x\widehat{f}_0(x)\,\mathrm{d}x$ is precisely equal to the data mean $\bar{x}$. When compared to traditional kernel methods it will be seen to smooth less, actually with an amount corresponding to a locally varying bandwidth of size $O(n^{-1/2})$, as opposed to the traditional optimal size $O(n^{-1/5})$. The latter does assume two derivatives of the underlying density, however, whereas the (11) estimator has been constructed directly from the data, without any further smoothness assumptions.

5 The Lorenz curve and the Gini index

Quantile functions are used in many spheres of theoretical and applied statistics. One such is that of econometric studies of income distributions, where information is often quantified and compared in terms of the so-called Lorenz curve (going back a hundred years, to Lorenz, 1905), along with various summary measures, like the Gini index; see e.g. Aaberge (2001) and Aaberge, Bjerve and Doksum (2005). This section considers nonparametric Bayes inference for such curves and indices.

When the distribution F of data is supported on the positive halfline, the *Lorenz curve* is defined as

$$L(y) = \int_0^y Q(u)\,\mathrm{d}u \Big/ \int_0^1 Q(u)\,\mathrm{d}u \quad \text{for } 0 \le y \le 1.$$

The numerator is also equal to $\int_0^{Q(y)} x\,\mathrm{d}F(x)$, and the denominator is simply equal to the mean μ of the distribution. It is in general convex, and is equal to the diagonal $L(y) = y$ if and only if the underlying distribution is concentrated in a single point (perfect equality of income).

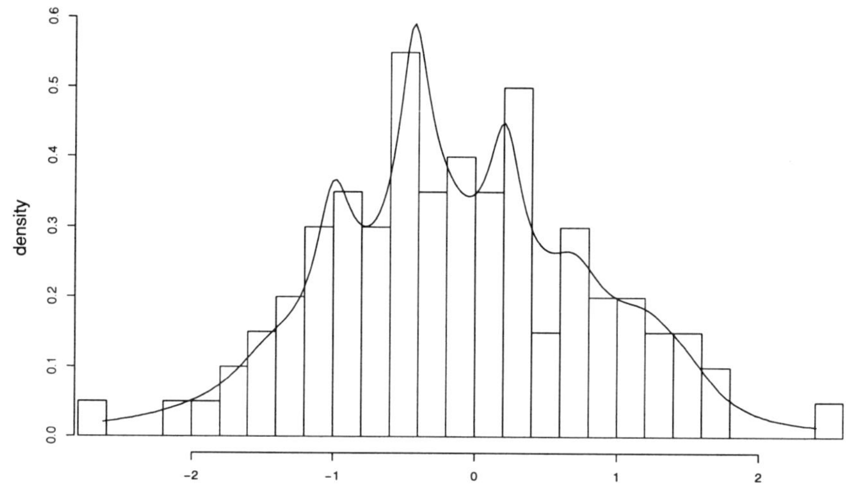

Figure 2 A histogram (with more cells than usual) over $n = 100$ data points from the standard normal, along with the automatic density estimator of (11).

Bayesian inference can now be carried out for L, for example through simulation of Q curves from the posterior distribution. A natural Bayes estimator takes the form

$$\widehat{L}_a(y) = \int_0^y \widehat{Q}_a(u)\,\mathrm{d}u \Big/ \int_0^1 \widehat{Q}_a(u)\,\mathrm{d}u,$$

stemming from keeping the weighted squared error loss function for Q, transforming the solution to L scale. Particularly interesting is the non-informative limit version

$$\widehat{L}_0(y) = \frac{\int_0^y \widehat{Q}_0(u)\,\mathrm{d}u}{\int_0^1 \widehat{Q}_0(u)\,\mathrm{d}u} = \Big\{n^{-1}\sum_{i=1}^n \mathrm{Be}(y; i, n-i+1)x_{(i)}\Big\}\Big/\bar{x} \quad \text{for } 0 \le y \le 1.$$

The *Gini index* is a measure of closeness of the L curve to the diagonal, i.e. the egalitarian case, and is defined as $G = 2\int_0^1 \{y - L(y)\}\,\mathrm{d}y$. With a Dirichlet prior for F and any weighted integrated squared error loss function for the quantile function, we get a Bayes estimator $\widehat{G}_a = 2\int_0^1 \{y - \widehat{L}_a(y)\}\,\mathrm{d}y$. The non-informative limiting version is of particular interest. Some algebra shows that $\widehat{G}_0 = 2\int_0^1 \{y - \widehat{L}_0(y)\}\,\mathrm{d}y$ may be expressed as

$$\widehat{G}_0 = 1 - 2\frac{1}{n}\sum_{i=1}^n \Big(1 - \frac{i}{n+1}\Big)\frac{x_{(i)}}{\bar{x}} = 2\frac{1}{n}\sum_{i=1}^n \frac{i}{n+1}\frac{x_{(i)}}{\bar{x}} - 1.$$

Its value may be supplemented with a credibility interval via posterior simulation of L curves.

6 Doksum's shift and Parzen's comparison

Assume data $X'_1, \ldots, X'_m$ come from the distribution G, independently of $X_1, \ldots, X_n$ from F. When inspecting such data there are various options for portraying, characterising and testing for differences between the two distributions.

Doksum (1974a) introduced the so-called *shift function*

$$D(x) = G^{-1}(F(x)) - x.$$

Its essential property is that $X + D(X)$ has the same distribution as X'. The shift function has a particularly useful role in situations with control and treatment groups. If the distributions of X and X' differ only in location, for example, then $D(x)$ is constant; if on the other hand G is a location-and-scale translation of F, then $D(x)$ is linear. Doksum (1974a) studied the natural nonparametric estimator $\widetilde{D}(x) = G_m^{-1}(F_n(x)) - x$, in terms of the empirical cumulative distributions F_n and G_m; see Section 7.3 below for its key large-sample properties. Here we describe how Bayesian inference can be carried out, starting with independent priors $F \sim \mathrm{Dir}(aF_0)$ and $G \sim \mathrm{Dir}(bG_0)$.

The posterior distribution at a fixed x is

$$K_{m,n}(t) = \Pr\{G^{-1}(F(x)) - x \le t \,|\, \mathrm{data}\} = \Pr\{F(x) \le G(x+t) \,|\, \mathrm{data}\},$$

which can be evaluated via numerical integration, using the Beta distributions involved. For the non-informative case,

$$\begin{aligned} K_{m,n}(t) &= \Pr\{\mathrm{Beta}(nF_n(x), n\bar{F}_n(x)) \le \mathrm{Beta}(mG_m(x+t), m\bar{G}_m(x+t))\} \\ &= \int_0^1 \mathrm{Be}(g, nF_n(x), n\bar{F}_n(x))\mathrm{be}(g, mG_m(x+t), m\bar{G}_m(x+t))\,dg. \end{aligned}$$

This can be used to compute the posterior median estimator $K_{m,n}^{-1}(\frac{1}{2})$, along with a pointwise credibility band, say $[K_{m,n}^{-1}(0.05), K_{m,n}^{-1}(0.95)]$. It follows from results of Section 7 that such a band will have frequentist coverage level converging to the required 90%, for each x, when the sample sizes grow.

We also provide formulae for the posterior mean and variance, for the non-informative case. These are found by first conditioning on F, viz.

$$\begin{aligned} \mathrm{E}\{G^{-1}(F(x)) \,|\, \mathrm{data}, F\} &= \sum_{j=1}^m \binom{m-1}{j-1} F(x)^{j-1}\bar{F}(x)^{m-j} x'_{(j)}, \\ \mathrm{E}\{G^{-1}(F(x))^2 \,|\, \mathrm{data}, F\} &= \sum_{j=1}^m \binom{m-1}{j-1} F(x)^{j-1}\bar{F}(x)^{m-j} (x'_{(j)})^2. \end{aligned}$$

Using Beta moment formulae this gives the Bayes estimator $\widehat{D}_0(x)$ as

$$\sum_{j=1}^{m}\binom{m-1}{j-1}\frac{\Gamma(n)}{\Gamma(nF_n)\Gamma(n\bar{F}_n)}\frac{\Gamma(nF_n+j-1)\Gamma(n\bar{F}_n+m-j)}{\Gamma(n+m-1)}x'_{(j)}-x,$$

writing F_n and $\bar{F}_n$ for $F_n(x)$ and $\bar{F}_n(x)$, while the posterior variance $\widehat{V}_0(x)$ can be found as

$$\sum_{j=1}^{m}\binom{m-1}{j-1}\frac{\Gamma(n)}{\Gamma(nF_n)\Gamma(n\bar{F}_n)}\frac{\Gamma(nF_n+j-1)\Gamma(n\bar{F}_n+m-j)}{\Gamma(n+m-1)}(x'_{(j)})^2 - \{\widehat{D}_0(x)+x\}^2.$$

The theory of Section 7 guarantees that the band $\widehat{D}_0(x) \pm 1.645\,\widehat{V}_0(x)^{1/2}$ has pointwise coverage level converging to 90%, for example, as the sample sizes increase.

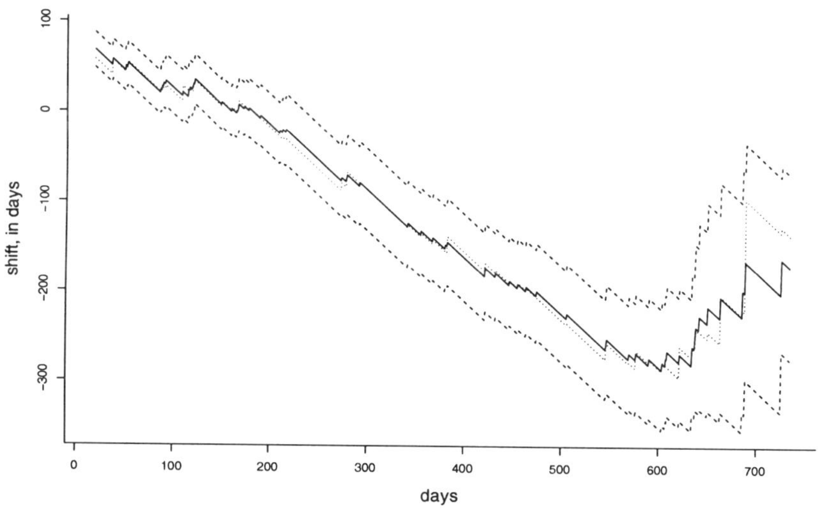

Figure 3 For the 65 guinea pigs in the control group and the 60 in the treatment group, we display the Bayes estimator [full line] of the shift function associated with the two survival distributions, alongside Doksum's sample estimator [dotted line]. Also given is the approximate pointwise 90% credibility band.

Doksum (1974a) illustrated his shift function using survival data of guinea pigs in Bjerkedal's (1960) study of the effect of virulent tubercle bacilli, with 65 in the control group and 60 in the treatment group, the latter receiving a dose of such bacilli. Here we re-analyse Bjerkedal and

Doksum's data, with Figure 3 displaying the Bayes estimate $\widehat{D}_0(x)$, seen there to be quite close to Doksum's direct estimate. Also displayed is the approximate 90% pointwise confidence band. The figure illustrates dramatically that the weaker pigs (those who tend to die early) will tend to have longer lives with the treatment, while the stronger pigs (those whose lives tend to be long) are made drastically weaker, i.e. their life lengths will decrease. This analysis agrees with conclusions in Doksum (1974a). For example, pigs with life expectancy around 500 days can expect to live around 200 days less if receiving the virulent tubercle bacilli in question.

Parzen (1979, 1982, 2002) has repeatedly advocated analysing and estimating the function $\pi(y) = G(F^{-1}(y))$, which he terms the *comparison distribution*. This function, or estimates thereof, may be plotted against the identity function $\pi_{\rm id}(y) = y$ on the unit interval; equality of the two distributions is equivalent to $\pi = \pi_{\rm id}$. See also Newton's interview with Parzen (2002, p. 372–374). We now consider nonparametric Bayesian estimation of the Parzen curve via independent Dirichlet process priors on F and G, with parameters respectively aF_0 and bG_0.

A formula for the posterior mean $\widehat{\pi}(y)$ may be derived as follows. Let $\widehat{G}_m = w'_m G_0 + (1 - w'_m)G_m$ be the posterior mean of G, in terms of $w'_m = b/(b+m)$ and the empirical distribution G_m for the m data points. Then $\widehat{\pi}(y)$ is the mean of $\mathrm{E}\{G(Q(y))\,|\,Q, \mathrm{data}\}$, i.e. the mean of $\widehat{G}_m(Q(y))$ given data, leading to

$$\begin{aligned}\widehat{\pi}(y) &= w'_m \mathrm{E}\{G_0(Q(y))\,|\,\mathrm{data}\} + (1-w'_m)\mathrm{E}\{G_m(Q(y))\,|\,\mathrm{data}\}\\ &= w'_m \int_0^1 \Pr\{G_0(Q(y)) > z\,|\,\mathrm{data}\}\,\mathrm{d}z\\ &\qquad + (1-w'_m)\frac{1}{m}\sum_{j=1}^m \Pr\{x'_j \le Q(y)\,|\,\mathrm{data}\}\\ &= w'_m \int_0^1 \mathrm{Be}(y; (aF_0 + nF_n)(G_0^{-1}(z)), (a\bar{F}_0 + n\bar{F}_n)(G_0^{-1}(z)))\,\mathrm{d}z\\ &\qquad + (1-w'_m)\frac{1}{m}\sum_{j=1}^m \mathrm{Be}(y; (aF_0+nF_n)(x'_j-), (a\bar{F}_0 + n\bar{F}_n)(x'_j-)),\end{aligned}$$

where the second term is explicit and the first not difficult to compute numerically. If there are no ties between the x'_j and the x_i points for the two samples, $(aF_0 + nF_n)(x'_j-)$ is the same as $(aF_0 + nF_n)(x'_j)$. For the non-informative case of a and b both going to zero, we have the particularly appealing estimator

$$\widehat{\pi}_0(y) = \frac{1}{m}\sum_{j=1}^m \mathrm{Be}(y; nF_n(x'_j-), n\bar{F}_n(x'_j-)).$$

Its derivative, which is an estimate of what Parzen terms the comparison density $g(F^{-1}(y))/f(F^{-1}(y))$, provided the densities $g = G'$ and $f = F'$ exist, is quite simply $(1/m)\sum_{j=1}^{m} \text{be}(y; nF_n(x'_j-), n\bar{F}_n(x'_j-))$. The posterior variance of $\pi(y)$ may also be calculated with some further efforts. For the non-informative case of $a = b = 0$, we find

$$\begin{aligned}\text{Var}\{\pi(y)\,|\,\text{data}\} &= \frac{1}{m+1}\widehat{\pi}_0(y)\{1-\widehat{\pi}_0(y)\} \\ &\quad + \frac{m}{m+1}\Big\{\frac{1}{m^2}\sum_{j,k}\text{Be}(y; nF_n(x'_{j,k}-), n\bar{F}_n(x'_{j,k}-)) - \widehat{\pi}_0(y)^2\Big\},\end{aligned}$$

in which $x'_{j,k} = \max(x'_j, x'_k)$.

It is seen that $\widehat{\pi}_0(y)$ provides a smoother alternative to the direct nonparametric Parzen estimator. The theory of Section 7 implies that the two estimators are asymptotically equivalent, and also that the simple credibility band $\widehat{\pi}_0(y) \pm 1.96\,\widehat{\text{sd}}(y)$, with $\widehat{\text{sd}}(y)$ the posterior standard deviation computed as above, is a band reaching 95% level coverage, in both the frequentist and Bayesian settings, as sample sizes grow.

Laake, Laake and Aaberge (1985) discussed relations between hospitalisation, as a measure of morbidity, and mortality. The patient material consisted of 367 consecutive admissions at hospitals in Oslo in 1980 (176 males and 191 females), while data on mortality in Oslo consisted of 6140 deaths (2989 males and 3151 females). Letting F be the distribution of age at hospitalisation and G the distribution of age at death, Laake, Laake and Aaberge suggested studying $\Lambda(y) = G^{-1}(y) - F^{-1}(y)$, a direct comparison of the two quantile functions. It is a close cousin of the Doksum curve in that $\Lambda(F(x)) = D(x)$.

We have re-analysed the data of Laake, Laake and Aaberge (1985, Table 1) using the Bayes estimator $\widehat{\Lambda}(y) = \widehat{Q}_G(y) - \widehat{Q}_F(y)$, with components as in (9). For our illustration, we 'made' continuous data from their table, by distributing the number of observations in question evenly over the required age interval; thus 12 and 17 observed hospitalised women in the age groups 50–54 and 55–59 gave rise to 12 and 17 Xs spread uniformly on the intervals $[49.5, 54.5]$ and $[54.5, 59.5]$, and so on. Figure 4 presents these curves, for women and for men separately, along with confidence band $\widehat{\Lambda}(y) \pm 1.96\,\widehat{\text{sd}}(y)$, where $\widehat{\text{sd}}(y)^2$ is the sum of the two variance estimates involved, computed as in (10). It follows from the theory of Section 7 that this band indeed has the intended approximate 95% confidence level at each quantile value y. The analysis shows that to the first order of approximation, and apart from noticeable deviations for the very young and the very old, age at hospitalisation and age at death are similar, with a constant shift between them, about seven years for women and six years for men.

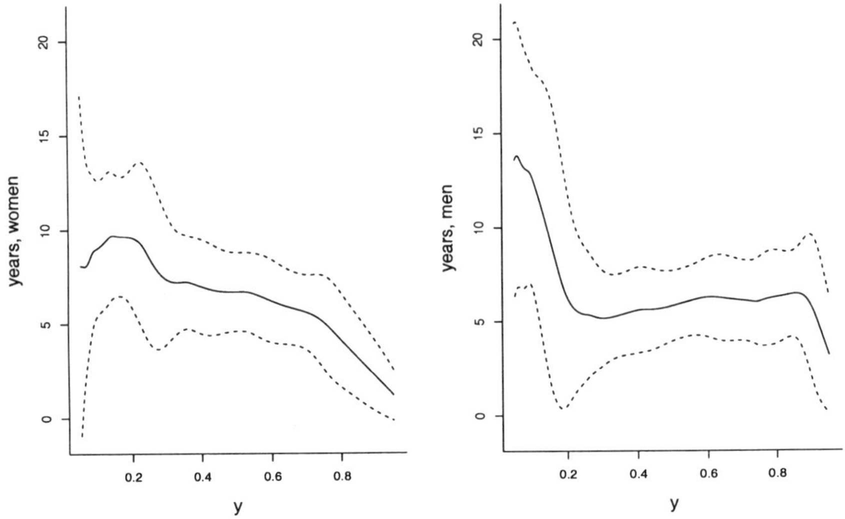

Figure 4 Estimated quantile difference $G^{-1}(y) - F^{-1}(y)$ between age at death distribution and age at hospitalisation distribution, along with pointwise 95% confidence bands, for women (left) and for men (right).

This interpretation is in essential agreement with conclusions reached by Laake, Laake and Aaberge.

7 Large-sample analysis

In this section we discuss large-sample behaviour of the estimation schemes we have developed, from both the Bayesian and frequentist perspectives.

7.1 *Nonparametric Bernshteĭn–von Mises theorems*

To set results reached below in perspective, it is useful first to recall some well-known results about the limiting behaviour of maximum likelihood and Bayes estimators, as well as about the posterior distribution, valid for general parametric models. Specifically, assume i.i.d. data $Z_1, \ldots, Z_n$ follow a parametric density $g(z, \theta)$, with $\theta_{\rm tr}$ the true parameter, and let $\widehat{\theta}_{\rm ml}$ and $\widehat{\theta}_B$ be the maximum likelihood and posterior mean Bayes estimator under a suitable prior $\pi(\mathrm{d}\theta)$. Then, under mild regularity conditions, discussed e.g. in Bickel and Doksum (2001, Ch. 5–6), four notable results are valid: (i) $\sqrt{n}(\widehat{\theta}_{\rm ml} - \theta_{\rm tr}) \to_d \mathrm{N}(0, J(\theta_{\rm tr})^{-1})$; (ii)

$\sqrt{n}(\widehat{\theta}_B - \widehat{\theta}_{\rm ml}) \to_p 0$; (iii) with probability one, the posterior distribution is such that $\sqrt{n}(\theta - \widehat{\theta}_B) \,|\, \text{data} \to_d \text{N}(0, J(\theta_{\rm tr})^{-1})$. Here $J(\theta)$ is the information matrix of the model, see e.g. Bickel and Doksum (2001, Ch. 6). With a consistent estimator $\widehat{J}$ of this matrix one may compute the approximation $\text{N}(\widehat{\theta}_{\rm ml}, n^{-1}\widehat{J})$ to the posterior distribution of θ. Result (iv) is that this simple method is first-order asymptotically correct, i.e. $\widehat{J}^{-1/2}(\theta - \widehat{\theta}_{\rm ml}) \,|\, \text{data}$ goes a.s. to $\text{N}(0, I)$, the implication being that one may approximate the posterior distribution without carrying out the Bayesian updating calculations as such. Results of the (iii) and (iv) variety are often called Bernshteĭn–von Mises theorems; see e.g. LeCam and Yang (1990, Ch. 7). Note that Bayes and maximum likelihood estimators have the same limit distribution, regardless also of the prior one starts out with, as a consequence of (ii).

Such statements and results become more complicated in non- and semiparametric models, and sometimes do not hold. There are situation when Bayes solutions do not match the natural frequentist estimators, and other situations where the posterior distribution goes awry, or have a limit different from that indicated by Bernshteĭn–von Mises heuristics; see e.g. Diaconis and Freedman (1986a, 1986b), Hjort (1986, 1996, 2003). For the present case of Dirichlet process priors there are no such surprises, however, as long as inference about F is concerned, as one may prove the following. Here the role of the maximum likelihood estimator is played by the empirical distribution F_n, with Bayes estimator (posterior mean) equal to $\widetilde{F}_n = (a/(a+n))F_0 + (n/(a+n))F_n$. Below, W^0 is a Brownian bridge, i.e. a Gaußian zero-mean process on $[0,1]$ with covariance structure $t_1(1-t_2)$ for $t \le t_2$.

Proposition 3. *Assume the Dirichlet process with parameter aF_0 is used for the distribution of i.i.d. data $X_1, X_2, \ldots$, and assume that the real generating mechanism for these observations is a distribution $F_{\rm tr}$. Then (i) the process $\sqrt{n}\{F_n(t) - F_{\rm tr}(t)\}$ converges to $W^0(F_{\rm tr}(t))$; (ii) the difference $\sqrt{n}(\widetilde{F}_n - F_n)$ goes to zero; and (iii) the posterior distribution process $V_n(t) = \sqrt{n}\{F(t) - \widetilde{F}_n(t)\} \,|\, \text{data}$ also converges, with probability one, to $W^0(F_{\rm tr}(t))$. The convergence is w.r.t. the Skorokhod topology in the space of right-continuous functions with left hand limits.*

Proof. The first result is classic and may be found in e.g. Billingsley (1968, Ch. 4). The second statement is immediate from the explicit representation of $\widetilde{F}_n$. Proving the third involves showing finite-dimensional convergence in distribution and tightness, as per the theory of convergence of probability measures laid out in e.g. Billingsley (1968).

To show finite-dimensional convergence we start with $t_1 < \cdots < t_m$ and work with differences $\Delta V_{n,j} = \sqrt{n}\{F(t_{j-1}, t_j] - \widetilde{F}_n(t_{j-1}, t_j]\}$. The vector of $D_j = F(t_{j-1}, t_j]$ has a Dirichlet distribution with parameters $(n+a)\widetilde{F}_n(t_{j-1}, t_j]$. Also, on a set Ω of probability one, both F_n and $\widetilde{F}_n$ tend uniformly to $F_{\rm tr}$, by the Glivenko–Cantelli theorem. Finishing this part of the proof is therefore more or less equivalent to the following lemma: If $(U_1, \ldots, U_m)$ is a Dirichlet distributed vector with parameters $(kp_1, \ldots, kp_m)$, where $p_1 + \cdots + p_m = 1$, then the vector with components $(k+1)^{1/2}(U_j - p_j)$ tends with growing k to a multinormal vector with mean zero and 'multinomial' covariance structure $p_i(\delta_{i,j} - p_j)$, writing $\delta_{i,j} = I_{\{i=j\}}$. Proving this can be done via Scheffé's theorem on convergence of densities, or more easily via the representation $U_j = G_j/(G_1 + \cdots + G_m)$ in terms of independent $G_j \sim \text{Gamma}(kp_j, 1)$, and for which one quickly establishes that $k^{1/2}(G_j/k - p_j)$ tends to a normal $(0, p_j)$.

It remains to demonstrate the almost sure tightness of V_n. For this purpose, take first (U, V, W) to be Dirichlet with parameter (kp, kq, kr), where $p + q + r = 1$. Then some fairly long calculations show that

$$\mathrm{E}(U-p)^2(V-q)^2 = \frac{pq}{(k+1)(k+2)(k+3)}\{k - (k-6)(p+q-3pq)\}.$$

Applying this to the posterior process, writing $V_n(s,t] = V_n(t) - V_n(s)$ and so on, shows that $\mathrm{E}\{V_n(s,t]^2 V_n(t,u]^2 \mid \text{data}\}$ is bounded by $3\widetilde{F}_n(s,t]\widetilde{F}_n(t,u]$, with the right hand side converging, under Ω, towards a quantity bounded by $3\,F_{\rm tr}(s,u]^2$. Tightness now follows from the proof of Theorem 15.6 (but not quite by Theorem 15.6 itself) in Billingsley (1968). □

The result above was also in essence proved in Hjort (1991), and is also related to large-sample studies of the Bayesian bootstrap, see e.g. Lo (1987). We also note that $(n+a+1)^{1/2}$ is a somewhat superior scaling, compared to $\sqrt{n}$, giving exactly matched first and second moments for the posterior process.

We further note that the above conclusions hold also when the strength parameter a of the prior is allowed to grow with n, as long as $a/\sqrt{n} \to 0$. In the more drastic case when $a = cn$, say, the frequentist and Bayesian schemes do not agree asymptotically, as $\widetilde{F}_n$ goes a.s. to $F_\infty = (c/(c+1))F_0 + (1/(c+1))F_{\rm tr}$. But the arguments regarding (iii) still go through, showing that the posterior distribution of $(n+a+1)^{1/2}(F - \widetilde{F}_n)$ tends a.s. to that of $W^0(F_\infty(\cdot))$.

7.2 *Behaviour of the posterior quantile process*

Here we aim at obtaining results as above for the quantile processes involved. For the quantiles, the natural frequentist estimator is F_n^{-1}, while several Bayesian schemes may be considered, including $\widetilde{F}_n^{-1}$ and the posterior mean function $\widehat{Q}_a(y)$ and its natural non-informative limit $\widehat{Q}_0(y)$.

Proposition 4. *Assume, in addition to conditions listed in Proposition 3, that the $F_{\rm tr}$ distribution has a positive and continuous density $f_{\rm tr}$, and let $Q_{\rm tr}(y)$ and $q_{\rm tr}(y) = 1/f_{\rm tr}(Q_{\rm tr}(y))$ be the true quantile and quantile density functions. Then (i) the process $\sqrt{n}\{F_n^{-1}(y) - Q_{\rm tr}(y)\}$ tends to $q_{\rm tr}(y)W^0(y)$; (ii) the difference $\sqrt{n}\{F_n^{-1}(y) - \widetilde{F}_n^{-1}(y)\}$ goes to zero in probability; and (iii) the posterior distribution process $\sqrt{n}\{Q(y) - \widetilde{F}_n^{-1}(y)\}\,|\,\text{data}$ converges a.s. to the same limit $q_{\rm tr}(y)W^0(y)$. The convergence takes place in each of the spaces $D[\varepsilon, 1-\varepsilon]$ of left-continuous functions with right-hand limits, equipped with the Skorokhod topology, where $\varepsilon \in (0, \frac{1}{2})$.*

Proof. The first result is again classic, see e.g. Shorack and Wellner (1986, Ch. 3). It is typically proven by tending to the uniform case first, involving say $F_{n,\rm unif}^{-1}(y)$, and then applying the delta method using the representation $F_n^{-1}(y) = Q_{\rm tr}(F_{n,\rm unif}^{-1}(y))$. Results (ii) and (iii) may be proven in different ways, but the apparently simplest route is via the method devised by Doss and Gill (1992), which acts as a functional delta method operating on the inverse functional $F \mapsto Q = F^{-1}$. We saw above that $\sqrt{n}\{F(t) - \widetilde{F}_n(t)\}\,|\,\text{data}$ tends a.s. to $V(t) = W^0(F_{\rm tr}(t))$. From a slight extension of Doss and Gill's Theorem 2, employing the set Ω of probability 1 encountered in the previous proposition, follows that $\sqrt{n}\{Q(y) - \widehat{F}_n^{-1}(y)\}\,|\,\text{data}$ must tend a.s. to the process $-V(Q_{\rm tr}(y))/f_{\rm tr}(Q_{\rm tr}(y))$, which is the same as $-q_{\rm tr}(y)W^0(y)$. This proves (iii), since by symmetry W^0 and $-W^0$ have identical distributions. Statement (ii) follows similarly from Doss and Gill (op. cit., Theorem 1), again with the slight extension to secure an 'almost sure' version rather than an 'in probability' version, since the process $\sqrt{n}(F_n - \widetilde{F}_n)$ has the zero process as its limit. □

Remark 2. We also note that $\sqrt{n}(\widehat{Q}_a - \widehat{Q}_0) \to_p 0$ follows, by the same type of arguments, starting from $\sqrt{n}(\widetilde{F}_n - F_n) \to_p 0$. In particular, different Bayesians using different Dirichlet process priors will all agree asymptotically. Also, the two estimators $\widehat{Q}_0$ (the Bernshteĭn smoothed quantiles) and F_n^{-1} (the direct quantiles) become equivalent for large samples, in the sense of $\sqrt{n}(\widehat{Q}_0 - F_n^{-1}) \to_p 0$. This also follows from work of Sheather and Marron (1990) about kernel smoothing of quantile functions; see also Cheng (1995). ■

An important consequence of the proposition is that the posterior variance of $\sqrt{n}(Q - F_n^{-1})$ tends to the variance of $q_{\rm tr}W^0$. This is valid for each Dirichlet strength parameter a, as $n \to \infty$. For $a = 0$, n times the posterior variance $\widehat{V}_0(y)$ of (10) converges a.s. to $q_{\rm tr}(y)^2 y(1-y)$. This fact, which may also be proved via results of Conti (2004), is among the ingredients necessary to secure that the natural confidence bands $\widehat{Q}_0 \pm z_0\,\widehat{V}_0^{1/2}$ have the correct limiting coverage level. This comment also applies to constructions in the following subsection.

7.3 *Doksum's shift and Parzen's comparison*

Here we first state results for the natural nonparametric estimators $\widetilde{D}(x)$ and $\widetilde{\pi}(y)$ of Doksum's shift function $D(x)$ and Parzen's comparison distribution, respectively, before we go on to describe the behaviour of their Bayesian cousins, introduced in Section 6. For data $X_1, \ldots, X_n$ from $F_{\rm tr}$ and $X'_1, \ldots, X'_m$ from $G_{\rm tr}$, let again F_n and G_m be the empirical distribution functions. We write $N = n + m$ and assume that $n/N \to c$ and $m/N \to 1 - c$ as the sample sizes increase. Here $F_{\rm tr}$ and $G_{\rm tr}$ are the real underlying distributions, for which we used Dirichlet process priors $\mathrm{Dir}(aF_0)$ and $\mathrm{Dir}(bG_0)$ in Section 6.

The Doksum estimator is $\widetilde{D}(x) = G_m^{-1}(F_n(x)) - x$. Some analysis, involving the frequentist parts of Propositions 3 and 4, shows that the $N^{1/2}\{\widetilde{D}(x) - D_{\rm tr}(x)\}$ process tends to

$$\begin{aligned}(G_{\rm tr}^{-1})'(F_{\rm tr}(x))\,&\{(1-c)^{-1/2}W_1^0(F_{\rm tr}(x)) + c^{-1/2}W_2^0(F_{\rm tr}(x))\}\\ &= \{c(1-c)\}^{-1/2}(G_{\rm tr}^{-1})'(F_{\rm tr}(x))W^0(F_{\rm tr}(x)),\end{aligned} \tag{12}$$

where $D_{\rm tr}(x) = G_{\rm tr}^{-1}(F_{\rm tr}(x)) - x$ and W_1^0 and W_2^0 are two independent Brownian bridges; these combine as indicated into one such Brownian bridge W^0. This result was given in Doksum (1974a), and underlies various methods for obtaining pointwise and simultaneous confidence bands for $D(x)$; see also Doksum and Sievers (1976).

Arguments used to reach the limit result above may now be repeated mutatis mutandis, in combination with the Bernshteĭn–von Mises results in Propositions 3–4, to reach

$$N^{1/2}\{D(x) - \widetilde{D}(x)\}\,|\,\mathrm{data} \to_d Z_D(x), \tag{13}$$

say, using Z_D to denote the limit process in (12). The convergence takes place in each Skorokhod space $D[a, b]$ over which the underlying densities $f_{\rm tr}$ and $g_{\rm tr}$ are positive, and holds with probability 1, i.e. for almost all sample sequences. Result (13) is valid for the informative case with a and b positive (but fixed) as well as for the limiting case where $F\,|\,\mathrm{data} \sim \mathrm{Dir}(nF_n)$ and $G\,|\,\mathrm{data} \sim \mathrm{Dir}(mG_m)$. It is also valid with $\widetilde{D}(x)$ replaced by either the posterior mean $\widehat{D}_0(x)$ or posterior median $K_{m,n}^{-1}(\frac{1}{2})$ estimators discussed in Section 6.

Similarly, the nonparametric Parzen estimator is $\widetilde{\pi}(y) = G_m(F_n^{-1}(y))$, and a decomposition into two processes shows with some analysis that

$N^{1/2}\{\widetilde{\pi}(y) - \pi_{\rm tr}(y)\}$ tends to the process

$$\begin{aligned} Z_P(y) &= \frac{1}{(1-c)^{1/2}} W_1^0(G_{\rm tr}(F_{\rm tr}^{-1}(y))) + \frac{1}{c^{1/2}} \frac{g_{\rm tr}(F_{\rm tr}^{-1}(y))}{f_{\rm tr}(F_{\rm tr}^{-1}(y))} W_0^2(y) \\ &= (1-c)^{-1/2} W_1^0(\pi_{\rm tr}(y)) + c^{-1/2} \pi'_{\rm tr}(y) W_2^0(y), \end{aligned} \tag{14}$$

with $\pi_{\rm tr}(y) = G_{\rm tr}(F_{\rm tr}^{-1}(y))$. For the case $F_{\rm tr} = G_{\rm tr}$, one has $\pi_{\rm tr}(y) = y$, and the limit result translates to the quite simple $(mn/N)^{1/2}(\widetilde{\pi} - \pi) \to_d W^0$. This provides an easy and informative way of checking and testing proximity of two distributions via the $\widetilde{\pi}$ plot. "Why aren't people celebrating these facts?", as says Parzen in the interview with Newton (2002, p. 373). Similarly worthy of celebrations, in the Bayesian camp, should be the fact that (14) has a sister parallel in the present context, namely that $N^{1/2}\{\pi(y) - \widehat{\pi}(y)\} \,|\, \text{data}$ tends to the same limit process as in (14). Here $\widehat{\pi}(y)$ can be the posterior median estimator or the posterior mean estimator found in Section 6.

8 Quantile regression

Consider the regression situation where certain covariates $(x_{i,1}, \ldots, x_{i,p})^{\rm t} = x_i$ are available for individual i, thought to influence the distribution of Y_i. Assume that $Y_i = \beta^{\rm t} x_i + \sigma\varepsilon_i$, where $\beta = (\beta_1, \ldots, \beta_p)^{\rm t}$ contains unknown regression parameters and $\varepsilon_1, \ldots, \varepsilon_n$ are independent error terms, coming from a scaled residual distribution F. Thus a prospective observation Y, with covariate information x, will have distribution $F(t \,|\, x) = F((t - \beta^{\rm t} x)/\sigma)$, conditional on (β, σ, F). Its quantile function becomes $Q(u \,|\, x) = \beta^{\rm t} x + \sigma\, Q(u)$, writing again Q for F^{-1}.

The problem to be discussed now is that of Bayesian inference for $Q(u \,|\, x)$, starting out with a prior for (β, σ, F). Take (β, σ) and F to be independent, with a prior density $\pi(\beta, \sigma)$ and a $\text{Dir}(aF_0)$ prior for F, where the prior guess F_0 has a density f_0. The posterior distribution of (β, σ, F) may then be described as follows. First, the posterior density of β can be shown to be

$$\pi(\beta, \sigma \,|\, \text{data}) = \text{const.}\, \pi(\beta, \sigma) \prod_{\text{distinct}} f_0((y_i - \beta^{\rm t} x_i)/\sigma),$$

where the product is taken over distinct values of $y_i - \beta^{\rm t} x_i$. This may be shown via techniques in Hjort (1986). Secondly, given data and (β, σ), Q acts as the posterior quantile process from a Dirichlet F with parameter $aF_0 + \sum_{i=1}^n \delta((y_i - \beta^{\rm t} x_i)/\sigma)$, with $\delta(z)$ denoting unit point mass at z; in particular, expressions for $\widehat{Q}_a(u \,|\, \beta, \sigma) = \text{E}\{Q(u) \,|\, \beta, \sigma, \text{data}\}$ may be written down using the results of earlier sections.

In combination, this gives for each x_0 an estimator for $Q(u\,|\,x_0)$ of the form

$$\begin{aligned}\widehat{Q}_a(u\,|\,x_0) &= \mathrm{E}\{\beta^{\mathrm{t}}x_0+\sigma Q(u)\,|\,\mathrm{data}\}\\ &= \widehat{\beta}^{\mathrm{t}}x_0+\mathrm{E}\{\sigma\widehat{Q}_a(u\,|\,\beta,\sigma)\,|\,\mathrm{data}\}\\ &= \widehat{\beta}^{\mathrm{t}}x_0+\int\sigma\widehat{Q}_a(y\,|\,\beta,\sigma)\pi(\beta,\sigma\,|\,\mathrm{data})\,\mathrm{d}\beta\,\mathrm{d}\sigma,\end{aligned}$$

where $\widehat{\beta}$ is the posterior mean of β. For the particular case of a tending to zero, this gives

$$\widehat{Q}_0(u\,|\,x_0)=\widehat{\beta}^{\mathrm{t}}x_0+\sum_{i=1}^{n}\binom{n-1}{i-1}u^{i-1}(1-u)^{n-i}\,e_i.$$

Here $e_i=\int(y-\beta^{\mathrm{t}}x)_{(i)}\pi(\beta\,|\,\mathrm{data})\,\mathrm{d}\beta$, where, for each β, $(y-\beta^{\mathrm{t}}x)_{(i)}$ is the result of sorting the n values of $y_j-\beta^{\mathrm{t}}x_j$ and then finding the ith ranked one. The simplest implementation might be to draw a large number of βs from the posterior density, and then for each of these sort the values of $y_j-\beta^{\mathrm{t}}x_j$. Averaging over all simulations then gives e_i as the posterior mean of $(y-\beta^{\mathrm{t}}x)_{(i)}$, for each $i=1,\ldots,n$, and in their turn $\widehat{Q}_0(u\,|\,x_0)$ for all x_0.

One may also give a separate recipe for making inference for Q, the residual quantile process. Other Bayesian approaches to quantile regression are considered in Kottas and Gelfand (2001) and Hjort and Walker (2006).

9 Concluding remarks

In our final section we offer some concluding comments, some of which might point to further problems of interest.

Other priors. There are of course other possibilities for quantifying prior opinions of quantile functions. One may e.g. start with a prior more general than or different from the Dirichlet process for F, like Doksum's (1974b) neutral to the right processes, or mixtures of Dirichlet processes, and attempt to reach results for the consequent quantile processes $Q=F^{-1}$. Another and more direct approach is via the versatile class of quantile pyramid processes developed in Hjort and Walker (2006). These work by first drawing the median $Q(\frac{1}{2})$ from a certain distribution; then the two other quartiles $Q(\frac{1}{4})$ and $Q(\frac{3}{4})$ given the median; then the three remaining octiles $Q(\frac{j}{8})$ for $j=1,3,5,7$; and so on. The Dirichlet process can actually be seen to be a special case of these pyramid constructions. While the treatment in Hjort and Walker leads to recipes which can handle the prior to posterior updating task for any quantile pyramid, this relies on simulation techniques of the McMC variety. Part of the contribution of the present chapter is that explicit formulae and characterisations are developed, partly obviating the need for such simulation work, for the particular case of the Dirichlet processes.

An invariance property. Our canonical Bayes estimator (9) was derived by starting with a $\mathrm{Dir}(aF_0)$ prior for F and then letting a go to zero. Extending the horizon beyond the simple i.i.d. setting, suppose for illustration that data are assumed to be of the form $X_i = \xi + \sigma Z_i$, with Z_i having distribution G. One may then give a semiparametric prior for the distribution $F(t) = G((t-\xi)/\sigma)$ of X_i, with a prior for (ξ, σ) and an independent $\mathrm{Dir}(aG_0)$ prior for G. This leads to a more complicated posterior distribution for $Q(y) = \xi + \sigma Q_G(y)$, say. But since G given data and the parameters is a Dirichlet with parameter $aG_0 + \sum_{i=1}^n \delta((x_i - \mu)/\sigma)$, results of Sections 2 and 3 give formulae for $\mathrm{E}\{Q(y) \mid \text{data}, \xi, \sigma\}$. For the non-informative case of $a = 0$,

$$\mathrm{E}\{Q(y) \mid \text{data}, \xi, \sigma\} = \xi + \sigma \sum_{i=1}^{n} \binom{n-1}{i-1} y^{i-1}(1-y)^{n-1} \frac{x_{(i)} - \xi}{\sigma}.$$

But the extra parameters cancel out, showing that the posterior mean is again the (9) estimator, which therefore is the limiting Bayes rule for rather wider classes of priors than only the pure Dirichlet. The argument goes through for each monotone transformation $X_i = a_\theta(Z_i)$ with a prior for (θ, G).

In situations where the Lorenz curve and Gini index are of interest, for example, one might think of data as $X_i = \theta Z_i$, with separate priors for θ and the distribution G of Z_i. The above argument shows that the θ information is not relevant for $Q(y) = \theta Q_G(y)$, when a is small, thus lending further support to the estimators $\widehat{L}_0$ and $\widehat{G}_0$ of Section 5.

Alternative proofs. There are other venues of interest towards proving Proposition 4 or other versions thereof. Johnson and Sim (2006) give a different proof of the large-sample joint normality of a finite number of posterior quantiles, including asymptotic expansions. Conti (2004) has independently of the present authors reached results for the posterior process $\sqrt{n}(Q - \widetilde{F}_n^{-1})$, partly using strong Hungárian representations. His approach gives results that are more informative than Proposition 4 concerning the boundaries, i.e. for y close to 0 and y close to 1, where our direct method works best on $D[\varepsilon, 1-\varepsilon]$ for a fixed small ε. Another angle is to exploit approximations to the Beta and Dirichlet distributions associated with the random F and turn these around to good approximations for Q. A third possibility of interest is to express the random posterior quantile process as $Q(y) = x_{(N(y))}$, with $N(y)$ the random process described in Section 2.4, climbing from 1 at zero to n at one. One may show that $\sqrt{n}\{N(y)/n - y\}$ tends to a Brownian bridge, and couple this with $Q(y) = Q_n(N(y)/n)$ to give yet another proof of the Bernshteĭn–von Mises part of Proposition 4.

Simultaneous confidence bands. In our illustrations we focussed on confidence bands with correct pointwise coverage. One may also construct simultaneous bands for the different situations, with some more work. For the Doksum shift function, in the frequentist setting, such simultaneous bands were constructed in Doksum (1974a), Doksum and Sieverts (1976) and Switzer (1976). To match this in the Bayesian setting, one might simulate a large number of $D(x)$ curves from the posterior process, and note the quantiles of the distribution of simulated $\max_{[a,b]} |D(x) - \widehat{D}_0(x)|$ across some interval $[a, b]$ of interest. Another method, using result (13), is to note that $N^{1/2} \max_{a \le x \le b} |D(x) - \widehat{D}_0(x)| \mid \text{data}$ tends in distribution to

$$\max_{a \le x \le b} |Z_D(x)| = \frac{1}{\{c(1-c)\}^{1/2}} \max_{F(a) \le v \le F(b)} \frac{|W^0(v)|}{g_{\rm tr}(G_{\rm tr}^{-1}(v))}.$$

With appropriate consistent estimation of the denumerator one might simulate the required quantile of the limiting distribution. Other bands evolve with alternative weight functions.

Further quantilian quantities. There are yet other statistical functions or parameters of interest that depend on quantile functions and that can be worked with using methods from our chapter. One such quantity is the total time on test statistic $T(u) = \int_0^{Q(u)} \{1 - F(x)\}\,{\rm d}x$. Doksum and James (2004) show how inference for T may be carried out via Bayesian bootstraps.

More informative priors for two-sample problems. In situations where the Doksum band contains a horizontal line it indicates that the shift function is nearly constant, which corresponds to a location translation from F to G, say $G(t) = F(t - \delta)$. For the Doksum–Bjerkedal data analysed in Figure 3 the band nearly contains a linear curve, which indicates a location-and-scale translation, say $G(t) = F((t - \delta)/\tau)$. The present point is that it is fruitful to build Bayesian prior models for such scenarios, linking F and G together, as opposed to simply assuming prior independence of F and G. One version is to take $F \sim {\rm Dir}(aF_0)$ and then $G(t) = F((t - \delta)/\tau)$ with a prior for (δ, τ). This leads to fruitful posterior models for (F, δ, τ).

Appendix: Various proofs

Relation between Beta cumulatives. Let ${\rm be}(\cdot; a, b)$ and ${\rm Be}(\cdot; a, b)$ denote the density and cumulative distribution of a Beta variable with parameters

(a,b). Then, by partial integration, for $b>1$,

$$\mathrm{Be}(c;a,b)-\mathrm{Be}(c;a+1,b-1)=\frac{\mathrm{be}(c;a+1,b)}{a+b}=\frac{\mathrm{be}(1-c;b,a+1)}{a+b}. \quad \text{(A1)}$$

Proof of Proposition 1. There are several ways in which to prove this, including analysis via Taylor type expansions of the (4) probabilities and their sum; see also Conti (2004). Here we briefly outline another and more probabilistic argument. The idea is to decompose the posterior distribution of F in two parts, corresponding to jumps $D_1,\ldots,D_n$ at the data points and a total probability $E=F(\mathbb{R}-\{x_1,\ldots,x_n\})$ representing all increments between the data points. Thus

$$F(t)=\sum_{i=1}^n D_i I\{x_{(i)}\le t\}+\sum_{i=1}^n E_i I\{x_{(i)}\le t\}=\widetilde{F}(t)+F^*(t),$$

say, with E_i the part of E corresponding to the window $(x_{(i-1)},x_{(i)})$ between data points. The point here is that $(D_1,\ldots,D_n,E)$ has a Dirichlet $(1,\ldots,1,a)$ distribution, with E becoming small in size as n increases. In fact, $E\le a/\sqrt{n}$ with probability at least $1-1/\sqrt{n}$. Thus $F=\widetilde{F}+F^*$ with $F-\widetilde{F}\le a/\sqrt{n}$, with high probability, and $Q=F^{-1}$ must with a high probability be close to $\widetilde{Q}=\widetilde{F}^{-1}$. But the latter has all its jumps exactly situated at the data points. ■

Proof of Proposition 2. We first recall that for any cumulative distribution function H on the real line,

$$\int_0^\infty x\,\mathrm{d}H(x)=\int_0^\infty\{1-H(x)\}\,\mathrm{d}x,\qquad \int_{-\infty}^0 x\,\mathrm{d}H(x)=-\int_{-\infty}^0 H(x)\,\mathrm{d}x.$$

These results can be shown using partial integration and the Fubini theorem, and hold in the sense that finiteness of one integral implies finiteness of the sister integral, and vice versa. These formulae are what is being used when we in Section 3 preferred formula (8) to (7).

With the above formulae and characterisations we learn that the finite existence of the posterior mean of $Q(y)$ hinges on the finiteness of the extreme parts $\int_c^\infty \mathrm{Be}(y;aF_0(x)+n,a\bar{F}_0(x))\,\mathrm{d}x$, for $c\ge x_{(n)}$, and $\int_{-\infty}^b \mathrm{Be}(1-y;a\bar{F}_0(x)+n,aF_0(x))\,\mathrm{d}x$, for $b\le x_{(1)}$. Using $\Gamma(v)=\Gamma(v+1)/v$ the first integral may be expressed as

$$\int_c^\infty \frac{\Gamma(a+n)a\bar{F}_0(x)}{\Gamma(aF_0(x)+n)\Gamma(a\bar{F}_0(x)+1)}\Big[\int_0^y u^{aF_0(x)+n-1}(1-u)^{a\bar{F}_0(x)-1}\,\mathrm{d}u\Big]\,\mathrm{d}x,$$

which is of the form $\int_c^\infty a\bar{F}_0(x)g(x)\,\mathrm{d}x$ for a bounded function g; hence this the integral is finite if and only if $\int_c^\infty\{1-F_0(x)\}\,\mathrm{d}x$ is finite. We may similarly show that the second integral is finite if and only if $\int_{-\infty}^b F_0(x)\,\mathrm{d}x$ is finite. These arguments are valid for any n, also for the no-sample prior case of $n=0$. This proves the proposition. ■

Acknowledgements

The authors gratefully acknowledge support and hospitality from the Department of Mathematics at the University of Oslo and the Istituto di Metodi Quantitativi at Bocconi University in Milano, at reciprocal research visits. Constructive comments from Dorota Dabrowska, Alan Gelfand, Pietro Muliere, Vijay Nair and Stephen Walker have also been appreciated.

References

1. Billingsley, P. (1968). *Convergence of Probability Measures.* Wiley, New York.
2. Bickel, P.J. and Doksum, K.A. (2001). *Mathematical Statistics: Basic Ideas and Selected Topics* (2nd ed.), Volume 1. Prentice Hall, Upper Saddle River, New Jersey.
3. Bjerkedal, T. (1960). Acquisition of resistance in guinea pigs infected with different doses of virulent tubercle bacilli. *American Journal of Hygiene* **72**, 132–148.
4. Cheng, C. (1995). The Bernstein polynomial estimator of a smooth quantile function. *Statistics and Probability Letters* **24**, 321–330.
5. Conti, P.L. (2004). Approximated inference for the quantile function via Dirichlet processes. *Metron* **LXII**, 201–222.
6. Diaconis, P. and Freedman, D.A. (1986a). On the consistency of Bayes estimates [with discussion]. *Annals of Statistics* **14**, 1–67.
7. Diaconis, P. and Freedman, D.A. (1986b). On inconsistent Bayes estimates of location. *Annals of Statistics* **14**, 68–87.
8. Doksum, K.A. (1974a). Empirical probability plots and statistical inference for nonlinear models in the two-sample case. *Annals of Statistics* **2**, 267–277.
9. Doksum, K.A. (1974b). Tailfree and neutral random probabilities and their posterior distributions. *Annals of Probability* **2**, 183–201.
10. Doksum, K.A. and Sievers, G.L. (1976). Plotting with confidence: Graphical comparisons of two populations. *Biometrika* **63**, 421–434.
11. Doksum, K.A. and James, L.F. (2004). On spatial neutral to the right processes and their posterior distributions. In *Mathematical Reliability: An Expository Perspective* (eds. R. Soyer, T.A. Mazzuchi and N.D. Singpurvalla), Kluwer International Series, 87–104.
12. Doss, H. and Gill, R.D. (1992). An elementary approach to weak convergence for quantile processes, with applications to censored survival data. *Journal of the American Statistical Association* **87**, 869–877.
13. Ferguson, T.S. (1973). A Bayesian analysis of some nonparametric problems. *Annals of Statistics* **1**, 209–230.

14. Ferguson, T.S. (1974). Prior distributions on spaces of probability measures. *Annals of Statistics* **2**, 615–629.

15. Hjort, N.L. (1986). Discussion contribution to P. Diaconis and D. Freedman's paper 'On the consistency of Bayes estimates', *Annals of Statistics* **14**, 49–55.

16. Hjort, N.L. (1991). Bayesian and empirical Bayesian bootstrapping. Statistical Research Report, University of Oslo.

17. Hjort, N.L. (1996). Bayesian approaches to non- and semiparametric density estimation [with discussion]. In *Bayesian Statistics 5*, proceedings of the Fifth International València Meeting on Bayesian Statistics (eds. J. Berger, J. Bernardo, A.P. Dawid, A.F.M. Smith), 223–253. Oxford University Press.

18. Hjort, N.L. (2003). Topics in nonparametric Bayesian statistics [with discussion]. In *Highly Structured Stochastic Systems* (eds. P.J. Green, S. Richardson and N.L. Hjort), Oxford University Press.

19. Hjort, N.L. and Walker, S.G. (2006). Quantile pyramids for Bayesian nonparametrics. *Annals of Statistics*, to appear.

20. Johnson, R.A. and Sim, S. (2007). Nonparametric Bayesian inference about percentiles. A chapter in *Advances in Statistical Modeling and Inference*, edited by V. Nair.

21. Kottas, A. and Gelfand, A. (2001). Bayesian semiparametric median regression modeling. *Journal of the American Statistical Association* **96**, 1458–1468.

22. LeCam, L. and Yang, G.L. (1990). *Asymptotics in Statistics.* Springer-Verlag, New York.

23. Lo, A.Y. (1987). A large-sample study of the Bayesian bootstrap. *Annals of Statistics* **15**, 360–375.

24. Lorenz, M.C. (1905). Methods of measuring the concentration of wealth. *Journal of the American Statistical Association* **9**, 209–219.

25. Laake, P., Laake, K. and Aaberge, R. (1985). On the problem of measuring the distance between distribution functions: Analysis of hospitalization versus mortality. *Biometrics* **41**, 515–523.

26. Newton, H.J. (2002). A conversation with Emanuel Parzen. *Statistical Science* **17**, 357–378. Correction, op. cit., 467.

27. Parzen, E. (1979). Nonparametric statistical data modeling [with discussion]. *Journal of the American Statistical Association* **74**, 105–131.

28. Parzen, E. (1982). Data modeling using quantile and density-quantile functions. *Some recent advances in statistics*, Symposium Lisbon 1980, 23–52.

29. Parzen, E. (2002). Discussion of Breiman's 'Statistical modeling: The two cultures'. *Statistical Science* **16**, 224–226.

30. Sheather, S.J. and Marron, J.S. (1990). Kernel quantile estimation. *Journal of the American Statistical Association* **80**, 410–416.

31. SHORACK, G.R. AND WELLNER, J. (1986). *Empirical Processes With Applications to Statistics.* Wiley, New York.

32. SWITZER, P. (1976). Confidence procedures for two samples. *Biometrika* **53**, 13–25.

33. AABERGE, R. (2001). Axiomatic characterization of the Gini coefficient and Lorenz curve orderings. *Journal of Economic Theory* **101**, 115–132. Correction, ibid.

34. AABERGE, R., BJERVE, S. AND DOKSUM, K.A. (2005). Lorenz, Gini, Bonferroni and quantile regression. Unpublished manuscript.

PART 8
Rank-Based Methods

Chapter 24

ASYMPTOTIC DISTRIBUTION THEORY OF EMPIRICAL RANK-DEPENDENT MEASURES OF INEQUALITY

Rolf Aaberge

Research Department
Statistics Norway, Oslo, NORWAY

E-mail: rolf.aaberge@ssb.no

A major aim of most income distribution studies is to make comparisons of income inequality across time for a given country and/or compare and rank different countries according to the level of income inequality. However, most of these studies lack information on sampling errors, which makes it difficult to judge the significance of the attained rankings. This chapter derives the asymptotic properties of the empirical rank-dependent family of inequality measures. A favorable feature of this family of inequality measures is that it includes the Gini coefficients, and that any member of this family can be given an explicit and simple expression in terms of the Lorenz curve. By relying on a result of Doksum (1974), it is easily demonstrated that the empirical Lorenz curve converges to a Gaussian process. This result forms the basis of the derivation of the asymptotic properties of the empirical rank-dependent measures of inequality.

Key words: Gini coefficient; Lorenz curve; Rank-dependent measures of inequality; Sampling errors.

1 Introduction

The standard practice in empirical analyses of income distributions is to make separate comparisons of the overall level of income (the size of the cake) and the distribution of income shares (division of the cake), and to use the Lorenz curve as a basis for analysing the distribution of income shares. [See e.g. Atkinson, Rainwater, and Smeeding (1995) who make cross-country comparisons of Lorenz curves allowing for differences between countries in level of income and Lambert (1993) for a discussion of applying

Lorenz dominance criteria as basis for evaluating distributional effects of tax reforms.] By displaying the deviation of each individual income share from the income share that corresponds to perfect equality, the Lorenz curve captures the essential descriptive features of the concept of inequality. [For a discussion of the normative aspects of Lorenz curve orderings, see Kolm (1969, 1976a, 1976b), Atkinson (1970), Yaari (1987, 1988) and Aaberge (2001).] When Lorenz curves do not intersect, it is universally acknowledged that the higher Lorenz curve displays less inequality than the lower Lorenz curve. This is due to the fact that the higher of two non-intersecting Lorenz curves can be obtained from the lower Lorenz curve by means of rank-preserving income transfers from richer to poorer individuals. However, since observed Lorenz curves normally intersect weaker ranking criteria than the dominance criterion of non-intersecting Lorenz curves are required. In this case one may either search for weaker dominance criteria, see e.g. Shorrocks and Foster (1978), Dardanoni and Lambert (1988), Lambert (1993) and Aaberge (2000b), or one may apply summary measures of inequality. The latter approach also offers a method for quantifying the extent of inequality in income distributions, which may explain why numerous alternative measures of inequality are introduced in the literature. The most well-known and widely used measure of inequality is the Gini coefficient, which is equal to twice the area between the Lorenz curve and its equality reference. However, to get a broader picture of inequality than what is captured by the Gini coefficient the use of alternative measures of inequality is required. By making explicit use of the Lorenz curve Mehran (1976), Donaldson and Weymark (1980,1983), Weymark (1981), Yitzhaki (1983) and Aaberge (2000a, 2001) introduce various generalized Gini families of inequality measures. Moreover, Aaberge (2000a) demonstrates that one of these families, called the Lorenz family of inequality measures, can be considered as the moments of the Lorenz curve and thus provides a complete characterization of the Lorenz curve. This means that the Lorenz curve can be uniquely recovered from the knowledge of the corresponding Lorenz measures of inequality, i.e. without loss of information examination of inequality in an income distribution can be restricted to application of the Lorenz measures of inequality. Note that a subclass of the extended Gini family introduced by Donaldson and Weymark (1980,1983) is uniquely determined by the Lorenz family of inequality measures. [See Aaberge 2000a for a proof.] Since the different alternative generalized families of inequality measures can be considered as subfamilies of Mehrans (1976) general family of rank-dependent measures of inequality it appears useful to consider the asymptotic properties of the empirical version of the general family of rank-dependent measures of inequality rather than to restrict to the empirical version of the Lorenz family of inequality measures. The plan of the paper is as follows. Section 2 provides formal definitions of the Lorenz curve and the family of rank-dependent measures of inequality and the corresponding non-parametric estimators. By relying on a result of Doksum (1974) it is demonstrated in Section 3.1 that the empirical Lorenz curve (regarded as a stochastic process) converges to a Gaussian process. This result forms the basis of the derivation of the asymptotic properties of the empirical rank-dependent measures of inequality that are presented in Section 3.2.

2 Definition and estimation of the Lorenz curve and rank-dependent measures of inequality

Let X be an income variable with cumulative distribution function F and mean μ. Let $[0,\infty\rangle$ be the domain of F where F^{-1} is the left inverse of F and $F^{-1}(0)\equiv 0$. The Lorenz curve L for F is defined by

$$L(u)=\frac{1}{\mu}\int_0^u F^{-1}(t)dt,\quad 0\leq u\leq 1.$$

Thus, the Lorenz curve $L(u)$ shows the share of total income received by the $100u$ per poorest of the population. By introducing the conditional mean function $H(\cdot)$ defined by

$$H(u)=E\left(X|X\leq F^{-1}(u)\right)=\frac{1}{u}\int_0^u F^{-1}(t)dt, 0\leq u\leq 1,$$

Aaberge (1982) found that the Lorenz curve can be written on the following form

$$L(u)=u\frac{H(u)}{H(1)}0\leq u\leq 1. \tag{1}$$

Let $X_1, X_2,\ldots X_n$ be independent random variables with common distribution function F and let be the corresponding empirical distribution function. Since the parametric form of F is not known, it is natural to use the empirical distribution function F_n to estimate F and to use

$$H_n(u)=\frac{1}{u}\int_0^u F_n^{-1}(t)dt, 0\leq u\leq 1$$

to estimate $H(u)$, where F_n^{-1} is the left inverse of F_n. Now replacing $H(u)$ by $H_n(u)$ in the expression (1) for $L(u)$, we get the empirical Lorenz curve

$$L_n(u)=u\frac{H_n(u)}{H_n(1)}, 0\leq u\leq 1. \tag{2}$$

To obtain an explicit expression for $H_n(u)$ and the empirical Lorenz curve, let $X_{(1)} \leq X_{(2)}, \leq \ldots \leq X_{(n)}$ denote the ordered $X_1, X_2, \ldots X_n$. For $u = i/n$ we have

$$H_n\left(\frac{i}{n}\right) = \frac{1}{i}\sum_{j=1}^{i} X_{(j)}, i = 1, 2, \ldots, n$$

and

$$L_n\left(\frac{i}{n}\right) = \frac{\sum_{j=1}^{i} X_{(j)}}{\sum_{j=1}^{n} X_j}, i = 1, 2, \ldots, n \tag{3}$$

which is the familiar estimate formula of the empirical Lorenz curve. As mentioned in Section 1 the ranking of Lorenz curves becomes problematic when the Lorenz curves in question intersect. For this reason and to be able to quantify the inequality in distributions of income it is common to apply summary measures of inequality. As justified in Section 1 it appears attractive to consider the family of rank-dependent measures of inequality introduced by Mehran (1976) and defined by

$$J_R(L) = 1 - \int_0^1 R(u)L(u)du \tag{4}$$

where R is a non-negative weight-function. [A slightly different version of J_R was introduced by Piesch (1975), whereas Giaccardi (1950) considered a discrete version of J_R. For alternative normative motivations of the J_R−family and various subfamilies of the J_R−family we refer to Donaldson and Weymark (1983), Yaari (1987,1988), BenPorath and Gilboa (1994) and Aaberge (2001). See also Zitikis (2002) and Tarsitano (2004) for a discussion on related families of inequality measures.] By inserting for the following two alternative subclasses R_1 and R_2 of R,

$$R_{1k}(u) = k(k+1)(1-u)^{k-1}, k > 0$$

and

$$R_{2k}(u) = (k+1)u^{k-1}, k > 0$$

we get the following subfamilies of the general rank-dependent family of inequality measures J_R,

$$G_k \equiv J_{R_{1k}}(L) = 1 - k(k+1)\int_0^1 (1-u)^{k-1}L(u)du, k > 0 \tag{5}$$

and

$$D_k \equiv J_{R_{2k}}(L) = 1 - (k+1)\int_0^1 u^{k-1}L(u)du, k > 0. \tag{6}$$

Note that $\{G_k : k > 0\}$ was denoted the extended Gini family and $\{D_k : k > 0\}$ the illfare-ranked single series Ginis by Donaldson and Weymark (1980). [See Zitikis and Gastwirth (2002) for a derivation of the asymptotic distribution of the empirical extended Gini family of inequality measures.] However, as mentioned in Section 1, Aaberge (2000a) proved that each of the subfamilies $\{D_k : k = 1, 2, \ldots\}$ (denoted the Lorenz family of inequality measures) and $\{G_k : k = 1, 2, \ldots\}$ provides a complete characterization of the Lorenz curve, independent of whether the distribution function F is defined on a bounded interval or not. Thus, any distribution function F defined on the positive halfline $\mathcal{R}^+$ can be specified by its mean and Lorenz measures of inequality even if some of the conventional moments do not exist. It follows directly from expressions (5) and (6) that the Gini coefficient defined by

$$G = 1 - 2\int_0^1 L(u)du$$

is included in the extended Gini family as well as in the Lorenz family of inequality measures. By replacing L by L_n in the expression (4) for J_R, we get the following estimator of J_R,

$$\hat{J}_R \equiv J_R(L_n) = 1 - \int_0^1 R(u)L_n(u)du. \tag{7}$$

For $R(u) = 2$, (7) gives the estimator of G as

$$\hat{G} = 1 - 2\int_0^1 L_n(u)du = 1 - \frac{2\sum_{i=1}^n \sum_{j=1}^i X_{(j)}}{(n+1)\sum_{j=1}^n X_j}. \tag{8}$$

[The asymptotic properties of the empirical Gini coefficient has been considered by Hoeffding (1948), Goldie (1977), Aaberge (1982), Zitikis (2002,2003) and Zitikis and Gastwirth (2002).]

3 Asymptotic distribution theory of the empirical Lorenz curve and empirical rank-dependent measures of inequality

As demonstrated by expressions (4) and (7), the rank-dependent measures of inequality and their empirical counterparts are explicitly defined in terms of the Lorenz curve and its empirical counterpart, respectively. Thus, in order to derive the asymptotic distribution of the empirical rank-dependent measures of inequality it is convenient to firstly derive the asymptotic properties of the empirical Lorenz curve. To this end we utilize the close formal connection between the shift function of Doksum (1974) and the

Lorenz curve. As an alternative to the approach chosen in this paper we can follow Zitikis (2002) by expressing the rank-dependent measures of inequality in terms of L-statistics and rely on asymptotic distribution results for $L-$statistics. [On general results for $L-$statistics see e.g. Chernoff et al. (1967), Shorack (1972), Stigler (1974) and Serfling (1980).] Note that Csörgő, Gastwirth and Zitikis (1998) have derived asymptotic confidence bands for the Lorenz and the Bonferroni curves without requiring the existence of the density f. Moreover, Davydov and Zitikis (2003, 2004) have considered the case where observations are allowed to be dependent. As demonstrated by Zitikis (1998) note that the Vervaat process proves to be a particularly helpful device in deriving asymptotic properties of various aggregates of empirical quantiles.

3.1 *Asymptotic properties of the empirical Lorenz curve*

Since F_n is a consistent estimate of $F, H_n(u)$ and $L_n(u)$ are consistent estimates of $H(u)$ and $L(u)$, respectively. Approximations to the variance of L_n and the asymptotic properties of L_n can be obtained by considering the limiting distribution of the process defined by

$$Z_n(u) = n^{\frac{1}{2}} \left[L_n(u) - L(u)\right]. \tag{9}$$

In order to study the asymptotic behavior of $Z_n(u)$ we find it useful to start with the process defined by

$$Y_n(u) = n^{\frac{1}{2}} \left[H_n(u) - H(u)\right] = \frac{1}{n} \int_0^u n^{\frac{1}{2}} \left(F_n^{-1}(t) - F^{-1}(t)\right) dt. \tag{10}$$

Assume that the support of F is a non-empty finite interval $[a, b]$. (When F is an income distribution, a is commonly equal to zero.) Then $Y_n(u)$ and $Z_n(u)$ are members of the space D of functions on $[0, 1]$ which are right continuous and have left hand limits. On this space we use the Skorokhod topology and the associated $\sigma-$field (e.g. Billingsley 1968, page 111). We let $W_0(t)$ denote a Brownian Bridge on $[0, 1]$, that is, a Gaussian process with mean zero and covariance function $s(1-t), 0 \leq s \leq t \leq 1$.

Theorem 1. *Suppose that F has a continuous nonzero derivative f on $[a,b]$. Then $Y_n(u)$ converges in distribution to the process*

$$Y_(u) = \frac{1}{u}\int_0^u \frac{W_0(t)}{f(F^{-1}(t))}dt. \tag{11}$$

Proof. It follows directly from Theorem 4.1 of Doksum (1974) that

$$n^{\frac{1}{2}}\left(F_n^{-1}(t) - F^{-1}(t)\right)$$

converges in distribution to the Gaussian process $W_0(t)/f\left(F^{-1}(t)\right)$. Using the arguments of Durbin (1973, section 4.4), we find that $Y(u)$ as a function of $\left(W_0(t)/f\left(F^{-1}(t)\right)\right)$ is continuous in the Skorokhod topology. The result then follows from Billingsley (1968, Theorem 5.1). □

The following result states that $Y(u)$ is a Gaussian process and thus that $Y_n(u)$ is asymptotically normally distributed, both when considered as a process, and for fixed u.

Theorem 2. *Suppose the conditions of Theorem 1 are satisfied. Then the process $uY(u)$ has the same probability distribution as the Gaussian process*

$$\sum_{j=1}^{\infty} q_j(u)Z_j$$

where $q_j(u)$ is given by

$$q_j(u) = \frac{2^{\frac{1}{2}}}{j\pi}\int_0^u \frac{\sin(j\pi t)}{f\left(F^{-1}(t)\right)}dt \tag{12}$$

and $Z_1, Z_2, \ldots$ are independent $N(0,1)$ variables.

Proof. Let

$$V_N(t) = \frac{2^{\frac{1}{2}}}{f\left(F^{-1}(t)\right)}\sum_{j=1}^{N}\frac{\sin(j\pi t)}{j\pi}Z_j$$

and note that

$$2\sum_{j=1}^{\infty}\frac{\sin(j\pi s)\sin(j\pi t)}{(j\pi)^2} = s(1-t), 0 \le s \le t \le 1. \tag{13}$$

Thus, the process $V_N(t)$ is Gaussian with mean zero and covariance function

$$\begin{aligned} cov\left(V_N(s), V_N(t)\right) &= \frac{2}{f\left(F^{-1}(s)\right)f\left(F^{-1}(t)\right)}\sum_{j=1}^{N}\frac{\sin(j\pi s)\sin(j\pi t)}{(j\pi)^2} \\ &\to cov\left(V(s), V(t)\right), \end{aligned}$$

where

$$V(t) = \frac{W_0(t)}{f\left(F^{-1}(t)\right)}.$$

In order to prove that $V_N(t)$ converges in distribution to the Gaussian process $V(t)$, it is, according to Hajek and Sidak (1967, Theorem 3.1.a, Theorem 3.1.b, Theorem 3.2) enough to show that

$$E[V_N(t) - V_N(s)]^4 \leq M(t-s)^2, 0 \leq s, t \leq 1,$$

where M is independent of N. Since for normally distributed random variables with mean 0,

$$EX^4 = 3[EX^2]^2,$$

we have

$$\begin{aligned}
E[V_N(t) - V_N(s)]^4 &= 3[var\left(V_N(t) - V_N(s)\right)]^2 \\
&= 3\{2var[\sum_{j=1}^{N} \frac{1}{j\pi}\left(\frac{\sin(j\pi t)}{f\left(F^{-1}(t)\right)} - \frac{\sin(j\pi s)}{f\left(F^{-1}(s)\right)}\right) Z_j]\}^2 \\
&= 3\{2\sum_{j=1}^{N}[\frac{1}{j\pi}\left(\frac{\sin(j\pi t)}{f\left(F^{-1}t\right)} - \frac{\sin(j\pi s)}{f\left(F^{-1}(s)\right)}\right)]^2\}^2 \\
&\leq 3\{2\sum_{j=1}^{\infty}[\frac{1}{j\pi}\left(\frac{\sin(j\pi t)}{f\left(F^{-1}(t)\right)} - \frac{\sin(j\pi s)}{f\left(F^{-1}(s)\right)}\right)]^2\}^2 \\
&= 3\{\frac{t(1-t)}{f^2\left(F^{-1}(t)\right)} + \frac{s(1-s)}{f^2\left(F^{-1}(s)\right)} - 2\frac{cov\left(W_0(s), W_0(t)\right)}{f\left(F^{-1}(s)\right) f\left(F^{-1}(t)\right)}\}^2.
\end{aligned}$$

Since $0 < f(x) < \infty$ on $[a, b]$, there exists a constant M such that

$$f\left(F^{-1}(t)\right) \geq M^{-\frac{1}{4}} \text{ for all } t \in [0, 1].$$

Then

$$E[V_N(t) - V_N(s)]^2 \leq 3M(t-s)^2(1-|t-s|)^2 \leq 3M(t-s)^2.$$

Hence $V_N(t)$ converges in distribution to the process $V(t)$. Thus, according to Billingsley (1968, Theorem 5.1)

$$\int_0^u V_N(t)dt = \sum_{j=1}^{N} q_j(u) Z_j$$

converges in distribution to the process

$$\int_0^u V(t)dt = \int_0^u \frac{W_0(t)}{f\left(F^{-1}(t)\right)}dt = uY(u).$$

□

Now, let h_j be a function defined by

$$h_j(u) = \frac{1}{u}[q_j(u) - q_j(1)L(u)] \tag{14}$$

where $q_j(u)$ is given by (12).

Theorem 3. *Suppose the conditions of Theorem 1 are satisfied. Then $Z_n(u)$ given by (9) converges in distribution to the Gaussian process*

$$Z(u) = \sum_{j=1}^{\infty} h_j(u)Z_j \tag{15}$$

where $Z_1, Z_2, \ldots$ are independent $N(0,1)$ variables and $h_j(u)$ is given by (14).

Proof. By combining (2), (9) and (10) we see that

$$Z_n(u) = \frac{1}{H_n(1)}[uY_n(u) - L(u)Y_n(1)]$$

where $Y_n(u)$ is given by (10). Now, Theorem 1 implies that the process

$$uY_n(u) - L(u)Y_n(1)$$

converges in distribution to the process

$$uY(u) - L(u)Y(1)$$

where $Y(u)$ is given by (11). Then, since $H_n(1)$ converges in probability to μ, Cramer-Slutskys theorem gives that $Z_n(u)$ converges in distribution to the process

$$\frac{1}{\mu}[uY(u) - L(u)Y(1)].$$

Thus, by applying Theorem 2 the proof is completed. □

In order to derive the asymptotic covariance functions of the processes $Y_n(u)$ and $Z_n(u)$, the following lemma is needed.

Lemma 1. *Suppose the conditions of Theorem 1 are satisfied. Then*

$$\sum_{i=1}^{\infty} q_i(u)q_i(v) = \tau^2(u) + \lambda(u,v), 0 \leq u \leq v \leq 1,$$

where $q_i(u)$ is defined by (12) and $\tau^2(u)$ and $\lambda(u,v)$ are given by

$$\tau^2(u) = 2\int_a^{F^{-1}(u)} \int_a^y F(x)\,(1-F(y))\,dxdy, 0 \leq u \leq 1 \qquad (16)$$

and

$$\lambda(u,v) = \int_{F^{-1}(u)}^{F^{-1}(v)} \int_a^{F^{-1}(u)} F(x)\,(1-F(y))\,dxdy, 0 \leq u \leq v \leq 1. \qquad (17)$$

Proof. Assume that $0 \leq u \leq v \leq 1$. From the definition of $q_i(u)$ we have that

$$\sum_{i=1}^{\infty} q_i(u)q_i(v) = \sum_{i=1}^{\infty} \int_0^v \int_0^u [\frac{2}{f\left(F^{-1}(t)\right) f\left(F^{-1}(s)\right)} \frac{\sin(i\pi t)\sin(i\pi s)}{(i\pi)^2}]dtds.$$

By applying Fubinis theorem (e.g. Royden 1963) and the identity (13) we get

$$\begin{aligned}
\sum_{i=1}^{\infty} q_i(u)q_i(v) &= \int_0^v \int_0^u [\frac{2}{f\left(F^{-1}(t)\right) f\left(F^{-1}(s)\right)} \sum_{i=1}^{\infty} \frac{\sin(i\pi t)\sin(i\pi s)}{(i\pi)^2}]dtds \\
&= 2\int_0^u \int_0^s \frac{t(1-s)}{f\left(F^{-1}(s)\right) f\left(F^{-1}(s)\right)} dtds \\
&\quad + \int_u^v \int_0^u \frac{t(1-s)}{f\left(F^{-1}(t)\right) f\left(F^{-1}(s)\right)} dtds \\
&= 2\int_a^{F^{-1}(u)} \int_a^y (F(x)(1-F(y))dxdy) \\
&\quad + \int_{F^{-1}(u)}^{F^{-1}(v)} \int_a^{F^{-1}(u)} F(x)\,(1-F(y))\,dxdy \\
&= \tau^2(u) + \lambda(u,v).
\end{aligned}$$

□

As an immediate consequence of Theorem 1, Theorem 2 and Lemma 1 we have the following corollary.

Corollary 1. *Under the conditions of Theorem 1, $Y_n(u)$ has asymptotic covariance function $\Theta^2(u,v)$ given by*

$$\Theta^2(u,v) = \frac{1}{uv}[\tau^2(u) + \lambda(u,v)], 0 < u \leq v \leq 1.$$

From Theorem 3 and Lemma 1 we get the next corollary.

Corollary 2. *Under the conditions of Theorem 1, $Z_n(u)$ has asymptotic covariance function $v^2(u,v)$ given by*

$$\begin{aligned} v^2(u,v) &= \frac{1}{\mu^2}[\tau^2(u) + \lambda(u,v) - L(u)\left(\tau^2(v) + \lambda(v,1)\right) \\ &\quad -L(v)\left(\tau^2(u) + \lambda(u,1)\right) + L(u)L(v)\tau^2(1)], 0 < u \le v \le 1. \end{aligned} \tag{18}$$

In order to construct confidence intervals for the Lorenz curve at fixed points, we apply the results of Theorem 3 and Corollary 2 which imply that the distribution of

$$n^{\frac{1}{2}}\frac{L_n(u) - L(u)}{v(u,u)}$$

tends to the $N(0,1)$ distribution for fixed u, where $v^2(u,u)$ is given by

$$v^2(u,u) = \frac{1}{\mu^2}[\tau^2(u) - 2L(u)\left(\tau^2(u) + \lambda(u,1)\right) + L^2(u)\tau^2(1)], 0 < u \le 1.$$

Before this result can be applied, we must estimate the asymptotic variance $v^2(u,u)$, i.e., we must estimate μ, L, τ^2 and λ. The estimates of μ and L are given by $\overline{X}$ and (3), respectively. Now, by introducing the statistics a_k and b_k defined by

$$a_i = \left(1 - \frac{k}{n}\right)\left(X_{(k+1)} - X_{(k)}\right)$$

and

$$b_k = \frac{k}{n}\left(X_{(k+1)} - X_{(k)}\right),$$

we obtain the following consistent estimates of τ^2 and λ,

$$\hat{\tau}^2\left(\frac{i}{n}\right) = 2\sum_{k=1}^{i-1}\left(a_k\sum_{i=1}^{l} b_l\right), i = 2,3,\ldots,n \tag{19}$$

and

$$\hat{\lambda}\left(\frac{i}{n},\frac{j}{n}\right) = \left(\sum_{k=1}^{j-1} a_k\right)\left(\sum_{l=1}^{i-1} b_l\right), i = 2,3,\ldots,n-1; j \ge i+1. \tag{20}$$

Thus, replacing μ, L, τ^2, and λ by their respective estimates in the expression (18) for v^2 we obtain a consistent estimate of v^2. To get an idea of how reliable $L_n(u)$ is as an estimate for $L(u)$, we have to construct a confidence band based on $L_n(u)$ and $L(u)$. Such a confidence band can be obtained from statistics of the type

$$K_n = n^{\frac{1}{2}} sup_{0 \le u \le 1} \frac{|L_n(u) - L(u)|}{\psi(L_n(u))}$$

where ψ is a continuous nonnegative weight function. By applying Theorem 3 and Billingsley (1968, Theorem 5.1), we find that K_n converges in distribution to

$$K = sup_{0 \le u \le 1} \left| \sum_{j=1}^{\infty} \frac{h_j(u)}{\psi(L(u))} Z_j \right|.$$

Let

$$T_m(u) = \sum_{j=1}^{m} \frac{h_j(u)}{\psi(L(u))} Z_j,$$

$$T(u) = \sum_{j=1}^{\infty} \frac{h_j(u)}{\psi(L(u))} Z_j$$

and

$$K'_m = sup_{0 \le u \le 1} |T_m(u)|.$$

Since T_m converges in distribution to T, we find by applying Billingsley (1968, Theorem 5.1) that K'_m converges in distribution to K. Hence, for a suitable choice of m and ψ, for instance $\psi = 1$, simulation methods may be used to obtain the distribution of K'_m and thus an approximation for the distribution of K.

3.2 *Asymptotic properties of the empirical rank-dependent family of inequality measures*

We shall now study the asymptotic distribution of the statistics $\hat{J}_R$ given by (7). Mehran (1976) states without proof that $n^{\frac{1}{2}} \left(\hat{J}_R - J_R \right)$ is asymptotically normally distributed with mean zero. The asymptotic variance, however, cannot be derived, as maintained by Mehran (1976), from Stigler (1974), Theorem 3.1). [See also Zitikis an Gastwirth (2002) on the asymptotic estimation of the $S-$Ginis, Zitikis (2003) on the asymptotic estimation of the $E-$Gini index and a more general discussion in Davydov and Zitikis (2004).] However, as will be demonstrated below Theorem 3 forms a helpful basis for deriving the asymptotic variance of $\hat{J}_R$. Let ω^2 be a parameter defined by

$$\omega^2 = \frac{1}{\mu^2}\Big\{2\int_0^1\int_0^v [\tau^2(u)+\lambda(u,v)]R(u)R(v)dudv \\ -2[\int_0^1 uR(u)du - J_R][\int_0^1 \left(\tau^2(u)+\lambda(u,1)\right)R(u)du] \\ +\tau^2(1)[\int_0^1 uR(u)du - J_R]^2\Big\} \tag{21}$$

Theorem 4. *Suppose the conditions of Theorem 1 are satisfied and $\omega^2 < \infty$. Then the distribution of*

$$n^{\frac{1}{2}}\left(\hat{J}_R - J_R\right)$$

tends to the normal distribution with zero mean and variance ω^2.

Proof. From (4), (7) and (9) we see that

$$n^{\frac{1}{2}}\left(\hat{J}_R - J_R\right) = -\int_0^1 R(u)Z_n(u)du.$$

By Theorem 3 we have that $Z_n(u)$ converges in distribution to the Gaussian process $Z(u)$ defined by (15). By applying Billingsley (1968, Theorem 5.1) and Fubinis theorem we get that $n^{\frac{1}{2}}\left(\hat{J}_R - J_R\right)$ converges in distribution to

$$-\int_0^1 R(u)Z(u)du = -\int_0^1 R(u)\left(\sum_{j=1}^{\infty} h_j(u)Z_j\right)du \\ = -\sum_{j=1}^{\infty}[\int_0^1 R(u)h_j(u)du]Z_j$$

where $Z_1, Z_2, \ldots$ are independent $N(0,1)$ variables and $h_j(u)$ is given by (14), i.e., the asymptotic distribution of $n^{\frac{1}{2}}\left(\hat{J}_R - J_R\right)$ is normal with mean zero and variance

$$\sum_{j=1}^{\infty}[\int_0^1 R(u)h_j(u)du]^2. \tag{22}$$

Then it remains to show that the asymptotic variance is equal to ω^2. Inserting (14) in (22), we get

$$\begin{aligned}\sum_{j=1}^{\infty}[\int_0^1 R(u)h_j(u)du]^2 &= \frac{1}{\mu^2}\sum_{j=1}^{\infty}[\int_0^1 R(u)\left(q_j(u)-q_j(1)L(u)\right)du]^2 \\ &= \frac{1}{\mu^2}\Big\{\sum_{j=1}^{\infty}[\int_0^1 R(u)q_j(u)du]^2 \\ &\quad -2[\int_0^1 R(u)L(u)du][\sum_{j=1}^{\infty} q_j(1)\int_0^1 R(u)q_j(u)du] \\ &\quad +[\sum_{j=1}^{\infty} q_j^2(1)][\int_0^1 R(u)L(u)du]^2\Big\}.\end{aligned}$$

In the following derivation we apply Fubinis theorem and the identity (13),

$$\begin{aligned}\sum_{j=1}^{\infty}[\int_0^1 R(u)q_j(u)du]^2 &= \sum_{j=1}^{\infty}\int_0^1\int_0^1 R(u)q_j(u)R(v)q_j(v)dudv \\ &= \int_0^1\int_0^1\Big[\int_0^v\int_0^u \frac{2}{f\left(F^{-1}(t)\right)f\left(F^{-1}(s)\right)} \\ &\quad \times\left(\sum_{j=1}^{\infty}\frac{\sin(j\pi t\sin(j\pi s)}{(j\pi)^2}\right)dtds\Big]R(u)R(v)dudv \\ &= 2\int_0^1\int_0^v\Big[2\int_0^u\int_0^s \frac{t(1-s)}{f\left(F^{-1}(t)\right)f\left(F^{-1}(s)\right)}\,dt\,ds \\ &\quad +\int_u^v\int_0^u \frac{t(1-s)}{f\left(F^{-1}(t)\right)f\left(F^{-1}(s)\right)}\Big]R(u)R(v)dudv \\ &= 2\int_0^1\int_0^v\Big[2\int_a^{F^{-1}(u)}\int_a^y F(x)\left(1-F(y)\right)dxdy \\ &\quad +\int_{F^{-1}(u)}^{F^{-1}(v)}\int_a^{F^{-1}(u)} F(x)\left(1-F(y)\right)dxdy\Big] \\ &\quad \times R(u)R(v)dudv \\ &= 2\int_0^1\int_0^v[\tau^2(u)+\lambda(u,v)]R(u)R(v)dudv\end{aligned}$$

where $\tau^2(u)$ and $\lambda(u,v)$ are given by (16) and (17), respectively. Similarly, we find that

$$\sum_{j=1}^{\infty} q_j(1)\int_0^1 R(u)q_j(u)du = \int_0^1[\tau^2(u)+\lambda(u,1)]R(u)du.$$

From Lemma 1 it follows that

$$\sum_{j=1}^{\infty} q_j^2(1) = \tau^2(1).$$

Finally, by noting that

$$\int_0^1 R(u)L(u)du = \int_0^1 uR(u)du - J_R,$$

the proof is completed. □

For $R(u) = 2$, Theorem 4 states that $\omega^2 = \gamma^2$, where γ^2 is defined by

$$\gamma^2 = \frac{4}{\mu^2}\Big\{2\int_0^1\int_0^v [\tau^2(u) + \lambda(u,v)]\, dudv - (1-G)\int_0^1 [\tau^2(u) + \lambda(u,1)]du \\ +\frac{1}{4}(1-G)^2\tau^2(1)\Big\}, \tag{23}$$

is the asymptotic variance of $n^{\frac{1}{2}}\hat{G}$. The estimation of γ^2 is straightforward. As in Section 2, we assume that the parametric form of F is not known. Thus, replacing F by the empirical distribution function F_n in expression (23) for γ^2, we obtain a consistent nonparametric estimator for γ^2. The current estimator is given by

$$\hat{\gamma}^2 = \frac{4}{\overline{X}^2}\Big\{\frac{2}{n^2}\sum_{j=2}^{n}\sum_{i=2}^{j}\hat{\tau}^2\left(\frac{i}{n}\right) + \frac{2}{n^2}\sum_{j=3}^{n}\sum_{i=2}^{j-1}\hat{\lambda}\left(\frac{i}{n},\frac{j}{n}\right) \\ -\frac{1}{n}\left(1-\hat{G}\right)[\sum_{i=2}^{n}\hat{\tau}^2\left(\frac{i}{n}\right) + \sum_{i=2}^{n-1}\hat{\lambda}\left(\frac{i}{n},1\right)] + \frac{1}{4}\left(1-\hat{G}\right)^2\hat{\tau}^2(1)\Big\}$$

where $\hat{\tau}^2, \hat{\lambda}$ and $\hat{G}$ and are given by (19), (20) and (8), respectively. Similarly, a consistent estimator for ω^2 is obtained by replacing τ^2, λ, μ and J_R by their respective estimates in the expression (21) for ω^2.

Acknowledgement

Part of this paper was written when the author was visiting ICER in Torino. The financial support and excellent working conditions provided by ICER are gratefully acknowledged. The author would also like to thank Dorota Dabrowska and an anonymous referee for helpful comments.

References

1. Aaberge, R. (1982): On the problem of measuring inequality. (In Norwegian). *Rapporter* **82/9**, Statistics Norway.
2. Aaberge, R. (2000a): Characterizations of Lorenz curves and income distributions. *Social Choice and Welfare*, **17**, 639-653.
3. Aaberge, R. (2000b): "Ranking intersecting Lorenz curves", Discussion Paper 271, *Statistics Norway.*
4. Aaberge, R. (2001): Axiomatic characterization of the Gini coefficient and Lorenz curve orderings. *Journal of Economic Theory*, **101**, 115-132.
5. Atkinson, A. B. (1970): On the measurement of inequality, *Journal of Economic Theory*, **2**, 244-263.
6. Atkinson, A. B., L. Rainwater and T. Smeeding (1995): Income distribution in OECD countries: The evidence from the Luxembourg Income Study (LIS), *Social Policy Studies* **18**, OECD, Paris.
7. Ben Porath, E. and I. Gilboa (1994): Linear measures, the Gini index, and the income-equality trade-off. *Journal of Economic Theory*, **64**, 443-467.
8. Billingsley, P. (1968): *Convergence of Probability Measures.* John Wiley & Sons, Inc., New York.
9. Chernoff, H., J. L. Gastwirth and M. V. Johns (1967): Asymptotic distribution of linear combinations of functions of order statistics with application to estimation. *Annals of Mathematical Statistics*, **38**, 52-72.
10. Csörgő, M., J. L. Gastwith and R. Zitikis (1998): Asymptotic confidence bands for the Lorenz and Bonferroni curves based on the empirical Lorenz curve, *Journal of Statistical Planning and Inference*, **74**, 65 - 91.
11. Dardanoni, V. and P. J. Lambert (1988): Welfare rankings of income distributions: A role for the variance and some insights for tax reforms. *Social Choice and Welfare*, **5**, 1-17.
12. Davydov, Y. and R. Zitikis (2003): Generalized Lorenz curves and convexifications of stochastic processes, *Journal of Applied Probability*, **40**, 906 - 925.
13. Davydov, Y. and R. Zitikis (2004): Convex rearrangements of random elements, Asymptotic Methods in Stochastics, Fields Institute communications, *American Mathematical Society*, **44**, 141 - 171.
14. Doksum, K. (1974): Empirical probability plots and statistical inference for nonlinear models in the two-sample case. *The Annals of Statistics*, **2** (2), 267-277.
15. Donaldson, D. and J. A. Weymark (1980): A single parameter generalization of the Gini indices of inequality. *Journal of Economic Theory*, **22**, 67-86.

16. DONALDSON, D. AND J. A. WEYMARK (1983): Ethically flexible indices for income distribution in the continuum. *Journal of Economic Theory*, **29**, 353-358.

17. DURBIN, J. (1973): Distribution Theory for Tests Based on the Sample Distribution Function. *Society for Industrial and Applied Mathematics*, Philadelphia.

18. GIACCARDI, F. (1950): Un criterio per la construzione di indici di conzentrazione. *Rivista Italiana di Demografia e Statistica*, **4**, 527-538.

19. GOLDIE, C. M. (1977).Convergence theorems for empirical Lorenz curves and their inverses. *Advances in Applied Probability*, **9**, 765-791.

20. HAJEK, J. AND Z. SIDAK (1967): *Theory of Rank Tests.* Academic Press, New York.

21. HOEFFDING, W. (1948): A Class of statistics with asymptotically normal distribution. *The Annals of Mathematical Statistics*, **19**, 293-325.

22. KOLM, S. CH. (1969): The optimal production of social justice. In J. Margolis and H. Guitton (eds.): *Public Economics.* Macmillan, New York/London.

23. KOLM, S. CH. (1976a): Unequal inequalities I, *Journal of Economic Theory*, **12**, 416-442.

24. KOLM, S. CH. (1976b): Unequal inequalities II, *Journal of Economic Theory*, **13**, 82-111.

25. LAMBERT, P. J. (1993): The Distribution and Redistribution of Income: A Mathematical Analysis. *Manchester University Press*, Manchester.

26. MEHRAN, F. (1976): Linear measures of inequality, *Econometrica*, **44**, 805-809.

27. PIESCH, W. (1975): *Statistische Konzentrationsmasse.* Mohr (Siebeck), Tbingen.

28. ROYDEN, H. (1963): *Real Analysis.* Macmillan, New York.

29. SERFLING, R. J. (1980): *Approximation Theorems of Mathematical Statistics.* John Wiley & Sons, Inc., New York.

30. SHORACK, G. R. (1972): Functions of order statistics. *The Annals of Mathematical Statistics*, **43** (2), 412-427.

31. SHORROCKS, A. F. and J.E. Foster (1978): Transfer sensitive inequality measures. *Review of Economic Studies*, **14**, 485-497.

32. STIGLER, S. M. (1974): Linear functions of order statistics with smooth weight functions. *The Annals of Statistics*, **2** (4), 676-693.

33. TARSITANO, A. (2004): A new class of inequality measures based on a ratio of L- Statistics, *Metron*, **LXII**, 137 - 160.

34. WEYMARK, J. (1981): Generalized Gini inequality indices. *Mathematical Social Science*, **1**, 409-430.

35. YITZHAKI, S. (1983): On an extension of the Gini inequality index. *International Economic Review*, **24**, 617-628.

36. YAARI, M. E. (1987): The dual theory of choice under risk. *Econometrica*, **55**, 95-115.

37. YAARI, M. E. (1988): A controversial proposal concerning inequality measurement. *Journal of Economic Theory*, **44**, 381-397.

38. ZITIKIS, R. (1998): The Vervaat process. In B. Szyszkowicz (ed): *Asymptotic Methods in Probability and Statistics.* North-Holland, Amsterdam.

39. ZITIKIS, R. (2002): Large Sample Estimation of a Family of Economic Inequality Indices, *Pakistan Journal of Statistics Special issue in honour of Dr. S. Ejaz Ahmad*, **18**, 225 - 248.

40. ZITIKIS, R. (2003): Asymptotic estimation of the E - Gini index, *Econometric Theory*, **19**, 587-601.

41. ZITIKIS, R. AND J. T. GASTWIRTH (2002): The asymptotic distribution of the S-Gini index, *Autralian and New Zealand Journal of Statistics*, **44**, 439 - 446.

Chapter 25

A MODIFIED KENDALL RANK-ORDER ASSOCIATION TEST FOR EVALUATING THE REPEATABILITY OF TWO STUDIES WITH A LARGE NUMBER OF OBJECTS

Tian Zheng and Shaw-Hwa Lo

Department of Statistics
Columbia University, New York, NY, U.S.A

E-mails: tzheng@stat.columbia.edu & slo@stat.columbia.edu

Different studies on the same objects under the same conditions often result in nearly uncorrelated ranking of the objects, especially when dealing with large number of objects.The problem arises mainly from the fact that the data contain only a small proportion of "interesting" or "important" objects which hold the answers to the scientific questions. This paper proposes a modified Kendall rank-order association test for evaluating the repeatability of two studies on a large number of objects, most of which are undifferentiated. Since the repeatability between two datasets is reflected in the association between the two sets of observed values, evaluating the extent and the significance of such association is one way to measure the strength of the signals in the data. Due to the complex nature of the data, we consider ranking association which is distribution-free. Using simulation results, we show that the proposed modification to the classic Kendall rank-order correlation coefficient has desirable properties that can address many of the issues that arise in current statistical studies.

Keywords: Repeatability; Kendall rank-order; Nonparametric association test; Truncated rank.

1 Introduction

Current technological developments in many scientific fields allow researchers to explore the research problems in more detail, on a larger scale, involving many possible factors, and in huge dimensions simultaneously. Such data, of unprecedentedly large sizes, provide new challenges to statisticians. In this paper, we discuss one of them: evaluating the repeatability of two studies that rank the same, large set of, objects. To avoid confusion with the statistical term "subject", we call the units of evaluation as *objects*.

Despite the huge amount of data collected, often only a small portion of the variables (factors) hold the keys to new scientific discovery or validation. Evaluating the importance of these variables (or factors) and ranking them accordingly has been a focus of much current statistical research. Such evaluation provides the basis for dimension reduction, model selection, feature selection, machine learning, etc. In this paper, for convenience of discussion, we refer to the variables (or factors, or so-called dimensions in some context) as *objects of evaluation*.

In most application, while the number of objects that need to be evaluated are frequently in the tens of thousands, only a few of them are relevant to the questions to be addressed. Thus, the datasets are enormous but contain only a small set of "interesting" or important objects. Typical examples of such scenarios can be easily found in microarray analysis. Only dozens or hundreds of genes are truly regulated due to the treatments under a single experiment, while the data contain expression levels from tens of thousands. Conventionally, genes of more interest are those that are highly expressed or those whose expression profiles correlate well with the treatments. The reliability of such studies is a major concern due to the fact that different studies on the same objects under the same condition usually result in different lists of genes that are highly expressed or highly regulated by the treatments. However, one expects the repeatability between two studies to be high if the scientific signal is real.

If the repeatability between two datasets is reflected in the association between the two sets of observed values, we can evaluate the extent and the significance of such association by measuring the strength of the signals in the data. Because of the complex nature of such data, nonparametric measures of association are more desirable since the distributions of the observed values are usually unknown. Thus, we study the repeatability between two studies through ranking association.

Consider a simple example using a microarry dataset (van 't Veer et al. 2002). In this study, gene expression levels of 24,479 biological oligonucleotides in samples from 78 breast cancer patients were measured (using 2-dye-hybridization experiments on DNA Microarrays). Among these 78 breast cancer patients, 44 remained disease-free for more than 5 years, while the other 34 patients developed metastases within 5 years (van 't Veer et al. 2002). The goal of this study was to identify genes that were associated with the risk of developing metastases and use these genes for a better prediction of disease outcomes. To illustrate the concepts and methods proposed in this paper, consider a small "experiment" using this data

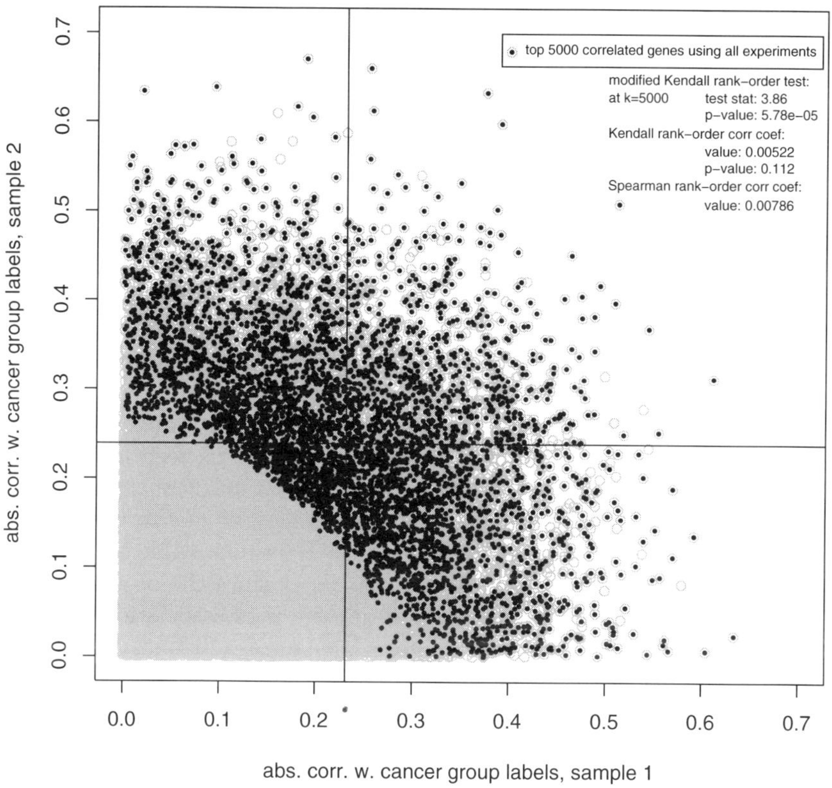

Figure 1 An example of two microarray experiment samples showing weak association.

(downloaded from the paper's web site). We randomly divide the data into two equal halves, each with 17 patients with poor prognosis (metastases within 5 years) and 22 patients with good prognosis (no metastases for more than 5 years). For convenience of discussion, we call these two halves: Sample 1 and Sample 2. For each gene, in either sample, correlation between the $\log_{10}$ gene expression ratios (the real gene expression versus the background gene expression) and the prognosis label (1=poor, 0=good) is calculated. In such a study, both positively and negatively correlated genes are regarded as important. Figure 1 shows the absolute correlation values from Sample 2 versus those of Sample 1. One can expect that the truly important genes will be strongly (positively or negatively) correlated with the prognosis label, in both Sample 1 and Sample 2. We should also expect that some genes will display strong association by chance in one of the two samples. The noisy pattern in Figure 1 shows a weak association between these two samples. Actually, the Kendall rank-order correlation coefficient of the 24,479 genes expression levels is 0.00522 (p-value = 0.112). Another interesting pattern is that, of the 5,000 top correlated genes selected based on the combined data of Sample 1 and Sample 2, only 17% demonstrated strong correlation with the prognosis label in both samples. It seems that the association between gene expressions and the cancer outcome are nearly random and uncorrelated between samples. However, the two samples have about 1,131 top genes in common if one simply select the top 5,000 ones with the highest absolute correlation values. Is this due to chance or true signal?

It is not surprising that the association on all objects is low since the unimportant or "uninteresting" objects should be similar and undifferentiated in the analysis, except for random noises, and lead to random rankings that result in the overall nearly uncorrelated pattern as seen in Figure 1. However, the main interest is to examine the repeatability of the top genes since their repeated appearance in the top list may be due to true signals related to the mechanism of breast cancer. It is impossible to compute measures of association using only the top ranks since the objects do not overlap completely. On the other hand, measures of association using all objects inevitably include many uninformative ones, and thus do not have power when dealing the situation where only the top few matter. Simply examining the number of overlapping top objects will result in loss of power as we will show later.

In this paper, we propose a modified Kendall rank-order association test to address this difficulty. Since it is based on rank-orders, it is a nonparametric test of association, which does not rely on any assumption of the quantitative evaluation scores of the two studies. Through simulation results, we show the the proposed modification to the classic Kendall rank-order correlation coefficient has desirable properties that will address the needs of many modern statistical studies. For instance, for the data in Figure 1, the p-value derived for the modified Kendall rank-order test is 5.78×10^{-5}, indicating the association of the top genes are significant and supporting the existence of true biological signals.

2 Method

This section studies the problem of repeatability through nonparametric tests of association between two sets of evaluations based on a same set of objects. To allow the tests to be distribution-free, the association is studied on rankings, not the actual observation or evaluation values.

2.1 *Notation*

Assume that n objects, $S_1, S_2, \ldots, S_n$ are under evaluation. Let X_i, $i = 1, \ldots, n$ and Y_i, $i = 1, \ldots, n$ be two independent rankings, in decreasing order, received by these samples. (Throughout, we discuss rankings in decreasing order unless otherwise noted.)

Denote α_i as the *true merit* of object S_i. The two sets of rankings are based on random observations and thus are random representations of the true ranking, Rank(α_i). Here, the notation Rank$(\cdot)$ is short for "the rank of ...". For convenience of formulation, we assume that

$$X_i = \text{Rank}(\alpha_i + \varepsilon_i), \qquad \text{and} \qquad Y_i = \text{Rank}(\alpha_i + \delta_i),$$

where ε_i's are independent random departures with distribution function F, δ_i's are independent random departures with distribution function G, and the ε's are independent with the δ's. Here, the α's and the distribution functions F, G are introduced only for convenience of discussion; they are neither assumed known nor used in the inference.

Without loss of generality, we assume that the objects S_1, ..., S_n are arranged in the order of their true merits, that is $\alpha_1 \geq \alpha_2 \geq \cdots \geq \alpha_n$. Under the null hypothesis that there is no difference among the objects in terms of their true merits, i.e., $\alpha_1 = \alpha_2 = \cdots = \alpha_n$, the ranking X, now reduced to the Rank(ε), would be independent of the ranking Y, Rank(δ). On the other hand, consider the extreme alternative where $\alpha_1 > \alpha_2 > \cdots > \alpha_n$, the ranking X and the ranking Y will then be positively correlated; the degree of correlation depends on the random variation of ε's and δ's. Thus, the correlation between ranking X and ranking Y can be used to measure the variation among the objects' merit, relative to the random error variation.

2.2 *Significance testing problem on ranking association*

In practice, we frequently need to separate out a small sample of objects with higher merit from the rest of the pool. If two rankings on the same objects are uncorrelated, the objects with high ranks may not be truly superior to the other objects. This is because it may be primarily due to chance that these objects are at the top of the list. On the other hand, in studies that involve evaluation of a large number of objects (eg. gene expression analysis), it is very common that the majority of the pool consists of objects that are quite similar if not identical in their true merits, and only few cases have higher, ordered true merits. Under such circumstances, the rankings are correlated for the few objects among the top, while the rankings are uncorrelated for the rest of the observations with undifferentiated merits, even when the random error variations are low. Such a low correlation among the lower ranks dilutes the overall "coordination" between the ranking X and the ranking Y. As a result, the overall correlation between X and Y would not be high. If, however, the objects are grouped into top ranks and low ranks (that is, higher rank value in a decreasing rank order), one would expect to observe a higher degree of correlation between the groupings according to the rankings X and Y than the correlation between the original rankings X and Y, reflecting the important true merits of the few top objects.

Here we consider the inference question of testing the null hypothesis $H_0 : \alpha_1 = \alpha_2 = \cdots = \alpha_n \equiv \alpha$ versus a *local* alternative $H_a : \exists 1 \leq k_0 \ll n$, s.t. $\alpha_1 \geq \alpha_2 \geq \cdots \geq \alpha_{k_0} > \alpha_{k_0+1} = \alpha_{k_0+2} = \cdots = \alpha_{n-1} = \alpha_n$. We propose to study a *modified Kendall rank-order test* which detects objects with true and high merits using the important association between the top ranks of two rankings and the association between the grouping of ranks, without effects from the noises in the lower ranks.

2.3 *Problems with the Kendall rank-order test*

We first examine the original *Kendall rank-order correlation coefficient* (Siegal and Castellan 1988, Chapter 9). It uses the number of agreements and disagreements defined as follows: consider all possible *pairs* of ranks (X_i, X_j) in which X_i is lower than X_j, if

- the corresponding Y_i is lower than Y_j, it is then an *agreement*;
- the corresponding Y_i is higher than Y_j, it is then an *disagreement*.

Using these two counts, the Kendall rank-order correlation coefficient is formulated as

$$T = \frac{\#\text{ agreements} - \#\text{ disagreements}}{\text{total number of pairs}} \tag{1}$$

Since the maximum possible values for the number of agreements and the number of disagreements are both the number of the total possible pairs, T ranges between -1 and 1, same as the conventional Pearson's coefficient of correlation.

It is easy to show that

$$\#\text{ agreements} = \sum_{i=1}^{n} \sum_{i \neq j} \mathbf{1}_{(X_i < X_j)} \mathbf{1}_{(Y_i < Y_j)}, \tag{2}$$

$$\#\text{ disagreements} = \sum_{i=1}^{n} \sum_{i \neq j} \mathbf{1}_{(X_i < X_j)} \mathbf{1}_{(Y_i > Y_j)}, \tag{3}$$

and, assuming there are no tied observations,

$$\#\text{ agreements} + \#\text{ disagreements} = n(n-1)/2.$$

Under the null hypothesis, E(# agreements) =E(# disagreements) = $\frac{1}{4}n(n-1)$. And the variance of the number of agreements can be shown to be $\frac{1}{16}(\frac{4n}{9}+\frac{10}{9})n(n-1)$ (see Siegel and Castellan, 1988).

It follows that, under the null hypothesis, $E(T) = 0$ and

$$\mathrm{Var}(T) = \mathrm{Var}\left(\frac{4(\#\text{ agreements}) - 2n(n-1)}{n(n-1)}\right) = \frac{2(2n+5)}{9n(n-1)}. \quad (4)$$

Conventionally, for $n > 10$, the significance of T is evaluated using an normal approximation (Siegel and Castellan 1988, Chapter 9). From (1) and (4), one can see that such tests of association based on the Kendall rank-order correlation coefficient is equivalent to a z-test for the number of agreements. By constructing the test through the number of agreements, the computation is reduced by half.

The test is reasonably powerful when the majority of the observations are random representations of a sequence of completely ordered (differentiated) true merits. However, when only a few objects have a sequence of ordered α values and the majority of the objects are similar in their α's, those undifferentiated objects will result in large numbers of random agreements and disagreements, and thus add substantial noise. As a result, the power of the test based on the Kendall rank-order correlation coefficients diminishes under local alternatives even when the local association exists and is strong.

Simulation results under two sets of local alternatives are shown in Figure 4. The powers of the Kendall rank-order correlation coefficients under different alternatives is the right ends of the curves. We see that the power is affected more when the number of important objects is smaller. More details and discussion on these examples are in Section 3.2 and the simulation details are discussed in Section 3.1.

2.4 *Problems with test based on number of overlapping top objects*

When only the top objects are of interest, it is natural to consider a test based on the number of overlapping top objects as a solution. Denote k as the number of top objects that are to be considered. The number of overlapping top objects between ranking X and ranking Y can then be defined as

$$O = \sum_{i=1}^{n} \mathbf{1}_{(X_i \le k)} \mathbf{1}_{(Y_i \le k)}.$$

It is easy to show that, under the null hypothesis,

$$\mathrm{E}(O) = \frac{k^2}{n}, \ \mathrm{Var}(O) = \frac{k^2}{n} + \frac{k^2(k-1)^2}{n(n-1)} - \frac{k^4}{n^2}.$$

Thus a simple test can be constructed using $\frac{O - \mathrm{E}(O)}{\sqrt{\mathrm{Var}(O)}}$, and the sampling distribution can be approximated by the standard normal distribution. The exact significance can also be evaluated based on the values of k and n using permutations.

The power of this test is satisfactory when the specification of k is close to the true number of important objects, under the local alternatives we consider in this paper. Actually, under such ideal situations, the performance of this test is nearly comparable to that of the modified Kendall rank-order association test, which is to be discussed in the next section. However, the performance of the test based on the number of overlapping top objects deteriorates dramatically when the specified value of k departs from the ideal specification k_0 (the true number of objects with higher true merits). Such a trend, found in the simulation studies, is shown in Figure 5. Details of this example will be discussed in Section 3.2.

2.5 *Modified Kendall rank-order association test*

We now propose a solution to the problem raised in Section 2.2 by evaluating the correlation between two rankings, while only a small number, $k \ll n$, of the top ranks are actually considered. The value of k is pre-specified, which can be based on the knowledge of k_0 or inferred from previous results regarding the data.

This modified Kendall rank-order test statistic is based on the truncated ranks defined as $X_i^c = \min(X_i, k)$. The number of agreements as defined in the Kendall rank-order correlation coefficient is now calculated based on these truncated ranks. The test statistic is then the number of agreements standardized by the mean and standard deviation of its sampling distribution under the null hypothesis. Similar to the formulation of (2) and (3), we have

$$\# \text{ agreements} = \sum_{i=1}^{n}\sum_{i\neq j} \mathbf{1}_{(X_i^c<X_j^c)}\mathbf{1}_{(Y_i^c<Y_j^c)} \tag{5}$$

$$= \sum_{i=1}^{n}\sum_{i\neq j} \mathbf{1}_{(\min(X_i,k)<\min(X_j,k))}\mathbf{1}_{((\min(Y_i,k)<\min(Y_j,k))}$$

$$\# \text{ disagreements} = \sum_{i=1}^{n}\sum_{i\neq j} \mathbf{1}_{(X_i^c<X_j^c)}\mathbf{1}_{(Y_i^c>Y_j^c)} \tag{6}$$

$$= \sum_{i=1}^{n}\sum_{i\neq j} \mathbf{1}_{(\min(X_i,k)<\min(X_j,k))}\mathbf{1}_{((\min(Y_i,k)>\min(Y_j,k))}$$

Under the null hypothesis, the ranking $X = \{X_1, \dots, X_n\}$ is a random permutation of $1, \dots, n$. Using (5) and (6), it is easy to show that

$$\mathrm{E}(\#\text{ agreements}) = \mathrm{E}(\#\text{ disagreements}) = \frac{1}{4}n(n-1)\left(1 - \frac{\binom{n-k+1}{2}}{\binom{n}{2}}\right)^2$$

$$= \frac{1}{4}n(n-1)\left(1 - \frac{(n-k+1)(n-k)}{n(n-1)}\right)^2. \tag{7}$$

The variance of the number of the agreements under the null hypothesis can also be calculated as follows:

$$\mathrm{Var}\left(\sum_{i=1}^{n}\sum_{j\neq i} \mathbf{1}_{(X_i^c<X_j^c)}\mathbf{1}_{(Y_i^c<Y_j^c)}\right) = \sum_{i=1}^{n}\sum_{j\neq i}\mathrm{Var}\left(\mathbf{1}_{(X_i^c<X_j^c)}\mathbf{1}_{(Y_i^c<Y_j^c)}\right)$$

$$+\sum_{i=1}^{n}\sum_{j\neq i}\sum_{(k,l)\neq(i,j)}\mathrm{Cov}\left(\mathbf{1}_{(X_i^c<X_j^c)}\mathbf{1}_{(Y_i^c<Y_j^c)},\ \mathbf{1}_{(X_k^c<X_l^c)}\mathbf{1}_{(Y_k^c<Y_l^c)}\right)$$

$$= n(n-1)\left\{\frac{1}{4}n(n-1)\left(1 - \frac{(n-k+1)(n-k)}{n(n-1)}\right)^2\right.$$

$$+(n-2)(n-3)\left(\frac{1}{4}\frac{\binom{k-1}{4}}{\binom{n}{4}} + \frac{1}{4}\frac{\binom{k-1}{3}\binom{n-k+1}{1}}{\binom{n}{4}} + \frac{1}{6}\frac{\binom{k-1}{2}\binom{n-k+1}{2}}{\binom{n}{4}}\right)^2$$

$$+(n-2)\frac{1}{6}\left(\frac{\binom{k}{3}}{\binom{n}{3}} + \frac{\binom{k-1}{2}\binom{n-k}{1}}{\binom{n}{3}}\right)^2$$

$$+(n-2)\frac{1}{9}\left(1 - \frac{\binom{n-k+1}{3}}{\binom{n}{3}}\right)^2$$

$$\left.-\frac{1}{16}(n^2-n)\left(1 - \frac{(n-k+1)(n-k)}{n(n-1)}\right)^4\right\}.$$

The modified Kendall rank-order test statistic is then defined as

$$T^c = \frac{\#\text{ agreements} - \text{E}(\#\text{ agreements})}{\sqrt{\text{Var}\,(\#\text{ agreements})}}, \tag{8}$$

where the sampling distribution can be approximated by the standard normal. The exact significance can also be evaluated based on the values of k and n under the null hypothesis through permutations.

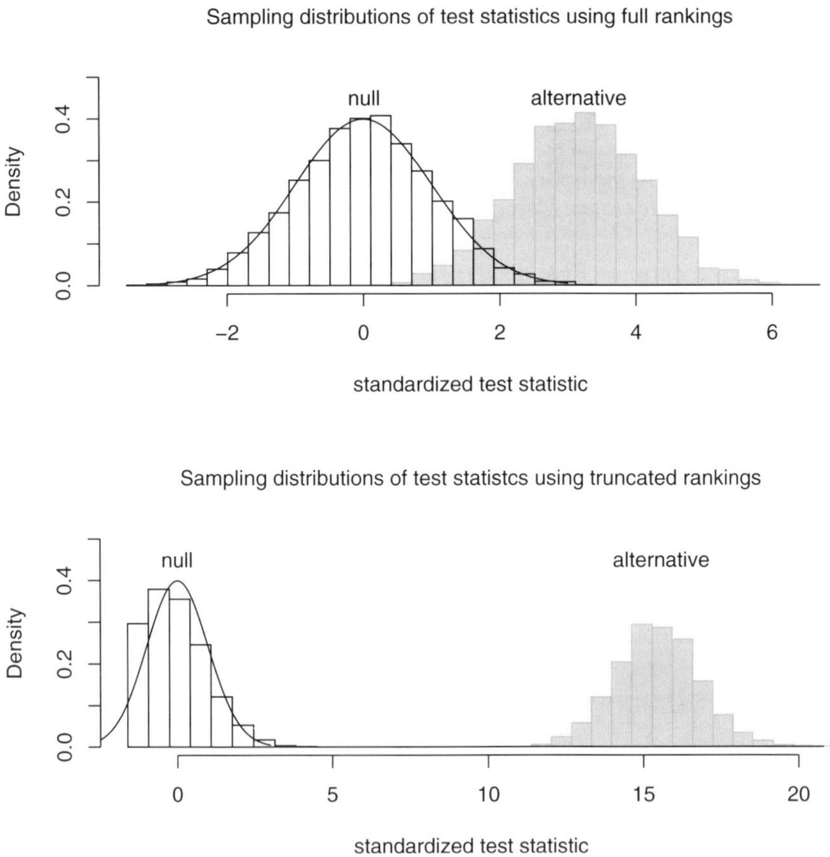

Figure 2 Sampling distributions using truncated rankings versus full rankings under the null and alternative hypotheses. Each distribution is based on 5000 simulations on 1000 objects. The alternative hypothesis used is specified as in Figure 3 with $\delta = 5$ and $k_0 = 50$. The smooth curves in the plots represent the standard normal distribution used in the approximation of p-values.

The advantage of the modified statistic is illustrated as in Figure 2. The sampling distributions were simulated under the same null and alternative hypotheses for the test statistics of the original Kendall rank-order correlation coefficient and the modified Kendall rank-order association test statistic. For the modified statistic, the sampling distribution under the alternative is well separated from the sampling distribution under the null hypothesis, while the distributions for the original statistic have a substantial overlap. It indicates that by focusing on the top ranks, as with the modified statistic, the signal becomes stronger because of the removal of a substantial amount of noise.

3 Simulations and Results

3.1 *Simulation models*

The local alternatives can take many forms of departures from the null hypothesis. For assessing the proposed method, we consider here a class of alternatives of the form as shown in Figure 3. As mentioned in Section 2.2, we assume the true merits α's are ordered. In each of the simulation model, a number, k_0, of α's are set to be higher than the rest of the objects that have identical merit values. For those that have higher merit values, we specifies the values to have linear increments, while the highest α value, α_1, was δ higher than the value of the undifferentiated objects. In other words, $\alpha_{i-1} - \alpha_i = \delta/k_0$, for $2 \leq i \leq k_0$ and $\alpha_{i-1} - \alpha_i = 0$ for $k_0 + 1 \leq i \leq n$. For convenience, F and G were taken to be normal distributions with mean 0 and standard deviation σ.

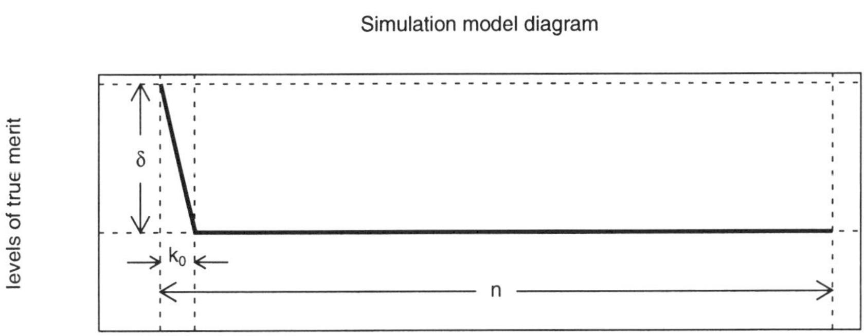

Figure 3 Alternative model used for simulations.

3.2 *Results*

For the class of alternatives that are considered in the simulation studies, the strength of the signal from the objects with higher α values depends on the elevation of merit, δ, and the noise standard deviation σ: The higher the ratio between δ and σ, the more distinct the top objects are from the rest of the evaluation pool. Without loss of generality, we fix σ to be 1 and vary only the value of δ.

In Figure 4, power performance of the modified Kendall rank-order test on simulated data with 500 objects are plotted at each possible truncation values, k, under two specifications of δ and different values of k_0. When $k = n$, the modified test becomes the original Kendall rank-order test. First, we observe that the power curves attain the peaks around the "right" specification of k, i.e., around the real value of k_0. Trimming too many objects (smaller values k) and too few (larger values of k) both result in loss of power. The most striking performance gain due for the modified rank-order test is observed when the signal is weaker and the number of the objects with higher merits is smaller (say, $\delta = 3$, k_0=10). As shown in this example, the original Kendall rank-order statistic has little power when the proportion of true signal is less than 5%, while the modified test maintains a power of higher than 70% for a reasonable range of k values around the true (unknown) value of k_0.

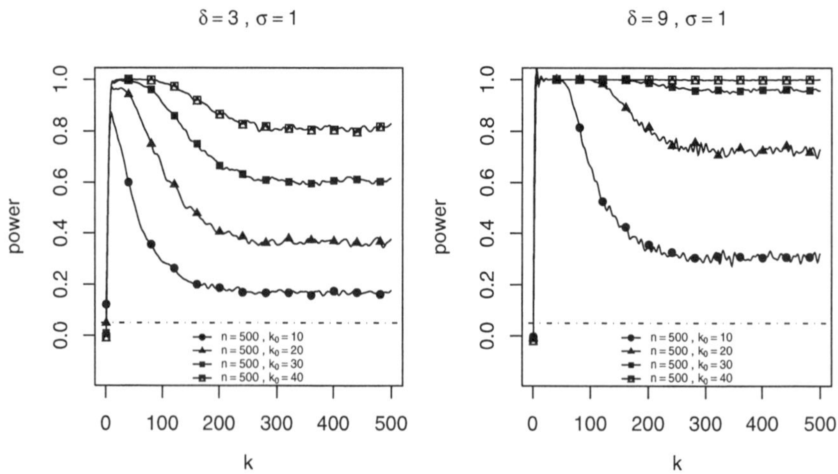

Figure 4 Power performance of the modified Kendall rank-order association test under two sets of alternatives. The power is estimated using 500 simulations. The models used for simulation are specified as in Figure 3 with $\delta = 3$ versus $\delta = 9$, while k_0 takes four different values.

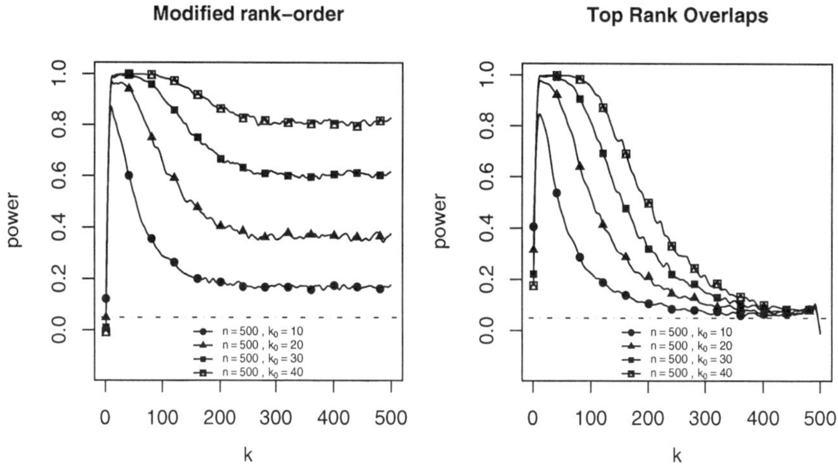

Figure 5 Power performance of the modified Kendall rank-order association test and the test based on the overlapping top objects. The power is estimated using 500 simulations. The model used for simulation is specified as in Figure 3 with $\delta = 3$ and $\sigma = 1$, while k_0 takes four different values.

Figure 5 shows the comparison of the power performance between the modified rank-order test and the test based on the number of overlapping top objects as described in Section 2.4. The models used for this comparison have $\delta = 3$ and $\sigma = 1$ with k_0 varying. For the test based on the number of overlapping top objects, the clearest pattern in Figure 5 is that the power curves drop dramatically as the specified truncation number k departs from the true value k_0. This method is not useful when the value k get closer to the total number of the objects, n, by definition. On the other hand, the signal of the top objects is better preserved and reflected by the modified rank-order test when k differs from k_0. This is due to the fact that the test using the overlap of top objects only reduces noises from the undifferentiated objects, while the modified rank-order test takes into account the informative order of the top ranks through the use of the truncated ranks and gain more power to detect the real signals.

4 Discussion and Conclusion

4.1 *Departure from normality*

The sampling distributions of the modified Kendall rank-order test statistic can be well approximated by normal distributions when k is not very small. However, when based on the truncated ranks, if k and n are both small (say $n \leq 30$, $k \leq 5$), the sampling distribution under the null hypothesis becomes more discrete. This is because, conditioning on the total number of agreements and disagreements combined, the distribution of the number of agreements is a binomial distribution with probability 0.5. The combined total number of agreements and disagreements equals the number of informative pairs of objects (those that have at least one falling into the top k ranks in both X and Y rankings). This number becomes very small when n and k/n are both small. Actually, the expected total number of agreements and disagreements under the null hypothesis is

$$
\begin{aligned}
&\mathrm{E}(\#\text{ agreements} + \#\text{ disagreements} \mid H_0) \\
&\qquad = \frac{n(n-1)}{2}\left(1 - \frac{(n-k)(n-k-1)}{n(n-1)}\right)^2 \\
&\qquad = \frac{(2n-k-1)^2 k^2}{2n(n-1)}.
\end{aligned}
$$

which is approximately $2k^2$ for most cases we consider in this paper, provided n is large and $k \ll n$. Thus, the test derived based on the normal approximation should be reasonable when $k \geq 10$ and $k \ll n$.

4.2 *Disagreements versus agreements*

The original Kendall rank-order correlation coefficient is calculated using both the number of agreements and disagreements while adjusting for the numbers of ties in both the X and Y rankings. For the truncated ranks, the mean and variance of such a score becomes complicated to evaluate. It is believed, though, that the number of disagreements should also contain useful information. This is an issue that remains to be addressed in the future.

4.3 *More general alternatives*

Other forms of departures from the null hypothesis can also take place in different applications. For instance, in the microarray data example considered in the introduction, if we look at the coefficient of correlation, we should look at both ends of the fully-sorted "merit" list. On the positive value end will be genes that are highly expressed in good prognosis patients but are suppressed in poor prognosis patients, while genes that expressed in good prognosis patients but not poor prognosis patients will have correlation values clustered around the lower end (large negative values). For a situation like this, a natural modification of the proposed method will be to change the right-truncation of the rank to center truncation, i.e., modifying the rank by assigning a single rank value to the mid-portion of the ordered observations.

4.4 *Computational considerations*

The computational complexity of the test statistic is the same as that of the original Kendall rank-order correlation coefficient, which is $O(n^2)$. For a large value of n, the sorting of the objects and the counting of the agreements can be time consuming. However, the amount of computation needed can be adequately handled by current computing power. For the example we have used in the introduction section, the calculation for 24,479 genes took about 3 minuets to finish using R on a Pentium-4 PC, which should be acceptable by current computing standards although this can be much improved using scientific computing languages such as C or Fortran. The mean and standard deviation for the normal approximation only depend on the specification of n and k, which is easy to calculate and can be prepared in advance.

4.5 *Conclusion*

In this paper, we proposed a modified Kendall rank-order association test for studying the repeatability of two studies on a large number of objects, most of which are undifferentiated. The method addresses new issues posed by the low signal-to-noise ratio of current data sets. The test statistic is intuitive, easy to implement, and fast to calculate. The exact sampling distribution can be derived using permutation simulations or conveniently approximated by normal distributions in most practical situations. Simulations on a class of general alternatives show substantial gains in power due to the proposed modification, compared to the original Kendall rank-order coefficient of correlation. Through the use of rank-order based on truncated ranks, the test statistic still manages to capture the informative order of the objects with higher merits while removes the noises from the undifferentiated objects. We believe this new test can find important applications in statistical studies that involves large number of objects for evaluation.

Acknowledgements

The research activities of these authors presented in this paper are, in part, supported by NIH grant R01 GM070789.

References

1. Siegel S. C. (1988) *Nonparametric statistics for the behavioral sciences*, 2nd Ed. McGraw-Hill.

2. van't Veer L. J., Dai H., van de Vijver M. J., He Y. D., Hart A. A., Mao M., Peterse H. L., van der Kooy K., Marton M. J., Witteveen A. T., Schreiber G. J., Kerkhoven R. M., Roberts C., Linsley P. S., Bernards R. and Friend S. H. (2002) Gene expression profiling predicts clinical outcome of breast cancer. *Nature* 415:530-6

Chapter 26

RANK PROCESS, STOCHASTIC CORRIDOR AND APPLICATIONS TO FINANCE

Ryozo Miura

Graduate School of International Corporate Strategy
Hitotsubashi University, Tokyo, JAPAN

E-mail: rmiura@ics.hit-u.ac.jp

Stock prices have been modeled in the literature as either discrete or continuous versions of geometric Brownian motions (GBM). This chapter uses rank statistics of the GBM to define a new exotic derivative called a stochastic corridor. This rank statistic measures how much time, during a given period, the stock prices stay below the price of a prefixed day. The properties of the stochastic corridor and its applications in finance are studied.

Key words: Rank Process; Stochastic Corridor; Look-back Option; Exotic Derivatives; Geometric Brownian Motion; Stock Prices

1 Introduction

In Mathematical Finance, stock prices in discrete time are typically assumed to follow lognormal distributions. More specifically, the logarithm of stock price at time t is assumed to be a sum of t independent identically distributed (iid) normal random variables. Functions of these stock prices are used to define *exotic derivatives*, especially the ones called *look-back options*. The most popular one is the arithmetic average; others include the maximum and the minimum (Goldman et al., 1979). Miura(1992) considered the alpha-quantile option, which is based on the quantile or order statistics of stock prices. The options based on Max, Min, and alpha-quantiles are called look-back options since, at the end of the time interval, we have to look back at all the stock prices that occurred during the time interval in order to compute the value of these statistics.

In this chapter, we will use a continuous-time framework which can be viewed as a limit of the discrete-time setting. Specifically, we will assume that the stock price $\{S_u,\ u \in [0,T]\}$ follows a geometric Brownian motion (GBM), i.e.,

$$S_u = S_0 e^{X_u} = S_0 e^{\mu u + \sigma W_u}, \text{ for } u \in [0,T], \tag{1}$$

where S_0 is the random initial price and W_u is a standard Brownian motion with zero mean.

Now, the *corridor* is defined as follows. For any fixed constant K, let

$$F(K) = \frac{1}{T}\int_0^T I\{S_u \le K\}du. \tag{2}$$

This is just a continuous-time version of the empirical process for stock prices during the time interval $[0,T]$. It is a measure of the proportion of time the stock prices stay below the given fixed value K during the time interval $[0,T]$. We will call it a fixed-level corridor or fixed corridor for short.

This quantity depends on the path of stock price, and it can be determined only at the end of the time interval. For example, consider an application in discrete-time setting such as the currency exchange rate derivatives. This statistic counts the proportion of days the exchange rate stays below the given fixed level K, and the *pay-off* (which is the value of the derivative at the time of exercise, or of expiration) of the derivative (contract) may promise to pay to the holder of the derivative the amount of money proportional to the statistic. This is called a *corridor option.* These corridors could also be used in principle for other applications such as weather derivatives to count the number of days where the daily-temperature stays below a fixed level.

Given $\alpha \in [0,1]$, the α-quantile of the process is defined as the quantity $m(\alpha)$ such that

$$\alpha = \frac{1}{T}\int_0^T I\{S_u \le m(\alpha)\}du. \tag{3}$$

Note that $F(m(\alpha)) = \alpha$. Then, $m(\alpha)$ is the level below which the stock price stays for $100\alpha-$percent of the time during the time interval $[0,T]$. Thus it could be viewed as a continuous-time version of "order statistics" of stock prices observed during the time interval $[0,T]$.

In this chapter, we define and study the properties of a new derivative called *stochastic corridor.* Specifically, consider a fixed day t with stock price S_t, which is random. Define the rank process

$$R(t) = \frac{1}{T}\int_0^T I\{S_u \le S_t\}du. \tag{4}$$

This has a similar interpretation as the fixed corridor except that the fixed value of K in the fixed corridor has been replaced by S_t which is stochastic. Note that $R(t)$ does not depend on S_0 since

$$I\{S_u \le S_t\} = I\{S_0 e^{X_u} \le S_0 e^{X_t}\} = I\{X_u \le X_t\}.$$

The fixed-level corridor and the stochastic-level corridor both can be used as payoff of derivatives.

The rest of the chapter is organized as follows. In Section 2, we briefly review the look-back options based on nonparametric statistics of stock prices observed during the time interval. There we also review the results for the probability distributions for non-parametric statistics such as α-quantiles and rank processes. In Section 3, a brief discussion is given of the risk-neutral measure in derivative pricing. In Section 4, we consider the stochastic corridor option and introduce a swap or an exchange contract between the stochastic corridor and a fixed corridor. We further introduce, in Section 5, an option to buy/sell the stochastic corridor by a fixed corridor. In Section 6, we define the forward starting corridors, and then we go on to discuss a swap and an option based on these which are forward starting versions of the spot starting ones in Sections 4 and 5. A specific feature of the forward starting stochastic corridor is that its probability distribution is independent of the starting stock price at the beginning of the future time interval since it is a rank statistic determined by the relative magnitude of the stock prices.

The contributions in this chapter are twofold. First, the look-back options defined in Sections 4, 5, and 6 are new results in the literature. Second, we develop technical results on how to calculate the prices of these derivatives.

2 Distribution Results

The following lemma plays a key role at several parts in this chapter where a calculation is encountered for an expectation of the nonparametric statistics such as a fixed corridor or a stochastic corridor. The proof can be found in Fujita(1997) or Fujita & Miura(2002,2004). See also the handbook Borodin & Salminen(2002) for the result without proof.

Lemma 1.
$P(W_t \in da, \int_0^t I\{W_s < 0\}ds \in du) = (\int_u^t (\frac{a}{2\pi\sqrt{s^3(t-s)^3}} e^{\frac{-a^2}{2(t-s)}} ds) da \quad du,$ *for* $a > 0$.
$P(W_t \in da, \int_0^t I\{W_s < 0\}ds \in du) = (\int_0^u (\frac{-a}{2\pi\sqrt{s^3(t-s)^3}} e^{\frac{-a^2}{2s}} ds) da \quad du$ *for* $a < 0$.

Consider now the distribution of the fixed corridor $F(K)$ or equivalently $TF(K)$. This was derived for the driftless case ($\mu = 0$) in Miura(1992) using the fact that the distribution can be reduced to that of $\int_0^T I\{W_u \leq 0\}du$. The Cameron-Martin theorem can then be used to get the result for the general $\mu \neq 0$ case.

Let $\mu = 0$ for now, and assume throughout that $S_0 < K$. Let

$$A = \frac{1}{\sigma}\log\left(\frac{K}{S_0}\right) \tag{5}$$

and

$$\tau = \inf\{u : W_u \geq A, 0 < u < T\}. \tag{6}$$

Note that $W_\tau = A$. It can be shown (Miura (1992)) that the probability distribution of $TF(K)$ is

$$G(x, K : 0, \sigma) = \int_0^x \frac{2}{\pi}\sin^{-1}((\frac{x-s}{1-s})^{\frac{1}{2}})\ h_A(s)ds$$

where h_A is the probability density of the stopping time τ.

Consider now the case with drift $\mu \neq 0$. By Cameron-Martin theorem, for any integrable function $h(\cdot)$,

$$\begin{aligned} E[h(TF(K))] &= E[h(\int_0^T I\{S_0 e^{\mu u + \sigma W_u} \leq K\}du)] \\ &= E[e^{\frac{\mu}{\sigma}W_T - (\frac{\mu}{\sigma})^2\frac{T}{2}} h(\int_0^T I\{W_u \leq \frac{1}{\sigma}\log\frac{K}{S_0}\}du)]. \end{aligned}$$

Define $Z_{u-\tau} = W_u - W_\tau)$ and recall that $A = \frac{1}{\sigma}\log\left(\frac{K}{S_0}\right)$. Now we can get the general distribution of $T\ F(K)$ in the $\mu \neq 0$ case as

$$\begin{aligned} G(x, K; \mu, \sigma) &= E[e^{\frac{\mu}{\sigma}W_T - (\frac{\mu}{\sigma})^2\frac{T}{2}} I\{\int_0^T I\{W_u \leq A\}du < x\}] \\ &= E[e^{\frac{\mu}{\sigma}W_T - (\frac{\mu}{\sigma})^2\frac{T}{2}} I\{(\tau + \int_\tau^T I\{Z_{u-\tau} \leq 0\}du) < x\}] \\ &= E[e^{\frac{\mu}{\sigma}A - (\frac{\mu}{\sigma})^2\frac{T}{2}} e^{\frac{\mu}{\sigma}Z_{T-\tau}} I\{(\tau + \int_0^{T-\tau} I\{Z_u \leq 0\}du) < x\}]. \end{aligned}$$

The last term can be calculated by using the joint probability density function of $(Z_{T-\tau}, \int_0^{T-\tau} I\{Z_u \leq 0\}du < x)$ as shown in Lemma 1 and the distribution of the first hitting time of τ.

From this, we get the distribution of the α-quantile $m(\alpha)$ as $\mathrm{P}\{m(\alpha) < y\} = 1 - G(T\alpha, y : \mu, \sigma)$. See also Akahori(1995). Fujita (1997) derived the joint probability density function of $(S_T, m(\alpha))$ in order to price a call option with payoff function $\max\{S_T - m(\alpha), 0\}$.

Let us turn to the distribution of the stochastic corridor $[TR(t)]$. Fujita and Miura (2004) noted that this rank statistic in the case $\mu = 0$, $\sigma = 1$ can be decomposed into a weighted sum of two independent random variables, each of which follows an arcsine law.

Let $\tilde{R}(t)$ denote the rank statistics for this case. Then,

$$\begin{aligned} T\tilde{R}(t) &= \int_0^T I\{W_u \leq W_t\}du \\ &= \int_0^t I\{W_u - W_t \leq 0\}du + \int_t^T I\{W_u - W_t \leq 0\}du \\ &= t\frac{1}{t}\int_0^t I\{Z_s \leq 0\}ds + (T-t)\frac{1}{T-t}\int_0^{T-t} I\{Z_s^* \leq 0\}du, \end{aligned}$$

where $Z_s = W_{t-s} - W_t$ and $Z_s^* = W_{t+s} - W_t$). Then, as shown in Fujita and Miura (2004), Lemma 1 and the Cameron-Martin theorem can be used to handle the general (μ, σ) case. More specifically, for the general case, for any integrable function $h(\cdot)$,

$$\begin{aligned} &E[h(TR(t))] \\ &= E[e^{\frac{\mu}{\sigma}W_T - (\frac{\mu}{\sigma})^2\frac{T}{2}}h(T\tilde{R}(t))] \\ &= \int_{-\infty}^{\infty}\left[\int_0^1 e^{\frac{\mu}{\sigma}x - (\frac{\mu}{\sigma})^2\frac{T}{2}}h(y)f_{(W_T, T\tilde{R}(t))}(x,y)dy\right]dx \\ &= E[e^{\frac{\mu}{\sigma}(Z^*_{T-t} - Z_t) - (\frac{\mu}{\sigma})^2\frac{T}{2}}h(\int_0^t I\{Z_s \leq 0\}ds + \int_0^{T-t} I\{Z_s^* \leq 0\}ds)] \end{aligned}$$

where $W_T = Z^*_{T-t} - Z_t$ in the equation on the righthand side above. Now the last equation above can be expressed as

$$\begin{aligned} &\iint_{-\infty<x_1<\infty, 0<y_1<T} {}_t\iint_{\infty<x_2<\infty, 0<y_2<t} e^{\frac{\mu}{\sigma}(x_1 - x_2) - (\frac{\mu}{\sigma})^2\frac{T}{2}} \\ &\quad h(y_2 + y_1)f_{(Z^*_{T-t}, \int_0^{T-t} I\{Z_s^* \leq 0\}ds)}(x_1, y_1)dy_1dx_1 \\ &\quad f_{(Z_t, \int_0^t I\{Z_s \leq 0\}ds)}(x_2, y_2)dy_2dx_2 \end{aligned}$$

Thus, it is enough to derive a joint probability distribution function (or density function) of $(W_T, \tilde{R}(t))$ rather than that of $(W_T, R(t))$ in order to calculate the above expectation. The joint densities of the decomposed variables $[f_{(Z^*_{T-t}, \int_0^{T-t} I\{Z_s^* \leq 0\}ds)}(x_1, y_1), f_{(Z_t, \int_0^t I\{Z_s \leq 0\}ds)}(x_2, y_2)]$ can be obtained from Lemma 1.

3 Derivative Pricing

In the mathematical theory of derivative pricing, the conditional expectation is taken for the stock price S_u under the so-called risk neutral probability measure (or equivalent martingale measure). Under this risk neutral measure Q, the newly defined stochastic process W^* (shifted version of the original Brownian motion W under the original measure P) behaves as a Brownian motion. This is done by an application of Girsanov's theorem as discussed briefly below (see, for example, Baxter and Rennie (1996) for details.)

Recall that $S_t = S_0 e^{\mu t + \sigma W_t}$ or $dS_t = S_t((\mu + \frac{1}{2}\sigma^2)dt + \sigma dW_t)$. Given a constant interest rate r, define $\gamma = \frac{\mu + \frac{1}{2}\sigma^2 - r}{\sigma}$ and $W^* = W_t + \gamma t$. Then, we can write $S_t = S_0 e^{rt - rt + \mu t + \sigma W_t} = S_0 e^{rt - \frac{1}{2}\sigma^2 t + \sigma(W_t + \gamma t)} = S_0 e^{rt - \frac{1}{2}\sigma^2 t + \sigma W_t^*}$. Q is then defined by

$$\frac{dQ}{dP} = e^{-\gamma W_T - \frac{1}{2}\gamma^2 T}$$

and under this Q, W^* is a Brownian motion. Further, $\{S_t\}$ behaves as

$$dS_t = S_t(rdt + \sigma dW_t^*).$$

In the following sections, all the conditional expectations are done in a derivative pricing framework. Hence, we assume that the conditional expectation taken for S is under Q.

4 Corridor Swap

The fixed corridor $F(K)$ and the stochastic corridor $R(t)$ both can be used separately as payoff of derivatives. Their prices at time 0 are given respectively by

$$e^{-rT} E_0[F_{K,T}^{r,\sigma}], \qquad e^{-rT} E_0[R_{t,T}^{r,\sigma}]$$

in a Black-Scholes market.

We go further to define a "swap" or an exchange of the two derivatives which requires appropriate choice of the value of K. The payoff of the swap contract is $F(K) - R(t)$. This price at time of the contract is zero so that we have, as usual,

$$0 = e^{-rT} E_0[\int_0^T I\{S_u \le K\}du - \int_0^T I\{S_u \le S_t\}du]$$

Thus, the constant K has to be chosen to satisfy the equation

$$E_0[\int_0^T I\{S_u \le K\}du] = E_0[\int_0^T I\{S_u \le S_t\}du].$$

Note that righthand-side is a non-negative bounded constant less than T, and the lefthand-side is a strictly increasing continuous function of K ranging from zero to T. So there must exist a constant K which satisfies the above equality.

It is necessary to have an explicit functional form of these expectations in order to obtain the numerical value of K. They can be obtained by using the distributional results in Section 2.

5 Corridor Option

It is possible to define Put-type and Call-type options using the fixed and stochastic corridors. Their pricing can be done in a straightforward manner since it does not require any further distributional results.

We define a corridor call option on the stochastic corridor with the fixed level corridor as its exercise value. The pay-off of the corridor call option is $V_{C,T} = \max(R(t) - F(K),\ 0)$. Similarly, the pay-off of the corridor put option is $V_{P,T} = \max(F(K) - R(t),\ 0)$. The prices of these Call and Put at time zero in the Black-Scholes model are given by $V_{C,0} = e^{-rT}E_0[V_{C,T}]$ and $V_{P,0} = e^{-rT}E_0[V_{P,T}]$ respectively. The expectation for Call option can be calculated as follows.

Theorem 1.

$$V_{C,0} = e^{-rT}E_0[B^{r,\sigma}] = e^{-rT}E_0[e^{\frac{r}{\sigma}W_T - (\frac{r}{\sigma})^2\frac{T}{2}}B^{0,\sigma}]$$

where

$$\begin{aligned} B^{r,\sigma} &= B_1^{r,\sigma} + B_2^{r,\sigma} \\ &= B_{1,1}^{r,\sigma} + B_{1,2}^{r,\sigma} + B_{2,1}^{r,\sigma} + B_{2,2}^{r,\sigma} \end{aligned}$$

and

$$B^{0,\sigma} = B_{1,1}^{0,\sigma} + B_{1,2}^{0,\sigma} + B_{2,1}^{0,\sigma} + B_{2,2}^{0,\sigma},$$

where

$$\begin{aligned} B^{\mu,\sigma} &= \max\{\int_0^T I\{S_u \le S_t\}du - \int_0^T I\{S_u \le K\}du, 0\} \\ &= \int_0^T I\{S_u \le S_t\}du \cdot I\{S_t > K\} - \int_0^T I\{S_u \le K\}du \cdot I\{S_t > K\} \\ &\triangleq B_1^{\mu,\sigma} + B_2^{\mu,\sigma} \end{aligned}$$

$$= \int_t^T I\{S_u \le S_t\}du \cdot I\{S_t > K\} - \int_t^T I\{S_u \le K\}du \cdot I\{S_t > K\}$$
$$+ \int_0^t I\{S_u \le S_t\}du \cdot I\{S_t > K\} - \int_0^t I\{S_u \le K\}du \cdot I\{S_t > K\}$$
$$\triangleq B^{\mu,\sigma}_{1,1} + B^{\mu,\sigma}_{1,2} + B^{\mu,\sigma}_{2,1} + B^{\mu,\sigma}_{2,2}$$

The terms above are calculated in the following lemmas.

Lemma 2.

$$E_0[B^{r,\sigma}_1] = E_0[\{\int_t^T I\{S_u \le S_t\}du - \int_t^T I\{S_u \le K\}du\} \cdot I\{S_t > K\}]$$
$$= E_0[e^{\frac{r}{\sigma}Z_{T-t}-(\frac{r}{\sigma})^2\frac{T-t}{2}} \int_0^{T-t} I\{Z_v \le 0\}dv] \cdot E_0[I\{W_t > A^{r,\sigma}\}]$$
$$- \int_{\{S_0 e^{rt+\sigma w} > K\}} \{E_0[e^{\frac{r}{\sigma}Z_{T-t}-(\frac{r}{\sigma})^2\frac{T-t}{2}} \int_0^{T-t} I\{Z_v \le rt$$
$$+\sigma w + \log\frac{K}{S_0}\}dv\,|W_t = w]\}n(w:0,t)dw$$

Proof. Define $A^{r,\sigma} = \frac{1}{\sigma}(\log\frac{K}{S_0} - rt))$, and $Z_{u-t} = W_u - W_t$. Note that W_t and $W_u - W_t$ are independent.

$$B^{r,\sigma}_{1,1} = \int_t^T I\{S_u \le S_t\}du \cdot I\{S_t > K\}$$
$$= \int_t^T I\{r(u-t) + \sigma(W_u - W_t) \le 0\}du \cdot I\{W_t > A^{r,\sigma}\}$$
$$= \int_0^{T-t} I\{ru + \sigma Z_u \le 0\}du \cdot I\{W_t > A^{r,\sigma}\}$$
$$B^{r,\sigma}_{1,2} = \int_t^T I\{S_u \le K\}du\} \cdot I\{S_t > K\}$$
$$= \int_t^T I\{S_0 e^{ru+\sigma W_u} \le K\}du\} \cdot I\{S_0 e^{rt+\sigma W_t} > K\}$$
$$= \int_t^T I\{r(u-t) + \sigma(W_u - W_t) \le -rt - \sigma W_t + \log\frac{K}{S_0}\}du$$
$$\times\ I\{W_t > A^{r,\sigma}\}$$
$$= \int_0^{T-t} I\{rv + \sigma Z_v \le rt + \sigma W_t + \log\frac{K}{S_0}\}dv \cdot I\{W_t > A^{r,\sigma}\}.$$
(where $v = u - t$)

In the last step, note that $\log\left(\frac{K}{S_t}\right) = -rt - \sigma W_t + \log\frac{K}{S_0} < 0$ in $\{W_t > A^{r,\sigma}\} = \{S_t > K\}$. Now,

$$\begin{aligned}
E_0[B_1^{r,\sigma}] &= E_0[B_{1,1}^{r,\sigma}] + E_0[B_{1,2}^{r,\sigma}] \\
&= E_0[e^{\frac{r}{\sigma}Z_{T-t}-(\frac{r}{\sigma})^2\frac{T-t}{2}} \int_0^{T-t} I\{Z_v \le 0\}dv] \cdot E_0[I\{W_t > A^{r,\sigma}\}] \\
&\quad - \int_{\{S_0 e^{rt+\sigma w} > K\}} \{E_0[e^{\frac{r}{\sigma}Z_{T-t}-(\frac{r}{\sigma})^2\frac{T-t}{2}} \int_0^{T-t} I\{Z_v \le \frac{1}{\sigma}(rt \\
&\quad +\sigma w + \log\frac{K}{S_0})\}dv \,|W_t = w]\,\}n(w:0,t)dw \\
&= E_0[e^{\frac{r}{\sigma}Z_{T-t}-(\frac{r}{\sigma})^2\frac{T-t}{2}} \int_0^{T-t} I\{Z_v \le 0\}dv] \cdot E_0[I\{W_t > A^{r,\sigma}\}] \\
&\quad - \int_{\{S_0 e^{rt+\sigma w} > K\}} \{E_0[e^{\frac{r}{\sigma}Z_{T-t}-(\frac{r}{\sigma})^2\frac{T-t}{2}}(\tau \\
&\quad + \int_\tau^{T-t} I\{Z_v \le 0\}dv) \,|W_t = w]\,\}n(w:0,t)dw,
\end{aligned}$$

where $\tau = \inf\{v : Z_v \ge A^*, 0 \le v \le T\}$, $A^* = \frac{1}{\sigma}(rt + \sigma w + \log\frac{K}{S_0})$ and $n(w:0,t)$ is the density of the normal distribution with mean zero and variance t. □

Lemma 3.

$$\begin{aligned}
E_0[B_2^{r,\sigma}] &= E_0[\{\int_0^t I\{S_u \le S_t\}du - \int_0^t I\{S_u \le K\}du\} \cdot I\{S_t > K\}] \\
&= E_0[e^{\frac{r}{\sigma}Z_t-(\frac{r}{\sigma})^2\frac{t}{2}} \int_0^t I\{0 \le Z_v\}dv \cdot I\{S_0 e^{Z_t} > K\}] \\
&\quad - \int_0^t E_0[e^{\frac{r}{\sigma}(Z_{t-s}+\frac{1}{\sigma}\log\left(\frac{K}{S_0}\right))-(\frac{r}{\sigma})^2\frac{t}{2}}(s + \int_0^{t-s} I\{Z_v \le 0\}dv) \\
&\quad \times I\{Z_{t-s} > 0\}\ |\tau = s]g(s)ds
\end{aligned}$$

Proof. We rely on Cameron-Martin theorem to reduce the calculations for the case $\mu \ne 0$ to that for the case $\mu = 0$.

$$
\begin{aligned}
E_0[B^{r,\sigma}_{2,1}] &= E_0[\{\int_0^t I\{S_u \le S_t\}du \cdot I\{S_t > K\}] \\
&= E_0[\int_0^t I\{S_0 e^{ru+\sigma W_u} \le S_0 e^{rt+\sigma W_t}\}du \cdot I\{S_0 e^{rt+\sigma W_t} > K\}] \\
&= E_0[\int_0^t I\{0 \le rv + \sigma Z_v\}dv \cdot I\{S_0 e^{rt+\sigma Z_t} > K\}] \\
&\quad \text{(where } Z_v = W_t - W_{t-v}, \text{ and note that } Z_t = W_t) \\
&= E_0[e^{\frac{r}{\sigma}Z_t - (\frac{r}{\sigma})^2 \frac{t}{2}} \int_0^t I\{0 \le Z_v\}dv \cdot I\{S_0 e^{Z_t} > K\}] \\
&= \iint_{\log(\frac{K}{S_0})<x<\infty, 0<y<t} e^{\frac{r}{\sigma}x - (\frac{r}{\sigma})^2 \frac{t}{2}} y f_{(Z_t, \int_0^t I\{Z_s \le 0\}ds)}(x,y)dydx.
\end{aligned}
$$

$$
\begin{aligned}
E_0[B^{r,\sigma}_{2,2}] &= E_0[\{\int_0^t I\{S_u \le K\}du\} \cdot I\{S_t > K\}] \\
&= E_0[\{\int_0^t I\{\frac{r}{\sigma}u + W_u \le \frac{1}{\sigma}\log\left(\frac{K}{S_0}\right)\}du\} \cdot I\{\frac{r}{\sigma}t + W_t > \frac{1}{\sigma}\log(\frac{K}{S_0})\}] \\
&= E_0[e^{\frac{r}{\sigma}W_t - (\frac{r}{\sigma})^2 \frac{t}{2}} \int_0^t I\{W_u \le \frac{1}{\sigma}\log\left(\frac{K}{S_0}\right)\}du \cdot I\{W_t > \frac{1}{\sigma}\log(\frac{K}{S_0})\}] \\
&= E_0[e^{\frac{r}{\sigma}W_t - (\frac{r}{\sigma})^2 \frac{t}{2}} (\tau + \int_\tau^t I\{W_u - W_\tau \le 0\}du) \cdot I\{W_t > \frac{1}{\sigma}\log(\frac{K}{S_0})\}] \\
&\quad \text{(where } \tau = \inf\{u : W_u > \frac{1}{\sigma}\log(K/S_0), 0 < u < t\}.)
\end{aligned}
$$

$$
\begin{aligned}
&= E_0[e^{\frac{r}{\sigma}(Z_{t-\tau} + \frac{1}{\sigma}\log\left(\frac{K}{S_0}\right)) - (\frac{r}{\sigma})^2 \frac{t}{2}} (\tau + \int_0^{t-\tau} I\{Z_v \le 0\}dv) \cdot I\{Z_{t-\tau} > 0\}] \\
&= \int_0^t E_0[e^{\frac{r}{\sigma}(Z_{t-s} + \frac{1}{\sigma}\log\left(\frac{K}{S_0}\right)) - (\frac{r}{\sigma})^2 \frac{t}{2}} (s + \int_0^{t-s} I\{Z_v \le 0\}dv) \\
&\quad \times I\{Z_{t-s} > 0\} \,|\tau = s]g(s)ds \\
&\quad \text{(where } Z_v = W_{\tau+v} - W_\tau, g(\cdot) \text{ is the probability density function of } \tau. \\
&\quad \text{Note } Z_{t-\tau} = W_t - W_\tau = W_t - \frac{1}{\sigma}\log\left(\frac{K}{S_0}\right)) \\
&= \int_{0<s<t} \iint_{0<x<\infty, 0<t-s} e^{\frac{r}{\sigma}(x + \frac{1}{\sigma}\log\left(\frac{K}{S_0}\right)) - (\frac{r}{\sigma})^2 \frac{t}{2}} (s+y) \\
&\quad \times f_{(Z_{t-s}, \int_0^{t-s} I\{Z_v < 0\}dv)(x,y)} g(s)dxdyds
\end{aligned}
$$

□

Put-Call Parity

For any random variables X and Y, we have an equality; $\max\{X-Y,0\}+Y = X+\max\{Y-X,0\}$. Applying this relation to our Call and Put regarding X and Y as our stochastic corridor $R(t)$ and fixed level corridor $F(K)$,

$$\max\left[\int_0^T I\{S_u \le S_t\}du - \int_0^T I\{S_u \le K\}du,\ 0\right] + \int_0^T I\{S_u \le K\}du$$
$$= \int_0^T I\{S_u \le S_t\}du + \max\left[\int_0^T I\{S_u \le K\}du - \int_0^T I\{S_u \le S_t\}du,\ 0\right]$$

Since the pay-off of the left hand side and the right hand side of the equation coincide, the prices at time zero of the derivatives corresponding to each side must be equal under the assumption that the market does not allow any arbitrage. Hence, using the linearity of expectation, we have (the price of corridor call)+(price of fixed corridor) =(the price of stochastic corridor)+(price of corridor put)

6 Forward Starting Corridor

Let $[T_0, T_1]$, $0 < T_0 < T_1$, be a future time interval where a corridor option counts the amount of time that the stock prices stay below a level, either fixed or stochastic. Now, the payoffs of forward starting fixed corridor and stochastic corridor are respectively,

$$F(K,(T_0,T_1)) = \int_{T_0}^{T_1} I\{S_u \le K\}du$$

and

$$R(t,(T_0,T_1)) = \int_{T_0}^{T_1} I\{S_u \le S_t\}du.$$

A contract is made at time 0 and the payoff is paid to the holder at time T_1. Then the prices of these options in the Black-Scholes model are

$$e^{-rT_1}E_0[\textstyle\int_{T_0}^{T_1} I\{S_u \le K\}du]$$

$$e^{-rT_1}E_0[\textstyle\int_{T_0}^{T_1} I\{S_u \le S_t\}du]$$

respectively.

As we saw in the previous section that the probability distribution of the stochastic corridor is independent of the value S_{T_0}, the value of the initial stock price in the future time interval $[T_0, T_1]$. This independence property may be expected to be useful in practice when they set a level for the fixed corridor. In order to decide a constant level K, it may be required in practice to have a certain idea or a prediction of overall level of stock prices during the future time interval $[T_0, T_1]$. Since it is not easy to make a prediction, it may be plausible sometimes to depend on a stochastic value to determine an overall level, for example S_{T_0}. Or there might be a special time point t during the future time interval $[T_0, T_1]$ that is suitable for making S_t the stochastic level for the stochastic corridor.

If one wants to compensate the result from the ambiguity of a suitable value of K with the difference between the two forward starting corridors, one can swap to exchange the forward starting fixed corridor with the forward starting stochastic corridor. The payoff of this swap is

$$\int_{T_0}^{T_1} I\{S_u \leq S_t\}du - \int_{T_0}^{T_1} I\{S_u \leq K\}du$$

or

$$\int_{T_0}^{T_1} I\{S_u \leq K\}du - \int_{T_0}^{T_1} I\{S_u \leq S_t\}du.$$

As in Section 3, we need to be able to determine a proper theoretical value of K which makes the price of the swap contract be zero at the time of the contract, i.e. at time 0. That is, K has to satisfy the equation

$$0 = e^{-rT_1} E_0[\int_{T_0}^{T_1} I\{S_u \leq K\}du - \int_{T_0}^{T_1} I\{S_u \leq S_t\}du]$$

In other words,

$$E_0[\int_{T_0}^{T_1} I\{S_u \leq K\}du] = E_0[\int_{T_0}^{T_1} I\{S_u \leq S_t\}du].$$

The above expectations are the conditional expectations taken under the condition that the value of S_0 is given. The existence of such a constant K is assured using the same argument as in the previous section.

The probability distribution of $\int_{T_0}^{T_1} I\{S_u \leq S_t\}du$ is S_{T_0}- independent and is the same as that of $\int_0^{T_1-T_0} I\{S_u \leq S_t\}du$. (See Fujita and Miura(2004)). However, the calculation for $E_0[\int_{T_0}^{T_1} I\{S_u \leq K\}du]$ requires some additional comments.

$$\begin{aligned} E_0[\int_{T_0}^{T_1} I\{S_u \leq K\}du\,|S_0] &= E_0[E_{T_0}[\int_{T_0}^{T_1} I\{S_u \leq K\}du\,|S_{T_0}]\,|S_0] \\ &= E_0[E_{T_0}[\int_{T_0}^{T_1} I\{e^{X_u - X_{T_0}} \leq \frac{K}{S_{T_0}}\}du\,|S_{T_0}]\,|S_0]. \end{aligned}$$

Since for any u in $[T_0, T_1]$, $(X_u - X_{T_0})$ and X_{T_0} or equivalently S_{T_0} are stochastically independent of each other, the expectation inside can be calculated with any given value of S_{T_0} and the result integrated with respect to the density function of S_{T_0}. So this does not involve a joint distribution function.

A call option with a payoff

$$\max\left[\int_{T_0}^{T_1} I\{S_u \leq S_t\}du - \int_{T_0}^{T_1} I\{S_u \leq K\}du, \quad 0\right],$$

is possible. Its price can be calculated in a similar way to that for the option for the spot starting corridor.

References

1. AKAHORI, J. (1995). "Some formulae for a new type of path-dependent option." Ann. Appl. Probab. 5. 383-388.
2. BAXTER, M. AND A. RENNIE. (1996) Financial Calculus; An Introduction to Derivative Pricing. Cambridge University Press.
3. BORODIN, A. N. AND P. SALMINEN. (2002). Handbook of Brownian Motion - Facts and Formulae, 2nd. edition p.256. (the first edition was published in1996). Birkhauser.
4. DASSIOS, A. (1995). "The distribution of the quantile of a Brownian motion with drift and the pricing of related path-dependent options." Ann. Appl. Probab. 5. 389-398.
5. EMBRECHT, P., ROGERS, L. C. G. AND YOR, M. (1995). "A proof of Dassios's representation of the α-quantile of Brownian motion with drift." Annals of Applied Probability. 5,757-767.
6. FUJITA, T. (1997). "On the price of α-percentile options." Working Paper Series #24, Faculty of Commerce Hitotsubashi University.
7. FUJITA, T. (2000). "A note on the joint distribution ofα,βpercentiles and its applications to the option pricing." Asia-Pacific Financial Markets. Vol.7(4), 339-344.
8. FUJITA, T. AND MIURA, R.(2002). "Edokko Options: A New Framework of Barrier Options." Asia-Pacific Financial Markets. Vol.9(2),December,141-151.
9. FUJITA, T. AND MIURA, R. (2006). " The distribution of Continuous Time Rank Process." Advances in Mathematical Economics,Vol.9,25-32. Springer Verlag.
10. GOLDMAN M. B., HOWARD B. S. AND GATTO, M. A. (1979) "Path Dependent Options ; Buy at the low, Sell at the high." Journal of Finance, Vol.34, pp.1111-1127.

11. Miura, R. (1992). " A note on look-back option based on order statistics." Hitotsubashi Journal of Commerce and Management. 27,15-28.

12. Miura, R. (2005). "Rank Process and Stochastic Corridor: Nonparametric Statistics of Lognormal Observations and Exotic Derivatives based on them." Working Paper #FS-2005-E-01, Graduate School of International Corporate Strategy, Hitotsubashi University, Tokyo, Japan.

13. Shreve, S. E. (2004). Stochastic Calculus for Finance II; Continuous-Time Models. Springer.

14. Yor, M. (1995)."The distribution of Brownian quantiles." Journal of Applied Probability. 2, 405-416.

PART 9

Monte Carlo and Resampling Methods

Chapter 27

CONDITIONAL MONTE CARLO BASED ON SUFFICIENT STATISTICS WITH APPLICATIONS

Bo Henry Lindqvist and Gunnar Taraldsen

Department of Mathematical Sciences
Norwegian University of Science and Technology, Trondheim, NORWAY

SINTEF Information and Communication Technology, Trondheim, NORWAY

E-mails: bo@math.ntnu.no & gunnar.taraldsen@sintef.no

We review and complement a general approach for Monte Carlo computations of conditional expectations given a sufficient statistic. The problem of direct sampling from the conditional distribution is considered in particular. This can be done by a simple parameter adjustment of the original statistical model if certain conditions are satisfied, but in general one needs to use a weighted sampling scheme. Several examples are given in order to demonstrate how the general method can be used under different distributions and observation plans. In particular we consider cases with, respectively, truncated and type I censored samples from the exponential distribution, and also conditional sampling for the inverse Gaussian distribution. Some new theoretical results are presented.

Key words: Sufficiency; Conditional distribution; Monte Carlo simulation; Pivotal statistic; Truncated exponential distribution; Type I censoring; Inverse Gaussian distribution.

1 Introduction

We consider a pair (X, T) of random vectors with joint distribution indexed by a parameter vector θ. Throughout the paper we assume that T is sufficient for θ compared to X, meaning that the conditional distribution of X given $T = t$ can be specified independent of θ [Bickel and Doksum (2001), Ch. 1.5, Lehmann and Casella (1998), Ch. 1.6]. Statistical inference is often concerned with conditional expectations of the form $E\{\phi(X)|T = t\}$, which will hence not depend on the value of θ. Applications include construction of optimal estimators, nuisance parameter elimination and goodness-of-fit testing.

Only in exceptional cases is one able to compute $E\{\phi(X)|T = t\}$ analytically. Typically this is not possible, thus leading to the need for approximations or simulation algorithms. Apparently because of the computational difficulties involved, methods based on conditional distributions given sufficient statistics are often not exploited in statistical applications. In fact, the literature is scarce even for the normal and multinormal distributions. Cheng (1984) used a result for Gamma-distributions to simulate conditional normal samples with given sample mean and sample variance, and then showed how to modify the idea to sample conditionally given the sufficient statistic for the inverse Gaussian distribution. Subsequently he extended the idea from his 1984 paper to derive a corresponding algorithm for the multivariate normal case [Cheng (1985)]. A related approach based on random rotations was recently suggested by Langsrud (2005). Lindqvist and Taraldsen (2005) derived a method for the multinormal distribution based on a parametrization via Cholesky-decompositions. Diaconis and Sturmfels (1998) derived algorithms for sampling from discrete exponential families conditional on a sufficient statistic.

Engen and Lillegaard (1997) considered the general problem of Monte Carlo computation of conditional expectations given a sufficient statistic. Their ideas were further developed and generalized in Lindqvist and Taraldsen (2005) [in the following referred to as LT (2005)] and in the technical report Lindqvist and Taraldsen (2001) where a more detailed measure theoretic approach was employed.

The present paper reviews basic ideas and results from LT (2005). The main purpose is to complement LT (2005) regarding computational aspects, examples and theoretical results. In particular we consider some new examples from lifetime data analysis with connections to work by Kjell Doksum [Bickel and Doksum (1969), exponential distributions; Doksum and Høyland (1992), inverse Gaussian distributions].

2 Setup and basic algorithm

Following LT (2005) we assume that there is given a random vector U with a known distribution, such that (X, T) for given θ can be simulated by means of U. More precisely we assume the existence of functions χ and τ such that, for each θ, the joint distribution of $(\chi(U, \theta), \tau(U, \theta))$ equals the joint distribution of (X, T) under θ. Let in the following $f(u)$ be the probability density of U.

Example 1. *Exponential distribution.* Suppose $X = (X_1, \ldots, X_n)$ are i.i.d. from the exponential distribution with hazard rate θ, denoted Exp(θ). Then $T = \sum_{i=1}^n X_i$ is sufficient for θ. Letting $U = (U_1, \ldots, U_n)$ be i.i.d. Exp(1) variables we can put

$$\chi(U, \theta) = (U_1/\theta, \ldots, U_n/\theta),$$
$$\tau(U, \theta) = \sum_{i=1}^n U_i/\theta.$$

Consider again the general case and suppose that a sample from the conditional distribution of X given $T = t$ is wanted. Since the conditional distribution by sufficiency does not depend on θ, it is reasonable to believe that it can be described in some simple way in terms of the distribution of U, and thus enabling Monte Carlo simulation based on U. A suggestive method for this would be to first draw U from its known distribution, then to determine a parameter value $\hat{\theta}$ such that $\tau(U, \hat{\theta}) = t$ and finally to use $X_t(U) = \chi(U, \hat{\theta})$ as the desired sample. In this way we indeed get a sample of X with the corresponding T having the correct value t. The question remains, however, whether or not $X_t(U)$ is a sample from the conditional distribution of X given $T = t$.

Example 1 (continued). For given t and U there is a unique $\hat{\theta} \equiv \hat{\theta}(U, t)$ with $\tau(U, \hat{\theta}) = t$, namely

$$\hat{\theta}(U, t) = \frac{\sum_{i=1}^n U_i}{t}.$$

This leads to the sample

$$X_t(U) = \chi\{U, \hat{\theta}(U, t)\} = \left(\frac{tU_1}{\sum_{i=1}^n U_i}, \ldots, \frac{tU_n}{\sum_{i=1}^n U_i} \right),$$

and it is well known [Aitchison (1963)] that the distribution of $X_t(U)$ indeed coincides with the conditional distribution of X given $T = t$.

The algorithm used in Example 1 can more generally be described as follows:

Algorithm 1. *Conditional sampling of X given $T = t$.*

(1) Generate U from the density $f(u)$.
(2) Solve $\tau(U, \theta) = t$ for θ. The (unique) solution is $\hat{\theta}(U, t)$.
(3) Return $X_t(U) = \chi\{U, \hat{\theta}(U, t)\}$.

The following so called pivotal condition, discussed and verified in LT (2005), ensures that Algorithm 1 produces a sample $X_t(U)$ from the conditional distribution of X given $T = t$. Note that uniqueness of $\hat{\theta}(U, t)$ in Step 2 is required.

The pivotal condition. Assume that $\tau(u, \theta)$ depends on u only through a function $r(u)$, where the value of $r(u)$ can be uniquely recovered from the equation $\tau(u, \theta) = t$ for given θ and t. This means that there is a function $\tilde{\tau}$ such that $\tau(u, \theta) = \tilde{\tau}\{r(u), \theta\}$ for all (u, θ), and a function $\tilde{v}$ such that $\tilde{\tau}\{r(u), \theta\} = t$ implies $r(u) = \tilde{v}(\theta, t)$. Note that in this case $\tilde{v}(\theta, T)$ is a pivotal quantity in the classical meaning that its distribution does not depend on θ.

Example 1 (continued). The pivotal condition is satisfied here with $r(U) = \sum_{i=1}^{n} U_i$. Thus Algorithm 1 is valid, as verified earlier by a direct method.

3 General algorithm for unique $\hat{\theta}(u, t)$

Algorithm 1 will in general not produce samples from the correct conditional distribution, even if the solution $\hat{\theta}(u, t)$ of $\tau(u, \theta) = t$ is unique. This was demonstrated by a counterexample in Lindqvist, Taraldsen, Lillegaard and Engen (2003). A modified algorithm can, however, be constructed. The main idea [LT (2005)] is to consider the parameter θ as a random variable Θ, independent of U, and with some conveniently chosen distribution π. Such an approach is similar to the one of Trotter and Tukey (1956), and this idea is also inherent in the approach of Engen and Lillegaard (1997).

The key result is that the conditional distribution of X given $T = t$ equals the conditional distribution of $\chi(U, \Theta)$ given $\tau(U, \Theta) = t$. This is intuitively obvious from the definition of sufficiency, which implies that this holds when Θ is replaced by any fixed value θ. Note, however, that independence of U and Θ is needed for this to hold. It follows that conditional expectations $E\{\phi(X)|T = t\}$ can be computed from the formula

$$E\{\phi(X)|T = t\} = E[\phi\{\chi(U, \Theta)\}|\tau(U, \Theta) = t]. \tag{1}$$

Assume in the rest of the section that the equation $\tau(u, \theta) = t$ has the unique solution $\hat{\theta}(u, t)$ for θ. Then $\theta = \hat{\theta}\{u, \tau(u, \theta)\}$ is an identity in θ

and u, and this fact together with (1) imply that

$$\begin{aligned} E\{\phi(X)|T=t\} &= E[\phi\{\chi(U,\Theta)\}|\tau(U,\Theta)=t] \\ &= E[\phi\{\chi(U,\hat{\theta}\{U,\tau(U,\Theta)\})\}|\tau(U,\Theta)=t] \\ &= E[\phi\{\chi(U,\hat{\theta}(U,t)\}|\tau(U,\Theta)=t]. \end{aligned}$$

Thus we need only the conditional distribution of U given $\tau(U,\Theta)=t$. Assuming this is given by a density $f(u|t)$, Bayes' formula implies that $f(u|t) \propto f(t|u)f(u)$, where $f(t|u)$ is the conditional density of $\tau(U,\Theta)$ given $U=u$ and $f(u)$ is the density of U. Now since U and Θ are independent, $f(t|u)$ is simply the density of $\tau(u,\Theta)$ which we in the following denote by $W_t(u)$. It should be stressed that $W_t(u)$ is the density of $\tau(u,\Theta)$ as a function of t, for each fixed u, while in the following it will usually be considered as a function of u. From this we get

$$E\{\phi(X)|T=t\} = \frac{E[\phi\{X_t(U)\}W_t(U)]}{E\{W_t(U)\}}, \tag{2}$$

where the denominator $E\{W_t(U)\} = \int W_t(u)f(u)du$ is merely the normalization of the conditional density $f(u|t)$. The formula shows that $W_t(u)$ acts as a weight function for a sample u from $f(u)$.

It follows from (2) that sampling from the conditional distribution in principle can be done by the following scheme:

Algorithm 2. *Weighted conditional sampling of X given $T=t$.*
Let Θ be a random variable and let $t \mapsto W_t(u)$ be the density of $\tau(u,\Theta)$.

(1) Generate V from a density proportional to $W_t(u)f(u)$.
(2) Solve $\tau(V,\theta)=t$ for θ. The (unique) solution is $\hat{\theta}(V,t)$.
(3) Return $X_t(V) = \chi\{V,\hat{\theta}(V,t)\}$.

The weight function $W_t(u)$ in the Euclidean case

Suppose that the vector X has a distribution depending on a k-dimensional parameter θ and that $T(X)$ is a k-dimensional sufficient statistic. Choose a density $\pi(\theta)$ for Θ and let $W_t(u)$ be the density of $\tau(u,\Theta)$. Since $\tau(u,\theta)=t$ if and only if $\theta=\hat{\theta}(u,t)$ it follows under standard assumptions that

$$W_t(u) = \pi\{\hat{\theta}(u,t)\}|\det\partial_t\hat{\theta}(u,t)| = |\frac{\pi(\theta)}{\det\partial_\theta\tau(u,\theta)}|_{\theta=\hat{\theta}(u,t)}. \tag{3}$$

The formula (2) can thus be written

$$E\{\phi(X)|T=t\} = \frac{\int \phi[\chi\{u,\hat{\theta}(u,t)\}]|\frac{\pi(\theta)}{\det\partial_\theta\tau(u,\theta)}|_{\theta=\hat{\theta}(u,t)}f(u)du}{\int |\frac{\pi(\theta)}{\det\partial_\theta\tau(u,\theta)}|_{\theta=\hat{\theta}(u,t)}f(u)du}, \tag{4}$$

and can be computed by simulation using a pseudo-sample from the distribution of U as will be explained in Section 6.1.

Example 2. *Truncated exponential lifetimes.* Let $X = (X_1, \ldots, X_n)$ be a sample from the exponential distribution with hazard rate θ, but assume now that X_i is an observation truncated at τ_i $(i = 1, \ldots, n)$, where the $\tau_i > 0$ are known numbers. This means that the distribution function of X_i is

$$F_i(x_i, \theta) = \frac{1 - e^{-\theta x_i}}{1 - e^{-\theta \tau_i}}, \; 0 \le x_i \le \tau_i, \; i = 1, \ldots, n. \tag{5}$$

As for the non-truncated exponential case in Example 1, the statistic $T = \sum_{i=1}^n X_i$ is sufficient for θ. Suppose we wish to consider the conditional distribution of X given $T = t$. It turns out to be convenient to extend the parameter set to allow θ to be any real number. Indeed, F_i defined in (5) is a c.d.f. for all real θ if we define $F_i(x_i, 0) = x_i/\tau_i$, $0 \le x_i \le \tau_i$, obtained by taking the limit as $\theta \to 0$ in (5).

Now a sample X can be simulated by ordinary inversion based on (5) using a sample $U = (U_1, U_2, \ldots, U_n)$ from the standard uniform distribution, denoted $\text{Un}[0, 1]$. This gives
$\chi(U, \theta) = (\eta_1(U_1, \theta), \ldots, \eta_n(U_n, \theta))$, $\tau(U, \theta) = \sum_{i=1}^n \eta_i(U_i, \theta)$ where

$$\eta_i(u_i, \theta) = \begin{cases} -\log\{1 - (1 - e^{-\theta\tau_i})u_i\}/\theta & \text{if } \theta \neq 0 \\ \tau_i u_i & \text{if } \theta = 0 \end{cases}.$$

The function $\eta_i(u_i, \theta)$ is strictly decreasing in θ, which follows since $F_i(x_i, \theta)$ is strictly increasing in θ. Consequently the solution $\hat{\theta}(u, t)$ of $\tau(u, \theta) = t$ is unique.

It turns out that the pivotal condition of Section 2 is not satisfied in the present case. Indeed, Lindqvist et al. (2003) studied the case $n = 2$ and found that Algorithm 1 does not produce the correct distribution. Thus we use instead Algorithm 2 and (4), for which we need to compute $|\partial_\theta \tau(u, \theta)|_{\theta = \hat{\theta}(u,t)}$. We obtain

$$|\partial_\theta \tau(u, \theta)|_{\theta = \hat{\theta}(u,t)} = \frac{1}{\hat{\theta}(u, t)} \left(t - \sum_{i=1}^n \frac{\tau_i u_i e^{-\hat{\theta}(u,t)\tau_i}}{1 - (1 - e^{-\hat{\theta}(u,t)\tau_i})u_i} \right).$$

In principle we can then use (4) with any choice of the density $\pi(\theta)$ for which the integrals exist. The simple choice of $\pi(\theta) = 1/|\theta|$ turns out to work well in this example and is in accordance with the discussion in Section 6.3 regarding the use of noninformative priors.

We close the example by noting that since $\theta = 0$ corresponds to the X_i being uniform, the target conditional distribution is that of n independent $\text{Un}[0, \tau_i]$ random variables given their sum. There seems to be no simple expression for this distribution, not even when the τ_i are equal.

4 The general case

Recall the basic idea described in Section 3 that conditional expectations $E\{\phi(X)|T = t\}$ can be computed from the formula (1) where we have introduced the random parameter Θ. In the general case, where there may not be a unique solution of $\tau(u, \theta) = t$, we compute (1) by conditioning on U in addition to $\tau(U, \Theta) = t$. This leads to the most general result of LT (2005) which states that

$$E\{\phi(X)|T = t\} = \frac{\int Z_t(u)W_t(u)f(u)du}{\int W_t(u)f(u)du}, \tag{6}$$

where $Z_t(u)$ is the conditional expectation of $\phi\{\chi(u, \Theta)\}$ given $\tau(u, \Theta) = t$ for fixed u, $W_t(u)$ is the density of the variable $\tau(u, \Theta)$ at t, for fixed u, and $f(u)$ is the density of U.

Thus our method essentially amounts to changing computations of conditional expectations of $\phi\{\chi(U, \theta)\}$ given $\tau(U, \theta) = t$ for fixed θ into the often much simpler problem of computing conditional expectations of $\phi\{\chi(u, \Theta)\}$ given $\tau(u, \Theta) = t$ for fixed u. Note the freedom to choose a suitable distribution π for Θ.

The formula (6) implies the following principal scheme for simulation of X given $T = t$.

Algorithm 3. *General weighted conditional sampling of X given $T = t$.* Let Θ be a random variable and let $t \mapsto W_t(u)$ be the density of $\tau(u, \Theta)$.

(1) Generate V from a density proportional to $W_t(u)f(u)$ and let the result be $V = v$.
(2) Generate Θ_t from the conditional distribution of Θ given $\tau(v, \Theta) = t$.
(3) Return $X_t(V) = \chi(V, \Theta_t)$.

4.1 *The general Euclidean case*

As in Section 3, suppose that the vector X has a distribution depending on a k-dimensional parameter θ and that $T(X)$ is a k-dimensional sufficient statistic. In this case, the equation $\tau(u, \theta) = t$ will typically have a finite number of solutions, where this number may vary as u varies. Define

$$\Gamma(u, t) = \{\hat{\theta} : \tau(u, \hat{\theta}) = t\}$$

and note that the density $t \mapsto W_t(u)$ of $\tau(u, \Theta)$ is now given by

$$W_t(u) = \sum_{\hat{\theta} \in \Gamma(u,t)} \frac{\pi(\hat{\theta})}{|\det \partial_\theta \tau(u, \theta)|_{\theta = \hat{\theta}}}. \tag{7}$$

Furthermore, the conditional distribution of Θ given $\tau(u,\Theta)=t$ is concentrated on $\Gamma(u,t)$ and is given by

$$Pr\{\Theta=\hat{\theta}\mid\tau(u,\Theta)=t\}=\frac{\pi(\hat{\theta})}{|\det\partial_\theta\tau(u,\theta)|_{\theta=\hat{\theta}}W_t(u)},\quad \hat{\theta}\in\Gamma(u,t). \tag{8}$$

The following formula generalizes the result (4):

$$\mathrm{E}\{\phi(X)|T=t\}=\frac{\int\sum_{\hat{\theta}\in\Gamma(u,t)}\phi(\chi(u,\hat{\theta}))\frac{\pi(\hat{\theta})}{|\det\partial_\theta\tau(u,\theta)|_{\theta=\hat{\theta}}}f(u)du}{\int\sum_{\hat{\theta}\in\Gamma(u,t)}\frac{\pi(\hat{\theta})}{|\det\partial_\theta\tau(u,\theta)|_{\theta=\hat{\theta}}}f(u)du}. \tag{9}$$

We note that the treatment of multiple roots of the equation $\tau(u,\theta)=t$ in the present context is similar to the treatment in Michael et al. (1976) in connection with generation of random variates from transformations with multiple roots. Formulas (7) and (8) can in fact together be considered as a multivariate generalization of equation 3 in Michael et al. (1976) [see also Taraldsen and Lindqvist (2005)].

The following two examples illustrate the use of Algorithm 3 and equation (9). In the first example $\Gamma(u,t)$ contains at most one value of θ, but may be empty. In the second example we may have an arbitrary number of elements in $\Gamma(u,t)$.

Example 3. *Type I censored exponential lifetimes.* Let n units with potential lifetimes $Y_1,Y_2,\ldots,Y_n$ be observed from time 0, but assume that the observation of the ith unit is censored at a given time $c_i>0$ $(i=1,\ldots,n)$. This means that we observe only $X_i=\min(Y_i,c_i)$. In the reliability terminology this is called Type I censoring. Suppose $Y_1,\ldots,Y_n$ are i.i.d. with distribution Exp(θ). Then the likelihood of $X_1,\ldots,X_n$ can be written $\theta^R\exp(-\theta S)$ where $R=\sum_i I(X_i<c_i)$ is the number of noncensored observations and $S=\sum_i X_i$ is the sum of all observations. Here $I(A)$ is the indicator function of the event A. Now $T=(R,S)$ is sufficient for θ, but note that a two-dimensional statistic is here sufficient for a one-dimensional parameter.

It should be remarked that the potential censoring times c_i are assumed known also for the units where $X_i<c_i$. For example this is the case if n machines, or patients in a medical study, are observed from possibly different starting points in time, and until a common terminal point. Let $c_1,\ldots,c_n$ be fixed, known numbers in the following.

As in Example 1, let $U=(U_1,\ldots,U_n)$ be a vector of n i.i.d. Exp(1) variables. We then simulate X for a given value of θ by means of $\chi(U,\theta)=(\eta_1(U_1,\theta),\ldots,\eta_n(U_n,\theta))$ where

$$\eta_i(u_i,\theta)=\min(u_i/\theta,c_i),\ i=1,\ldots,n.$$

Thus $T = (R,S)$ is simulated by $\tau(U,\theta) = (\gamma(U,\theta),\psi(U,\theta))$ where $\gamma(U,\theta) = \sum_i I(U_i/\theta < c_i)$ and $\psi(U,\theta) = \sum_i \eta_i(U_i,\theta)$.

We now show how to find the functions $W_t(u)$ and $Z_t(u)$ needed in (6). First we show that the equation $\tau(u,\theta) = t$ has at most one solution for θ for fixed u, t, but may have none. Let the observed value of the sufficient statistic, $t = (r, s)$, be fixed with $0 < r \leq n$, $0 < s < \sum_i c_i$. Then consider the equations $\gamma(u,\theta) = r$, $\psi(u,\theta) = s$ for a given u. Since $\psi(u,\theta)$ is strictly decreasing in θ, from $\sum_i c_i$ to 0, there is a unique $\hat{\theta}$ which satisfies $\psi(u,\hat{\theta}) = s$. However, this $\hat{\theta}$ may not solve $\gamma(u,\theta) = r$. In the cases where indeed $\gamma(u,\hat{\theta}) = r$, put $K(u,t) = 1$ and put $K(u,t) = 0$ otherwise. If $K(u,t) = 1$ then define $I(u,t) = \{i_1,\ldots,i_r\}$ to be the set of indices i for which $u_i/\hat{\theta} < c_i$. With this notation we can express the solution $\hat{\theta}$ when $K(u,t) = 1$ as

$$\hat{\theta}(u,t) = \frac{\sum_{i\in I(u,t)} u_i}{s - \sum_{i\notin I(u,t)} c_i}.$$

Next, choose a density $\pi(\theta)$ for $\theta > 0$, for example $\pi(\theta) = 1/\theta$ in accordance with Example 2. We then find the density $W_t(u) \equiv W_{(r,s)}(u)$ of $\tau(u,\Theta)$ to be

$$\begin{aligned} W_t(u)\,ds &= \pi\{\theta : \gamma(u,\theta) = r, s \leq \psi(u,\theta) \leq s + ds\} \\ &= \begin{cases} 0 & \text{if } K(u,t) = 0 \\ \hat{\theta}(u,t)^2\pi(\hat{\theta}(u,t))ds/\sum_{i\in I(u,t)} u_i & \text{if } K(u,t) = 1. \end{cases} \end{aligned}$$

Further, $Z_t(u)$ is the conditional expectation of $\phi\{\chi(u,\Theta)\}$ given $\tau(u,\Theta) = t$. This is easily found since the conditional distribution of Θ given $\tau(u,\Theta) = t$ is a one-point mass at $\hat{\theta}(u,t)$ if $K(u,t) = 1$ and can be arbitrarily chosen otherwise. Formula (9) therefore gives

$$E\{\phi(X)|T = t\} = \frac{E\{K(U,t)\phi[\chi\{U,\hat{\theta}(U,t)\}]W_t(U)\}}{E\{K(U,t)W_t(U)\}}.$$

The choice $\pi(\theta) = 1/\theta$ yields the simple weight function

$$W_t(u) = (s - \sum_{i\notin I(u,t)} c_i)^{-1}, \tag{10}$$

valid when $K(u,t) = 1$.

An important special case is when the c_i are all equal. In this case $W_t(u)$ in (10) does not depend on u and we can sample directly from the conditional distribution of X given $T = t$ for fixed t by drawing u until $K(u,t) = 1$ and then using $X_t(u) = \chi\{u,\hat{\theta}(u,t)\}$.

Example 4. *Inverse Gaussian distributed lifetimes.* Let $X = (X_1, \ldots, X_n)$ be a sample from the inverse Gaussian distribution with density

$$f(x; \mu, \phi) = \sqrt{\frac{\mu\phi}{2\pi x^3}} \exp\left(-\frac{\mu\phi}{2x} - \frac{\phi x}{2\mu} + \phi\right), \quad x > 0 \tag{11}$$

[Seshadri (1999), p. 2] where $\mu, \phi > 0$ are parameters. Denote this distribution by $\mathrm{IG}(\mu, \phi)$. Note that a more common parametrization uses μ together with $\lambda = \mu\phi$, but the one used in (11) is more convenient for our purposes as will become clear below. Doksum and Høyland (1992) considered models for accelerated life testing experiments which were based on the inverse Gaussian distribution. In the present example we shall consider conditional sampling given the sufficient statistic, which may have several interesting applications in this connection.

A sufficient statistic is given by [Seshadri (1999), p. 7]

$$T = (T_1, T_2) = \left(\sum_{i=1}^{n} X_i, \sum_{i=1}^{n} 1/X_i\right).$$

Since μ is a scale parameter in (11) we can simulate from $\mathrm{IG}(\mu, \phi)$ by first simulating from $\mathrm{IG}(1, \phi)$ and then multiplying the result by μ. We shall use the method suggested by Michael et al. (1976) which seems to be easier than ordinary inversion since there is no closed form expression for the inverse cumulative distribution function.

Let U_i be $\mathrm{Un}[0,1]$ and V_i be χ_1^2 for $i = 1, \ldots, n$, where all variables are independent. Here χ_1^2 means the chi-square distribution with 1 degree of freedom. Let

$$\begin{aligned} W_i &= 1 - (2\phi)^{-1}\left(\sqrt{V_i^2 + 4\phi V_i} - V_i\right), \\ Z_i &= (1 + W_i)^{-1}. \end{aligned}$$

Then [Michael et al. (1976)] the variables

$$\eta(U_i, V_i, \phi) = I(U_i \le Z_i)\, W_i + I(U_i > Z_i)\, (1/W_i)$$

are distributed as $\mathrm{IG}(1, \phi)$ and hence

$$\chi(U, V, \mu, \phi) = (\mu\eta(U_1, V_1, \phi), \ldots, \mu\eta(U_n, V_n, \phi))$$

is a simulated sample of size n from $\mathrm{IG}(\mu, \phi)$. Here $U = (U_1, \ldots, U_n)$, $V = (V_1, \ldots, V_n)$. Moreover, we simulate $T = (T_1, T_2)$ by

$$\begin{aligned} \tau(U, V, \mu, \phi) &= (\tau_1(U, V, \mu, \phi), \tau_2(U, V, \mu, \phi)) \\ &= \left(\sum_{i=1}^{n} \mu\eta(U_i, V_i, \phi), \sum_{i=1}^{n} (1/\mu)(1/\eta(U_i, V_i, \phi))\right). \end{aligned}$$

In order to compute conditional expectations or to sample from the conditional distribution of X given $T = t$ we need to solve the equations $\tau(u, v, \mu, \phi) = t = (t_1, t_2)$ with respect to μ and ϕ. This can be done by first solving the equation

$$\tau_1(u, v, \mu, \phi) \cdot \tau_2(u, v, \mu, \phi) = t_1 t_2, \tag{12}$$

which is free of μ. It turns out that the solution for ϕ is not necessarily unique. In fact, the number of roots is finite but may vary with (u, v). However, for each root found for ϕ we can easily solve for μ using $\tau_1(u, v, \mu, \phi) = t_1$. It should be noted that the functions $\eta(u_i, v_i, \phi)$ are discontinuous in ϕ due to the indicator functions involved in their definition. However, the discontinuities are easy to calculate, and the functions behave smoothly as functions of ϕ between them. This simplifies the solution of the equation (12) and enables rather straightforward computation of $W_t(u)$ in (7). A possible choice of the density π is to put $\pi(\mu, \phi) = 1/(\mu\phi)$ since Jeffreys' priors for, respectively, known ϕ and known μ are $1/\mu$ and $1/\phi$ (see Section 6.3 for the use of Jeffreys' priors in the present context). The desired simulations and computations can thus be performed by the methods of the present section.

As mentioned in the introduction, Cheng (1984) presented a method for simulation of conditional distributions in the case of inverse Gaussian distributed samples. His method is based on a subtle decomposition of chi-squared random variates and appears to be somewhat simpler than the method presented here.

4.2 *The discrete case*

Suppose that both X and T have discrete distributions, while the parameter space is a subset of the k-dimensional Euclidean space. In this case the sets $\Gamma(u, t)$ are usually sets with positive Lebesgue measure. These may in many cases be found explicitly, so that $W_t(u) = Pr\{\tau(u, \Theta) = t\}$ can be computed directly. In some instances, however, the set $\Gamma(u, t)$ is difficult to find. For such cases Engen and Lillegaard (1997) suggest replacing π by a discrete measure, such as the counting measure on a grid of points in the parameter space.

A thorough treatment of the discrete case is given in LT (2005), including an example with logistic regression.

5 On the distribution of $\hat{\theta}(U, t)$

Consider again the case when $\tau(u, \theta) = t$ has the unique solution $\hat{\theta}(u, t)$. For computational reasons it may be desirable to have some knowledge of the probability distribution of $\hat{\theta}(U, t)$ as a function of U.

Note first that for the case when θ is one-dimensional and T is stochastically increasing in θ, Lillegaard and Engen (1999) used the variates $\hat{\theta}(U, t)$ to derive exact confidence intervals for θ. More precisely they showed that one obtains an exact $(1 - 2k/(m+1))$-confidence interval for θ by sampling $m + 1$ values of $\hat{\theta}(U, t)$ and then using the interval from the kth smallest to the kth largest of them. They called this method conditional parametric bootstrapping. Their result can be rephrased to say that the interval between the $\alpha/2$ and $1 - \alpha/2$ percentiles of the distribution of $\hat{\theta}(U, t)$ is an exact $1 - \alpha$ confidence interval for θ. In fact, the distribution of $\hat{\theta}(U, t)$ is in this case a fiducial distribution in the sense of Fisher [Wilks (1962), p. 370]. This suggests that, under given standard conditions, the distribution of $\hat{\theta}(U, t)$ should at least asymptotically be comparable to that of a decent estimator of θ, for example the maximum likelihood estimator.

This turns in fact out to be true under reasonable conditions. A rough argument for the extended case where θ and T are k-dimensional can be given as follows. Suppose that we have $U = (U_1, U_2, \ldots, U_n)$ where we shall consider the case where $n \to \infty$. Furthermore, assume that the parametrization is such that $\theta = E\{T\} = E\{\tau(U, \theta)\}$. Lehmann and Casella (1998, p. 116) calls this the mean value parametrization. In this case T is itself an unbiased estimator of θ, and is the maximum likelihood estimator if the underlying model is an exponential family [Lehmann and Casella (1998), p. 470]. Our basic assumption for the following derivation is that

$$n^{1/2}(\tau(U, \theta) - \theta) \xrightarrow{d} N_k(0, \Sigma(\theta))$$

as $n \to \infty$ for some positive definite matrix $\Sigma(\theta)$. This is satisfied in the exponential family case, where $\Sigma(\theta)$ is the inverse Fisher information matrix.

Now we consider a fixed value of t and define $\hat{\theta}(U, t)$ to be the unique solution of $\tau(U, \theta) = t$. Assume furthermore that we can show that $\hat{\theta}(U, t) \to t$ in probability as $n \to \infty$. In this case, for any U,

$$\begin{aligned} t &= \tau(U, \hat{\theta}(U, t)) \\ &= \tau(U, t) + \partial_\theta \tau(U, \theta)|_{\theta = \tilde{\theta}}(\hat{\theta}(U, t) - t), \end{aligned}$$

where $\tilde{\theta}$ is between $\hat{\theta}(U, t)$ and t in the sense that each component of $\tilde{\theta}$ is a convex combination of the corresponding components of $\hat{\theta}(U, t)$ and t.

Hence

$$n^{1/2}(\hat{\theta}(U,t)-t) = (\partial_\theta \tau(U,\theta)|_{\theta=\tilde{\theta}})^{-1} n^{1/2}(t-\tau(U,t))$$

and provided $\partial_\theta \tau(U,\theta)|_{\theta=\tilde{\theta}} \xrightarrow{p} I$ (where I is the identity matrix) we have

$$n^{1/2}(\hat{\theta}(U,t)-t) \to N_k(0,\Sigma(t)) \tag{13}$$

The requirement that $\partial_\theta \tau(U,\theta)|_{\theta=\tilde{\theta}} \xrightarrow{p} I$ is typical in asymptotic results related to estimating equations, see for example Welsh (1996, Section 4.2.4) and Sørensen (1999) for sufficient conditions. The reason for the limit I above is that $E\{\tau(U,\theta)\} = \theta$. We will not pursue this further here, since the methods we derive are meant for use in non-asymptotic inference.

The conclusion is that for a large class of models $\hat{\theta}(U,t)$ has the same asymptotic distribution as T under the parameter value $\theta = t$. Thus in multiparameter exponential models we conclude that $\hat{\theta}(U,t)$ (under given conditions) has the same asymptotic distribution as the maximum likelihood estimator for θ. Note that by the invariance property of the maximum likelihood estimator and of $\hat{\theta}(U,t)$ (see Section 6.3) this holds under any parametrization.

Finally we can reinterpret our result (13) to say that conditionally on T, $n^{1/2}\{\hat{\theta}(U,T)-T\}$ has the same limiting distribution as $n^{1/2}(T-\theta)$. This result is analogous to asymptotic results for bootstrapping (Bickel and Freedman, 1981), in which the $\hat{\theta}(U,T)$ are replaced by bootstrapped statistics.

6 Computational aspects

6.1 *Monte Carlo computation of conditional expectations*

A basic idea of our approach is that expectations of functions of U, such as (4) and (9), can be evaluated by Monte Carlo simulation. Basically, we can estimate $E\{h(U)\}$ by $(1/m)\sum_{i=1}^m h(u_i)$ where $u_1,\ldots,u_m$ is a computer generated pseudo sample from the distribution of U. The literature on Monte Carlo simulation [for example Ripley (1987)] contains various methods for improving on this naive approach of estimating $E\{h(U)\}$.

6.2 *Choice of simulation method for* (X,T)

Our approach relies on the functions $(\chi(U,\theta),\tau(U,\theta))$ chosen for simulation of (X,T) in the original model. There is usually no unique way of selecting a simulation method. In the simulation of inverse Gaussian variables in Example 4 it would be possible, for example, to use ordinary inversion based on the cumulative distribution function, or even to use simulation of Wiener processes as described in Chhikara and Folks (1989). Each simulation scheme would give a different solution technique for handling the conditional distributions.

6.3 *Choice of the density π. Jeffreys' prior*

For a given setup in terms of $(\chi(U,\theta), \tau(U,\theta))$ we need to specify a density $\pi(\theta)$, except when conditions for using Algorithm 1 are fulfilled. In practice the effectiveness of an algorithm is connected to variation in the $W_t(u)$ which should be small or at best absent. For example, in order to minimize this variation in the case of formula (4), the density $\pi(\theta)$ should be chosen so that $\pi\{\hat{\theta}(u,t)\}$ is similar to $|\text{det}\partial_\theta \tau(u,\theta)|_{\theta=\hat{\theta}(u,t)}$.

Under the pivotal condition (Section 2) we may always choose π so that $W_t(u)$ does not depend on u. As a simple illustration, consider the simple pivotal case where θ is one-dimensional and $\tau(u,\theta) = r(u)\theta$. This means that T/θ is a pivotal quantity. Assume that the parametrization is such that $E\{r(U)\} = 1$ so that we have the mean value parametrization. In this case $\hat{\theta}(u,t) = t/r(u)$ so $\partial_t \hat{\theta}(u,t) = 1/r(u) = \hat{\theta}(u,t)/t$. Hence we get $W_t(u)$ in (3) constant in u by choosing $\pi(\theta) = 1/\theta$. Assuming that T is the maximum likelihood estimator of θ then under regularity conditions the Fisher-information is given by $1/\text{Var}\{\tau(U,\theta)\} \propto 1/\theta^2$, so $1/\theta$ is Jeffreys' prior here. As another illustration it is shown in LT (2005) that in the case where X is a sample from $N(\mu,\sigma)$ we obtain constant $W_t(u)$ by choosing $\pi(\mu,\sigma) = 1/\sigma$, which is the standard improper, noninformative prior for this case.

In fact there are reasons to choose improper, noninformative priors, such as Jeffreys' prior, also in general for the distribution π. Consider in particular a one-to-one reparametrization from a k-dimensional parameter θ to the k-dimensional ξ defined by $\theta = h(\xi)$. We then define $\tau_h(u,\xi) = \tau(u,h(\xi))$ from which it follows that the $\hat{\xi}(u,t)$ which solves the equation $\tau_h(u,\xi) = t$ satisfies $\hat{\theta}(u,t) = h\{\hat{\xi}(u,t)\}$. Now let J be the $k \times k$-matrix with elements $J_{ij} = \partial h_i(\xi)/\partial \xi_j$. Then we can write (3) as

$$W_t(u) = \pi[h\{\hat{\xi}(u,t)\}]\ |\text{det}J\ \text{det}\partial_t \hat{\xi}(u,t)|.$$

This shows that if we change the parametrization, then the weights $W_t(u)$ are unchanged provided we change π by the ordinary change of variable formula for densities. Thus a consistent principle for choosing π should have this property of invariance under reparametrizations. It is well known that Jeffreys' prior (Jeffreys, 1946) has this property, and there seems to be reasons why in fact Jeffreys' prior distribution is a reasonable candidate for general use.

6.4 *Direct sampling from the conditional distributions*

Algorithm 1 describes how to sample from the conditional distribution of X given $T = t$ under special conditions. Sampling from the conditional distribution using Algorithms 2 or 3 may, however, in general be difficult since the normalizing constant of the density $W_t(u)f(u)$ may not be easily available. Rejection sampling can be used if we are able to bound $W_t(u)$ from above. Looking at (3) we find that a possible way of doing this is to seek a positive function $\rho(\theta)$ such that for all u we have

$$|\text{det}\partial_\theta \tau(u,\theta)|_{\theta=\hat{\theta}(u,t)} \geq \rho\{\hat{\theta}(u,t)\}.$$

In this case we can put $\pi(\theta) = \rho(\theta)$ to get $W_t(u) \leq 1$ for all u. Then we may simulate V in Step 1 by first drawing a $U = u$ and then accepting it with probability $W_t(u)$.

A possible method for sampling without bounding $W_t(u)$ is by means of the SIR-algorithm of Rubin (1988). In the case of Algorithm 2 this method can be described as follows:

First sample $u_1, \ldots, u_m$ independently from the density $f(u)$. Then define $w_i = W_t(u_i)$ for $i = 1, \ldots, m$ and let F_m denote the discrete probability measure which assigns probability $w_i / \sum_{i'=1}^{m} w_{i'}$ to u_i. Then F_m converges to the desired conditional distribution as $m \to \infty$. Hence for m large enough we can obtain independent samples in Step 1 of Algorithms 2 and 3 by sampling from F_m.

Samples in Step 1 of Algorithms 2 and 3 can also be obtained by using the independence sampler based on the Metropolis-Hastings algorithm [Tierney (1994)], but this leads to dependent samples from the conditional distribution.

References

1. Aitchison, J. (1963). Inverse distributions and independent gamma distributed products of random variables. *Biometrika* **50** 505-508.

2. Bickel, P. J. and Doksum, K. A. (1969). Tests for monotone failure rate based on normalized spacings. *Ann. Math. Statist.* **40** 1216-1235.

3. Bickel, P. J. and Doksum, K. A. (2001). *Mathematical Statistics: Basic Ideas and Selected Topics, 2nd Edition, Vol I.* Prentice-Hall, Upper Saddle River, NJ.

4. Bickel, P. J. and Freedman, D. A. (1981). Some asymptotic theory for the bootstrap. *Ann. Statist.* **9** 1196-1217.

5. Cheng, R. C. H. (1984). Generation of inverse Gaussian variates with given sample mean and dispersion. *Appl. Stat.* **33** 309-316.

6. CHENG, R. C. H. (1985). Generation of multivariate normal samples with given sample mean and covariance matrix. *J. Statist. Comput. Simul.* **21** 39-49.

7. CHHIKARA, R.S. AND FOLKS, J. L. (1989). *The inverse Gaussian distribution. Theory, Methodology, and Applications.* Marcel Dekker, New York.

8. DIACONIS, P. AND STURMFELS, B. (1998). Algebraic algorithms for sampling from conditional distributions. *Ann. Statist.* **26** 363-397.

9. DOKSUM, K. A. AND HØYLAND, A. (1992). Models for variable-stress accelerated life testing experiments based on Wiener processes and the Inverse Gaussian distribution. *Technometrics* **34** 74-82.

10. ENGEN, S. AND LILLEGAARD, M. (1997). Stochastic simulations conditioned on sufficient statistics. *Biometrika* **84** 235-240.

11. JEFFREYS, H. (1946). An invariant form for the prior probability in estimation problems. *Proc. Roy. Soc. A* **186** 453-461.

12. LANGSRUD, Ø. (2005). Rotation tests. *Statistics and Computing* **15** 53-60.

13. LEHMANN, E. L. AND CASELLA, G. (1998). *Theory of Point Estimation.* Springer-Verlag, New York.

14. LILLEGAARD, M. AND ENGEN, S. (1999). Exact confidence intervals generated by conditional parametric bootstrapping. *J. Appl. Stat.* **26** 447-459.

15. LINDQVIST, B. H. AND TARALDSEN, G. (2001). Monte Carlo conditioning on a sufficient statistic. Statistics No. 9/2001, Dep. of Math. Sciences, Norwegian University of Science and Technology, Trondheim. Available as *http://www.math.ntnu.no/preprint/statistics/2001/S9-2001.ps.*

16. LINDQVIST, B. H. AND TARALDSEN, G. (2005). Monte Carlo conditioning on a sufficient statistic. *Biometrika* **92** 451-464.

17. LINDQVIST, B. H., TARALDSEN, G., LILLEGAARD, M. AND ENGEN, S. (2003). A counterexample to a claim about stochastic simulations. *Biometrika* **90** 489-490.

18. MICHAEL, J. R., SCHUCANY, W. R. AND HAAS, R. W. (1976). Generating random variates using transformations with multiple roots. *Am. Stat.* **30** 88-90.

19. RIPLEY, B. (1987). *Stochastic Simulation.* Wiley, New York.

20. RUBIN, D.B. (1988). Using the SIR algorithm to simulate posterior distributions (with discussion). In *Bayesian Statistics, Vol. 3* (Bernardo et al., eds.) 395-402. Oxford University Press, Oxford.

21. SESHADRI, V. (1999). *The Inverse Gaussian Distribution. Statistical Theory and Applications.* Lecture Notes in Statistics 137. Springer, New York.

22. SØRENSEN, M. (1999). On asymptotics of estimating functions. *Brazilian J. Prob. Statist.* **13** 111 - 136.

23. TARALDSEN, G. AND LINDQVIST, B. H. (2005). The multiple roots simulation algorithm, the inverse Gaussian distribution, and the sufficient conditional Monte Carlo method. Statistics No. 4/2005, Dep. of Math. Sciences, Norwegian University of Science and Technology, Trondheim. Available as *http://www.math.ntnu.no/preprint/statistics/2005/S4-2005.pdf.*

24. TIERNEY, L. (1994). Markov chains for exploring posterior distributions (with discussion). *Ann. Statist.* **22** 1701-1762.

25. TROTTER AND TUKEY (1956). Conditional Monte Carlo for normal samples. In *Symposium on Monte Carlo Methods* (H. A. Meyer, ed.) 64-79. Wiley, New York.

26. WELSH, A. H. (1996). *Aspects of Statistical Inference.* Wiley, New York.

27. WILKS, S. S. (1962). *Mathematical Statistics.* Wiley, New York.

Chapter 28

FURTHER EXPLORATIONS OF LIKELIHOOD THEORY FOR MONTE CARLO INTEGRATION

Augustine Kong, Peter McCullagh, Xiao-Li Meng, and Dan L. Nicolae

deCode Genetics, Reykjavik, ICELAND

Department of Statistics, The University of Chicago, Chicago, IL, U.S.A.

Department of Statistics, Harvard University, Cambridge, MA, U.S.A.

Department of Statistics, The University of Chicago, Chicago, IL, U.S.A.

E-*mails: kong@decode.is, pmcc@galton.uchicago.edu, meng@stat.harvard.edu & nicolae@galton.uchicago.edu*

Monte-Carlo estimation of an integral is usually based on the method of moments or on an estimating equation. Recently, Kong et al. (2003) proposed a likelihood based theory, which puts Monte-Carlo estimation of integrals on a firmer, less *ad hoc*, basis by formulating the problem as a likelihood inference problem for the baseline measure with simulated observations as data. In this paper, we provide further exploration and development of this theory. After an overview of the likelihood formulation, we first demonstrate the power of the likelihood-based method by presenting a universally improved importance sampling estimator. We then prove that the formal, infinite-dimensional Fisher-information based variance calculation given in Kong et al. (2003) is asymptotically the same as the sampling based "sandwich" variance estimator. Next, we explore the gain in Monte Carlo efficiency when the baseline measure can be parameterized. Furthermore, we show how the Monte Carlo integration problem can also be dealt with by the method of empirical likelihood, and how the baseline measure parameter can be properly profiled out to form a profile likelihood for the integrals only. As a byproduct, we obtain four equivalent conditions for the existence of unique maximum likelihood estimate for mixture models with known components. We also discuss an apparent paradox for Bayesian inference with Monte Carlo integration.

Keywords: Bridge Sampling; Fisher information; Importance sampling; Mixture model; Normalizing constant; Profile likelihood.

1 Introduction and overview

1.1 *The need for computing normalizing constants*

Let $\{q_\theta\}$ be a family of unnormalized probability/density functions on some sample space Γ, and let P_θ be the corresponding probability measure, $P_\theta(A) = \int_A q_\theta(x)\mu(dx)/c(\theta)$, for any measurable $A \subset \Gamma$. Here $c(\theta)$ is the normalizing constant, and μ is the dominating measure. Whereas the normalizing constant is not required for many sampling methods, particularly Markov chain Monte Carlo (MCMC) methods (e.g. Gilks et al., 1996), it is a central quantity in many statistical and scientific problems. In physics, it is known as the partition function, often estimated via MCMC (e.g. Bennett, 1976; Ceperley, 1995; Voter, 1985). In genetics, many likelihoods are computed using the following identity

$$p(Y_{\text{mis}} \,|\, Y_{\text{obs}}, \theta) = \frac{L(\theta \,|\, Y_{\text{obs}}, Y_{\text{mis}})}{L(\theta \,|\, Y_{\text{obs}})},$$

where $L(\theta|Y_{\text{obs}})$ is the likelihood of interest, and $L(\theta \,|\, Y_{\text{obs}}, Y_{\text{mis}})$ is the complete-data likelihood, typically easier to evaluate because of its simpler structure. With the help of sophisticated MCMC algorithms that simulate from $p(Y_{\text{mis}} \,|\, Y_{\text{obs}}, \theta)$, computing $L(\theta \,|\, Y_{\text{obs}})$ becomes a problem of estimating the normalizing constant $c(\theta) = L(\theta \,|\, Y_{\text{obs}})$ of the unnormalized $f(Y_{\text{mis}} \,|\, Y_{\text{obs}}, \theta)$, namely, $q_\theta(Y_{\text{mis}}) = L(\theta \,|\, Y_{\text{obs}}, Y_{\text{mis}})$, using our generic notation (here Y_{obs} is fixed). Methods such as importance sampling and bridge sampling are then used to estimate $L(\theta \,|\, Y_{\text{obs}})$ (e.g. Ott, 1976; Geyer and Thompson, 1992; Irwin et al., 1994; Jensen and Kong, 1999; Stephens and Donnelly, 2001; Thompson, 2000).

In statistics, besides the obvious need for $c(\theta)$ in computing the likelihood, the computation of a Bayes factor is precisely a problem of computing normalizing constants. These likelihood and Bayesian computation problems are the major reasons for the recent interest in this topic, particularly given the increased complexity of Bayesian models (e.g. Geyer, 1994; Gelfand and Dey, 1994; Newton and Raftery, 1994; Chib, 1995; Verdinelli and Wasserman, 1995; Meng and Schilling, 1996; Meng and Wong, 1996; Chen and Shao, 1997b,a; DiCiccio et al., 1997 Gelman and Meng, 1998; Johnson, 1999; Chib and Jeliazakov, 2001; Meng and Schilling, 2002). In all these problems, the quantities of interest are either ratios of normalizing constants (e.g., likelihood ratios) or can be formulated as such. Even for computing the normalizing constant for a single model, for computational efficiency, we can estimate its value relative to the known value of the normalizing constant of a simple approximation to the model (DiCiccio et al., 1997; Meng and Schilling, 2002). Thus we focus on estimating a set of normalizing constants modulo a common positive multiple. By extending

$\{q_\theta\}$ to include integrable but not necessarily non-negative functions, the formulation also covers general Monte Carlo integrations.

1.2 *A review of the likelihood theory*

An intriguing aspect of Monte Carlo (MC) integration is that there is no obvious (non-trivial) lower bound on the Monte Carlo variance with a *given* simulation size. This is, of course, not surprising, because it is well-known that the variance of the importance sampling estimator approaches zero as the distance between the target density and trial density approaches zero. Also, in MC integration problems, we know all quantities in $c(\theta) = \int_\Gamma q_\theta(x)\, \mu(dx)$, so there appears to be no inference problem to speak of. This has led to several quandaries in the general attempts to model MC simulated data just as real data; see Meng (2005).

To address this problem, Kong et al. (2003) proposed to treat the baseline measure, μ, as the unknown parameter; additional arguments on why this is a natural strategy are given in Meng (2005). Specifically, let $q_1, \ldots, q_k$ be real-valued non-negative functions on Γ, and let μ be any non-negative measure on Γ. We are interested in estimating c_r/c_s, where $c_r = \int_\Gamma q_r(x)\, d\mu$ is assumed to be positive and finite. The simulated data are $n_r > 0$ independent samples from the rth weighted distribution

$$P_r(dx) = c_r^{-1} q_r(x)\, \mu(dx). \tag{1}$$

There may be additional functions $q_r, r = k+1, \ldots, k+m$ for which (the relative value of) $c_r = \int_\Gamma q_r(x)\, \mu(dx)$ must also be estimated, and these functions need not be non-negative. Also, as a theoretical device, by extending q_r to be a joint density of dependent draws, the formulation covers the practical situation where draws are realizations of a Markov chain.

Under the model of Kong et al. (2003), the parameter space is the set of all non-negative measures on Γ, but our interest lies in the $k + m$ linear functionals

$$c_r = \int_\Gamma q_r(x)\, d\mu, \quad r = 1, \ldots, k+m. \tag{2}$$

Since the simulated data are n independent pairs: $(y_1, x_1), \ldots, (y_n, x_n)$, where the labels $y_i \in \{1, \ldots, k\}$ are determined by the simulation design and $x_i \sim P_{y_i}$, the full likelihood for μ is

$$L(\mu; X) = \prod_{i=1}^{n} P_{y_i}(\{x_i\}) = \prod_{i=1}^{n} \mu(\{x_i\})\, c_{y_i}^{-1} q_{y_i}(x_i). \tag{3}$$

Here we have assumed that Γ is countable; the uncountable case is discussed in Section 1.3. Re-parameterizing in terms of the canonical parameter $\theta(x) = \log \mu(\{x\})$, the log likelihood for θ, except for a constant, is

$$\sum_{i=1}^{n} \theta(x_i) - \sum_{s=1}^{k} n_s \log c_s(\theta) = n \int_{\Gamma} \theta(x)\, d\hat{P} - \sum_{s=1}^{k} n_s \log c_s(\theta), \qquad (4)$$

where $\hat{P}$ is the standard empirical measure which puts $1/n$ mass at each observed data point. The maximum likelihood estimate of μ is given by

$$\hat{\mu}(dx) = \frac{n\hat{P}(dx)}{\sum_{s=1}^{k} n_s \hat{c}_s^{-1} q_s(x)}, \qquad (5)$$

where $\hat{c}_s$ is the MLE of c_s, which are obtained (up to a proportionality constant) as the solution of the first k equations of

$$\hat{c}_r = \int_{\Gamma} q_r(x)\, d\hat{\mu} = \sum_{i=1}^{n} \frac{q_r(x_i)}{\sum_{s=1}^{k} n_s \hat{c}_s^{-1} q_s(x_i)}, \quad r = 1, \ldots, k+m. \qquad (6)$$

Note that this set of equations has a unique solution (up to a multiplicative constant) if and only if the set of values $\{q_r(x_i) \geq 0, i = 1, \ldots, n; r = 1, \ldots, k\}$ satisfies the "connected" condition of Vardi (1985), which we assume. We also remark that in the above formulation, the labels $\{y_1, \ldots, y_n\}$ play no role because they are not a part of the minimum sufficient statistic (Vardi, 1985). However, such label information is crucial for the "warp transformation" formulation, as discussed in Section 4.2.

1.3 *Uncountable sample spaces*

While the likelihood theory given in Kong et al. (2003) can be formally extended to cases where Γ is uncountable, the definition of θ becomes problematic because it requires the existence of a dominating measure ν on Γ such that the logarithmic derivative

$$\theta(x) = \log\left(\frac{d\mu}{d\nu}(x)\right)$$

is well-defined on Γ. That is to say, the parameter space is restricted to the set of measures on Γ that are absolutely continuous with respect to ν. This construction is unsatisfactory when Γ is uncountable. The difficulty is that if μ is Lebesgue measure on $\mathbb{R}$, the "MLE" $\hat{\mu}$ is atomic, and thus not in the parameter space as described. In fact, there does not exist on $\mathbb{R}$ a common dominating measure ν such that $\hat{\mu} \ll \nu$ for all possible estimates $\hat{\mu}$.

The problem is more of a mathematical technicality than a practical obstacle, as equation (6) is clearly well-defined whether or not Γ is countable. Nevertheless, it is of some interest to acknowledge the problem, to offer a resolution and to explore the consequences. First, we define $(\Gamma, \mathcal{A})$

as a measure space, in which $\mathcal{A}$ is a σ-algebra of subsets sufficiently rich to include all singletons of Γ. Second, we assume that the functions $q_s(x)$ are $\mathcal{A}$-measurable. Finally, the parameter space $\mathcal{M}$ is taken to be the set of all non-negative measures defined on $\mathcal{A}$. The likelihood at $\mu \in \mathcal{M}$, $L(\mu; X)$, is still given by (3). Note here that we define likelihood through its original form using the *probability* of the observed event $\{x_1, \ldots, x_n\}$, not through any *density* function, which is not suitable here as there is no single dominating measure for all elements in $\mathcal{M}$.

From (3), it is clear that if $\mu(\{x_i\}) = 0$ or $c_{y_i} = \infty$ for at least one i, then $L(\mu; X) = 0$. Furthermore, for each $\mu \in \mathcal{M}$ such that $\mu(\{x_i\}) > 0$ for all $i = 1, \ldots, n$ and $\mu(\Gamma \backslash \{x_1, \ldots, x_n\}) > 0$, we define a $\tilde{\mu} \in \mathcal{M}$ so that $\tilde{\mu}(\{x_i\}) = \mu(\{x_i\})$ for all $i = 1, \ldots, n$, but $\tilde{\mu}(\Gamma \backslash \{x_1, \ldots, x_n\}) = 0$. Recall that $q_s(x) > 0$ for all $x \in \Gamma$, it is then evident that for each $s \in \{1, \ldots, k\}$,

$$\tilde{c}_s = \int_\Gamma q_s(x) d\tilde{\mu} < \int_\Gamma q_s(x) d\mu = c_s,$$

so that $L(\tilde{\mu}; x) > L(\mu; x)$. Therefore, as far as MLE is concerned, we can concentrate on measures with support on $\{x_1, \ldots, x_n\}$. This effectively implies that we can proceed as if Γ were countable.

Regardless of whether Γ is countable or not, the real power of the likelihood-based method is that any usable knowledge about μ can (and should) be used to form a sub-model for estimating μ, and hence to improve MC efficiency for the resulting estimates of the $c's$. The next section details such an exercise in the context of importance sampling.

2 A universal improvement for importance sampling

2.1 *Symmetrized importance sampling*

While the formulation and results in Section 1 cover the most general bridge sampling (Meng and Wong, 1996), the case with $k = 1$ is of special interest, because it corresponds to the importance sampling approach via

$$\gamma \equiv \frac{c_2}{c_1} = \int_\Gamma \frac{q_2(x)}{q_1(x)} [c_1^{-1} q_1(x)] d\mu = \int_\Gamma \frac{q_2(x)}{q_1(x)} dP_1. \tag{7}$$

Here we assume $\mathcal{S}_2 \subset \mathcal{S}_1$, where $\mathcal{S}_r$ is the support of q_r, and we typically set $c_1 = 1$ because the trial density P_1 is completely known. We highlight this common application to emphasize that many current MC integrations can be improved upon because they needlessly ignore usable symmetry properties in the baseline measure (e.g., Lebesgue measure), which can be captured easily by a sub-model under the likelihood formulation.

Specifically, let $\mathcal{G}$ be a compact group acting on Γ in such a way that μ is invariant: $\mu(gA) = \mu(A)$ for each $g \in \mathcal{G}$. The sub-model is the one where the parameter space consists only of measures that are invariant under $\mathcal{G}$. The log likelihood function (4) simplifies because $\theta(x) = \theta(gx)$ for each $g \in \mathcal{G}$. The MLE of μ is still given by (5) and (6), but with q_s and $\hat{P}$ replaced respectively by their group averages $\bar{q}_s(x) = \text{ave}_{g\in\mathcal{G}}\, q_s(gx)$ and $\hat{P}^{\mathcal{G}}(A) = \text{ave}_{g\in\mathcal{G}}\, \hat{P}(gA)$.

To illustrate, the sub-model shows that (7) can be *symmetrized* by group-averaging:

$$\gamma = \frac{c_2}{c_1} = \int_\Gamma \frac{\bar{q}_2(x)}{\bar{q}_1(x)}\, d\bar{P}_1. \tag{8}$$

Consequently, the usual importance sampling estimator

$$\hat{\gamma}_n = \frac{1}{n}\sum_{i=1}^{n} \frac{q_2(x_i)}{q_1(x_i)} \equiv \frac{1}{n}\sum_{i=1}^{n} w(x_i), \tag{9}$$

where $\{x_1, \ldots, x_n\}$ are draws from P_1, is replaced by

$$\hat{\gamma}_n^{\mathcal{G}} = \frac{1}{n}\sum_{i=1}^{n} \frac{\bar{q}_2(x_i)}{\bar{q}_1(x_i)} \equiv \frac{1}{n}\sum_{i=1}^{n} w^{\mathcal{G}}(x_i). \tag{10}$$

Because of (8), $\hat{\gamma}_n^{\mathcal{G}}$ is unbiased for γ as long as $\mathcal{S}_2^{\mathcal{G}} \subseteq \mathcal{S}_1^{\mathcal{G}}$ (where $\mathcal{S}_r^{\mathcal{G}}$ is the support of $\bar{q}_r$), which is a weaker requirement than $\mathcal{S}_2 \subseteq \mathcal{S}_1$. Therefore the group average improves (or at least does no harm to) the robustness of importance sampling in the sense of providing more assurance of having enough support in the trial density.

There are several ways to see why (10) is more efficient than (9). First, (9) is an "extreme" special case of (10) with $\mathcal{G}$ the identity transformation, while the original analytic integration/summation over Γ in (7) can be viewed as the other extreme case of (10) with a group rich enough such that each of the $w^{\mathcal{G}}(x_i)$ in (10) is exactly the target value γ. The latter can be most easily seen when Γ is finite, where we can take $\mathcal{G}$ be the full permutation group on Γ. By using a suitable, much smaller sub-group, (10) takes advantage of our ability to do partial analytic and/or numerical summation, and then uses Monte Carlo to deal with the rest. In contrast, (9) relies entirely on MC simulation to estimate γ.

Second, we can view the group transformation as "reparametrizing" the sample space into a set of *orbits* and a *cross-section* that indexes the orbits. Group averaging then analytically integrates over each orbit, and leaves only the integration over the cross-section to MC simulation. For example, suppose that $\Gamma = \mathbb{R}^d$, μ is Lebesgue, and $\mathcal{G}$ is the orthogonal group. Then group averaging is the same as the $d-1$ dimensional integration over all the angles in polar coordinates. This analytic averaging, if feasible, makes (10) more efficient than (9), because it effectively "Rao-Blackwellizes" out the

$d-1$ angle coordinates, and thus (10) becomes a one-dimensional MC integral over the radius. Note that the correct "Rao-Blackwellization" carried out by the sub-model, as shown in Section 2.2, averages individual q_r, not the ratios $w = q_2/q_1$. The latter does not in general provide a consistent estimator because P_1 is usually not invariant under $\mathcal{G}$.

Third, group averaging increases the overlap among the underlying "densities" (in quotes as some "densities" can be negative), and thus reduces the variability in the importance-sampling weight w. That is, (10) is more efficient than (9) for any compact group $\mathcal{G}$. Asymptotically, this is a consequence of the Fisher information results for general k and m, obtained in Section 3, because a sub-model necessarily possesses larger Fisher information than the full model for the same set of parameters. For $k = m = 1$, namely the importance sampling, in Section 2.2 we prove this for any finite sample size by showing that (10) is the Rao-Blackwell projection of (9) given the minimum sufficient statistic, i.e., the cross-section.

2.2 *A theoretical comparison*

The following theorem establishes that averaging over a larger group necessarily reduces the variance of (10) for each sample size, under the independence assumption, because of the usual "Rao-Blackwellization."

Theorem 1. *Suppose $\mathcal{G}_2 \subset \mathcal{G}_1$ are two finite groups and μ is $\mathcal{G}_1$-invariant. Let $\mathcal{G}x = \{gx : g \in \mathcal{G}\}$ be the $\mathcal{G}$-orbit of x and*

$$w_j(x) = \frac{\text{ave}_{g\in\mathcal{G}_j}\, q_2(gx)}{\text{ave}_{g\in\mathcal{G}_j}\, q_1(gx)} = \frac{\int_{t\in\mathcal{G}_j x} q_2(t)\, d\mu}{\int_{t\in\mathcal{G}_j x} q_1(t)\, d\mu}, \quad j = 1, 2.$$

If $\{x_1, \ldots, x_n\}$ is an i.i.d. sample from P_1, then

$$\text{Var}(\hat{\gamma}_n^{\mathcal{G}_1}) \le \text{Var}(\hat{\gamma}_n^{\mathcal{G}_2}), \quad \forall\, n \ge 1, \tag{11}$$

where the equality holds if and only if there exists $\Omega \subset \mathcal{S}_1$, the support of P_1, such that $P_1(\Omega) = 1$ and,

$$\forall\, x \in \Omega, \quad w_2(gx) = w_2(x), \quad \text{for all } g \in \mathcal{G}_1. \tag{12}$$

Proof. We first prove that for $X \sim P_1$,

$$E\left[w_2(X) |\, \mathcal{G}_1 x\right] = w_1(x). \tag{13}$$

The left-hand side of (13) is

$$\frac{\int_{t\in\mathcal{G}_1 x} w_2(t) q_1(t)\, d\mu}{\int_{t\in\mathcal{G}_1 x} q_1(t)\, d\mu} = \frac{\int_{t\in\mathcal{G}_1 x} w_2(t) \left[\text{ave}_{g\in\mathcal{G}_2}\, q_1(gt)\right] d\mu}{\int_{t\in\mathcal{G}_1 x} q_1(t)\, d\mu}, \tag{14}$$

where the equality holds because $w_2(t)$ and μ are $\mathcal{G}_2$-invariant and $\mathcal{G}_2 \subset \mathcal{G}_1$. The right side of (14) is $w_1(x)$ because $w_2(t)\,[\mathrm{ave}_{g\in\mathcal{G}_2}\, q_1(gt)] = \mathrm{ave}_{g\in\mathcal{G}_2}\, q_2(gt)$, and

$$\int_{t\in\mathcal{G}_1 x} \mathrm{ave}_{g\in\mathcal{G}_2}\, q_2(gt)\, d\mu = \mathrm{ave}_{g\in\mathcal{G}_2} \int_{t\in\mathcal{G}_1 x} q_2(gt) d\mu = \int_{t\in\mathcal{G}_1 x} q_2(gt) d\mu,$$

where the last equality follows from $\int_{t\in\mathcal{G}_1 x} q_2(gt) d\mu$ being $\mathcal{G}_2$-invariant.

It follows from (13) that $\hat{\gamma}_n^{\mathcal{G}_1} = \sum_i w_1(x_i)/n$ is the Rao-Blackwell projection of $\hat{\gamma}_n^{\mathcal{G}_2}$ when $\{x_1, \ldots, x_n\}$ are i.i.d. because

$$E\left[\hat{\gamma}_n^{\mathcal{G}_2} \,|\, \mathcal{G}_1 x_1, \ldots, \mathcal{G}_1 x_n\right] = \hat{\gamma}_n^{\mathcal{G}_1}.$$

Consequently, (11) holds, with equality if and only if there exists $\tilde{\Omega} \subset \mathcal{S}_1$ with $P_1(\tilde{\Omega}) = 1$ such that

$$\forall x \in \tilde{\Omega}, \quad w_2(x) = w_1(x). \tag{15}$$

To prove that (15) implies (12), let $\mathcal{B} = \cup_{g\in\mathcal{G}_1}\{gx :\ x \in \mathcal{S}_1\backslash\tilde{\Omega}\}$. Then $P_1(\mathcal{B}) = 0$ because $\mathcal{G}_1$ is finite. Let $\Omega = \mathcal{S}_1\backslash\mathcal{B} \subset \tilde{\Omega}$. Then $P_1(\Omega) = 1$. Furthermore, if $x \in \Omega$, then $gx \in \Omega \subset \tilde{\Omega}$ for any $g \in \mathcal{G}_1$. Consequently, for any $x \in \Omega$, since $w_1(x)$ is $\mathcal{G}_1$-invariant, (15) implies $w_2(gx) = w_1(gx) = w_1(x) = w_2(x)$ for any $g \in \mathcal{G}_1$, which is (12).

To prove the converse, we first note that for any x,

$$\mathrm{ave}_{g\in\mathcal{G}_1}\, q_r(gx) = \mathrm{ave}_{g_1\in\mathcal{G}_1}\{\mathrm{ave}_{g_2\in\mathcal{G}_2}\, q_r(g_2 g_1 x)\}, \quad r = 1, 2, \tag{16}$$

which implies

$$\mathrm{ave}_{g\in\mathcal{G}_1}\, q_2(gx) = \mathrm{ave}_{g_1\in\mathcal{G}_1}\{w_2(g_1 x)\, \mathrm{ave}_{g_1\in\mathcal{G}_2}\, q_1(g_2 g_1 x)\}. \tag{17}$$

Consequently, if (12) holds, then (17) becomes

$$\mathrm{ave}_{g_1\in\mathcal{G}_1}\{w_2(x)\, \mathrm{ave}_{g_2\in\mathcal{G}_2}\, q_1(g_2 g_1 x)\}$$

for any $x \in \Omega$, which together with (16) implies

$$\mathrm{ave}_{g\in\mathcal{G}_1}\, q_2(gx) = w_2(x)\, \mathrm{ave}_{g\in\mathcal{G}_1}\, q_1(gx).$$

This establishes (15) when we take $\tilde{\Omega} = \Omega$. □

This theorem provides a theoretical confirmation that our ability to carry out the $\mathrm{ave}_{\mathcal{G}_1}$ operator analytically, in addition to our ability to evaluate $\mathrm{ave}_{\mathcal{G}_2}$, always helps to reduce the MC error unless the group averages under $\mathcal{G}_1$ are already invariant under $\mathcal{G}_2$, in the sense of (12).

2.3 *Practical implications*

A consequence of (11)–(12) is that we can improve on the standard MC estimators such as (9) by using convenient choices of $\mathcal{G}$ for which (10) dominates (9). For the most common applications of (9) in statistics, where $\Gamma = \mathbb{R}^d$ and μ is the Lebesgue measure, we can always take the two-element group $\mathcal{G}_O = \{I_d, -I_d\}$, where I_d is the $d \times d$ identity matrix. For any problem where (9) can be implemented, we can implement (10) with $\mathcal{G} = \mathcal{G}_O$:

$$\hat{\gamma}_n^{\mathcal{G}_O} = \frac{1}{n}\sum_{i=1}^{n} \frac{q_2(x_i) + q_2(-x_i)}{q_1(x_i) + q_1(-x_i)}, \tag{18}$$

where $q_r(x) = 0$ if x is outside the support of $q_r, r = 1, 2$. By (11)–(12), $\text{Var}(\hat{\gamma}_n^{\mathcal{G}_O}) < \text{Var}(\hat{\gamma}_n)$ unless $q_2(-x)/q_1(-x) = q_2(x)/q_1(x)$ for almost all $x \in \mathcal{S}_1$.

In fact, even when (9) fails to provide a consistent estimator because $\mathcal{S}_2 \not\subset \mathcal{S}_1$, (18) can still be consistent as it requires the weaker assumption $\mathcal{S}_2^{\mathcal{G}} \subseteq \mathcal{S}_1^{\mathcal{G}}$, namely, the support of $q_1(x) + q_1(-x)$ covers that of $q_2(x) + q_2(-x)$. As an extreme example, consider $d = 1$, $c_1 = 1$, $\mathcal{S}_2 = \mathbb{R}$ but $\mathcal{S}_1 = [0, +\infty)$. Then (9) will only estimate $\int_0^{+\infty} q_2(x)\, d\mu$. By contrast, because $q_1(-x_i) = 0$ for $x_i \sim P_1$, (18) becomes

$$\frac{1}{n}\sum_{i=1}^{n} \frac{q_2(x_i)}{q_1(x_i)} + \frac{1}{n}\sum_{i=1}^{n} \frac{q_2(-x_i)}{q_1(x_i)}, \tag{19}$$

which correctly estimates

$$\int_0^{+\infty} q_2(x)\, d\mu + \int_0^{+\infty} q_2(-x)\, d\mu = \int_{-\infty}^{+\infty} q_2(x) d\mu.$$

Upon recognizing the support problem of q_1, one would apply (9) twice to form (19) to estimate $\int_{\mathbb{R}} q_2(x)\, d\mu$, whereas (18) achieves this automatically. This illustrates the robustness of (18), or other versions of (10), over (9) in dealing with the well-known "tail" problem with importance sampling, because it can greatly reduce biases caused by lack of support in the trial density. The requirement of making more function evaluations by (18) or (10) is often a negligible premium for its greater efficiency *and* robustness, especially in comparisons with the expense of making the MC draws. In fact, for cases like (19) the corrected importance sampling requires the same number of function evaluations as implementing (18).

The $\mathcal{G}_O$ group is only one of many that can be used to improve efficiency. For example, one can replace $-I_d$ in $\mathcal{G}_O$ by any of the $2^d - 2$ other diagonal matrices where the diagonal elements are either 1 or -1. Each of these groups represents reflections with respect to some of the d axes, and which one is optimal depends on q_2 and q_1. One advantage of using $\mathcal{G}_O$ is that it automatically symmetrizes $\bar{q}_r$ on each one-dimensional subspace. While the comparisons of (10) among non-nested groups can be mathematically complicated, intuitively $\mathcal{G}_O$ is a good "default" choice compared to other two-element reflection groups. If d is not too large, then we can and should consider using the full reflection group consisting of all 2^d diagonal matrices with diagonal elements ± 1, which is superior to any of its sub-group as guaranteed by Theorem 2.1. Further improvement is also possible by using different reflection points/axes for different distributions, as investigated in Meng and Schilling (2002); see Section 4.2.

As briefly mentioned in Kong et al. (2003), there are some similarities between the group averaging method with the *importance link function* (ILF) method of MacEachern and Peruggia (2000). The key of the ILF method is to construct a finite number of 1–1 and onto importance link functions $g_i, i = 1, \ldots, I$ with domain $B_i \subset \Gamma_0$, where Γ_0 is a subset of Γ, such that $\{T_i \equiv g_i(B_i), i = 1, \ldots, I\}$ forms a partition of Γ. Thus, integration on Γ can be estimated from the integral on each T_i via importance sampling using draws from a trial density concentrated on Γ_0. This is a very effective strategy to deal with a common problem in MCMC where the draws "get stuck" in part of the space, say Γ_0, but one needs to estimate integrals on the whole space Γ. Indeed, MacEachern and Peruggia (2000) proposed to use this method for handling reducible chains. In this regard, group averaging achieves the same goal and provides a more systematic way to construct link functions. If Γ_0 is a cross-section of the orbits, then $\{g\Gamma_0, g \in \mathcal{G}\}$ automatically form a partition of Γ and (10) will be the same as the ILF estimator using the same $\mathcal{G}$. Estimator (19) is such an example with $g_1(x) = x$ and $g_2(x) = -x$. In addition, the group formulation makes it clear that Γ_0 needs to contain at least a cross-section in order for the support of the $\bar{P}_1$ to cover Γ. When Γ_0 is richer than a cross-section, the group averaging estimator (10) is more efficient than the ILF estimator using the same $\mathcal{G}$ because the latter is generally not the Rao-Blackwell projection given the cross-section. As a trade-off, the ILF estimator does not require $\{g_i, i = 1, \ldots, I\}$ to be a group, when they are constructed based on information other than symmetries in the baseline measure.

3 Asymptotic covariance matrix

3.1 *Formal Fisher information calculation*

Apart from the special case with $k = 1$ (and $m \geq 1$), the exact calculation of the variance of the MLE of c is not tractable. However, we can obtain the asymptotic covariance matrix via the usual Fisher information calculation, at least formally. The following result (provided in Kong et al., 2003), based on the concept of *Fisher information measure* (e.g. McCullagh, 1999), was introduced to deal with Fisher information "matrix" of infinite order, countably or uncountably.

Specifically, the Fisher information measure for $\theta = \log \mu$ is

$$n\mathcal{I}(A, B) = \sum_{r=1}^{k} n_r \big(P_r(A \cap B) - P_r(A)P_r(B)\big),$$

where P_r is the distribution in (1). When Γ is countable, we also use $\mathcal{I}$ (without argument) to denote the $|\Gamma| \times |\Gamma|$ density matrix of the Fisher information measure; thus $\mathcal{I}$ is the usual Fisher information matrix, although of countably infinite order. The asymptotic covariance matrix of $\hat{\theta}$ is the (generalized) inverse Fisher information matrix, $n^{-1}\mathcal{I}^{-}$, and the asymptotic covariance matrix of $d\hat{\mu}$ is $n^{-1} d\mu(x)\, d\mu(y)\, \mathcal{I}^{-}(x, y)$, where $\mathcal{I}^{-}(x, y)$ is the (x, y) element of $\mathcal{I}^{-}$, as indexed by $\Gamma \times \Gamma$. From expression (6) for $\hat{c}_r$, we find that the asymptotic covariance of $\log \hat{c}$ is given by

$$\operatorname{cov}(\log \hat{c}_r, \log \hat{c}_s) = n^{-1} \int_{\Gamma \times \Gamma} \mathcal{I}^{-}(x, y)\, dP_r(x)\, dP_s(y), \quad 1 \le r, s \le k + m. \tag{20}$$

Before we proceed further, we remark that so far as contrasts of $\log \hat{c}$ are concerned, two variance matrices V, V' are equivalent if $a^\top V b = a^\top V' b$ for all contrast vectors a, b. In other words, $a^\top (V - V')b = 0$, so we may add to V any (symmetric) matrix W such that $a^\top W b = 0$ without affecting the value $a^\top (V + W)b$, the covariance of two contrasts $a^\top \log \hat{c}$ and $b^\top \log \hat{c}$, where $\hat{c} = (\hat{c}_1, \ldots, \hat{c}_{k+m})$. The set $\mathcal{W}$ of such symmetric matrices is the set $1x^\top + x1^\top$, which is a vector subspace of dimension k. The set of symmetric matrices that are equivalent to V is the coset $V + \mathcal{W}$ of symmetric $k \times k$ matrices. Not all elements of this coset need be positive definite. Cosets of this sort arise naturally as the set of symmetric generalized inverses of a non-invertible symmetric matrix A whose kernel is $\mathbf{1}$, the set of constant vectors. In particular, if $A1 = 0$ then $AWA = 0$ for each $W \in \mathcal{W}$. If A^- is a generalized inverse of A, i.e. $AA^-A = A$, then $A^- + W$ is also a generalized inverse of A for each $W \in \mathcal{W}$. If A^- is symmetric and $\ker(A) = \mathbf{1}$, the coset $A^- + \mathcal{W}$ is the set of symmetric generalized inverses of A. In this paper, whenever A is symmetric, A^- will be restricted to be a symmetric generalized inverse, and any equality between generalized inverses is interpreted in the sense of equivalence. The use of such generalized inverses makes it possible to express the asymptotic covariance matrix of $\log \hat{c}$, which is singular, in a symmetric form without the awkwardness and asymmetry associated with fixing an arbitrary component of $\hat{c}$.

3.2 Deriving the matrix version

While expression (20) is extendable to cases where Γ is uncountable, for the case where Γ is finite or countably infinite, we can obtain the usual matrix form. Specifically, let $P_{\text{mix}} = \sum_{r=1}^{k} f_r P_r$ be the mixture probability where $f_r = n_r/n$, $r = 1, \ldots, k$. Then the matrix $\mathcal{I}$ is given by

$$\mathcal{I} = D - \mathcal{P}_k F \mathcal{P}_k^\top,$$

where $D = \text{diag}\{P_{\text{mix}}(\{x\})\}$, $\mathcal{P}_k$ is a $|\Gamma| \times k$ matrix with rth column given by $P_r(\{x\})$ for $x \in \Gamma$ with x as the row index, and $F = \text{diag}\{f_1, \ldots, f_k\}$. Provided that P_{mix} is strictly positive on Γ, the matrix $\mathcal{I}$ has kernel equal to $\mathbf{1}$. Let $O_k = \mathcal{P}_k^\top D^{-1} \mathcal{P}_k$ and let $(F^{-1} - O_k)^-$ be a generalized inverse. Then the matrix

$$\mathcal{I}^- = D^{-1} + D^{-1}\mathcal{P}_k\big(F^{-1} - O_k\big)^- \mathcal{P}_k^\top D^{-1} \tag{21}$$

is a generalized inverse of $\mathcal{I}$. Thus, writing $\hat{c}^{(k)} = (\hat{c}_1, \ldots, \hat{c}_k)$, asymptotically,

$$n \,\text{cov}(\log \hat{c}^{(k)}) = \mathcal{P}_k^\top \mathcal{I}^- \mathcal{P}_k = O_k + O_k(F^{-1} - O_k)^- O_k, \tag{22}$$

which involves only symmetric matrices of order k. The inverse asymptotic variance of $\log \hat{c}^{(k)}$, i.e. the asymptotic precision, is $n(O_k^- - O_k^- O_k F O_k O_k^-)$.

Similarly, for $\hat{c}^{(k+m)} = (\hat{c}_1, \ldots, \hat{c}_{k+m})$, asymptotically we have

$$n \;\text{cov}(\log \hat{c}^{(k+m)}) = \begin{pmatrix} O_k + O_k L_k O_k & O_{m,k}^\top + O_k L_k O_{m,k}^\top \\ O_{m,k} + O_{m,k} L_k O_k & O_m + O_{m,k} L_k O_{m,k}^\top \end{pmatrix}, \tag{23}$$

where $L_k = (F^{-1} - O_k)^-$, $O_m = \mathcal{P}_m^\top D^{-1} \mathcal{P}_m$, $O_{m,k} = \mathcal{P}_m^\top D^{-1} \mathcal{P}_k$, with $\mathcal{P}_m$ the counterpart of $\mathcal{P}_k$ but for $r = k+1, \ldots, k+m$. Note that for $f = (f_1, \ldots, f_k)^\top$, we have $O_k f = 1$ and $(F^{-1} - O_k) f = 0$, so $F^{-1} - O_k$ is singular. For each vector $\alpha \in \mathbb{R}^k$, $(F^{-1} - O_k)^- + \alpha f^\top + f \alpha^\top$ is also a symmetric generalized inverse. But the choice of α has no effect on the variance of any contrast in expression (23).

We remark in passing that from (21) $\mathcal{I}^- \geq D^{-1}$ in the sense of Löwner ordering, from which we obtain the asymptotic inequality,

$$\text{Var}(\log(\frac{\hat{c}_r}{\hat{c}_s})) \geq n^{-1} \int_\Gamma \Big(\frac{dP_r(x)}{dP_{\text{mix}}(x)} - \frac{dP_s(x)}{dP_{\text{mix}}(x)}\Big)^2 P_{\text{mix}}(dx), \quad \forall 1 \leq r, s \leq k+m.$$

3.3 Estimating equation "sandwich" version

A technical difficulty with the Fisher information approach arises when Γ is uncountable. In such cases, the log density estimate $\hat{\theta}$ is generally inconsistent in the sense of pointwise convergence and thus the Fisher information approach presented in Section 3.1 can only be viewed as a formal calculation, suggestive but not rigorous. In this section we show the formula (20) or equivalently (23) are indeed correct even when Γ is uncountable.

First, equation (6) gives an estimating equation for $\log \hat{c}^{(k)}$ via

$$\sum_{i=1}^{n} \left. \frac{\partial \log[P(x_i|y_i;c)]}{\partial \log c} \right|_{c=\hat{c}^{(k)}} = 0, \tag{24}$$

where

$$P(x|y;c) = \frac{f_y q_y(x) c_y^{-1}}{\sum_{s=1}^{k} f_s q_s(x) c_s^{-1}}. \tag{25}$$

Applying the standard "sandwich" approach, albeit on the quotient space $\log c \in \mathbb{R}^k/\mathbf{1}$, we obtain, asymptotically

$$n \operatorname{cov}(\log \hat{c}^{(k)}) = \tilde{\mathcal{I}}^{-} V \tilde{\mathcal{I}}^{-},$$

where, denoting by E_r and cov_r, the expectation and variance under P_r,

$$\tilde{\mathcal{I}} = \sum_{r=1}^{k} f_r \mathrm{E}_r \left[- \left. \frac{\partial^2 \log[P(x|y;c)]}{(\partial \log c)(\partial \log c)^\top} \right|_{c=\hat{c}^{(k)}} \right] = F O_k F - F,$$

and

$$V = \sum_{r=1}^{k} f_r \operatorname{cov}_r \left[\left. \frac{\partial \log[P(x|y;\log c)]}{\partial \log c} \right|_{c=\hat{c}^{(k)}} \right] = F O_k F - F O_k F O_k F.$$

Therefore, asymptotically,

$$n \operatorname{cov}(\log \hat{c}^{(k)}) = (I - O_k F)^{-} (O_k - O_k F O_k)(I - O_k F)^{-\top}, \tag{26}$$

where $A^{-\top}$ denotes $(A^{-1})^\top$.

To show that (22) and (26) are equivalent, we only need to show that for a particular choice of the generalized inverse, they are equivalent. A convenient choice is the Moore-Penrose generalized inverse A^+ for any matrix A, which is unique and satisfies $AA^+A = A$, $A^+AA^+ = A^+$, and A^+A and AA^+ are symmetric. Using this choice, it can be shown that both (22) and (26) are equivalent to $(I - O_k F)^+ O_k$. And thus for computing the variance of any contrast of $\log \hat{c}^{(k)}$, (22) and (26) are equivalent. The extended version (23) can be derived directly in similar way, although the algebra is a bit more involved. In this derivation, a key is to observe that the f_y in (25) plays no role in (24) because $\log f_y$ is a constant. Thus we can effectively remove f_y from (25), which would then allow the y index extended to include $y = k+1, \ldots, k+m$ (for which $f_y = 0$).

3.4 A numerical example

For a numerical example, we take $k = 3$, μ unit Poisson, $n_1 = n_2 = n_3$, and $q_r(x) = r^x$ so that P_r is Poisson with mean r. Using (22), we find that

$$O_3 = \begin{pmatrix} 1.426 & 0.958 & 0.616 \\ 0.958 & 1.038 & 1.004 \\ 0.616 & 1.004 & 1.380 \end{pmatrix}, \qquad \mathcal{P}_3^\top \mathcal{I}^- \mathcal{P}_3 = \begin{pmatrix} 1.576 & 0.949 & 0.475 \\ 0.949 & 1.039 & 1.012 \\ 0.475 & 1.012 & 1.513 \end{pmatrix}.$$

The variance of any contrast $\log(\hat{c}_r/\hat{c}_s)$ is remarkably little affected by the relative allocation frequencies n_r/n in the design. For example, if the relative frequencies are $f = (0.2, 0.3, 0.5)^\top$ we find

$$O_3 = \begin{pmatrix} 1.733 & 1.085 & 0.656 \\ 1.085 & 1.051 & 0.936 \\ 0.656 & 0.936 & 1.176 \end{pmatrix}, \qquad \mathcal{P}_3^\top \mathcal{I}^- \mathcal{P}_3 = \begin{pmatrix} 2.431 & 1.452 & 0.772 \\ 1.452 & 1.250 & 1.002 \\ 0.772 & 1.002 & 1.200 \end{pmatrix}.$$

The asymptotic variances of $\big(\log(\hat{c}_1/\hat{c}_2), \log(\hat{c}_1/\hat{c}_3), \log(\hat{c}_2/\hat{c}_3)\big)$ are thus $(0.716, 2.138, 0.528)/n$ and $(0.775, 2.086, 0.446)/n$ for the first and second allocation respectively.

In the sense of minimizing the average variance of pairwise contrasts, the relative frequencies in the optimal allocation are approximately $(0.0, 0.8, 0.2)$, with no observations from P_1. But the average variance achieved by this allocation is only 8% less than the average variance in the design with equal weights. Further, the inferior design with equal weights may be superior for interpolation or extrapolation, i.e. for estimating ratios $\log(c_r/c_s)$ with r and/or s in $\{k+1, \ldots, k+m\}$.

The term *bridge sampling* has been used by Meng and Wong (1996) and Gelman and Meng (1998), in connection with the practice of sampling from intermediate distributions $P_2, \ldots, P_{k-1}$ in order to estimate the ratio c_1/c_k more accurately. Since the optimal allocation in the preceding example puts weight zero on P_1, the optimal bridge is in fact a cantilever, supported entirely on P_2, P_3. While the term 'bridge sampling' has a certain metaphoric appeal, this example indicates that it can be misleading to interpret the structural stability of the bridge as evidence of its statistical efficiency.

3.5 Asymptotic covariance matrix from sub-models

Since the estimating equation obtained from a sub-model is the same as (6) except that q_r is replaced by a group-average $\bar{q}_r$, $r = 1, \ldots, k+m$, the general covariance formula (23) is obviously applicable with the same replacement. In notation, this replacement is signified with $\bar{O}$ in place of O in (22) and other formulas as necessary. To see the potential gain in efficiency by a sub-model, consider a case where Γ is countable, μ is the counting measure, and $k = 2$. Let A be a finite subset of Γ, and let $\mathcal{G}_A$ be the permutation group on A. Then averaging P_1 and P_2 over A effectively makes both of them uniform on A. The asymptotic variance of $\log \gamma^{\mathcal{G}_A}$, where $\gamma^{\mathcal{G}_A} = \hat{c}_2^{\mathcal{G}_A}/\hat{c}_1^{\mathcal{G}_A}$, from (22) is (also see Meng and Wong, 1996)

$$\mathrm{Var}(\log \hat{\gamma}^{\mathcal{G}_A}) = \frac{1}{nf_1f_2}(\bar{o}_{12}^{-1} - 1), \tag{27}$$

where $\bar{o}_{12}$ is the off-diagonal element of $\bar{O}_2$, that is,

$$\bar{o}_{12} \equiv \int_\Gamma \frac{d\bar{P}_1(x)}{d\bar{P}_{\mathrm{mix}}(x)} \frac{d\bar{P}_2(x)}{d\bar{P}_{\mathrm{mix}}(x)} \bar{P}_{\mathrm{mix}}(dx) \geq \int_A \frac{d\bar{P}_1(x)}{d\bar{P}_{\mathrm{mix}}(x)} \frac{d\bar{P}_2(x)}{d\bar{P}_{\mathrm{mix}}(x)} \bar{P}_{\mathrm{mix}}(dx). \tag{28}$$

Using the fact that $\bar{P}_1$ and $\bar{P}_2$ are uniform on A, (27)-(28) yields

$$\mathrm{Var}(\log \hat{\gamma}^{\mathcal{G}_A}) \leq \frac{1}{n_1}\frac{P_1(A^c)}{P_1(A)} + \frac{1}{n_2}\frac{P_2(A^c)}{P_2(A)}. \tag{29}$$

Consequently, the variance decreases with the increase of mass of A under both P_1 and P_2, and it approaches zero as A approaches Γ. This is not surprising because as A approaches Γ, the difficulty of summing over A, as required by the $\mathrm{ave}_{g\in\mathcal{G}_A}$ operator, approaches that of the original summation problem we try to avoid. Putting it differently, the choice and especially the size of A models what we consider to be usable information and computationally feasible. In implementing this sub-model estimator, we do not need to actually perform the permutation because for any $x \in A$, $\bar{P}_r(\{x\}) = \sum_{w\in A} P_r(\{w\})/|A|$, so the computation is linear in $|A|$.

As an illustration, let $\Gamma = \{0, 1, 2, \ldots, \}$, μ counting measure, $q_r(x) = (r\lambda)^x/x!$, $r = 1, 2$, and thus P_r is Poisson with mean $r\lambda$. Take $A = A_k = \{0, 1, \ldots, k\}$. Then $P_r(A_k^c) \leq (r\lambda)^{k+1}/(k+1)!$, and thus the right-hand side of (29) goes to zero rapidly as k goes to infinity. Figure 1 gives the relative variance of $\log(\hat{c}_1^{A_k}/\hat{c}_2^{A_k})$ verse the same estimator but without permutation (i.e., using $k = 0$), based on the asymptotic variance formula (27). It is seen that the size of the set of values of λ, the difference between the means, that show much improvement increases with k. This is expected as when k is suitably large compared to λ and 2λ, A_k will cover a substantial amount of mass under both P_1 and P_2, and thus the group averaging will be significantly better than the original un-permuted one. This also suggests that we can choose other A's, such as the union of two neighborhoods (not necessarily overlapping) of the mean/mode of each distribution, that may lead to even more efficient estimator with the same $|A|$. The key message here is that an effective strategy for increasing overlap between two distributions is to make both of them as close to uniform as possible.

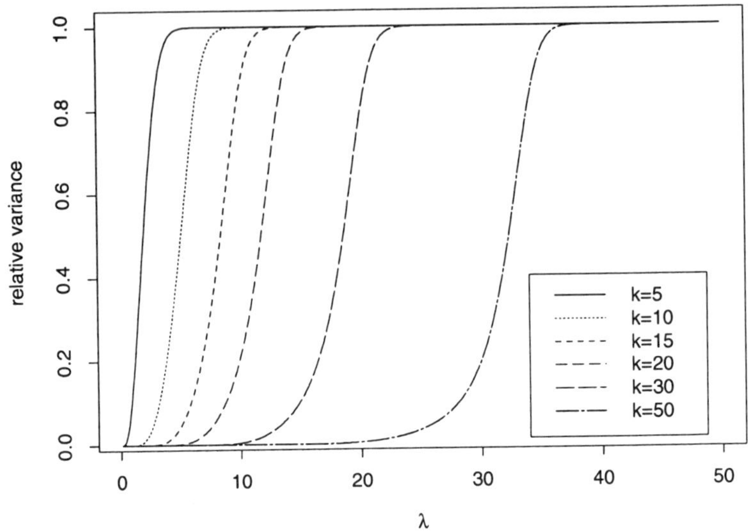

Figure 1 Variance of sub-model MLEs relative to that of MLE for the Poisson example of Section 3.5 (here k is the group size).

4 Modeling with additional information

4.1 *Parameterizing baseline measure*

An extreme form of additional information arises when we can parameterize the baseline measure μ up to a finite set of unknown parameters. Although this is of little practical interest (as it effectively assumes that we can perform integrations analytically or numerically once we are given the values of the unknown parameters), it provides a framework for examining the maximum possible gains by using additional information on μ. We give here two examples to illustrate two possible scenarios.

Example 1. Let $k = 2$, $\Gamma = \{0, 1, 2, \dots, \}$, $q_r(x) = r^x$, and suppose that μ is known to be a distribution in the Poisson family, but with unknown mean λ. Direct calculation shows that, for $r = 1, 2$, $c_r = e^{\lambda(r-1)}$, and that $\lambda_r = r\,\lambda$ is the mean of the distribution P_r. Thus $\xi \equiv \log(c_2/c_1) = \lambda_2 - \lambda_1 = \lambda$ may be estimated by the moment estimator $\hat{\xi}_{MNT} = \bar{X}_2 - \bar{X}_1$, or more efficiently by the MLE $\hat{\xi}_{MLE} = \sum_{i=1}^{n} X_i/(n_1 + 2n_2)$. The MLE is minimum-variance, unbiased and with variance

$$\mathrm{Var}(\hat{\xi}_{MLE}) = \frac{\lambda}{n}\frac{1}{f_1 + 2f_2}, \tag{30}$$

where $f_i = n_i/n$. Note the efficiency of $\hat{\xi}_{MNT}$ relative to $\hat{\xi}_{MLE}$ is $[9 + 2(\sqrt{f_2/f_1} - \sqrt{f_1/f_2})^2]^{-1} \leq 1/9$, achieved when $f_1 = f_2 = 1/2$.

To see the loss of efficiency from not parameterizing μ, we compute the asymptotic variance of the semi-parametric MLE $\hat{\xi}_{SMLE}$ from Section 1, which is also the optimal bridge sampling estimator (Meng and Wong, 1996). This variance is given by (27) (with the original o_{12} in the place of $\bar{o}_{12}$), where for our current problem, o_{12} is a function of λ given by

$$o_{12}(\lambda) = \sum_{x=0}^{\infty} \frac{(2\lambda)^x e^{-2\lambda}}{x!(f_1 + f_2 2^x e^{-\lambda})}.$$

Figure 2 plots the asymptotic efficiency of $\hat{\xi}_{SMLE}$ relative to $\hat{\xi}_{MLE}$ as a function of λ, where $f_1 = f_2 = 1/2$. It is seen that the relative efficiency is always below one and it approaches zero as $\lambda \to \infty$. This is expected because as the difference in means, $\lambda = \lambda_2 - \lambda_1$, increases, $\mathrm{Var}(\hat{\xi}_{MLE})$ goes up linearly in λ as seen in (30), but $\mathrm{Var}(\hat{\xi}_{SMLE})$ goes up exponentially in λ. The latter can be seen by using the inequality (8.4) of Meng and Wong (1996, p. 850), which states that when $f_1 = f_2 = 1/2$,

$$H_{12}^2 \leq o_{12} \leq H_{12} \equiv \int_{\Gamma} \sqrt{\frac{dP_1(x)}{d\mu(x)} \frac{dP_2(x)}{d\mu(x)}} \mu(dx), \tag{31}$$

where H_{12} determines the Hellinger distance between P_1 and P_2: $\sqrt{2(1 - H_{12})}$. Since for our current problem $H_{12} = e^{-(1.5-\sqrt{2})\lambda}$, we have

$$\frac{4}{n}\left(e^{(3-2\sqrt{2})\lambda} - 1\right) \geq \mathrm{Var}(\hat{\xi}_{SMLE}) \geq \frac{4}{n}\left(e^{(1.5-\sqrt{2})\lambda} - 1\right).$$

The phenomenon that the variance of the optimal bridge sampling estimator (when $k = 2$) goes up exponentially with the difference in means was also reported in Meng and Wong (1996), which suggests that it is crucial to increase the overlap between the P_1 and P_2, using methods such as those given in Meng and Schilling (2002), in order to improve the efficiency of bridge sampling estimators. It is also interesting to note that the parameterized MLE, $\hat{\xi}_{MLE}$, resembles the behavior of some estimators from *path sampling*, which is bridge sampling with infinitely many bridges, in the sense that the latter can also have variances that go up linearly in the difference in the means (Gelman and Meng, 1998).

Example 2. Let $\Gamma = \mathbb{R}^+$, $q_r(x) = e^{-\beta_r x}$, where $0 \leq \beta_1 < \beta_2 < \cdots < \beta_k$ are known, and $\mu(dx) = e^{-\rho x}dx$ for some unknown $\rho > 0$. Then $c_r = 1/(\beta_r + \rho)$, $r = 1, \ldots, k$, and P_r is exponential with mean c_r. The model is thus inverse linear (McCullagh and Nelder, 1989, chap. 2) with one unknown parameter ρ. The sample ratio, $\bar{X}_2/\bar{X}_1$ is a consistent, but not

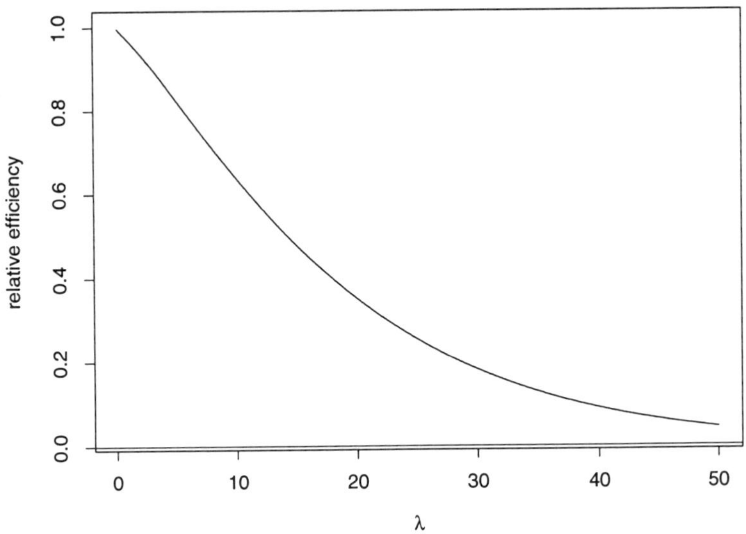

Figure 2 Relative efficiency of the semi-parametric MLE versus MLE for the Example 1 of Section 4.1.

fully efficient, estimate of the ratio c_2/c_1. The MLE of ρ is $\hat{\rho} = \max(0, \tilde{\rho})$, where

$$\sum_{r=1}^{k} \frac{f_r}{\beta_r + \tilde{\rho}} = \bar{X},$$

which has a unique solution in $(-\beta_1, \infty)$. The Fisher information is

$$I(\rho) = n \sum_{r=1}^{k} \frac{f_r}{(\beta_r + \rho)^2} = n \sum_{r=1}^{k} f_r c_r^2. \tag{32}$$

To investigate the gain of efficiency by parameterizing, we consider the case of $k = 2$. From (32), the asymptotic variance of $\hat{\xi}_{MLE} = \log(\beta_2 + \hat{\rho}) - \log(\beta_1 + \hat{\rho})$ is

$$\mathrm{Var}(\hat{\xi}_{MLE}) = \frac{1}{n} \frac{(c_1 - c_2)^2}{f_1 c_1^2 + f_2 c_2^2}.$$

On the other hand, $H_{12} = 2\sqrt{c_1 c_2}/(c_1 + c_2)$, and thus by (27) and (31), when $f_1 = f_2 = 1/2$,

$$\frac{1}{n}\left[\frac{2(\sqrt{c_1} - \sqrt{c_2})^2}{\sqrt{c_1 c_2}}\right] \le \mathrm{Var}(\hat{\xi}_{SMLE}) \le \frac{1}{n}\left[\frac{(c_1 - c_2)^2}{c_1 c_2}\right].$$

Consequently, the asymptotic relative efficiency of $\hat{\xi}_{SMLE}$ is bounded by

$$\min\left\{\frac{(\sqrt{c_1}+\sqrt{c_2})^2\sqrt{c_1c_2}}{(c_1^2+c_2^2)},\ 1\right\} \geq \frac{\mathrm{Var}(\hat{\xi}_{MLE})}{\mathrm{Var}(\hat{\xi}_{SMLE})} \geq \frac{2c_1c_2}{c_1^2+c_2^2}. \tag{33}$$

Consider the case where $\beta_1 = 1$ and $\beta_2 = \beta > 1$. Then the lower bound on the asymptotic efficiency in (33) has a minimum value $2\beta/(1+\beta^2)$, achieved when $\rho = 0$. Thus, unlike Example 1, in this example the gain in efficiency from using $\hat{\xi}_{MLE}$ may not be significant. For example, when $\beta = 2$, the asymptotic relative efficiency of $\hat{\xi}_{SMLE}$ is at least 80%, irrespective the value of ρ. This is due to substantial overlap between exponentials with mean $(1+\rho)^{-1}$ and with mean $(2+\rho)^{-1}$, regardless of the value of ρ. Of course, when $\beta \to \infty$, the asymptotic efficiency of $\hat{\xi}_{SMLE}$ tends to zero by (33), because the exponential with mean $(\beta+\rho)^{-1}$ becomes concentrated at zero, and thus has little overlap with the exponential with mean $(1+\rho)^{-1}$.

4.2 *Using label information*

As we derived in Section 3, the asymptotic covariance is determined by the design matrix F and the overlap measure matrix $O = \{o_{rs},\ 1 \leq r, s \leq k+m\}$, where

$$o_{rs} = \int_\Gamma \frac{dP_r(x)}{dP_{\mathrm{mix}}(x)}\frac{dP_s(x)}{dP_{\mathrm{mix}}(x)} P_{\mathrm{mix}}(dx). \tag{34}$$

Generally speaking, the more overlap as measured by O, the more accurate the MLE of $\log \hat{c}$, where $c = \{c_1, \ldots, c_{k+m}\}$. When Γ has certain topological structure, we can consider transformations to "warp" P_r's into similar shapes (and locations) in such a way that the transformations do not alter the normalizing constants. For example, suppose the dominating measure μ is Lebesgue. Then we can consider transforming each $x_r \sim P_r$ via an one-to-one transformation $g_r(x)$. That is, (2) is replaced by

$$c_r = \int_\Gamma q_r(g_r^{-1}(x))J_r(x)\,d\mu, \tag{35}$$

where J_r is the Jacobian for g_r^{-1}. Since we know both g_r and J_r, (35) is simply (2) with q_r replaced by $\tilde{q}_r = q_r(g_r^{-1}(x))J_r(x)$, and thus we can proceed as before. However, with appropriate choices of g_r, the corresponding new $\tilde{P}_r$'s can have substantially more overlap than the original P_r's, as measured by O. Therefore, this *warping* technique can help greatly to reduce the Monte Carlo error. We emphasize that the label information is crucial for such a procedure, in contrast to the likelihood formulation given in Section 1, where we do not assume any additional knowledge (e.g., the topological structures of P_r's, including the support Γ).

Meng and Schilling (2002) provide extensive empirical evidence on the effectiveness of this warping strategy. They considered first, second, and third order warping transformations, which correspond to location shift, scale/rotation matching, and symmetrization. The first two orders of warping can be summarized by an affine transformation $g_r(x) = S_r(x - m_r)$, where m_r and S_r can be (i) estimated from the draws from P_r (e.g., sample mean and precision) or (ii) determined analytically from the known q_r (e.g., its mode and the square root of the negative Hessian matrix at the mode). For the third order warping, Meng and Schilling (2002) suggested using mixtures of a density with its various reflections to eliminate skewness. This is mathematically equivalent to group averaging when the reflections used in the mixture form a group.

An unresolved problem with these label-specific transformations is that we currently do not have a model-based way to choose, for example, the transformation parameters $\{m_r, S_r\}$. The difficult is that in order to let the model properly estimate $\{m_r, S_r\}$ such that the warped distributions will be close to each other, we need to build such a requirement into the model. While we can estimate the distributional summaries of each P_r using the draws from it, the likelihood function based on these data do not contain direct information on how to transform these sampling distributions together. This appears to be another interesting and challenging problem in modeling our inability, namely the inability to analytically maximize overlap, as measured by O, among the underlying distributions.

5 Using the profile likelihood approach

5.1 *Profiling the empirical likelihood*

In this section, we show that the likelihood (4) can be partially maximized to produce the same results as given in Section 2 and Section 3. This empirical likelihood approach not only yields a profile likelihood for c, but also provides another explanation for why the "retrospective likelihood" for c studied in Geyer (1994) is only first-order correct, as demonstrated before (Kong et al., 2003, Section 6).

In the empirical likelihood approach, we treat both $\theta = \{\theta_1, \ldots, \theta_n\}$, where $\theta_i = \theta(x_i)$, and $c = \{c_1, \ldots, c_k\}$ as parameters. Because of (4), the profile likelihood of c is defined by

$$l(c) = \max_{\theta \in \Theta(c)} \left(\sum_{i=1}^{n} \theta_i - \sum_{s=1}^{k} n_s \log c_s \right), \tag{36}$$

where

$$\Theta(c) = \{\theta : \sum_{i=1}^{n} e^{\theta_i} q_r(x_i) c_r^{-1} \equiv \sum_{i=1}^{n} P_r(\{x_i\}) = 1, \quad r = 1, \dots, k.\}. \tag{37}$$

The equality constraint in (37) is motivated by the discussion given in Section 1.3, where it is shown that in maximum likelihood calculations, we can restrict ourselves to measures with support on $\{x_1, \dots, x_n\}$.

Before we proceed, we need to set conditions to guarantee that $\Theta(c)$ is not empty. To do so, we let

$$\mathcal{W}(c) = \{W = (w_1, \dots, w_k) : \sum_{s=1}^{k} w_s = 1 \;\&\; p_{\text{mix}}(x_i|W) > 0, \forall 1 \le i \le n\}, \tag{38}$$

where $p_{\text{mix}}(x|W) \equiv \sum_{s=1}^{k} w_s p_s(x)$, with $p_s = q_s c_s^{-1}$ denoting the normalized density. Note that (38) does not require nonnegative w_i's, but only that $p_{\text{mix}}(x_i|W)$ is positive for all i, i.e. we allow negative "weights", as long as the corresponding mixture density is positive. Clearly $\mathcal{W}(c)$ is convex, and is non-empty because $(k^{-1}, \dots, k^{-1}) \in \mathcal{W}(c)$ under our sample design.

Intuitively, because any two densities with respect to the same dominating measure cannot dominate each other in order to integrate to one, we need the following necessary condition for $\Theta(c)$ to be non-empty:

No Dominance Condition *There does not exist a $W \in \mathcal{W}(c)$ and a $1 \le t \le k$ such that*

$$p_{\text{mix}}(x_i|W) \ge p_t(x_i), \quad \text{for all } i = 1, \dots, n, \tag{39}$$

and where the inequality is strict for at least one i.

It is interesting that this intuitive necessary condition turns out to be also sufficient, as seen from the following theorem, proved in the Appendix. The $\mathcal{A}(c)$ set used in the theorem is the collection of all $(a_1, \dots, a_k)$ such that $\sum_{s=1}^{k} a_s = 0$ and $p_{\text{mix}}(x_i|A) \ge 0$ for all $i = 1, \dots, n$ with the inequality being strict for at least one i. Note that the conditions (I), (IV) and (V) were considered before (Tan, 2004), but (II) and (III) appear to be new.

Theorem 2. *Assuming $Q_{n\times k} = \{q_j(x_i)\}$ is of rank k, the following five conditions are equivalent:*

(I) *$\Theta(c)$ is non-empty;*
(II) *The "No Dominance Condition" holds;*
(III) *$\mathcal{A}(c)$ is empty;*
(IV) *$\mathcal{W}(c)$ is a bounded convex set;*

(V) *The equations*

$$\sum_{i=1}^{n} \frac{p_r(x_i)}{p_{\text{mix}}(x_i|W)} = n, \quad \text{for all} \quad r = 1, \ldots, k, \tag{40}$$

have a unique solution in the interior of $\mathcal{W}(c)$.

We remark here that while conditions (III) and (IV) are geometrically appealing, condition (V) seems to be most convenient for practical purposes, because we can check numerically the existence of the solution to (40). We also remark that when $c_r \propto \hat{c}_r$, where $\{\hat{c}_r, r = 1, \ldots, k\}$ is the solution of (6), $W = (f_1, \ldots, f_k)$ satisfies (40), where $f_r = n_r/n$. Therefore, we know that there exists at least one c such that $\Theta(c)$ is not empty.

For $k = 2$, the bounds given by (IV) can be established explicitly because condition (II) implies that the following two sets

$$N_1 = \{i : p_1(x_i) - p_2(x_i) < 0\} \quad \text{and} \quad N_2 = \{i : p_1(x_i) - p_2(x_i) > 0\}$$

are non-empty. Consequently, it is easy to verify directly that $\mathcal{W}(c)$ consists of all (w_1, w_2) such that $w_1 + w_2 = 1$ and

$$\max_{i \in N_2} \left\{ \frac{-p_2(x_i)}{p_1(x_i) - p_2(x_i)} \right\} < w_1 < \min_{i \in N_1} \left\{ \frac{-p_2(x_i)}{p_1(x_i) - p_2(x_i)} \right\}.$$

5.2 *The computation of the profile likelihood*

Equipped with Theorem 2, we can now proceed to find a computable expression for the profile likelihood $l(c)$ of (36). We first observe that the constraint in (37) is equivalent to

$$\sum_{s=1}^{k} w_s \left(\sum_{i=1}^{n} e^{\theta_i} q_s(x_i) c_s^{-1} \right) \equiv \sum_{i=1}^{n} e^{\theta_i} p_{\text{mix}}(x_i|W) = 1, \tag{41}$$

for any $W \in \mathcal{W}(c)$. Taking logarithms on both sides of (41), and then applying Jensen's inequality to the log function, we obtain that

$$\sum_{i=1}^{n} \theta_i \leq -\sum_{i=1}^{n} [\log p_{\text{mix}}(x_i|W) + \log n], \tag{42}$$

for any $\theta \in \Theta(c)$ and $W \in \mathcal{W}(c)$. This implies that

$$\max_{\theta \in \Theta(c)} \sum_{i=1}^{n} \theta_i \leq -\max_{W \in \mathcal{W}(c)} \sum_{i=1}^{n} [\log p_{\text{mix}}(x_i|W) + \log n]. \tag{43}$$

We now show that the above inequality is actually an equality, and hence the maximization needed to compute the profile likelihood $l(c)$ is equivalent to maximizing the log-likelihood for the unknown weights W under the mixture model as given by $p_{\text{mix}}(x|W)$, with $\mathcal{W}(c)$ as the parameter space. This happens because equation (40) is just the normal equation for the maximum likelihood estimate (MLE) of W under this mixture model. Let us denote with $W(c)$ the MLE under this mixture model (recall we assume condition (V) here), namely the unique solution of (40), and let

$$\theta_i(c) = -[\log p_{\text{mix}}(x_i|W(c)) + \log n]. \tag{44}$$

Evidently, this choice of θ makes (43) equality. Furthermore, (40) implies $\theta(c) \in \Theta(c)$. Consequently, $\theta(c)$ is the maximizer in (36), and therefore

$$l(c) = -\sum_{i=1}^{n} \log p_{\text{mix}}(x_i|W(c)) - n\log n - \sum_{s=1}^{k} n_s \log c_s. \tag{45}$$

Now because $W(c)$ is the solution of (40), $W(c) \in \mathcal{W}(c)$ and c must satisfy

$$T_r(W(c), c) \equiv \sum_{i=1}^{n} \frac{q_r(x_i)c_r^{-1}}{\sum_{s=1}^{k} w_s(c)q_s(x_i)c_s^{-1}} = n, \quad r = 1, \ldots, k. \tag{46}$$

We remark here that the above derivation is similar to the maximization approach used for finding MLE with control variates, as investigated in Tan (2003a, 2004) and Meng (2005). A key advantage of (45) is that it provides a direct "marginal likelihood" for c, which can be treated as a likelihood to be used in Bayesian inference for c when we have reliable prior information on it; see Section 6.1.

5.3 *Computing the MLE and the observed Fisher information*

To maximize $l(c)$, we first identify its stationary point(s). We therefore calculate

$$\frac{\partial l(c)}{\partial \log c_r} = c_r\left[-\sum_{i=1}^{n} \frac{\sum_{s=1}^{k} p_s(x_i)\frac{\partial w_s(c)}{\partial c_r} - p_r(x_i)c_r^{-1}w_r(c)}{\sum_{s=1}^{k} p_s(x_i)w_s(c)} - \frac{n_r}{c_r}\right] = nw_r(c) - n_r, \tag{47}$$

where the last equality is due to (46) and $\sum_{r=1}^{k} w_r(c) = 1$. Consequently, any stationary point c must satisfy $w_r(c) = n_r/n = f_r, r = 1, \ldots, k$. By the uniqueness of the solution $W(c)$ for (40), we can conclude that c must be the solution of $T_r(f, c) = n$ for all $r = 1, \ldots, k$, which is exactly (6) for $r = 1, \ldots, k$. Under the "connectivity" assumption of Vardi (1985), which we always assume, (6) has a unique solution (up to a multiplicative constant), which is the MLE of c under the likelihood (4). Since $l(c)$ is the profiled likelihood derived from (4), it is clear that the very same $\hat{c}$ also maximizes $l(c)$ in (45), as it should.

It is interesting to observe that if we let $W = f$ in (42) and (44), but *without* realizing that the resulting θ from (44) may not satisfy (37), we would have arrived at a "profile" log-likelihood (45) with $W = f$, as in Geyer (1994). This would be exactly the wrong log-likelihood obtained by the retrospective argument – see (6.1) of Kong et al. (2003). In other words, the wrong likelihood is the same as the incorrectly "profiled" likelihood without realizing the strong compatibility requirement between W and c, as in (46). However, because of (47), this incorrectly "profiled" log-likelihood does provide the correct MLE as it coincides with the correct profile likelihood when $c = \hat{c}$ from (6) since $W(\hat{c}) = f$.

The incorrect "profile" likelihood does not provide the correct second order inference, as demonstrated in Kong et al. (2003). The correct asymptotic covariance for $\log \hat{c}$ can be estimated by the inverse of the observed Fisher information from the profile likelihood (45) (Murphy and Van Der Vaart, 1999), namely,

$$\hat{\mathcal{I}} = -\left[\frac{\partial^2 l(c)}{\partial \log c (\log c)^\top}\right]\Bigg|_{c=\hat{c}} = -n\, \frac{\partial W(c)}{\partial \log c}\Bigg|_{c=\hat{c}}, \tag{48}$$

where the last equality is due to (47). By (46), we have

$$\left[\frac{\partial T(W,c)}{\partial W}\right]\left[\frac{\partial W(c)}{\partial \log c}\right] = -\frac{\partial T(W,c)}{\partial \log c}, \tag{49}$$

where $T = (T_1, \ldots, T_k)^\top$. Using $W(\hat{c}) = f$, it is easy to check that for any $r, s \in \{1, \ldots, k\}$,

$$\frac{\partial T_r}{\partial w_s}\Bigg|_{c=\hat{c}} = -\hat{o}_{rs} \quad \text{and} \quad \frac{\partial T_r}{\partial \log c_s}\Bigg|_{c=\hat{c}} = \hat{o}_{rs} f_s - \delta_{\{r=s\}}, \tag{50}$$

where

$$\hat{o}_{rs} = \int_\Gamma \left[\frac{dP_r(x)}{dP_{\text{mix}}(x)}\right]\left[\frac{dP_s(x)}{dP_{\text{mix}}(x)}\right] \hat{P}_{\text{mix}}(dx),$$

which is the sample version of (34). It follows from (48)-(50) that $\hat{O}_k \hat{\mathcal{I}}/n = I - \hat{O}_k F$, which implies

$$n\hat{\mathcal{I}}^- = (I - \hat{O}_k F)^+ \hat{O}_k,$$

which is equivalent to (22) and (26), except with $\hat{O}_k$ estimating O_k.

6 A paradox and some future work

6.1 *A Bayesian paradox?*

Given the success of likelihood based methods, it is natural to ask the question, "what about Bayesian methods?" Indeed, it seems so obvious that Bayesian methods should be particularly useful for dealing with simulated data, since the usual dispute of the correctness of prior information no longer exists. For example, by mathematical inequalities, such as Jensen's inequality or Cauchy-Schwartz, we may know for certain that a normalizing constant c is between two known values, a and b. Surely such prior information can and should be used in our MC integration. But how? Because the model parameter is the baseline measure μ, to put a prior on μ that respects $a \leq c = \int_\Gamma q(x)\mu(dx) \leq b$ would require similar or even greater analytic effort than what is needed to calculate c analytically. In other words, in order to carry out the Bayesian method, we need more effort than what is needed to solve the original problem.

One, of course, could try to use the profile likelihood as given in (45) to conduct Bayesian inference. This is certainly a topic worth investigating, particularly with respect to the question of trading computational efficiency with statistical efficiency because the computation of (45) is not cost-free (but at least it is numerically feasible). Nevertheless, from a philosophical point of view, profile likelihood is not an legitimate Bayesian approach, which finds marginal likelihood via integration, not maximization/profiling. Therefore, statistical inference for MC integration appears to be an ultimate paradox for Bayesian inference, because it appears that Bayesian methods can solve (at least in theory) every other inference problem except for their own computational problems (as Bayesian methods rely heavily on MC integration for implementation). Or as Kong et al. (2003) put it "This computational black hole, an infinite regress of progressively more complicated models, is an unappealing prospect, to say the least."

6.2 *Some future work*

Two important challenges are to extend the models to include the estimation of label-specific transformations, and to use effectively information on dependence structure (e.g., auto-correlation), as in an MCMC setting. There has been no progress regarding the first, but the empirical evidence provided in Meng and Schilling (2002) from the use of "warp transformation" methods suggest that the gain of efficiency by appropriately modeling the label-specific transformations can be substantial. As for the second, several papers (Tan, 2003b, 2004, 2006) show great promise. In particularly, the use of the kernel functions under the likelihood modeling in a Gibbs sampling setting, or more generally with Metropolis-Hastings algorithms can lead to estimates of normalizing constants with a n^{-1} rate of convergence, instead of the usual $n^{-1/2}$ rate (Kong et al., 2003; Tan, 2003a).

Both of these extensions can therefore have substantial practical consequences, as well as provide theoretical insight into how we should balance analytical, numerical, and simulation efforts in effective use of MC methods. As discussed before (van Dyk and Meng, 2001, rejoinder), model selection with simulated data has a different goal than with real data, because the key question is not which model is approximately true — all models that can link the simulated data to our estimand are known and true. The goal is rather to select a model that provides an effective compromise between computational complexity, human effort, and statistical efficiency. Our submodeling via group averaging was guided by this goal, and we look forward to further explorations of this method and to development of other methods that will help to achieve the same goal.

Appendix: Proof of Theorem 2

(I)⇒ (II) We prove by contradiction. Suppose (II) is false. Then there exists a $W \in \mathcal{W}(c)$ and t that satisfy (39). It follows that for any $\theta \in \Theta(c)$, we will reach the following contradiction:

$$1 = \sum_{i=1}^{n} e^{\theta_i} p_{\text{mix}}(x_i|W) > \sum_{i=1}^{n} e^{\theta_i} p_t(x_i) = 1,$$

where the first and last equality are due to $\theta \in \Theta(c)$ and $W \in \mathcal{W}(c)$, and the inequality is a consequence of (39).

(II)⇒ (III) If there exists an $A \in \mathcal{A}(c)$, then it is easy to see that $A+e_1 \in \mathcal{W}(c)$, where $e_1 = (1, 0, \ldots, 0)$, and that (39) is satisfied for this element of $\mathcal{W}(c)$ and for $t = 1$, with the inequality being strict for at least one i. This contradicts assumption (II).

(III)⇒ (IV) We again prove by contradiction. Suppose there exists a sequence $W^{(m)} \in \mathcal{W}(c)$ such that it is unbounded as $m \to \infty$. For any m, let r_m be the index such that $|W_{r_m}^{(m)}| = \max\{|W_r^{(m)}|, r = 1, \ldots, k\}$. Because r_m can only take k values, as $m \to \infty$, there is a subsequence such that r_m takes the same value, say, $r_m = 1$.

Along this subsequence $|W_1^{(m)}| \to \infty$, and $a_r^{(m)} \equiv W_r^{(m)}/|W_1^{(m)}| \in [-1, 1]$ for any $r \geq 1$. Therefore we can choose a subsequence such that the limit of $a_r^{(m)}$ exists for all r. Denote this limit by $a = (a_1, \ldots, a_k)$. Then it is clear from $\sum_r W_r^{(m)} = 1$ that

$$\sum_r a_r = \lim_{m\to\infty} \sum_r \frac{W_r^{(m)}}{|W_1^{(m)}|} = \lim_{m\to\infty} \frac{1}{|W_1^{(m)}|} = 0. \tag{51}$$

Furthermore, from $p_{\text{mix}}(x_i|W^{(m)}) > 0$, we obtain that

$$p_{\text{mix}}(x_i|A) = \lim_{m\to\infty} \frac{p_{\text{mix}}(x_i|W)}{|W_1^{(m)}|} \geq 0, \quad i = 1, \ldots, n. \tag{52}$$

If we can prove that at least one inequality in (52) is strict, then (51) and (52) allow us to conclude that $a \in \mathcal{A}(c)$, which contradicts assumption (III). To prove this, suppose $p_{\rm mix}(x_i|A) = 0$ for all $i = 1, \ldots, n$. It follows that $P_{n\times k} a^\top = 0$, where $P_{n\times k} = \{p_j(x_i)\}$ is the $n \times k$ matrix of the normalized density values. Since $P_{n\times k} = Q_{n\times k}\,\mathrm{diag}\{c_1^{-1}, \ldots, c_k^{-1}\}$ and therefore it is of full rank k under our assumption that $Q_{n\times k}$ is of full rank k, we can conclude that $a = 0$. But this is impossible because $|a_1| = 1$ by our construction.

(IV) ⇒ (V) Because $\mathcal{W}(c)$ is bounded, all its boundaries are determined by W such that $p_{\rm mix}(x_i|W) = 0$. Therefore,

$$f(W) = \sum_{i=1}^{n} \log p_{\rm mix}(x_i|W),$$

which is a concave function on $\mathcal{W}(c)$, must be maximized at an interior point of $\mathcal{W}(c)$ because at any of these boundaries $f(W) = -\infty$. Since this interior point, labeled by $W(c)$, must be a stationary point, by the method of Lagrange multiplier, it must satisfy (40). To prove this solution is unique (and hence $f(W)$ has only one stationary point, the global maximizer), let us suppose there are two solutions, W_1 and W_2, that satisfy (40). This implies that, by summing up the left-hand side of (40) with respect to either weight and then summing up the two sums,

$$\sum_{i=1}^{n} \left[\frac{p_{\rm mix}(x_i|W_1)}{p_{\rm mix}(x_i|W_2)} + \frac{p_{\rm mix}(x_i|W_2)}{p_{\rm mix}(x_i|W_1)} - 2 \right] = 0. \tag{53}$$

Using the fact that $a + a^{-1} \geq 2$ for any $a > 0$, where the equality holds if and only if $a = 1$, we can conclude from (53) that $p_{\rm mix}(x_i|W_1) = p_{\rm mix}(x_i|W_2)$ for all $i = 1, \ldots, n$, namely $P_{n\times k}(W_1 - W_2) = 0$. It follows immediately that $W_1 = W_2$ because $P_{n\times k}$ is of rank k.

(V) ⇒ (I) Let $W(c)$ be the unique solution of (40), and define $\theta(c)$ as in (44). Then this $\theta(c)$ satisfies the constraints required by (37) and hence $\Theta(c)$ must be non-empty.

References

1. Bennett, C. H. (1976). Efficient estimation of free energy differences from Monte Carlo data. *Journal of Computational Physics* **22** 245–268.
2. Ceperley, D. M. (1995). Path integrals in the theory of condensed helium. *Reviews of Modern Physics* **67** 279–355.

3. CHEN, M. H. AND SHAO, Q. M. (1997a). Estimating ratios of normalizing constants for densities with different dimensions. *Statistica Sinica* **7** 607–630.
4. CHEN, M. H. AND SHAO, Q. M. (1997b). On Monte Carlo methods for estimating ratios of normalizing constants. *The Annals of Statistics* **25** 1563–1594.
5. CHIB, S. (1995). Marginal likelihood from the Gibbs output. *Journal of the American Statistical Association* **90** 1313–1321.
6. CHIB, S. AND JELIAZKOV, I. (2001). Marginal likelihood from the Metropolis-Hastings output. *Journal of the American Statistical Association* **96** 270–281.
7. DICICCIO, T. J., KASS, R. E., RAFTERY, A. and WASSERMAN, L. (1997). Computing Bayes factors by combining simulation and asymptotic approximations. *Journal of the American Statistical Association* **92** 903–915.
8. GELFAND, A. E. AND DEY, D. (1994). Bayesian model choice: asymptotics and exact calculations. *Journal of the Royal Statistical Society, Series B* **56** 501–514.
9. GELMAN, A. AND MENG, X.-L. (1998). Simulating normalizing constants: From importance sampling to bridge sampling to path sampling. *Statistical Science* **13** 163–185.
10. GEYER, C. J. (1994). Estimating normalizing constants and reweighting mixtures in Markov chain Monte Carlo. Tech. Rep. 568, School of Statistics, University of Minnesota.
11. GEYER, C. J. AND THOMPSON, E. A. (1992). Constrained Monte Carlo maximum likelihood for dependent data (with discussion). *Journal of the Royal Statistical Society B* **54** 657–699.
12. GILKS, W. R., RICHARDSON, S. and SPIEGELHALTER, D. J. (1996). *Markov chain Monte Carlo in Practice.* Chapman & Hall, London.
13. IRWIN, M., COX, N. J. AND KONG, A. (1994). Sequential imputation for multilocus linkage analysis. *Proc. Natl. Acad. Sci. USA* **91** 11684–11688.
14. JENSEN, C. S. AND KONG, A. (1999). Blocking Gibbs sampling for linkage analysis in large pedigrees with many loops. *American Journal of Human Genetics* 885–901.
15. JOHNSON, V. (1999). Posterior distributions on normalizing constants. Tech. rep., Duke University.
16. KONG, A., MCCULLAGH, P., MENG, X.-L., NICOLAE, D. AND TAN, Z. (2003). A statistical theory for Monte Carlo integration (with discussion). *Journal of the Royal Statistical Society, Series B* **65** 585–618.
17. MACEACHERN, S. AND PERUGGIA, M. (2000). Importance link function estimation for Markov chain Monte Carlo methods. *Journal of Computational and Graphical Statistics* **9** 99–121.

18. McCullagh, P. (1999). Quotient spaces and statistical models. *Canadian Journal of Statistics* **27** 447–456.

19. McCullagh, P. and Nelder, J. A. (1989). *Generalized linear models (Second edition).* Chapman & Hall, London.

20. Meng, X.-L. (2005). Discussion: Computation, survey, and inference (discussion of "Qausi Monte Carlo and control variates" by Hickernell, Lemieux, and Owen). *Statistical Science* **20** 21–28.

21. Meng, X.-L. and Schilling, S. (1996). Fitting full-information item factor models and an empirical investigation of bridge sampling. *Journal of the American Statistical Association* **91** 1254–1267.

22. Meng, X.-L. and Schilling, S. (2002). Warp bridge sampling. *The Journal of Computational and Graphical Statistics* **11** 552–586.

23. Meng, X.-L. and Wong, W. H. (1996). Simulating ratios of normalizing constants via a simple identity: A theoretical explanation. *Statistica Sinica* **6** 831–860.

24. Murphy, S. A. and Van Der Vaart, A. W. (1999). Observed information in semi-parametric models. *Bernoulli* **5** 381–412.

25. Newton, M. A. and Raftery, A. E. (1994). Approximate Bayesian inference and the weighted likelihood bootstrap (with discussion). *Journal of the Royal Statistical Society, Series B* **56** 3–48.

26. Ott, J. (1979). Maximum likelihood estimation by counting methods under polygenic and mixed models in human pedigrees. *American Journal of Human Genetics* **31** 161–175.

27. Stephens, M. and Donnelly, P. (2001). Inference in molecular population genetics (with discussion). *Journal of the Royal Statistical Society B* **62** 605–655.

28. Tan, Z. (2003a). *A likelihood approach for Monte Carlo integration.* Ph.D. thesis, The University of Chicago, Dept. of Statistics.

29. Tan, Z. (2003b).Monte Carlo integration with Markov chain. Tech. Rep. 20, Johns Hopkins Biostatistics.

30. Tan, Z. (2004). On a likelihood approach for Monte Carlo integration. *Journal of the American Statistical Association* **99** 1027–1036.

31. Tan, Z. (2006). Monte Carlo integration with acceptance-rejection. *Journal of Computational and Graphical Statistics* (to appear).

32. Thompson, E. A. (2000). *Statistical Inference from Genetic Data.* NSF-CBMS Regional Conference Series in Probability and Statistics, Institute of Mathematical Statistics, Hayward, California.

33. van Dyk, D. A. and Meng, X.-L. (2001). The art of data augmentation. *Journal of Computational and Graphical Statistics (with discussion)* **10** 1–111.

34. Vardi, Y. (1985). Empirical distributions in selection bias models. *The Annals of Statistics* **13** 178–203.

35. Verdinelli, I. and Wasserman, L. (1995). Computing Bayes factors using a generalization of the Savage-Dickey density ratio. *Journal of the American Statistical Association* **90** 614–618.

36. Voter, A. F. (1985). A Monte Carlo method for determining free-energy differences and transition state theory rate constants. *J. Chem. Phys.* **82(4)** 1890–1899.

Chapter 29

CONFIDENCE NETS FOR CURVES

Tore Schweder

Department of Economics
University of Oslo, Oslo, NORWAY

E-mail: tore.schweder@econ.uio.no

A confidence distribution for a scalar parameter provides confidence intervals by its quantiles. A confidence net represents a family of nested confidence regions indexed by degree of confidence. Confidence nets are obtained by mapping the deviance function into the unit interval. For high-dimensional parameters, product confidence nets, represented as families of simultaneous confidence bands, are obtained from bootstrapping utilizing the abc-method. The method is applied to Norwegian personal income data.

Key words: Abc-method; Bootstrapping; Confidence cure; Likelihood; Simultaneous confidence; Quantile regression; Personal income.

1 Introduction

Confidence intervals, confidence regions and p-values are the prevalent concepts for reporting inferential results in applications, although Bayesian posterior distributions are increasingly used. In 1930, R.A. Fisher challenged the Bayesian paradigm of the time, and proposed fiducial distributions to replace posterior distributions based on flat priors. When pivots are available, fiducial distributions follow, and are usually termed confidence distributions (Efron 1998, Schweder and Hjort 2002). The cdf of a confidence distribution could also be termed a p-value function. Neyman (1941) showed the connection between his confidence intervals and fiducial distributions. Exact confidence distributions are only available in simple models, but approximate confidence distributions might be found through simulation.

The general concept of confidence distribution is difficult in higher dimension. For vector parameters one must therefore settle for a less ambitious construct to capture the inferential uncertainty. I propose to use the *confidence net.* A confidence net is a stochastic function from parameter space to the unit interval with level sets representing simultaneous confidence regions. One important method to construct a confidence net is to map the deviance function into the probability interval such that the distribution of the transformed deviance at the true value is uniformly distributed. That the confidence net evaluated at the true value is uniformly distributed is actually the defining property of confidence nets.

Confidence net for a scalar parameter was introduced by Birnbaum (1961) under the name 'confidence curve'. The method has been been repeatedly proposed under different names (Bender, Berg and Zeeb 2005) but has only found sporadic use in applied work. I will use the term 'confidence net' to keep it apart from the estimated curve.

Curves such as correlation curves (Bjerve and Doksum 1993) are often represented by its ordinate values at a finite number of argument values, and with a method to connect neighboring values by a piece of continuous curve. I regard the set of values defining the curve as a parameter, usually of high but finite dimension. A simultaneous confidence band for the curve is a product confidence region for the vector parameter. It has the shape of a box or a rectangle in parameter space. A nested family of product confidence regions indexed by their coverage probabilities, constitutes the level sets of a confidence net for the curve. Such confidence nets can be constructed from families of point-wise confidence bands by adjusting the nominal levels to be simultaneous coverage probabilities. Beran (1988) developed this construction into a theory of balanced simultaneous confidence sets. My confidence nets for curves are essentially variants of his method.

Confidence nets for curves might be hard to develop analytically except in special cases. Bootstrapping or other simulation techniques are generally more useful. The abc-method of Efron (1987) leads to confidence nets for scalar parameters which are easily combined to a product confidence net for the curve.

In the next section I discuss confidence nets in general, and give some examples. Then I discuss confidence nets for curves, and show how they might be found by bootstrapping and the abc-method. In the final section I study personal income in Norway by quantile regression curves with associated confidence nets. I am particularly interested in the 5% upper quantile of income on capital as a function of wage for given age. I will use the abc-method on a vector of 29 components.

2 Confidence distributions and confidence nets

The setup is the familiar, with data X being distributed over a measurable space according to a parameterized distribution P_θ. The parameter space is Euclidean of finite dimension.

Assume first θ to be scalar, and that a pivot $piv(\theta, X)$, increasing in θ and with continuous cdf F, is available. The probability transformed pivot $C(\theta; X) = F(piv(\theta, X))$ is then the cdf of a confidence distribution for θ. For any observed value of the data X, $C(\theta; X)$ is in fact a cdf in θ representing the confidence distribution inferred from the data, and for any value of θ, $C(\theta; X)$ is uniformly distributed on the unit interval when $X \sim P_\theta$. These two properties are basic for confidence distributions (Schweder and Hjort (2002)).

Let $C^{-1}(p; X)$ be the confidence quantile. Since C is uniformly distributed at the true value, $P_\theta(C(\theta; X) \leq p) = P_\theta(\theta \leq C^{-1}(p; X)) = p$. The interval $\left(C^{-1}(\alpha; X);\quad C^{-1}(\beta; X)\right)$ is thus a confidence interval of degree $\beta - \alpha$ for all choices of $0 \leq \alpha \leq \beta \leq 1$. The distribution represented by C therefore distributes confidence over interval statements concerning θ. This is the reason for the name *confidence distribution.*

To ease notation, X will often be suppressed. Whether $C(\theta)$ is a stochastic element or a realization, and whether it is a function of θ or a value, should be clear from the context. Similarly, $L(\theta)$, $D(\theta)$, and $N(\theta)$ denotes likelihood function, deviance function, and confidence net (to be defined below) respectively.

Let

$$N(\theta) = 1 - 2\min\{C(\theta), 1 - C(\theta)\} = |1 - 2C(\theta)|. \tag{1}$$

At each level $1 - \alpha$, the level set $\left(C^{-1}(\alpha/2);\quad C^{-1}(1 - \alpha/2)\right)$ of N is a tail-symmetric confidence interval. I will call N a tail-symmetric *confidence net* for θ. The concept of confidence net is not confined to scalar parameters. Confidence nets share with pivots the property of having a constant distribution at the true value of the parameter:

Definition 1. A stochastic function N from parameter space to the unit interval is a confidence net if for each θ, $N(\theta; X)$ is uniformly distributed on the unit interval when $X \sim P_\theta$.

The level sets $\{\theta : N(\theta; X) \leq \alpha\}$ are clearly confidence regions for θ. N is called a confidence net since its level sets constitute a net in the mathematical sense.

A confidence distribution is only partially characterized by a related confidence net. This is the case in one dimension, and even more so in higher dimensions. There are actually many confidence nets stemming from the same confidence distribution. In the scalar case for example, there are various tail-asymmetric confidence nets such as $N_s(\theta) = 1 - \min\{C(\theta)/s, (1-C(\theta))\,/\,(1-s)\}$ for s between zero and one. Here N_s has level sets $\left(C^{-1}(s\,(1-\alpha));\quad C^{-1}(1-(1-s)\,(1-\alpha))\right)$ which are tail-asymmetric for $s \neq 1/2$.

Note that $C(\theta) = N_0(\theta)$ and $1 - C(\theta) = N_1(\theta)$ both are confidence nets with extreme tail skewness. They are one-sided confidence nets.

A confidence distribution in one dimension might be displayed by its cdf, its density $c(\theta) = C'(\theta)$, or often preferably by a confidence net which usually would be tail-symmetric. Other distributions such as Bayesian posteriors might also be displayed by (confidence) nets rather than by their densities.

Example 1. Let $X > 0$ be exponentially distributed with mean $\theta > 0$. Then $Y = -X/\theta$ is exponentially distributed on the negative half-axis, and is thus a pivot with cdf $exp(y)$, $y < 0$. The probability transformed pivot is the confidence cdf $C(\theta) = \exp(-\theta X)$. Figure **??** shows representations of the confidence distribution. The five realizations of the confidence net in the lower right panel cross the vertical line at the true value $\theta = 1$ at uniformly distributed levels.

With $\widehat{\theta}$ the maximum likelihood estimator,

$$D(\theta) = -2\ln(L(\theta)\,/L(\widehat{\theta})) \tag{2}$$

is the deviance function. The deviance gives rise to confidence nets, also when θ is a vector parameter. So do other suitable objective functions, such as the profile deviance. The following proposition is trivial.

Proposition 1. *If the (profile) deviance evaluated at the true value, $D(\theta)$, has continuous cumulative distribution function F_θ, then*

$$N(\theta) = F_\theta(D(\theta))$$

is a confidence net.

In regular cases the null distribution F_θ is asymptotically independent of the parameter, and it is approximately the chi-square distribution with degrees of freedom equal to the dimension of θ.

Since the deviance is invariant, the confidence net based on the deviance is invariant. But the deviance might be biased in the sense that the maximum likelihood estimator $\widehat{\theta}$ is biased. I prefer to define bias in terms of the median rather than the mean, to make the notion of no bias invariant to monotonous transformations.

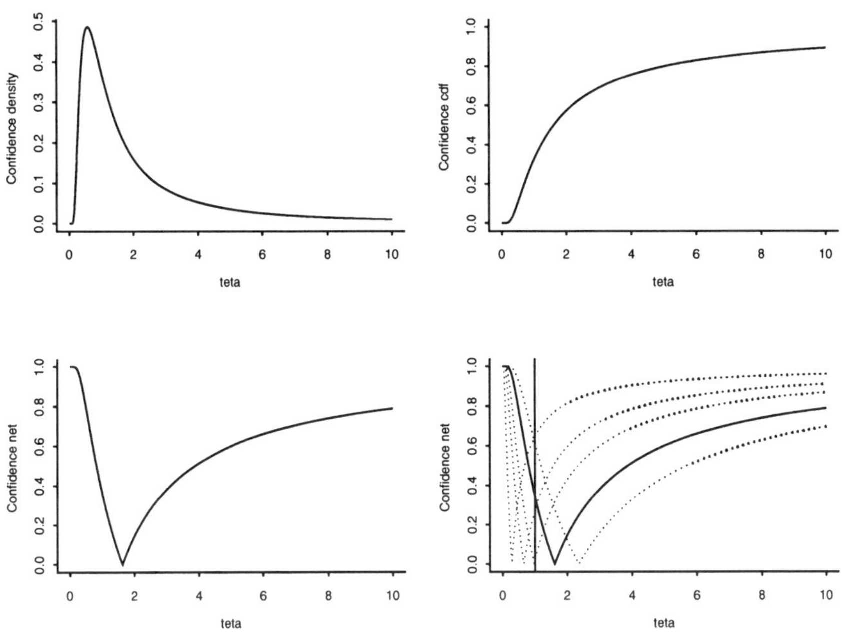

Figure 1 Confidence density (upper left), confidence cdf (upper right), tail-symmetric confidence net (lover left) for data $X = 1.12$ drawn from the standard exponential distribution, and also four other realizations of N based on data from the same distribution (lower right).

Monotonicity and median might be defined in several ways when the dimension is higher than one. For vector parameters representing curves it is natural to define these notions component-wise, as we shall do.

Definition 2. A confidence net N is unbiased when the point estimator $\widetilde{\theta} = \arg\min N(\theta)$ is median unbiased in each component.

Writing m for the vector of component medians, let $b(\theta) = m\left(\widehat{\theta}\right)$ for the maximum likelihood estimator $\widehat{\theta}$. With b invertible $\widetilde{\theta} = b^{-1}\left(\widehat{\theta}\right)$ is median-unbiased and it minimizes $D(b(\theta))$. With F_θ the cdf of $D(b(\theta))$, $N(\theta) = F_\theta(D(b(\theta)))$ is a bias-corrected confidence net.

Example 2. The maximum likelihood estimator $\widehat{\sigma}$ of σ is badly biased in the Neyman-Scott example of highly stratified normal data. Let $X_{ij} \sim N(\mu_i, \sigma^2)$ $i = 1, \cdots, n$ $j = 1, 2$. The maximum likelihood estimator is $\widehat{\sigma}^2 = \frac{1}{4n}\sum_{i=1}^{n}(X_{i1}-X_{i2})^2$ and the profile deviance is $D(\sigma) = 2n\left(\widehat{\sigma}^2/\sigma^2 - \ln\left(\widehat{\sigma}^2/\sigma^2\right) - 1\right)$. $\sigma^2/\widehat{\sigma}^2 \sim 2n/\chi^2_n$ is a pivot yielding a confidence distribution and a confidence net. From the pivotal distribution, $b(\sigma) = m(\widehat{\sigma}) = \sigma\sqrt{m(\chi^2_n)/(2n)}$. The null distribution of $D(b(\sigma))$ is that of $2n\left(\chi^2_n/m\left(\chi^2_n\right) - \ln\left(\chi^2_n/m\left(\chi^2_n\right)\right) - 1\right)$, which is free of σ and is easily obtained by simulation. The null distribution is actually very nearly that of $2\chi^2_1$ rather than that of χ^2_1. Figure 2 shows both the confidence net based on the bias-corrected profile deviance and the tail-symmetric confidence net from the confidence distribution based on the pivot, for $n = 20$ and $\widehat{\sigma} = 0.854$. The null distribution of the bias-corrected profile deviance was based on a sample from the χ^2_n of size only 500 to enable the dashed line to be seen. The two nets are practically identical when using the exact distribution of $D(b(\sigma))$.

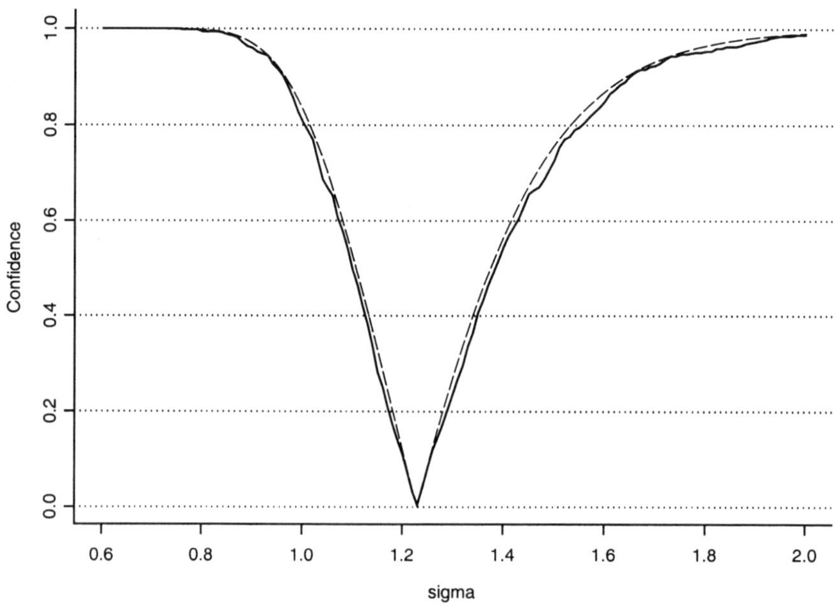

Figure 2 Exact tail-symmetric confidence net for σ in the Neyman-Scott example (dashed line), and approximate confidence net based on the bias-corrected deviance (solid line, 500 simulation replicates), $n = 20$ and $\widehat{\sigma} = 0.854$.

Example 3. Consider a situation with two variance parameters, σ_1^2 and σ_2^2, with independent estimators distributed proportional to chi-square distributions with 9 and 4 degrees of freedom respectively. We are primarily interested in $\psi = \sigma_2/\sigma_1$. The maximum likelihood estimator $\widehat{\psi}$ has median 0.95ψ while $\widehat{\sigma}_1$ has median $0.96\sigma_1$. Applying median bias-correction to each component separately, the deviance function yields the confidence net that is contoured in Figure 3. The maximum likelihood estimates behind this net are $\widehat{\sigma}_1 = 1$ and $\widehat{\psi} = 2$. The distribution of the two-parametric null deviance is independent of the parameter, and is found by simulation. It is nearly chi-square 2. The confidence net based on the bias-corrected profile deviance for ψ, with null distribution nearly that of $1.08\chi_1^2$, is shown in the lower panel. The exact tail-symmetric confidence net based on the F-distribution is also plotted, but is practically identical to that based on the profile deviance.

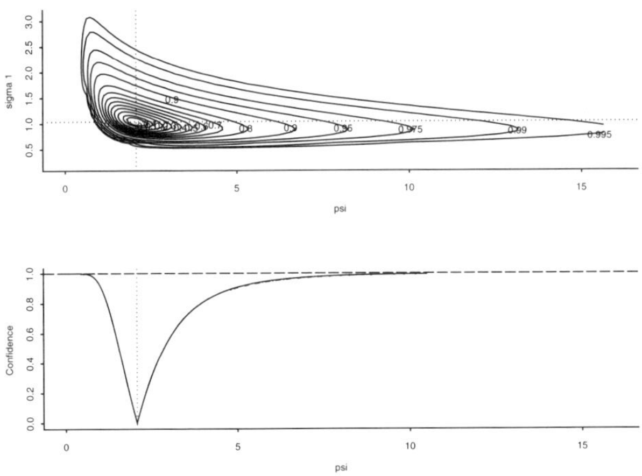

Figure 3 Confidence net for $(\psi = \sigma_2/\sigma_1, \sigma_1)$ (upper panel) for two normal samples. In lower panel, confidence net for ψ based on the bias-corrected profile deviance. The median unbiased estimates are indicated by dotted lines. Null distributions found by simulation with 100,000 replicates.

My experience is that confidence nets in one dimension based on median bias-corrected profile deviances almost perfectly agree with exact confidence nets based on the same statistic when the latter exist.

3 Confidence nets for curves

For non-parametric curves, represented by a vector of ordinates θ, simultaneous confidence bands are product confidence regions for θ. Product confidence nets are therefore desirable for capturing the inferential uncertainty in curve estimates.

For parametric curves, the confidence bands might be obtained from simultaneous confidence regions for the basic parameter. In the linear normal model for example, the method of Scheffé (1959) for multiple comparison is based on elliptical confidence nets. A curve constructed as an indexed set of linear parameters, is obtained from the elliptical confidence net dating back to Working and Hotelling (1929). If the curve represents all the linear functions that can be constructed from an r-dimensional linear parameter, the product confidence net for the curve is equivalent to the r-dimensional elliptical confidence net for the parameter.

Tukey's method of simultaneous inference for pair-wise differences (Scheffé 1959) might be regarded as a problem of developing a product confidence net for the vector of pair-wise differences, and could be solved by simulation as outlined below when the Studentized range distribution is of questionable validity.

Nair (1984) constructed simultaneous confidence bands for survival functions by suitably expanding the set of point-wise intervals. Beran (1988) went a step further by constructing simultaneous product confidence sets from what he called a root. The root is essentially a collection of confidence nets for the components of the vector parameter. For given nominal confidence coefficient for all the point-wise intervals, their product set is a box-shaped confidence set of a degree that depends on the dependence structure and other particulars. This simultaneous confidence degree is often found by simulation. Beran's method is then simply to chose nominal degree to obtain the desired simultaneous degree. It leads to balance in the sense that the simultaneous confidence set has the same marginal coverage probability for each component of the parameter seen in isolation.

My product confidence net for a vector parameter is simply representing the collection of Beran's balanced product sets. Let the curve be represented by the vector parameter θ of dimension T and with generic component θ_t. Let $N_t(\theta_t)$ be the one dimensional confidence net for θ_t. This point-wise confidence net is typically found from a confidence distribution, or from a marginal, conditional or profile deviance function. N_t is uniformly distributed at the true value of the parameter. The confidence net for θ I am looking for is the product net $(K(N_t(\theta_t)) \quad t = 1, \cdots, T)$, written $(K(N_t(\theta_t)))$, for a suitable transformation $K : (0,1) \to (0,1)$ called the *adjustment function.*

Definition 3. An increasing transform K turns a set of point-wise confidence nets $N_t(\theta_t)$ into a balanced product confidence net for $\theta =$

$(\theta_1, \cdots, \theta_T)$,

$$N(\theta) = (K(N_t(\theta_t))),$$

if

$$P_\theta (K(N_t(\theta_t)) \leq \alpha \text{ for } t = 1, \cdots, T) = P_\theta \left(\max_{t=1\cdots T} N_t(\theta_t) \leq K^{-1}(\alpha) \right) = \alpha$$

for all θ and $0 < \alpha < 1$.

Analogous to Tukey's problem, which is solved by the Studentized range distribution, my problem is to find the distribution with cdf K of the maximum null net $\max_{t=1\cdots T} N_t(\theta_t)$. Only in rare cases is it possible to solve this problem analytically.

The net $N(\theta)$ is balanced in the sense of Beran (1988) since each of its confidence regions is the product of intervals with the same nominal degree of confidence across the coordinates.

Example 4. Continue the previous example. With $\psi = \sigma_2/\sigma_1$, $\theta = (\psi, \sigma_1)$ is not much of a curve, but is used to illustrate the Chinese box structure of product confidence nets. A product confidence net has rectangular level sets shown in Figure 4. The root for this product net consists of the two confidence nets $N(\sigma_1) = |1 - 2F_{\nu_1}(\nu_1\widehat{\sigma}_1^2/\sigma^2)|$ and $N(\psi) = |1 - 2F_{\nu_2\nu_1}\left(\left(\widehat{\psi}/\psi\right)^2\right)|$. Here, F_{ν_1} and $F_{\nu_2\nu_1}$ are the cdfs of the appropriate chi-square distribution and the F-distribution respectively. The adjustment function K is now the cdf of $\max\{|1 - 2F_{\nu_1}(X)|, |1 - 2F_{\nu_2\nu_1}(Y/X)|\}$ where X and Y are independent chi square distributed with ν_1 and ν_2 degrees of freedom respectively. Figure 4 shows also the adjustment function K together with the approximate adjustment function determined by the simple Bonferroni method.

Balance is not always desirable for product confidence nets. Some components of the parameter might be of more interest than others, and, for these, narrower projected nets are required on the expense of wider projected nets for less interesting components. A practical weighting scheme is provided by component-specific transformations of the form

$$K_t(c) = K\left(c^{1/w_t}\right).$$

Here, w_t is a weight of interest in component θ_t. With K being the cdf of $\max_{t=1\cdots T} N_t(\theta_t)^{1/w_t}$, $N(\theta) = (K_t(N_t(\theta_t)))$ is indeed a confidence net for θ.

I return to a balanced product confidence net based on a root of pointwise confidence nets. When the latter is obtained by way of simulation, the

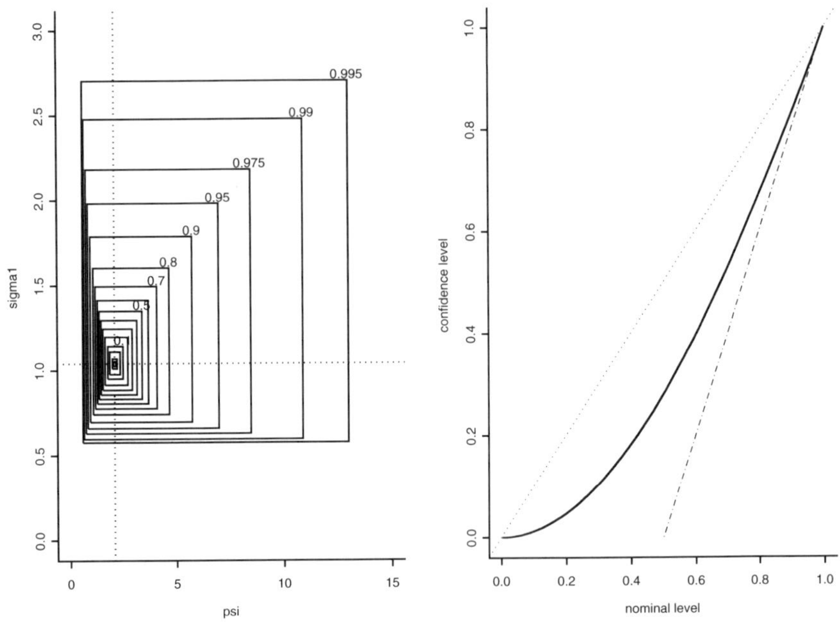

Figure 4 Level sets of a product confidence net for $\psi = \sigma_2/\sigma_1$ and σ_1 (left panel), and K (solid line, right panel) with the diagonal and also the simple Bonferroni adjustment function (dashed line). $\nu_1 = 9$ and $\nu_2 = 4$. 100,000 simulations.

adjustment function K can be estimated from the same set of simulated data sets. Let N_t^* be the one dimensional confidence net for θ_t based on a random set of data simulated with $\theta = \theta^0$. When the simulation is done by non-parametric bootstrapping, the estimate $\widehat{\theta}$ based on the observed data serves as reference value, $\theta^0 = \widehat{\theta}$. Maximizing the value of the net at the reference value, across components, leaves us with $\max_{t=1\cdots T} N_t^*\left(\theta_t^0\right)$. The adjustment function K is then simply the cdf of this random variable.

Since the point-wise confidence nets are transformed to a common scale, e.g. to have uniform null distribution, the distribution of the maximum null net will be independent of the true value of θ^0 to a good approximation when the correlation structure in the null nets varies little with the parameter.

Exact confidence nets are available for the individual components only if pivots are available. Unfortunately, models with exact pivots are rare. Under regularity conditions, approximate pivots are however available for large data. Beran (1988) suggests to use bootstrapping to obtain such approximate pivots as input to his construction, and he develops first order asymptotic results which apply to the direct simulation approach and the bootstrapping approach based on Efron's adjusted percentile method for constructing confidence intervals.

Efron (1987) introduced acceleration and bias corrected percentile intervals for one dimensional parameters, see also Schweder and Hjort (2002) who term them abc intervals. The idea is that on some transformed but monotonous scale Γ, $\widehat{\gamma} = \Gamma\left(\widehat{\theta}\right)$ is normally distributed with mean $\gamma - b\,(1 + a\gamma)$ and variance $(1 + a\gamma)^2$, and with $\gamma = \Gamma(\theta)$. With a value for the acceleration constant a, which might take some effort to find, and for the bias constant $b = \Phi^{-1}\left(H(\widehat{\theta})\right)$, the tail-symmetric abc-net is

$$\begin{aligned} N_{abc}(\theta;\widehat{\theta}) &= |1 - 2C_{abc}(\theta)|, \\ C_{abc}(\theta;\widehat{\theta}) &= \Phi\left(\frac{\Phi^{-1}\left(H(\theta)\right) - b}{1 + a\left(\Phi^{-1}\left(H(\theta)\right) - b\right)} - b\right), \end{aligned} \tag{3}$$

where H is the cdf of the bootstrap distribution for the estimator $\widehat{\theta}$ assumed here to be based on $B = \infty$ replicates. The scale transformation is related to H as

$$\Gamma(s) = (1 + a\widehat{\gamma})\left\{\Phi^{-1}\left(H(s)\right) - b\right\} + \widehat{\gamma}. \tag{4}$$

The confidence adjustment of the product of these point-wise abc-nets is easily obtained, as explained in the following proposition.

Proposition 2. *The balanced product confidence net for* $\theta = (\theta_1, \cdots, \theta_T)$ *obtained from point-wise abc-nets* $N^t_{abc}(\theta_t;\widehat{\theta}_t)$ *given by (3) from non parametric bootstrapping of the data, is*

$$N_{abc}(\theta) = \left(K\left(N^t_{abc}\left(\theta_t\right)\right)\right)$$

where K *is the cdf of*

$$V = \max_t |1 - 2H_t(\theta^*_t)|, \tag{5}$$

and H_t *is the cdf of the bootstrapped component estimates* θ^*_t.

Proof. The basis for non parametric bootstrapping is that the bootstrapped curve estimate θ^* has nearly distribution $P_{\widehat{\theta}}$ when $\widehat{\theta}$ has distribution P_θ. The max null-net distribution of the unadjusted product net is thus found from the joint distribution of $C_{abc}(\widehat{\theta}_t;\theta^*_t)$. Consider a given component t. Using (4) for Γ_t and utilizing the invariance property of confidence distributions,

$$C_{abc}(\widehat{\theta}_t;\theta^*_t) = C_{abc}(\widehat{\gamma}_t;\gamma^*_t) = \Phi\left(\frac{\widehat{\gamma}_t - \Gamma_t(\theta^*_t)}{1 + a_t\widehat{\gamma}} - b_t\right) = 1 - H_t\left(\theta^*_t\right).$$

This proves that the adjustment function K indeed is given by (5). □

As mentioned, the same set of bootstrapped curve estimates can be used to calculate the point-wise abc-nets and the adjustment function K. This is correct to first order when the curve is of finite dimension and the assumptions behind the abc method holds for each component.

For non-parametric curves, or parameters of infinite dimension, the transformation K is not directly available. Beran (1988) showed however that sampling from the infinite index set solves the problem, at least asymptotically. For regression curves over compact support, Claeskens and Van Keilegom (2003) found asymptotically correct simultaneous confidence sets based on bootstrapping.

4 Application to Norwegian income data

Statistics Norway surveyed income and wealth for a random sample of individuals in 2002. I consider males 18 years and above, and only look at yearly income on capital Y by yearly income from all other sources X, controlled for age A. Income is measured in Norwegian crowns, and all figures are before tax. Sample size is 22496. I am interested in the upper quantiles in the conditional distribution of Y given X and A, particularly in how the 95% quantile varies with X when controlling for the effect of age. I assume an additive quantile regression function of the form

$$Q_p(Y|X,A) = h(X) + g(A) + error.$$

Q_p is the p-quantile function, here acting on the conditional distribution of Y given X and A. The smooth curve h is represented by the vector $(h(x_1), \cdots, h(x_{29}))$ for x_t the $t/30$ quantile in the marginal distribution of X. Similarly, g is represented by a 29-dimensional vector.

The model is fitted by a simplified version of the backfitting algorithm of Yu and Lu (2004) as follows. Let $bin(x_t)$ be the bin for X containing the value x_t. Similarly for $bin(a_t)$. The iteration alternates between updating h and g by taking quantiles of adjusted values of Y for values in appropriate bins, and starts with $g = g_0 = 0$. In the i-th iteration, first $h_i(x_t) = Q_p(Y - g_{i-1}(A)|X \in bin(x_t))$ $t = 1, \cdots, 29$. Then $g_i^1(a_t) = Q_p(Y - h_i(X)|A \in bin(a_t))$, which is median shifted to $g_i(a_t) = g_i^1(a_t) - Q_{.5}(g_i^1)$, also for $t = 1, \cdots, 29$. Since the sample size is large, I used no smoothing across bins in this simplified algorithm.

Using $p = .95$ the backfitting algorithm converges quickly, see Figure 5. I thus settle for 6 iterations in the estimation. The curve estimator to be bootstrapped is $h = s(h_6)$ where s is the smoothing spline with 10 degrees of freedom found by gam in Splus. This degree of smoothing was chosen to allow the rather sharp bend in h come through (Figure 5).

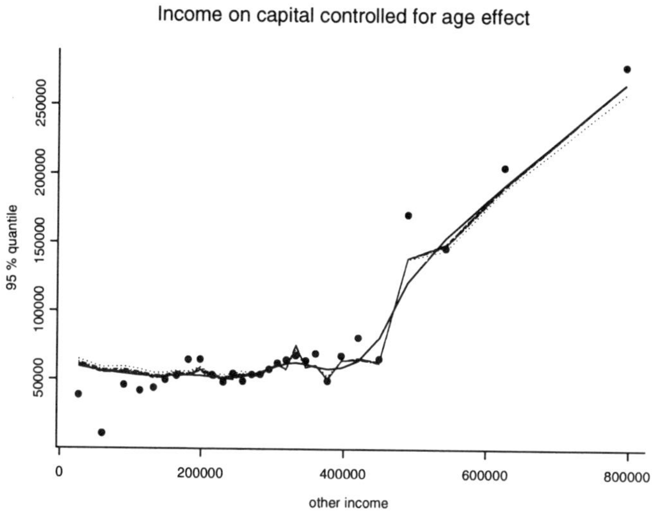

Figure 5 Points are 95% quantiles of capital income in bins by other income, regardless of age. The iterates of the algorithm of Yu and Lu (2004) are shown by different line types. They converge quickly to the rugged curve appearing to be solid. The smooth solid curve is the smooth of the 6th iteration, $df = 10$.

A non-parametric bootstrap experiment with 1000 replicates was carried out. The bootstrap results were first studied for each component separately to see whether there seems to be acceleration in the standard deviation on the transformed scale on which the bootstrap values are normally distributed. If, on the other hand, the assumption $a = 0$ can be made, the level sets of the abc-confidence nets are particularly simple to compute. After some trial and error I found that $\sqrt[10]{\widehat{h}(x_t)}$ is reasonably normally distributed for each component. The question is then whether the standard deviation in bootstrap estimates transformed to this scale is close to constant when plotted against the transformed estimate across the components.

The lower right panel of Figure 6 shows the scatter over the 29 values of x of marginal standard deviation in bootstrapped samples versus curve estimate, both at the 10th root scale. On this scale, standard deviation increases only slightly with curve estimate. From simple regression the slope is estimated to be 0.06. This small value allows the acceleration to be neglected, and $a = 0$ is assumed in (3). Figure 6 also shows that little bias correction is needed (upper left panel), and that the variance on the transformed scale indeed is relatively stable (lower left panel). The normal probability plot of the bootstrapped $\sqrt[10]{\widehat{h}(332000)}$ in the upper right panel

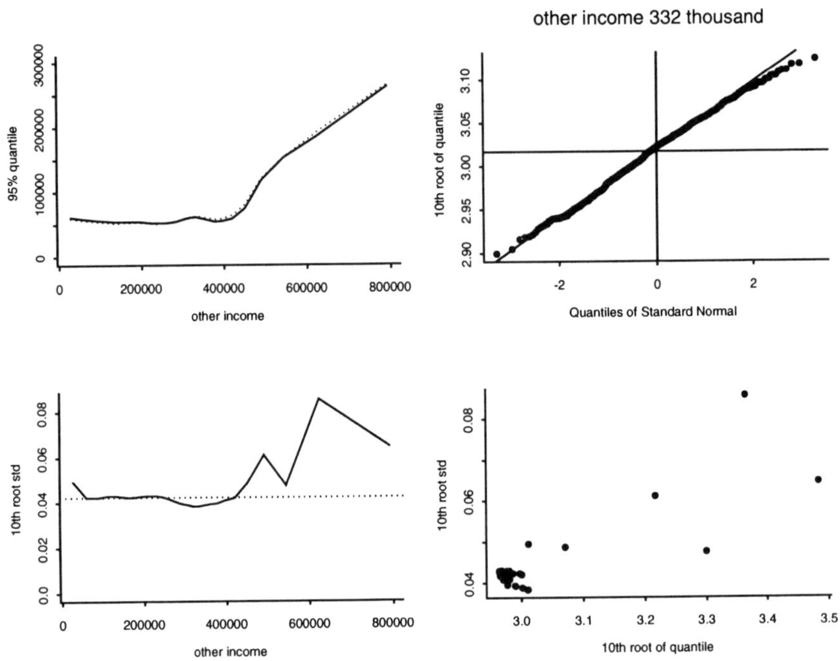

Figure 6 Curve estimate, bias corrected by the abc method (solid line) and uncorrected estimate (dotted line) in upper left panel. Standard deviation of 10th root of bootstrapped estimates in lower left panel. Upper right panel: normal probability plots of tenth root of curve estimator at a typical level of other income. Lower right panel shows the scatter of marginal standard deviation in the bootstrap samples versus curve estimate, both at the 10th root scale.

Table 1 Nominal pointwise confidence needed to obtain the abc net and the simple Bonferroni net at given simultaneous confidence. The default smoothing by gam in Splus is denoted by default df.

	.50	.75	.90	.95	.99
abc : $K^{-1}(\alpha)$ *df* $= 10$	.908	.964	.988	.994	.998
abc : $K^{-1}(\alpha)$ default *df*	.816	.926	.976	.990	.998
Bonferroni : $1-(1-\alpha)/29$	.983	.991	.997	.998	1.00

shows that this bootstrap distribution is close to normal. This is the case also at the other values of x.

The max null net (5) has distribution with cdf K found from the same set of 1000 bootstrap replicates as that used for the point-wise abc-nets.

Bjerve and Doksum (1993) used the Bonferroni method to construct a simultaneous confidence band. This construction leads to a competing

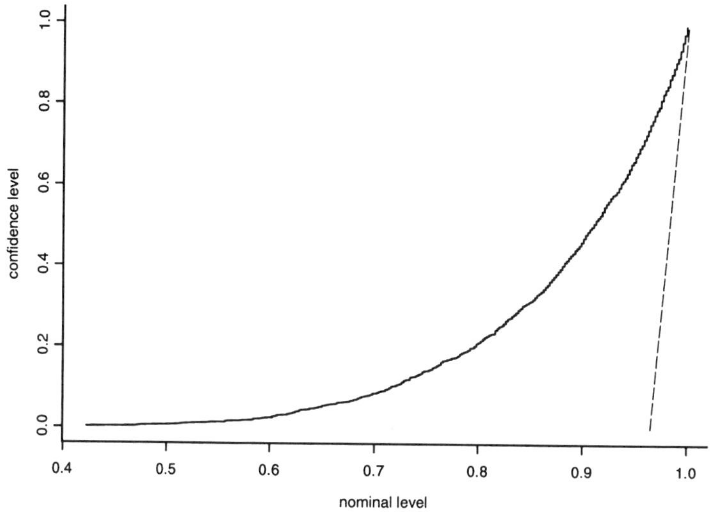

Figure 7 Adjustment function K for the abc product net (solid curve) and the simple Bonferroni adjustment (dashed line).

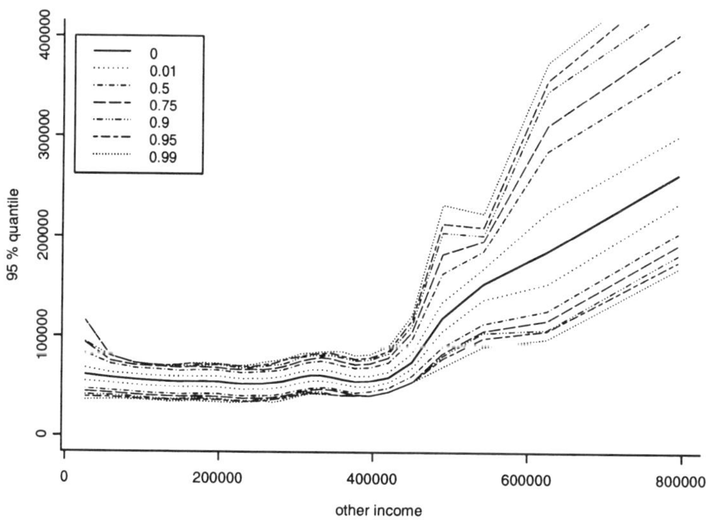

Figure 8 Level sets for the abc product confidence net for the quantile regression curve of income on capital by other income, controlling for age. Levels given in the legend.

confidence net. Figure 7 compares the Bonferroni adjustment with the abc product net adjustment. The latter method is more conservative, as is well known, and by a considerable degree. Table 1 spells this out numerically. It also shows adjustment results under more smoothing of the non-parametric quantile regression curve. The smoothed curve in Figure 5 has 10 degrees of freedom. With heavier smoothing there is more autocorrelation in the curve estimator, and the adjustment curve is lifted up.

The resulting confidence net is shown in Figure 8. One conclusion is that the 95% quantile regression of capital income on other income, controlling for the effect of age, is nearly flat up to other income 400000 crowns (slightly less than a professor's wage), and is then nearly linearly increasing. I cannot explain this strong signal in the data, and will pass the question to the economists.

Acknowledgments

Data from Statistics Norway's Survey of Income and Wealth were made available to me in anonymized form by Norwegian social science data services.

References

1. Beran, R. 1988. Balanced simultaneous confidence sets. *J. Amer. Statis. Assoc.* **83**: 679-686.
2. Bender R., Berg G., Zeeb H. 2005. Tutorial: Using confidence curves in medical research. *Biometrical Journal.* **47**: 237-247
3. Birnbaum, A. 1961. Confidence curves: an omnibus technique for estimation and testing statistical hypotheses. *J. Amer. Statis. Assoc.* **56**: 246-249.
4. Bjerve, S. and Doksum, K. 1993. Correlation curves: measures of association as functions of covariate values. *The Annals of Statistics.* **21**:890-902.
5. Claeskens, G. and Van Keilegom, I. 2003. Bootstrap confidence bands for regression curves and their derivatives. *The Annals of Statistics.* **31**: 1852-1884.
6. Efron, B. 1987. Better bootstrap confidence intervals (with discussion). *J. Amer. Statis. Assoc.* **82**, 171-200.
7. Efron, B. 1998. R.A. Fisher in the 21st century (with discussion). *Statistical Science*, **13**:95-122.
8. Fisher, R. A. 1930. Inverse probability. *Proc. Cambridge Phil. Society.* **26**: 528-535.

9. NAIR, V. N. 1984. Confidence bands for survival functions with censored data: a comparative study. *Technometrics* **26**: 265-275.
10. NEYMAN, J. 1941. Fiducial argument and the theory of confidence intervals. *Biometrika* **32**: 128-150.
11. SCHEFFÉ, H. 1959. *The Analysis of Variance.* Wiley
12. SCHWEDER, T. AND HJORT, N. L. 2002. Confidence and likelihood. *Scandinavian Journal of Statistics.* **29**: 309-332.
13. WORKING, H. AND HOTELLING, H. 1929. Application of the theory of error to the interpretation of trends. *J. Amer. Stat. Assoc.*, Mar. Suppl.: 73-85.
14. YU, K. AND LU, Z. 2004. Local linear additive quantile regression. *Scandinavian Journal of Statistics.* **31**: 333-346.

PART 10

Constrained Inference

Chapter 30

DENSITY ESTIMATION BY TOTAL VARIATION REGULARIZATION

Roger Koenker and Ivan Mizera

Department of Economics
U. of Illinois, Champaign, IL, U.S.A

Department of Mathematical and Statistical Sciences,
U. of Alberta, Edmonton, Alberta, CANADA

E-mails: rkoenker@uiuc.edu & mizera@stat.ualberta.ca

L_1 penalties have proven to be an attractive regularization device for nonparametric regression, image reconstruction, and model selection. For function estimation, L_1 penalties, interpreted as roughness of the candidate function measured by their total variation, are known to be capable of capturing sharp changes in the target function while still maintaining a general smoothing objective. We explore the use of penalties based on total variation of the estimated density, its square root, and its logarithm – and their derivatives – in the context of univariate and bivariate density estimation, and compare the results to some other density estimation methods including L_2 penalized likelihood methods. Our objective is to develop a unified approach to total variation penalized density estimation offering methods that are: capable of identifying qualitative features like sharp peaks, extendible to higher dimensions, and computationally tractable. Modern interior point methods for solving convex optimization problems play a critical role in achieving the final objective, as do piecewise linear finite element methods that facilitate the use of sparse linear algebra.

Key words: Density estimation; Penalized likelihood; Total variation; Regularization.

1 Introduction

The appeal of pure maximum likelihood methods for nonparametric density estimation is immediately frustrated by the simple observation that maximizing log likelihoods,

$$\sum_{i=1}^{n} \log f(X_i) = \max_{f \in \mathcal{F}}! \tag{1}$$

over any moderately rich class of densities, $\mathcal{F}$, yields estimators that collapse to a sum of point masses. These notorious "Dirac catastrophes" can be avoided by penalizing the log likelihood

$$\sum_{i=1}^{n} \log f(X_i) - \lambda J(f) = \max_{f \in \mathcal{F}}! \tag{2}$$

by a functional J that imposes a cost on densities that are too rough. The penalty regularizes the original problem and produces a family of estimators indexed by the tuning parameter λ.

Penalized maximum likelihood methods for density estimation were introduced by Good (1971), who suggested using Fisher information for the location parameter of the density as a penalty functional. Good offered a heuristic rationale of this choice as a measure of the sensitivity of the density to location shifts. The choice has the added practical advantage that it permits the optimization to be formulated as a convex problem with the (squared) L_2 penalty,

$$J(f) = \int (\sqrt{f}\,')^2 dx. \tag{3}$$

In subsequent work Good and Gaskins (1971) found this penalty somewhat unsatisfactory, producing estimates that sometimes "looked too straight." They suggested a modified penalty that incorporated a component penalizing the *second* derivative of $\sqrt{f}$ as well as the first. This component has a more direct interpretation as a measure of curvature and therefore as a measure of roughness of the fitted density.

Eschewing a "full-dress Bayesian approach," Good and Gaskins refer to their methods as a "Bayesian approach in mufti." Ideally, penalties could be interpreted as an expression of prior belief about the plausibility of various elements of $\mathcal{F}$. In practice, the justification of particular penalties inevitably has a more heuristic, ad-hoc flavor: data-analytic rationality constrained by computational feasibility. While penalties may be applied to the density itself rather than to its square root, a possibility briefly mentioned in Silverman (1986), a more promising approach considered by Leonard (1978) and Silverman (1982) replaces $\sqrt{f}$ by $\log f$ in the penalty term. When the second derivative of $\log f$ is penalized, this approach privileges exponential densities; whereas penalization of the third derivative of $\log f$ targets the normal distributions.

The early proposals of Good and Gaskins have received detailed theoretical consideration by Thompson and Tapia (1990) and by Eggermont and LaRiccia (2001), who establish consistency and rates of convergence. A heuristic argument of Klonias (1991) involving influence functions suggests that penalized likelihood estimators perform automatically something similar in effect to the "data sharpening" of Hall and Minnotte (2002) – they take mass from the "valleys" and distribute it to the "peaks." Silverman (1984) provides an nice link between penalty estimators based on the rth derivative of $\log f$ and adaptive kernel estimators, and he suggests that the penalty approach achieves a degree of automatic adaptation of bandwidth without reliance on a preliminary estimator. Taken together this work constitutes, we believe, a convincing *prima facie* case for the regularization approach to density estimation.

From the computational point of view, all these proposals, starting from those of Good, can be formulated as convex optimization problems and therefore are in principle efficiently computable. However, the practice has not been that straightforward, and widely accessible implementations may still not be always available. In particular, the earlier authors thinking in terms of techniques for minimization of quadratic functionals might still view the constraints implied by the fact that the optimization must be performed over f that are densities as a computational pain. Penalization of $\sqrt{f}$ or $\log f$ is often motivated as a practical device circumventing the nonnegativity constraint on f; penalizing the logarithm of the density as noted by Silverman (1982), offers a convenient opportunity to eliminate the constraint requiring the integral of f to be 1. In contrast to these advantages, penalizing the density f itself requires a somewhat more complicated strategy to ensure the positivity and integrability of the estimator. In any case, the form of the likelihood keeps the problem nonlinear; hence iterative methods are ultimately required. Computation of estimators employing the L_2 penalty on $(\log f)''$ has been studied by O'Sullivan (1988). An implementation in R is available from the package **gss** of Gu (2005). Silverman's (1982) proposal to penalize the third derivative of $\log f$, thereby shrinking the estimate toward the Gaussian density, has been implemented by Ramsay and Silverman (2002).

The development of modern interior-point methods for convex programming not only changes this outlook – convex programming works with constraints routinely – but also makes various other penalization proposals viable. In what follows, we would like to introduce several new nonparametric density estimation proposals involving penalties formulated in terms of total variation. Weighted sums of squared L_2 norms are replaced by weighted L_1 norms as an alternative regularization device. Squaring penalty contributions inherently exaggerates the contribution to the penalty of jumps and sharp bends in the density; indeed, density jumps and piecewise linear bends are impossible in the L_2 framework since the penalty evaluates them as "infinitely rough." Total variation penalties are happy to tolerate such jumps and bends, and they are therefore better suited to identifying discrete jumps in densities or in their derivatives. This is precisely the property that has made them attractive in imaging applications.

From a computational perspective, total-variation based penalties fit comfortably into modern convex optimization setting. Exploiting the inherent sparsity of the linear algebra required yields very efficient interior point algorithms. We will focus our attention on penalizing derivatives of $\log f$, but other convex transformations can be easily accommodated. Our preliminary experimentation with penalization of $\sqrt{f}$ and f itself did not seem to offer tangible benefits.

Total-variation penalties also offer natural multivariate generalizations. Indeed, we regard univariate density estimation as only a way station on a road leading to improved multivariate density estimators. To this end, the fact that penalty methods can easily accommodate qualitative constraints on estimated functions and their boundary values is an important virtue. One of our original motivations for investigating total variation penalties for density estimation was the ease with which qualitative constraints – monotonicity or log-concavity, for instance – could be imposed. In this context it is worth mentioning the recent work of Rufibach and Dümbgen (2004) who show that imposing log-concavity *without any penalization* is enough to regularize the univariate maximum likelihood estimator, and achieve attractive asymptotic behavior.

Total variation penalties for nonparametric regression with scattered data have been explored by Koenker, Ng and Portnoy (1994), Mammen and van de Geer (1997), Davies and Kovac (2001, 2004) and Koenker and Mizera (2002, 2004). Total variation has also played an important role in image processing since the seminal papers of Mumford and Shah (1989), and Rudin, Osher, and Fatemi (1992).

We begin by considering the problem of estimating univariate densities, and then extend the approach to bivariate settings.

2 Univariate Density Estimation

Given a random sample, $X_1, \dots, X_n$ from a density f_0, we will consider estimators that solve,

$$\max_f \{ \sum_{i=1}^n \log f(X_i) - \lambda J(f) \mid \int_\Omega f = 1 \}, \tag{4}$$

where J denotes a functional intended to penalize for the roughness of candidate estimates, $\mathcal{F}$, and λ is a tuning parameter controlling the smoothness of the estimate. Here the domain Ω may depend on *a priori* considerations as well as the observed data.

We propose to consider roughness penalties based on total variation of the transformed density and its derivatives. Recall that the total variation of a function $f : \Omega \to \mathcal{R}$ is defined as

$$\bigvee_{\Omega}(f) = \sup \sum_{i=1}^{m} |f(u_i) - f(u_{i-1})|,$$

where the supremum is taken over all partitions, $u_1 < \ldots < u_m$ of Ω. When f is absolutely continuous, we can write, see e.g. Natanson (1974, p.259),

$$\bigvee_{\Omega}(f) = \int_{\Omega} |f'(x)| dx.$$

We will focus on penalizing the total variation of the first derivative of the log density,

$$J(f) = \bigvee_{\Omega}((\log f)') = \int_{\Omega} |(\log f)''|,$$

so letting $g = \log f$ we can rewrite (3) as,

$$\max_{g} \{ \sum_{i=1}^{n} g(X_i) - \lambda \bigvee_{\Omega}(g') \mid \int_{\Omega} e^g = 1 \}. \tag{5}$$

But this is only one of many possibilities: one may consider

$$J(f) = \bigvee_{\Omega}(g^{(k)}),$$

where $g^{(0)} = g$, $g^{(1)} = g'$, etc., and g may be $\log f$, or $\sqrt{f}$, or f itself, or more generally $g^{\kappa} = f$, for $\kappa \in [1, \infty]$, with the convention that $g^{\infty} \equiv e^g$. Even more generally, linear combinations of such penalties with positive weights may be considered. From this family we adopt $\kappa = \infty$ and $k = 1$; see Sardy and Tseng (2005) for $\kappa = 1$ and $k = 0$. In multivariate settings $g^{(k)}$ is replaced by $\nabla^k g$, as described in the next section.

As noted by Gu (2002), even for L_2 formulations the presence of the integrability constraint prevents the usual reproducing kernel strategy from finding exact solutions; some iterative algorithm is needed. We will adopt a finite element strategy that enables us to exploit the sparse structure of the linear algebra used by modern interior point algorithms for convex programming.

Restricting attention to f's for which $\log(f)$ is piecewise linear on a specified partition of Ω, we can write $J(f)$ as an ℓ_1 norm of the second weighted differences of f evaluated at the mesh points of the partition. More explicitly, let Ω be the closed interval $[x_0, x_m]$ and consider the partition $x_0 < x_1 < \cdots < x_m$ with spacings $h_i = x_i - x_{i-1}, \quad i = 1, \cdots m$. If $\log(f(x))$ is piecewise linear, so that

$$\log(f(x)) = \alpha_i + \beta_i x \qquad x \in [x_i, x_{i+1}),$$

then

$$J(f) = \bigvee_{\Omega}((\log f)') = \sum_{i=1}^{m} |\beta_i - \beta_{i-1}| = \sum_{i=1}^{m} |(\alpha_{i+1} - \alpha_i)/h_{i+1} - (\alpha_i - \alpha_{i-1})/h_i|,$$

where we have imposed continuity of f in the last step. We can thus parameterize functions $f \in \mathcal{F}$ by the function values $\alpha_i = \log(f(x_i))$, and this enables us to write our problem (3) as a linear program,

$$\max\{\sum_{i=1}^{n} \alpha_i - \lambda \sum_{j=1}^{m} (u_j + v_j) | D\alpha - u + v = 0,\ (\alpha, u, v) \in \mathbb{R}^n \times \mathbb{R}_+^{2m}\} \quad (6)$$

where D denotes a tridiagonal matrix containing the h_i factors for the penalty contribution, and u and v represent the positive and negative parts of the vector $D\alpha$, respectively.

An advantage of parameterization of the problem in terms of $\log f$ is that it obviates any worries about the non-negativity of $\hat{f}$. But we have still neglected one crucial constraint. We need to ensure that our density estimates integrate to one. In the piecewise linear model for $\log f$ this involves a rather awkward nonlinear constraint on the α's,

$$\sum_{j=1}^{m} h_i \frac{e^{\alpha_i} - e^{\alpha_{i-1}}}{\alpha_i - \alpha_{i-1}} = 1.$$

This form of the constraint cannot be incorporated directly in its exact form into our optimization framework, nevertheless its approximation by a Riemann sum on a sufficiently fine grid provides a numerically satisfactory solution.

2.1 *Data Augmentation*

In the usual Bayesian formalism, the contribution of the prior can often be represented as simple data augmentation. That is, the prior can be interpreted as what we would believe about the model parameters if we had observed some "phantom data" whose likelihood we could evaluate. This viewpoint may strain credulity somewhat, but under it the penalty, $J(f)$, expresses the belief that we have "seen" m observations on the second differences of $\log f$ evaluated at the x_i's, *all zero*, and independent with standard Laplacian density, $\frac{1}{2}e^{-|x|}$. The presence of λ introduces a free scale parameter that represents the strength of this belief. Data dependent strategies for the choice of λ obviously violate Bayesian orthodoxy, but maybe condoned by the more pragmatic minded.

Pushing the notion of Bayesian virtual reality somewhat further, we may imagine observing data at new x_i values. Given that our estimated density is parameterized by its function values at the "observed" x_i values, these new values introduce new parameters to be estimated; these "phantom observations" contribute nothing to the likelihood, but they do contribute to the penalty term $J(f)$. But by permitting $\log f$ to bend at the new x_i points in regions where there is otherwise no real data, flexibility of the fitted density is increased. In regions where the function $\log f$ is convex, or concave, one large change in the derivative can thus be broken up into several smaller changes, without affecting the total variation of its derivative. Recall that the total variation of a monotone function on an interval is just the difference in the values taken at the endpoints of the interval.

Rather than trying to carefully select a few x_i values as knots for a spline representation of the fitted density, as described in Stone, Hansen, Kooperberg, and Truong (1997), all of the observed x_i are retained as knots and some virtual ones are thrown in as well. Shrinkage, controlled by the tuning parameter, λ, is then relied upon to achieve the desired degree of smoothing. The use of virtual observations is particularly advantageous in the tails of the density, and in other regions where the observed data are sparse. We will illustrate the use of this technique in both univariate and bivariate density estimation in the various examples of subsequent sections.

Example 1. Several years ago one of us, as a class exercise, asked students to estimate the density illustrated in Figure 1(a), based on a random sample of 200 observations. The density is a mixture of three, three-parameter lognormals:

$$f_1(x) = \sum_{i=1}^{3} w_i\phi(\log((x-\gamma_i-\mu_i)/\sigma_i))/(\sigma_i(x-\gamma_i)), \tag{7}$$

where ϕ denotes the standard normal density, $\mu = (0.5, 1.1, 2.6)$, $\gamma = (.0.4, 1.2, .2.4)$, $\sigma = (0.2, 0.3, .0.2)$, and $w = (0.33, 0.33, 0.33)$. In the figure we have superimposed the density on a histogram of the original data using an intentionally narrow choice of binwidth.

The most striking conclusion of the exercise was how poorly conventional density estimators performed. With one exception, none of the student entries in the competition were able to distinguish the two tallest peaks, and their performance on the lower peak wasn't much better. All of the kernel estimates looked very similar to smoother of the two kernel estimates displayed in Figure 1(b). This is a fixed-bandwidth Gaussian kernel estimate with bandwidth chosen by Scott's (1992) biased cross-validation criterion as implemented in R and described by Venables and Ripley (2002). The other kernel estimate employs Scott's alternative unbiased cross-validation bandwidth, and clearly performs somewhat better. Gallant and Nychka's (1987) Hermite series estimator also oversmooths when the order of the estimator is chosen with their BIC criterion, but performs better when AIC order selection is used, as illustrated in Figure 1(c). In Figure 1(d) we illustrate two variants of the most successful of the student entries based on the logspline method of Kooperberg and Stone (1991): one constrained to have positive support, the other unconstrained. Figure 1(e) illustrates two versions of the logspline estimator implemented by Gu (2002). Finally, Figure 1(f) illustrates two versions of a total variation penalty estimator; both versions employ a total variation penalty on the derivative of $\log f$, and use in addition to the 200 sample observations, 300 "virtual observations" equally spaced between 0 and 25. These estimators were computed with the aid of the MOSEK package of E. D. Andersen, an implementation for MATLAB of the methods described in Andersen and Ye (1998). The penalty method estimators all perform well in this exercise, but the kernel and Hermite series estimators have difficulty coping with the combination of sharp peaks and smoother foothills.

3 Bivariate Density Estimation

In nonparametric regression piecewise linear fitting is often preferable to piecewise constant fitting. Thus, penalizing total variation of the gradient, ∇g, instead of total variation of g itself, is desirable. For smooth functions we can extend the previous definition by writing,

$$\bigvee_{\Omega} \nabla g = \int_{\Omega} \|\nabla^2 g\|, \tag{8}$$

where $\|\cdot\|$ can be taken to be the Hilbert-Schmidt norm, although other choices are possible as discussed in Koenker and Mizera (2004). This penalty is closely associated with the thin plate penalty that replaces $\|\nabla^2 g\|$ with $\|\nabla^2 g\|^2$. The latter penalty has received considerable attention, see e.g. Wahba (1990) and the references cited therein. We would stress, however, that as in the univariate setting there are important advantages in taking the square root.

For scattered data more typical of nonparametric regression applications, Koenker and Mizera (2004) have proposed an alternative discretiza-

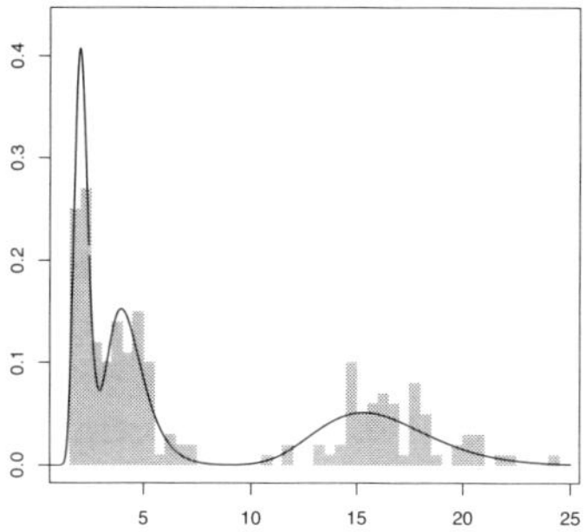

(a) Histogram

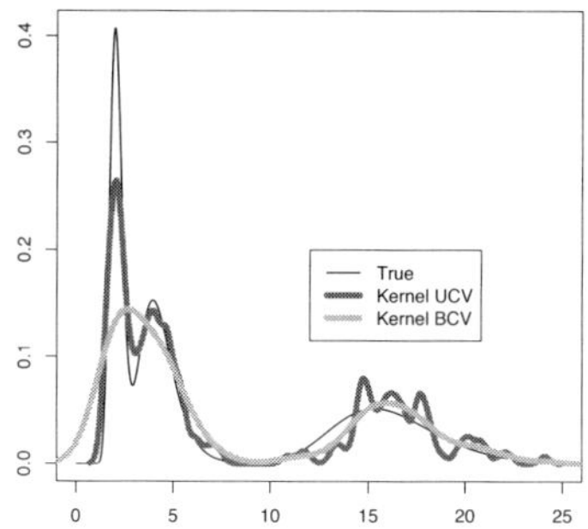

(b) Kernel

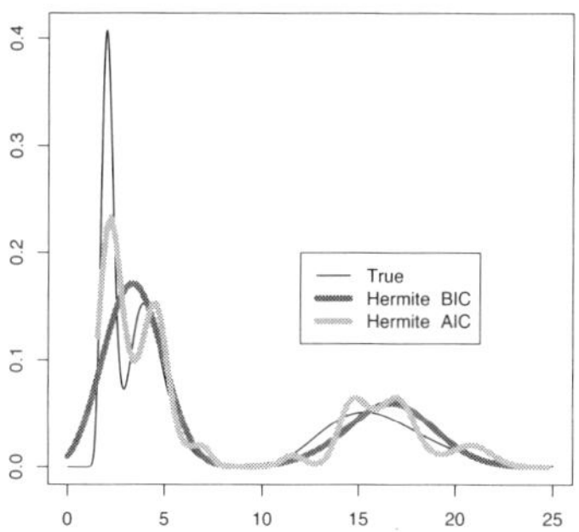

(c) Hermite

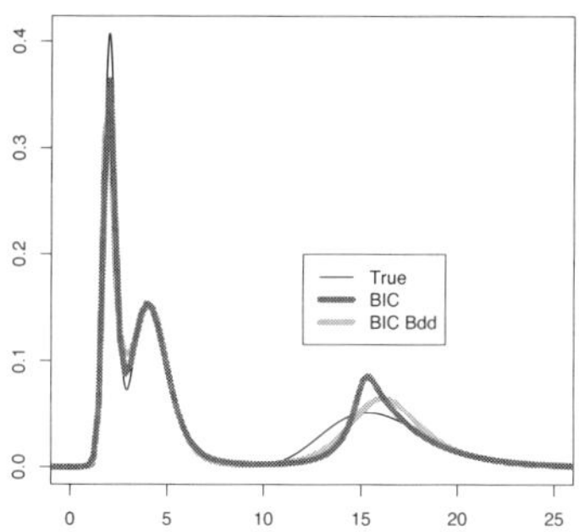

(d) Logspline

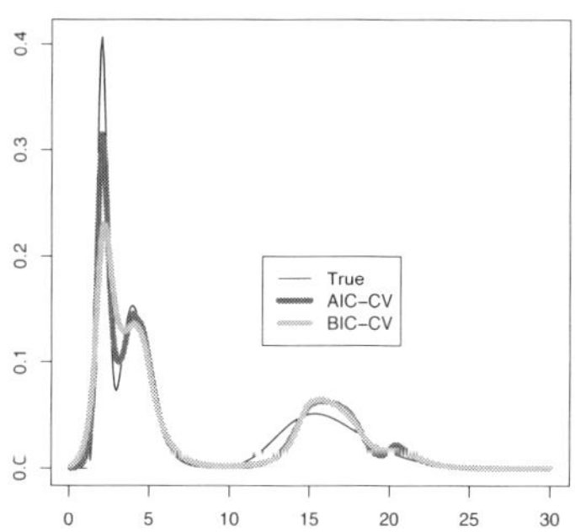

(e) Gulog

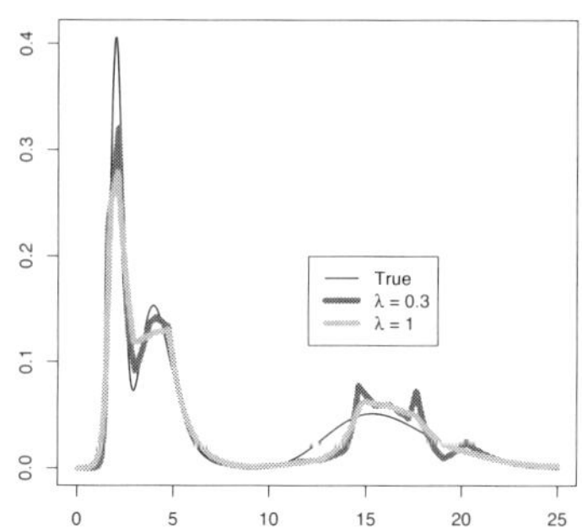

(f) TVlog

Figure 1 Comparison of Estimates of the 3-Sisters Density.

tion of the total variation penalty based on continuous, piecewise-linear functions defined on triangulations of a convex, polyhedral domain. Following Hansen, Kooperberg, and Sardy (1998), such functions are called triograms. The penalty (7) can be simplified for triograms by summing the contributions over the edges of the triangulation,

$$\bigvee_{\Omega} \nabla g = \sum_k \|\nabla g^+_{e_k} - \nabla g^-_{e_k}\| \, \|e_k\|. \tag{9}$$

Each edge is associated with two adjacent triangles; the contribution of the edge is simply the product of the Euclidean norm of the difference between the gradients on the two triangles multiplied by the length of the edge. The interiors of the triangles, since they are linear, contribute nothing to the total variation, nor do the vertices of the triangulation. See Koenker and Mizera (2004) for further details.

Choice of the triangulation is potentially an important issue especially when the number of vertices is small, but numerical stability favors the classical Delaunay triangulation in most applications. Hansen and Kooperberg (2002) consider sequential (greedy) model selection strategies for choosing a parsimonious triangulations for nonparametric regression without relying on a penalty term. In contrast, Koenker and Mizera (2004) employ the total variation penalty (8) to control the roughness of the fit based on a much more profligate triangulation. As in the univariate setting it is often advantageous to add virtual vertices that can improve the flexibility of the fitted function.

Extending the penalized triogram approach to bivariate density estimation requires us, as in the univariate case, to make a decision about what is to be penalized? We will focus exclusively on total variation penalization of the log density with the understanding that similar methods could be used for the density itself or another (convex) transform.

Given independent observations $\{x_i = (x_{1i}, x_{2i}) : i = 1, \cdots, n\}$ from a bivariate density $f(x)$, let $g = \log f$, and consider the class of penalized maximum likelihood estimators solving

$$\max_{g \in \mathcal{G}} \sum_{i=1}^{n} g(x_i) - \lambda J(g),$$

where J is the triogram penalty, given by (8). The set $\mathcal{A}$ consists of triogram densities: continuous functions from a polyhedral convex domain Ω to $\mathbb{R}_+$, piecewise linear on a triangulation of Ω and satisfying the condition,

$$\int_{\Omega} e^g = 1.$$

It follows that $\log f$ can be parameterized by its function values at the vertices of the triangulation. As in the univariate case, adding virtual vertices is advantageous especially so in the region outside the convex hull of the observed data where they provide a device to cope with tail behavior.

Example 2. To illustrate the performance of our bivariate density estimator, we consider the density

$$\begin{aligned} f_2(x_1, x_2) &= f(x_2|x_1)f(x_1) \\ &= 2\phi(2(x_2 - \sqrt{x_1})) \cdot f_1(x_1), \end{aligned}$$

where f_1 is the univariate test density given above. Two views of this density can be seen in the upper panels of Figure 2. There is one very sharp peak and two narrow "fins". In the two lower panels we depict views of a fitted density based on 1000 observations. The tuning parameter λ is taken to be 2, and the fit employs virtual observations on a integer grid over the rectangle $\{[0, 30] \times [0, 6]\}$.

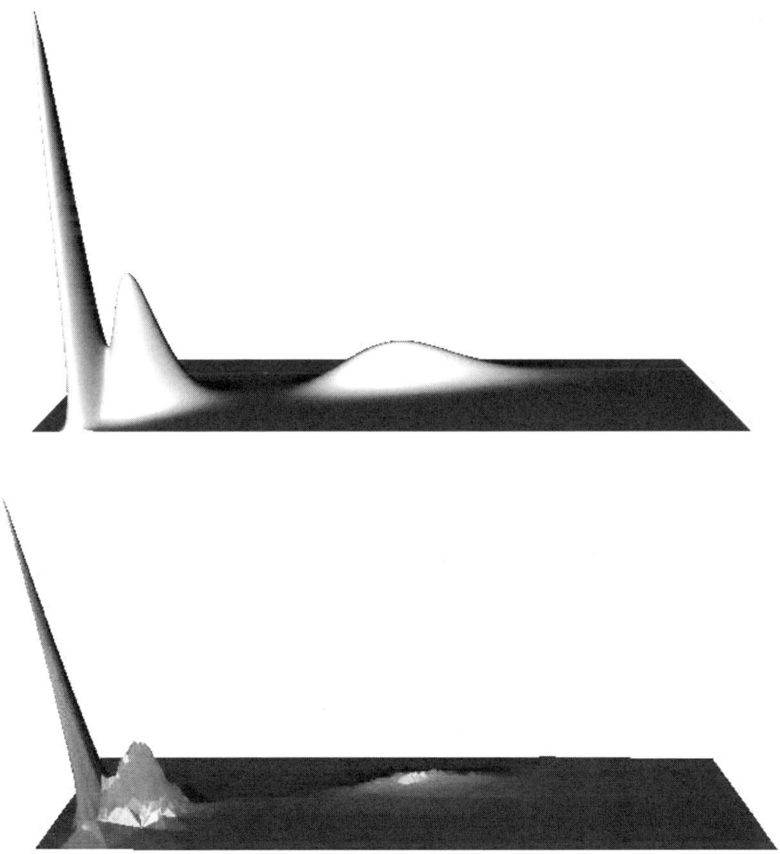

Figure 2 Bivariate 3-Sisters Density and an Estimate.

4 Duality and Regularized Maximum Entropy

An important feature of convex optimization problems is that they may be reformulated as dual problems, thereby often offering a complementary view of the problem from the other side of the looking glass. In addition to providing deeper insight into the interpretation of the problem as originally posed, dual formulations sometimes yield substantial practical benefits in the form of gains in computational efficiency. In our experience, the dual formulation of our computations exhibits substantially better performance than the original penalized likelihood formulation. Execution times are about 20 percent faster and convergence is more stable. We will show in this section that total variation penalized maximum likelihood density estimation has a dual formulation as regularized form of maximum entropy estimation.

As we have seen already, piecewise linear log density estimators can be represented by a finite dimensional vector of function values

$$\alpha_i = g(x_i) \quad i = 1, \cdots, m,$$

evaluated at knot locations, $x_i \in \Omega$. These points of evaluation can be sample observations or "virtual" observations, or a mixture of the two. They may be univariate, bivariate, or in principle, higher dimensional. We approximate our integral by the Riemann sum,

$$\int_{\Omega} e^g \approx \sum_{i=1}^{m} c_i e^{\alpha_i},$$

a step that can be justified rigorously by introducing points of evaluation on a sufficiently fine grid, but is also motivated by computational considerations. Provisionally, we will set the tuning parameter $\lambda = 1$, so our primal problem is,

$$\max\{\delta^\top \alpha - \|D\alpha\|_1 \mid \sum_i c_i e^{\alpha_i} = 1\}. \qquad (P)$$

In the simplest case the vector $\delta \in \mathbb{R}$ is composed of zeros and ones indicating which elements of α correspond to sample points and thus contribute to the likelihood term. In the case that the x_i are *all* virtual, chosen to lie on a regular grid, for example, we can write, $\delta = B1_n$, where B denotes an m by n matrix representing the n sample observations expressed in terms of the virtual points, e.g. using barycentric coordinates.

The integrability constraint can be conveniently incorporated into the objective function using the following discretized version of a result of Silverman (1982).

Lemma 1. *$\hat{\alpha}$ solves problem (P) if and only if $\hat{\alpha}$ maximizes,*

$$R(\alpha) = \delta^\top \alpha - \|D\alpha\|_1 - n\sum_i c_i e^{\alpha_i}.$$

Proof. Note that any differential operator, D, annihilates constant functions, or the vector of ones. Thus, evaluating R at $\alpha^* = \alpha - \log \sum c_i e^{\alpha_i}$, so $\sum c_i e^{\alpha_i^*} = 1$, we have

$$R(\alpha^*) = R(\alpha) + n\sum_i c_i e^{\alpha_i} - n \log \sum_i c_i e^{\alpha_i} - 1,$$

but $t - \log t \geq 1$, for all $t > 0$ with equality only at $t = 1$. Thus, $R(\alpha^*) \geq R(\alpha)$, and it follows that $\hat{\alpha}$ maximizes R if and only if $\hat{\alpha}$ maximizes R subject to $\sum_i c_i e^{\alpha_i} = 1$. This constrained problem is equivalent to (P). □

Introducing the artificial barrier vector β, the penalty contribution can be reformulated slightly, and we can write (P) as,

$$\max_{\alpha,\beta}\{\delta^\top \alpha - 1^\top \beta - \sum_i c_i e^{\alpha_i} \mid D\alpha \leq \beta, \quad -D\alpha \leq \beta\}.$$

We seek to minimize the Lagrangian,

$$\begin{aligned} L(\alpha, \beta, \nu_1, \nu_2) &= \delta^\top \alpha - 1^\top \beta - n\sum c_i e^{\alpha_i} + \nu_1^\top (D\alpha - \beta) + \nu_2^\top (-D\alpha - \beta) \\ &= (\delta + D^\top (\nu_1 - \nu_2))^\top \alpha - (1 - \nu_1 - \nu_2) 1^\top \beta - n\sum c_i e^{\alpha_i}, \end{aligned}$$

subject to the feasibility constraints,

$$\gamma \equiv \delta + D^\top (\nu_1 - \nu_2) \geq 0, \ \nu_1 + \nu_2 = 1, \ \nu_1 \geq 0, \text{ and } \nu_2 \geq 0.$$

Now, differentiating the Lagrangian expression with respect to the α_i's yields

$$\frac{\partial L}{\partial \alpha_i} = \delta_i - d_i^\top (\nu_1 - \nu_2) - c_i e^{\alpha_i} = 0, \ i = 1, \cdots, m.$$

Convexity assures that these conditions are satisfied at the unique optimum:

$$f_i \equiv (\delta_i - d_i^\top (\nu_1 - \nu_2))/c_i = e^{\alpha_i} \quad i = 1, \cdots, m,$$

so we can rewrite our Lagrangian problem with $C = \text{diag}(c)$ as

$$\min\{\sum c_i f_i \log f_i \mid f = C^{-1}(\delta + D^\top y) \geq 0. \|y\|_\infty \leq 1\}.$$

Reintroducing the tuning parameter λ we obtain the final form of the dual problem.

Theorem 1. *Problem (P) has equivalent dual formulation*

$$\max\{-\sum c_i f_i \log f_i \mid f = C^{-1}(\delta + D^\top y) \geq 0, \ \|y\|_\infty \leq \lambda\}. \qquad (D)$$

Remarks

(1) We can interpret the dual as a maximum entropy problem regularized by the ℓ_∞ constraint on y with added requirement that an affine transformation of the vector of dual variables, y, lies in the positive orthant.
(2) The ℓ_∞ constraint may be viewed as a generalized form of the tube constraint associated with the taut string methods of Davies and Kovac (2004). In the simplest setting, when total variation of the log density itself, rather than its derivative, is employed as a penalty for univariate density estimation, D is a finite difference operator and the dual vector, y, can be interpreted as a shifted estimate of the distribution function constrained to lie in a band around the empirical distribution function. In more general settings the geometric interpretation of the constraints on the dual vector, y, in terms of the sample data is somewhat less clear. See also Koenker and Mizera (2006).
(3) The weights c_i appearing in the objective function indicate that the sum may be interpreted as a Riemann approximation to the entropy integral. Expressing the problem equivalently as the maximization of

$$\sum_i c_i f_i \log \frac{c_i}{c_i f_i} + \log n$$

we arrive at an interpretation in terms of the Kullback-Leibler divergence, $\mathcal{K}(\phi, \nu)$, of the probability distribution $\phi = (c_i f_i)$, corresponding to the estimated density f, from the probability distribution $\nu = n(c_i)$, corresponding to the density uniform over Ω. Thus, our proposal can be interpreted in terms of regularized minimum distance estimation,

$$\min\{\mathcal{K}(\phi, \nu) | \phi = (\delta + D^T y) \geq 0, \|y\|_\infty < \lambda\},$$

a formulation not entirely surprising in the light of our knowledge about maximum likelihood estimation. The choice of the uniform "carrier" density could be modified to obtain exponentially tilted families as described in Efron and Tibshirani (1996).

(4) Density estimation methods based on maximum entropy go back at least to Jaynes (1957). However, this literature has generally emphasized imposing exact moment conditions, or to use the machine learning terminology, "features," on the estimated density. In contrast, our dual problem may be viewed as a regularized maximum entropy approach that specifies "soft" feature constraints imposed as inequalities. Dudík, Phillips, and Schapire (2004) consider a related maximum entropy density estimation problem with soft feature constraints. Donoho, Johnstone, Hoch, and Stern (1992) consider related penalty methods based on entropy for a class of regression type imaging and spectroscopy problems, and show that they have superior performance to linear methods based on Gaussian likelihoods and priors.

5 Monte-Carlo

In this section we report the results of a small Monte-Carlo experiment designed to compare the performance of the TV penalized estimator with three leading competitors:

TS The taut string estimator of Davies and Kovac (2005) using the default tuning parameters embedded in the function `pmden` of their R package `ftnonpar`.

Kucv The fixed bandwidth kernel density estimator implemented by the function `density` in the R `stats` package, employing Scott's (1992) "unbiased cross validation" bandwidth selection.

Kbcv The fixed bandwidth density estimator as above, but using Scott's biased cross-validation bandwidth.

For purposes of the Monte-Carlo, automatic selection of λ for the TV estimator was made according to the following recipe. Estimates were computed at the fixed λ's, $\{.1, .2, \ldots, .9, 1.0\}$, using virtual observations on a grid, $\mathcal{G}$, of 400 points equally spaced on $[-4, 4]$. For each of these estimates the Kolmogorov distance between the empirical distribution function of the sample, $\hat{F}_n$, and the smoothed empirical, $\tilde{F}_{n,\lambda}$, corresponding to the density estimate

$$\kappa(\lambda) \equiv K(\hat{F}_n, \tilde{F}_{n,\lambda}) = \max_{x_i \in \mathcal{G}} |\hat{F}_n(x_i) - \tilde{F}_{n,\lambda}(x_i)|$$

was computed. Based on preliminary investigation, $\log \kappa(\lambda)$ was found to be approximately linear in $\log \lambda$, so we interpolated this log-linear relationship to find the λ that made $\kappa(\lambda)$ approximately equal to the cutoff $c_\kappa = .3/\sqrt{n}$. The value .3 was chosen utterly without any redeeming the-

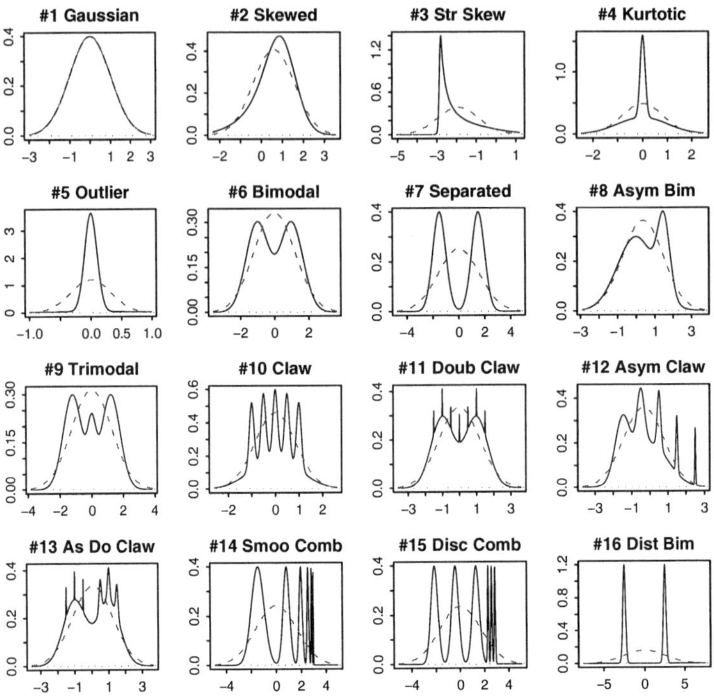

Figure 3 The Marron and Wand candidate densities.

oretical justification. In rare cases for which this interpolation fails, i.e., $\hat{\lambda} \notin [.1, 1]$, we use $\hat{\lambda} = \max\{\min\{\hat{\lambda}, 1\}, .1\}$.

As candidate densities, we use the familiar Marron and Wand (1992) normal mixtures illustrated in Figure 1. Random samples from these densities were generated in with the aid of the R `nor1mix` package of Mächler (2005). All computations for the taut string and kernel estimators are conducted in *R*; computations for the TV estimator are made in matlab with the aid of the `PDCO` function of Saunders (2004) as described above using the sample data generated from *R*.

Three measures of performance are considered for each of the 16 test densities. Table 1 reports the proportion replications for which each method obtained the correct identification of the number of modes of the true density. Table 2 reports median MIAE (mean integrated absolute error), and Table 3 reports median MISE (mean integrated squared error).

Clearly, the taut-string estimator performs very well in identifying unimodal and well separated bimodal densities, but it has more difficulties

Table 1 The proportion of correct estimates of the number of modes for the Marron-Wand densities: Sample size, $n = 500$ and replications $R = 1000$.

Distribution	TV	TS	K-ucv	K-bcv
MW 1	0.303	1.000	0.690	0.863
MW 2	0.304	1.000	0.354	0.456
MW 3	0.169	1.000	0.000	0.059
MW 4	0.152	1.000	0.000	0.176
MW 5	0.345	1.000	0.000	0.000
MW 6	0.634	0.329	0.718	0.937
MW 7	0.716	1.000	0.678	0.880
MW 8	0.522	0.067	0.279	0.592
MW 9	0.472	0.013	0.434	0.292
MW 10	0.680	0.528	0.000	0.001
MW 11	0.000	0.000	0.006	0.000
MW 12	0.010	0.014	0.017	0.000
MW 13	0.172	0.001	0.003	0.000
MW 14	0.122	0.021	0.000	0.014
MW 15	0.101	0.078	0.000	0.038
MW 16	0.772	1.000	0.000	1.000

Table 2 Median Integrated Absolute Error: Sample size, $n = 500$ and number of replications $R = 1000$.

Distribution	TV	TS	K-ucv	K-bcv
MW 1	0.109	0.166	0.089	0.082
MW 2	0.109	0.173	0.099	0.092
MW 3	0.130	0.218	0.191	0.200
MW 4	0.143	0.212	0.199	0.202
MW 5	0.120	0.177	0.150	0.140
MW 6	0.110	0.187	0.105	0.104
MW 7	0.127	0.204	0.120	0.116
MW 8	0.113	0.187	0.116	0.124
MW 9	0.120	0.204	0.118	0.132
MW 10	0.190	0.289	0.190	0.348
MW 11	0.144	0.193	0.118	0.117
MW 12	0.149	0.262	0.182	0.274
MW 13	0.186	0.214	0.146	0.143
MW 14	0.208	0.295	0.222	0.279
MW 15	0.173	0.311	0.224	0.248
MW 16	0.148	0.201	0.140	1.279

Table 3 Median Integrated Squared Error: Sample size, $n = 500$ and number of replications $R = 1000$.

Distribution	TV	TS	K-ucv	K-bcv
MW 1	0.0039	0.0074	0.0021	0.0018
MW 2	0.0042	0.0088	0.0028	0.0024
MW 3	0.0096	0.0468	0.0162	0.0280
MW 4	0.0117	0.0293	0.0163	0.0202
MW 5	0.0241	0.0577	0.0220	0.0183
MW 6	0.0037	0.0090	0.0029	0.0027
MW 7	0.0052	0.0121	0.0041	0.0037
MW 8	0.0042	0.0095	0.0041	0.0050
MW 9	0.0042	0.0104	0.0037	0.0043
MW 10	0.0163	0.0393	0.0137	0.0468
MW 11	0.0056	0.0101	0.0045	0.0043
MW 12	0.0066	0.0225	0.0115	0.0223
MW 13	0.0194	0.0136	0.0073	0.0071
MW 14	0.0238	0.0310	0.0174	0.0276
MW 15	0.0481	0.0334	0.0168	0.0231
MW 16	0.0054	0.0349	0.0145	0.5596

with the multimodal cases. Unbiased cross-validation is generally inferior to biased cross-validation from a mode identification viewpoint, producing too rough an estimate and therefore too many modes.

Unbiased CV has quite good MIAE performance. Not surprisingly, it does best at the normal model, but it is somewhat worse than our TV estimator for distributions 3, 4, 5, 14, and 15. In the other cases the performance is quite comparable. The biased CV kernel estimator is consistently inferior in MIAE except at the normal model. It fails spectacularly for the sharply bimodel density number 16. The TV estimator is not too bad from the MIAE perspective, consistently outperforming the taut-string estimator by a substantial margin, and very competitive with the kernel estimators except in the strictly Gaussian setting. Results for MISE are generally similar to those for MIAE.

6 Prospects and Conclusions

Total variation penalty methods appear to have some distinct advantages when estimating densities with sharply defined features. They also have attractive computational features arising from the convexity of the penalized likelihood formulation.

There are many enticing avenues for future research. There is considerable scope for extending the investigation of dual formulations to other penalty functions and other fitting criteria. It would also be valuable to explore a functional formulation of the duality relationship. The extensive literature on covering numbers and entropy for functions of bounded variation can be deployed to study consistency and rates of convergence. And inevitably there will be questions about automatic λ selection. We hope to be able to address some of these issues in subsequent work.

Acknowledgement. This research was partially supported by NSF grant SES-02-40781, and by the Natural Sciences and Engineering Research Council of Canada. The authors want to thank Zuzana Benešová for help with the computer processing of figures.

References

1. ANDERSEN, E. AND Y. YE (1998): "A computational study of the homogeneous algorithm for large-scale convex optimization," *Computational Optimization and Applications*, 10, 243–269.
2. DAVIES, P. L., AND A. KOVAC (2001): "Local extremes, runs, strings and multiresolution," *The Annals of Statistics*, 29, 1–65.
3. DAVIES, P. L., AND A. KOVAC (2004): "Densities, Spectral Densities and Modality," *The Annals of Statistics*, 32, 1093–1136.
4. DAVIES, P. L., AND A. KOVAC (2005): "ftnonpar: Features and Strings for Nonparametric Regression," `http://cran.R-project.org`.
5. DONOHO, D. L., I. M. JOHNSTONE, J. C. HOCH, AND A. S. STERN (1992): "Maximum entropy and the nearly black object," *Journal of the Royal Statistical Society, Series B*, 54, 41–67.
6. DUDÍK, M., S. PHILLIPS,] AND R. SCHAPIRE (2004): "Performance Guarantees for Regularized Maximum Entropy Density Estimation," in *Proceedings of the 17th Annual Conference on Computational Learning Theory*, ed. by J. Shawe-Taylor, and Y. Singer.
7. EFRON, B., AND R. TIBSHIRANI (1996): "Using Specially Designed Exponential Families for Density Estimation," *The Annals of Statistics*, 24, 2431–2461.
8. EGGERMONT, P. AND V. LARICCIA (2001): *Maximum Penalized Likelihood Estimation.* Springer-Verlag.
9. GALLANT, A. R. AND D. W. NYCHKA (1987): "Semi-nonparametric maximum likelihood estimation," *Econometrica*, 55, 363–390.
10. GOOD, I. J. (1971): "A nonparametric roughness penalty for probability densities," *Nature*, 229, 29–30.
11. GOOD, I. J. AND R. A. GASKINS (1971): "Nonparametric roughness penalties for probability densities," *Biometrika*, 58, 255–277.

12. Gu, C. (2002): *Smoothing spline ANOVA models.* Springer-Verlag.

13. Gu, C. (2005): "gss: An R Package for general smoothing splines," R package version 0.9-3, http://cran.R-project.org.

14. Hall, P. and M. C. Minnotte (2002): "High order data sharpening for density estimation," *Journal of the Royal Statistical Society, Series B*, 64, 141–157.

15. Hansen, M. and C. Kooperberg (2002): "Spline Adaptation in Extended Linear Models," *Statistical Science*, 17, 2–51.

16. Hansen, M., C. Kooperberg, and S. Sardy (1998): "Triogram Models," *Journal of the American Statistical Association*, 93, 101–119.

17. Jaynes, E. (1957): "Information theory and statistical mechanics," *Physics Review*, 106, 620–630.

18. Klonias, V. K. (1991): "The influence function of maximum penalized likelihood density estimators," in *Nonparametric Functional Estimation and Related Topics*, ed. by G. Roussas. Kluwer.

19. Koenker, R. and I. Mizera (2002): "Comment on Hansen and Kooperberg: Spline Adaptation in Extended Linear Models," *Statistical Science*, 17, 30–31.

20. Koenker, R. and I. Mizera (2004): "Penalized triograms: total variation regularization for bivariate smoothing," *Journal of the Royal Statistical Society, Series B*, 66, 145–163.

21. Koenker, R. and I. Mizera (2006): "The alter egos of the regularized maximum likelihood density estimators: deregularized maximum-entropy, Shannon, Rényi, Simpson, Gini, and stretched strings," *Prague Stochastics 2006*, ed. by M. Hušková and M. Janžura. Matfyzpress.

22. Koenker, R., P. Ng, and S. Portnoy (1994): "Quantile Smoothing Splines," *Biometrika*, 81, 673–680.

23. Kooperberg, C. and C. J. Stone (1991): "A Study of Logspline Density Estimation," *Computational Statistics and Data Analysis*, 12, 327–347.

24. Leonard, T. (1978): "Density estimation, stochastic processes and prior information," *Journal of the Royal Statistical Society, Series B*, 40, 113–132.

25. Mächler, M. (2005): "nor1mix: Normal (1-d) Mixture Models Classes and Methods," R package version 1.0-5, http://cran.R-project.org.

26. Mammen, E., and S. van de Geer (1997): "Locally Adaptive Regression Splines," *The Annals of Statistics*, 25, 387–413.

27. Marron, J. S. and M. P. Wand (1992): "Exact mean integrated squared error," *The Annals of Statistics*, 20, 712–736.

28. Mumford, D. and J. Shah (1989): "Optimal approximations by piecewise smooth functions and associated variational problems," *Communications on Pure and Applied Mathematics*, 42, 577–684.

29. NATANSON, I. (1974): *Theory of Functions of a Real Variable.* Ungar.

30. O'SULLIVAN, F. (1988): "Fast computation of fully automated log-density and log-hazard estimators," *SIAM Journal on Scientific and Statistical Computing*, 9, 363–379.

31. RAMSAY, J. O. AND B. W. SILVERMAN (2002): *Applied Functional Data Analysis.* Springer, New York.

32. RUDIN, L., S. OSHER, AND E. FATEMI (1992): "Nonlinear total variation based noise removal algorithms," *Physica D*, 60, 259–268.

33. RUFIBACH, K. AND L. DÜMBGEN (2004): "Maximum Likelihood Estimation of a Log-Concave Density: Basic Properties and Uniform Consistency," preprint.

34. SARDY, S. AND P. TSENG (2005): "Estimation of nonsmooth densities by total variation penalized likelihood driven by the sparsity L_1 information criterion," `http://statwww.epfl.ch/people/sardy`.

35. SAUNDERS, M. (2004): "PDCO convex optimization software (MATLAB)," `http://www.stanford.edu/group/SOL/soft-ware.html`.

36. SCOTT, D. W. (1992): *Multivariate density estimation: theory, practice, and visualization.* John Wiley & Sons.

37. SILVERMAN, B. W. (1982): "On the estimation of a probability density function by the maximum penalized likelihood method," *The Annals of Statistics*, 10, 795–810.

38. SILVERMAN, B. W. (1984): "Spline smoothing: The equivalent variable kernel method," *The Annals of Statistics*, 12, 898–916.

39. SILVERMAN, B. W. (1986): *Density estimation for statistics and data analysis.* Chapman & Hall.

40. STONE, C., M. HANSEN, C. Kooperberg, AND Y. TRUONG (1997): "Polynomial splines and their tensor products in extended linear modeling," *The Annals of Statistics*, 25, 1371–1470.

41. THOMPSON, J. R. AND R. A. TAPIA (1990): *Nonparametric function estimation, modeling, and simulation.* SIAM.

42. VENABLES, W. N. AND B. D. RIPLEY (2002): *Modern applied statistics with S.* Springer-Verlag.

43. WAHBA, G. (1990) *Spline models for observational data*, CBMS-NSF Regional Conference Series in Applied Mathematics. SIAM.

Chapter 31

A NOTE ON THE BOUNDED NORMAL MEAN PROBLEM

Jianqing Fan and Jin-Ting Zhang

Department of Operations Research and Financial Engineering
Princeton University Princeton, NJ, U.S.A.

Department of Statistics and Applied Probability
National University of Singapore, SINGAPORE

E-mails: jqfan@princeton.edu & stazjt@nus.edu.sg

The bounded normal mean problem has important applications in nonparametric function estimation. It is to estimate the mean of a normal distribution with mean restricted to a bounded interval. The minimax risk for such a problem is generally unknown. It is shown in Donoho, Liu and MacGibbon (1990) that the linear minimax risk provides a good approximation to the minimax risk. We show in this note that a better approximation can be obtained by a simple truncation of the minimax linear estimator and that the minimax linear estimator is itself inadmissible. The gain of the truncated minimax linear estimator is significant for moderate size of the mean interval, where no analytical expression for the minimax risk is available. In particular, we show that the truncated minimax linear estimator performs no more than 13% worse than the minimax estimator, comparing with 25% for the minimax linear estimator.

Keywords: Bounded normal mean; Minimax risk; Quadratic loss.

1 Introduction

The minimax estimation of the bounded normal mean problem has been extensively studied in the literature. It is closely related to the minimax density estimation and minimax nonparametric regression problems. This connection has been convincingly demonstrated in seminal work by Sacks and Strawderman (1982), Ibragimov and Hasminskii (1984), Donoho and Liu (1991) Fan (1993), Brown and Low (1991,1996) Nussbaum (1996), among others. A sharp bound in the bounded normal mean problem would provide a sharp minimax bound for the corresponding nonparametric problems and a better procedure for estimating the bounded normal mean would lead to a better procedure in the nonparametric problems. See for example Donoho et al. (1990), Donoho and Liu (1991) and Brown and Low (1991,1996). In fact, the nonparametric regression model, nonparametric density estimation problem and the Gaussian white noise models are asymptotically equivalent in the sense of Le Cam's deficiency distance, as demonstrated in Brown, Cai, Low and Zhang (2002) and Grama and Nussbaum (2002). Actually, the lower bound of the best nonlinear estimator for nonparametric regression model in Fan (1993) and Chen (2003) is derived by using the bound from the bounded normal mean problem.

The problem considered here is to estimate the mean of a normal distribution with a known variance under the quadratic loss when the mean lies in a given bounded interval. No generality is lost if we further assume that the known variance is 1 and the given interval is symmetric about the origin. That is, we may assume the observation X comes from $N(\theta, 1)$ with $\theta \in [-\tau, \tau]$ for some $\tau > 0$.

Let $\delta(X)$ be an arbitrary estimator for θ based on X. Its risk based on the quadratic loss is defined as $R(\delta, \theta) = E(\delta(X) - \theta)^2$, which will be used throughout this paper. The minimax estimator $\delta_N(X)$ is the one that minimizes the maximum risk $\sup_{\theta\in[-\tau,\tau]} R(\delta, \theta)$ among all possible estimators and the minimax risk is

$$\rho_N(\tau) = \inf_{\delta} \sup_{\theta\in[-\tau,\tau]} R(\delta, \theta) = \sup_{\theta\in[-\tau,\tau]} R(\delta_N, \theta).$$

It follows from Ghosh (1964) that there exists a unique symmetric least favorable prior, which concentrates its mass on a finite number of points and gives a minimax Bayes estimator. Casella and Strawderman (1981) showed that for small τ, i.e., $0 < \tau \leq 1.05$, this least favorable prior is a two-point one which puts its mass on the endpoints of the mean interval while for $1.4 \leq \tau \leq 1.6$, the least favorable three-point priors can be constructed for each individual τ. When $\tau \to \infty$, the limiting behavior of the least favorable priors was obtained by Bickel (1981) and the similar results can be found in Levit (1980). Unfortunately, it remains open how to construct the least favorable finite-point prior for a general $\tau > 0$ and hence $\rho_N(\tau)$ is generally unknown.

Several authors have proposed alternative procedures to approximate $\rho_N(\tau)$. Gatsonis, MacGibbon and Strawderman (1987) studied the properties of the Bayes estimator under the uniform prior. Donoho et al. (1990) showed that the linear minimax risk is not far from the minimax risk. It is well-known that the minimax linear estimator is

$$\delta_L(X) = (1+\tau^2)^{-1}\tau^2 X,$$

and the linear minimax risk is

$$\rho_L(\tau) = \inf_{\delta \text{ linear}} \sup_{\theta\in[-\tau,\tau]} R(\delta,\theta) = (1+\tau^2)^{-1}\tau^2.$$

The linear minimax risk serves as an obvious upper bound for $\rho_N(\tau)$. Let

$$\mu_{LN}(\tau) = \rho_L(\tau)/\rho_N(\tau) \quad \text{and} \quad \mu^*_{LN} = \sup_{\tau>0} \mu_{LN}(\tau).$$

It is noted from Donoho et al. (1990) that

$$\lim_{\tau\to 0} \mu_{LN}(\tau) = 1, \quad \text{and} \quad \lim_{\tau\to\infty} \mu_{LN}(\tau) = 1, \tag{1}$$

and

$$1 \le \mu_{LN}(\tau) \le \mu^*_{LN} \le 1.25.$$

Nevertheless, the above bound can be easily improved upon by observing that $\delta_L(X)$ may lie outside the interval $[-\tau,\tau]$ and hence $\delta_L(X)$ is inadmissible. An obvious improvement for $\delta_L(X)$ can be made by a simple truncation. That is, we define the truncated minimax linear estimator as

$$\delta_T(X) = \begin{cases} -\tau, & \delta_L(X) < -\tau, \\ \delta_L(X), & \delta_L(X) \in [-\tau,\tau], \\ \tau, & \delta_L(X) > \tau. \end{cases}$$

Define further

$$\rho_T(\tau) = \sup_{\theta\in[-\tau,\tau]} R(\delta,\theta),$$

and

$$\mu_{LT}(\tau) = \rho_L(\tau)/\rho_T(\tau) \quad \text{and} \quad \mu_{TN}(\tau) = \rho_T(\tau)/\rho_N(\tau). \tag{2}$$

It is easy to show that $\rho_N(\tau) \le \rho_T(\tau) \le \rho_L(\tau)$. Therefore, $1 \le \mu_{LT}(\tau) \le \mu_{LN}(\tau)$ and $1 \le \mu_{TN}(\tau) \le \mu_{LN}(\tau)$. Then by (1), we have

$$\begin{aligned}\lim_{\tau\to 0} \mu_{LT}(\tau) &= \lim_{\tau\to 0} \mu_{TN}(\tau) = \lim_{\tau\to 0} \mu_{LN}(\tau) = 1, \\ \lim_{\tau\to\infty} \mu_{LT}(\tau) &= \lim_{\tau\to\infty} \mu_{TN}(\tau) = \lim_{\tau\to\infty} \mu_{LN}(\tau) = 1.\end{aligned}$$

Thus, we gain little for small and large $\tau \in (0, \infty)$ by using $\delta_T(X)$ and $\delta_N(X)$ instead of $\delta_L(X)$. Nonetheless, the gain is large for moderate τ. By defining $\mu_{LT}^* = \sup_{\tau>0} \mu_{LT}(\tau)$ and $\mu_{TN}^* = \sup_{\tau>0} \mu_{TN}(\tau)$, we have the following theorem.

Theorem 1. *For any $\tau \in (0, \infty)$, the following inequalities hold:*
(a) $1 \leq \mu_{LT}(\tau) \leq \mu_{LT}^* \leq 1.23;$
(b) $1 \leq \mu_{TN}(\tau) \leq \mu_{TN}^* \leq 1.13;$
(c) $1.22 \leq \mu_{LT}^* \leq 1.23;$
(d) $1.22 \leq \mu_{LN}^* \leq 1.25.$

A consequence of the above theorem is that we provide a much sharper bound for $\rho_N(\tau)$:

$$1.13^{-1}\rho_T(\tau) \leq \rho_N(\tau) \leq \rho_T(\tau),$$

namely, the truncated minimax linear estimator performs at most within 13% away from the best possible one. This marks a considerable improvement for the bounds based on the minimax linear estimator:

$$1.25^{-1}\rho_L(\tau) \leq \rho_N(\tau) \leq \rho_L(\tau).$$

The above results have direct implications on minimax theory for other nonparametric problems. Following the explicit construction and the proofs in Fan (1992,1993), we can also improve the minimax efficiency of the best linear estimator constructed in Fan (1992) and Chen (2003) for nonparametric regression problem. The resulting truncated local polynomial estimator has minimax efficiency loss at most 13%.

The rest of this note is organized as follows. In Section 2, some properties of the truncated minimax linear estimator are investigated. In Section 3, we first outline a numerical approach to calculate the lower bounds of the minimax risk for moderate τ. The resulting lower bounds together with the maximum risk of $\delta_L(X)$ and $\delta_T(X)$ obtained in Section 2 for many selected τ are tabulated and so are their ratios. We then give a detailed proof of Theorem 1 there. In Section 4, we briefly discuss approximating the minimax risk over a hyperrectangle by using $\delta_T(X)$ instead of $\delta_L(X)$.

2 Truncated Minimax Linear Estimator

In this section, we investigate some properties of $R(\delta_T, \theta)$, the risk function of the truncated minimax linear estimator $\delta_T(X)$. An immediate result from the definition of δ_T is that for any $\theta \in [-\tau, \tau]$, $R(\delta_T, \theta) \leq R(\delta_L, \theta)$. That is,

Lemma 1. *δ_T dominates δ_L and hence δ_L is inadmissible.*

Some simple calculation leads to the exact expression of $R(\delta_T, \theta)$, which allows a quick and accurate evaluation of $\rho_T(\tau), \mu_{LT}(\tau)$ and μ^*_{LT}.

Lemma 2. *The risk function of $\delta_T(X)$ can be expressed in terms of the standard normal distribution function $\Phi(x)$ and its density $\phi(x)$ as*

$$\begin{aligned} R(\delta_T, \theta) &= \{\alpha^2 + (1-\alpha)^2\theta^2\}\{\Phi(\tau/\alpha - \theta) + \Phi(\tau/\alpha + \theta) - 1\} \\ &+ (\tau+\theta)^2\{1 - \Phi(\tau/\alpha + \theta)\} + (\tau-\theta)^2\{1 - \Phi(\tau/\alpha - \theta)\} \\ &- \alpha\{\tau + (\alpha-2)\theta\}\phi(\tau/\alpha - \theta) - \alpha\{\tau - (\alpha-2)\theta\}\phi(\tau/\alpha + \theta), \end{aligned}$$

for $\theta \in [-\tau, \tau]$ where $\alpha = (1+\tau^2)^{-1}\tau^2$.

From the above lemma, it is easy to see that $R(\delta_T, -\theta) = R(\delta_T, \theta)$. That is, $R(\delta_T, \theta)$ is symmetric in θ. To study the ratio $\mu_{LT}(\tau)$ defined in (2), we need knowledge about $\rho_T(\tau) = \sup_{\theta\in[-\tau,\tau]} R(\delta_T, \theta)$. By the symmetry of $R(\delta_T, \theta)$, we have

$$R'_\theta(\delta_T, \theta) = -R'_\theta(\delta_T, -\theta),$$

where $R'_\theta(\delta_T, \theta) = \frac{\partial R(\delta_T, \theta)}{\partial \theta}$. A trivial result is

Lemma 3. $R'_\theta(\delta_T, 0) = 0$.

This implies that $\theta = 0$ is a local extrema of the risk function $R(\delta_T, \theta)$. However, the closed form for other local extrema is not available and a numerical approach is needed to find the maximum risk due to the complicated form of the derivative function $R'_\theta(\delta_T, \theta)$. Figure 1 illustrates the risk curves for some selected τ's. It can be seen that for small τ, $\tau < \tau_0$, say, the maximum risk $\rho_T(\tau)$ is attained at $\theta = \pm\tau$ but this is not the case when τ is larger than τ_0. Numerical calculation shows that τ_0 is about 2.175. To have an impression of $\rho_T(\tau)$, we computed its values at a sequence of τ's having lag .2 over the interval $[.2, 8.6]$ using a numerical approach. These values are tabulated in Column 3 of Table 1 of next Section. They suggest that $\rho_T(\tau)$ is a non-decreasing function of τ over the interval $[0.2, 8.6]$.

We can show easily that $\rho_L(\tau)$ and $\rho_N(\tau)$ are non-decreasing functions of τ. Similarly, it is expected that $\rho_T(\tau)$ is a non-decreasing function of $\tau \in (0, \infty)$. This can also be seen in Figure 2 where the maximum risk curves $\rho_L(\tau), \rho_T(\tau)$ and the lower bound curve of $\rho_N(\tau)$ are displayed. For further discussion, we also tabulated $\rho_L(\tau)$ and $\mu_{LT}(\tau)$ in Columns 2 and 5 of Table 1.

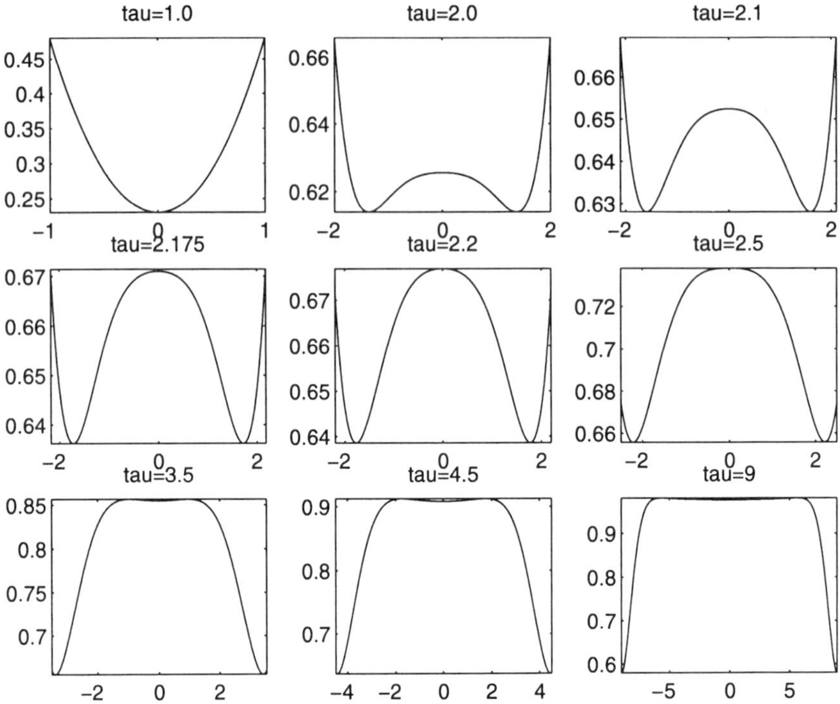

Figure 1 Risk curves of $\delta_T(X)$ for some selected τ's. When $\tau < \tau_0$, the maximum risk is attained at the endpoints; when $\tau_0 < \tau < \tau_1$, the maximum risk is attained at the origin; and when $\tau > \tau_1$, the maximum risk is attained at two symmetric points which are neither the origin nor the endpoints. Here τ_0 and τ_1 are some fixed constants.

3 Minimax Risk

It is quite challenging to obtain $\rho_N(\tau)$ or its sharp lower bounds, especially for moderate τ although the exact value or sharp lower bounds of $\rho_N(\tau)$ can be used to assess how close the linear minimax estimator $\delta_L(X)$ and the truncated minimax linear estimator $\delta_T(X)$ are from the minimax estimator $\delta_N(X)$. For small τ, say $0 < \tau \leq 1.05$, formula for computing exact $\rho_N(\tau)$ is available. In fact, Donoho et al. (1990) showed that

$$\rho_N(\tau) = \tau^2 e^{-\tau^2/2} \int \phi(t)/\cosh(\tau t)dt, \quad 0 < \tau \leq 1.05, \tag{3}$$

where $\phi(t)$ is the standard normal density function. This is the Bayes risk for the least favorable two-point prior based on the work of Casella and Strawderman (1981). For moderate and large τ, say $\tau > 1.05$, computing exact $\rho_N(\tau)$ is usually difficult if not impossible. As a result, sharp lower

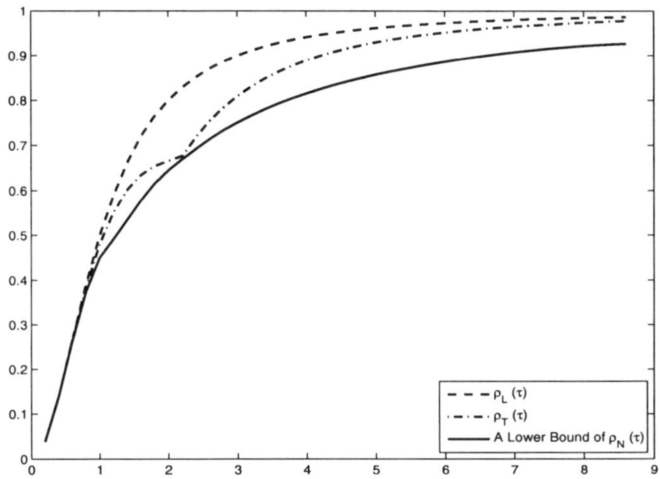

Figure 2 Maximum risk (or lower bound) curves for the estimators $\delta_L(X), \delta_T(X)$ and $\delta_N(X)$.

bounds for $\rho_N(\tau)$ are often computed instead. For large τ, Donoho et al. (1990) employed the following inequality:

$$\rho_N(\tau) \geq 1 - \sinh(\tau)/\{\tau \cosh(\tau)\}, \tag{4}$$

for sharp lower bounds. To calculate sharp lower bounds of $\rho_N(\tau)$ for moderate τ, however, one has to rely on the implicit characteristic of $\rho_N(\tau)$ as the maximum of Bayes risks:

$$\rho_N(\tau) = \sup_{\pi \in \Pi} \rho_\pi(\tau), \tag{5}$$

where Π denotes all prior distributions, $\pi(\theta)$, supported on $[-\tau, \tau]$ and $\rho_\pi(\tau)$ the associated Bayes risk. By the well-known Brown's identity, we have

$$\rho_\pi(\tau) = \inf_\delta \int_{-\tau}^{\tau} R(\delta, \theta)\pi(\theta)d\theta = 1 - \int_{-\infty}^{+\infty} f'_\pi(x)^2/f_\pi(x)dx, \tag{6}$$

where $f_\pi(x)$ is the marginal density of X, i.e.,

$$f_\pi(x) = \int_{-\tau}^{\tau} \pi(\theta)\phi(x-\theta)d\theta. \tag{7}$$

It follows that for any $\pi \in \Pi$, ρ_π provides a lower bound for $\rho_N(\tau)$ and sharper lower bounds are provided by better choices of priors. Since the interval $[-\tau, \tau]$ is symmetric, it is natural to consider only the symmetric priors. For a symmetric prior π over the interval $[-\tau, \tau]$, simple calculation yields that $f_\pi(-x) = f_\pi(x)$ and $f'_\pi(-x) = -f'_\pi(x)$. It then follows that

$$\rho_\pi(\tau) = 1 - 2\int_0^\infty f'_\pi(x)^2/f_\pi(x)dx. \tag{8}$$

Following Casella and Strawderman (1981), we just consider symmetric finite-point priors for π. Let s denote a nonnegative integer. We define $\pi_{2s}(\theta)$ as a symmetric $(2s)$-point prior by:

$$\pi_{2s}(\theta) = \begin{cases} p_i, & \text{if} \quad \theta = \pm\theta_i, i = 1, 2, \cdots, s, \\ 0, & \text{else,} \end{cases}$$

where $0 < \theta_1 < \theta_2 < \cdots < \theta_s = \tau$ and $2\sum_{i=1}^s p_i = 1$. Similarly, we define $\pi_{2s+1}(\theta)$ as a symmetric $(2s+1)$-point prior by:

$$\pi_{2s+1}(\theta) = \begin{cases} p_0, & \text{if} \quad \theta = 0, \\ p_i, & \text{if} \quad \theta = \pm\theta_i, i = 1, 2, \cdots, s, \\ 0, & \text{else,} \end{cases}$$

where $0 < \theta_1 < \theta_2 < \cdots < \theta_s = \tau$ and $p_0 + 2\sum_{i=1}^s p_i = 1$ with $p_0 > 0$ since if $p_0 = 0$, $\pi_{2s+1}(\theta)$ reduces to a symmetric $(2s)$-point prior. For a given $\tau \in (0, \infty)$, by suitably choosing s, θ_i, and p_i, we can obtain some good symmetric finite-point priors and hence obtain some sharp lower bounds for $\rho_N(\tau)$ and the corresponding good upper bounds for $\mu_{LN}(\tau)$ and $\mu_{TN}(\tau)$, by employing the formulas (5)-(8).

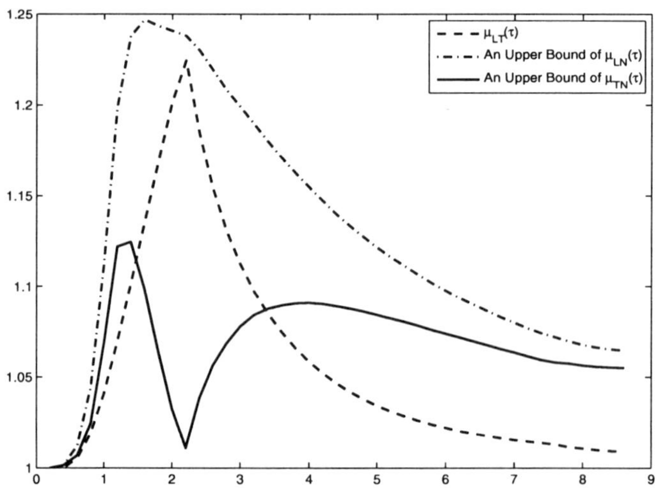

Figure 3 Maximum risk ratio (or upper bound) curves. It can been seen that the gain is considerable for moderate τ by using $\delta_T(X)$ instead of $\delta_L(X)$.

Table 1 Maximum risks (or lower bound) of $\delta_L(X), \delta_T(X)$ and $\delta_N(X)$.

τ	$\rho_L(\tau)$	$\rho_T(\tau)$	$\rho_N(\tau) \geq$	$\mu_{LT}(\tau)$	$\mu_{TN}(\tau) \leq$	$\mu_{LN}(\tau) \leq$
0.2	0.0385	0.0385	0.0385	1.0000	1.0000	1.0000
0.4	0.1379	0.1379	0.1377	1.0000	1.0015	1.0015
0.6	0.2647	0.2633	0.2615	1.0053	1.0069	1.0122
0.8	0.3902	0.3829	0.3737	1.0191	1.0246	1.0442
1.0	0.5000	0.4804	0.4496	1.0408	1.0685	1.1121
1.2	0.5902	0.5521	0.4921	1.0690	1.1219	1.1993
1.4	0.6622	0.6015	0.5349	1.1009	**1.1245**	1.2380
1.6	0.7191	0.6338	0.5768	1.1346	1.0988	**1.2467**
1.8	0.7642	0.6540	0.6146	1.1685	1.0641	1.2434
2.0	0.8000	0.6658	0.6448	1.2016	1.0326	1.2407
2.2	0.8288	0.6769	0.6695	**1.2244**	1.0111	1.2379
2.4	0.8521	0.7196	0.6929	1.1841	1.0385	1.2298
2.6	0.8711	0.7549	0.7146	1.1539	1.0564	1.2190
2.8	0.8869	0.7843	0.7341	1.1308	1.0684	1.2081
3.0	0.9000	0.8090	0.7506	1.1125	1.0778	1.1990
3.2	0.9110	0.8302	0.7658	1.0973	1.0841	1.1896
3.4	0.9204	0.8483	0.7800	1.0850	1.0876	1.1800
3.6	0.9284	0.8639	0.7929	1.0747	1.0895	1.1709
3.8	0.9352	0.8774	0.8045	1.0659	1.0906	1.1625
4.0	0.9412	0.8892	0.8151	1.0585	1.0909	1.1547
4.2	0.9464	0.8994	0.8249	1.0523	1.0903	1.1473
4.4	0.9509	0.9084	0.8341	1.0468	1.0891	1.1400
4.6	0.9549	0.9163	0.8424	1.0421	1.0877	1.1335
4.8	0.9584	0.9233	0.8501	1.0380	1.0861	1.1274
5.0	0.9615	0.9295	0.8573	1.0344	1.0842	1.1215
5.2	0.9643	0.9350	0.8640	1.0313	1.0822	1.1161
5.4	0.9668	0.9399	0.8700	1.0286	1.0803	1.1113
5.6	0.9691	0.9444	0.8759	1.0262	1.0782	1.1064
5.8	0.9711	0.9483	0.8814	1.0240	1.0759	1.1018
6.0	0.9730	0.9519	0.8865	1.0222	1.0738	1.0976
6.2	0.9746	0.9552	0.8912	1.0203	1.0718	1.0936
6.4	0.9762	0.9580	0.8956	1.0190	1.0697	1.0900
6.6	0.9776	0.9606	0.8998	1.0177	1.0676	1.0865
6.8	0.9788	0.9629	0.9038	1.0165	1.0654	1.0830
7.0	0.9800	0.9651	0.9075	1.0154	1.0635	1.0799
7.2	0.9811	0.9670	0.9109	1.0146	1.0614	1.0768
7.4	0.9821	0.9687	0.9141	1.0138	1.0594	1.0741
7.6	0.9830	0.9704	0.9171	1.0130	1.0581	1.0719
7.8	0.9838	0.9726	0.9197	1.0115	1.0575	1.0697
8.0	0.9846	0.9741	0.9221	1.0108	1.0564	1.0678
8.2	0.9853	0.9755	0.9240	1.0100	1.0557	1.0663
8.4	0.9860	0.9768	0.9256	1.0094	1.0553	1.0653
8.6	0.9867	0.9779	0.9268	1.0090	1.0551	1.0646

For example, when $0 < \tau \leq 1.05$, the two-point prior, which puts equal mass on the endpoints of the mean interval (i.e., $s = 1$, $\theta_1 = \tau$ and $p_1 = 1/2$), is the appropriate one while for each $1.05 < \tau \leq 2$, a three-point prior with well chosen p_0 $[\theta_1 = \tau, p_1 = (1-p_0)/2]$ is desired; see Casella and Strawderman (1981) for details. For large τ, four-point, five-point or higher-order finite-point priors are generally needed to get good priors and hence good lower bounds for $\rho_N(\tau)$. Donoho et al. (1990) gave some lower bounds of $\rho_N(\tau)$ for many selected $\tau \in [.42, 4.2]$. To show Theorem 1, we need to calculate more lower bounds over a wider interval, $[.2, 8.6]$ say. For this purpose, a numerical approach has been developed and we omit it here for brevity. The resulting lower bounds for $\rho_N(\tau)$ for a sequence of τ's having lag .2 over the interval $[.2, 8.6]$ are listed in Column 4 of Table 1. The associated upper bounds of the $\mu_{LN}(\tau)$ and $\mu_{TN}(\tau)$ are listed in Columns 6 and 7 of Table 1 respectively.

Based on Table 1, we can show that Theorem 1 is valid over the interval $[.2, 8.6]$. From Columns 5-7 of Table 1, it is easy to see that $\mu^*_{LT} \approx 1.2244, \mu^*_{TN} \leq 1.1245$ and $\mu^*_{LN} \leq 1.2467$. Since $\mu_{LT}(\tau), \mu_{TN}$ and $\mu_{LN}(\tau)$ are continuous functions of τ, at least theoretically, we can estimate or predict them at any $\tau \in [.2, 8.6]$ using numerical interpolation so that we can have more accurate estimate of μ^*_{LT}, and more accurate upper bounds of μ^*_{TN} and μ^*_{LN}. In fact, by interpolation, we evaluated $\mu_{LT}(\tau)$, the upper bounds of $\mu_{TN}(\tau)$ and $\mu_{LN}(\tau)$ for a sequence of τ's having lag .02 over the interval $[.2, 8.6]$ and found that $\mu^*_{LT} \approx 1.2251, \mu^*_{TN} \leq 1.1292$ and $\mu^*_{LN} \leq 1.2467$. By these and Figure 3 where the ratio curve $\mu_{LT}(\tau)$, the upper bound curves of the ratio curves $\mu_{TN}(\tau)$ and $\mu_{LN}(\tau)$ are depicted as dashed, dashdotted and solid curves respectively, we can safely conclude that Parts (a)-(c) of Theorem 1 are valid for $\tau \in [.2, 8.6]$. The validity of Part (d) of Theorem 1 over the interval $[.2, 8.6]$ follows from

$$\begin{aligned}
\mu^*_{LN} &= \sup_{\tau>0} \frac{\rho_L(\tau)}{\rho_N(\tau)} = \sup_{\tau>0} \left\{ \frac{\rho_L(\tau)}{\rho_T(\tau)} \frac{\rho_T(\tau)}{\rho_N(\tau)} \right\} \\
&\geq \sup_{\tau>0} \frac{\rho_L(\tau)}{\rho_T(\tau)} = \mu^*_{LT} \approx 1.2251,
\end{aligned}$$

by noticing that we always have $\rho_T(\tau) \geq \rho_N(\tau)$.

It remains to show Theorem 1 holds for $\tau \in (0, .2] \cup [8.6, \infty)$. To this end, it is sufficient to show that $\mu_{LN}(\tau) \leq 1.13$ for $\tau \in (0, .2] \cup [8.6, \infty)$ since we always have $\mu_{LT}(\tau) \leq \mu_{LN}(\tau)$ and $\mu_{TN}(\tau) \leq \mu_{LN}(\tau)$ for $\tau \in (0, \infty)$. Now for $\tau \in (0, .2]$, by (3), we have

$$\begin{aligned}
\mu_{LN}(\tau) &= \rho_L(\tau)/\rho_N(\tau) = (1+\tau^2)^{-1}\tau^2/\{\tau^2 e^{-\tau^2/2} \int \phi(t)/\cosh(\tau t)dt\} \\
&\leq e^{\tau^2/2}\{\int \phi(t)/\cosh(\tau t)dt\}^{-1} \leq e^{.2^2/2}\{\int \phi(t)/\cosh(.2t)dt\}^{-1} \\
&\leq 1.04,
\end{aligned}$$

as desired. The second to the last inequality follows from the monotonicity of $e^{\tau^2/2}\{\int \phi(t)/\cosh(\tau t)dt\}^{-1}$ (see Donoho et al. 1990 for details). While for $\tau \geq 8.6$, by (4), we have

$$\begin{aligned}\mu_{LN}(\tau) &= \rho_L(\tau)/\rho_N(\tau) \leq (1+\tau^2)^{-1}\tau^2/\{1-\sinh(\tau)/(\tau\cosh(\tau))\} \\ &\leq (1+8.6^2)^{-1}8.6^2/\{1-\sinh(8.6)/(8.6\cosh(8.6))\} \\ &\leq 1.12,\end{aligned}$$

as desired. This completes the proof of Theorem 1.

4 Minimax Risk Over Hyperrectangles

Suppose we have a sequence of observations:

$$Y_i = \theta_i + \epsilon_i, \quad i = 1, 2, \cdots,$$

where ϵ_i are i.i.d $N(0, \sigma^2)$ with a given σ. Assume further that $\theta = (\theta_1, \theta_2, \cdots)$ lies in Θ, a hyperrectangle defined as

$$\Theta(\tau) = \{\theta : |\theta_i| \leq \tau_i, i = 1, 2, \cdots\}$$

for some given sequence of τ_i, tending to zero. Such a kind of Gaussian white noise model is closely related to the nonparametric regression model. See for example Brown and Low (1996), Nussbaum (1996), Brown et al. (2002) and Grama and Nussbaum (2002). Let $X_i = Y_i/\sigma$. Then $X_i \sim N(\theta_i/\sigma, 1)$. One wishes to estimate the vector θ by $\hat{\theta}$ with small quadratic risk $R(\hat{\theta}, \theta) = E\|\hat{\theta} - \theta\|^2$. Let

$$\hat{\theta}_N = \sigma\Big(\delta_N(X_1), \delta_N(X_2), \cdots\Big),$$

where $\delta_N(X_i) = \delta_N(Y_i/\sigma)$ is the minimax estimator of θ_i/σ. It is shown in Donoho et al. (1990) that $\hat{\theta}_N$ is a minimax estimator of θ with the minimax risk

$$\rho_N^*(\sigma) = \inf_{\hat{\theta}} \sup_{\theta\in\Theta} E\|\hat{\theta} - \theta\|^2 = \sup_{\theta\in\Theta} E\|\hat{\theta}_N - \theta\|^2 = \sigma^2 \sum \rho_N(\tau_i/\sigma).$$

Unfortunately, the minimax procedure $\delta_N(\cdot)$ is unknown to us. Instead, Donoho et al. (1990) consider a replacement of $\delta_N(\cdot)$ by the known procedure $\delta_L(\cdot)$, leading to the following minimax linear estimator

$$\hat{\theta}_L = \sigma\Big(\delta_L(Y_1/\sigma), \delta_L(Y_2/\sigma), \cdots\Big).$$

Such an estimator has a maximum risk

$$\rho_L^*(\sigma) = \sup_{\theta\in\Theta} E\|\hat{\theta}_L - \theta\|^2 = \sigma^2 \sum \rho_L(\tau_i/\sigma).$$

In Donoho et al. (1990), it is shown that

$$\rho_N^*(\sigma) \le \rho_L^*(\sigma) \le \mu_{LN}^* \rho_N^*(\sigma) \le 1.25\rho_N^*(\sigma).$$

Since δ_T dominates δ_L, a better estimator for the vector θ is

$$\hat{\theta}_T = \sigma\Big(\delta_T(Y_1/\sigma), \delta_T(Y_2/\sigma), \cdots\Big),$$

which has the maximum risk

$$\rho_T^*(\sigma) = \sup_{\theta\in\Theta} E\|\hat{\theta}_T - \theta\|^2 = \sigma^2 \sum \rho_T(\tau_i/\sigma).$$

By Theorem 1, we can easily show that

Theorem 2. $\rho_N^*(\sigma) \le \rho_T^*(\sigma) \le \mu_{TN}^* \rho_N^*(\sigma) \le 1.13\rho_N^*(\sigma)$.

Thus, the truncated minimax linear estimator $\hat{\theta}_T$ is a better estimator than $\hat{\theta}_L$, and a much closer bound for the minimax risk is obtained.

References

1. Bickel, P.J. (1981). Minimax estimation of the mean of a normal parameter space is restricted. *Ann. Statist.*, **9** 1301-1309.
2. Brown, L. D. and Low, M. G. (1991). Information inequality bounds on the minimax risk (with an application to nonparametric regression). *Ann. Statist.*, **19**, 329-337.
3. Brown, L. D. and Low, M. G. (1996). Asymptotic equivalence of nonparametric regression and white noise. *Ann. Statist.*, **24**, 2384-2398.
4. Brown, L. D., Cai, T. T., Low, M. G., and Zhang, C.H. (2002). Asymptotic equivalence theory for nonparametric regression with random design. *Ann. Statist.*, **30**, 688–707.
5. Casella, G. and Strawderman, W. E.(1981). Estimating a bounded normal mean. *Ann. Statist.*, **9**, 870-878.
6. Chen, K. (2003). Linear minimax efficiency of local polynomial regression smoothers. *J. Nonparametr. Stat.*, **15**, 343–353.
7. Donoho, D. L. and Liu, R. C.(1991). Geometrizing rate of convergence III. *Ann. Statist.*, **19**, 633-667.
8. Donoho, D. ., Liu, R. C. and MacGibbon, B. (1990). Minimax risk over hyperrectangles and implications. *Ann. Statist.*, **18**, 1416-1437.
9. Fan, J. (1992). Design-adaptive nonparametric regression. *J. Ameri. Statist. Assoc.* **87**, 998-1004.
10. Fan, J. (1993). Local linear regression smoothers and their minimax efficiency. *Ann. Statist.*, **21**, 196-216.

11. Gatsonis, C., MacGibbon, B. and Strawderman, W. E. (1987). On the estimation of a restricted normal mean. *Statist. Probab. Lett.* **6**, 21-30.

12. Ghosh, M. N. (1964). Uniform approximation of minimax point estimates. *Ann. Math. Statist.* **35**, 1031-1047.

13. Grama, I. and Nussbaum, M. (2002). Asymptotic equivalence for non-parametric regression. *Math. Methods Statist.*, **11**, 1–36.

14. Ibragimov, I. A. and Hasminskii, R. Z.(1984). Nonparametric estimation of the value of a linear functional in Gaussian white noise. *Theory Probab. Appl.*,**29**,19-32.

15. Levit, B. Ya.(1980). On the second order asymptotically minimax estimates. *Theory Probab. Appl.* **25**, 561-576.

16. Nussbaum, M. (1996). Asymptotic equivalence of density estimation and Gaussian white noise. *Ann. Statist.*, **24**, 2399-2430.

17. Sacks, J. and Strawderman, W. E. (1982). Improvements on linear minimax estimates. In *Statistical Decision Theory and Related Topics III*(S. S. Gupta and J. O. Berger, eds.). **2**, 287-304. Academic, New York.

Chapter 32

ESTIMATION OF SYMMETRIC DISTRIBUTIONS SUBJECT TO A PEAKEDNESS ORDER

Javier Rojo and José Batún-Cutz

Department of Statistics
Rice University, Houston, TX, U.S.A.

Department of Statistics and Probability
CIMAT, Guanajuato, Gto MÉXICO

E-mails: jrojo@stat.rice.edu & batun@cimat.mx

A distribution function F is more peaked about a point a than the distribution G is about the point b if $F((x+a)^-) - F(-x+a) \geq G((x+b)^-) - G(-x+b)$ for every $x > 0$. The problem of estimating symmetric distribution functions F or G, or both, under this constraint is considered in this paper. It turns out that the proposed estimators are projections of the empirical distribution function onto suitable convex sets of distribution functions. As a consequence, the estimators are shown to be strongly uniformly consistent. The asymptotic distribution theory of the estimators is also discussed.

Keywords: Stochastic ordering; Projection; Symmetry; Weak convergence; Linkage analysis.

1 Introduction

The concept of dispersion permeates the theory and applications of statistics. Doksum (1969) defined a tail-ordering between distributions F and G by requiring that $F^{-1}G(x) - x$ be nondecreasing. When F and G are continuous and strictly increasing, this concept is easily seen to be equivalent to the dispersive order defined by $F^{-1}(u) - F^{-1}(v) \leq G^{-1}(u) - G^{-1}(v)$ for all $0 \leq u \leq v \leq 1$. This order will be denoted by $F <_d G$. Doksum (1969) utilized this concept to study power properties of rank tests. In particular, he showed that the power of certain rank tests is isotonic with respect to this order. Rojo (1995b, 1999) considered the problem of estimating the quantile function F^{-1} and the distribution function F when $F <_d G$, and the asymptotic theory of the resulting estimators was delineated. Rojo and Wang (1994) also showed that the power of tests based on L-statistics is isotonic with respect to the dispersive order. For other properties of the dispersive order see Bickel and Lehmann (1979).

This paper considers a different concept of dispersion proposed by Birnbaum (1948). A distribution function F is said to be more peaked about the point a than the distribution function G is about the point b if, for all $x \geq 0$,

$$F((x+a)^-) - F(-x+a) \geq G((x+b)^-) - G(-x+b), \qquad (1)$$

where $h(x^-) = \lim_{\epsilon \downarrow 0} h(x-\epsilon)$. When (1) holds, we write $F >_p G$. If F and G are symmetric about the point 0, then the condition (1) is equivalent to

$$\begin{array}{ll} F(x^-) \geq G(x^-) & \text{for } x \geq 0 \\ F(x) \leq G(x) & \text{for } x < 0. \end{array} \qquad (2)$$

Note that (1) is equivalent to requiring that $|X-a|$ be stochastically smaller than $|Y-b|$, where $X \sim F$ and $Y \sim G$ respectively. Although, in general, $F <_d G \not\Rightarrow F >_p G$ and $F >_p G \not\Rightarrow F <_d G$, when F and G are symmetric and continuous, it can be seen that $F <_d G \Rightarrow F >_p G$.

As a way to motivate the use of the concept of peakedness in applications, we briefly discuss some international studies on Body Mass Index (BMI) of populations of African descent. These studies have shown that as populations migrate West, the distribution of the BMI for the various populations increases in dispersion. This is the case, for example, for a series of studies conducted by Loyola University in Chicago, (see, *e.g.*, Rotimi *et al.* (1995) and Colilla *et al.* (2000)), under the general umbrella of "The International Collaborative Study on Hypertension in Blacks". These studies "followed" populations of Nigerian descent as they migrated from Africa to the Caribbean, eventually settling in the United States. The studies compared, in particular, the male and female populations from Nigeria with African-American male and female populations of Nigerian descent in the Chicago suburb of Maywood, Illinois. The comparison was based on anthropometric variables of which BMI was of particular interest. The results of these studies demonstrated quite decisively that the sample distributions for the BMI of the Nigerian populations are more "peaked" than the BMI sample distributions for their African-American counterparts.

In the context of identifying genes linked to a specific phenotype, sib-paired data illustrates very clearly that the distribution functions of sib-pair differences are symmetrically distributed, and when the candidate gene is

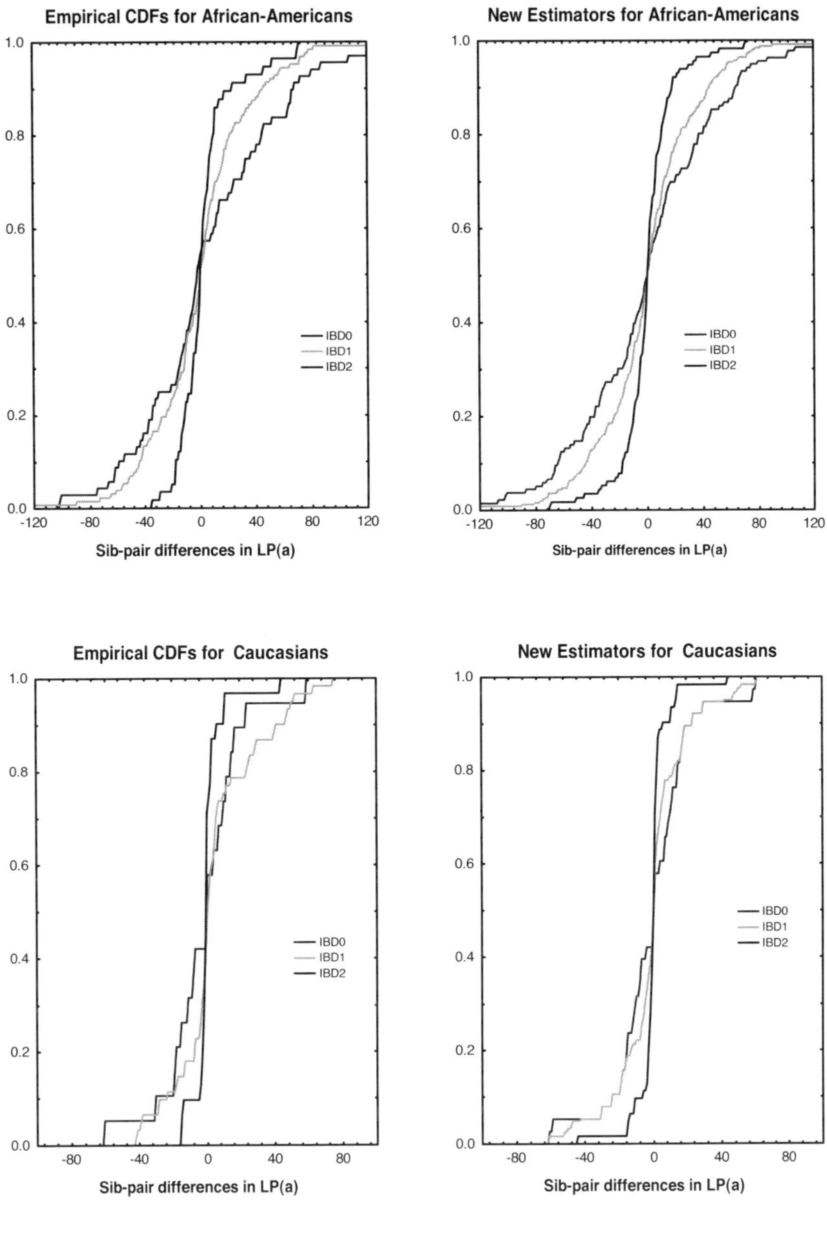

Figure 1 Empirical CDF and new estimators for the Lipoprotein data

linked to the phenotype of interest, the cumulative distributions of the differences within sib pairs are ordered by peakedness. The assumption of symmetry of sib-pair differences may be justified, for example, under a rather general assumption on the distribution of the sib-pair phenotypes (X, Y). That is, if $(X-\mu_X, Y-\mu_Y)$ has the same distribution as $(\mu_X - X, \mu_Y - Y)$, as it happens under the assumption of a bivariate normal distribution, and if the means μ_X and μ_Y are equal, then the sib-pair differences are symmetrically distributed. Figure 1 shows the empirical distribution functions for plasma Lipoprotein (a) differences within sib-pairs for a sample of African-American, and a sample of Caucasian individuals from the Dallas metroplex area. The sib-pairs have been separated into three groups: Those sharing zero alleles identical by descent (IBD); those sharing one allele IBD; and those sharing two alleles IBD. It can be seen from these plots that the assumptions of symmetry and peakedness are almost satisfied. A closer look, however, shows that there are several violations of these assumptions. We will return to this example at the end of the paper where we will illustrate our estimators by applying them to this example. Thus, the concept of peakedness arises in connection with many interesting applications.

The organization of the paper is as follows: Sections 2 and 3 consider the problem of estimating F (or G), when G (or F) is known. Several estimators are proposed. These estimators are shown to be projections of the empirical distribution function onto appropiate convex sets of distribution functions. As a result, the strong uniform convergence of the estimators is easily obtained. In fact, a stronger result holds in the one-sample problem. It can be shown that the resulting estimators are pointwise closer to the true distribution function than the empirical distribution is to the true distribution function in the sense that $|F_{ni}(x)-F(x)|+|F_{ni}(-x^-)-F(-x)| \le |F_n(x)-F(x)|+|F_n(-x^-)-F(-x)|$, where F_{ni} and F_n represent the new estimators and the empirical distribution function respectively. The weak convergence of these estimators is also considered. Section 4 deals with the case where both F and G are unknown, and estimators are provided which are shown to be strongly uniformly consistent and their asymptotic theory is discussed. Finally, section 5 discusses the example on sib-pair lipoprotein data.

2 One-sample problem

Rojo *et al.* (2006) considered the general problem of estimating F and G under the peakedness restriction but without the assumption of symmetry. Estimators were proposed which turned out to be strongly uniformly consistent. However, the estimators presented here take full advantage of the symmetry assumption and are not special cases of the procedures presented in Rojo *et al.* (2006). Moreover, they are simple to compute and can be interpreted as projections of the empirical distribution function.

Let $X_1, ..., X_n$ be a random sample from the distribution function F with $F >_p G$. To fix ideas, the case when G is a known distribution function is considered first. Suppose that F and G are symmetric distribution functions satisfying (2). It is clear that the empirical distribution function F_n does not necessarily satisfy conditions (1) and (2) and therefore estimators which satisfy these conditions may be needed. One possible way to proceed is to "project" the empirical distribution function F_n onto the convex set of distribution functions satisfying (1) and (2). There are several possible ways of doing this. Using results of Schuster (1973, 1975), F_n can be first projected onto the convex set of symmetric distributions to obtain, say, F_n^*. Subsequently, using ideas in Rojo and Ma (1996), F_n^* can be projected onto the convex set of distribution functions which satisfy (1) and (2) to obtain F_n^{**}. Of course, it is also possible to "project" first onto the convex set of distribution functions satisfying (1) and then onto the convex set of distribution functions satisfying (1) and (2). Doing this, produces two estimators whose finite sample and asymptotic properties are considered in this paper.

Schuster (1975) considered a functional of a distribution function F defined by

$$\Phi_1(F(x)) = \frac{1}{2}\left(F(x) + 1 - F(-x^-)\right). \tag{3}$$

This provides a distribution function $\Phi_1(F(x))$ that is symmetric about 0, but does not necessarily satisfy (2). Since (1) is equivalent to the stochastic order of the absolute values of X and Y, where $X \sim F$ and $Y \sim G$, using ideas similar to those presented in Rojo and Ma (1996), consider the following functional of the empirical distribution function

$$\Phi_2(F_n(x), G(x)) = \begin{cases} \max\{F_n(x), G(x)\} & x > 0 \\ \min\{F_n(x), G(x)\} & x < 0. \end{cases} \tag{4}$$

This gives a distribution function that satisfies (2), but is not necessarily symmetric.

The operators Φ_1 and Φ_2 can be applied to the empirical distribution function to obtain the following two estimators

$$F_{n,1}(x) = \Phi_1(\Phi_2(F_n(x), G(x)))$$

and

$$F_{n,2}(x) = \Phi_2\left(\Phi_1(F_n(x)), G(x)\right).$$

It turns out that both $F_{n,1}$ and $F_{n,2}$ are distribution functions which are symmetric about 0 and satisfy (2), as can be seen in the next lemma.

Lemma 1. *The distribution functions $F_{n,1}$ and $F_{n,2}$ are symmetric about 0 and satisfy (2).*

Proof. The distribution function $F_{n,1}$ is symmetric by definition. Let us prove that it satisfies (2). Let $x \geq 0$. Then

$$\begin{aligned} F_{n,1}(x) &= \frac{1}{2}\left[\Phi_2\left(F_n(x), G(x)\right) + 1 - \Phi_2\left(F_n(-x^-), G(-x^-)\right)\right] \\ &\geq \frac{1}{2}\left[G(x) + 1 - G(-x^-)\right] \geq G(x), \end{aligned}$$

where the last inequality follows from the symmetry of G about 0. The proof for $x < 0$ is analogous. Thus $F_{n,1}$ satisfies (2). Now for $F_{n,2}$, let $x \geq 0$. Then

$$\begin{aligned} F_{n,2}(x) &= \max\left\{\Phi_1(F_n(x)), G(x)\right\} \\ &= \max\left\{1 - \Phi_1(F_n(-x^-)), 1 - G(-x^-)\right\} \\ &= 1 - \min\left\{\Phi_1(F_n(-x^-)), G(-x^-)\right\} \\ &= 1 - \Phi_2\left(\Phi_1(F_n(-x^-)), G(-x^-)\right) \\ &= 1 - F_{n,2}(-x^-), \end{aligned}$$

where the second identity follows from the symmetry of $\Phi_1(F_n(x))$. The proof is analogous for $x < 0$. Thus $F_{n,2}$ is symmetric about 0. It is easy to check that $F_{n,2}$ satisfies the condition (2). □

It can be seen that the estimators $F_{n,1}$ and $F_{n,2}$ are "projections" of the empirical distribution function onto appropriate convex sets of distribution functions. To make this precise, let $\mathcal{F}_G$ be the set of distribution functions that satisfy (2) but are not necessarily symmetric. Then, the functions F, $\Phi_2(F_n, G)$, $F_{n,1}$ and $F_{n,2}$ are elements of $\mathcal{F}_G$. Under the norm $\|h\|_p = \{\int (h(x))^p du(x)\}^{1/p}$, for all p, and consequently under the norm $\|h\|_\infty = \sup_{-\infty<x<\infty} |h(x)|$, $\Phi_2(F_n, G)$ is the projection of F_n onto the set $\mathcal{F}_G$. In fact a stronger statement holds.

Theorem 1. *Let Φ_2 be defined by (4). Then $|\Phi_2(F_n(x), G) - F_n(x)| \leq |H(x) - F_n(x)|$ for all x and all $H \in \mathcal{F}_G$.*

Proof. Let $H \in \mathcal{F}_G$. If $x > 0$, then $|\Phi_2(F_n(x), G(x)) - F_n(x)| = 0$ when $F_n(x) \geq G(x)$. On the other hand, if $F_n(x) < G(x)$, then

$$|\Phi_2(F_n(x), G(x)) - F_n(x)| = G(x) - F_n(x) \leq H(x) - F_n(x),$$

since $G(x) \leq H(x)$. For $x < 0$, $|\Phi_2(F_n(x), G(x)) - F_n(x)| = 0$ when $F_n(x) \leq G(x)$; and when $F_n(x) > G(x)$

$$|\Phi_2(F_n(x), G(x)) - F_n(x)| = F_n(x) - G(x) \leq F_n(x) - H(x) \leq |H(x) - F_n(x)|. \qquad \square$$

As a consequence of Theorem (1) the following Corollary is obtained.

Corollary 1. *Let Φ_2 be defined by (4) and let F_n denote the empirical distribution function. Then* $\|\Phi_2(F_n, G) - F_n\|_p = \inf_{H \in \mathcal{F}_G} \|H - F_n\|_p$.

Let $\mathcal{S}$ be the set of all the distribution functions symmetric about 0. Applying the operator Φ_1 to the empirical distribution function F_n gives a projection of F_n onto $\mathcal{S}$, as stated in the following theorem.

Theorem 2. *Let $\Phi_1(.)$ be defined by (3). Then* $\|\Phi_1(F_n) - F_n\|_\infty = \inf_{H \in \mathcal{S}} \|H - F_n\|_\infty$

Proof. The proof is in Schuster (1973), Theorem 1. $\square$

It follows from these results that the estimator $F_{n,1}$ is obtained by first projecting the empirical distribution function onto the convex set of symmetric distributions and then projecting the resulting symmetric distribution onto the set of distributions satisfying (1) and (2). Similar arguments apply to $F_{n,2}$. In general $F_{n,1}(x) \neq F_{n,2}(x)$. In fact $F_{n,1} <_p F_{n,2}$ as demostrated in the following result.

Theorem 3. *Let $F_{n,1} = \Phi_1(\Phi_2(F_n, G))$ and $F_{n,2} = \Phi_2(\Phi_1(F_n), G))$. Then $F_{n,1} <_p F_{n,2}$.*

Proof. Suppose first that $x \geq 0$. Since $F_n(x) \leq \Phi_2(F_n(x), G(x))$ and $\Phi_2(F_n(-x^-), G(-x^-)) \leq F_n(-x^-)$, it follows that

$$F_n(x) + 1 - F_n(-x^-) \leq \Phi_2(F_n(x), G(x)) + 1 - \Phi_2(F_n(-x^-), G(-x^-)),$$

and

$$\Phi_1\left(F_n(x)\right) \leq \Phi_1(\Phi_2(F_n(x), G(x))) = F_{n,1}(x).$$

From the last inequality, we obtain $F_{n,2}(x) \leq F_{n,1}(x)$.

Now consider $x < 0$. In this case, we have the inequalities

$$F_n(x) \geq \Phi_2(F_n(x), G(x)) \text{ and } \Phi_2(F_n(-x^-), G(-x^-)) \geq F_n(-x^-).$$

Then,

$$\Phi_2(F_n(x), G(x)) + 1 - \Phi_2(F_n(-x^-), G(-x^-)) \leq F_n(x) + 1 - F_n(-x^-).$$

Thus, $F_{n,1}(x) \leq \Phi_1\left(F_n(x)\right)$. The desired result is obtained from the last inequality. $\square$

Since both $F_{n,1}$ and $F_{n,2}$ are symmetric, $F_{n,1}(0) = F_{n,2}(0)$ with probability 1, and as a result of the strong uniform consistency of both estimators, $F_{n,1}$ and $F_{n,2}$ are asymptotically equivalent. For finite n, it is still possible for $F_{n,1}$ and $F_{n,2}$ to be equal with positive probability.

Corollary 2.

a) *For* $x \geq 0$, $F_{n,1}(x) = F_{n,2}(x)$ *when* $G(x) \leq \min\{F_n(x), 1 - F_n(-x^-)\}$ *or* $G(x) \geq \max\{F_n(x),\ 1 - F_n(-x^-)\}$.

b) *For* $x < 0$, $F_{n,1}(x) = F_{n,2}(x)$ *when* $G(x) \leq \max\{F_n(x), 1 - F_n(-x^-)\}$ *or* $G(x) \geq \min\{F_n(x),\ 1 - F_n(-x^-)\}$.

Proof.

a) Let $x \geq 0$. If $G(x) \leq \min\{F_n(x), 1 - F_n(-x^-)\}$ then $G(x) \leq \Phi_1(F_n(x))$, and

$$\Phi_1(F_n(x)) \leq F_{n,1}(x) \leq F_{n,2}(x) = \Phi_1(F_n(x)).$$

Next suppose that $G(x) \geq \max\{F_n(x), 1 - F_n(-x^-)\}$. From $\Phi_2(F_n(x), G(x)) = G(x)$, and $G(x) \geq \Phi_1(F_n(x))$ we obtain

$$F_{n,1}(x) = \Phi_1(G(x)) = G(x) = F_{n,2}(x).$$

b) Let $x < 0$. If $G(x) \leq \max\{F_n(x), 1 - F_n(-x^-)\}$, then

$$G(x) \leq \Phi_1(F_n(x)), \Phi_2(F_n(x), G(x)) = G(x),$$

and $F_{n,1}(x) = G(x) = F_{n,2}(x)$. Consider now the case $G(x) \geq \min\{F_n(x), 1 - F_n(-x^-)\}$. From $\Phi_2(F_n(x), G(x)) = F_n(x)$, and $G(x) \geq \Phi_1(F_n(x))$, we obtain

$$F_{n,1}(x) = \Phi_1(F_n(x)) = F_{n,2}(x).$$

□

Note that $\Phi_1(F_n(x))$ is unbiased, and $\Phi_2(\Phi_1(F_n(x), G(x))) = \max(\Phi_1(F_n(x), G(x))$ for $x > 0$, while $\Phi_2(\Phi_1(F_n(x), G(x))) = \min(\Phi_1(F_n(x), G(x))$ for $x < 0$. Thus, F_{n2} is positively biased for $x > 0$ and it is negatively biased for $x < 0$. Similarly, it is easy to show that $\Phi_1(\Phi_2(F_n(x), G(x))) \geq \Phi_1(F_n(x))$ for $x > 0$, while $\Phi_1(\Phi_2(F_n(x), G(x))) \leq \Phi_1(F_n(x))$ for $x < 0$. Thus, both estimators, $F_{n,1}$ and $F_{n,2}$, are biased. Expressions for the expectation of both $F_{n,1}$ and $F_{n,2}$ may be obtained following ideas of Rojo and Ma (1996), and they are given in Batun (2005).

3 Asymptotic theory

The asymptotic theory of the estimators $F_{n,1}$ and $F_{n,2}$ is considered in this section. Their strong uniform consistency will be seen to be an immediate consequence of their being projections of the empirical distribution function in a way that is made precise later. As it turns out, both $F_{n,1}$ and $F_{n,2}$ are uniformly closer to F than the empirical is to F, in the sense that $|F_{ni}(x)-F(x)|+|F_{ni}(-x^-)-F(-x)| \leq |F_n(x)-F(x)|+|F_n(-x^-)-F(-x)|$, for $i = 1,2$. When $F(x_0) \neq G(x_0)$, $F_{n,1}$ and $F_{n,2}$, suitably normalized, have asymptotic normal distributions. However, when $F(x_0) = G(x_0)$, the asymptotic distributions of $F_{n,1}$ and $F_{n,2}$ are somewhat more complex.

The first result in this section is the key for showing the strong uniform consistency.

Theorem 4. *Let F and G be symmetric distribution functions with $F >_p G$, and G known. Let F_n be the empirical distribution function based on the random sample $X_1, ..., X_n$ from F. Then, for $i = 1, 2$,*

$$|F_{n,i}(x) - F(x)| \leq \frac{1}{2} |F_n(x) - F(x)| + \frac{1}{2} \left|F_n(-x^-) - F(-x^-)\right|.$$

Proof.

Let $x \geq 0$. Then,

$$\begin{aligned}
|F_{n,1}(x) - F(x)| &= |\Phi_1(\Phi_2(F_n(x), G(x))) - \Phi_1(\Phi_2(F(x), G(x)))| \\
&\leq \frac{1}{2} |\Phi_2(F_n(x), G(x)) - \Phi_2(F(x), G(x))| \\
&\quad + \frac{1}{2} \left|\Phi_2(F_n(-x^-), G(-x^-)) - \Phi_2(F(-x^-), G(-x^-))\right| \\
&\leq \frac{1}{2} |F_n(x) - F(x)| + \frac{1}{2} \left|F_n(-x^-) - F(-x^-)\right|,
\end{aligned}$$

and

$$\begin{aligned}
|F_{n,2}(x) - F(x)| &= |\max\{\Phi_1(F_n(x)), G(x)\} - \max\{\Phi_1(F(x)), G(x)\}| \\
&\leq \max\{|\Phi_1(F_n(x)) - \Phi_1(F(x))|, 0\}.
\end{aligned}$$

Since $|\Phi_1(F_n(x)) - \Phi_1(F(x))| = \frac{1}{2} |F_n(x) - F_n(-x^-) - F(x) + F(-x^-)|$ the result follows. A similar argument shows the result for $x < 0$. □

The strong uniform consistency of $F_{n,1}$ and $F_{n,2}$ follows from this theorem and is stated as the next corollary.

Corollary 3. *Under the assumptions of Theorem 4,*

(i) $\|F_{n,i} - F\|_\infty \leq \|F_n - F\|_\infty$ *for* $i = 1, 2$.

(ii) *For distribution functions F and H define $L(F,H) = V(\|F-H\|_\infty)$, where $V(x) \geq 0$ for all $x \geq 0$. Then F_n is inadmissible as an estimator of F with respect to all loss functions $V(.)$.*

Next, we turn our attention to the asymptotic distribution of $F_{n,1}$ and $F_{n,2}$. Clearly $F_{n,1}$ and $F_{n,2}$ are correlated but we only consider their marginal asymptotic distributions. To avoid technicalities, we consider the case of F and G continuous. Suppose first that $x > 0$ and $F(x) > G(x)$ so that $F(-x) < G(-x)$. Then,

$$\begin{aligned}\sqrt{n}(F_{n,1}(x) - F(x)) &= \frac{\sqrt{n}}{2} \max\{F_n(x) - F(x), G(x) - F(x)\} \\ &\quad - \frac{\sqrt{n}}{2} \min\left\{F_n(-x^-) - F(-x^-), G(-x) - F(-x)\right\}.\end{aligned}$$

Since $G(x) - F(x) < 0$, with $G(-x) - F(-x) > 0$, eventually with probability one,

$$\sqrt{n}(F_{n,1}(x) - F(x)) = \frac{\sqrt{n}}{2}(F_n(x) - F(x)) - \frac{\sqrt{n}}{2}(F_n(-x^-) - F(-x)).$$

Since $\{\sqrt{n}(F_n(x) - F(x)) : -\infty < x < \infty\}$ converges weakly to a mean zero Gaussian process with covariance function $F(s)(1-F(t))$, $s \leq t$, it follows that $\sqrt{n}(F_{n,1}(x) - F(x)) \xrightarrow{D} N(0, \frac{\overline{F}(x)}{2}(2F(x)-1))$.

The same argument applies to $x < 0$ and shows that $\sqrt{n}(F_{n,1}(x) - F(x)) \xrightarrow{D} N(0, \frac{F(x)}{2}(1-2F(x)))$. The asymptotic distribution of $F_{n,2}$ follows in a similar manner. Thus, we obtain the following theorem.

Theorem 5. *Let F and G be continuous symmetric distribution functions with $F(x) > (<)\ G(x)$ for $x > (<)\ 0$. Then, for $i = 1, 2$ and all x,*

$$\sqrt{n}(F_{n,i}(x) - F(x)) \xrightarrow{D} N(0, \frac{F(-|x|)}{2}(2F(|x|) - 1)).$$

The asymptotic distribution of $F_{n,i}$, $i = 1, 2$, is somewhat more involved if $F(x_0) = G(x_0)$ for some $x_0 > 0$. In this case, $\sqrt{n}(F_{n,i}(x_0) - F(x_0))$ converges in distribution to the random variable $(\max(Z_1, 0) - \min(Z_2, 0))/2$, where (Z_1, Z_2) is a zero-mean bivariate normal distribution function with $Var(Z_1) = Var(Z_2) = F(x_0)\overline{F}(x_0)$ and $Corr(Z_1, Z_2) = \overline{F}(x_0)/F(x_0)$. In particular, the asymptotic distribution of $\sqrt{n}(F_{n,i}(x_0) - F(x_0))$ assigns positive probability to zero equal to $P(Z_1 \leq 0, Z_2 \geq 0)$.

Although the processes $\{\sqrt{n}(F_{n,i}(x) - F(x)) : \infty < x < \infty\}$, $i = 1, 2$, are correlated, their marginal asymptotic theory is easily handled and it turns out that it is the same for both processes. If $F(x) > G(x)$ for all $x > 0$, it can be shown that the finite dimensional distributions of the process $\{\sqrt{n}(F_{n,i}(x) - F(x)) : -\infty < x < \infty\}$, $i = 1, 2$, converge weakly to the finite dimensional distributions of a mean zero Gaussian process $\{\tilde{B}(x) : -\infty < x < \infty\}$ with covariance function

$$Cov(\tilde{B}(x), \tilde{B}(y)) = \begin{cases} \frac{1}{2}\bar{F}(y)(F(x) - F(-x)), & |y| > |x| \\ \frac{1}{2}F(x)(F(-y) - F(y)), & |y| < |x|. \end{cases} \tag{5}$$

We follow the classical approach to weak convergence as presented by, for example, Billingsley (1968). Thus, the convergence of the finite dimensional distributions, coupled with the tightness of the process $\sqrt{n}(F_{n,i}(x) - F(x)) : -\infty < x < \infty\}$ $i = 1, 2$, yields the following result.

Theorem 6. *Under the assumptions of Theorem 5 the process* $\{\sqrt{n}(F_{n,i}(x) - F(x)) : -\infty < x < \infty\}$, $i = 1, 2$, *converges weakly to a zero-mean Gaussian process* $\{\tilde{B}(x) : -\infty < x < \infty\}$ *with covariance function (5).*

Proof. The proof follows directly from results in Rojo and Ma (1996) that show that $\sqrt{n}\max(F_n(x) - F(x), G(x) - F(x))$ and hence $\sqrt{n}\min(F_n(x) - F(x), G(x) - F(x))$ converge weakly to a mean zero Gaussian proces. Since $F_{n,i}$, $i = 1, 2$, are continuous functions of these processes with respect to $\|.\|_\infty$, the result follows. □

Next suppose that the functions F and G are distinct but overlap on an interval $(x_0, x_0 + \delta)$, $\delta > 0$. Of course, by symmetry, they also overlap on $(-x_0 - \delta, x_0)$, and in fact there may be other intervals where they overlap, but this is not important as it is enough that there is one "overlap" interval to show the lack of tightness of the process of interest. Thus $F(x) = G(x)$ for $x \in (x_0, x_0 + \delta)$ for some x_0 and some δ, while $F(x) \neq G(x)$ for, say, $x = x_0 + \delta$. It can be seen, as in Rojo (1995a), that the sequence of processes $\{\sqrt{n}(F_{ni}(x) - F(x)) : -\infty < x < \infty\}$ is not tight and hence cannot converge weakly. The lack of tightness follows from the fact that for $x \in (x_0, x_0 + \delta)$, $\sqrt{n}(F_{ni}(x) - F(x))$ converges in distribution to $(\max(Z_1, 0) - \min(Z_2, 0))/2$, where (Z_1, Z_2) has a mean-zero bivariate normal distribution, while for $x = x_0 + \delta$, the asymptotic distribution of $\sqrt{n}(F_{ni}(x) - F(x))$ is given by Theorem 5, and therefore condition (15.8) in Theorem 15.3 in Billingsley (1968) is not satisfied and, hence, tightness fails.

Finally, let $F(x) = G(x)$, for all x. This condition arises under the hypothesis that F and G are equal in "peakedness". The asymptotic behavior of the process $\{\sqrt{n}(F_{ni}(x) - F(x)) : -\infty < x < \infty\}$, for $i = 1, 2$ can be used to construct tests for the null hypothesis. We have,

$$\begin{aligned} \sqrt{n}(F_{n1}(x) - F(x)) &= \frac{1}{2}\max\{\sqrt{n}(F_n(x) - F(x)), 0\} \\ &\quad -\frac{1}{2}\min\{\sqrt{n}(F_n(-x^-) - F(-x)), 0\} \end{aligned}$$

for $x > 0$, and

$$\begin{aligned}\sqrt{n}(F_{n1}(x) - F(x)) &= \frac{1}{2}\min\{\sqrt{n}(F_n(x) - F(x)), 0\} \\ &\quad -\frac{1}{2}\max\{\sqrt{n}(F_n(-x^-) - F(-x)), 0)\},\end{aligned}$$

for $x < 0$. The process $\{\sqrt{n}(F_{n1}(x) - F(x)) : -\infty < x < \infty\}$, converges weakly to $\{H(x) : -\infty < x < \infty\}$,

$$H(x) = \begin{cases} \frac{1}{2}\max(Z(x), 0) - \frac{1}{2}\min(Z(-x), 0) & x > 0, \\ \\ \frac{1}{2}\min(Z(x), 0) - \frac{1}{2}\max(Z(-x), 0) & x < 0, \end{cases} \tag{6}$$

where $\{Z(x) : -\infty < x < \infty\}$ represents the weak limit of the empirical process $\{\sqrt{n}(F_n(x) - F(x)) : -\infty < x < \infty\}$. Similar arguments apply to the process defined by $\{\sqrt{n}(F_{n2}(x) - F(x)) : -\infty < x < \infty\}$. These results are summarized in the following theorem.

Theorem 7. *Let F and G be continuous distributions functions with $F >_p G$.*

(i) *When $F(x) = G(x)$ for all x, the processes $\{\sqrt{n}(F_{ni}(x) - F(x)) : -\infty < x < \infty\}$, $i = 1, 2$ converge weakly to the process $\{H(x) : -\infty < x < \infty\}$ defined by (6).*

(ii) *Suppose that the functions F and G are distinct but overlap on an interval $(x_0, x_0 + \delta)$, $\delta > 0$, for some x_0 and some $\delta > 0$. Then, the sequences $\{\sqrt{n}(F_{ni}(x) - F(x)) : -\infty < x < \infty\}$, $i = 1, 2$, are not tight and hence cannot converge weakly.*

4 The case when both F and G are unknown

Up to this point the distribution function G has been assumed to be known. In this section we consider the problem of estimating F and G when $F >_p G$ and independent random samples $X_1, ..., X_n \sim F$ and $Y_1, ..., Y_m \sim G$ are available. Several possibilities for estimating F and G, subject to the peakedness order (1), based on the operators Φ_1 and Φ_2 arise in this case. We consider one possible approach and briefly discuss others. For that purpose, let F_n and G_m be the respective empirical distributions, and consider the estimators

$$F^1_{nm}(x) = \Phi_1(\Phi_2(F_n(x), G^*_m(x))) \tag{7}$$

and

$$F^2_{nm}(x) = \Phi_2(\Phi_1(F_n(x)), G^*_m(x)), \tag{8}$$

where $G_m^*(x) = \Phi_1(G_m(x))$, the symmetrized version of G_m, and $\Phi_1(x)$ and $\Phi_2(x)$ are defined by (3) and (4). That is, $F_{nm}^1(x)$ ($F_{nm}^2(x)$) is defined in the same way as $F_{n,1}$ ($F_{n,2}$), except that $G(x)$ is replaced by $G_m^*(x)$. As it was the case with $F_{n,1}$ and $F_{n,2}$, F_{nm}^1 and F_{nm}^2 define symmetric distributions that satisfy (1) with G_m^* now playing the role of G. That is, G is estimated by G_m^*, while F is estimated by F_{nm}^1 or F_{nm}^2. It follows that $F_{nm}^i >_p G_m^*$, $i = 1, 2$.

Before discussing the asymptotic theory of the estimators defined through (7) and (8), note that there are other ways to obtain estimators of F and G that are symmetric and satisfy (1). For example, we could start by symmetrizing both F_n and G_m to obtain F_n^* and G_m^* and then apply (7) and (8) to get F_{nm}^{1*} and G_{nm}^{2*}. Of course the result of this will be that there is no difference between F_{nm}^{1*}, G_{nm}^{2*}, and F_{nm}^2. However, one could also define an operator

$$\Phi_2^*(F_n(x), G_m(x)) = \begin{cases} \min\{F_n(x), G_m(x)\} & x \geq 0 \\ \max\{F_n(x), G_m(x)\} & x < 0, \end{cases} \tag{9}$$

and estimate F by $\Phi_1(F_n(x))$ and G as follows:

$$G_{nm}^1(x) = \Phi_1(\Phi_2^*(F_n^*(x), G_m(x))), \tag{10}$$

and

$$G_{nm}^2(x) = \Phi_2^*(\Phi_1(F_n^*(x)), G_m(x)). \tag{11}$$

The main difference between the approach that estimates G by G_m^* and then uses (7) and (8) to estimate F, and the approach that estimates F by F_n^* and then uses (10) and (11) to estimate G, is that, in the former, G_m^* is unbiased for G with smaller variance than G_m and both F_{nm}^1 and F_{nm}^2 are biased for F, while the latter approach provides an unbiased estimator for F and biased estimators G_{nm}^i, $i = 1, 2$ for G. Attention will be focused on the asymptotic theory of F_{nm}^i, $i = 1, 2$, but the results apply almost verbatim to the estimators G_{nm}^i, $i = 1, 2$.

The strong uniform consistency of F_{nm}^1 and F_{nm}^2, defined by (7) and (8) respectively, follows directly from the following result which is analogous to Theorem 4.

Theorem 8. *Let F and G satisfy the assumptions of Theorem 4, and let F_n and G_m denote the empirical distribution functions based on $X_1, ..., X_n$ and $Y_1, ..., Y_m$. Then, for $i = 1, 2$,*

$$|F_{nm}^i(x) - F(x)| \leq \frac{1}{2} \max\{|F_n(x) - F(x)|, |G_m^*(x) - G(x)|\}$$
$$+ \frac{1}{2} \max\{|F(-x) - F_n(-x^-)|, |G_m^*(-x^-) - G(-x)|\}.$$

Proof. For $x \geq 0$

$$
\begin{aligned}
|F^1_{nm}(x) - F(x)| &= \frac{1}{2}|\max\{F_n(x), G^*_m(x)\} + 1 - \min\{F_n(-x^-), G^*_m(-x^-)\} \\
&\quad - \max\{F(x), G(x)\} - 1 + \min\{F(-x), G(-x)\}| \\
&\leq \frac{1}{2}\max\{|F_n(x) - F(x)|, |G^*_m(x) - G(x)|\} \\
&\quad + \frac{1}{2}\max\{|F_n(-x^-) - F(-x)|, |G^*_m(-x^-) - G(-x)|\}.
\end{aligned}
$$

For $x < 0$, the argument is similar, and the proof for F^2_{nm} follows almost verbatim. □

Since every term in the right of the last inequality converges almost surely to zero as $n, m \to \infty$, it follows that, pointwise, both $F^1_{nm}(x)$ and $F^2_{nm}(x)$ converge, with probability one, to $F(x)$ as $n, m \to \infty$. Since,

$$\|F^i_{nm} - F\|_\infty \leq \max\{\|F_n - F\|_\infty, \|G_m - G\|_\infty\}, \text{ for } i = 1, 2,$$

the strong uniform consistency of F^1_{nm} and F^2_{nm} follows.

Corollary 4. *Let F and G be as in Theorem 8. Then*

$$\|F^1_{nm} - F\|_\infty \to 0 \qquad \text{and} \qquad \|F^2_{nm} - F\|_\infty \to 0$$

as $n, m \to \infty$, with probability one.

As in the one-sample case, both estimators F^1_{nm} and F^2_{nm} have the same asymptotic distribution. Therefore, we will initially concentrate our attention on F^1_{nm}. Under the assumptions of Theorem 5, $F(x) > G(x)$ for every $x > 0$. Therefore, for arbitrary $x > 0$ let $\delta(x) = F(x) - G(x) > 0$. Henceforth, for ease of notation, we will write δ instead of $\delta(x)$. Write

$$
\begin{aligned}
&k_{n,m}(F^1_{nm}(x) - F(x)) \\
&\quad = \frac{k_{n,m}}{2}\{\max\{F_n(x) - F(x), G^*_m(x) - F(x)\} \\
&\quad - \min\{F_n(-x^-) - F(-x), G^*_m(-x^-) - F(-x)\}\} \\
&\quad = \frac{k_{n,m}}{2}\{\max\{F_n(x) - F(x), G^*_m(x) - G(x) - \delta\} \\
&\quad - \min\{F_n(-x^-) - F(-x), G^*_m(-x^-) - G(-x) + \delta\}\}, \qquad (12)
\end{aligned}
$$

where $k_{n,m} \to \infty$ as $n, m \to \infty$. Suppose also that $\frac{k_{n,m}}{\sqrt{n}} \to c$ for some $0 < c < \infty$, and that $\frac{k_{n,m}}{\sqrt{m}}$ remains bounded. For example, one may take $k_{n,m} = \sqrt{n}$ or $k_{n,m} = \sqrt{n+m}$ with $\frac{n}{m} \to c^*$, for some $0 < c^* < \infty$. It follows from (12) that

$$k_{n,m}(F^1_{nm}(x)-F(x)) =$$
$$\frac{1}{2}\{\max\{\frac{k_{n,m}}{\sqrt{n}}\{\sqrt{n}(F_n(x)-F(x))\}, \frac{k_{n,m}}{\sqrt{m}}\{\sqrt{m}(G^*_m(x)-G(x))\}-k_{n,m}\delta\}$$
$$-\min\{\frac{k_{n,m}}{\sqrt{n}}\{\sqrt{n}(F_n(-x^-)-F(-x)\}, \frac{k_{n,m}}{\sqrt{m}}\{\sqrt{m}(G^*_m(-x^-)-G(-x))\}$$
$$+k_{n,m}\delta\}\}. \tag{13}$$

Since $\delta > 0$, it follows from (13) that

$$\lim_{n,m\to\infty} P\{k_{n,m}(F^1_{nm}(x)-F(x)) \le t\} =$$
$$\lim_{n,m\to\infty} P\{\frac{k_{n,m}}{2\sqrt{n}}\{\sqrt{n}(F_n(x)-F(x))+\sqrt{n}(F(-x)-F_n(-x^-))\} \le t\}.$$

When $x < 0$ similar arguments yield the same result. It is now clear that when $F(x) > G(x)$ for $x > 0$, the asymptotic distribution of F^1_{nm}, and that of F^2_{nm} as well, is the same as the asymptotic distribution of $F_{n,1}$ and $F_{n,2}$ in the one-sample case. These results are summarized in the following theorem.

Theorem 9. *Let F and G be as in Theorem 5. Let $k_{n,m} \to \infty$ as $n, m \to \infty$, with $\frac{k_{n,m}}{\sqrt{n}} \to c$, and $\frac{k_{n,m}}{\sqrt{m}} \to c^*$ for some $0 < c, c^* < \infty$. Then, for $i = 1, 2$, as $n, m \to \infty$,*

$$\sqrt{n}(F^i_{nm}(x)-F(x)) \xrightarrow{D} N(0, \frac{c^2F(-|x|)}{2}(2F(|x|)-1)).$$

More precisely, $\lim_{n,m\to\infty} P\{k_{n,m}(F^i_{nm}(x)-F(x)) \le t\} = \Phi(\frac{t}{\sigma})$ *where* $\sigma^2 = c^2F(\frac{-|x|}{2})(2F(|x|)-1)$.

The weak convergence of the processes $\{\sqrt{n}(F^i_{nm}(x)-F(x)) : -\infty < x < \infty\}$, $i = 1, 2$, as $n, m \to \infty$, in the case that $F(x) > G(x)$ for all $x > 0$, may be obtained as in the one-sample case. The basic idea behind the proof is that when $F(x) > G(x)$, for all $x > 0$, as $n, m \to \infty$, eventually, with high probability,

$$\sqrt{n}(F^i_{nm}(x)-F(x)) = \frac{1}{2}\{\sqrt{n}(F_n(x)-F(x))+\sqrt{n}(F(-x)-F_n(-x^-))\}$$

and, therefore, the weak limit of the processes $\{\sqrt{n}(F^i_{nm}(x)-F(x)), -\infty < x < \infty\}$, $i = 1, 2$, is the same as the weak limit in the one-sample problem. This is made precise in the following theorem which considers only the case of $k_{n,m} = \sqrt{n}$. The results, however, hold for general $k_{n,m}$ satisfying the conditions of Theorem 9.

Theorem 10. *Let F and G be as in Theorem 5 and let $\frac{n}{m} \to c$, for some $0 < c < \infty$. Then, the processes $\{\sqrt{n}(F^i_{nm}(x) - F(x)), -\infty < x < \infty\}$, $i = 1, 2$, converge weakly to the zero-mean Gaussian process with covariance function given by (5).*

Proof. Only the proof for the weak convergence of $\{\sqrt{n}(F^2_{nm}(x) - F(x)), -\infty < x < \infty\}$, is provided here since the arguments to prove the weak convergence of $\{\sqrt{n}(F^1_{nm}(x) - F(x)), -\infty < x < \infty\}$ are very similar. Note that, for $x > 0$,

$$\begin{aligned} F^2_{nm}(x) - F(x) &= \max\{F^*_n(x), G^*_m(x)\} - F(x) \\ &= \frac{1}{2}\{F_n(x) - F(x) + F(-x) - F_n(-x^-)\} \\ &\quad + \max\{0, G^*_m(x) - F^*_n(x)\} \end{aligned} \tag{14}$$

where $F^*_n(x) = \frac{1}{2}\{F_n(x) + 1 - F(-x^-)\}$, the symmetrized version of F_n. Similarly, for $x < 0$,

$$\begin{aligned} F^2_{nm}(x) - F(x) &= \min\{F^*_n(x), G^*_m(x)\} - F(x) \\ &= \frac{1}{2}\{F_n(x) - F(x) + F(-x) - F_n(-x^-)\} \\ &\quad + \min\{0, G^*_m(x) - F^*_n(x)\}. \end{aligned} \tag{15}$$

It follows from (14) and (15) that, to prove the weak convergence of $\{\sqrt{n}(F^2_{nm}(x) - F(x)), -\infty < x < \infty\}$ to the zero-mean Gaussian process with covariance function given by (5), it suffices to show that

$$\begin{aligned} &\sup_x |\sqrt{n}\{(F^2_{nm}(x) - F(x)) - \frac{1}{2}\{F_n(x) - F(x) + F(-x) - F_n(-x^-)\}\}| \\ &= \sqrt{n}\max\{\sup_{x\geq 0}\max\{0, (G^*_m(x) - F^*_n(x))\}, \sup_{x<0}|\min\{0, (G^*_m(x) - F^*_n(x))\}|\} \end{aligned}$$

converges to zero in probability. It is easy to see, using a symmetry argument, that

$$\begin{aligned} P\{\sup_x |\sqrt{n}\{(F^2_{nm}(x) - F(x))\frac{1}{2}\{F_n(x) - F(x) + F(-x) - F_n(-x^-)\}\}| > \varepsilon\} \\ \leq 2P\{\sup_{x\geq 0}\sqrt{n}\max\{0, G^*_m(x) - F^*_n(x)\} > \varepsilon\}. \end{aligned} \tag{16}$$

Thus, it is enough to show that $P\{\sup_{x\geq 0}\sqrt{n}\max\{0, G^*_m(x) - F^*_n(x)\} > \varepsilon\} \to 0$ as $n, m \to \infty$. Recall that $F(x) > G(x)$ for all $x > 0$. Choose $\beta_n \downarrow 0$, with $\sqrt{n}\beta_n \to \infty$, and let

$$k_n = \inf\{y > 0 : F(y) - G(y) = \beta_n\} \quad \text{and}$$

$$k^*_n = \sup\{y > k_n : \inf_{k_n \leq x \leq y}(F(x) - G(x)) = \beta_n\}.$$

Then $k_n \downarrow 0$ and $k_n^* \to \infty$, and $F(x) - G(x) \geq \beta_n$ in $[k_n, k_n^*]$. Consider now

$$P\{\sup_{x\geq 0} \sqrt{n} \max\{0, G_m^*(x) - F_n^*(x)\} > \varepsilon\} =$$

$$P\{\max\{ \sup_{0\leq x\leq k_n} \sqrt{n} \max\{0, G_m^*(x) - F_n^*(x)\}, \sup_{k_n<x\leq k_n^*} \sqrt{n} \max\{0, G_m^*(x) - F_n^*(x)\}, \sup_{x>k_n^*} \sqrt{n} \max\{0, G_m^*(x) - F_n^*(x)\} > \varepsilon\}. \quad (17)$$

It is easy to see that the right side of (17) is bounded above by

$$P\{ \sup_{0\leq x\leq k_n} \sqrt{n}(G_m^*(x) - F_n^*(x)) > \varepsilon\} + P\{ \sup_{k_n<x\leq k_n^*} \sqrt{n}(G_m^*(x) - F_n^*(x) > \varepsilon\} + P\{ \sup_{x>k_n^*} \sqrt{n}(G_m^*(x) - F_n^*(x)) > \varepsilon\}.$$

Thus, it suffices to show that each of these terms goes to zero as $n, m \to \infty$. Consider first the second term and note that

$$\begin{aligned}
&P\{ \sup_{k_n<x\leq k_n^*} \sqrt{n}(G_m^*(x) - F_n^*(x)) > \varepsilon\} \\
&= P\{ \sup_{k_n<x\leq k_n^*} \sqrt{n}(G_m^*(x) - G(x) + G(x) - F(x) + F(x) - F_n^*(x)) > \varepsilon\} \\
&\leq P\{ \sup_{k_n<x\leq k_n^*} \sqrt{n}\{G_m^*(x) - G(x) + F(x) - F_n^*(x)\} > \varepsilon + \sqrt{n}\beta_n\} \\
&\leq P\{\sqrt{\frac{n}{m}} \left\|\sqrt{m}(G_m^*(x) - G(x))\right\|_\infty > \varepsilon + \sqrt{n}\beta_n\} \\
&\quad + P\{\left\|\sqrt{n}(F_n^*(x) - F(x))\right\|_\infty > \varepsilon + \sqrt{n}\beta_n\}.
\end{aligned}$$

Since $\sqrt{\frac{n}{m}}$ remains bounded while $\sqrt{n}\beta_n \to \infty$, it follows that $P\{ \sup_{k_n\leq x\leq k_n^*} \sqrt{n}(G_m^*(x) - F_n^*(x)) > \varepsilon\}$ converges to zero as $n, m \to \infty$.

Now consider

$$\begin{aligned}
&P\{ \sup_{0\leq x\leq k_n} \sqrt{n}(G_m^{**}(x) - F_n^*(x)) > \varepsilon\} \\
&= P\{ \sup_{0\leq x\leq k_n} \frac{\sqrt{n}}{2}\{2((G_m^*(x) - \frac{1}{2}) - (G(x) - \frac{1}{2})) \\
&\quad + 2\,(G(x) - F(x)) + 2((F(x) - \frac{1}{2}) - (F_n^*(x) - \frac{1}{2}))\} > \varepsilon\} \\
&\leq P\{ \sup_{0\leq x\leq k_n} \sqrt{n}\{2((G_m^*(x) - \frac{1}{2}) - (G(x) - \frac{1}{2}))\} > \varepsilon\} \\
&\quad + P\{ \sup_{0\leq x\leq k_n} \sqrt{n}\{2((F_n^*(x) - \frac{1}{2}) - (F(x) - \frac{1}{2}))\} > \varepsilon\}.
\end{aligned}$$

Now note that $2(G(x)-\frac{1}{2}) = P(|Y_i| \le x)$ and that $2(G_m^*(x)-\frac{1}{2})$ is the empirical distribution function defined as $\frac{1}{m}\sum_{i=1}^m I_{\{|Y_i|\le x\}}$. Therefore,

$$\{\sqrt{m}(2(G_m^*(x)-\frac{1}{2})-2(G(x)-\frac{1}{2})), 0 \le x < \infty\}$$

is the empirical process associated with the sequence $\{|Y_i|\}_{i=1}^\infty$. Similarly,

$$\{\sqrt{n}(2(F_n^*(x)-\frac{1}{2})-2(F(x)-\frac{1}{2})), 0 \le x < \infty\}$$

is the empirical process associated with $\{|X_i|\}_{i=1}^\infty$. Therefore, with $\sqrt{\frac{m}{n}}$ bounded away from 0, say $\sqrt{\frac{m}{n}} > A$ for some $A > 0$ and all m, n,

$$P\{\sqrt{\frac{n}{m}} \sup_{0\le x\le k_n} \sqrt{m}\{2((G_m^*(x)-\frac{1}{2})-(G(x)-\frac{1}{2}))\} > \varepsilon\} \le \frac{2(G(k_n)-\frac{1}{2})}{\varepsilon^2 A^2}.$$

Since $k_n \downarrow 0$, $G(k_n) \to \frac{1}{2}$ and thus the above probability goes to zero. Similar arguments show that

$$P\{\sup_{0\le x\le k_n} \sqrt{n}\{2((F_n^*(x)-\frac{1}{2})-(F(x)-\frac{1}{2}))\} > \varepsilon\} \le \frac{2(F(k_n)-\frac{1}{2})}{\varepsilon^2} \to 0,$$

and

$$P\{\sqrt{\frac{n}{m}} \sup_{x\ge k_n^*} \sqrt{m}\{2((G_m^*(x)-\frac{1}{2})-(G(x)-\frac{1}{2}))\} > \varepsilon\} \le \frac{2(1-G(k_n^*))}{\varepsilon^2 A^2} \to 0,$$

and

$$P\{\sup_{x\ge k_n^*} \sqrt{n}\{2((1-F_n^*(x))-(F(x)-\frac{1}{2}))\} > \varepsilon\} \le \frac{2(1-F(k_n^*))}{\varepsilon^2} \to 0,$$

where the last two results follow since $k_n^* \to \infty$. □

5 Example

This section illustrates the estimators defined in Section 4 using a data set on sib-pairs to study the linkage between apolipoprotein(a) and levels of Lipoprotein(a). Various studies have been conducted, e.g. Mooser *et al* (1997) and Boerwinkle *et al* (1992), to examine the relationship between levels of Lipoprotein(a), or Lp(a), and a polymorphic glycoprotein (apolipoprotein(a)), in African-American and Caucasian populations. In linkage analysis studies with sib-pair data, assuming that the locus of interest is biallelic, sib-pairs are divided into those sharing 0 alleles Identical By Descent (IBD); those sib-pairs sharing 1 allele IBD; and those sib-pairs sharing 2 alleles IBD. The idea is that if the locus of interest, (in this case the polymorphic glycoprotein apolippoprotein(a)), has a genetic effect on the phenotype of interest, (in this case levels of Lipoprotein(a)), siblings in group 0 IBD will be less similar, than those siblings in group 1 IBD, and these siblings in turn will be less similar than those siblings in group 2 IBD, where similarity is measured, for example, by the spread of the distributions of the pairwise phenotypic differences of the siblings within groups. This idea is made precise in the Haseman-Elston model (Haseman and Elston (1972)), which tests for linkage by regressing the siblings squared phenotypic differences on the proportion of alleles IBD, and then tests for the slope parameter being zero.

In this example we illustrate our estimators with a data set that comes from a study conducted at the University of Texas Southwestern Medical School. The data represent the differences in Lp(a) for pairs of siblings from a population of African-American and Caucasian families in the Dallas metroplex area. The empirical distribution functions for the sib-pair differences have been plotted in Figure 1, after grouping the sib-pairs by number of alleles identical by descent at the locus of interest – the Apo(a) gene in the present context. Under the usual model of bivariate normality of the sib-pair phenotypes, and equal mean levels of Lp(a) for each member of the pair, the differences should appear to be symmetrically distributed about 0. The data shown in Figure 1 seems to support this observation, except that various violations to this restriction can be observed. Note that under the model of bivariate normality with equal marginals, a common assumption in the quantitative trait linkage analysis literature, the pair (X, Y) is exchangeable and therefore, $X - Y$ has the same distribution as $Y - X$, thus giving rise to the symmetry of the phenotypic differences.

On the other hand, the studies mentioned at the beginning of this section, have also confirmed that the Apo(a) gene is linked to the levels of Lp(a). Therefore, the data should show isotonic peakedness with respect to the number of alleles IBD. This notion is also observed in Figure 1, although as it was the case with the assumption of symmetry, there are violations of the condition of peakedness. This is most noticeable in the Caucasian population where the IBD 1 empirical distribution function, which should lie within the curves for IBD 0 and IBD 2 falls below the IBD 0 curve for values of Lp(a) between 35 and 70. The estimators discussed earlier in Section 4 of the manuscript were applied to both data sets of sib-pair differences to obtain estimators which satisfy the restrictions of symmetry and peakedness. It can be observed from the plots on the right side of Figure 1, that the major modifications made in the case of the African-American population, consisted of adjustments to obtain the symmetry of the estimators as the peakedness restriction was already satisfied by the empirical distribution function. On the other hand, the modifications in the case of the Caucasian population were more substantial and adjusted for both the lack of symmetry and the lack of order in peakedness.

Acknowledgements. Javier Rojo's work was generously supported by grants NSF-DMS 0532346, NSF-DMS 0244184, NSF-DMS 0552590, and NSA 06101701. The second author acknowledges support by project No. 39017 of Conacyt, México. Rudy Guerra and Jonathan Cohen kindly made accessible the data on sibling pairs and their research was supported through NIH grant RO1 HL-53917. The authors also acknowledge the many helpful suggestions by Dorota Dabrowska.

References

1. Batun-Cutz, J. (2005). On some problems in nonparametric inference. Unpublished Ph.D. dissertation, Centro de Investigación en Matemáticas, Guanajuato, México.
2. Bickel, P. J. and Lehmann, E. L. (1979). Descriptive statistics for nonparametric models. IV. Spread. In *Contributions to Statistics, Jaroslav Hájek Memorial Volume*, ed. J. Jureckova, Dordrecht: Riedel, pp. 33-40.
3. Billingsley, P. (1968). *Convergence of Probability Measures,* Wiley, New York.
4. Birnbaum, Z. W. (1948). On random variables with comparable peakedness. *The Annals of Mathematical Statistics* 34, 1593-1601.
5. Boerwinkle, E., Leffert, C. C., Lin, J., Lackner, C., Chiesa, G. and Hobbs, H. H. (1992). Apolipoprotein(a) gene accounts for greater than 90% of the variation in plasma lipoprotein(a) concentrations. *The Journal of Clinical Investigation*, Vol. 90, 52-60.
6. Colilla, S., Rotimi, C., Cooper, R., Goldberg, J. and Cox, N. (2000). Genetic inheritance of body mass index in African-American and African families. *Genetic Epidemiology*, Vol. 18, 360-376.
7. Doksum, K. A. (1969). Star-shaped transformations and the power of rank tests. *The Annals of Mathematical Statistics* 40, 1167-1176.
8. Haseman, J. K. and Elston, R. C. (1972). The investigation of linkage between a quantitative trait and a marker locus. *Behavioral Genetics*, Vol. 2, 3-19.
9. Mooser, V., Scheer, D., Marcovina, S. M., Wang, J., Guerra, R., Cohen, J., and Hobbs, H. H. (1997). The Apo(a) gene is the major determinant of variation in plasma Lp(a) levels in African Americans. *American Journal of Human Genetics*, 61(2), 402-417.
10. Rojo, J. and He, G. Z. (1991). On the estimation of stochastically ordered survival functions, *Journal of Statistical Computation and Simulation*, Vol. 55, 1-51.

11. ROJO, J. AND WANG, J. (1994). Test based on L-statistics to test the equality in dispersion of two probability distributions, *Statistics and Probability Letters.* 21, 107-113.

12. ROJO, J. (1995a). On the weak convergence of certain estimators of stochastically ordered survival functions. *J. Nonparam. Statist.*, Vol. 4, 349-363.

13. ROJO, J. (1995b). Nonparametric quantile estimation under order constraints, *J. Nonparam. Statist.*, 5, 185-200.

14. ROJO, J. AND MA, Z. (1996). On the estimation of stochastically ordered survival functions. *Journal of Statistical Computation and Simulation*, Vol 55, 1-21.

15. ROJO, J. (1999). On the estimation of a survival function under a dispersive order constraint, *J. Nonparam. Statist.* 11, 107-135.

16. ROJO, J., BATUN-CUTZ, J. L. AND DURAZO, R. (2006). Inference under peakedness restrictions. *Statistica Sinica, To appear.*

17. ROTIMI, C. N., COOPER, C. S., ATAMANI, S. L., OSOTIMEHIN, B., KADIRI, S., MUNA, W., KINGUE, S., FRASER, H. AND MCGEE, D. (1995). Distribution of anthropometric variables and the prevalence of obesity in populations of west African origin: The International Collaboration Study on Hypertension in Blacks (ICSHIB), *Obesity Research*, Vol. 3, 95S-105S.

18. SCHUSTER, E. (1973). On the goodness-of-fit problem for continuous symmetric distributions. *Journal of the American Statistical Association,* Vol. 68, No. 343, 713-715.

19. SCHUSTER, E. (1975). Estimating the distribution function of a symmetric distribution. *Biometrika,* 62, 3, 631-635.

20. SCHUSTER, E. (1987). Identifying the closest symmetric distribution or density function. *The Annals of Statistics*, Vol. 15, 865-874.

Subject Index